East Asian Monsoon

WORLD SCIENTIFIC SERIES ON METEOROLOGY OF EAST ASIA

Vol. 1: East Asia and Western Pacific Meteorology and Climate: Selected Papers of the Fourth Conference
edited by C.-P. Chang, Guoxiong Wu, Ben Jou & C. Y. Lam

Vol. 2: East Asian Monsoon
edited by C.-P. Chang

Vol. 3: Observation, Theory, and Modeling of Atmospheric Variability Selected Papers of Nanjing Institute of Meteorology Alumni in Commemoration of Professor Jijia Zhang
edited by Xun Zhu, Xiaofan Li, Ming Cai, Shuntai Zhou, Yuejian Zhu, Fei-Fei Jin, Xiaolei Zou & Minghua Zhang

WORLD SCIENTIFIC SERIES ON METEOROLOGY OF EAST ASIA

Vol. 2

East Asian Monsoon

edited by

C.-P. Chang

Naval Postgraduate School,
Monterey, California, USA

World Scientific

NEW JERSEY • LONDON • SINGAPORE • BEIJING • SHANGHAI • HONG KONG • TAIPEI • CHENNAI

Published by

World Scientific Publishing Co. Pte. Ltd.
5 Toh Tuck Link, Singapore 596224
USA office: 27 Warren Street, Suite 401–402, Hackensack, NJ 07601
UK office: 57 Shelton Street, Covent Garden, London WC2H 9HE

British Library Cataloguing-in-Publication Data
A catalogue record for this book is available from the British Library.

EAST ASIAN MONSOON

ISBN 981-238-769-2

Printed in Singapore by Mainland Press

PREFACE

The term "Asian summer monsoon" most frequently refers to the heavy summer rainfall season in South Asia, particularly around India. To some in the western world, "East Asian summer monsoon" is a less familiar term. This is partly because seasonal mean maps have a large area of high rainfall in South Asia during boreal summer, while rainfall over a comparably sized area in East Asia is significantly lower. In fact, the East Asian summer monsoon has complex space and time structures that are distinct from the South Asian summer monsoon. It covers both subtropics and midlatitudes and its rainfall tends to be concentrated in elongated rain belts that stretch for many thousands of kilometers and affect China, Japan, Korea, and the surrounding seas. The narrow meridional scale and the tendency of alternating stationary and northward progression stages make the seasonal averaged rainfall in East Asia less concentrated than in South Asia. However, the impacts of floods and droughts on human lives and economics during to East Asian summer monsoon are just as great because the finer intraseasonal space-time structure, coupled with the narrow rivers, is more sensitive to interannual variations. For example, a flood event can result from heavier-than-normal rainfall during the normal subseasonal rain period or from sustained rainfall outside of the normal period or both.

The East Asian summer monsoon is also closely linked with the West Pacific summer monsoon. Both are part of the global climate system and are affected by El Nino – Southern Oscillation (ENSO) and surface temperature variations in the western Pacific and surrounding oceans, the tropospheric biennial oscillation, and the South Asian summer monsoon. In addition, typhoons in the western North Pacific are most active during the East Asian summer monsoon. They may be considered a component of the East Asian summer monsoon as they contribute substantial amounts of rainfall and have major impacts on the region.

The East Asian winter monsoon, or "Asian winter monsoon", may be a more familiar weather system to the western world. Its circulation encompasses a very large meridional domain. Cold air outbreaks emanate from the Siberian high and penetrate deeply into the equatorial region, where the center of maximum rainfall is found in the Maritime Continent region. This center has long been recognized as a major planetary scale heat source that provides a significant amount of the energy that drives the global circulation during boreal winter. The East Asian winter monsoon is directly connected with the Australian summer monsoon by vigorous circulations extending across the equator, and both are affected by ENSO and other oscillations in the Pacific and Indian Ocean. In addition, a number of global teleconnections associated with the East Asian winter monsoon have recently been found. These include a correlation of temperature and precipitation anomalies over North America with fluctuations in the East Asian jet stream, and an association of winter storm development as far away as Europe with precursor convection over the Maritime Continent.

Because of its impacts on nearly one third of the world population and on the global climate system including effects on climate change, the study of East Asian monsoon has received increased attention in both East Asian countries and in the United States. This book presents reviews of the recent research on the subject. The book is organized into five parts with three chapters each: East Asian summer and winter monsoon, interannual variations, general circulation modeling, synoptic and mesoscale processes, and interactions with other circulations.

The idea of this book originated when the World Meteorological Organization/ Commission on Atmospheric Science Working Group on Tropical Meteorology Research, under the past chair, Dr. Greg Holland, and the present chair, Prof. Lianshou Chen, organized the International Panel for East Asian Monsoon under the leadership of Prof. Shiyan Tao. The book would not have been possible without the efforts of the many reviewers. Besides the members of the Editorial Board of the World Scientific Series on Meteorology of East Asia who served as reviewers, the book benefited tremendously from the following individuals who provided critical reviews of individual chapters: Drs. Hanna Annamalai, Johnny Chan, Russ Elsberry, Bill Frank, Bob Haney, Harry Hendon, Brian Hoskins, Dick Johnson, T. N. Krishnamurti, Bill Kuo, Bill Kyle, Mark Lander, Peggy LeMone, Jerry Meehl, Takio Murakami, Julia Slingo, Wei-Kuo Tao, Mingfang Ting, Chuck Wash, and Song Yang. The technical editing was performed by Ms. Hway-Jen Chen. This work was supported in part by the National Science Foundation under Grant ATM-0101135, the National Oceanic and Atmospheric Administration under Grant NA01AANRG0011, and the Office of Naval Research Marine Meteorology Program.

Chih-Pei Chang
Naval Postgraduate School
Monterey, California

CONTENTS

IV. Synoptic and Mesoscale Processes

V. Interactions with Other Circulations

Part I

East Asian Summer and Winter Monsoon

1. SEASONAL MARCH OF THE EAST-ASIAN SUMMER MONSOON

YIHUI DING
National Climate Center
China Meteorological Administration
No.46 Zhongguancun Nan Da Jie
Haidian District, Beijing 100081, PRC
E-mail: yhding@public.bta.net.cn; dingyh@cma.gov.cn

This chapter presents a comprehensive review of the seasonal march of the East Asian summer monsoon. First, the earliest onset of the summer monsoon over the South China Sea (SCS) and the Indo-China Peninsula occurring around mid-May or even earlier, is discussed in terms of low-level and upper-level wind, OLR and rainfall fields. A sudden change in these fields before and after the onset is revealed. This is characterized by a wind switch from low-level easterlies and high-level westerlies to low-level westerlies and high-level easterlies, the rapid growth of convection and increase in rainfall, and the eastward retreat of the subtropical high. The onset of the summer monsoon over the Indo-China Peninsula and the SCS is preceded by the development of circulation features and convective activity in the tropical East Indian Ocean and the Bay of Bengal. These are characterized by the development of a twin cyclone crossing the equator, the rapid acceleration of low-level westerlies, and a significant increase in both the areal extent and the intensity of convection.

Climatologically, the seasonal march of the summer monsoon displays distinct stepwise northward and northeastward advances. Over East Asia, two abrupt northward jumps and three stationary periods have been identified while over the western North Pacific three stages in the onset and advance of the summer monsoon have been identified. The noted East Asian monsoon rainy seasons, including the pre-summer rainy season over South China, Meiyu/Baiu over eastern China and Japan, and Changma over Korea occur normally during the stationary periods that are imbedded in the northward advance of the summer monsoon.

Major physical processes and mechanisms responsible for the onset and seasonal progress are discussed with special emphasis on three controlling factors and related physical mechanisms. The land-sea thermal contrast and the effect of the elevated heat source of the Tibetan Plateau are identified as pre-conditions for the abrupt onset of the Asian summer monsoon over the SCS and Indo-China Peninsula through the rapid reversal of the meridional temperature gradient. The arrival of the intra-seasonal oscillations (ISO) provides a triggering mechanism, along with the several phase-locking wet ISO phases. The intrusion of mid-latitude troughs into the northern SCS and central and northern Indo-China Peninsula is also seen to be another triggering mechanism that induces the convective activity through the release of potential instability, thus enhancing the monsoon trough there through the feedback process of meso-scale convective systems.

1. Introduction

The East-Asian summer monsoon is a main component of the Asian summer monsoon system and plays an important role in the weather and climate over the surrounding areas as well as on a global scale. For example, the shift of the major seasonal rain belt and its unusual behavior that may lead to occurrence of extensive drought/flood disasters in East

Asia is greatly affected by the East-Asian summer monsoon (Ding 1992). Recent events include the two-month long-persistent excessively heavy rainfall/floods over the Yangtze-Huaihe River Basins during the 1991 Meiyu season (Ding 1993) and the prolonged, unprecedented heavy rainfall/floods over the Yangtze River Basins during the 1998 Meiyu season. The former brought about a serious loss of about 100 billion Chinese Yuan and numerous lives, and the latter caused an economic loss of at least 180 billion Chinese Yuan. Both events were closely related to the unusual activities of the summer monsoon in East Asia. (Ding 1993; National Climate Center 1998). To gain a better insight into the abnormal behavior of the East Asian summer monsoon, it is necessary to illustrate the seasonal march of the East-Asian summer monsoon from a climatological perspective. It has long been realized that there is a close relationship between the arrival of rainy seasons in different sub-regions and the seasonal march of the summer monsoon in East Asia (Ding 1992; Qian and Lee 2002).

The rainy season in China as well as in East Asia generally begins with the onset of the summer monsoon in the South China Sea (SCS) and ends with its withdrawal. This is the reason why in recent years, the study of the SCS summer monsoon has become a focus in field of the Asian summer monsoon and climate variability. The major seasonal rain belt over East Asia moves from low to middle and high latitudes as the summer monsoon develops and advances northward. Thereafter the summer monsoon retreats southward rapidly, thus leading to the end of the major precipitation season in summer in China as well as East Asia (Guo and Wang 1981; Ninomiya and Murakami 1987; Qian and Lee 2000; Wang and Lin 2002). The early or late arrival of the summer monsoon, rapid or slow northward movement and its intensity variation directly influence the temporal and spatial distributions of the major monsoonal rain belt in summer in East Asia, and the occurrence of drought/flood during the rainy seasons (National Climate Center 1998). Recently, Lau and Weng (2000) have reported teleconnection patterns that link major U.S. summer droughts and floods to variability in the Asian summer monsoon, mainly in the SCS and the West Pacific. Therefore, the study of the East Asian summer monsoon, especially the summer monsoon over the SCS, not only has a regional implication, but also has a global effect.

The above-described scientific rationale contributes to the scientific motivation behind launching the South China Sea Monsoon Experiment (SCSMEX) in 1998. A four-month field observation phase, which covered the period from May 1 to August 31, 1998, is one of the core components of the SCSMEX. Field observations consisted of an atmospheric observation network, oceanographic observation network, air-sea interface observation network and satellite observation network, which included upper-air radiosondes, surface observations, dual Doppler and digitized weather radars, scientific research ships, Aerosondes, satellite observations, oceanic boundary layers and flux measurement, integrated sounding system (ISS), radiation measurements, ATLAS moorings, drifting buoys, acoustic Doppler current profilers (ADCP), towed profiling (CTD) and air-borne expendable bathythermograph (AXBT). Many countries and regions were involved in the enhanced and intensive observations of the SCSMEX, including 10 provinces in mainland China, and regions of Taiwan, Hong Kong and Macao; USA, Australia, Thailand, Vietnam, Malaysia,

Singapore, Brunei, Indonesia and Philippines. The SCSMEX has designed the large-scale observation region and intensive observation domain. The former includes a large Asian-West Pacific region (70°E - 150°E, 10°S - 40°N) with the focus on conventional observations; the latter is centered on the SCS and its surrounding regions (95°E - 130°E, 10°S - 30°N). Two intensive flux arrays at nearly meso-scale within the intensive observation domain are further designed to carry out upper air sounding, surface observations, ISS, Aerosondes, dual-Doppler radar, radiation measurements, TRMM satellite observation, boundary layer flux and oceanic integrated observations (Figure 1).

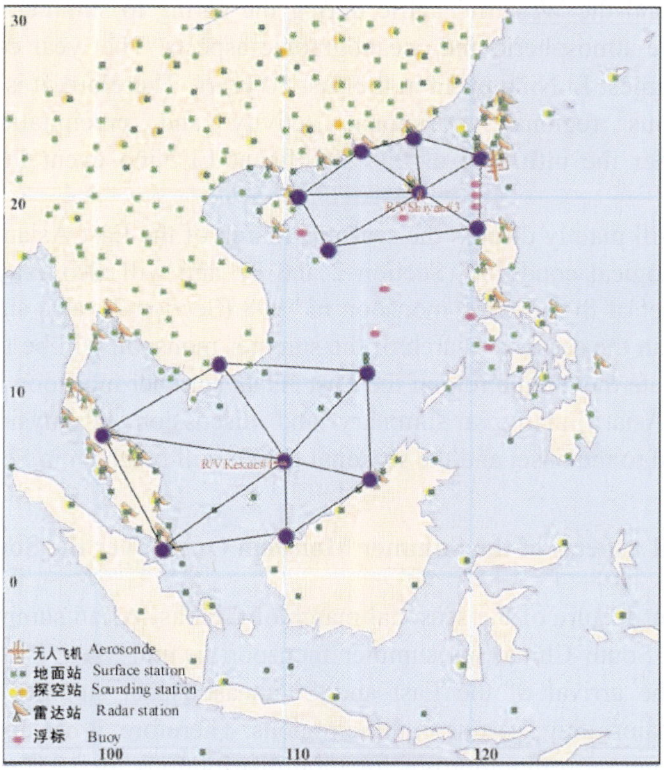

SCSMEX观测网

Figure 1. The network of the SCSMEX field observation in May-August, 1998 (SCSMEX Project Office 1998).

The enhanced phase of field observation for the SCSMEX spanned the period from May 1 to August 31, 1998, during which there were two intensive observation periods (IOP), May 5-25 and June 5-25. The first IOP focused on monitoring the onset of the SCS monsoon and its sudden seasonal change as well as the effect upon precipitation in South China and Southeast Asia, especially the monsoon trough over the northern part of the SCS and the meso-scale convective system. The second IOP focused on monitoring atmospheric and oceanic conditions over the SCS during the mature phase and northward migration of the

East Asian monsoon, and its effect upon precipitation in the Yanzgtze River Basin, the Korean Peninsula and Japan. Under the support of participating countries/regions, the four-month field observations have been completed successfully (SCSMEX Project Office 1998).

A wealth of information and data has been obtained on the early monsoon onset over the northern part of the SCS, the evolution and the northward propagation of the monsoon to the Yangtze River Basin in the context of 1997/1998 El Niño events, thus providing an excellent basis for further research. As recently revealed by some investigators (McPhaden 1999), an anomalous seasonal evolution can be found over the Indian Ocean, maritime continent, the South China Sea and the Western Pacific during the spring to summer season of 1998 in terms of large-scale atmospheric and oceanographic aspects. That year corresponded to the post phase of the largest El Niño event in the past 50 years. Therefore, it is very interesting to examine anomalous regional monsoonal activity and precipitation fields during SCSMEX/IOP under the influence of this significant El Niño event in Section 3 of this Chapter.

This chapter will mainly discuss the seasonal march of the East-Asian summer monsoon under the climatological condition (Section 2 and 4), and will also examine characteristic features of the onset of the summer monsoon in 1998 (Section 3), as a significant case. The moisture transport in the seasonal march of the summer monsoon will be further discussed in Section 5 to elucidate the unique role of the East Asian summer monsoon in the precipitation process in East Asia. Finally, a summary and discussion of physical processes and mechanisms related to the onset and the seasonal march will be given in Section 6.

2. Climatological Aspects of the Summer Monsoon Onset over the South China Sea

The most prominent feature of the seasonal march of the East-Asian summer monsoon is the rapid onset of the South China Sea summer monsoon in mid-May. The onset of the SCS monsoon marks the arrival of the East and Southeast Asian summer monsoon and the beginning of the major rainy seasons in these regions. Therefore, it has a significant regional implication. The long-term monthly precipitation distributions for 1951-1981 over the SCS and its surrounding area have shown that most of the stations undergo a sudden amplification of precipitation amounts after May 1, the month of the onset of the SCS summer (Ding *et al.* 1996). The climatological high reflective cloudiness (HRC) data also show that after the onset of the summer monsoon over the SCS in the middle of May, the HRC abruptly increases, with a relatively early occurrence in the northern and southern SCS. However, there has not been a truly unified, agreed-upon definition on the onset date nor a dynamically consistent viewpoint on the reason for the earliest onset of the SCS summer monsoon (Lu and Chan 1999; Li and Zeng 2002). Some investigators have defined the onset of the SCS summer monsoon in terms of the sudden change in prevailing wind fields (Yan 1997). Others have defined the onset using the monsoon precipitation increase exceeding a certain threshold value (Tao and Chen 1987; Matsumoto 1997; Wang and Lin 2002). Lau and Yang (1997) used 6 mm day^{-1} precipitation amount derived from GPI (Global Precipitation Index) as a characteristic value to define the onset date of the summer monsoon in Southeast Asia and the South China Sea. The significant changes in cloudiness and outgoing longwave radiation (OLR) measured with satellite observations have been also used as an indicator to define the onset dates (Hirasawa *et al.* 1995; Xie *et al.* 1996). The active convection is sometimes defined as an onset condition when an OLR value of less than 230 W m^{-2} is observed. Recent studies have tended to combine the wind and precipitation variables (or OLR data) to define the onset date (Chan *et al.* 2000; Qian and Lee 2000). In the paper by Ding and Liu (2001), a similar but somewhat different definition of the onset of the SCS summer monsoon is used: (1) the low-level wind at 850 hPa becomes westerly or southwesterly, but must be sustained for at least 5 days and attain a maximum wind speed greater than 5 m s^{-1}. This condition will exclude those cases of transient fluctuation of the westerly wind caused by some moving tropical weather systems or events. This defined change in wind direction mainly reflects the seasonal transition of large-scale circulation features; (2) a significant and persistent increase of precipitation and active convection should accompany this wind change. No threshold value of precipitation is used due to the dearth of precipitation observations in the SCS region. It should be pointed out that most of the definitions of monsoon onset are somewhat arbitrary. However, the rapid switch of prevailing wind direction and its sustenance for a certain period, by definition, is the most essential among these criteria. Based on numerous previous works, an estimated averaged onset period is the 4th pentad of May. It should be pointed out that the range of interannual variation of averaged onset dates is quite large, lasting from the 1st pentad of May to the 2nd pentad of June (Figure 2). The early or late onset might be closely related to the effect of ENSO events or other processes. In this chapter, this problem will not be addressed in more detail.

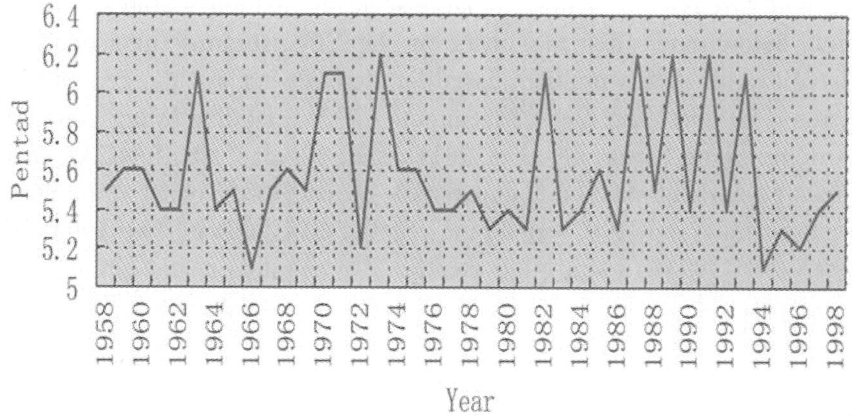

Figure 2. Interannual variability of onset pentads of the SCS summer monsoon (based on data of He *et al.* 2001).

Fig. 3 shows the area-averaged (10°N - 20°N, 110°E - 120°E) annual courses of climatological mean CMAP precipitation, OLR, BT and the 850 hPa zonal wind for 1980-1995 (Qian and Lee 2000). Similar time series have been produced by other authors. It can be seen that all indicators show an abrupt change around the fourth pentad of May (16-20

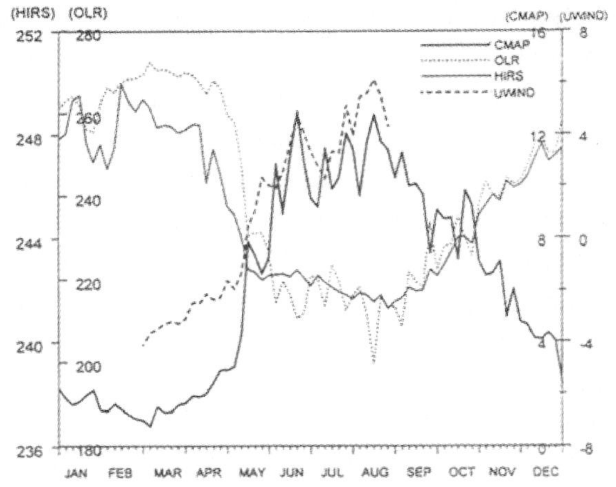

Figure 3. Climatological time series of CMAP (Climate Precipitation Center Merged Analysis of Precipitation) precipitation (mm day^{-1}), BT (Brightness Temperature) (K), OLR (W m^{-2}) and 850 hPa zonal wind (m s^{-1}) averaged over the South China Sea area of (10°N - 20°N, 110°E - 120°E) for 1980-1995 (Qian and Lee 2000).

May), with an increase in CMAP precipitation from 3mm day^{-1} in the 3rd pentad to 8 mm day^{-1} in the 4th pentad, the decrease in OLR from 247 W m^{-2} to 230 W m^{-2} and the decrease in BT (see the curve of HIRS) from 245 to 243 K. Thus, the thermodynamic criteria for the onset of the SCS summer monsoon, 6 mm day^{-1}, 230 W m^{-2} and 244 K are chosen for thresholds of the daily precipitation rate, OLR and BT, respectively. The criterion for wind is the change in direction of the zonal wind from easterly to westerly. According to these criteria, it can be seen from Figure 3 that the SCS summer monsoon starts from mid-May and ends early in November.

Below, we will discuss the change in large-scale circulation features over the Asian monsoon region before and after the onset of the SCS summer monsoon. Numerous investigators have examined this problem based on the large-scale wind, geopotential height and OLR patterns (Lau and Yang 1997; Matsumoto 1997; Fong and Wang 2001; Wang and Lin 2002). Among them, Fong *et al.* (2001) used 40-year NCEP/NCAR datasets to characterize the large-scale circulation features in the process of the onset of the SCS summer monsoon, while Lau and Yang (1997) used nine years of GPI dataset (1986-1994) and NASA DAO reanalysis (1985-1993) to study the change in precipitation and circulation patterns during the onset process of the SCS summer monsoon. The length of the datasets for precipitation, wind and OLR in a similar study by Matsumoto is about 9-12 years. Therefore, the latter two studies seem to use shorter data length, so that their results are less representative from the climatological viewpoint. The discussion to be given below will be mainly based on the results derived by Fong and Wang (2001), and Ding and Sun (2001) who used 21-year long NCEP reanalysis datasets. From Figures 4-7, it can be seen that a dramatic change occurs from the pentad of May 11-15 to the pentad of May 16-20 for these fields.

The southwesterlies rapidly expand from the equatorial East Indian Ocean region, across the Indo-China Peninsula, down to most of the South China Sea (Figure 4(a)-(d)). At the same time, the OLR values significantly decrease from 240W m^{-2} to values below 240 W m^{-2}

during this short transition period (Figure 5), implying that convective clouds and precipitation abruptly develop over the SCS during the onset process. Yan (1997) and Ding *et al.* (1996) have indicated that explosive development of high clouds over the SCS is an important event for the onset of the SCS summer monsoon, heralding the end of the dry season and the arrival of the wet season in this region. It is very interesting to examine the figures for differences of 850 hPa wind and OLR patterns between the pentad of May 21-25 (after the onset) and the pentad of May 6-10 (before the onset) (Figure 4 (e) and Figure 5 (e)). The most significant change of the low-level wind pattern is the acceleration and eastward extension of tropical westerlies from the tropical East Indian Ocean to the central and southern SCS. The Somali jet upstream also undergoes a considerable intensification. From the northern part of the Bay of Bengal to the northern SCS, a wind shear line with two cyclonic circulations embedded is generated. This fact indicates the development of the monsoon trough which is connected with the tailing part of mid-latitude frontal systems. Therefore, the onset of the summer monsoon in the SCS seems to be considered as a regional demonstration of the rapid seasonal intensification of the whole Asian summer monsoon. Correspondingly, the most significant change in the OLR pattern is also seen in the Arabian Sea, the tropical East Indian Ocean and the Bay of Bengal, and the SCS and the tropical West Pacific. These changes reflect abrupt enhancement of cloud and rainfall in these regions. Among them, the change in the SCS is most marked. Another sudden change is the rapid weakening and eastward retreat of the subtropical high over the West Pacific from the Indo-China Peninsula and the SCS (Figure 6). At the same time, a trough over the Bay of Bengal continuously extends southward and deepens. In Figure 6(e), the most significant changes in the 500 hPa height pattern are found over the Bay of Bengal and the SCS. The deepening of the upper trough over the Bay of Bengal is closely related to the development of a monsoon trough with the cyclonic circulation at low-levels, which greatly favors local development of intensive convective activity as well as the acceleration and eastward propagation of low-level westerlies in the tropical East Indian Ocean. Now it is not clear which one, eastward extension of low-level southwesterlies or the eastward retreat of the subtropical high, is the primary cause for leading to large-scale abrupt changes in the above chain of events. Fong and Wang (2001) pointed out that the establishment of the SCS summer monsoon during the pentad of May 16-20 is only confined to the lower troposphere, with a ridge of the subtropical high and upper-level southeasterlies or northeasterlies still dominating the SCS, showing that the main body of the subtropical high has not significantly changed. Therefore, they concluded that the onset of the low-level monsoon in the SCS initially forces the subtropical high to move eastward, whereas the feedback process of convective and meso-scale activities over the Indo-China peninsula and the SCS possibly makes an important contribution to the final withdrawal out of the SCS of the subtropical high at a later stage. Intensive vertical transports of heat produced by vigorous convective activity may significantly heat the upper troposphere (Figure 9) and induce high-level out flow (Figure 8), thus leading to decrease of pressure at low-level or the development of a trough region in the lower troposphere. This kind of the feedback mechanism should be further studied with numerical simulation and theoretical analysis. Such a situation is quite similar to the breaking of the continuous subtropical high belt in the Northern Hemisphere around the region of the Bay of Bengal due to a local development of the monsoon trough when the pre-summer monsoon begins in late April or early May.

The most salient feature of the 200 hPa wind patterns is the significant development and northward movement of the South Asian high over the eastern part of the Indo-China Peninsula. Before the onset of the SCS summer monsoon, the South-Asian high is located in the southern part of the Indo-China Peninsula, and has a weaker intensity (Figure 7(a)). Thereafter, this high moves toward the northwest and significantly intensifies (Figure 7(b)-(d)). The upper-level westerly jet and the easterly jet on either flank of the high accelerate, thus leading to intensification of upper level divergence and convective activity. One may clearly see the development of an anomalous upper-anticyclonic circulation over

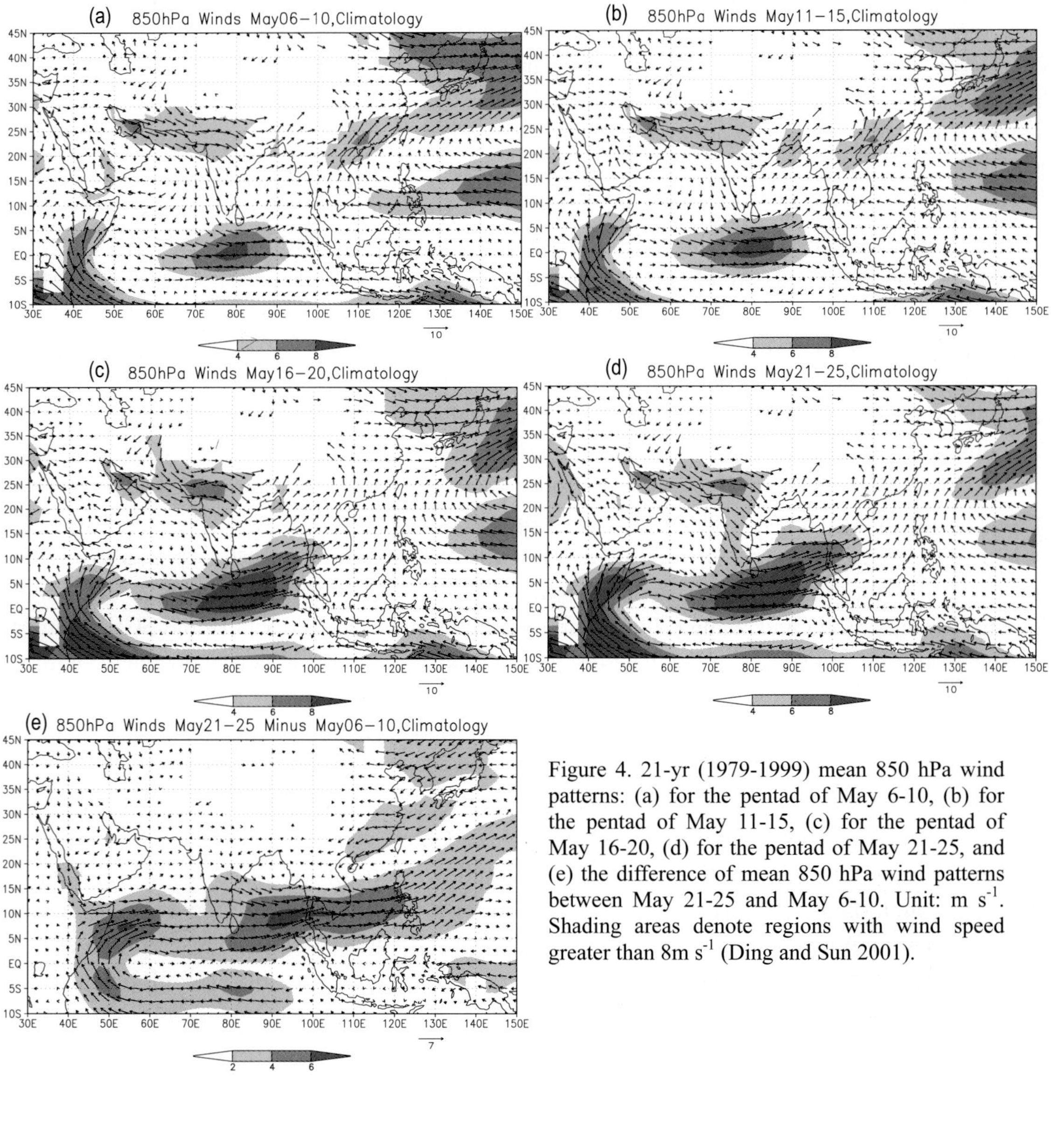

Figure 4. 21-yr (1979-1999) mean 850 hPa wind patterns: (a) for the pentad of May 6-10, (b) for the pentad of May 11-15, (c) for the pentad of May 16-20, (d) for the pentad of May 21-25, and (e) the difference of mean 850 hPa wind patterns between May 21-25 and May 6-10. Unit: m s^{-1}. Shading areas denote regions with wind speed greater than 8m s^{-1} (Ding and Sun 2001).

East Asia during the process of the onset of the SCS summer monsoon (Figure 7(e)). Figure 8 shows the change in the pentad-mean 200 hPa velocity potential and divergent wind patterns. Before the onset (Figure 8 (a), (b)), the major divergent center is located in the tropical West Pacific. During the pentad of May 15-20 when the onset of the SCS summer monsoon takes place, a major divergent center rapidly moves over the central and southern SCS, with marked divergent airflow found in these regions. This pattern of upper level divergence lends some support to the idea of a feedback mechanism driven by enhanced convective activity. From the heating pattern during this period (Figure 9 (right column)), it can be seen that this major outflow region corresponds to an extensive area of the heat source (Q1$>$0) in these regions.

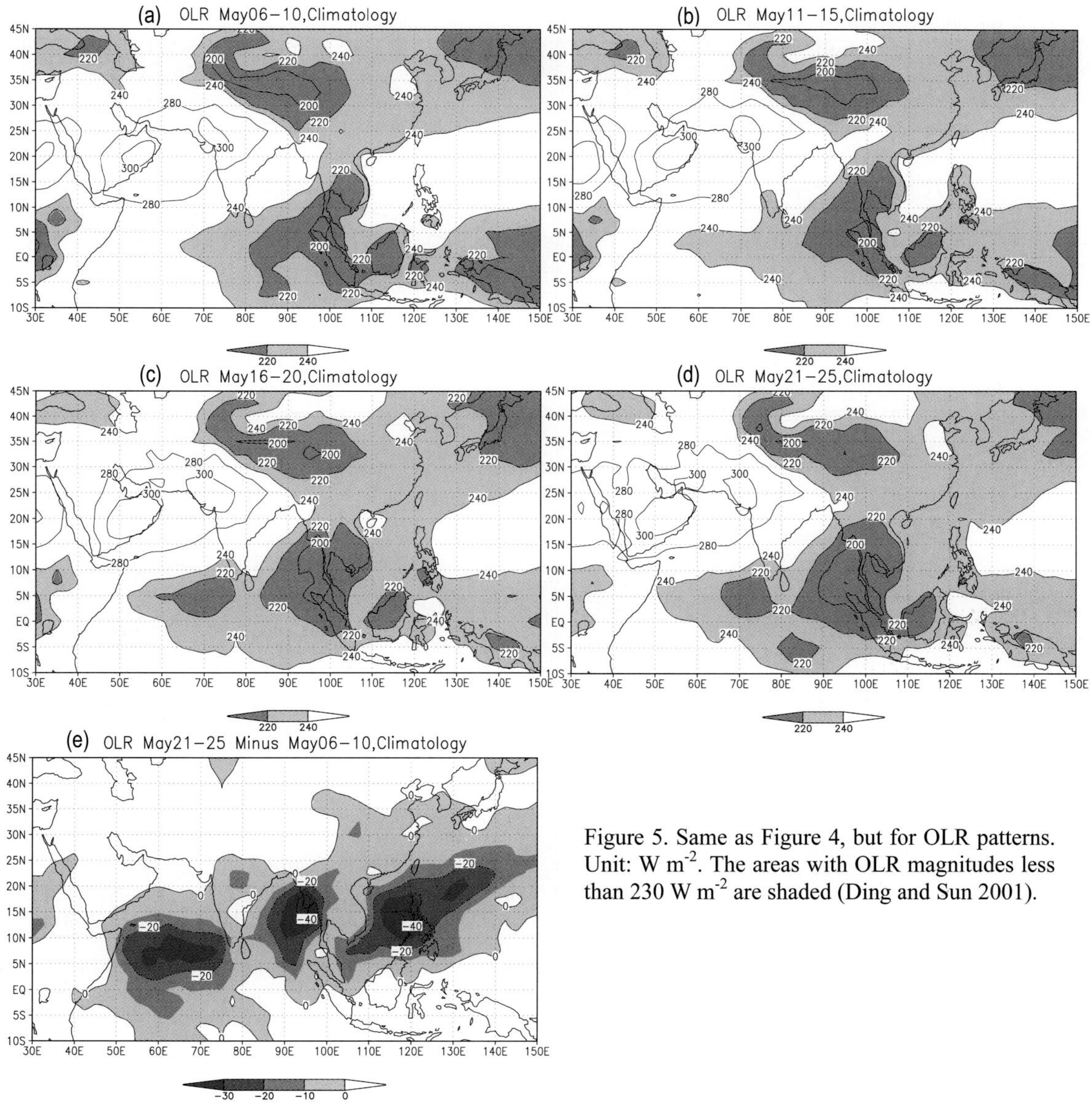

Figure 5. Same as Figure 4, but for OLR patterns. Unit: W m^{-2}. The areas with OLR magnitudes less than 230 W m^{-2} are shaded (Ding and Sun 2001).

Other investigators have obtained more or less similar results. Lau and Yang (1997) have shown that before the onset, low-level easterlies prevail over the SCS under the persistent influence of the subtropical high over the West-Pacific, and after the onset the subtropical high retreats eastwards by about 50° longitude from the Indo-China Peninsula while the low-level westerly flow from the equatorial Indian Ocean extends into the SCS. Strong southerly flow over the SCS and South China is observed due to a switch of the local circulation from anticyclonic to cyclonic over the SCS (see also Figure 4(d)). At this time, although the summer monsoon has not begun in India, the westerlies to its south over the Indian Ocean are well established. The changes in the upper tropospheric circulation associated with the onset of the SCS summer monsoon is the intensification of the large-scale

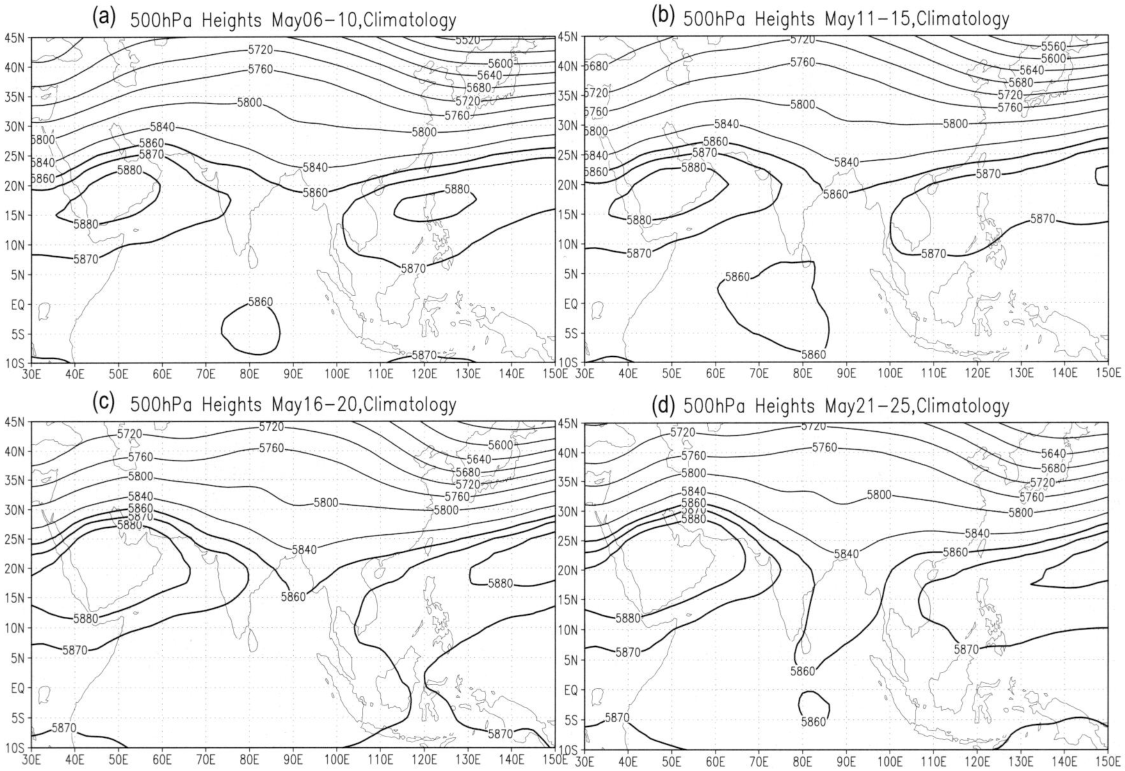

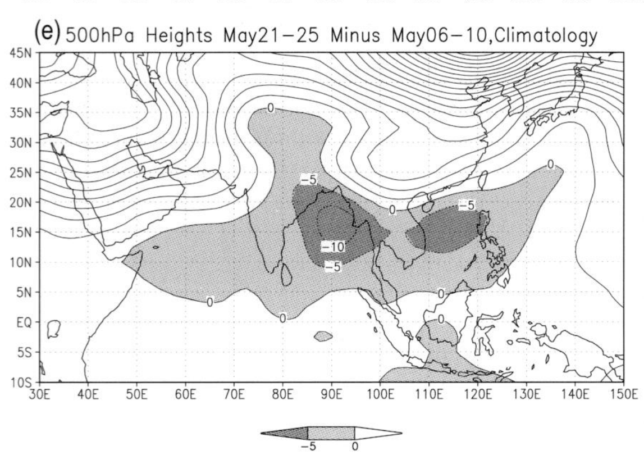

Figure 6. Same as Figure 4, but for geopotential height distributions at 500 hPa. Unit: Geopotential meters (gpm). The areas of high pressure center are shaded (Ding and Sun 2001).

South-Asian anticyclone concentrated over the Indo-China Peninsula region. The strong northeasterly outflow emanating from the SCS after the onset reflects the pronounced strengthening of the upper-level easterlies near 20°N. It is also due to the rapid development and northward advance of the upper-level anticyclone (see also Figure 7). The analyses of rainfall, wind and OLR fields by Matsumoto (1997) indicated that the summer rainy season over the inland Indo-China Peninsula begins from late April under the effect of the mid-latitude westerlies. After that, the summer monsoon circulation covers both the Indo-China Peninsula and the central SCS in mid-May when vigorous convective activity over the northern SCS begins to enhance abruptly. Therefore, he also believed that mid-May is an important turning point for the onset of the summer monsoon over Indo-China Peninsula and the SCS.

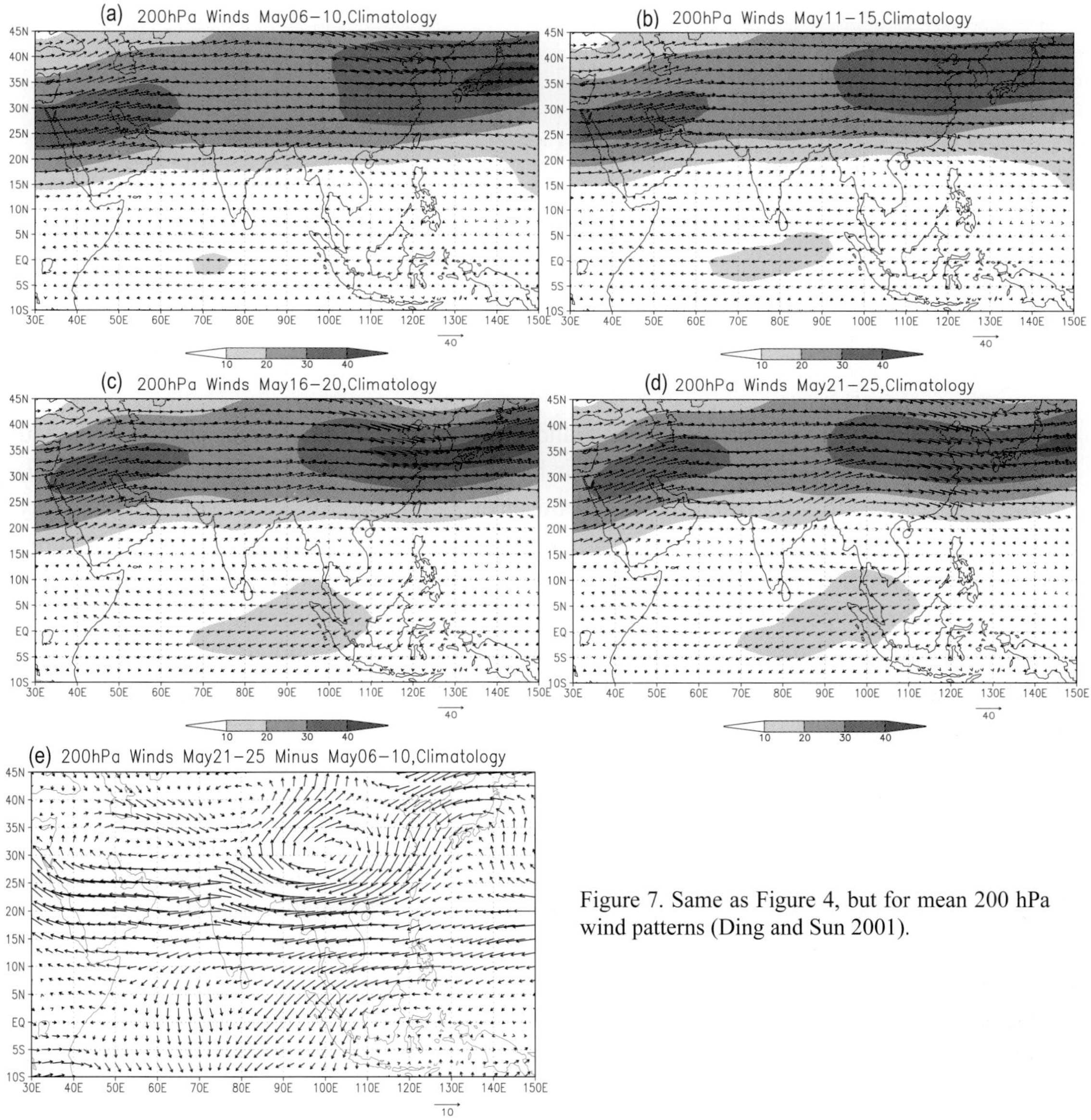

Figure 7. Same as Figure 4, but for mean 200 hPa wind patterns (Ding and Sun 2001).

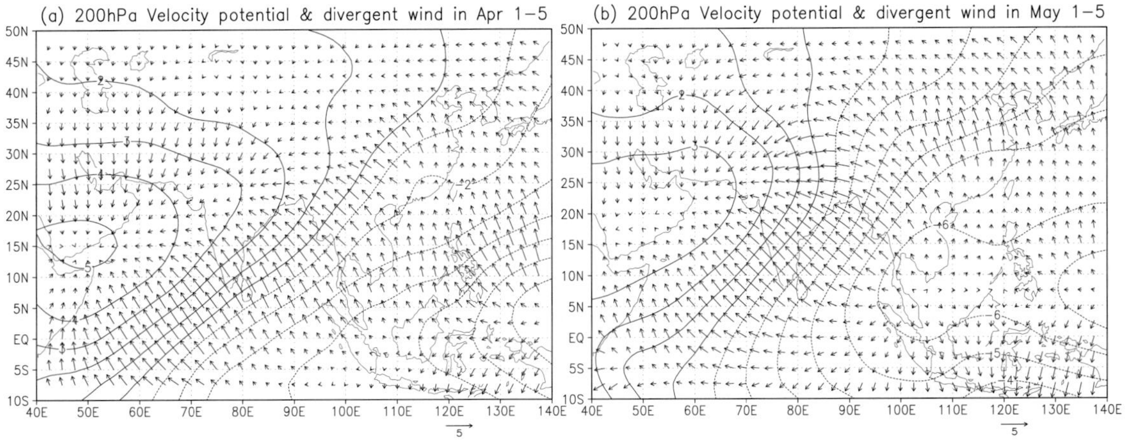

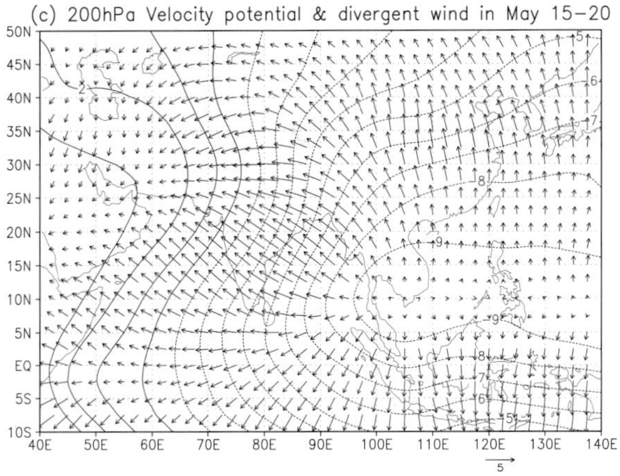

Figure 8. Pentad mean 200 hPa velocity potential (10^6 s^{-1}) and divergent wind (m s^{-1}) for (a) April 1-5, (b) May 1-5 and (c) May 15-20 (Zhang *et al.* 2004).

The changes of monsoonal convection and rainfall are consistent with changes of wind and circulation features. Based on the analysis by Zhang *et al.* (2004) in this regard (Figure 9), on the first pentad of May when the monsoon onset is confined to south of 12.5°N, heavy rainfall in excess of 10mm day^{-1} occur over the southwestern part of the Indo-China Peninsula and South China. This implies that the rainfall belt is of mid-latitude origin, or at least of the hybrid type of tropical and frontal rainfall (left column in Figure 9). In the next two pentads(see also Figure 5(a) and (b)), as the tropical monsoon advances over the whole of the southern Indo-China Peninsula, heavy rainfall regions correspondingly expand and persist. Note that at this time there is no precipitation observed in the central and northern SCS. An abrupt change in precipitation patterns occurs in the pentad of May 16-20, with extensive areas of precipitation rapidly covering the whole SCS, indicating the onset of the SCS summer monsoon. During several pentads after the onset, significant precipitation over the SCS region is maintained. As the monsoonal airflow intensifies, the precipitation in South China becomes heavier. It is interesting that two significant precipitation maxima greater than 10 mm day^{-1} are found over South China and the Indo-China Peninsula, respectively. Notice that these two maxima are connected with each other to form an elongated northeast-southwest oriented rain belt. At this time, the precipitation maximum in South China is likely produced due to the convergence of cold air from mid-latitudes in East Asia

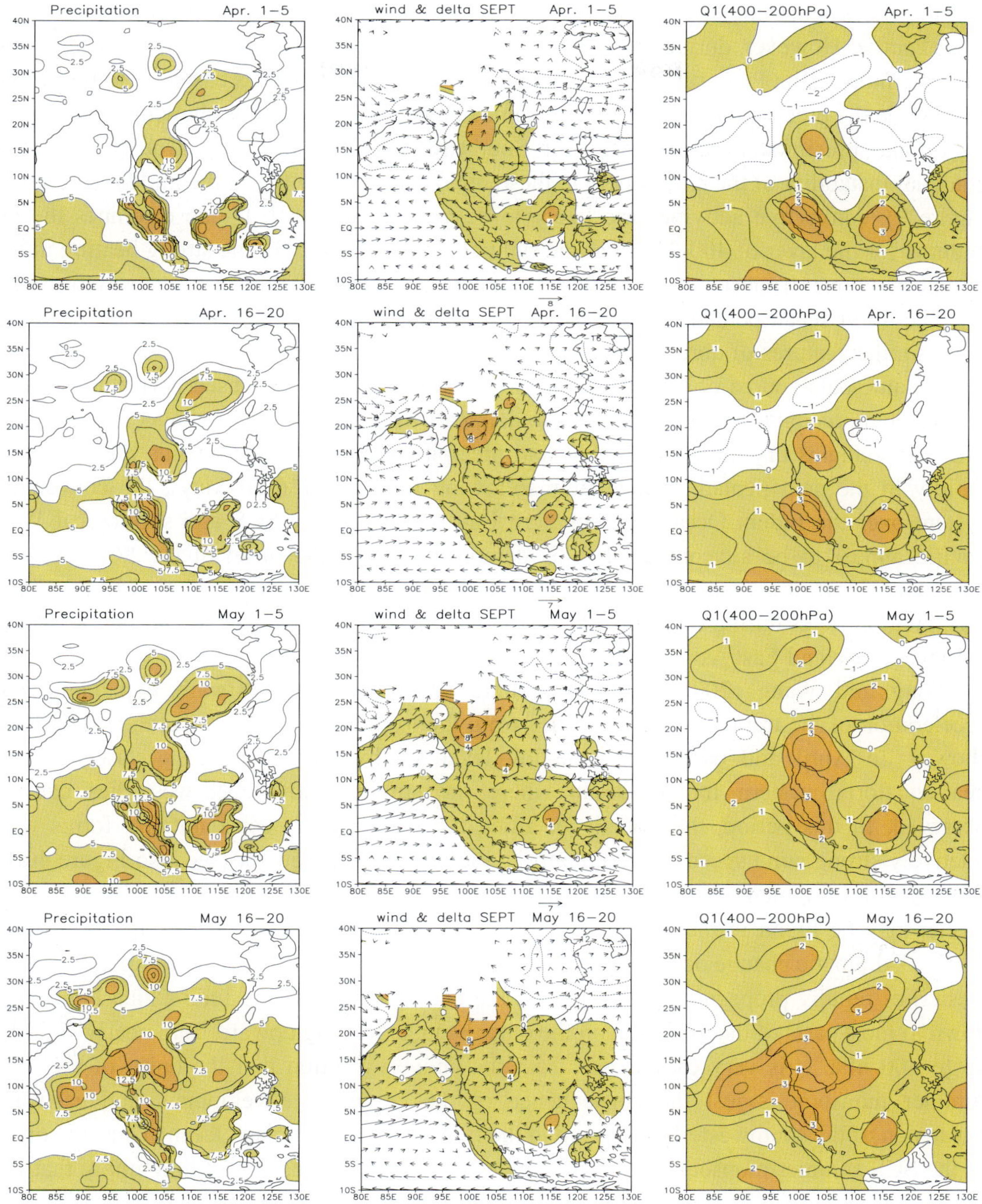

Figure 9. Evolution of precipitation rate (mm day^{-1}) (left clumn), 1000-850 hPa mean wind (arrows, m s^{-1}) and $\Delta\theta_{se}$ (θ_{se} (1000-700 hPa)$-\theta_{se}$ (600-300 hPa) contour, °C) (middle column), and mean Q_1 (°C day^{-1}) over 400-200 hPa (right column) for May 1-5 (top) to May 16-20 (bottom). Light and dark shadings in the left column denote the values $>$5 mm day^{-1} and $>$10mm day^{-1}, that in middle column $>$0°C day^{-1} and $>$2°C day^{-1}, respectively. The distribution over regions higher than 2000 m is excluded in the middle column. NCEP reanalysis data set for 1979-1995 is used in this analysis and Fig. 10 (Zhang *et al*. 2004).

and the sub-tropical moist southwesterly flow to the west of the sub-tropical high over the West Pacific. However, during the pentad of the onset (May 16-20), the tropical precipitation over the Indo-China Peninsula is greatly enhanced and rapidly extends eastward to the whole SCS while the precipitation rate in South China considerably decreases, showing that tropical monsoonal precipitation dominates over Southeast Asia and the SCS. Due to the great potential instability of the atmosphere in these regions (positive $\Delta\theta_{se}$, in the middle column of Figure 9), convective activities are likely to be greatly enhanced. If one examines the Q_1 patterns for the layer of 400-200 hPa (the right column of Figure 9), the strong heating in the middle and upper troposphere is found over an extensive area of Southeast Asia and the SCS.

The 40-yr mean pentad precipitation maps obtained by Fong and Wang (2001) showed quite similar features of abrupt change around the pentad of May 16-20, possibly due to use of the same data source of NCEP Reanalysis. Lau and Yang (1997) used 8-yr pentad mean GPI (global precipitation index) patterns to delineate the precipitation or convection changes. The SCS remains dry until 11-15 May, and precipitation and convection fills the SCS very rapidly in the next pentad. Immediately following the onset (16-20 May), the convection system over the Indian Ocean, the southern Bay of Bengal, the SCS and subtropical West Pacific appear to be all connected. This demonstrates that the onset of the SCS summer monsoon is part of a chain of events of abrupt change in large-scale circulation features and precipitation fields during the seasonal transition from spring to summer in the Bay of Bengal, Southeast Asia and East Asia.

Fong and Wang (2001) clearly illustrate the process of eastward expanding and acceleration of the tropical southwesterlies from the tropical East Indian Ocean (Figures not shown). In April and early May, the low-level southwesterlies develop slowly in the near-equatorial eastern Indian Ocean where the strongest low-level westerlies can be observed. In this period, the Somali jet along the East African coast initially forms with relatively strong wind speed observed only to the south of the equator (see also Figure 4(a)). The early development of the near-equatorial strong westerlies in the equatorial East Indian Ocean may be found in the longitudinal range of 80°E - 90°E (Matsumoto and Murakami 2002). A dramatic change occurs from the 27th to the 28th pentad (Fong and Wang 2001), with westerlies rapidly propagating eastward, on the one hand, across the Indo-China Peninsula, down to most parts of the SCS, and on the other hand, extending upward from nearly 700 hPa up to 400 hPa. It is very easy to see from the above evolution that the onset of the SCS summer monsoon is just the consequence of the explosive development and eastward expansion of the southwesterly monsoon in the tropical Eastern Indian Ocean. Thus, the onset of the SCS summer monsoon is also of explosive nature. The above fact may imply that the summer monsoon in the tropical East Indian Ocean, Indo-China Peninsula and the SCS is of the same summer monsoon system, with a single wind maximum steadily situated in the lower- troposphere over the tropical East Indian Ocean. In contrast, the wind maximum of the Indian southwest monsoon has not been established in this time period. Therefore, the Indian southwest monsoon is in a sense different from the SCS summer monsoon. As indicated previously, an interesting problem associated with the eastward expansion of the southwest monsoon is the relative importance of the subtropical high in the onset of the SCS summer monsoon. According to Fong and Wang (2001), as the southwest monsoon rapidly propagates eastward and extends upward, the main body of the subtropical high still dominates the middle and upper troposphere, with the southeasterlies prevailing above 700 hPa over much of the SCS region. Therefore, in this process, its whole entity mainly shows a weakening intensity, rather than the large-scale retreat out of the SCS region. That implies that the eastward retreat of the subtropical high might be a passive process driven by a feedback process produced by active convection.

Based on the above analysis, the chain of significant events during the onset of the SCS summer monsoon may be identified below:

- the development of a cross-equatorial current in the equatorial East Indian Ocean (80°E - 90°E) and off the Somali coast, and the rapid seasonal enhancement of heat sources over the Indo-China Peninsula, South China, Tibetan Plateau, neighboring areas;
- the acceleration of low-level westerly wind in the tropical eastern Indian ocean;
- the development of a monsoon depression or cyclonic circulation and the breaking of the continuous subtropical high belt around the Bay of Bengal;
- the eastward expansion of tropical southwest monsoon from the tropical East Indian Ocean;
- the arrival of the rainy season in the regions of Bay of Bengal and Indo-China Peninsula with involvement of impacts from mid-latitudes;
- further eastward expansion of the southwesterly monsoon into the SCS region;
- the significant weakening and eastward retreat of the main body of the subtropical high, and eventual onset of the SCS summer monsoon with convective clouds, rainfall, low-level southwesterly wind and upper-level northeasterly wind suddenly developing in this region.

Based on the changes in wind and OLR patterns, Zhang *et al.* (2004) have further studied the advance of the Asian summer monsoon onset in the tropical region south of 20°N. They define the onset of the Asian summer monsoon with the following two conditions: (1) Establishment of a zonal vertical shear, with low-level (*e.g.* 850 hPa) westerlies and upper-level (*e.g.* 200 hPa) easterlies and (2) the pentad mean OLR less than 240 W m^{-2}. Figure 10 presents their results. During the first pentad of May (Figure 10(a)), the summer monsoon is established only over Sumatra. In the next two pentads (Figures 10(b), (c)), the tropical monsoon advances up to the land bridge, first establishing itself over the southwestern Indo-China Peninsula and then expanding to the entire southern peninsula. During the pentad of May 16-20 (Figure 10(d)), the build up of the summer monsoon is observed over the central Indo-China Peninsula. At the same time, the onset location extends into the central and southern SCS, accompanied by a rainfall rate of >5 mm day^{-1} over the entire SCS. In the next pentad, onset expands quickly and almost covers the entire SCS (Figure 10(f)). On the other hand, the Asian summer monsoon also advances northeastward to the Indian monsoon region from the near-equatorial East Indian Ocean and the Indo-China Peninsula starting from mid-May (Figure 10(d)). Earliest onset of the Asian summer monsoon in this region may be observed over the southern tip of the Indian subcontinent. In early June (Figures 10(g), (f)), the Asian summer rapidly advances northwestward, arriving in

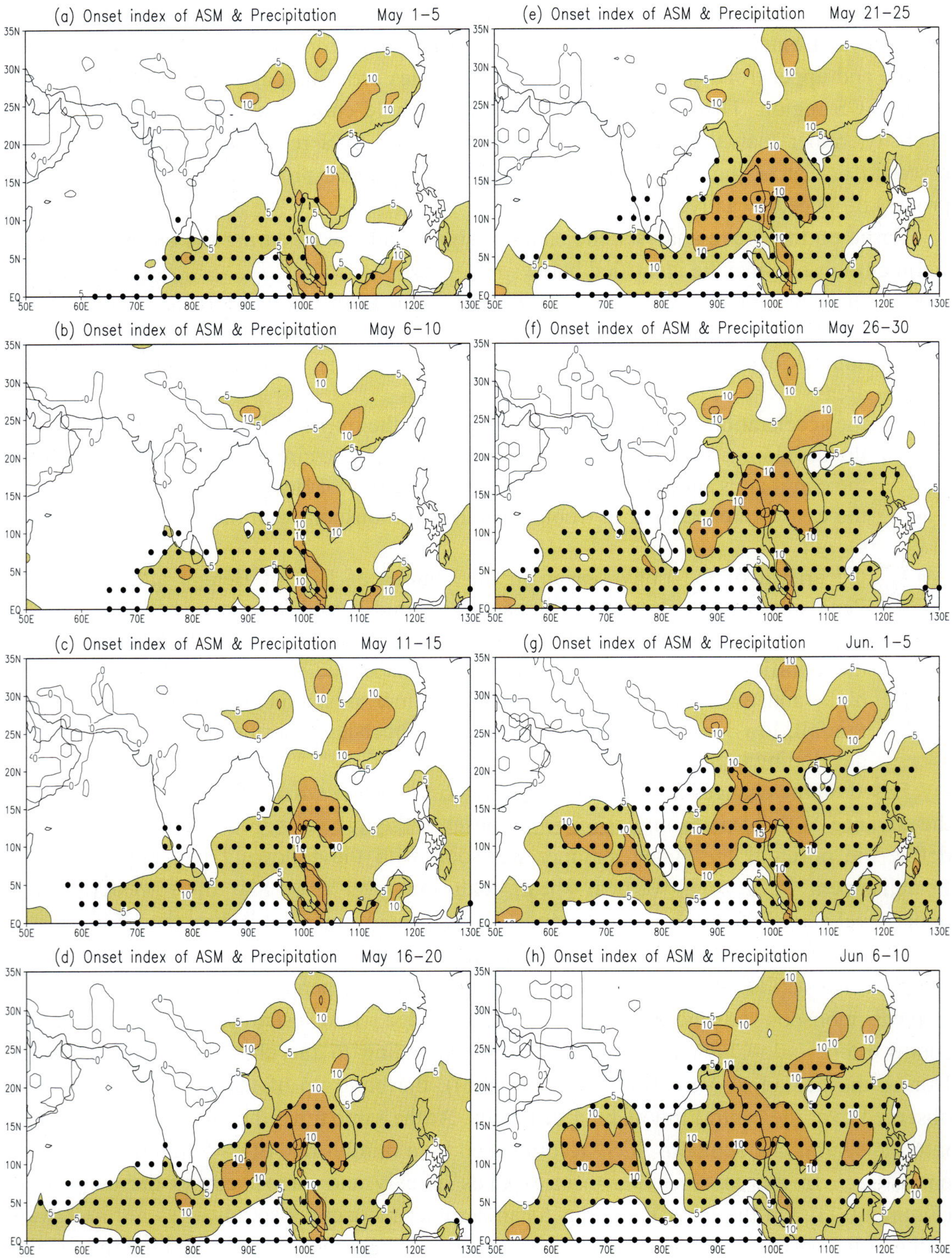

Figure 10. Climatological pentad-averaged precipitation rates (mm day^{-1}) for May1-5(a) to June 6-11(h) in sequence. Light and dark shadings indicate precipitation regions greater than 5mm day^{-1} and 10mm day^{-1}, respectively. The black dots represent the location of onset of the summer monsoon (Zhang *et al*. 2002).

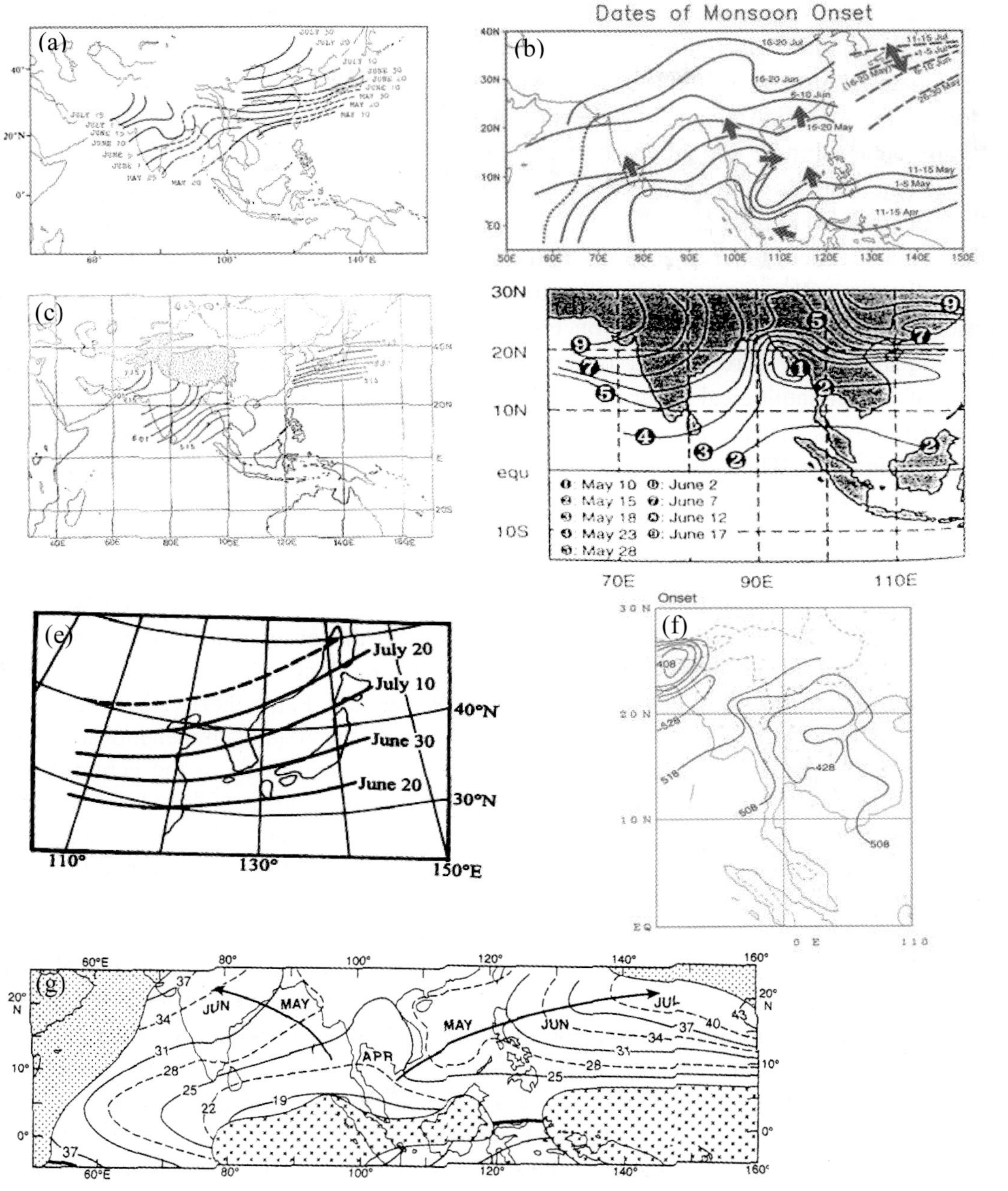

Figure 11. The climatological dates of the onset of the Asian summer monsoon defined by various investigators: (a) Tao and Chen (1987), (b) Lau and Yang (1997), (c) Tanaka (1992), (d) Webster *et al.* (1998), (e) Kim (1992), (f) Matsumato (1997) and (g) Wang and Fan (1999) ((a)-(c)taken from Wang and Ho 2002).

the central Indian subcontinent. Meanwhile, the onset over the Arabian Sea and the western coast of the Indian subcontinents is observed, due mainly to the enhancement of the cross-equatorial airflow off the Somali coast and the development of the onset vortex in the central and northern Arabian Sea (Krishnamurti *et al.* 1981; Ding 1981). This date is generally believed to be normal onset dates for the Indian summer monsoon. So, the onset of the East Asian and Southeast summer monsoon, and the Southern Asian summer monsoon is closely interrelated in the context of the Asian summer monsoon. However, the earliest onset of the Asian summer monsoon occurs over the Indo-China Peninsula and the SCS.

Finally, we will summarize the climatological dates of the onset of the Asian summer monsoon in different regions based on Figure 10 and the results of various investigators, which are shown in Figure 11. It is to some extent difficult to obtain a unified and consistent picture of the climatological onset dates of the Asian summer monsoon in different regions due to differences in data, monsoon indices and definitions of the monsoon onset used in these investigations. Thus, the method to summarize these results here is to characterize the common features identified by various authors, with the major discrepancies indicated in an appropriate manner. The whole onset process of the Asian summer monsoon may be divided into four stages:

(1) Stage 1. The earliest onset is often observed in the central Indo-China Peninsula late in April and early in May, but in some cases, the onset may first begin in the southern part or the western part of the Indo-China Peninsula (Lau and Yang 1997; Matsumoto 1997; Webster *et al.* 1998; Wang and Fan 1999). However, some investigators have argued that the earliest onset of the Asian summer monsoon may occur in the Bay of Bengal, Myanma (Wu and Zhang 1998) or the central South China Sea (Chen *et al.* 2000). Therefore, the earliest onset of the Asian summer monsoon may occur over an extensive area around the Indo-China Peninsula, with the central Indo-China Peninsula as the most frequent region. In addition, the precipitation patterns at this stage are quite well modulated by mid-latitude effects.

(2) Stage 2. This stage is characterized by the areal extension of the summer monsoon, advancing northward up to the Bay of Bengal and eastward down to the SCS. This event usually is very abrupt or rapid, with most frequent occurrence during the period from mid to late May. The onset of the summer monsoon over the SCS just occurs at this stage. Numerous investigators have illustrated the major features of flow and rainfall patterns at this stage (Lin 1987; Tanaka 1992, 1994; Lau and Yang 1997; Chan *et al.* 2000). This stage is in good correspondence to the first stage of the three distinct stages of onset of the Asian summer monsoon (Wu and Wang 2001). They charctierize this stage with the following abrupt changes of circulation and rainfall: the large increase of rain rate, height decrease at 500 hPa due to a sudden northeastward retreat of the subtropical high over West Pacific, the development of cyclonic wind change at 850 hPa over the SCS, increase of the zonal winds at 850 hPa along the zonal bands of 5°N - 10°N and from the Bay of Bengal to the SCS, the anticyclonic wind development over the Tibetan Plateau and increase of the upper-level divergence and the sudden growth of deep convection. During this stage (the first dekad of June), the presummer rainy season in South China (also including

the Meiyu in Taiwan) reaches its peak.

(3) Stage 3. This stage is well known for the onset of the Indian summer monsoon and the arrival of the East-Asian rainy season such as the Meiyu over the Yangtze River Basin and the Baiu season in Japan. These significant events usually occur during the period from the second dekad to the first dekad of June (Ninomiya and Murakami 1987; Tanaka 1994; Ding 1994). The earlier onsets of Changma in southern Korean Peninsula may be observed in the third dekad of June for some years (Oh *et al.* 1997).

(4) Stage 4. The summer monsoon at this stage can advance up to North China, the Korean Peninsula and even North Japan in the first or second dekads of July (Ding 1994; Tanaka 1994; Oh *et al.* 1997). Starting from the second half of July, the rainy season in Northeast China begins with the summer monsoon prevailing in this region. A detailed description of the advance and retreat of the summer monsoon in East Asia will be given in Section 4.

3. The Onset of the East Asian Summer Monsoon during the 1998 SCSMEX

The onset process of the SCS summer monsoon in 1998 began with development of a twin cyclone around the oceanic region near Sri Lanka (Lau *et al.* 2000; Ding and Liu 2001). The pentad mean streamline patterns at 850 hPa over the large-scale observation region of the SCSMEX in early May suggested that a depression near Sri Lanka first began to develop and at the same time another cyclone occurred on the other side of the equator in this region, thus forming a twin cyclone straddling the equator. In between, the equatorial westerly wind appeared over the equatorial East Indian Ocean in the second pentad of May and accelerated (not shown). The SCS region was controlled by the easterlies to the south of the subtropical high over the West Pacific. In the third pentad of May (Figure 12(a)), the significant change in circulation features upstream of the SCS was characterized with eastward and northward movement of the Sri Lanka depression toward the Bay of Bengal and further acceleration of the equatorial Indian westerlies, but their effect did not reach the Indo-China Peninsula at this time. Over most of the SCS and northern part of the Indo-China Peninsula, the easterlies at low-level and middle-levels prevailed due to dominance of the subtropical high in those regions. The westerlies were only observed in South China and the northern Indo-China Peninsula. This kind of low-level westerlies is defined as the subtropical monsoonal westerlies due to their origin to the north of the subtropical high. Therefore, during this pentad the summer monsoon of tropical origin did not set in over the SCS and the Indo-China Peninsula. Note that the subtropical westerlies were separated from the equatorial East Indian westerlies with a region of easterlies over the Malaysian Peninsula and eastern Bay of Bengal. In the fourth pentad, the tropical southwesterly wind to the west of the subtropical high and equatorial East Indian westerly winds started to accelerate and dominated over the northern part of the SCS (Figure 12(b)). The main body of the subtropical high over the West Pacific that had controlled over the SCS moved northeastward rapidly. It can be found that this process of eastward propagation of the tropical monsoon occurred nearly at the same time as the eastward withdrawal of the subtropical high and significant southward intrusion of cold air from mid-latitudes into South China and the northern SCS (Fig. 12(c)). The low-level northerlies over the China mainland rapidly developed and intruded into the northern SCS accompanied by the southward passage of a cold front and temperature drop (Chan *et al.* 2000).

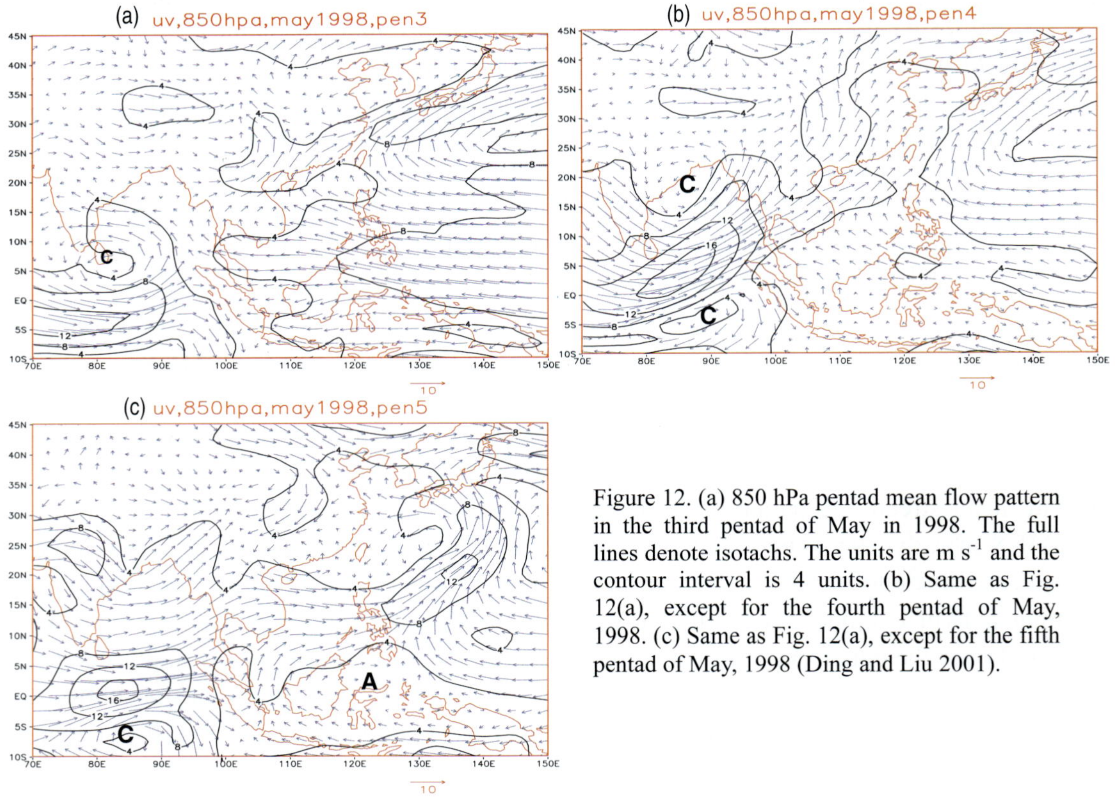

Figure 12. (a) 850 hPa pentad mean flow pattern in the third pentad of May in 1998. The full lines denote isotachs. The units are m s^{-1} and the contour interval is 4 units. (b) Same as Fig. 12(a), except for the fourth pentad of May, 1998. (c) Same as Fig. 12(a), except for the fifth pentad of May, 1998 (Ding and Liu 2001).

Figure 13 presents the daily area-averaged precipitation time series in the northern and central SCS. It can be seen that the significant increase of precipitation occurred after May 15, 1998 with daily precipitation amount exceeding 6 mm day^{-1}. Therefore, by the definition discussed in Section 2, the summer monsoon may be considered to set in initially over the northern SCS (e.g. Lau *et al.* 2000; Chan *et al.* 2000). Note that the considerable increase of rainfall amount in the central SCS occurred a little later, around May 17, 1998, indicating a trend of southward propagation of the major rainfall process. The onset of the summer monsoon over the Indo-China Peninsula was also observed in this pentad. This event of the near-synchronous onset of the northern SCS and the Indo-China Peninsula is quite different from the climatological condition, as described in Section 2.

It is remarkable from the above process that the early development of the summer southwesterly monsoon in the northern SCS was a large-scale event rather than the localized phenomena, with the tropical southwesterly air current dominating over the extensive region from the Bay of Bengal *via* the Indo-China Peninsula to the coastal areas in southeastern

China, although the effect from the mid-latitudes also may be obviously perceived. The twin-cyclone event at this time was still evident, with another weak cyclonic vortex being maintained near the equatorial East Indian Ocean. In the fifth pentad of May, the subtropical high fully retreated out of the SCS and the strong tropical westerly air current at low-level dominated over the Bay of Bengal, Indo-China Peninsula and the whole SCS, symbolizing a full onset of the summer monsoon over the whole SCS (Figure 12(c)). A shear line in the low-level wind field or the monsoon trough, in which are embedded some vortices, became very evident. In the sixth pentad of May, the southwesterly wind continued to develop and prevailed over the middle and northern parts of the SCS (not shown). Under such condition, the monsoon trough associated with the low-level wind shear line moved to coastal areas of South China and the concentrating intense rainfalls during the pre-summer season were observed there.

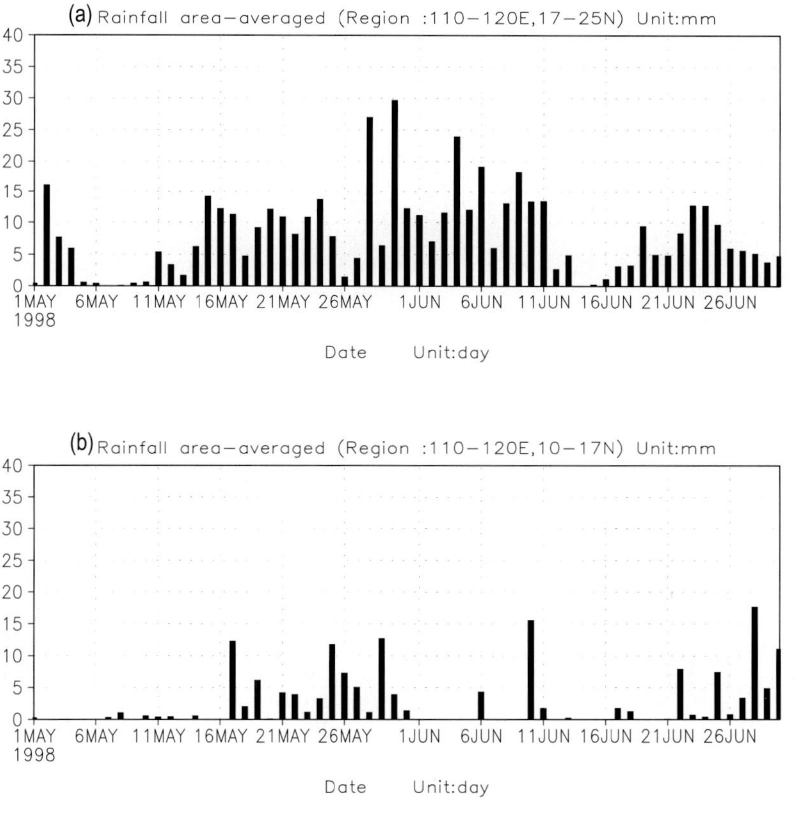

Figure 13. Area-averaged daily precipitation time series for (a) the northern SCS (110°E - 120°E, 17°N - 25°N) from May 1-June 30, 1998 and (b) same as Fig. 13(a), but for the central SCS (110°E - 120°E, 10°N - 17°N). The unit is mm day^{-1} (Ding and Liu 2001).

Corresponding to the 850 hPa circulation situation, the pentad-to-pentad evolution of the wind patterns at 200 hPa also showed a significant change before and after the onset of the SCS summer monsoon. In the third pentad of May (Figure 14(a)), the South Asian high formed and gradually dominated over the Indo-China Peninsula. The center of the high was located over the region (~10°N) to the south of the Bay of Bengal and a deep trough was situated to the east of the high northwesterly air currents to the east of the anticyclone were prominently observed over the northern part of the SCS. The upper trough was closely related to an intrusion of cold air into the northern SCS. In the fourth pentad of May (Figure 14(b)), the South Asian high moved eastward slightly and the deep trough downstream moved out of the SCS. The whole SCS was completely controlled by the northwesterly and northeasterly air currents to the east of the anticyclone, indicating that the SCS was mainly under the influence of the tropical upper-level circulation. This matched the timing of the monsoon onset over the northern part of the SCS. In the fifth pentad of May, with enhancement and northward movement of the center of the anticyclone over the region from the Indo-China Peninsula to the SCS, the northeasterly or northerly air currents to the east of the South Asian high entirely prevailed over the SCS. Corresponding to the low-level southwesterly air current, this implied the full onset of the summer monsoon over the whole SCS (Figure 14c).

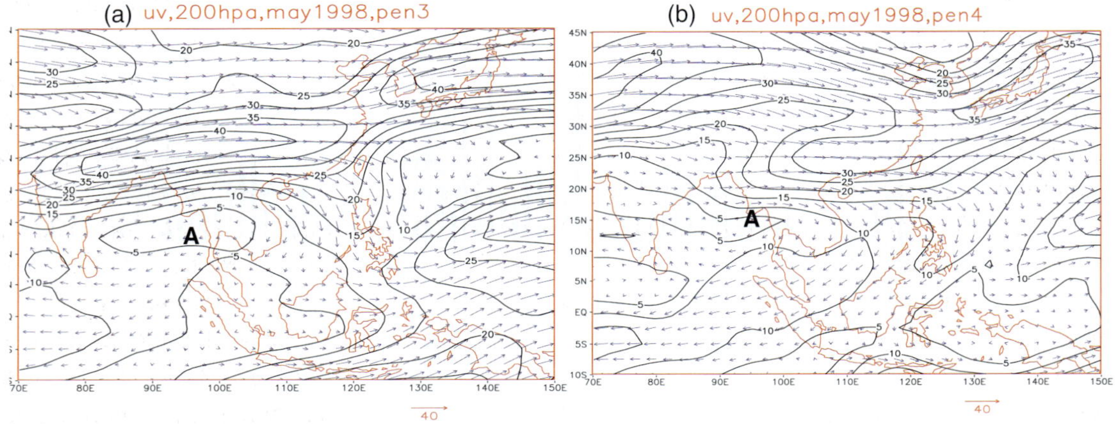

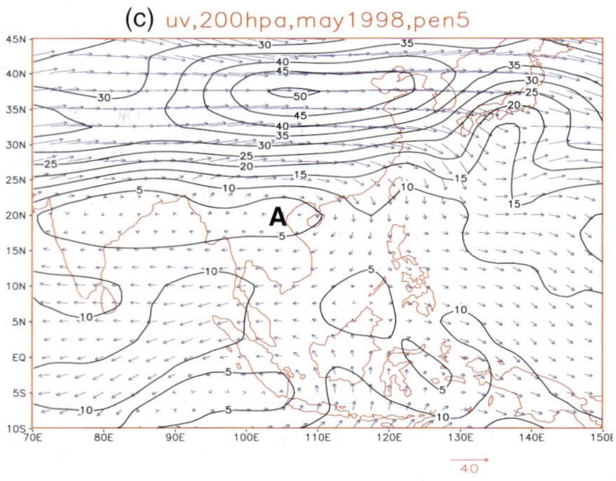

Figure 14. Same as Figure 12, except for 200 hPa flow patterns in the third (a), fourth (b) and fifth (c) pentad of May in 1998. The unit is m s^{-1} and the contour interval is 5 units (Ding and Liu 2001).

Based on the pentad mean OLR fields, no convection was observed over the equatorial region of the Indian Ocean in the first pentad of May (not shown), with only the Maritime Continent covered with local convection. The SCS was also cloud-free (high OLR values). In the second pentad of May (not shown), an obvious center of low OLR value appeared over the equatorial Indian Ocean region to the south of Sri Lanka, indicating the occurrence of strong convection, in correspondence to the development of a monsoonal depression there, as described above. In the third pentad of May (Figure 15(a)), the zone of OLR low value over the equatorial East Indian Ocean began to extend gradually eastward, and the strong convection also grew rapidly over the Maritime Continent. Subsequently, a continuous convective zone near the equator formed over the region from the equatorial Indian Ocean to the Maritime Continent. Over the most of the Indo-China Peninsula and the SCS, there was no significant convective activity. In the fourth pentad of May (Figure 15(b)), the zone of low OLR value coming from the equatorial East Indian Ocean merged with the low OLR band coming from the China mainland, and further strengthened in the northern part of the SCS, thus leading to rapid growth of the convective activity over the northern part of the SCS. One may see that a wide band of low OLR value reflecting vigorous convection activities was concurrently located in the northern SCS and the western Indo-China Peninsula and the Bay of Bengal. However, the northeastern portion of the convective cloud band was connected with the frontal cloud band of mid-latitude origin. This fact might imply that the earliest onset of the summer monsoon over the northern SCS is greatly affected by the activity of cold air from mid-latitudes and of the hybrid nature of air of mid-latitude and tropical origins. Figure 16(a) may clearly show the processes of southward intrusion of a cold wave before the onset of the SCS summer monsoon. It can be seen that a strong cold upper trough dipped southward during the second pentad. In the third pentad, the cold air intruded further southward. This was a process of strong cold wave in East Asia. Around May 16, the leading front of cold air reached the northern SCS, in good agreement with the onset date of the summer monsoon in this region. Therefore, the cold air could play an important role in triggering the onset. Another southern intrusion of cold air occurring on May 26 is also evident. Each southward intrusion of cold waves was concurrently accompanied by the similar southern extension of the upper jet stream in westerlies from mid-latitudes (Figure 16(b)). Chan *et al.* (2000) also emphasized the importance of a band of relatively cool air originating north of 35°N which propagated southward and reached South China coast by the onset day. In this pentad, no convection was observed over the central and southern SCS. In the fifth pentad of May, the center of high OLR value over the SCS gradually retreated northeastward. The vigorous convection associated with the area of low OLR value was already established in most of the SCS region. This was also a sign of the full onset of the SCS monsoon (Figure 15(c)). Therefore, the OLR field may be used as an indicator to characterize the process of the onset of the SCS summer monsoon as previously indicated (Xie *et al.* 1996).

The above analysis has shown that the onset of the Asian summer monsoon for this year occurred earliest over the SCS, with two phases identified. The first onset of the summer monsoon over the northern part of the SCS occurring in the fourth pentad was of hybrid nature, which was greatly affected by the interaction between the weather systems at middle latitudes and the tropical monsoon air current. Johnson and Ciesielski (2001) have recently studied the characteristics of the onset of the summer monsoon over the northern SCS utilizing observation from the May - June 1998 SCSMEX. They have shown that the onset occurred in the northern SCS in mid-May with a rapid increase in deep convection over a week to ten-day period, thus producing the significant vertical transports of heat and moisture. During the undisturbed pre-onset period (May 6-12) (Figure 17), there is upper-level convergence, low-level divergence and deep subsidence, consistent with the mostly clear skies and high values of OLR. Q_1 (the apparent heat source) is negative at all levels with values in excess of -2.5K day^{-1} in the upper troposphere. The profiles of divergence and vertical motion during the onset (May 16-22) and post-onset periods (June 3-9) are dramatically different from the pre-onset period. Low-level convergence, upper-level divergence and strong upward motion occur during both periods. Deep convergence extends to 300-500 hPa, Q_1 and Q_2 (the apparent moisture sink) profiles are characteristic of deep convection during these periods, with a peak heating rate of 5 K day^{-1} located at about 400 hPa and a considerable separation of the two curves, suggesting more vigorous deep convection at that time.

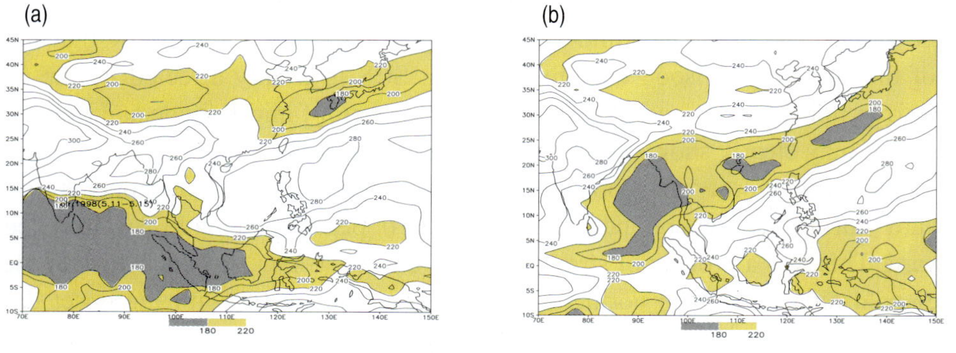

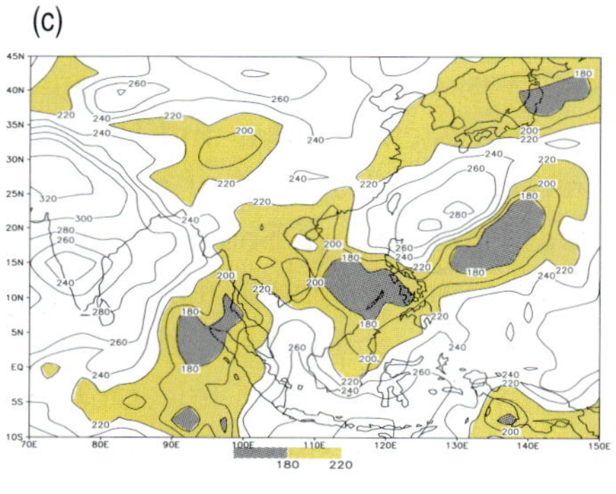

Figure 15. Same as Figure 14, except for OLR patterns in the third (a), fourth (b) and fifth pentad (c) of May in 1998. The units are W m^{-2} and the contour interval is 20 units. The areas of low OLR values below 220 W m^{-2} are shaded (Ding and Liu 2001).

The second large-scale monsoon onset in the fifth pentad of May was basically of tropical nature, being induced by rapid changes in tropical circulation features. Before the onset of the SCS monsoon, a monsoon depression over Sri Lanka and a near-equatorial low vortex over the equatorial East Indian Ocean formed as a twin cyclone. With the subtropical high retreating eastward and the depression over Sri Lanka advancing northward up to the Bay of Bengal region, the equatorial westerly wind from the Indian Ocean greatly accelerated, extended eastward and converged with the southwesterly air current to the west of the subtropical high over the SCS, thus bringing about onset of the monsoon over the SCS. Consequently, the cloud and rain areas suddenly developed, rapidly covering the whole SCS region, and began to have the transition toward the condition of the summer monsoon. The

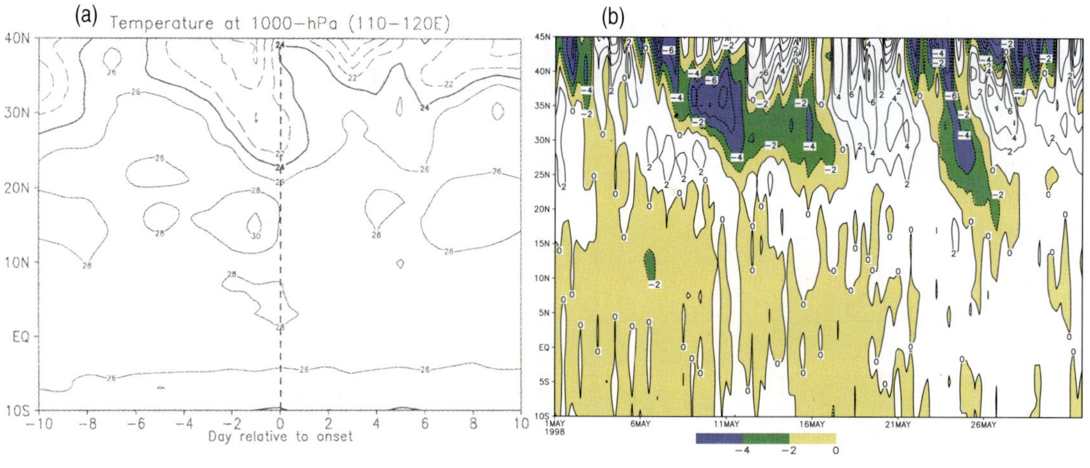

Figure 16. (a) Latitude-time cross-section of 850 hPa temperature departure averaged for longitudinal range of 110°E - 120°E for May of 1998. The departure is obtained from subtracting the monthly mean temperature of May 1998 from the daily temperature. The unit is °C (Chan *et al.* 2000). (b) Same as 16(a), but for 200 hPa U-component for May of 1998. Westerlies (solid lines) are shaded. Dashed lines represent easterlies. The unit is m s^{-1} (Ding and Liu 2001).

onset of the SCS monsoon was significantly affected by the southwesterly air current from the equatorial Indian Ocean, in particular for the second phase (the full SCS onset), which was closely correlated with development and intensification of the Somali jet. Thus, onset of the 1998 SCS monsoon occurred under the combined effect of large-scale circulation features characterized by the eastward retreat out of the SCS of the subtropical high over West Pacific, and large-scale acceleration and eastward extension of the low level equatorial westerlies from the Indian Ocean. The latter may in turn be closely associated with the land (the Asian continent)-ocean temperature gradient or thermal contrast, especially the elevated heating effect of the Tibetan Plateau (Li and Yanai 1997; Ueda and Yasunari 1998; Ding and Sun 2001) that will been elucidated from study of the climatological aspect in Section 6.

It should be pointed out that the process of the two-stage onset of the SCS summer monsoon is not only seen in 1998, but also can be found in other years (Lau *et al.* 2000; Zhang *et al.* 2000; Ding and Sun 2001). In fact, based on the climatological study of dates of the onset of the SCS summer monsoon for 1953-1999 (Zhang *et al.* 2002), 13 years out of a total of 47 years may fall into the type of the two stage onset (1957, 1968, 1969, 1972, 1973, 1981, 1982, 1983, 1984, 1992, 1993, 1997, 1998). This accounts for 27.6% of the SCS summer monsoon onsets. Sometimes the first onset of the SCS summer monsoon initially occurs in the southern SCS and then gradually extends northward. This type of the first onset in the southern SCS accounts for 15% (7 years). Although this onset also assumes a two-stage process, it is rather gradual, without a distinct transition from the first stage to the second stage. The type of the full onset of the SCS summer monsoon at one time accounts for 57.4% (27 years). Therefore, the process of the first onset of the summer monsoon over the northern SCS is to some extent representative of a major monsoon onset process in the SCS, even from the climatological viewpoint. Based on their study and other studies (Yan 1997), the

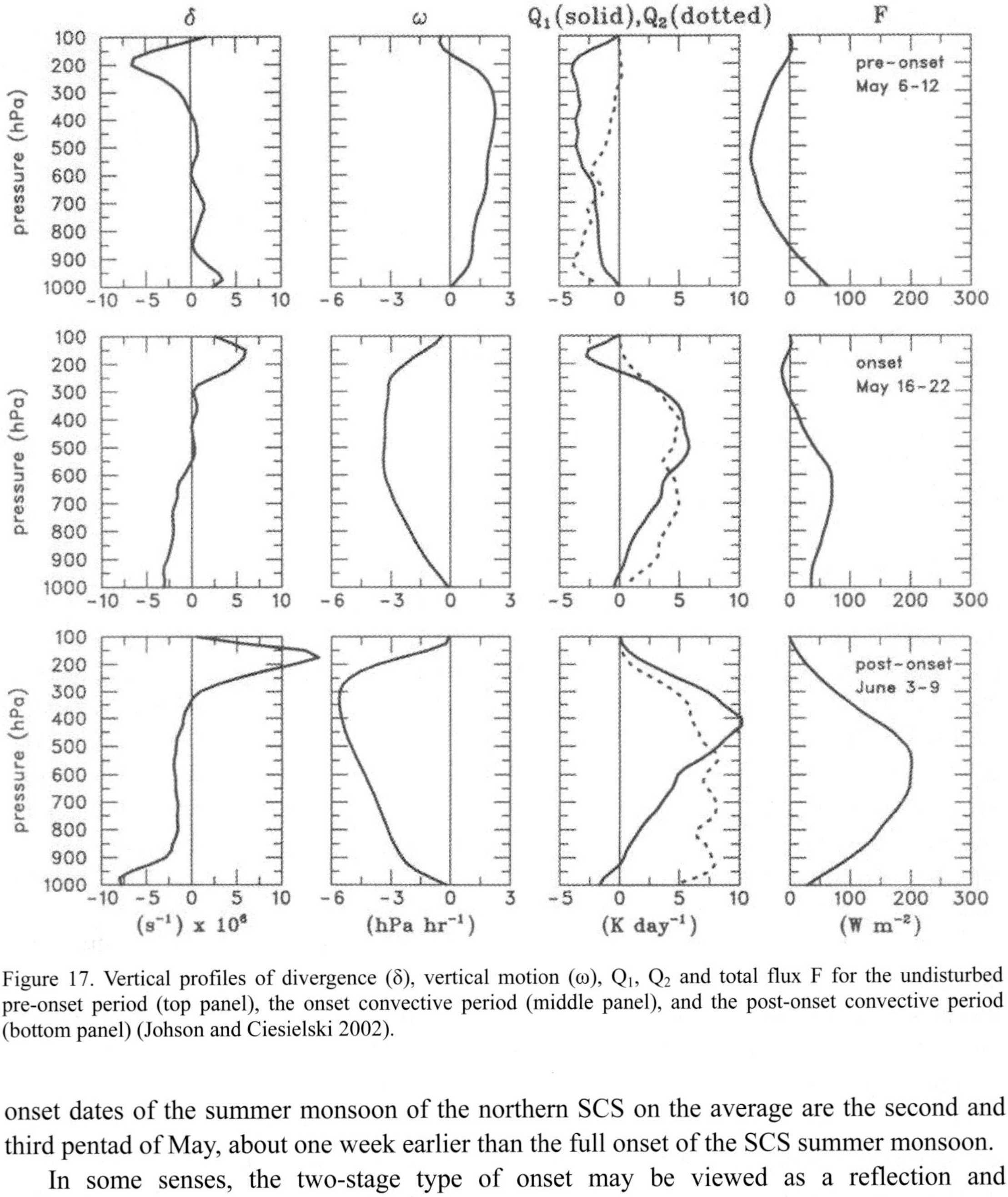

Figure 17. Vertical profiles of divergence (δ), vertical motion (ω), Q_1, Q_2 and total flux F for the undisturbed pre-onset period (top panel), the onset convective period (middle panel), and the post-onset convective period (bottom panel) (Johson and Ciesielski 2002).

onset dates of the summer monsoon of the northern SCS on the average are the second and third pentad of May, about one week earlier than the full onset of the SCS summer monsoon.

In some senses, the two-stage type of onset may be viewed as a reflection and demonstration of the evolutional process of seasonal adjustment of the circulation features over East Asia, tropical Indian Ocean and West Pacific Ocean. The subtropical high first retreats in the northern SCS, with the main body still dominating the central and southern SCS, thus providing a favorable pre-condition of the development of low-level westerlies in northern SCS region. If a twin cyclone develops in the near-equatorial East Indian Ocean (near Sri Lanka), the equatorial westerlies are accelerated and extend northeastward into the

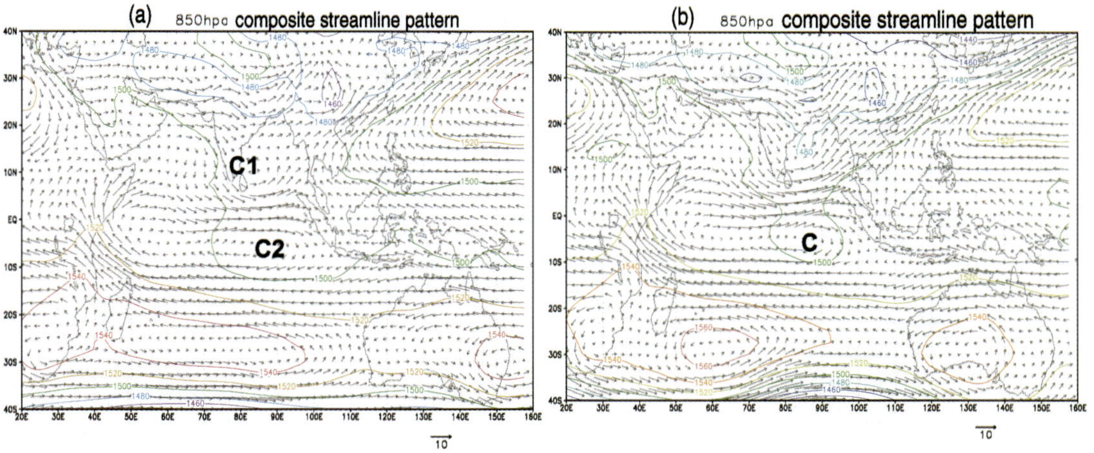

Figure 18. The composite 850 hPa streamline pattern averaged for 13 cases for one pentad before the onset of the summer monsoon over the northern SCS (a) and during the onset of the summer monsoon over the northern SCS (b). 13 cases used for average include the following years: 1957, 1968, 1969, 1972, 1973, 1981, 1982, 1983, 1984, 1992, 1993, 1997 and 1998 (Zhang *et al.* 2000).

Indo-China Peninsula and the northern SCS, thus greatly enhancing the low-level westerlies there. Therefore, the second pre-condition of the onset of the northern SCS summer monsoon is the development of a twin cyclone over equatorial East Indian Ocean. The climatological and case study both have shown that this pre-condition generally exists (Lau *et al.* 1998, 2000; Zhang *et al.* 2002). For 13 cases of the first monsoon onset of the northern SCS out of 47 years from 1953 to 1999, the twin cyclones over the equatorial East Indian Ocean were always observed, although their developmental processes may have some differences. Figure 18 is the composite 850 hPa wind field for 13 cases of the first onset of the northern SCS summer monsoon, showing the existence of the twin cyclone (C1 and C2) straddling the equator over the tropical East Indian Ocean and significant acceleration of equatorial westerlies in this region, which can extend northeastward up to the northern SCS. The third pre-condition for the first monsoon onset of the northern SCS is the southward intrusion of strong cold air accompanied by an obvious cold front. As pointed out by Chan *et al.* (2000), this process from mid-latitudes may be a triggering mechanism for the monsoon onset in the northern SCS, especially for rapid amplification of regional precipitation and convection. This problem will be discussed in more detail in Section 6.

4. Seasonal Advance and Retreat of the East-Asian Summer Monsoon

The seasonal advance and retreat of the summer monsoon in East Asia behaves in a stepwise way, not in continuous way. When the summer monsoon advances northward, it undergoes three standing stages and two stages of abrupt northward shifts. In this process, as does the monsoonal airflow, the monsoon rain belt and its associated monsoon air mass also demonstrate a similar northward movement. These stepwise northward jumps are closely related to seasonal changes in the general circulation in East Asia, mainly the seasonal

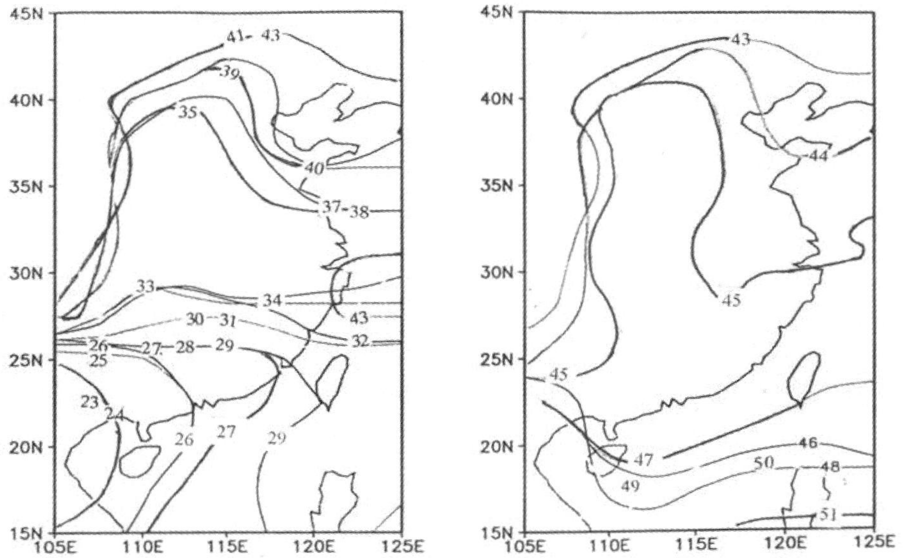

(a) 23rd - 43rd pentad (21 April - 3 August)

(b) 43rd - 51st pentad (3 August - 12 September)

Figure 19. The 40-yr mean pentad-to-pentad variations of the leading zone of the summer monsoon in East Asia from the 23rd to the 51st pentad. (a) From the 23rd to the 43rd pentad (21 April to 3 August) and (b) from the 43rd to the 51st pentad (3 August to 12 September) (Fong and Wang 2001).

evolution of the planetary frontal zone, the westerly upper-level jet stream and the subtropical high over the West Pacific. Recently, Wang and Lin (2002) have studied the large-scale onset, peak and withdrawal of the Asian monsoon rainy season, and have identified two phases in the evolution process. The first phase begins with the rainfall surges over the South China Sea in mid-May, which establishes a planetary-scale monsoon rainband extending from the South Asian marginal seas (the Arabian Sea, the Bay of Bengal, and the SCS) to the subtropical western North Pacific (WNP). The second phase of the Asian monsoon onset is characterized by the synchronized initiation of the Indian rainy season and the Meiyu/Baiu in early to mid-June. The peak rainy seasons tend to occur primarily in three stepwise phases, in late June over the Meiyu/Baiu regions, the northern Bay of Bengal and the vicinity of the Philippines; in late July over India and northern China; and in mid-August over the tropical WNP. Their results are basically in agreement with earlier ones, but with a perspective of the planetary scale Asian monsoon system. Below, we will focus our discussion on the East Asian region.

Generally, the isopleth of θ_{se}=340K at 850 hPa may be used to characterize the thermodynamic property of the air mass of the summer monsoon, and the 850 hPa wind fields are used to define the leading zone of the summer monsoonal airflow (Ding 1994). Recently, Fong and Wang (2001) used a slightly different definition to characterize the advance and retreat of the summer monsoon in East Asia. They used the southwesterly wind with velocity greater than 2.5 m s^{-1} and θ_{se}=335K at 850 hPa to represent the leading zone of the summer monsoon. Figure 19 shows the pentad-to-pentad variation of the leading zone of the summer monsoon defined by them. It can be seen that the northward shift of the leading zone of the summer monsoon begins with the onset of the South China Sea summer monsoon (26 to 30 May). These leading zones show spatial clustering in three regions during the northward movement: South China, the Yangtze River Basin and North China, reflecting the stationary feature of the leading zone of the summer monsoon. In between, the northward movement of the leading zone is very rapid, with two significant northward jumps identified. The first jump from South China to the Yangtze River Basin occurs in the 33rd to 34th pentad (10-19 June), which is consistent with the timing and the location of the Meiyu rainy season, and the second northward jump from the Yangtze River Basin to North China occurs in the first ten days of July, thus bringing about the beginning of the rainy season in North China. In the 41st pentad (20-24 July), the leading zone of the summer monsoon advances up to Northeast China. This is the northernmost location where the East Asian summer monsoon may reach.

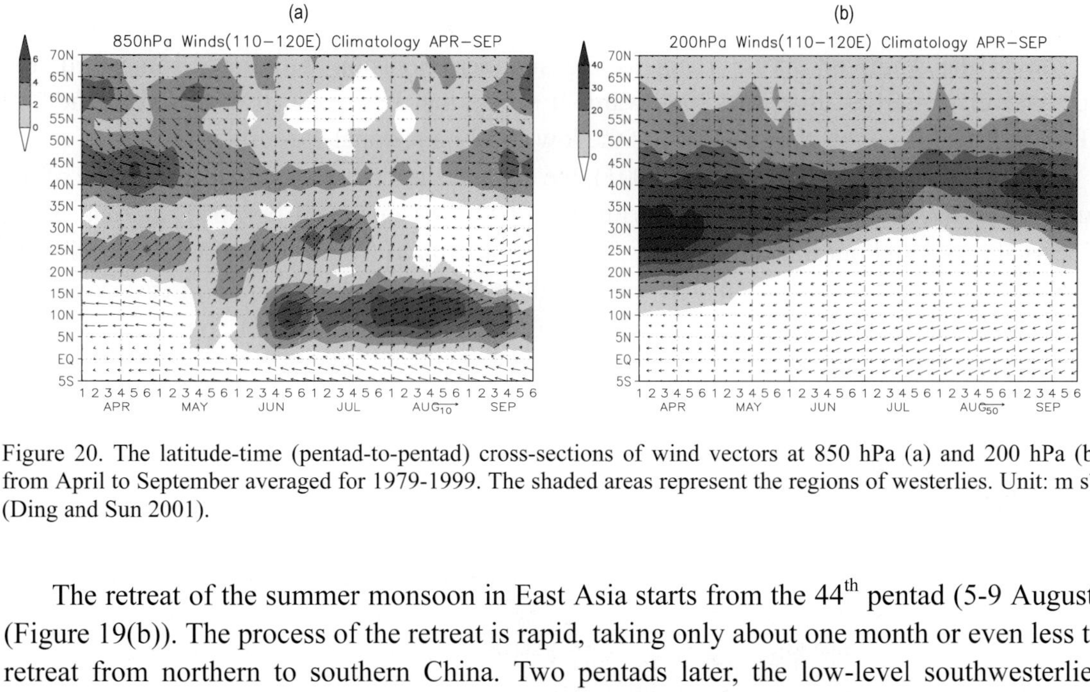

Figure 20. The latitude-time (pentad-to-pentad) cross-sections of wind vectors at 850 hPa (a) and 200 hPa (b) from April to September averaged for 1979-1999. The shaded areas represent the regions of westerlies. Unit: m s^{-1} (Ding and Sun 2001).

The retreat of the summer monsoon in East Asia starts from the 44th pentad (5-9 August) (Figure 19(b)). The process of the retreat is rapid, taking only about one month or even less to retreat from northern to southern China. Two pentads later, the low-level southwesterlies disappear in the region to the south of the Yangtze River Basin. Early in September, the leading zone of the summer monsoon quickly withdraws southward to the northern part of the South China Sea and then is stationary there, marking the end of the summer monsoon in East Asia. Matsumoto (1997) also indicated that the monsoon westerlies are already replaced by easterlies in the northern part of the South China Sea in early September, while monsoonal westerlies are still dominant over the Indo-China Peninsula, until late October when the summer monsoon fully ends over Southeast Asia. So, the life cycle of the activity of the summer monsoon in East Asia is about four months from April to September. It can also be seen clearly that the monsoonal westerlies at 850 hPa assume two northward shifts (the first ten days of June and around mid-July) and three stationary periods (South China and the northern South China Sea from mid May to early June, 25°N - 30°N for mid-June to middle July and 40°N - 45°N from the last ten days of July to mid-August). Figure 20 shows pentad-to-pentad 850 hPa (a) and 200 hPa (b) wind vector variations from April to September. It can be seen clearly that the monsoonal westerlies at 850 hPa assume two northward shifts (the first ten days of June and around mid-July) and three stationary periods. In the low latitude (from the equator to 15°N), an independent southwesterlies system is maintained. On the other hand, the mid-latitude westerlies can be observed at the latitudinal zone of 35°N - 50°N, with significant weakening in intensity from mid-June to mid-August when the summer monsoon advances northward. So, in East Asia, three separate low-level wind systems actually exist in summer, and their interaction sets a large-scale stage for occurrence of many significant weather and climate events in East Asia. The wind field at 200 hPa also demonstrates a significant seasonal variation, mainly in the intensity and location of the westerly jet. Early in May, a northward jump may be observed. When the summer monsoon advances to its northernmost latitude, the westerly jet also has a maximum northward extension and the weakest intensity while the upper-level easterlies dominate the extensive region to the south of 30°N. Then, as the summer monsoon withdraws southward, the upper-level westerlies also shift southward. It seems that the seasonal variation of the 200 hPa wind field is not so abrupt as that of the low-level monsoonal westerlies.

As indicated by Fong and Wang (2001), the advance and retreat of the leading zone of the summer monsoon is quite consistent with that of the seasonal rain belt in East Asia. Ding (1994) also pointed out this consistency. The rainy season in China generally begins with the onset of the summer monsoon and ends with its withdrawal. Rainfall intensity and its variation are closely related to the fluctuations of the summer monsoon. When the summer monsoon advances northward, its leading zone and the monsoon rain belt correspondingly

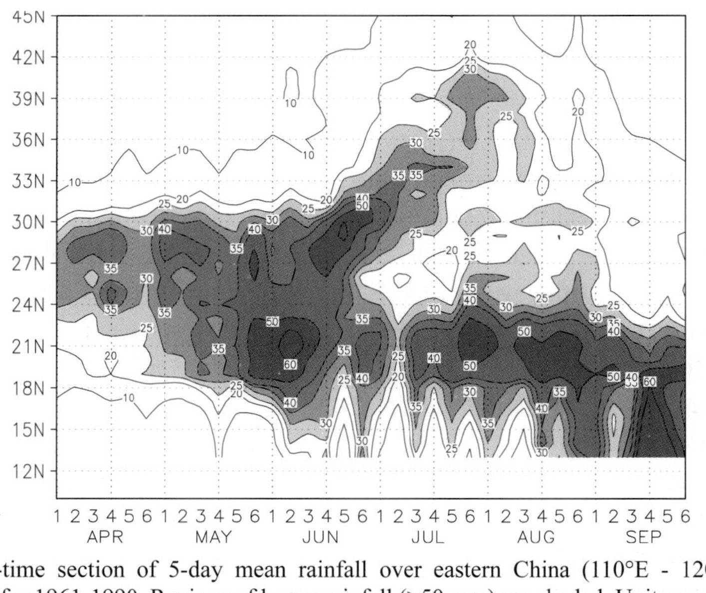

Figure 21. Latitude-time section of 5-day mean rainfall over eastern China (110°E - 120°E) from April to September averaged for 1961-1990. Regions of heavy rainfall (>50 mm) are shaded. Unit: mm (Sun 2002).

moves from a low-latitude to mid-and high latitudes. In this process, as does the leading zone of the monsoonal airflow, the monsoon rain belt undergoes three standing stages and two stages of abrupt northward jump. Based on the time-latitude cross-section of 5-day rainfall amount for eastern China (Figure 21) (Sun 2002), the most conspicuous feature is the monsoon onset between 18°N and 25°N as indicated by the steep rise in precipitation starting from the first 10-day period of May. This rainy episode is so-called pre-summer rainy season in South China, Hong Kong and Taiwan. This characteristic feature has been also studied by other investigators (Guo and Wang 1981; Lau *et al.* 1988). The first standing stage of the major rain belt generally continues into the first 10-day period of June, and afterwards it rapidly shifts to the valley of the Yangtze River. This second stationary phase initiates the Meiyu rainy season in central China. The time span of the season on the average lasts for one month (10 June-10 July). The wind and thermal fields in the Meiyu region are usually characterized by a low-pressure trough (the so-called the East Asian summer monsoon trough), a weak stationary front at surface, significant horizontal wind shear across the front and frequent occurrence of prolonged heavy rainfall. The Baiu in Japan and Changma in Korea also occur in a similar situation, but with a regional difference in locations, timing and duration which will be discussed later. The Meiyu rain belt is usually quasi-zonally oriented near 30°N over the eastern China and tilts east-northeastward toward Korea and Japan. The quasi-stationary front (the Meiyu front in China, Baiu front in Japan and Changma front in Korea) often extends from a cyclonic low center west-southwestward as the center itself moves eastward or east-northeastward. The heaviest rainfall is mostly associated with eastward-moving meso- to synoptic scale disturbances along the front. Numerous investigators have studied the development mechanism of this kind of the East Asian summer monsoon (Meiyu) disturbances and their contribution to the Meiyu rainfall (Tao and Ding 1981; Ninomiya and Akiyama 1992; Chang *et al.* 1998, 2000).

From mid-July, the rain belt rapidly moves over North China and in August farther over Northeast China, the northernmost position of summer monsoon rainfall. This standing stage of the rain belt causes the rainy season in the northern China that generally lasts for one month. In the early or middle part of August the rainy season of North China comes to end, with the major monsoon rain belt disappearing. From the end of August to early September the monsoon rain belt moves back to South China again. At this time, most of the eastern part of China is dominated by a dry spell. Note that from mid-July, the second rainy season predominates over South China, with a gap of a time period of about 20 days or one month between the pre-summer rainy season and this rainy season, that is mainly caused by typhoons, the movement of the ITCZ and other tropical disturbances in the monsoonal airflow. Around the beginning of October, the rainy season over eastern China more or less ends and rapidly makes way for the winter monsoon. Guo and Wang (1981), with analysis of climatological patterns of summer rainfall by use of a longer record of precipitation in China (1950-1979) similarly pointed out that the summer monsoon rain belt assumed three relatively stable periods, with the first period (the pre-summer rainy season in South China) having the longest persistence, the third period (the rainy season in North China) the second longest persistence and the second period (the Meiyu season) the shortest.

The Baiu season in Japan and the Changma season in Korea also occur during the seasonal northward advance of the summer monsoon. As indicated by Ninomiya and Muraki (1986), the Baiu in Japan begins in early June when rainfall in Okinawa reaches its peak. In the last ten days of June, the rainfall peak moves to the western and southern parts of Japan. Then the rainfall peak further moves northward in the first ten days of July. North of 40°N, no rainfall peaks associated with the Baiu can be observed. So, the Baiu season in Japan mainly lasts from early June to mid- July almost concurrently with the occurrence of the Meiyu in China.

The rainy season in Korea, the so-called Changma, accompanied with a belt-like peak rainfall zone, begins with the influence of the quasi-stationary convergence zone between the tropical maritime airmass from the south, and both continental and maritime polar airmasses

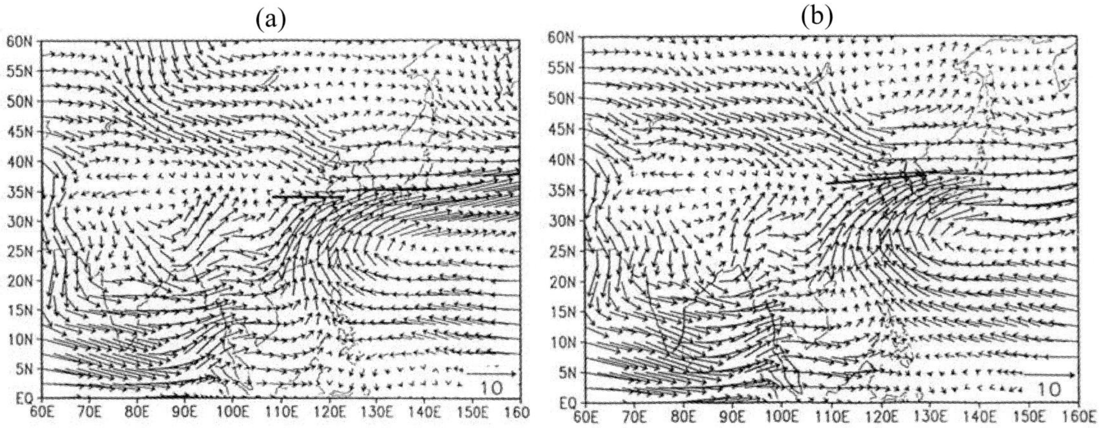

Figure 22. Climatological mean wind fields at the 700 hPa: (a) for 2-6 July and (b) 7-11 July. Bold-solid line indicates the convergence line or zone. A scale of the wind (Unit: m s^{-1}) is given in the lower right corners (Qian and Lee 2000).

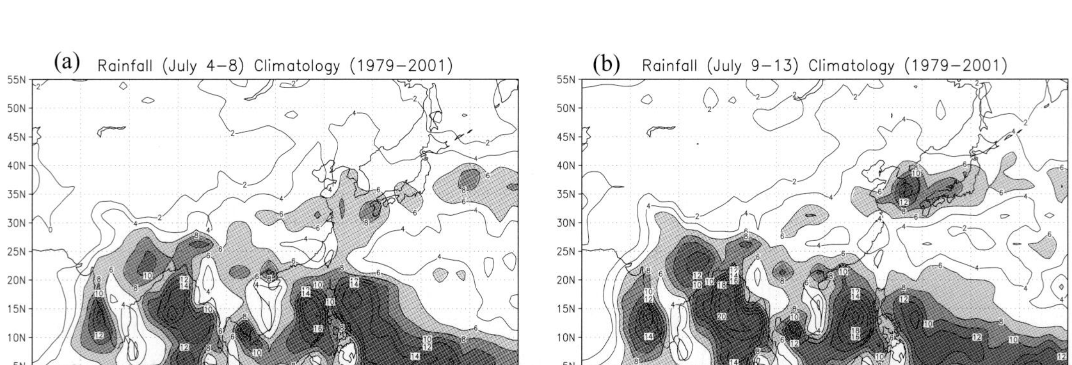

Figure 23. Climatological mean precipitation fields (1979-2001). CAMP dataset is used here. (a) 4-8 July and (b) 9-13 July. Unit: mm day^{-1} (Sun 2002).

from the north (Oh *et al.* 1997). Figure 22 clearly depicts this convergence zone at 700 hPa (Qian and Lee 2000) (Figure 22(a)). During 2-7 July, the convergence zone or line of both southwesterly and northwesterly winds is located in the Huaihe River Basin between the lower Yangtze River and the lower Yellow River. The major rainfall area is not observed over the Korean Peninsula (Figure 23(a)). However, in the next pentad (7-11 July) when the convergence zone migrates northward (Figure 22(b)), wind and intensive rainfall with 12 mm day^{-1} start to dominate the Korean Peninsula and Japan (Figure 23(b)), which is evidently separated from the extensive area of tropical precipitation. Therefore, based on the precipitation peak and lower tropospheric circulation features, the onset date of the Northeast Asia summer monsoon or Changma rainy season can be determined as the period of the 37th to 39th pentad (late June—mid-July), with a significant interannual variability. Therefore, the Changma is a shorter monsoonal rainy season, with mean Changma period being 20 days to one month long.

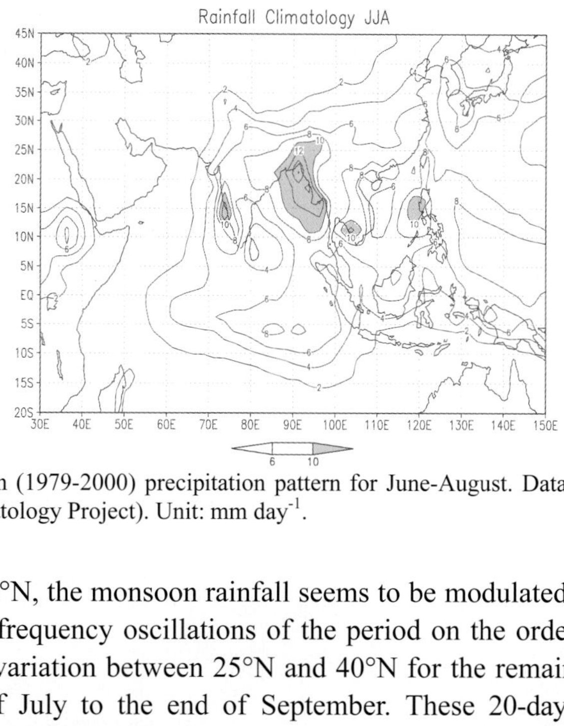

Figure 24. Long-term mean (1979-2000) precipitation pattern for June-August. Data sets are taken from GPCP (Global Precipitation Climatology Project). Unit: mm day^{-1}.

To the south of 30°N, the monsoon rainfall seems to be modulated by a 30-50 day period oscillation while low-frequency oscillations of the period on the order of 20 days appear to dominate the rainfall variation between 25°N and 40°N for the remaining part of the season from the later part of July to the end of September. These 20-day fluctuations appear to originate around 25°N and propagate northward to 40°N (Lau *et al.* 1988). Shi and Chao (1983) documented the presence of this low-frequency oscillation based on a power-spectrum analysis of the wind field and some thermodynamic quantities at Beijing (40°N). This indicates that this characteristic oscillation of the monsoon regime may propagate up to very high latitude.

The 40-day rainfall fluctuations over East Asia are most likely related to the global 40-50 day oscillation found by Madden and Julian. Lau *et al.* (1988) suggest that abrupt changes or "sudden jumps" by the major rainfall belt are related to the phase-locking between an intraseasonal oscillation such as the 40-day mode and 20-day mode and the seasonal cycle. This unique process leads to wave-like progress from south to north of monsoon rainfall over East Asia from April to September, as described above. This problem will be elaborated on further in Section 6.

5. The Climatological Aspect of the Moisture Transport of the East-Asian Summer Monsoon

The East-Asian summer monsoon plays a crucial role in moisture transport and supply for precipitation in this region. In addition, the moisture transport in this region is a requirement of the moisture and water balance over the Asian monsoon region. This section will discuss the climatological aspects of moisture transport and supply of the East Asia summer monsoon under the context of its seasonal march.

Figure 24 is the 21-yr mean (1979-2000) precipitation pattern for June to August produced with GPCP data sets. In the Asian summer monsoon region, the maximum precipitation is observed in the Bay of Bengal. The western part of the Indian Peninsula, the Indo-China Peninsula and the southern part of the South China Sea are also maximum precipitation regions, but with much less areal extent. In East Asia, especially in East China, western Japan and the Korean Peninsula, the precipitation is quite significant, but with much lower rainfall amount than in the Indian and Southeast Asian monsoon regions. Therefore, from the viewpoint of the mean precipitation field, the convection and precipitation in the East Asian summer monsoon region is not so vigorous and abundant as in the Indian summer monsoon region.

Figure 25 represents the moisture source and sink distribution in the period June-August, indicating that most of the Asian monsoon region is a moisture sink, *i.e.*, precipitation has a

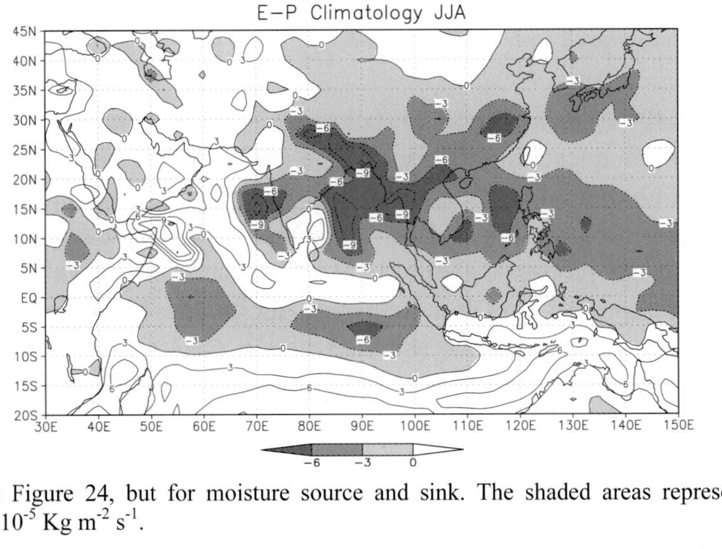

Figure 25. Same as Figure 24, but for moisture source and sink. The shaded areas represent the regions of moisture sink. Unit: 10^{-5} Kg m^{-2} s^{-1}.

surplus over evaporation. The major regions of moisture sink are located in the Bay of Bengal, the western Indian Peninsula, the South China Sea and South China, consistent with the maximum precipitation regions as described in Figure 24. The regions of moisture source are located in the Arabian Sea, the coastal area of East Africa and the southern Indian Ocean. Therefore, much moisture supply available for precipitation in the Asian summer monsoon region comes from the Southern Hemisphere, implying the special importance of the interhemispheric moisture transport. Figure 26 further confirms this fact that the most remarkable moisture channel originates in the southern Indian Ocean, and crossing the equator near the Somali coastal region, further flows in the Arabian Sea, the Bay of Bengal and the South China Sea. From there, the moisture transport flows northward into East Asia.

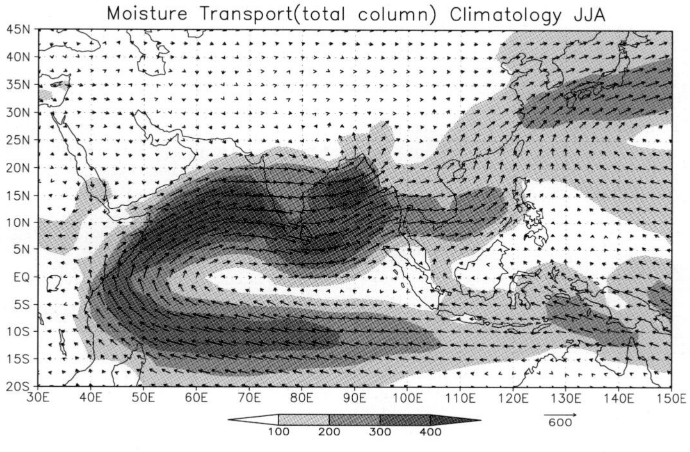

Figure 26. Same as Figure 24, but for vertically integrated (1000-300 hPa) moisture transport pattern. Shaded areas denote strong transport region. Unit: Kg m^{-1} s^{-1}.

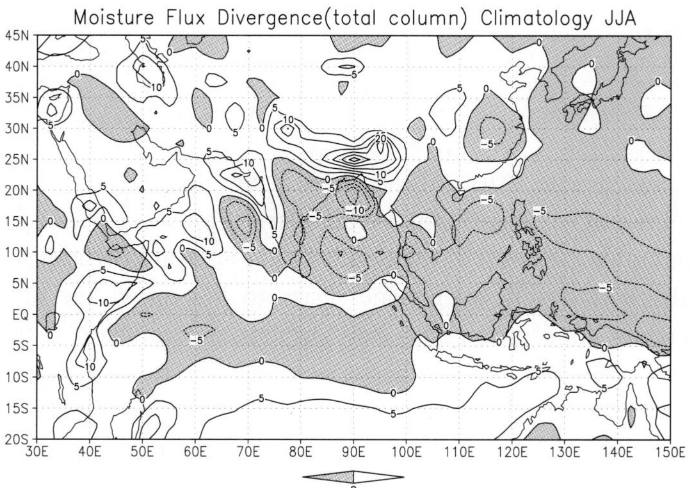

Figure 27. Same as Figure 24, but for vertically integrated (1000-300 hPa) moisture flux divergence. Shaded areas are convergence regions. Unit: 10^{-5} Kg m^{-2} s^{-1}.

Another secondary moisture channel comes from the southern and western periphery of the subtropical high over the West-Pacific. These two moisture channels merge into a single one in the South China Sea and the East Asian region. The pattern of the moisture flux divergence (Figure 27) is very similar to the patterns of precipitation (Figure 24) and moisture source and sink (Figure 25). The strongest moisture convergence is located in the Bay of Bengal, most of the Indian Peninsula, the Indo-China Peninsula, the South China Sea, East China and Northeast Asia. Simmonds *et al.* (1999), utilizing the ECWMF data sets, have analyzed the climatological fields of moisture transport and divergence in the Asian monsoon region. Their results are basically similar to those described here.

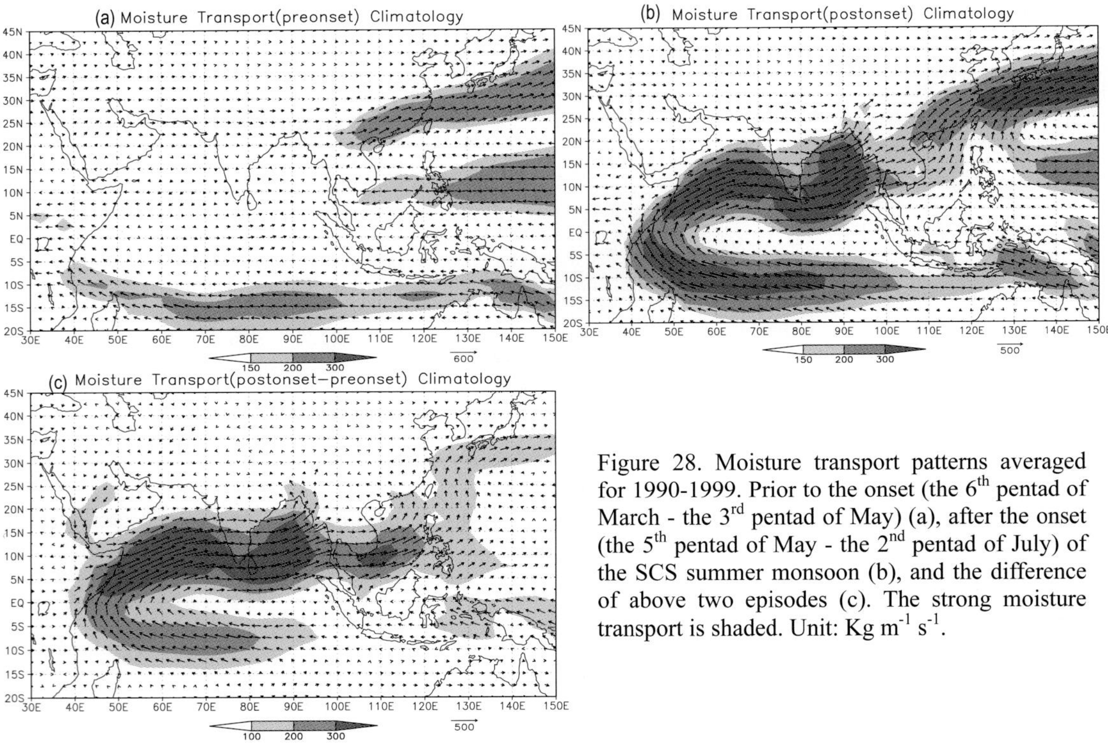

Figure 28. Moisture transport patterns averaged for 1990-1999. Prior to the onset (the 6th pentad of March - the 3rd pentad of May) (a), after the onset (the 5th pentad of May - the 2nd pentad of July) of the SCS summer monsoon (b), and the difference of above two episodes (c). The strong moisture transport is shaded. Unit: Kg m^{-1} s^{-1}.

Below we will discuss the effect of the onset of the South China Sea summer monsoon on moisture transport and regional moisture budget in East Asia. From Figure 28, it can be seen that before and after the onset of the South China Sea summer monsoon, the patterns of moisture flux undergo a significant change. Prior to the onset, the moisture transport is generally weak, mainly concentrating on the northern and southern part of the subtropical high. The moisture in the southern Indian Ocean cannot flow northward into the Asian monsoon region. After the onset, the moisture coming from the Southern Hemisphere flows into the Asian region and forms a continuous moisture channel as illustrated in Figure 26. The most remarkable change occurs in the South and Southeast Asian monsoon regions (See Figure 28(c)). Similarly, the significant change in moisture flux divergence field is also noted (Figure 29).

It is very interesting to study regional moisture budgets prior to and after the onset of the SCS summer monsoon (Figure 30). Before the onset of the SCS summer monsoon, the interhemispheric moisture transport is rather weak and even southward. The northward moisture transport across the northern boundaries of various regions is generally weak, except for the regions of the Indo-China Peninsula and the SCS. The moisture sinks occur in the regions of Bay of Bengal, the Indo-China Peninsula and South China, where the enhanced precipitation may be observed. After the onset the whole picture of the moisture transport and budget rapidly changes and becomes well-organized. The cross-equatorial flow has its maximum moisture transport in the western part of the equatorial Indian Ocean. The second maximum moisture transport is located in the equatorial East Indian Ocean. In the South Asian and Southeast Asian monsoon regions, one may see consistent eastward moisture transport, all the way to the SCS. The moisture sinks from the Indian Peninsula to the SCS are consistent with the major observed precipitation regions, with the Bay of Bengal having the maximum.

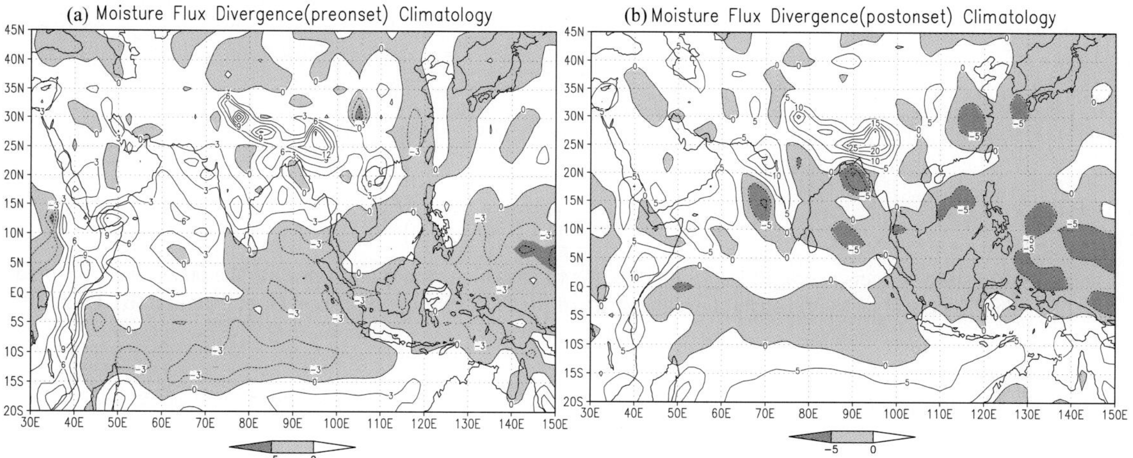

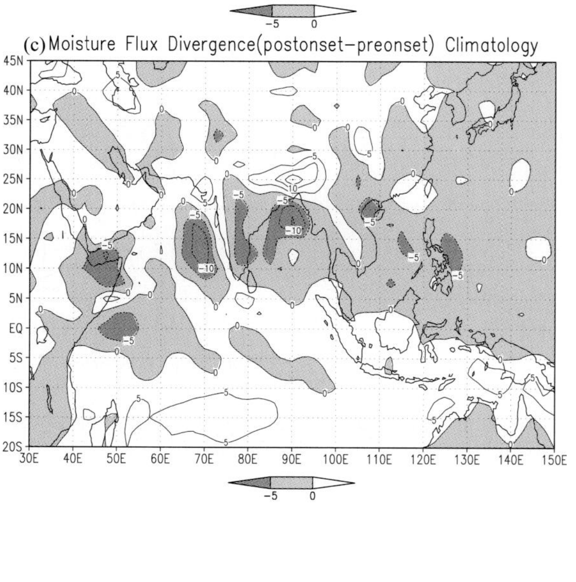

Figure 29. Same as Figure 28, but for moisture flux divergence field. Convergence areas are shaded. Unit: 10^{-5} Kg m^{-2} s^{-1}.

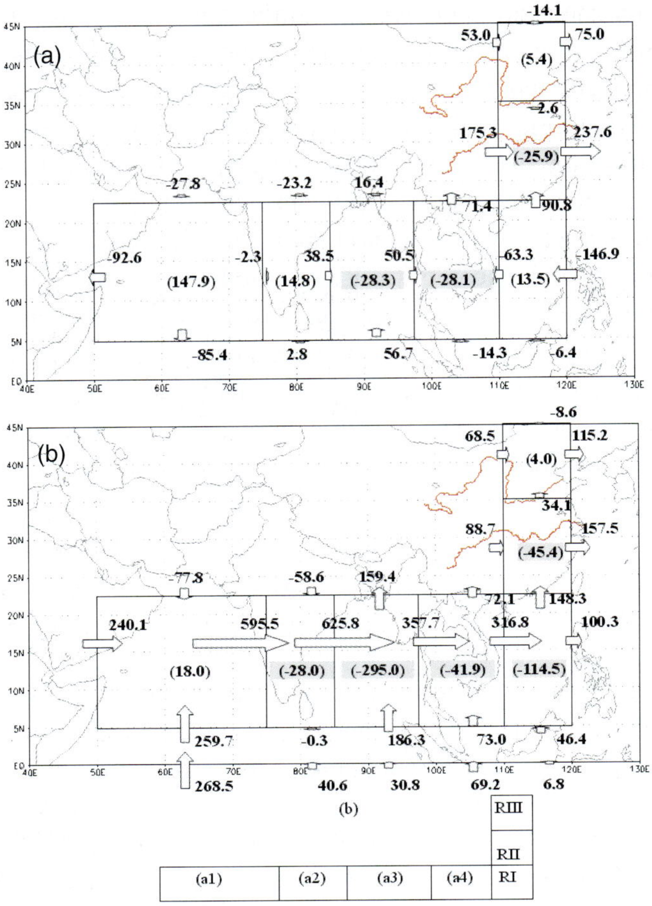

Figure 30. Moisture budgets averaged for 1990-1999 for various monsoon regions prior to the onset (the 1st pentad of April - the 2nd pentad of May) (a) and after the onset of the SCS summer monsoon (June-August) (b). Unit: 10^6 Kg s^{-1}.

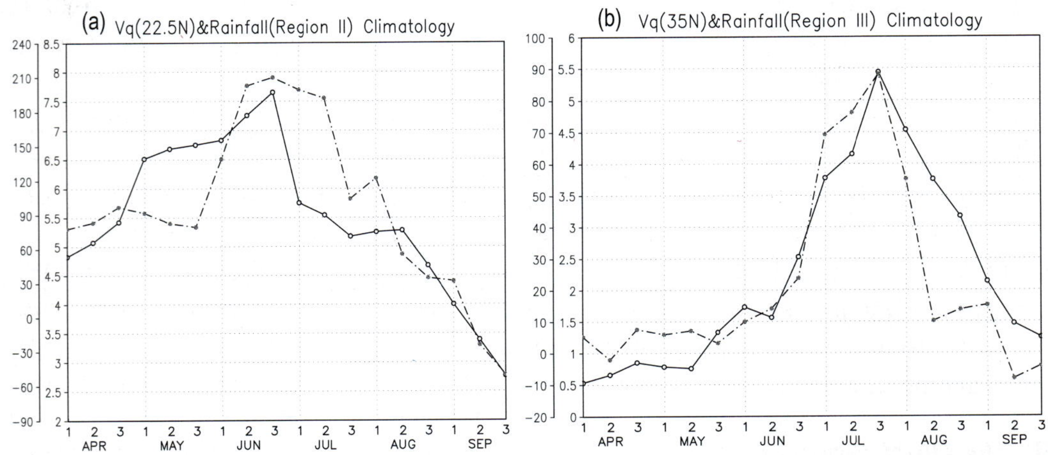

Figure 31. Comparisons of the pentad mean precipitation (Unit: mm day^{-1}) and the meridional moisture transport (Unit: Kg m^{-1} s^{-1}) for April-September averaged for the period of 1990-1999: (a) for the Region RII and (b) for the region RIII.

The northward moisture transport *via* the northern boundaries has its maximum in the region of the Bay of Bengal. The SCS takes the second place. But, if one combines together the moisture transport coming from the Indo-China Peninsula and the SCS, the northern moisture transport into the East Asian region will come to 22.3×10^6 Kg s^{-1}, which obviously exceeds the northward transport *via* the Bay of Bengal. Note the moisture budget for the RII (East China). The moisture supply in this region mainly comes from the SCS. This is consistent with the results obtained by Simmonds *et al.* (1999) and implies the critical role of the moisture transport from the SCS in the precipitation in East Asia. From Figure 31(a), one may see a close relationship between the pentad mean precipitation amount and northward moisture transport *via* 22.5°N (the northern boundary of the SCS region). The moisture coming from the SCS may further be transported to North China and be fed into precipitation systems there. Figure 31(b) shows a very nice relationship between the pentad mean precipitation amount and the northward moisture transport across the northern boundary (35°N) of region RII.

6. Summary and Discussion of the Physical Processes and Mechanisms Related to the Onset and the Seasonal March of the East Asian Summer Monsoon

In the preceding sections, we have discussed the climatological aspects of the onset and the seasonal march of the East Asian summer monsoon in terms of large-scale circulation and rainfall. Four characteristic features can be summarized from these discussions.

(I) The Earliest Onset of the Asian Summer Monsoon Occurs in Most of Cases in the Central and Southern Indo-China Peninsula.

The summer monsoon sets in over these regions early in May, or even earlier at the end of April. Then, with the eastward retreat of the subtropical high over the West Pacific and rapid acceleration and eastward propagation of the low-level westerlies in the tropical East Indian Ocean, the summer monsoon starts to break out during the period from May 16-20 in the SCS. In some years, as exemplified by the detailed analysis of the 1998 case in this study, the onset of the Asian summer monsoon may occur nearly synchronously around mid-May in the Indo-China Peninsula and the SCS. The initial onset of the summer monsoon in the SCS is often observed in its northern part, with the trigger of the synoptic systems coming from the mid-latitudes. After the onset of the Asian summer monsoon, one may see the two preferred routes of advance, one propagating northwestward to arrive in the Indian Peninsula, thus leading to the onset of the Indian monsoon from early June to early July, but almost several weeks or one month later than the onset over the SCS. Another propagates northward and northeastward to East Asia (China, Korea and Japan) and the Northwest Pacific. The physical factors and processes underlying this onset and march will be discussed later.

(II) The Onset Process over the SCS and the Indo-China Peninsula is very Abrupt, with Dramatic Changes of Large-Scale Circulation and Rainfall Occurring during a Quite Short Time Period of about One Week.

After this sudden onset, low-level easterlies and upper-level westerlies rapidly switch to westerlies and easterlies, respectively. At the same time, the dry season which lasted for the cold season rapidly changes into the wet season, indicating the earliest arrival of the summer monsoon rainy season in the Asian - western North Pacific monsoon region. This suddenness is clearly illustrated in Figure 32. Over the SCS, the major precipitation belt is steadily located in the zonal band of 15°S - 5°N before mid-May. Another rain belt located in South China (20°N - 28°N) corresponds to the pre-summer rainy season there. Around mid-May the near-equatorial rain belt suddenly moves northward and merges with the South China rain belt. It can be seen from Figure 32(b) that this process is accomplished during a quite short time period. In contrast, over the Indian longitudes (Figure 32(a)) this onset process is more gradual even though a large increase in rainfall in this region may still be observed. This suddenness of the onset process in the SCS has been well documented by numerous investigators with both climatological and case studies (see Section 2 and 3).

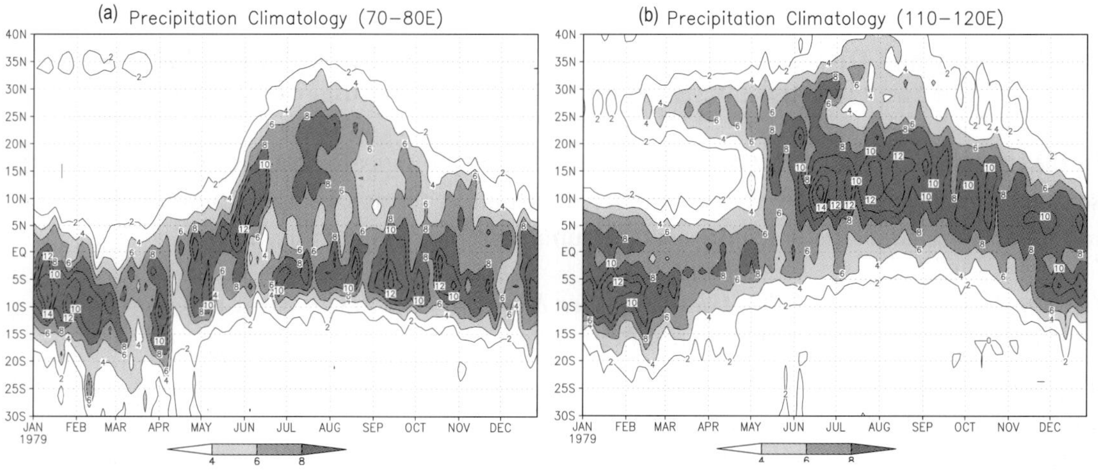

Figure 32. Latitude-time cross-sections of mean precipitation (1979-2001) along 70°E - 80°E (a) and 110°E - 120°E (b). The CMAP precipitation dataset is used here. Unit: mm day^{-1} (Sun 2002).

(III) The Climatological Summer Monsoon Onset and Seasonal March Displays a Distinct Stepwise Northward and Northeastward Advance.

Over East Asia, two abrupt northward jumps and three stationary periods have been identified in the process of the seasonal march of the summer monsoon while over the western North Pacific three stages of onset and advance of the summer monsoon have been identified (Wu and Wang 2001). These stepwise onset and advance in the two major monsoon regions appear to be nearly concurrent. The monsoon rain commences over the region from the Indo-China Peninsula-the SCS-Philippines during the period from early May to mid-May, then it extends abruptly to the Yangtze River Basin, and western and southern Japan, and the southwestern Philippine Sea in early to mid-June and finally penetrates to North China, Korea and part of Japan, and the topical western West Pacific (around 150°E and 17.5°N). In association, three abrupt changes have been identified in the large-scale circulation features. In particular, the western North Pacific subtropical high demonstrates a sudden eastward or quick northward displacement and the monsoon trough pushes abruptly eastward or northward at the stepwise onset and seasonal march.

(IV) The Onset of the Summer Monsoon over the Indo-China Peninsula and the SCS is Preceded by Development of Circulation Features and Convective Activity in the Tropical East Indian Ocean and the Bay of Bengal that is Characterized by the Development of a Twin Cyclone Crossing the Equator, the Rapid Acceleration of Low-Level Westerlies and Significant Increase of Convective Activity in both Areal Extent and Intensity.

The evidence implies that the onset of the summer monsoon over the Indo-China Peninsula and the SCS is closely related to upstream development and its eastward propagation. A recent study by Matuamoto and Murakami (2002) views these developmental processes as a component of climatological monsoon transition from one hemisphere to another. During spring and early summer over the East Indian Ocean is a prominent symmetric Rossby-type basic flow, which serves as a medium for convection to propagate northward, although not in a well organized manner, from SIOM (South Indian Ocean Monsoon) to SEAM (Southeast Asian Monsoon). The convection center remains weak until the 28th pentad (May 16-20). At this time period the monsoon trough or ITCZ returns to Southeast Asia from the South Indian Ocean due to the effective sensible heating. Sudden amplification occurs after the 31st pentad (June 1-5). Thus, they consider this pentad as the onset phase of the Southeast Asian monsoon.

Finally, we will discuss the physical processes and mechanisms. From different perspectives, a number of investigators have studied the possible contributing factors and physical mechanisms which may be used to explain the onset and the seasonal march of the East Asian summer monsoon. Recently, Hung and Yanai (2002) have examined the factors contributing to the onset of the Australian summer monsoon. Four major factors contributing to the onset are identified by them: (1) land-sea thermal contrast, (b) barotropic instability, (c) arrival of Madden-Julian oscillation (MJO), and (4) intrusion of mid-latitude troughs. In the Asian monsoon region, likewise, the thermal contrast due to differential heating between land and sea in the process of seasonal march of solar radiation acts as a seasonal precondition for the onset. However, in contrast to the Australian monsoon region, the Asian monsoon is not only forced by the thermal effect of land-sea contrast, but also by the elevated heat source produced by the huge massif of the Tibetan Plateau (Yeh and Gao 1979; Ding 1992). Based on the estimate of heat budget made by Yeh and Gao (1979) using 1961-1990 station data, the total energy supplied by the Tibetan Plateau has its maximum in late spring and early summer, with a peak occurring in May. This heat flux from the surface to the atmosphere has its maximum contribution from the sensible heat. Thus, the atmosphere over the Tibetan Plateau in May and June becomes the strongest atmospheric heat source in a year, and has abnormally high temperature with the warmest region in July and August found in the region of the longitudinal range of 50°E - 110°E. It is very interesting that during the transition season from spring to summer, the warming in this region occurs earlier than in other zones of the same latitude. In March, the increase in thickness (500-300 hPa) is also evident and attains its maximum in May and June (Yeh and Gao 1979), preceding the onset of the Asian summer monsoon in timing. For the MONEX year (the year of 1979) when a better observational coverage of upper-air radiosondes was available, the estimates of the heat budget over the Tibetan Plateau showed that the most significant heating ($Q_1 > 0$) occurred during late May and early June of that year, just before onset of the summer monsoon (Murakami and Ding 1982; Luo and Yanai 1984). All of these studies have well documented the thermal forcing of land-sea contrast, especially the Tibetan Plateau and its surrounding areas, on the onset of the Asian summer monsoon.

Next, one may naturally ask why the earliest onset occurs in the Indo-China peninsula and the SCS, rather than in other locations. The study by He *et al.* (1987) made an initial attempt to provide some evidence to address this important problem by using the data of 1979. They found that a sudden temperature increase over the eastern Plateau and the central China plain (85ºE - 115°E) occurred during the period from 6 May to 15 May. At the same time, the reversal of the meridional temperature gradient first occurred over the longitudes east of 85°E and then over the longitudes west of 85°E. The two stages of the reversal of the temperature gradient (as well as the geopotential height gradient) coincide with the two stages of the onset of the low-level southwesterlies and organized rains over the Bay of Bengal and the Arabian Sea. The dominant role played by the temperature increases over the land areas including the plateau in this reversal has been further documented by the works of Wu and Wang (2001) and Zhang *et al.* (2004). Figure 33 also clearly shows occurrence of the first reversal of the temperature gradient in the entire troposphere in the longitudinal range of 90°E - 120°E. At the end of April, the positive meridional temperature gradient is confined in the layer below 700 hPa west of 105°E, with the maxima over India and the Arabian Peninsula (Figure 33(a)). In the upper troposphere, only a slight reversal is observed over the Bay of Bengal and the Indo-China Peninsula. The westerly winds prevail above 400 hPa and the easterlies prevail below 400 hPa. In the following pentads, the area of positive meridional temperature gradient at high levels expands and penetrates downward with time and finally merges with a layer of weak positive meridional temperature gradient in the lower troposphere in mid-May (Figure 33(b)-(e)). This result suggests that the reversal of the meridional temperature gradient in the entire troposphere is accomplished first over the Indo-China Peninsula. This process of reversal clearly propagates from the high to the lower troposphere. At the same time, the upper-level westerlies and low-level easterlies over the Indo-China Peninsula and the SCS are replaced by easterlies and westerlies, respectively.

Wu and Wang (2001) also pointed out that the change of the wind direction or the vertical shear (200-850 hPa) (Figure 34(a)) can be explained by the reversal of the meridional temperature gradient (Figure 34(b)). The meridional temperature gradient averaged over the layer of 850-200 hPa reverses first over the Indo-China Peninsula because the atmosphere heats up more quickly over the land than over the ocean. The thermal advection of the warm air from the Tibetan Plateau in relation to the westerly winds at middle and upper levels before the onset is also important. The latent heat released by the pre-summer or spring

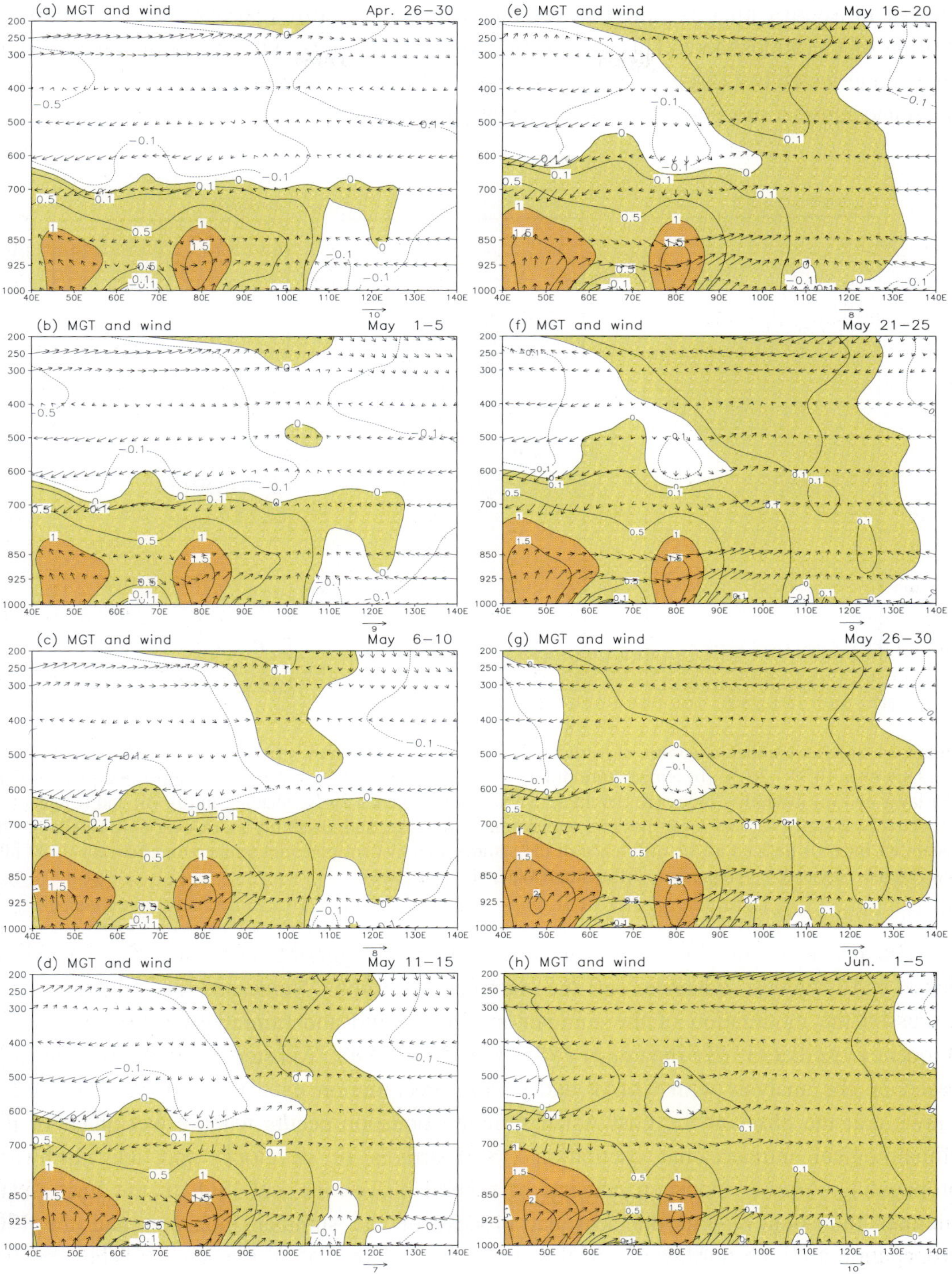

Figure 33. Longitude-pressure cross-sections of pentad mean meridional temperature gradient (°C per 2.5 latitude) and wind (m s^{-1}) over 5°N - 20°N from (a) April 26-30 to (h) June 1-5 in sequence. Light and dark shadings denote the positive meridional temperature gradient and greater than 1°C per latitude, respectively (Zhang *et al.* 2004).

rainfall in South China and the Indo-China Peninsula possibly make some contribution to heating of the atmosphere. This view is supported by the development of the zone of high convective instability (Figure 34(c)). As a result, the easterly vertical shear and the onset of the Asian summer monsoon develops first along Southeast Asian longitudes.

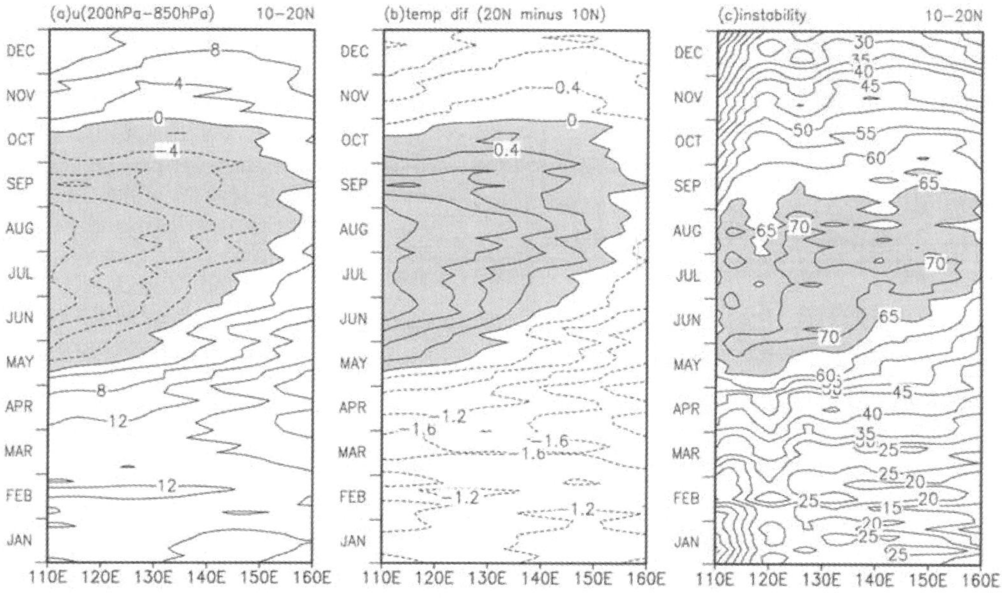

Figure 34. (a) Hovemoller diagrams of vertical shear of zonal wind (m s^{-1}) between a 200 hPa and 850 hPa averaged over 10°N - 20°N, (b) temperature difference (°C) between 20°N and 10°N averaged over the 850-200 hPa layer and (c) instability index (K/1000hPa) averaged over 10°N - 20°N. Shading in (a), (b) and (c) denotes, respectively, easterly vertical shear, positive temperature difference, and instability index over 65K/1000hPa. The instability index is defined as the difference of the saturated equivalent potential temperature between 1000 hPa and 700 hPa (divided by the pressure difference) (Wu and Wang 2001).

The arrival of the MJO oscillation is likely to be a triggering mechanism for the sudden onset and stepwise advances of the summer monsoon. Numerous investigators have discussed the modulation of the summer monsoon wind and rainfall by a 30-60 day low frequency oscillation (Krishnamurti 1985; Lau *et al.* 1988; Webster *et al.* 1998). Recently, based on the analysis with CMAP and station precipitation data, Qian *et al.* (2002) have shown that the onset of the East Asian summer monsoon occurs when a wet phase of the climatological intraseasonal oscillation (ISO) arrives or develops, and the northward propagating summer monsoon consists of several phase-lock wet ISO. In the East Asian summer monsoon region, the seasonal process of the summer monsoon and the ISO propagation are both northward and they are interconnected at all the stages of the seasonal march and in all the subregions of East Asia. Using the 1998 data derived from the SCSMEX, Chan *et al.* (2002) have found that the activity of the SCS summer monsoon for this year was mainly controlled by the 30-60 day oscillation, but modified by the 10-20 day mode. Therefore, their result further supports the previous findings that two preferred modes of ISO

(30-60 day and 10-20 day modes) in the East Asian monsoon region may play an important role in modulation of the summer monsoon activity. In recent years, Wang and Xu (1997) have identified four cycles of statistically significant climatological intraseasonal oscillation (CISO) from May to October in the Northern Hemisphere summer monsoon regions. CISO here represents a broad time scale oscillation spectrum from 20 to 60 days with the most prominent peak occurring at 30 days, the second peak at about 50-60 days and a weaker peak at 20-25 days. The first cycle has a peak wet phase in mid-May that starts the monsoon over the SCS and Philippines. Its dry phase in late May and early June brings the pre-monsoon dry weather over the regions of western North Pacific summer monsoon (WNPSM), Meiyu/Baiu, and Indian summer monsoon (ISM). The wet phase of Cycle II peaking in mid-July marks the onsets of WNPSM and ISM, and the end of Meiyu. The wet phase of Cycle III peaking in mid-August benchmarks the height of WNPSM, which was followed by a conspicuous dry phase propagating westward and causing the second break of WNPSM (in early September) and ISM (in mid-September). The wet phase of Cycle IV represents the last active WNPSM and withdrawal of ISM in mid-October.

In the western North Pacific (WNP), the seasonal progress of the summer monsoon has little linkage with that of the East Asian monsoon. In this region, the seasonal progress is mainly northeastward and the ISO propagates northwestward from the equator or develops locally (Wu and Wang 2001). The seasonal progress is gradual and the large-scale background is established first in the west (the SCS) and then in the east (WNP). Thus the first onset occurs only in the west part (in the SCS around pentad 27-28 (May 11-20)) where the seasonal cycle reaches its transition. The second onset in the middle (in the western Philippine Sea around 135°E and 10°N around the pentad 34 (June 15-19)) occurs when a second wet phase of the climatological ISO arrives and at the same time the large-scale background transition is established. The third onset (around 150°E and 17.5°N around pentad 40-41 (July 15-24)) develops under similar situations with the seasonal northeastward progress of the large-scale background and the dominant period of the ISO about 30 days in the WNP. The onset displays stepwise northwestward advance with a time interval of about one month between the consecutive stages of the onset.

Though the climatological ISO is often the primary reason for the sudden onset, the onset is paced by the seasonal evolution of large-scale circulation and thermodynamics that determines the direction of the onset advance. With the large-scale background established by the seasonal evolution, the arrival of several one-after-another ISO wet phases triggers the development of deep convection. Due to the seasonal regulation, the ISO has a tendency to be phase-locked with respect to the calendar year so that the climatological onset displays multiple stages. The stepwise march of the onset is observed each year (Wu and Wang 2001).

In the northeastward march of the onset of monsoon rain in the WNP, the possible effect of the SST (the sea surface temperature) is also indicated by Wu and Wang (2001). With the northeastward migration of high SST in the SCS and WNP, the thermal contrast changes with the warm center developing in the SCS before mid-May and in the Philippine Sea after mid-May. The related thermal contrast induces low-level moisture convergence and results in the increase of the low-level moisture and convective instability. The westerlies are accelerated to the southwest of the warm SST center. Those changes favor the northeastward march of the monsoon trough and the high convective instability zone. The relative role of the land-sea thermal contrast and the SST change may differ between the SCS and WNP monsoon onset because of the difference of the geographic location. For the SCS monsoon onset, both the land-sea thermal contrast and SST change are important. Because the land lies directly to the north of the SCS, the land-sea thermal contrast can affect the wind changes over the SCS through the change of the meridional temperature gradient. The role of the SST change can be felt through the surface heat fluxes and the turbulent mixing in the atmospheric boundary layer. These processes can modify the lower-level winds and the static stability. For the WNP monsoon onset, due to the lack of geographically fixed land to the north, the land-sea thermal contrast may play a less important role.

The MJO alone is not sufficient to trigger the onset of the summer monsoon in some years and some regions. In such cases, the mid-latitude events (troughs and ridges) may play a substantial role in the monsoon onset (Davidson *et al.* 1983; Chang and Chen 1995; Hung and Yanai 2002). However, very few investigators have studied the physical processes and mechanisms of triggering the onset by the intrusion of mid-latitude troughs or frontal systems in detail. Ding and Liu (2001) summarized the possible triggering mechanisms in their study on the effect of change in circulation features at mid-latitudes on the onset of the northern SCS summer monsoon based on various previous studies: (1) lifting effect to release the existing convectively potential instability for occurrence of convection and precipitation; (2) accelerating the low-level northeasterly wind with enhancing the meridional pressure gradient to increase the shear vorticity and cyclonic circulation of wind shear line; (3) enhancing the baroclinicity due to increase of horizontal temperature gradient, thus providing some amount of available potential energy for development of disturbances or meso-scale systems in the frontal zone; (4) exciting the growth of extensive convective cloud systems, which is a favorable environment for development of meso-scale systems in the low-level wind shear zone between northeasterly and southwesterly winds and associated low troughs which may force the subtropical high to retreat southward and eastward through some kind of feedback process.

As a matter of fact, these theoretical considerations may well apply to interpretation of the first onset of the northern SCS summer monsoon in a similar way. In 1998, a strong mid-latitude trough dipped southward before the onset of the SCS summer monsoon. Around May 16, the associated leading front of cold air reached the northern SCS, in good agreement with the onset date of the summer monsoon in this region (*see* Figure 16). Accompanying the continuous southward excursion of this cold wave, the major rainfall zone moved southward, covering the central and southern SCS. At the same time, the similar southern extension of the upper jet stream in westerlies was observed and the low-level westerlies were rapidly accelerated through coupling the upper-level westerlies and low-level westerlies or the downward transport of the westerly wind momentum. Thus, the mid-latitude events are an important triggering mechanism for the onset of the SCS summer monsoon through enhancing the convection and rainfall as well as the low-level westerlies in this region. Chan *et al.* (2000) also emphasized the importance of southward intrusion of cold air from mid-latitudes to trigger the onset of the SCS summer monsoon. Its role is to lift the warm, moist and unstable air to release the convective available potential energy (CAPE), when the atmospheric convective instability is already established before the onset through the heat and moisture transport by the low-level tropical or subtropical southwesterlies.

The triggering mechanism of mid-latitude events in the East Asian summer monsoon region is unique. It is quite different from the effect of southward intrusion of cold air from the mid-latitudes in the South Asian summer monsoon region. Several studies have indicated that the anomalous development and long-persistence of the blocking high over the European region may delay the onset of the Indian summer monsoon (Sikka 1981; Murakami and Ding 1982; Ding and Murakami 1984). This is because the persistent blocking high can force the upper air westerlies to split into two branches, the southern branch of which can flow southward to the region to south of the Himalaya foothills, thus leading to the southward intrusion of surface cold highs into the central and northern parts of India which is a unfavorable condition for the monsoon trough to move northward. In addition, these studies have also shown that the onset will be late if the upper-level westerlies prevailing over the northern India are anomalously strong during the transition (May and early June).

The impact from mid-latitudes may be observed not only for the onset of the East Asian summer monsoon, but also for all stages of its seasonal progress. The Meiyu/Baiu rainy season occurring during mid June-early July when the East Asian summer monsoon rapidly moves from South China and the SCS to the Yangtze River Basin and southern and western Japan is produced by the persistent interaction of cold air, and moist and warm monsoonal airflow in these regions. The continuous southward intrusion of cold air and accompanying frontal systems (the so-called Meiyu/Baiu front) is excited by the development and prevailing of blocking highs in the mid-and high latitudes over Eurasia. The dual blocking high situation, one located over the Ural Mountains and another located over the Okhotsk Sea, is the most favorable situation for prolonged Meiyu/Baiu heavy rainfall (Ding 1991; Zhang and Tao 1998; Wu 2002). Numerous studies have addressed this problem that will not be discussed here in detail. As the East Asian summer monsoon further moves northward (mid July-mid August), this marked interaction of cold air and the monsoonal flow occurs in North China and Northeast China as well as the Korean peninsula and Japan in a similar way, thus bringing about the significant rainy seasons there. So, one of the main differences between the Indian and East Asian summer monsoon is the different effect of mid-latitudes events. Due to significant interaction between the mid-latitude impacts and the summer monsoon in East Asia at all the stages of the seasonal march of the summer monsoon, the climate prediction including the seasonal prediction in this part of the world monsoon regions region is very difficult. This complex interaction needs to be further studied in the future.

Acknowledgements

I would like to thank the staff of National Climate Center, China Meteorological Administration for their help in preparing this manuscript. All figures in Section 5 were provided by Dr. Sun Ying.

References

Chan, J. C. L., Y. G. Wang, and X. J. Xu, 2000: Dynamic and thermodynamic characteristics associated with the onset of the 1998 South China Sea summer monsoon. *J. Meteor. Soc. Japan*, **78**, 367-380.

______, W. X. Ai, and J. J. Xu, 2002: Mechanisms responsible for the maintenance of the 1998 South China Sea summer monsoon. *J. Meteor. Soc. Japan*, **80**, 1103-1113.

Chang, C.-P., and G. T. J. Chen, 1995: Tropical circulation associated with southwest monsoon onset and westerly surges over the South China Sea. *Mon. Wea. Rev.*, **123**, 3254-3267.

______, S. C. Hou, H. C. Kuo, and G. T. J. Chen, 1998: The development of an intense East Asian summer monsoon disturbance with strong vertical coupling. *Mon. Wea. Rev.*, **126**, 2692-2712.

______, L. Yi, and G. T. J. Chen, 2000: A numerical simulation of vortex development during the 1992 East Asian summer monsoon onset using the Navy's regional model. *Mon. Wea. Rev.*, **128**, 1604-1631.

Chen, L. X., W. Li, P. Zhao, and S. Y. Tao, 2000: On the process of summer monsoon onset over East Asia. *Climate and Envir. Res.*, **5**, 345-355 (in Chinese).

Davidson, N. E., J. L. McBride, and B. J. McAvaney, 1983: The onset of the Australian monsoon during winter MONEX: synoptic aspects. *Mon. Wea. Rev.*, **111**, 496-516.

Ding, Y. H., 1981: A case study of formation and structure of a depression over the Arabian Sea. *Chinese J. Atmos. Sci.*, **5**, 267-280 (in Chinese).

______, and T. Murakami, 1984: Wind and temperature change over Eurasia during active, break and withdrawal periods of the Indian summer monsoon in 1979. *Acta Meteorologica Sinica*, **42**, 198-210 (in Chinese).

______, 1991: *Advanced Synoptic Meteorology*. China Meteorological Press, 792pp (in Chinese).

______, 1992: Effects of the Qinghai-Xizang (Tibet) Plateau on the circulation features over the plateau and its surrounding areas. *Adv. in Atmos. Sci.*, **9**, 112-130.

______, 1992: Summer monsoon rainfalls in China. *J. Meteor. Soc. Japan*, **70**, 373-396.

______ (Ed.), 1993: *A study on the Excessively Heavy rainfall over the Yangtze-Huaihe River Basin in 1991*. China Meteorological Press, 2-4 (in Chinese), 254pp.

______, 1994: *Monsoons over China.* Kluwer Academic Publishers, Dordrecht/Boston/London, 419pp.

______, Q. Y. Wang, and J. Y. Yan, 1996: Some climatological aspects of the summer monsoon over the South China Sea. *Atmospheric Circulation to Global Change*. IAP, ed., Chinese Academy of Sciences, China Meteorological press, 329-339.

______, and Y. Sun, 2001: A study on anomalous activities of East Asian summer monsoon during 1999. *J. Meteor. Soc. Japan*, **79**, 1119-1137.

______, and Y. J. Liu, 2001: Onset and the evolution of the summer monsoon over the South China Sea during SCSMEX field experiment in 1998. *J. Meteor. Soc. Japan*, **79**, 255-276.

______, and Y. Sun, 2001: Some aspects of climatology of the Asian summer monsoon. *Acta Meteorological Sinica. (*Submitted)

Fong, S. K., and A. Y. Wang (Eds.), 2001: *Climatological Atlas for Asian Summer Monsoon*. Macau Meteorological and Geophysical Bureau and Macau Foundation, pp318.

Guo, Q. Y., and J. Q. Wang, 1981: The distribution of precipitation in China during the summer monsoon period for recent 30 years. *Acta Geographica Sinica*, **36**, 187-195 (in Chinese).

He, H. Y., J. W. McGinnis, Z. S. Song, and M. Yanai, 1987: Onset of the Asian summer monsoon in 1979 and the effect of the Tibetan Plateau. *Mon. Wea. Rev.*, **115**, 1966-1995.

He, J. H., Y. H. Ding, H. Gao, and H. M. Xu, (Eds.), 2001: *Determination of Onset Dates of the South China Sea Summer Monsoon and Monsoon Index.* China Meteorological Press, 123pp.

Hirasawa, N., K. Kato, and T. Takeda, 1995: Abrupt change in the characteristics of the cloud zone in subtropical East Asia around middle of May. *J. Meteor. Soc. Japan*, **73**, 221-239.

Hung, C.-W., and M. Yanai, 2002: Factors contributing to the onset of the Australian summer monsoon. *Quart. J. Roy. Meteor. Soc.*. (Submitted)

Johnson, R. H., and P. E. Ciesielski, 2002: Characteristics of the 1998 summer monsoon onset over the northern South China Sea. *J. Meteor. Soc. Japan*, **80**, 561-578.

Kim, K. S., 1992: *Encyclopedia of Meteorology.* Hwangmunsa, Seoul, Korea, 735pp.

Krishnamurti, T. N., P. Ardanuy, Y. Ramanathan, and R. Pasch, 1981: The onset-vertex of the summer monsoon. *Mon. Wea. Rev.*, **109**, 344-363.

______, 1985: Summer monsoon experiment—A review. *Mon. Wea. Rev.*, **113**, 1590-1626.

Lau, K. M., G. J. Yang, and S. H. Shen, 1988: Seasonal and intraseasonal climatology of summer monsoon rainfall over East Asia. *Mon. Wea. Rev.*, **116**, 18-37.

______, and S. Yang, 1997: Climatology and interannual variability of the Southeast Asian summer monsoon. *Adv. in Atmos. Sci.*, **14**, 141-162.

______, H. T. Wu, and S. Yang, 1998: Hydrological process associated with the first transition of the Asian summer monsoon: A pilot study. *Bull. Amer. Meteor. Soc.*, **79**, 1871-1882.

______, Y. H. Ding, J. T Wang, R. Johnson, T. Keenan, R. Cifelli, J. Gerlach, O. Thiele, T. Rikenbach, S. C. Tsay, and P. H. Lin, 2000: A report of the field operation and early results of the South China Sea Monsoon Experiment (SCSMEX). *Bull. Amer. Meteor. Soc.*, **81**, 1261-1270.

______, and H. Y. Weng, 2000: Remote forcing of summertime U.S. droughts and floods by the Asian monsoon. *WCRP/GEWEX News*, **10(2)**, 5-6.

Li, C., and M. Yanai, 1996: The onset and interannual variability of the Asian summer monsoon in relation to land-sea thermal contrast. *J. Climate*, **9**, 358-375.

Li, C. Y., and Z. X. Long, 2001: An index for the South China Sea monsoon and its interannual variability. *Determination of Onset Dates of the South China Sea Summer Monsoon and Monsoon Index.* He, J. H., Y. H. Ding, H. Gao and H. H. Xu, Eds., China Meteorological Press, 123pp.

Li. J. P., and Q. C. Zeng, 2002: A unified monsoon index. *Geophys. Res. Lett.*, **29**, 1151-1154.

Lin, Z. G., 1987: A further study of advance and retreat of summer monsoon in East China. Collected Papers, **No. 10**, Chinese Academy of Meteorological Sciences, 24-31 (in Chinese).

Liu, E., and J. C. L. Chan, 1999: A unified monsoon index for South China. *J. Climate*, **12**, 2375-2385.

Luo, H. B., and M. Yanai, 1984: The large-scale circulation and heat sources over the Tibetan Plateau and surrounding areas during the early summer of 1979, part II: Heat and moisture budgets. *Mon. Wea. Rev.*, **112**, 966-989.

Matsumoto, J., 1997: Seasonal transition of summer rainy season over Indo-China and adjacent monsoon region. *Adv. in Atmos. Sci.*, **14**, 231-245.

______, and T. Murakami, 2002: Seasonal migration of monsoon between the Northern and Southern Hemisphere as revealed from equatorially symmetric and asymmetric OLR data. *J. Meteor. Soc. Japan*, **80**, 419-437.

McPhaden, M. J., 1999: Genesis and evolution of the 1997-1998 El Nino. *Science*, **283**, 950-953.

Murakami, T., and Y. H. Ding, 1982: Wind and temperature changes over Eurasia during the early summer of 1979. *J. Meteor. Soc. Japan*, **60**, 183-196.

National Climate Center/China Meteorological Administration, 1998: *Heavy Flooding and Climate Anomalies in China in 1998.* China Meteorological Press, **2-4**, 62-63 (in Chinese).

Ninomiya, K., and H. Muraki, 1986: Large scale circulation over East Asia during Baiu period of 1979. *J. Meteor. Soc. Japan*, **64**, 409-429.

______, and T. Murakami, 1987: The early summer rainy season (Baiu) over Japan. *Monsoon Meteorology*. C. P. Chang and T. N. Krishnamurti, Eds, Oxford University Press, 93-121.

______, K., and T. Akiyama, 1992: Multiscale feature of Baiu, the summer monsoon of Japan and the East Asia. *J. Meteor. Soc. Japan*, **70**, 467-495.

Oh, J. H., W. T. Kwon, and S. B. Ryoo, 1997: Review of the researches on Changma and future observational study (KORMEX). *Adv. in Atmos. Sci.*, **14**, 207-222.

Qian, W. H., and D. K. Lee, 2000: Seasonal march of Asian summer monsoon. *Intl J. Climatol.*, **20**, 1371-1378.

______, H.-S. Kang, and D.-K. Lee, 2002: Distribution of seasonal rainfall in East Asian monsoon region. *Theor. and Appl. Climatol.*, **73**, 151-168.

SCSMEX Project Office, 1998: The operational summary of the South China Sea Monsoon Experiment (SCSMEX), 3-4 (in Chinese). Published by SCSMEX Operations Center.

Shi, S. W., and J. M. Chao, 1983: The summer monsoon in North China. *Proc. Symposium on the Summer Monsoon in South and East Asia.* 1-5, October, 1982, Kunming, China. Yunnan People's Press, 72-85 (in Chinese).

Sikka, D. R., 1981: An appraisal of the onset of summer monsoon over India in the light of MONEX-79 data. Results of Summer MONEX Field Phase Research (part B), FGGE Operation Report, **No. 9**, ISCU/WMO, 87-95.

Simmonds, I., D. Bi, and P. Hope, 1999: Atmospheric water vapor flux and its association with rainfall over China in summer. *J. Climate*, **12**, 1351-1367.

Sun, Y., 2002: A study of 1998 anomalous summer monsoon activity and its mechanism. Ph.D thesis. 280pp. Available from National Climate Center, CMA, Beijing (in Chinese).

Tanaka, M., 1992: Intraseasonal oscillation and the onset and retreat dates of the summer monsoon over the east, southeast and western north Pacific region using GMS high cloud amount data. *J. Meteor. Soc. Japan*, **20**, 613-629.

______, 1994: The onset and retreat dates of the Austral summer monsoon over Indonesia, Australia and New Guinea. *J. Meteor. Soc. Japan*, **72**, 255-267.

Tao, S. Y., and L. X. Chen, 1997: A review of recant research of the East Asian summer monsoon in China. *Monsoon Meteorology*, C. P. Chang and T. N. Krishnamurti, Eds., Oxford University Press, 60-92.

______, and Y.-H. Ding, 1981: Observational evidence of the influence of the Qinhai Xizang (Tibet) plateau on the occurrence of heavy rain and severe convective storms in China. *Bull. Amer. Meteor. Soc.*, **62**, 23-30.

Ueda, H., and T. Yasunari, 1998: Role of warming over the Tibetan Plateau in early onset of the summer monsoon over the Bay of Bengal and the South China Sea. *J. Meteor. Soc. Japan*, **76**, 1-12.

Wang, B., and X. H. Xu, 1997: Northern Hemisphere summer monsoon singularities and climatological intraseasonal oscillation. *J. Climate*, **10**, 1071-1085.

______, and Z. Fan, 1999: Choice of South Asian summer monsoon indices. *Bull. Amer. Meteor. Soc.*, **80**, 630-638.

______, and Lin Ho, 2002: Rainy season of the Asian-Pacific summer monsoon. *J. Climate*, **15**, 386-398.

Webster, P. J., V. O. Magana, T. N. Palmer, J. Shukla, R. A. Tomas, M. Yanai, and T. Yasunari, 1998:

Monsoons: Process, predictability, and the prospects for prediction. *J. Geophys. Res.*, **103**, 14451-14510.

______, and S. Yang, 1992: Monsoon and ENSO: Selectively interactive system. *Quart. J. Roy. Meteor. Soc.*, **118**, 877-926.

Wu, G. X., and Y. S. Zhang, 1998: Tibetan Plateau forcing and the timing of the monsoon onset over South Asia and the South China Sea. *Mon. Wea. Rev.*, **126**, 913-927.

Wu, R., and B. Wang, 2001: Multi-stage onset of the summer monsoon over the western North Pacific. *Climate Dyn.*, **17**, 277-289.

Wu, R. G., 2002: A mid-latitude Asian circulation anomaly pattern in boreal summer and its connection with the Indian and East Asian summer monsoons. *Intl. J. Climatol.*, **22**, 1879-1895.

Xie, A., Lu, X., and Q. Ye, 1996: The climate characteristics of summer monsoon onset over the South China Sea. *The Recent Advances in Asian Monsoon Research*. He Jinhai, Ed., China Meteorological Press, 132-142.

Yan, J. Y., 1997: Climatological characteristics on the onset of the South China Sea southwest monsoon. *Acta Meteorologica Sinica*, **55**, 174-186.

Yeh, D. Z., and Y. X. Gao, 1979: *Meteorology over the Tibetan Plateau.* China Science Press, 278pp (in Chinese).

Zhang, Q. Y., and S. Y. Tao, 1998: Influence of Asian mid-high latitude circulation on East Asian summer rainfall. *Acta Meteorologica Sinica*, **56**, 199-211 (in Chinese).

Zhang, X. Z., J. L. Li, J. Y. Yan, and Y. H. Ding, 2002: Circulation features and index of the South China Sea summer monsoon onset. *Climatic Envir. Res.* **7**, 321-331.

Zhang, Z. Q., C. L. J. Chan, and Y. H. Ding, 2004: Characteristics, evolution and mechanisms of the summer monsoon onset over Southeast Asia. *Intl. J. Climatol.* (Accepted)

2. THE EAST ASIA WINTER MONSOON

JOHNNY C. L. CHAN
Laboratory for Atmospheric Research
Department of Physics and Materials Science
City University of Hong Kong
Hong Kong, China
E-mail: Johnny.chan@cityu.edu.hk

CHONGYIN LI
Institute of Atmospheric Physics
Chinese Academy of Sciences
Beijing, China
E-mail: lcy@lasg.iap.ac.cn

This chapter mainly focuses on the characteristics of the East Asia winter monsoon (EAWM). An examination of the climatology of the boreal winter in Asia shows that the EAWM results from the development of a cold-core high over the Siberia-Mongolia region. The movement of this cold air southward produces pressure surges and temperature drops across the Asian continent. Two types of such surges can be identified: the northerly surge (NS) and the easterly surge (ES). The initiation of the NS begins with the eastward passage of a polar jet streak west of Lake Balkhash. The eastward migration of this jet streak over the Siberia-Mongolia region intensifies a cold high there, which eventually leads to a southward outpour of the cold air in the lower troposphere. Such a push of the cold air then excites gravity waves that propagate across the South China Sea, which results in convection over the maritime continent. On the other hand, an ES is apparently the consequence of an initially eastward and then southeastward migration of a cold pool that splits off from a quasi-stationary high-pressure system over the Siberia-Mongolia region due to the passage of a 500-hPa ridge over the region. As the low-level anticyclone moves to the east coast of China, it initiates a southward surge of cool air and strong winds along the coast, resembling a coastal Kelvin wave. Its strength is usually much less than that of the NS. Other than these surges, a significant effect of the EAWM is the explosive development of low-pressure systems over the East China Sea as the cold air moves off the continent and over the warm water, which results from the strong baroclinity between the cold air from the continent and warm air over the ocean, and the subsequent potential instability, rising motion and latent heat release. The last section of the chapter discusses intraseasonal, interannual and interdecadal variations of the EAWM, which can be related to similar oscillations in other planetary-scale circulation features. These include the 10-20-day oscillation, the Madden-Julian Oscillation, the polar vortex, the El Niño/Southern Oscillation, sea-surface temperature anomalies in the North Pacific, the North Atlantic Oscillation, and the East Asia summer monsoon. Furthermore, "two-way" interactions between the EAWM and some of these oscillations have also been found.

1. Introduction

During the boreal winter, the planetary-scale circulation in the Northern Hemisphere is dominated by strong baroclinic systems in the mid to high latitudes. These include not only the cold high-pressure systems over the continents but also the intense low-pressure systems off the continents. Of the two major high-pressure systems, the one over Eurasia (commonly referred to as the Siberian High) is generally more intense, with surface pressure at the centre often reaching over 1040 hPa (see section 2). The snow cover over the Tibetan Plateau also creates a horizontal temperature gradient in the mid troposphere that is not found over the American continent. This, together with the topography of the Plateau, forces the migration of the cold air from the Siberian High to take on certain routes. The atmospheric flow associated with the eastward and southward movement of this cold air is generally refereed to as the East Asia winter monsoon (EAWM).

In this chapter, various characteristics of the EAWM will be discussed. The climatology of the boreal winter in Asia will first be introduced in section 2 to provide a background for understanding the evolution of the EAWM. The synoptic-scale aspects of the EAWM will then be explored in section 3. One major trait of the EAWM is the so-called cold-air "surges" that affect most parts of eastern and southern China, as well as the South China Sea (SCS). These surges carry a large amount of energy with them. They not only dominate the weather over China and southeast Asia, but also affect convection over the maritime continent (Chang and Lau 1980; see also Chapter 3), the Southern Hemisphere (SH) monsoon (Davidson *et al.* 1983) and the genesis of tropical cyclones in the SH (Love 1985). In addition, these surges have been linked to the development of the El Niño/Southern Oscillation (ENSO) event (Li 1990; Xu and Chan 2001). Because of their importance in various aspects of the planetary-scale circulation, the synoptic discussion in section 3 will be based on the different types of surges.

Other than the surges, a significant effect of the EAWM is the explosive development of low-pressure systems over the East China Sea as the cold air moves off the continent and over the warm water. Such development will be discussed in section 4.

Various researchers have identified significant temporal variations in the strength of the EAWM on time scales from intraseasonal to interannual and interdecadal. A documentation of such variations will be given in section 5. Possible linkages of these variations with other oscillations in the atmosphere (e.g. ENSO) or atmospheric and oceanographic conditions will also be discussed. A summary of the EAWM characteristics will be presented in section 6 together with some unresolved issues that need to be addressed in the future.

2. Climatology of the East Asia Winter

2.1. *Surface Conditions*

Boreal winter traditionally refers to the months of December, January and February. However, a dramatic transition apparently occurs between October and November from the surface temperature distribution. In October, the strongest horizontal temperature gradient lies in a zone oriented southwest to northeast across China and then in a nearly east-west direction from the Korean Peninsula eastward (Fig. 1a). The colder air to the southwest of

China is related to the higher terrain in the eastern foothills of the Tibetan Plateau (hereafter simply referred to as the Plateau). By November, this cold frontal zone has pushed significantly southward to ~ 25°N in the west and ~ 30°N in the east (Fig. 1b). Offshore, this zone has reached the southern part of the Korean Peninsula and southern Honshu. Between December and February, the zone of maximum horizontal temperature gradient remains at ~ 23°N over the continent (Figs. 1c, d and e). Notice that the temperature over much of central China, and extending into the Korean Peninsula and Japan, is quite uniform. By March, the cold air begins to retreat and the zone of maximum temperature gradient shifts to north of ~ 25°N.

The surface pressure distribution also shows the significant changes that take place between October and November. In October, pressure is the highest over Siberia, northwestern China and western Mongolia (Fig. 2a). Then, in November, a pressure "surge" pushes southward, with the 1020-hPa isobar extending to ~ 25°N (Fig. 2b). Pressure is the highest over western Mongolia and continues to increase, reaching a maximum of > 1040 hPa in January (Fig. 2d). This high-pressure system is referred to as the Siberia-Mongolia High (SMH). Meanwhile, further increases in surface pressure occur over southern China, with the 1020-hPa isobar reaching the northern tip of Taiwan in December (Fig. 2c) and the south China coast by January (Fig. 2d). As the SMH begins to weaken in February, the 1020-hPa isobar also retreats northward (Fig. 2e) and by March, the main belt of high pressure only extends to the northeastern part of China (Fig. 2f).

Corresponding to the surface pressure distributions, the surface winds are generally anticyclonic over the continent (Fig. 3). Two important features are of particular importance. First, the northerly to northeasterly winds over the continent south of the centre of the anticyclonic are highly ageostrophic (cf. the corresponding pressure distributions in Fig. 2) especially in the months of December through February (Figs. 3c–e). This will be discussed further in section 3 where detailed features of the northerly surge will be presented. Second, the winds from the Taiwan Strait to the South China Sea (SCS) are rather strong, ranging from 8 m s^{-1} in October to over 10 m s^{-1} from November through January (Figs. 3b–d). These rather strong northeasterly winds can come from two sources. One is the continuation of the northerly surge into the SCS, and the other is the easterly surge that originates from the coast of East China and propagates southward along the coast of China. This latter type of surge will also be discussed in detail in section 3.

2.2. *Upper-Level Features and Cross-Sections*

A very important feature of the cold air associated with the SMH is its shallow nature. In mid winter (January), while the maximum pressure at the surface (Fig. 2d), and the maximum geopotential heights at 1000 hPa (Fig. 4a) and 925 hPa (Fig. 4b), can clearly be seen, the 850-hPa geopotential height distribution simply shows a ridge over the area (Fig. 4c), and the ridge becomes even less obvious at 700 hPa (Fig. 4d). Similar situations are observed in other months (not shown). Indeed, a latitudinal cross-section through the centre of the SMH along 95°E shows the geopotential height surface to be essentially flat above 900 hPa (Fig. 5).

The shallow nature of the SMH does not mean that the upper level is not important. In mid winter (December through February), the polar jet at 500 hPa has its entrance region near the centre of the SMH (Fig. 6a). It extends southeastwards and merges over the Sea of Japan with the subtropical jet. Notice that the entrance region of the subtropical jet is to the

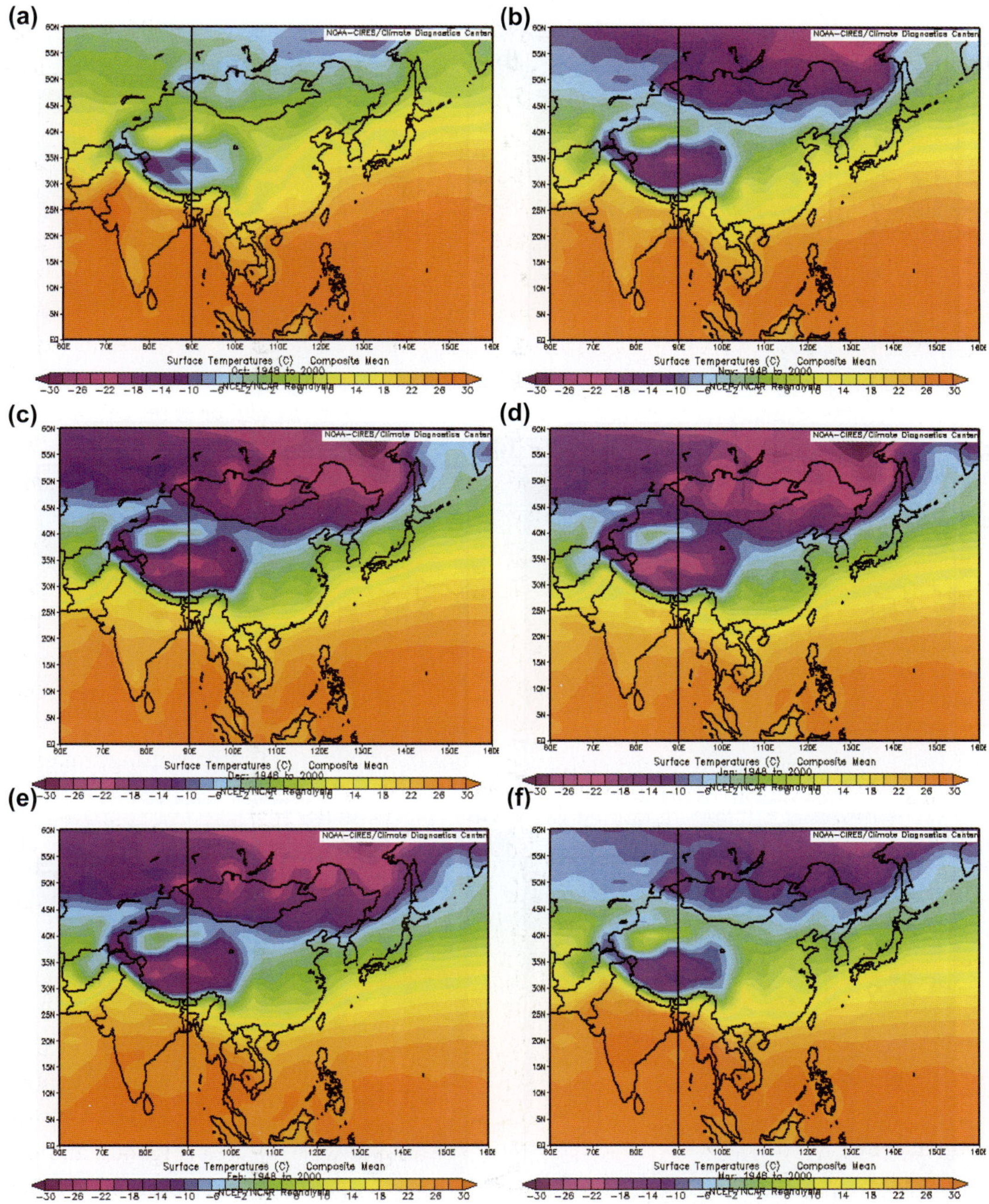

Figure 1. Mean monthly surface temperature over the Asian region in (a) October, (b) November, (c) December, (d) January, (e) February, and (f) March for the period 1948–2000 based on the NCAR/NCEP Reanalysis dataset. Contour interval: 2°C.

southwest of the Plateau. The 200-hPa flow is similar except for the absence of the polar jet (Fig. 6b). It is also important to note that the subtropical high south of the subtropical jet provides a channel for the divergent flow from the convection over the maritime continent (Chang and Lau 1980). These upper-level southeasterlies and southwesterlies couple with the low-level northeasterlies (see Fig. 3) to form the Hadley circulation in the wintertime (Chang

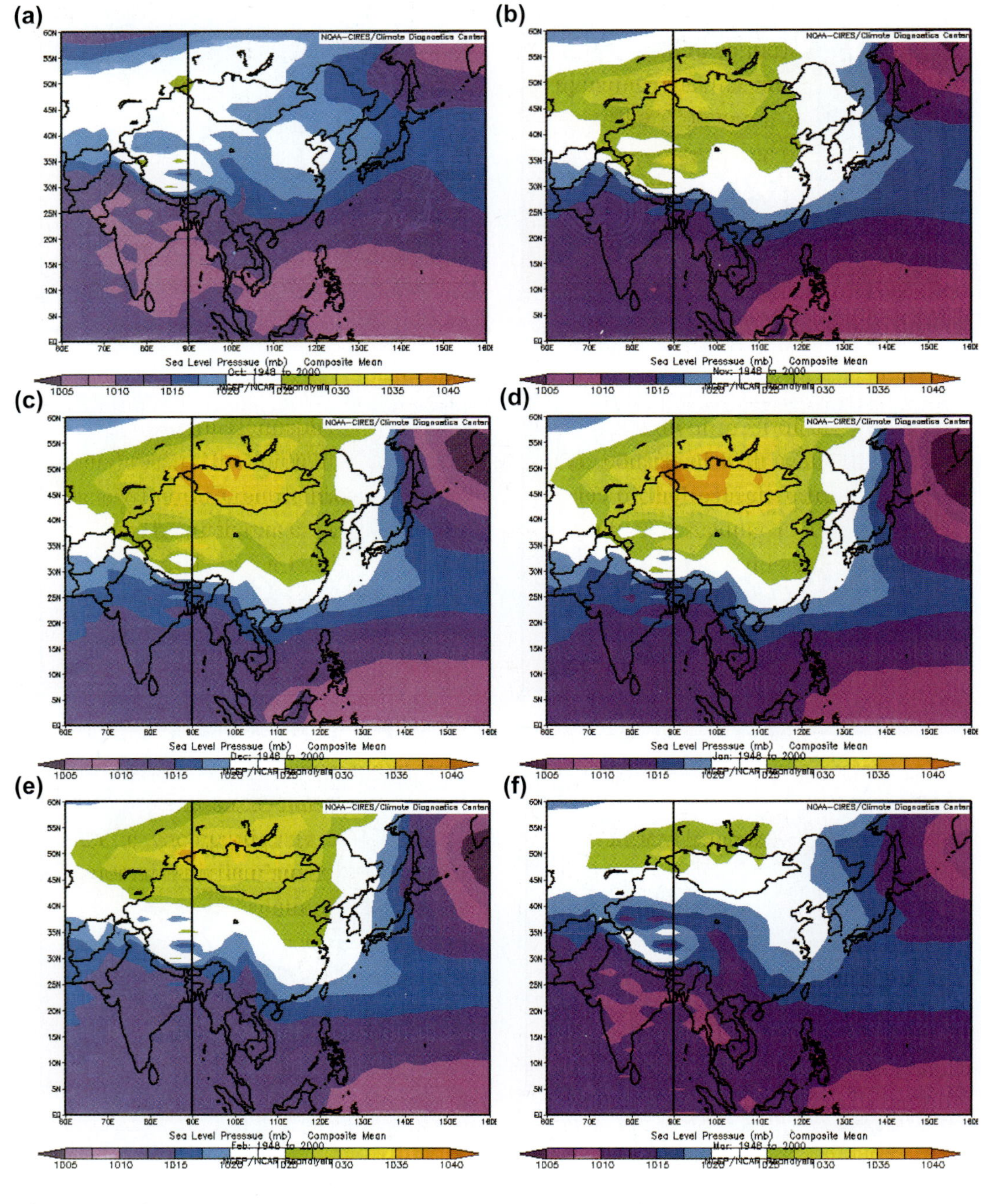

Figure 2. As in Figure 1 except for the mean-sea-level pressure. Contour interval: 2.5 hPa.

and Lau 1980). The absence of the polar jet at 200 hPa arises from the fact that the tropopause in that region is below 200 hPa (Fig. 7). Notice also that the atmosphere in that region below 700 hPa is very stable.

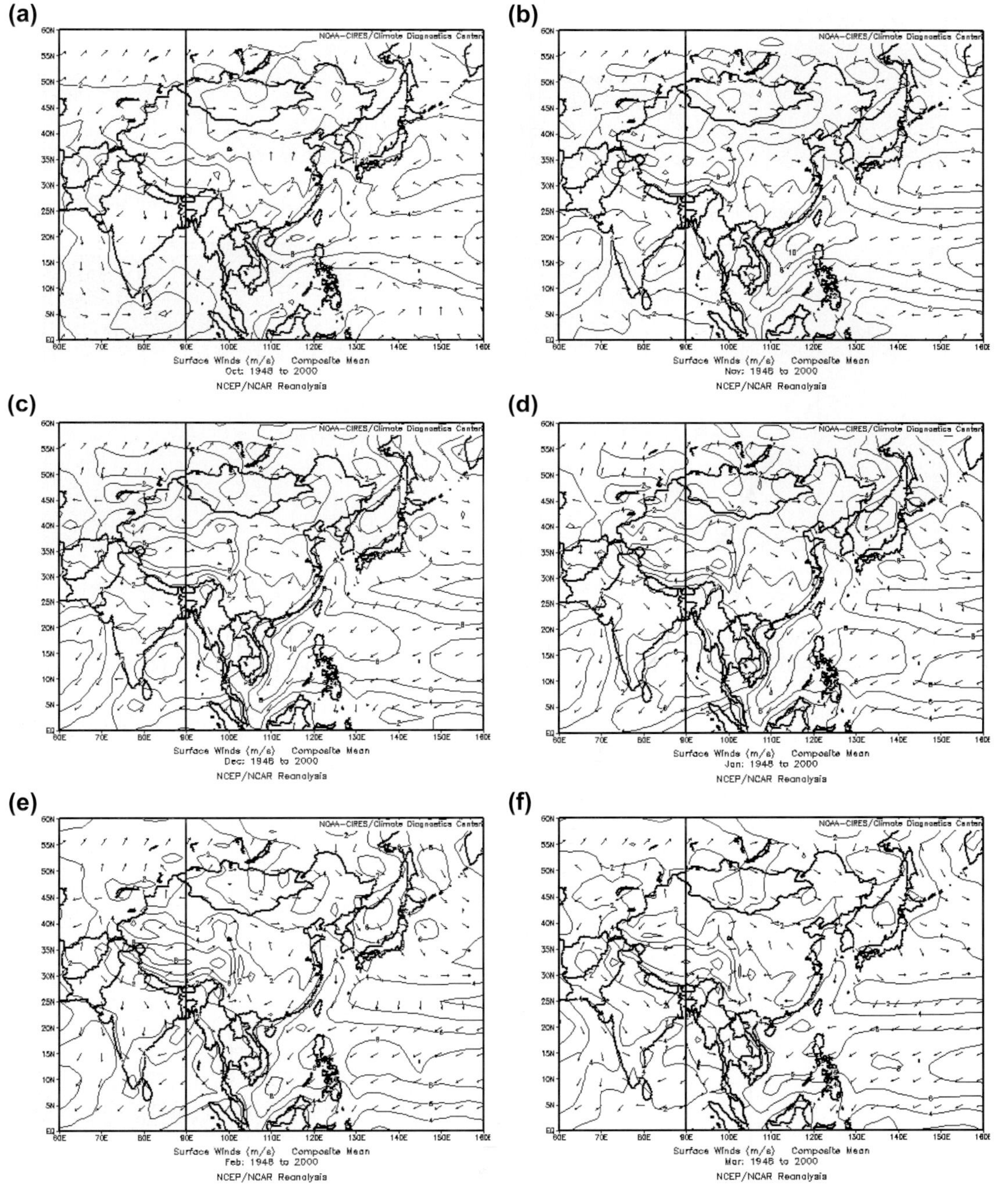

Figure 3. As in Figure 1 except for the surface wind. Contour interval of isotachs: 2 m s^{-1}.

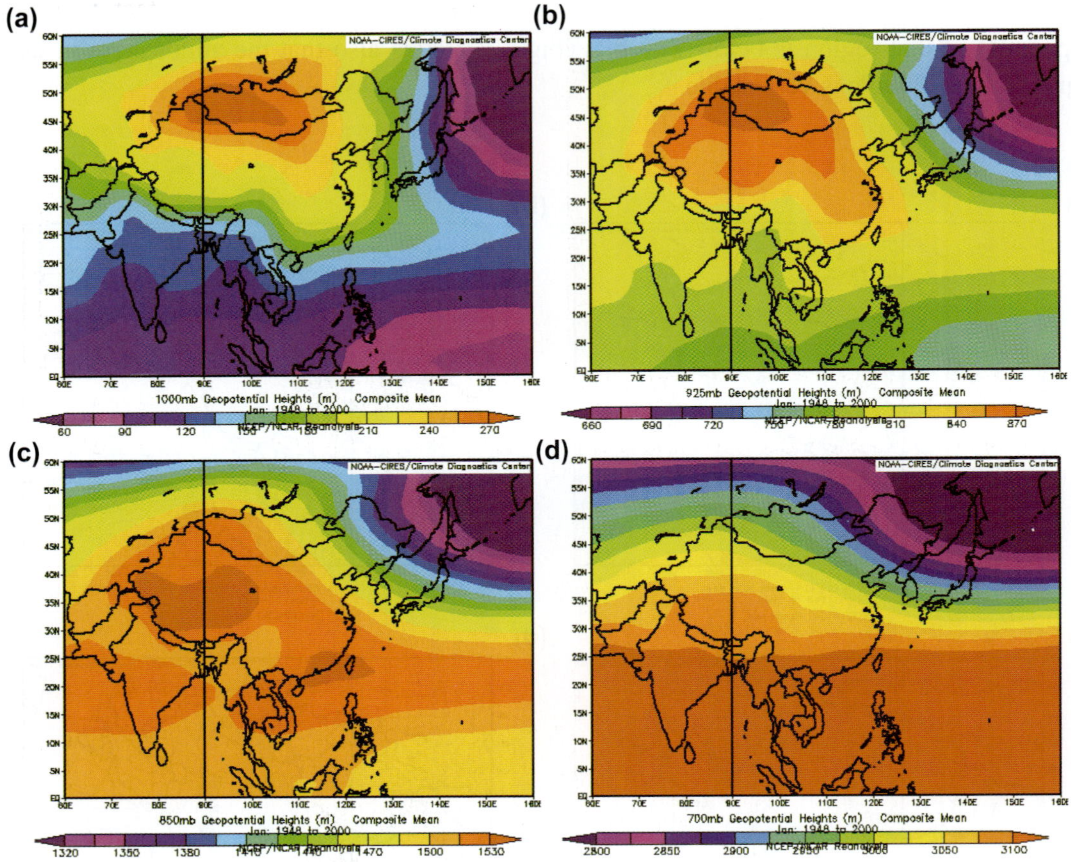

Figure 4. Mean monthly geopotential heights in January at (a) 1000 hPa, (b) 925 hPa, (c) 850 hPa and (d) 700 hPa for the period 1948–2000 based on the NCAR/NCEP Reanalysis dataset. Contour interval: (a)–(c) 15 gpm, (d) 25 gpm.

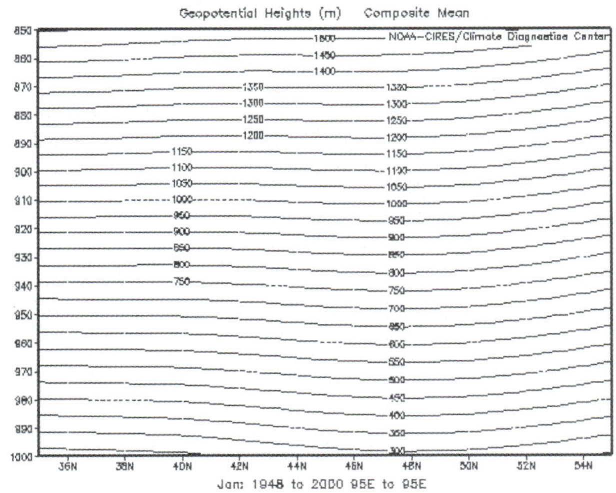

Figure 5. Latitude (35°N –55°N) – height (1000–850 hPa) cross-section along 95°E of monthly mean geopotential heights in January for the period 1948–2000 based on the NCAR/NCEP Reanalysis dataset.

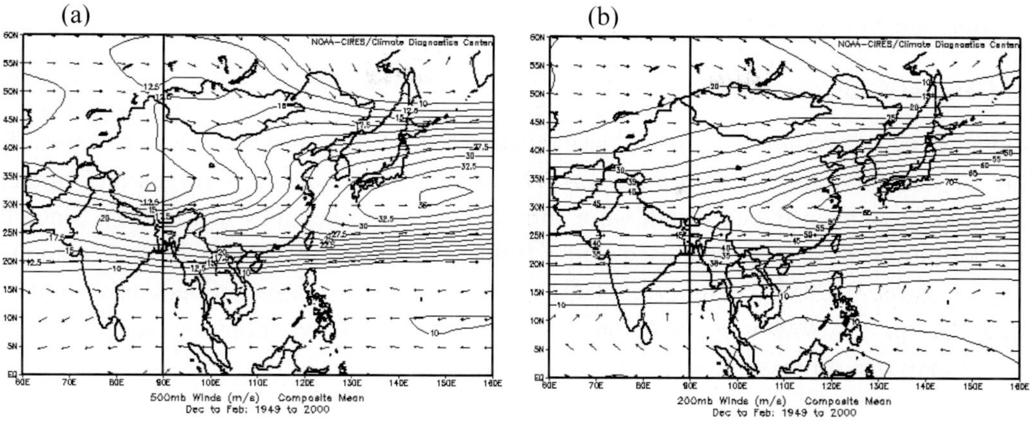

Figure 6. Mean winds between December and February of the following year at (a) 500 hPa and (b) 200 hPa for the period 1949–2000 based on the NCAR/NCEP Reanalysis dataset. Contour interval of isotachs: (a) 2.5 m s^{-1}, (b) 5 m s^{-1}.

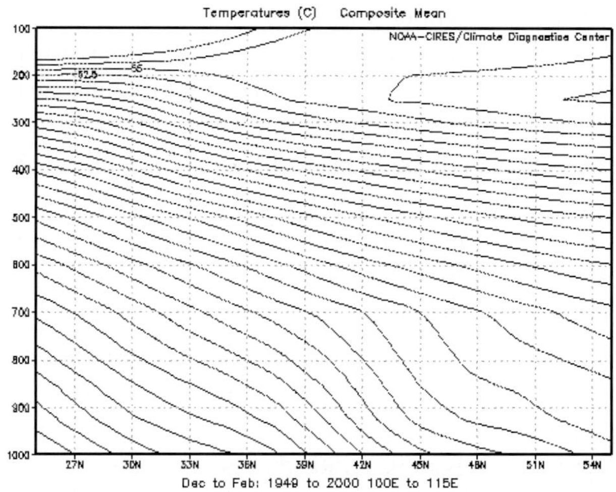

Figure 7. Latitude (25°N – 55°N) – height (1000–100 hPa) cross-section averaged between the longitude band 100°E – 115°E of mean temperature between December and February of the following year for the period 1948–2000 based on the NCAR/NCEP Reanalysis dataset.

2.3. *Summary*

To summarize, the EAWM results from the development of a cold-core high over the Siberia-Mongolia region. The movement of this cold air southward, apparently related to the polar and subtropical jet streams, produces pressure surges and temperature drops across the Asian continent. Associated with these surges are the strong low-level northeasterlies that extend all the way to the South China Sea. The return flow at the upper levels completes the Hadley circulation, which feeds back to the subtropical jet over the western North Pacific.

3. Monsoon Surges

3.1. *General Considerations*

As mentioned in section 1, a main characteristic of the EAWM is the cold-air surges. Thus, many of the previous studies of the EAWM have focused on such surges and numerous definitions have been made, which were summarized in Boyle and Chen (1987). Since then, more definitions have been proposed [Ding 1990a; Qiu *et al.* 1992; Ding 1994; Wu and Chan 1995 (hereafter WC95); Zhang *et al.* 1997]. Ding (1990b) has also defined active and break periods of the EAWM while Compo *et al.* (1999) examined monsoon surges based on spectral analyses of pressure, wind and temperature fluctuations.

Most of the definitions include temporal changes of one or more meteorological parameters at certain locations or over certain regions, which can include surface pressure rises, strengthening of surface or low-level winds, and/or temperature drops. Since different studies focused on different regions, it would be difficult to reach a universal definition or consensus. However, a general agreement is that these surges have the most dramatic effects in the tropics, in initiating convection over the maritime continent and in the subsequent feedback to the subtropical jet aloft. It seems therefore appropriate to choose a definition that focuses on what happens over South China and the South China Sea. For this reason, the definition of WC95 is adopted and their results, together with those in a later paper (Wu and Chan 1997a; hereafter WC97), will be used to illustrate the physical processes associated with EAWM surges.

In fact, the results presented in WC95 and WC97 represent the most detailed analyses of the evolution of the East Asia winter monsoon surges, with results that are consistent with many previous case studies. Another crucial reason for choosing their results is their separation of two types of surges: northerly and easterly. Most traditional, and even the latest, studies of monsoon surges focus on the former while the easterly surge is generally not discussed. However, the results of WC95 and WC97 suggest that in order to have a complete description of the EAWM surges, both types need to be examined.

In the remainder of this section, these two surges will be described in detail, first from a synoptic climatology perspective. The physical processes involved will then be discussed.

3.2. *Northerly Surges*

3.2.1. Definition

WC95 based on meteorological parameters measured in Hong Kong to define the occurrence of a northerly surge (NS). They labeled the day when an NS reached Hong Kong to be day 0, which must satisfy the following criteria:

(a) the daily mean temperature T is maximum on day 0 and the decrease in T from day 0 to day 2 must be $\geq 2°C$; or T (day 0) – T (day 1) $\geq 2°C$ and T (day 2) < T (day 1);

(b) the lowest daily mean temperature during the five days between T (day –2) and T (day 2) is less than the monthly normal;

(c) any one of the days after day –1 has northerly mean wind speed > 29 km h^{-1};

(d) a discontinuity in the prevailing wind direction: the wind direction on any day prior to day 0 or day 1 must differ from that on any one day after day 0 or day 1 by at least 60°.

These conditions basically combine those specified by others or are qualitatively consistent with their definitions (e.g. Chu 1978; Chang and Lau 1980; Lau *et al.* 1983; Lau and Lau 1984; Ding 1990a; Qiu *et al.* 1992; Ding 1994; Zhang *et al.* 1997). They highlight the importance of not only the temperature drop and strengthening of the northerly winds, but also the change in wind direction.

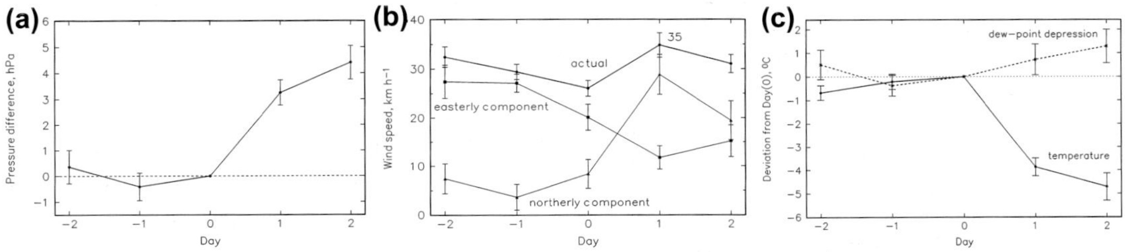

Figure 8. Composite variations of surface meteorological parameters measured in Hong Kong during the passage of northerly surges from two days before (day –2) to two days after (day 2). Day 0 is the day the NS arrives at Hong Kong. (a) Deviation of mean-sea-level pressure from day 0, (b) actual, easterly and northerly components of the surface wind, and (c) surface temperature (solid) and dew-point depression (dashed) relative to day 0. The vertical bars in all diagrams indicate the standard errors of the data within the sample. (Adapted from WC95)

3.2.2. Local Variations

Based on these definitions, WC95 identified 15 independent (at least a 3-day separation between surges) NS cases for three winters (October 1988–March 1991). The time variations of surface meteorological parameters over Hong Kong show a sharp rise of ~3 hPa on day 1 relative to day 0 (Fig. 8a), which is consistent with the strengthening of the northerly wind component to ~ 30 km h^{-1} (Fig. 8b) and the temperature drop of ~4°C (Fig. 8c). Note also the decrease in the surface moisture content from the variation in dew-point depression (Fig. 8c) and the significant weakening of the easterly component (Fig. 8b), which follows condition (d) in the definition. These two latter changes are often not emphasized in most other studies.

3.2.3. Surface Synoptic-Scale Features

The composited station data over Asia for these 15 NS cases show an increase in pressure gradient over the Siberia-Mongolia (SM) region, reaching over 1045 hPa on day (–1) (Fig. 9b). Although the centre of the SMH does not exhibit substantial movement, a dramatic equatorward shift in isobars over mainland China can be seen (cf. Figs. 9a with other panels in Fig. 9). The 1026-hPa isobar migrates from northern China (~40°N) on day –2 southward to reach a latitude of ~25°N by day 1 (Fig. 9d), which gives a southward speed of ~ 7 m s^{-1}. This value is roughly consistent with the estimate by Chang *et al.* (1983) for their second stage of the surge, the first stage being a gravity-wave type propagation that will be discussed in the next sub-section. Associated with this pressure surge is the surge of cold air, which

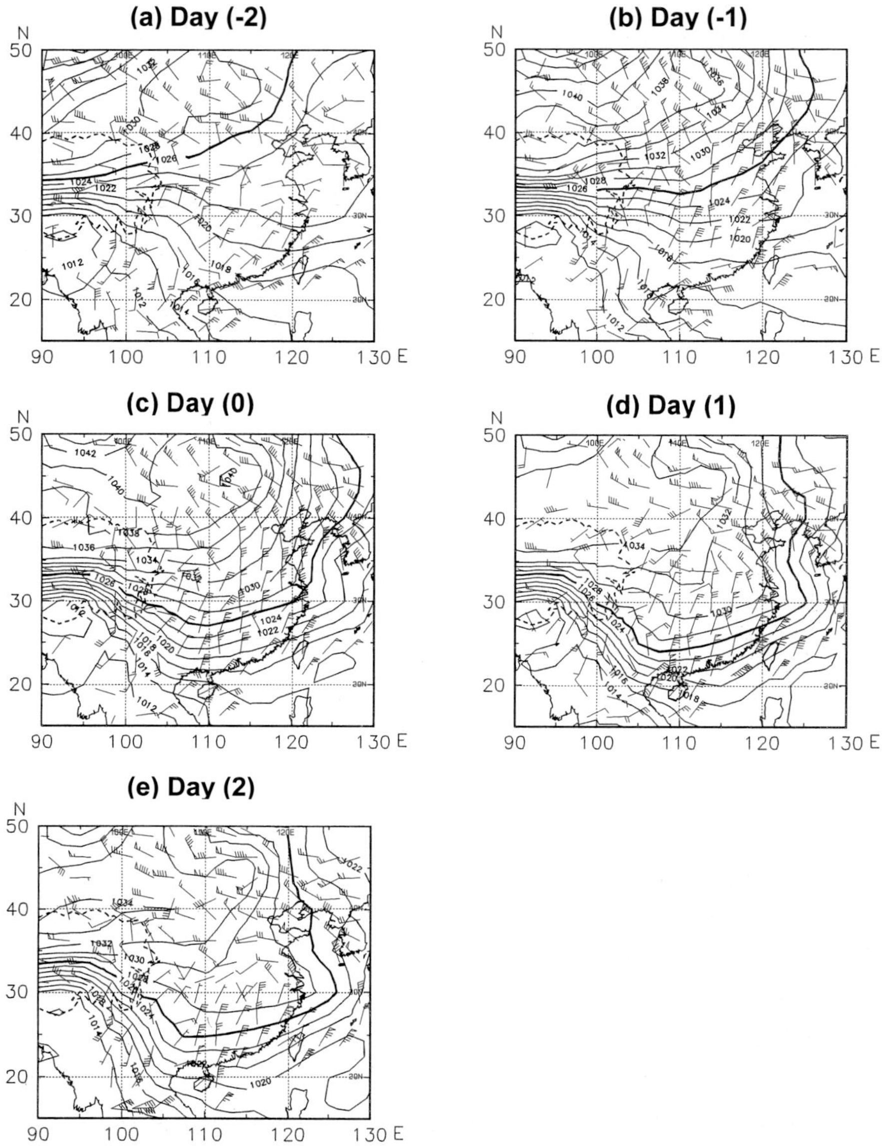

Figure 9. Composite mean-sea-level pressure distribution (contour interval: 2 hPa) and surface winds (magnitudes magnified by 10 and wind barbs plotted in the standard convention) during the passage of northerly surges from two days before (day –2) to two days after (day 2). Day 0 is the day the NS arrives at Hong Kong. The 1026-hPa isobar is thickened to highlight the southward push of the cold air. (Adapted from WC95)

leads to significant temperature drops over the region (Fig. 10), with the maximum cooling starting from northern China in day –1 to the south China coast by day 1.

Another important feature in Fig. 9 is the cross-isobaric flow of the surface wind, its acceleration clearly evident especially when the pressure surge reaches south China and the SCS (Figs. 9c–e). The strong winds over the SCS influence the sea-surface temperatures (SSTs) there–see chapter 3 for details.

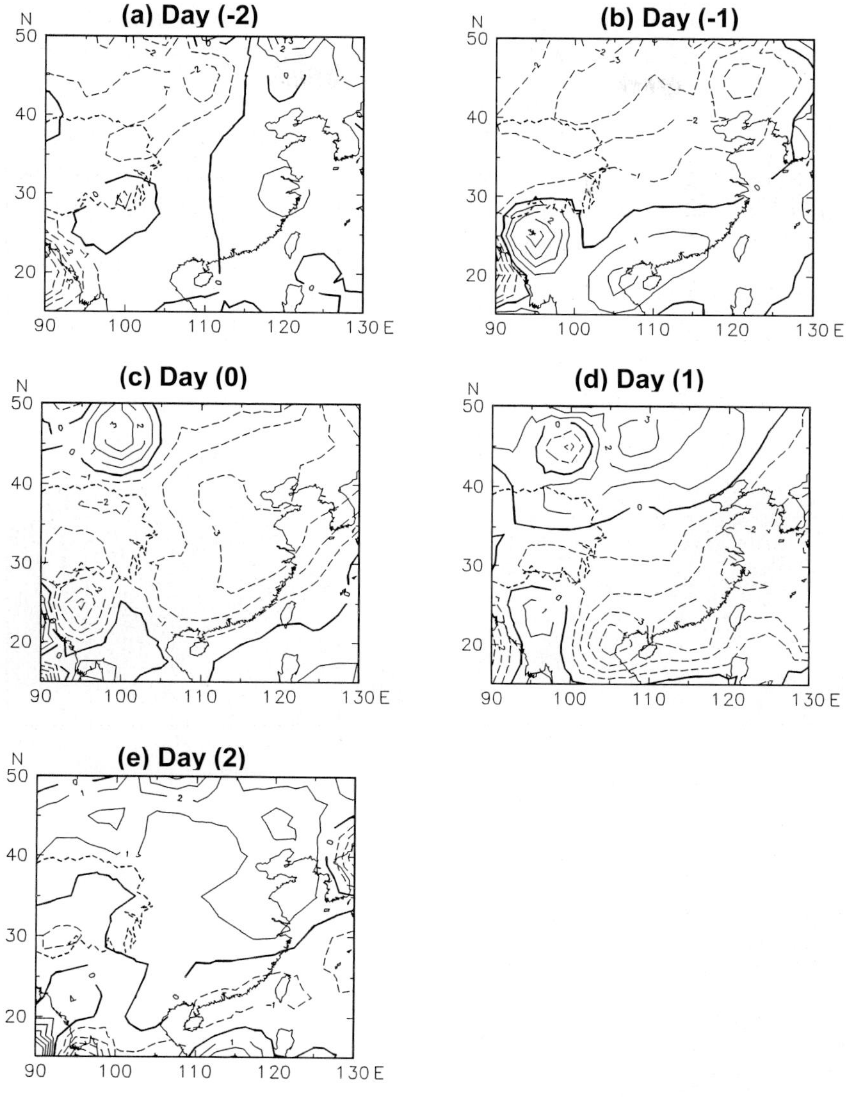

Figure 10. Composite 24-h surface temperature change during the passage of northerly surges from two days before (day −2) to two days after (day 2). Day 0 is the day the NS arrives at Hong Kong. Dashed lines indicate a decrease in temperature during the past 24 h. The thickened line is the 0°C change. (Adapted from WC95)

The apparent stationarity of the SMH found by WC95 may appear to be inconsistent with the results of Ding and Krishnamurti (1987) and those of Zhang *et al.* (1997) who were able to identify tracks of anticyclones coming out of the SMH (Figs. 11a and b respectively). Such an apparent discrepancy may be explained as follows. The surface wind fields in Fig. 9b show an abrupt change in wind direction at around 25°N, which is characteristic of a frontal boundary (see also Fig. 10b for the temperature change). It can thus be interpreted that a high-pressure centre is likely to be present to the north of this boundary. The likely scenario is that the main center of the SMH remains quasi-stationary but pockets of cold air propagate out of this main center in response to an upper-level forcing (see section 3.2.6). Each of these pockets shows up as an anticyclone that migrates eastward and/or southward, and hence the results from the various studies are consistent with one another.

In Fig. 9, the composite surface pressure at the centre of the SMH is ~1045 hPa. However, for individual cases, a surface pressure as high as 1085 hPa has been reported (Zhao and Ding 1991) and values of > 1050 hPa are not uncommon (Zhang *et al.* 1997; Zhang and Chen 1999). The evolution and dynamics of the SMH will be explored in section 3.2.7.

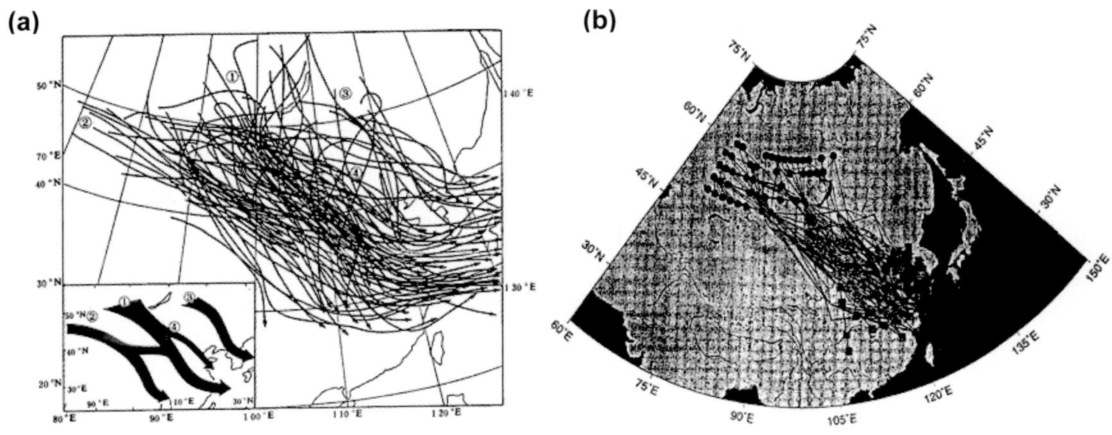

Figure 11. (a) Trajectories of anticyclones originating from the SMH crossing into China for the five winters (December to February) of 1980–1984. The inset on the bottom left gives a schematic of these tracks (from Ding and Krishnamurti 1987). (b) Trajectories of surface anticyclones associated with cold surges (defined by Zhang *et al.* 1997) for the period October to April of 1979/80–1994/95. The origin of the cold air and the end of the cold surges are denoted by circles and squares respectively. (from Zhang *et al.* 1997)

3.2.4. Gravity Wave Characteristics

Based on previous observational studies of the EAWM, Lim and Chang (1981) performed a theoretical study on the nature of the NS and suggested that the pressure surge could represent a transient, gravity-wave like motion due to a pressure-wind imbalance, which then allows a fast propagation of energy from the midlatitudes to the tropics. To test this hypothesis, Chang *et al.* (1983) carefully examined meteorological data at various stations near and within the SCS for NS events that occurred during the Winter MONEX. By plotting the time series of surface pressure and dew point, they were able to identify two stages of the surge for over half of the cases studied, the first one being a pressure rise (called the "edge") and the second one a significant drop in dew point (the "front"). The example in Fig. 12 shows how the edge and front times were identified. By comparing the edge times at various stations, they estimated that the edge (pressure surge) travels at a speed of ~ 40 m s^{-1}, which is much faster than the advective speed (generally ~ 10 m s^{-1}) and is therefore suggestive of a

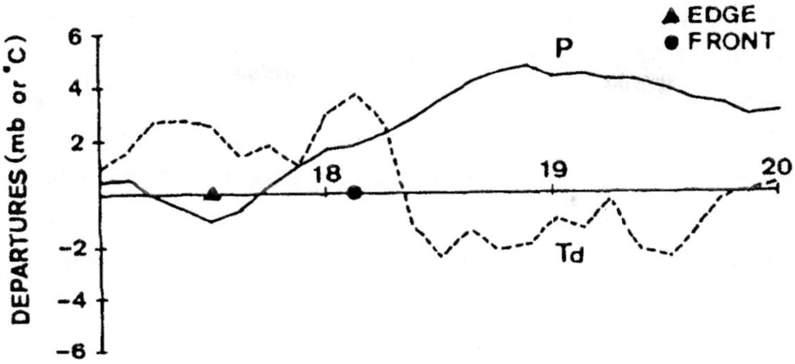

Figure 12. Example of how the edge and front times of a surge are defined based on the time series of pressure (solid) and dew point (dashed) at Taipei. The departures are from seasonal averages. (from Chang *et al.* 1983)

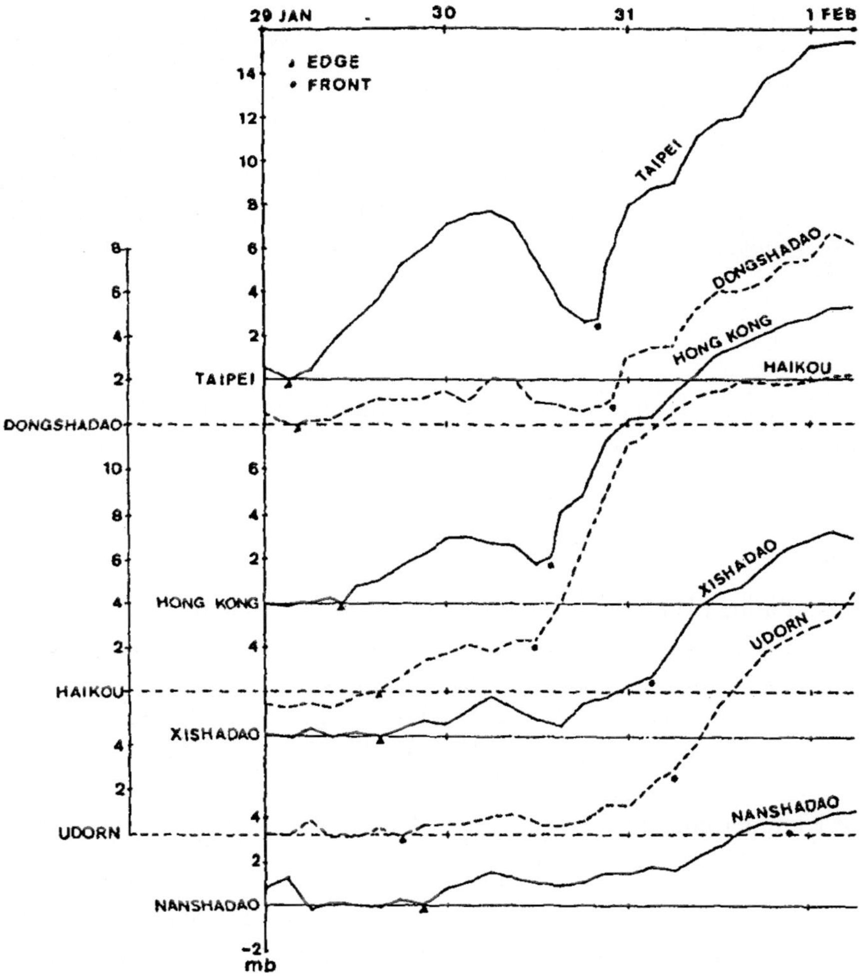

Figure 13. Pressure time series at a number of stations during the passage of an NS between 29 January and 1 February 1979. The scale for each station is shown at the respective zero line. (from Chang *et al.* 1983)

gravity-wave like character. An example of this propagation is given in Fig. 13. The same NS also produced a drop in dew point but the speed of the front was only ~ 10 m s^{-1}. More recent studies by Chen and Huang (1989) and Zhou (1989) that included ship reports over the SCS further confirmed the gravity-wave characteristics of the pressure surge over the SCS.

Lu and Zhu (1990) showed, using numerical simulations and a theoretical analysis, that the cold surges mainly result from a topographic gravity waves with a phase speed of ~ 30 m s^{-1}. In the following, a simple theoretical analysis is presented to illustrate their results.

The linear shallow-water equation without a basic flow can be written as

$$\begin{aligned} &\frac{\partial u}{\partial t} - f_0 v = -\frac{\partial \phi}{\partial x} \\ &\frac{\partial v}{\partial t} + f_0 u = -\frac{\partial \phi}{\partial y} \\ &\frac{\partial \phi}{\partial t} + g\frac{\partial (H-h)}{\partial x} + g\frac{\partial (H-h)}{\partial y} = 0 \end{aligned} \tag{1}$$

where u and v refer to the velocity components in the x and y directions respectively; ϕ the geopotential, g the acceleration of gravity, and H the mean depth of the fluid. The topographic altitude h is assumed to have the following form:

$$h = h(x) = h_0 - bx \tag{2}$$

where $0 \le x \le x_0$, and b is slope of the topography in the x direction.

Assuming the wave solution of (1) has the form

$$(u, v, \phi) = (U, V, \Phi)e^{i(my - \omega t)} \tag{3}$$

and after substituting (3) into (1), the following equation is obtained:

$$\zeta\frac{d^2K}{d\zeta^2} + (1-\zeta)\frac{dK}{d\zeta} + \left[\frac{\omega^2 - f_0^2}{2gmb} - \frac{f_0}{2\omega} - \frac{1}{2}\right]K = 0 \tag{4}$$

This is the Kummer equation. The function $K(\zeta) = \Phi e^{\zeta/2}$ is called the Kummer function, and $\zeta = 2m[x - (h_0 - H)/b]$. Under the condition

$$K|_{\zeta\to\infty} = O(\zeta^m) \tag{5}$$

the eigenvalues for (4) are

$$\frac{\omega^2 - f_0^2}{2gmb} - \frac{f_0}{2\omega} - \frac{1}{2} = n, \qquad n = 0, 1, 2 \cdots \tag{6}$$

and the angular frequency **ω** can be represented by

$$\omega_{1,2} \approx \pm\sqrt{f_0^2 + 2mgb\left(n + \frac{1}{2}\right)} \tag{7}$$

which are inertia-gravitational waves that propagate southwards/northwards, and

$$\omega_3 \approx -\frac{mgbf_0}{f_0^2 + 2mgb\left(n + \frac{1}{2}\right)} \tag{8}$$

which is a topographic gravity wave [since it disappears when there is no topography (b = 0)] with a lower frequency. Applying the typical values of $b = 0.015$, $f_0 = 10^{-4}$ s^{-1} and m ($= 2\pi/L_y$) = 3.14 x 10^{-6} m^{-1}, the topographic gravity wave has a southward phase speed c_3 = - 30 m s^{-1}, which is similar to the result based on observational data.

3.2.5. Lower-Tropospheric Features

The geopotential height distributions at 1000 hPa are rather similar to those at the surface, with the center of the SMH anchored to the southwest of Lake Baikal (Fig. 14a). The average height value at the center increases from ~280 m at day –1 to ~310 m at day 0 (not shown). The cold pocket that moves out of the SMH can best be seen at 850 hPa (Fig. 14b) where a

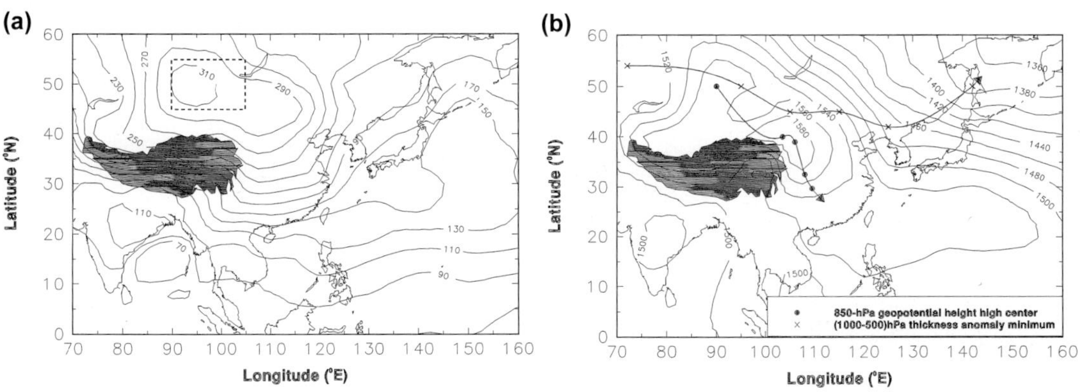

Figure 14. Composite geopotential height (gpm) at (a) 1000 hPa and (b) 850 hPa of an NS on day 0. In (a), the dashed box defines the region of the SMH. In (b), the solid curves indicate the track of the (local) maximum 850-hPa height center from 1200 UTC on day –1 to 1200 UTC on day 1 (bottom track, dots every 12 h) and the 1000–500 hPa (local) minimum thickness anomaly from day –3 to day 2 (top track, crosses every 12 h). Blackened area indicates the Tibetan Plateau with altitude greater than 3 km. (Adapted from WC97)

southeastward trajectory is apparent, which is consistent with the result of Ding and Krishnamurti (1987) and Zhang *et al.* (1997). The location of the 1000–500 hPa minimum thickness shown in Fig. 14b also suggests that the lower tropospheric air mass migrates eastward, which follows a trough passage at 500 hPa (see section 3.2.6). Such a vertical tilt is consistent with the development of a baroclinic wave (Holton 1992).

3.2.6. Upper-Tropospheric Features

The propagation of the jet stream is best identified from the zonal wind variations at 500 hPa (Fig. 15). At day –3, zonal winds over the Tarim Basin strengthen , which corresponds to the passage of the polar jet from upstream near Lake Balkhash (45°N, 75°E) (not shown). This jet streak propagates eastward and then merges with the subtropical jet by day –1 (Fig. 15a) to form the East Asian jet that continues to migrate eastward to Japan by day 0 and day 1 (Figs. 15b and c). This configuration is similar to the climatological pattern shown in Fig. 6a. The jet further strengthens on day 2 and day 3 due to a return flow from the equatorial region [not shown; see Chang and Lau (1980); Ding and Xiao (1992) or WC97 for a discussion of this feature], which is basically an enhancement of the Hadley cell leading to northward transport of angular momentum so that the subtropical jet or East Asian jet is strengthened (see also Palmén and Newton 1969).

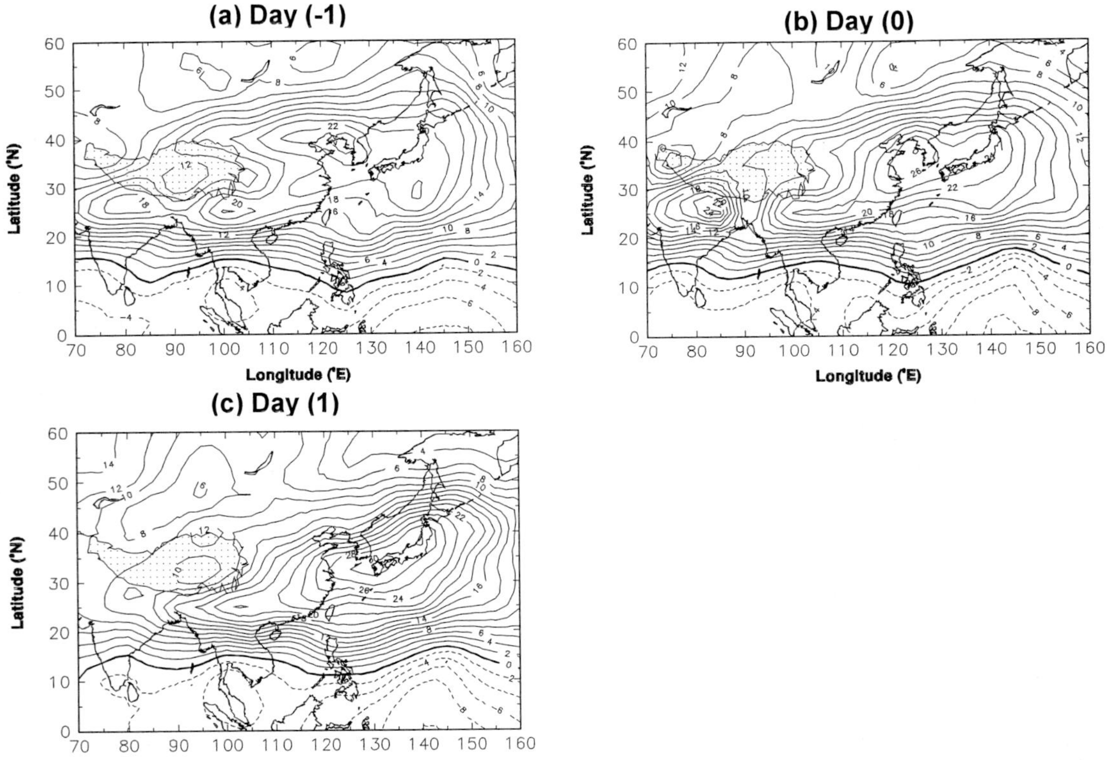

Figure 15. Composite 500-hPa zonal wind speed of an NS from day –1 to day 1. Stippled region indicates the Tibetan Plateau with altitude greater than 3 km. (Adapted from WC97)

This eastward propagation of the polar jet streak is consistent with the migration of the center of the cold anomaly shown in Fig. 14b. Thus, while the East Asian jet reaches a maximum on day 0 or day 1, which is consistent with the result of Chang and Lau (1980), the strengthening of the polar and subtropical jets appear to follow the establishment and southward push of the cold air. This in fact agrees with Bluestein's (1992) comment that the migration of cold or warm air could be responsible for the meandering of the jet stream through the thermal wind relationship.

3.2.7. The Siberia-Mongolia High

Since the NS originates from the SMH, it is of interest to examine the synoptic-scale features associated with this high-pressure system. Different authors have chosen different areas to be the boundary of the SMH (see e.g. Ding *et al.* 1991; Ding 1994; WC97). Here the results from WC97 will be used to illustrate the dynamics involved. Hence, the area selected has the following boundaries: 45°N – 55°N, 90°E – 105°E (marked in Fig. 14a).

The passage of an upper-level shortwave trough over the SMH is obvious from the time-pressure cross-section of relative vorticity (Fig. 16a) and geopotential height (Fig. 16b) averaged over the SMH domain. Strong cyclonic vorticity is observed at 300 hPa at around day –2 to day –1 (Fig. 16a) while the decrease in geopotential height appears to occur a little earlier (Fig. 16b). Below ~ 700 hPa, the geopotential height anomaly increases from ~day –1, which is accompanied by strong divergence (Fig. 16c). The increase in geopotential height anomaly and positive divergence suggests that the SMH is intensifying. Prior to this, the level of maximum divergence is actually at the upper levels, with weak convergence at the low levels. This implies that between days –3 and –4, strong rising motion occurs in the mid troposphere. This can also be inferred by examining the vertical distribution of absolute vorticity advection (Fig. 16d). During this time, absolute vorticity advection increases with height in the mid to upper troposphere. A similar pattern is observed in the potential temperature advection variations (not shown – see WC97). Then, as the shortwave trough passes, the vertical variation of vorticity advection reverses sign, which implies sinking motion. Notice that near day –1 and day 0, while the strongest vertical gradient of vorticity gradient is in the mid troposphere, the divergence extends all the way up to the upper troposphere (Fig. 16c). This means that the sinking motion does not result from the passage of the upper-level shortwave. Rather, it is induced by the strong divergence near the surface. Then at day 1, the anticyclone upstream of the shortwave (Fig. 16a) leads to an increase in geopotential height (Fig. 16b), which forces convergence through a deep layer (Fig. 16c) and a subsequent warming (not shown) and the weakening of the cold pool.

3.2.8. Physical Processes Associated with the Evolution of the NS

The previous sub-sections have discussed in detail the various features associated with the evolution of the NS. Based on these features, the physical processes involved in such an evolution may be described as follows. The initiation of the NS begins with the eastward passage of a polar jet streak west of Lake Balkhash. As the jet streak passes over the SMH domain, the thermally indirect transverse circulation at the exit region of the jet, coupled with the circulation ahead of the trough in a developing baroclinic wave, forces strong rising motion over the SMH, which leads to strong cooling and an intensification of the SMH (Fig. 17a). With the continued eastward propagation of the jet streak, the SMH domain is then in the entrance region of the jet so that subsidence now occurs over the main part of the domain, which is enhanced by the anticyclonic advection to the backside of the baroclinic trough (Fig. 17b). The sinking motion therefore builds up the SMH to its maximum intensity. An outpour of the cold air in the lower troposphere then occurs, which leads to further upper-level convergence and sinking motion. As this cold dome pushes southward, the polar jet also migrates in tandem and merges with the subtropical jet stream to form the East Asian jet. The subsidence warming eventually over-compensates radiational cooling, and leads to warming of the column. Together with the discharge of the cold air, the SMH then weakens. In the lower troposphere, the southward push of the cold air then excites gravity waves that propagate across the SCS. Convection breaks out over the maritime continent and the northward outflow from the deep convection then causes a second strengthening of the East Asian jet in the vicinity of Japan. This then completes the cycle of an NS.

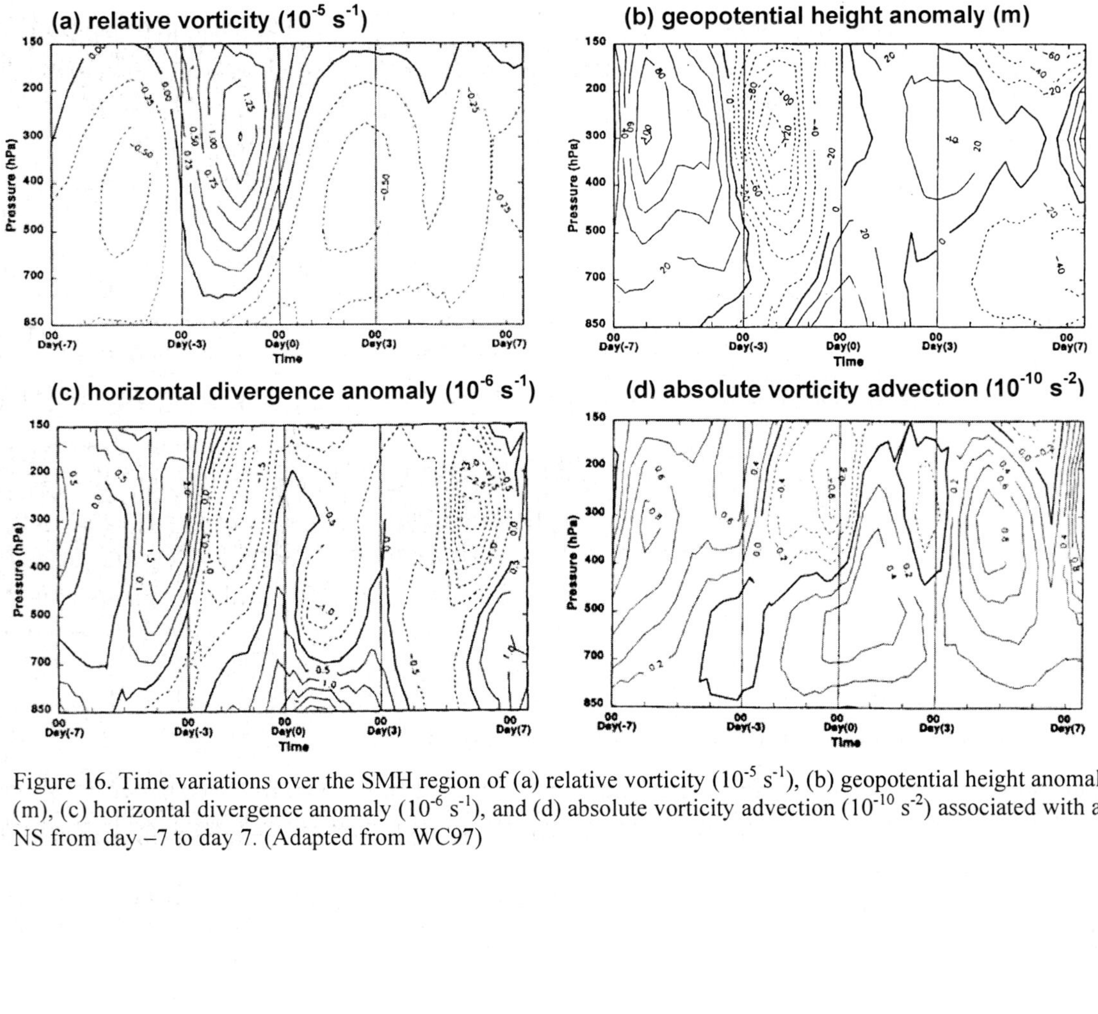

Figure 16. Time variations over the SMH region of (a) relative vorticity (10^{-5} s^{-1}), (b) geopotential height anomaly (m), (c) horizontal divergence anomaly (10^{-6} s^{-1}), and (d) absolute vorticity advection (10^{-10} s^{-2}) associated with an NS from day –7 to day 7. (Adapted from WC97)

While this physical description is mainly extracted from the discussion in WC97, many aspects of the evolution of the SMH and its relationship to cold surges determined by other researchers are similar. For example, Qiu (1988) and Xie *et al.* (1992) all emphasized the importance of ascending cooling in intensifying the cold dome. In two separate case studies, Lu *et al.* (1983) and Gao *et al.* (1992) both related this rising motion to the entrance and exit regions of the polar jet streak. The variations of the vorticity and geopotential height anomalies shown in Figs. 15a and b are also consistent with the composite study results of Ding *et al.* (1991).

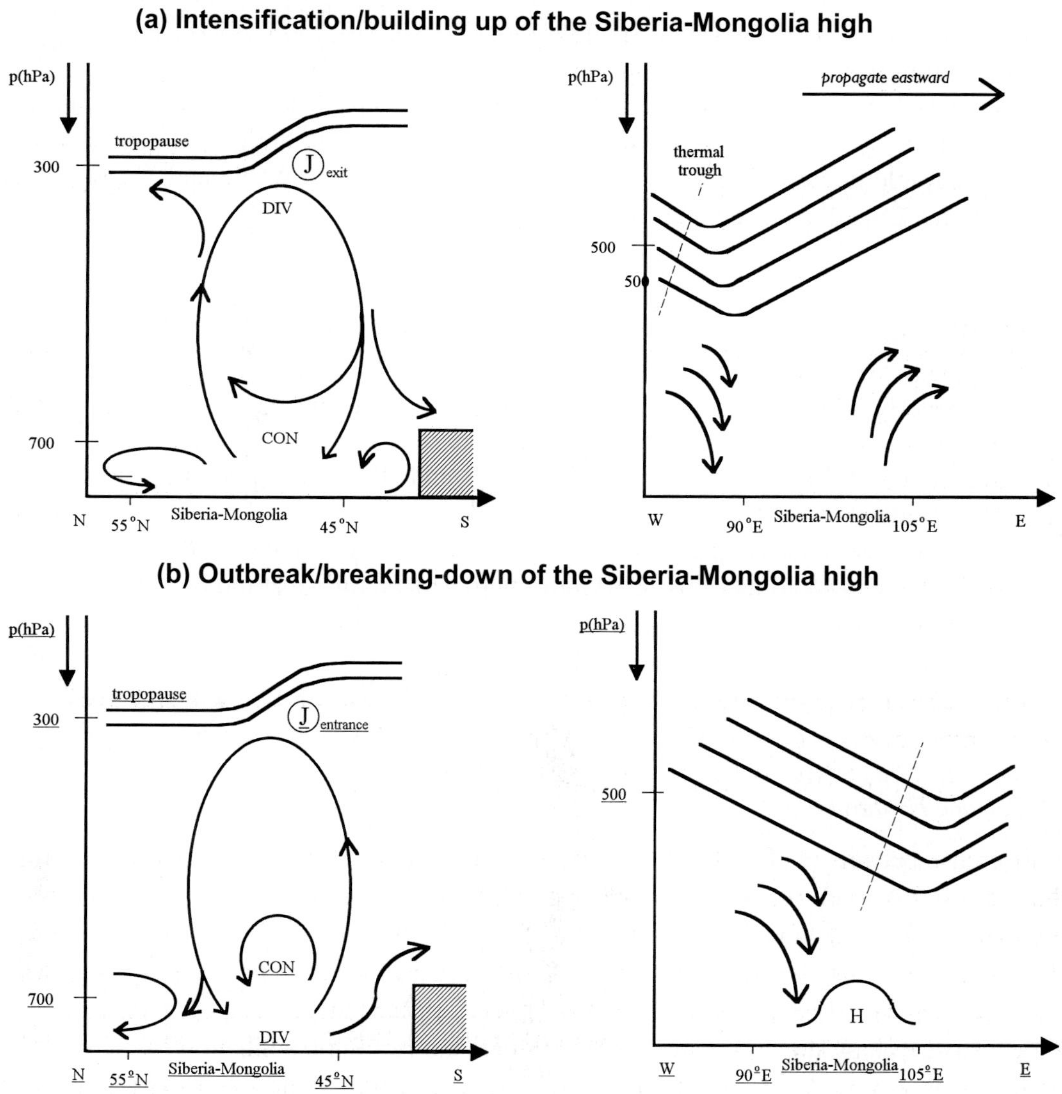

Figure 17. Schematic showing the evolution of the SMH in an NS at the stages of (a) intensification, and (b) outbreak. Left panels show the transverse circulation associated with the jet (marked with J and circled) and the (a) exit and (b) entrance regions. Right panels show the secondary circulation associated with the passage of the height and thermal trough when (a) the shortwave is heading toward, and (b) away from the SMH region. (Adapted from WC97)

Thus, the physical processes presented in this section are likely to be largely valid. With the availability of various reanalysis datasets, further investigations into the veracity of such a description should be carried out in the future, especially with regards to the dynamic and thermodynamic processes involved. Further, the role of topography (the SMH being situated in the Tarim Basin) in containing the cold air during the intensifying stage of the SMH should be examined.

3.3. *Easterly Surges*

As mentioned in section 3.1, the easterly surge (ES) is generally not discussed in most of the literature on the EAWM because relative to the NS, its main effect appears to be primarily in the strengthening of the winds and is quite localized. However, the ES represents a type of surge that is also initiated as a result of the movement of a cold dome and is therefore part of the EAWM. Further, it presents some interesting phenomena along the East China coast that are of considerable interest. Therefore, in this section, results from WC95 and WC97 on the evolution of the ES will be discussed.

3.3.1. Definition

Since the main trait of an ES is in the strengthening of the easterly winds (Morrice 1973), WC95 defined an ES as follows:

(a) the easterly wind component at Hong Kong must be greater than that of the monthly normal value by at least 5 km h^{-1} on day 0; and

(b) the mean wind speed is maximum on day 0 and is strictly increasing (decreasing) before (after) day 0.

For the three-year data sample of WC95, the mean increase of the easterly component at Hong Kong from day –2 to day 0 is ~25 km h^{-1} with a standard deviation of ~ 5 km h^{-1}. On day 0, the mean wind is almost exactly easterly with a average magnitude of 43 km h^{-1}. Throughout the ES, the northerly component is < 10 km h^{-1}.

Not much temperature drop can be identified in the case of ES, with the maximum decrease being less than 1°C at all locations along the coast (WC95).

3.3.2. Surface Features

The difference between an ES and an NS in the surface pressure distribution is in the location of the high-pressure cell center (cf. Figs. 9 and 18). In the case of ES, on day –1, the center is located near Dahingganling to the northwest of the Yellow Sea (Fig. 18a), with the mean central pressure of only around 1032 hPa. It then spreads southeastward and reaches the Yellow Sea by day 0 (Fig. 18b). Notice that there appears to be a relative maximum in surface pressure along the east coast of China. Indeed, a careful examination of the pressure distribution on day 0 shows an obvious coastal ridging from Ningpo to Hong Kong (Fig. 18e). This appears to be similar to the ridging phenomenon found by Holland and Leslie (1986) along the Australian coast. By day 1, the horizontal pressure gradient has relaxed and the center of the high pressure moved to the Korean Peninsula (Fig. 18c), and the surge begins to subside (Fig. 18d).

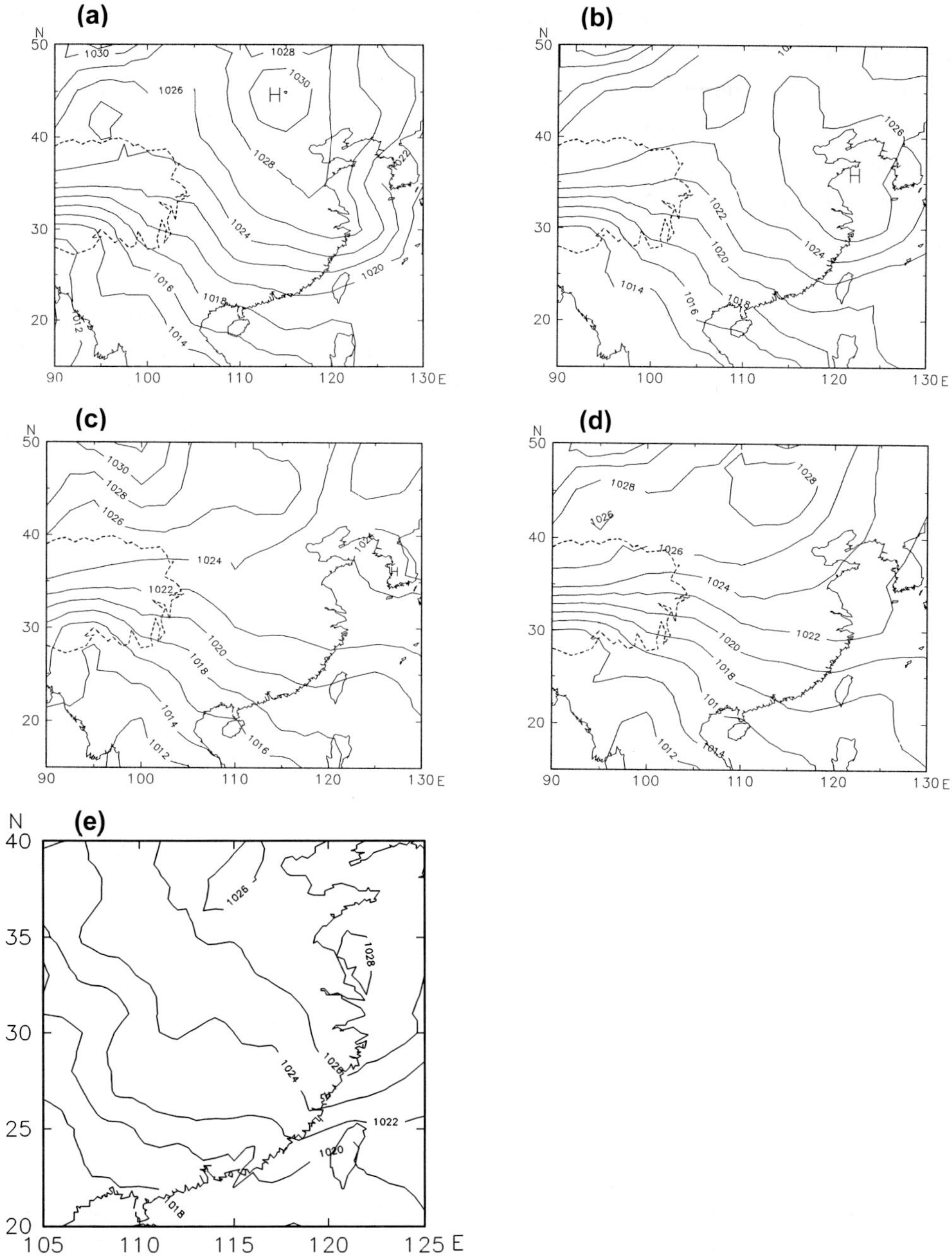

Figure 18. Composite mean-sea-level pressure distribution (contour interval: 2 hPa) during the passage of easterly surges from (a) one day before (day –1), (b) day 0, (c) day 1 to (d) day 2. Day 0 is the day the ES arrives at Hong Kong. (e) is the same as (b) except plotted with higher resolution and on the morning of day 0. (Adapted from WC95)

The coastal ridging is apparently related to a coastal Kelvin-wave-type propagation along the east coast of China, as can be seen from the wind speed anomaly at various coastal stations (Fig. 19). The wind first reaches a maximum at Dachen Dao about one and a half days before day 0. Moving along the coast southward, the maximum wind at Xiamen occurs about one day later. However, further southward propagation is not evident. Nevertheless, the winds over Hong Kong reach the maximum around day 0, by definition, although the easterlies apparently weaken rapidly so that at Yanjiang to the west of Hong Kong, not much increase in wind speed is observed.

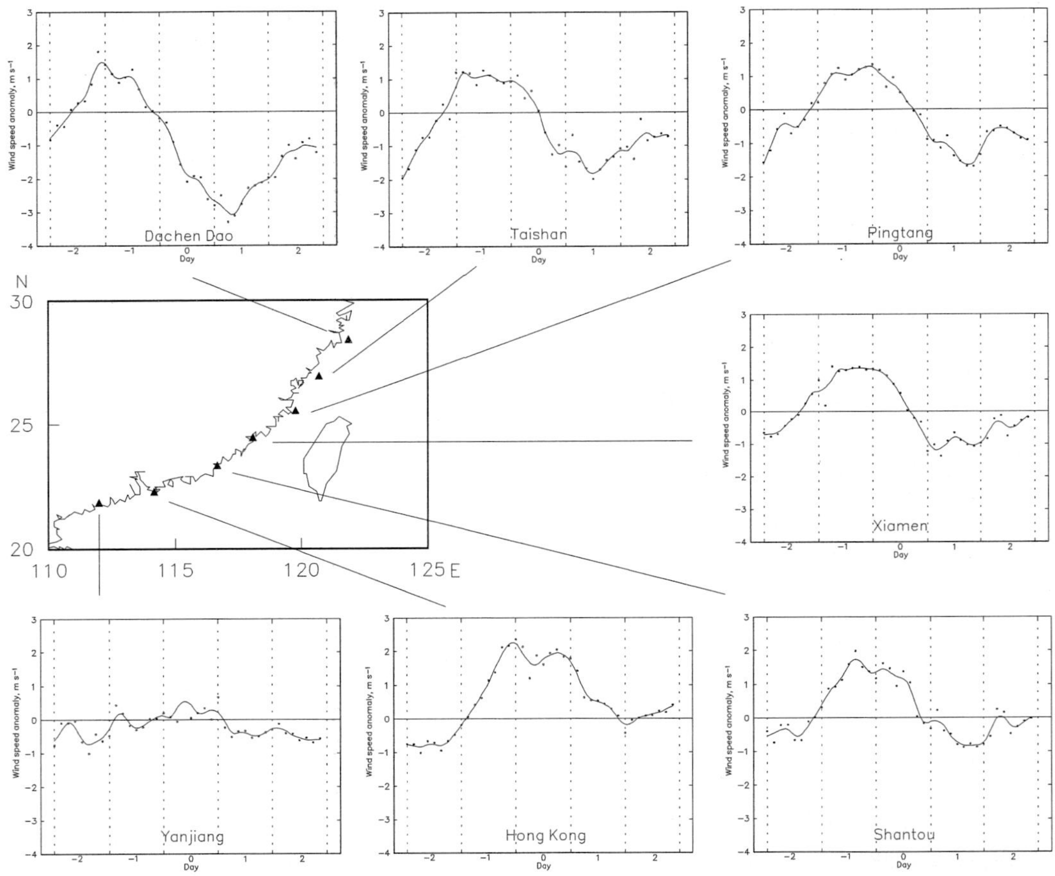

Figure 19. Wind speed anomalies for seven selected coastal stations during an ES. The time interval is 3 h between the dots on each graph. (Adapted from WC95)

3.3.3. Features at Other Tropospheric Levels

The origin of the surface high pressure is best identified from examining the 1000-hPa geopotential height distribution (Fig. 20). On day –3, the geopotential height of the SMH is ~ 290 gpm (Fig. 20a), which is slightly weaker than that for the NS. Then a clear split can be seen on day –2 (Fig. 20b), with the high over Dahingganling migrating southeastward (Fig. 20c), finally reaching the Shandong Peninsula by day 0.

At 850 hPa, a broad anticyclone is seen over central China from day –2 (Fig. 21a). Notice that the center of this anticyclone is actually to the *south* of that at 1000 hPa, which indicates that the high is actually warm. Apparently, the subsidence warming associated with the Dahingganling high over-compensates the radiational cooling and cold advection. By day 0 (Fig. 21b), the 850-hPa flow is actually southerly over the mainland of China, which further causes warming in northern China (not shown – see WC97) and weakens the high. Note that the ridge axis at 850 hPa is almost oriented zonally along the south China coast on day 1 (Fig. 21c), which would bring in warm and moist air from the SCS. Indeed, an ES is generally associated with cloudy or even drizzly conditions over south China (Morrice 1973).

According to WC97, the trigger of the split of the high from the SMH is still not clear. They found the passage of a 500-hPa ridge over the SMH domain from about day –3 (Fig. 22), in sharp contrast to the NS case in which a shortwave passage leads to the intensification of the SMH. However, the physical mechanisms involved were not investigated. In addition, WC97 found that the jet stream is actually weakening in the case of ES and they suggested that perhaps barotropic instability might be responsible for the migration of the high due to a

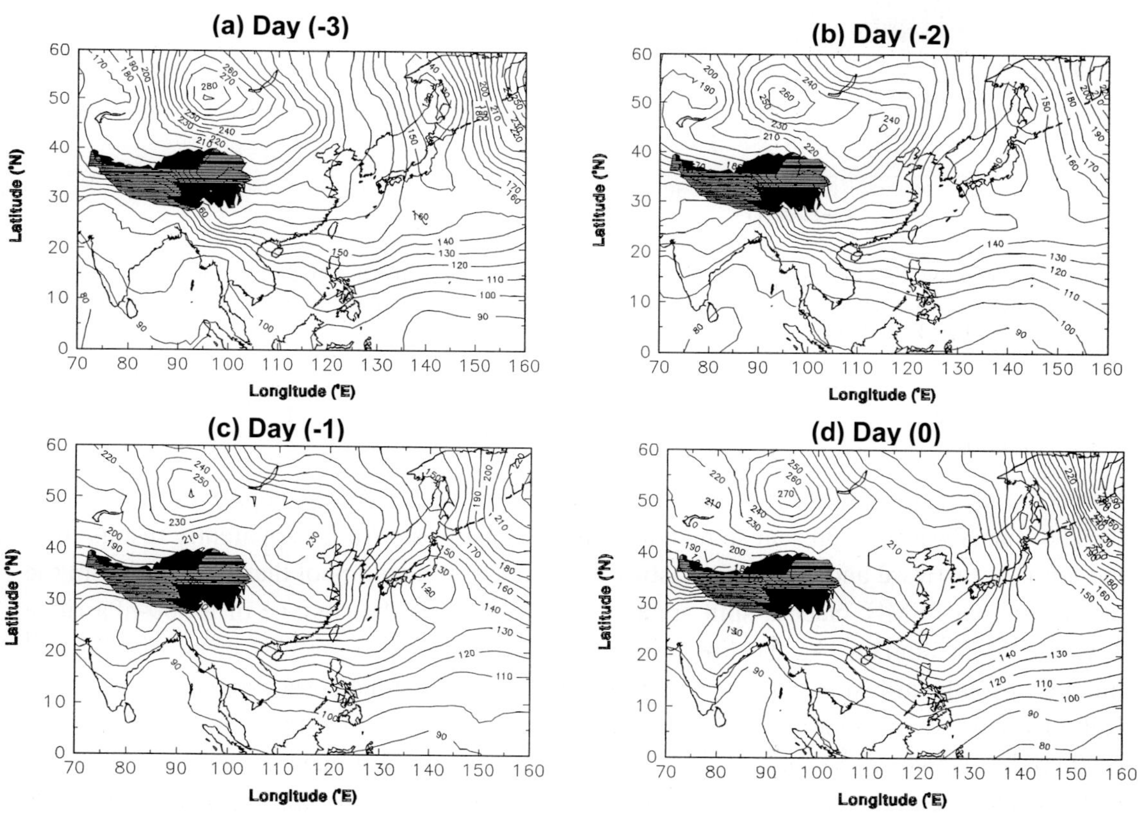

Figure 20. Composite 1000-hPa geopotential heights (gpm) during an ES from day –3 to day 0. Contour interval: 10 gpm. Blackened area indicates the Tibetan Plateau with altitude greater than 3 km. (Adapted from WC97)

conversion of kinetic energy of the upper westerlies to that of the disturbance. However, this proposed mechanism needs to be investigated further before it can be substantiated.

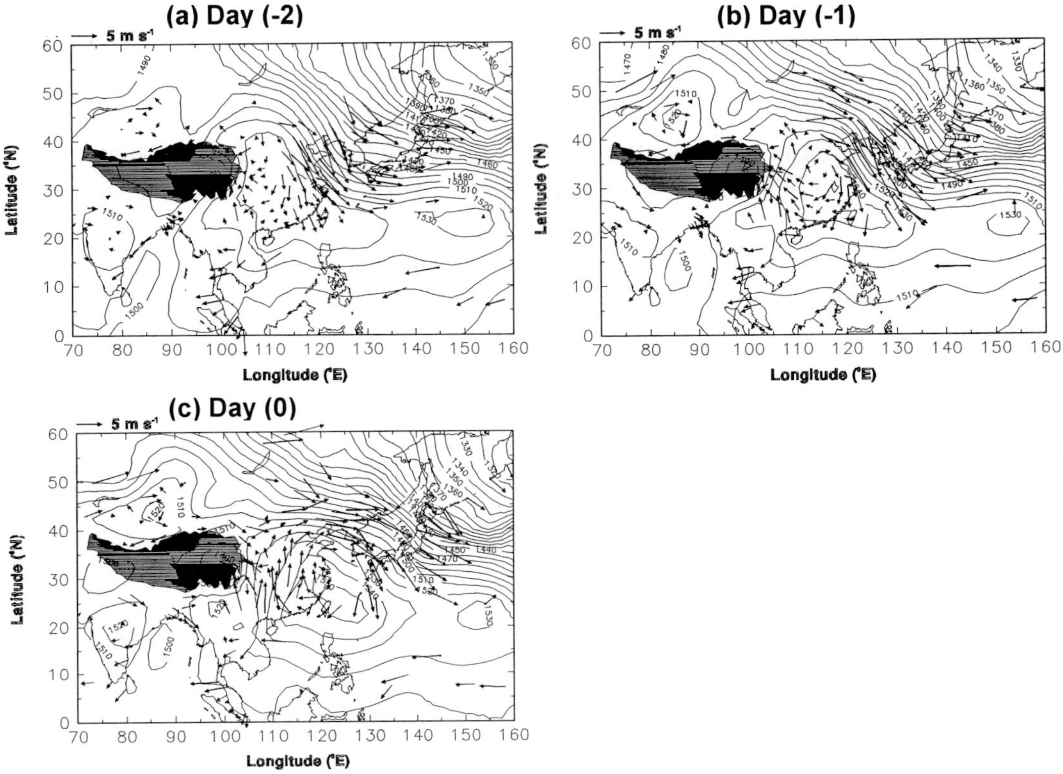

Figure 21. Composite 850-hPa geopotential heights (gpm) and winds (scale indicated on top left corner of each panel) during an ES on (a) day –2, (b) day 0 and (c) day 1. Blackened area indicates the Tibetan Plateau with altitude greater than 3 km. (Adapted from WC97)

3.3.4. Summary

While the changes in the atmosphere associated with the ES do not appear as dramatic as those of the NS, it does present some interesting observations. An ES apparently results from the initially eastward and then southeastward migration of a cold pool that splits off from the quasi-stationary SMH. Such a split appears to be related to the passage of a 500-hPa ridge over the SMH domain. As the low-level anticyclone moves to the east coast of China, it initiates a southward surge of cool air and strong winds along the coast, resembling a coastal Kelvin wave. By the time the wave reaches the south China coast, the winds strengthen to > 40 km h^{-1}. However, because the center of the anticyclone continues to move eastward, the winds cannot be maintained and therefore an ES is a short-lived surge. In addition, because of the shallow nature of the high, it weakens not only due to subsidence warming but also warm advection from the south.

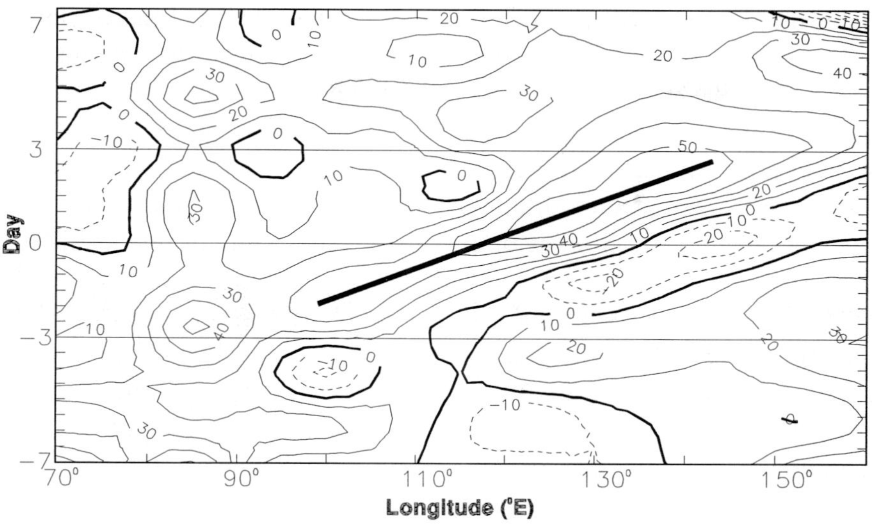

Figure 22. Hovmöller diagram of the 500-hPa geopotential height anomalies (gpm) averaged between 25°N and 35°N associated with an ES. The solid line indicates the passage of the ridge. (Adapted from WC97)

4. Explosive Cyclogenesis off the East Asian Continent

In addition to the monsoon surges described in the last section, an important feature of the winter monsoon is the often-explosive development of low-pressure systems off the East Asian coast associated with the eastward movement of the cold air off the continent. As the cold air encounters the relative warm ocean over the East China Sea, and the Kuroshio Current further east, a strong baroclinic zone is established. Rapid cyclogenesis often occurs. A similar situation occurs over the Atlantic as cold air moves off the North American continent and over the Gulf Stream (Sanders and Gyakum 1980).

In this section, a brief discussion of the climatology of such explosive cyclogenesis events will be given, followed by the physical processes involved in the development of such events.

4.1. *Climatology*

Different researchers studied different cyclones within different periods and therefore one must be careful when establishing a general climatology of cyclones associated with the EAWM. Asai *et al.* (1988) showed three principal cyclone tracks in the East Asia region (Fig. 23). The group of cyclones that originate from Mongolia and subsequently move across northeastern China to the Sea of Okhotsk is associated with the polar jet, and hence is generally not considered as being related to the EAWM. Thus, these can be eliminated from the discussion. Interested readers can consult the paper by Asai *et al.* (1988). The group of cyclones along ~ 30°N is obviously related to the EAWM based on the previous discussion.

Ding and Krishnamurti (1987) and Zhang *et al.* (1997) found tracks of surface anticyclones that originate from the SMH crossing the Yellow Sea and the Korean Peninsula (see Figs. 11a and b). Therefore, the middle group of cyclones should also be included. It must be cautioned, however, that in counting the number of cyclones, some researchers do not separate those with different origins so that the climatology among different researchers can be quite different. To be consistent with the theme of this chapter, which is on the EAWM, only those cyclones that develop from anticyclones originating from the SMH will be considered.

In most studies, a cyclone is counted if, based on sea-level pressure analyses with a 2-hPa interval, the low-pressure system has at least one surface closed isobar and generally lasts for at least over one day (Hanson and Long 1985; Asai *et al.* 1988; Yi and Ding 1993). A cyclone that deepens more than 24 hPa in 24 h is defined as an explosive cyclone, which follows the definition of Sanders and Gyakum (1980). Yi and Ding (1993) found that during the period 1973–88, about 20% of the EAWM-related cyclones went through the explosively deepening phase. In the following, only these explosive events will be discussed.

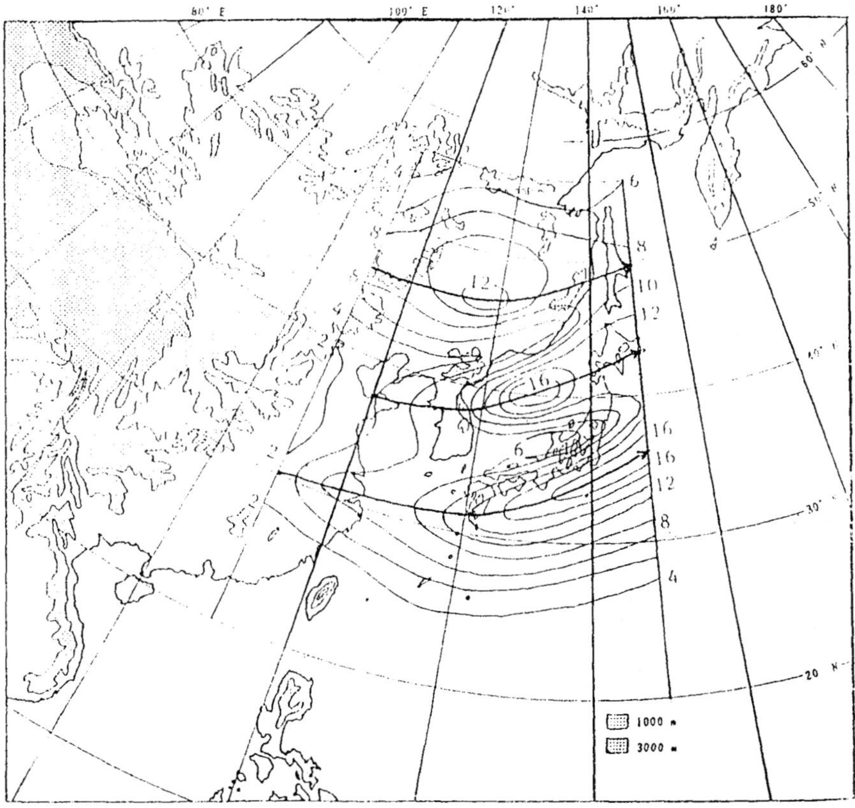

Figure 23. Frequency distribution of cyclones during the period 1961–1985. Thick solid lines and arrows indicate principal cyclone tracks. (from Asai *et al.* 1988)

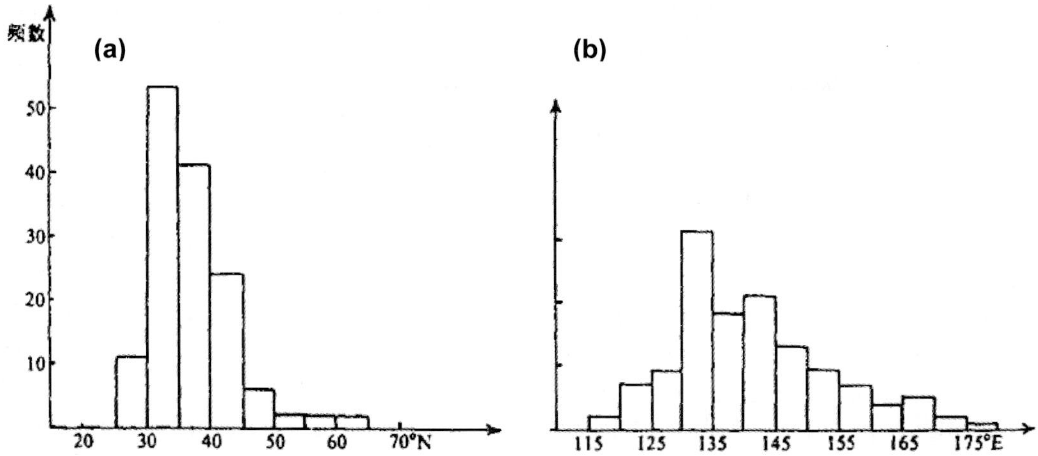

Figure 24. (a) latitudinal (averaged between 115°E – 180°), and (b) longitudinal (averaged between 20°N – 70°N) distributions of the number of explosive cyclones originating from China at the initial time of their explosive deepening phase during the period 1973–1988. (from Yi and Ding 1992)

The latitudinal and longitudinal frequency distributions of explosive cyclones suggest a concentration near the latitude band of 30°N – 35°N and the longitude band of 130°E – 135°E (Fig. 24). This result suggests that while cyclones mostly develop off the East Asian coast (Hanson and Long 1985), they generally do not deepen explosively until they move near the Kuroshio Current south of Japan where the SST gradient is apparently the strongest. The distribution given in Fig. 24b for the area west of 130°E, which gives about 1 explosive cyclone per year, is consistent with that of Dong and Li (1989) based on the data from 1966–85. Note, however, the average number of explosive cyclones is around 10 per year. If the origin of the cyclones is not considered, the number of explosive events over the entire western North Pacific reaches about 20 per year (Ouyang *et al.* 1990), which implies that half of the explosive cyclones are associated with the EAWM.

While it might be expected that the explosive events occur mostly in winter, the monthly distribution of the absolute maximum actually occurs in March (Fig. 25). The total frequencies of occurrence for the winter (December to February) and spring (March to May) months are about equal. The distributions of the total number of cyclones around similar regions found by Hanson and Long (1985) and Asai *et al.* (1988) also give a maximum frequency in March and April (not shown).

Yi and Ding (1993) also classified the intensity of explosive cyclones based on the scale proposed by Sanders (1986). They found that west of 130°E, strong cyclones (> 1.8 Bergeron[a]) occur only about once every 5 years (Table 1), but the frequency of occurrence

---

[a] 1 Bergeron = 24 hPa/24 h x ($\sin \phi / \sin 60°$), where ϕ is the latitude of the center of the cyclone (Sanders 1986).

between 130°E and 180° is close to 1.5 per year. Nevertheless, most of the cyclones, irrespective of their location, are in the moderate to weak categories.

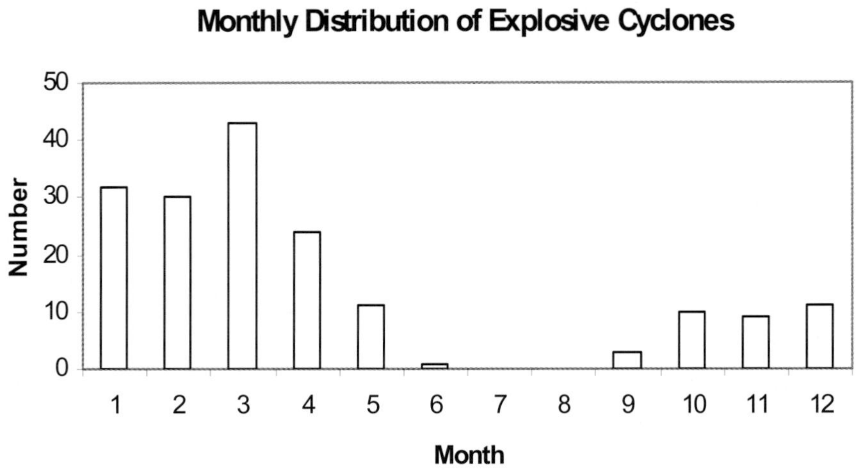

Figure 25. As in Figure 24 except for the monthly distribution. (drawn from the data in Yi and Ding 1992)

Table 1. Number and percent of total (in parentheses) of explosive cyclones grouped by intensity (in units of Bergeron) and region. (constructed from the data in Yi and Ding 1992)

Region	**Strong (> 1.8)**	**Moderate (1.3-1.8)**	**Weak (1.0-1.2)**
west of 130°E	3 (15%)	10 (50%)	7 (35%)
130°E - 180°E	22 (13%)	77 (45%)	73 (42%)

4.2. *Physical Processes*

Most of the observational or numerical investigations of the physical processes associated with the genesis and development of explosive cyclones are based on case studies. Even when many cases were included (e.g. Li and Ding 1989; Sun and Gao 1993), the diagnoses were performed on one or two specific cyclones. Although no two cyclones are exactly identical, it is possible to identify some common characteristics and processes that various researchers have determined to be important from their case studies.

Nitta and Yamamoto (1974) pointed out that oceanic cyclones can be classified into two types, (1) the "polar low" (Reed 1979; Mullen 1979) type with a horizontal scale of < 1000 km, and (2) the intermediate-scale cyclone with a horizontal scale of 1000–2000 km. The low-pressure systems generated off the East Asian coast usually fall within the first category although they may not have all the characteristics of the polar low. Intermediate-scale cyclones tend to form further east to the south of Japan. They tend to be shallower systems (Chen *et al.* 1985) and do not deepen explosively. Therefore, the focus in this section will be on the former type of cyclones.

In all the studies, latent heat release is identified as the crucial factor in determining whether a cyclone can deepen explosively. In a numerical study of the explosive development of a cyclone over the Kuroshio Current, Xu and Zhou (1999) found that by turning off latent heating, the deepening of the simulated cyclone was much less. Chen and Dell'Osso (1987) obtained similar results in a case study of a cyclone that deepened ~ 20 hPa in 24 h. Without latent heating, the modeled cyclone remained a shallow feature. The diagnostic analyses of Li and Ding (1989) and Yi and Ding (1992) also found the magnitude of latent heating to be the maximum during the explosive developing phase of these cyclones.

The latent heat release necessary for the development of these cyclones is generally attributed to the presence of strong rising motion. Mechanisms that are apparently responsible for causing the ascent include low-level convergence due to sensible heating of the boundary layer, upper-level divergence and positive vorticity advection due to the passing of a shortwave trough, vertical differential temperature advection, and secondary circulation at the exit region of a passing subtropical jet streak (e.g. Li and Ding 1989; Yi and Ding 1992; Sun and Gao 1993). Horizontal advection of heat and moisture into the low-pressure system by an ageostrophic low-level jet has also been identified as a mechanism that increases the instability of the atmosphere to allow the conversion of available potential energy to eddy kinetic energy of the cyclone.

The latent heating not only provides the energy necessary for the development of the system, but can also feed back to amplify either the shortwave trough (Xu and Zhou 1999) or the jet streak circulation (Chen and Dell'Osso 1987). It also enhances the rising motion and thus creates a positive feedback process, and hence an explosive development.

In summary, explosive cyclogenesis basically results from the strong baroclinity associated with the cold air moving off the East Asian coast and meeting the relative warm air over the East China Sea where the sea surface temperature is ~ 20°C (Hanson and Long 1985). Warm air also comes from the western flank of the subtropical high. The strong baroclinity then leads to potential instability, rising motion and latent heat release. As the upper-level shortwave trough passes, vorticity and thermal advection as well as the secondary circulation associated with the exit region of the jet streak all contribute towards an enhancement of the rising motion. Latent heating also feeds back to this process and thus explosive cyclogenesis occurs. It should also be noted that in most numerical experiments, sensible heating is found to play only a minor role.

Latent heating is also identified to be the most important factor in the development of the medium-scale cyclones, at least for an oceanic cyclone that developed during the Air-Mass Transformation Experiment in 1975. In fact in this case, the incipient stage of this cyclone was apparently one of the small-scale cyclones described above (Chen *et al.* 1985). The initial development was due to heat flux from the Kuroshio Current that destabilized the polar air mass coming off the East Asian continent. The subsequent rising motion led to latent heat release, which then caused the formation of a low-pressure system at the surface. Further release of latent heat then led to the development of the medium-size cyclone. While latent heating was found to be important in the development, the surface heat and moisture sources also appear to be crucial (Chen *et al.* 1983; Chang *et al.* 1987).

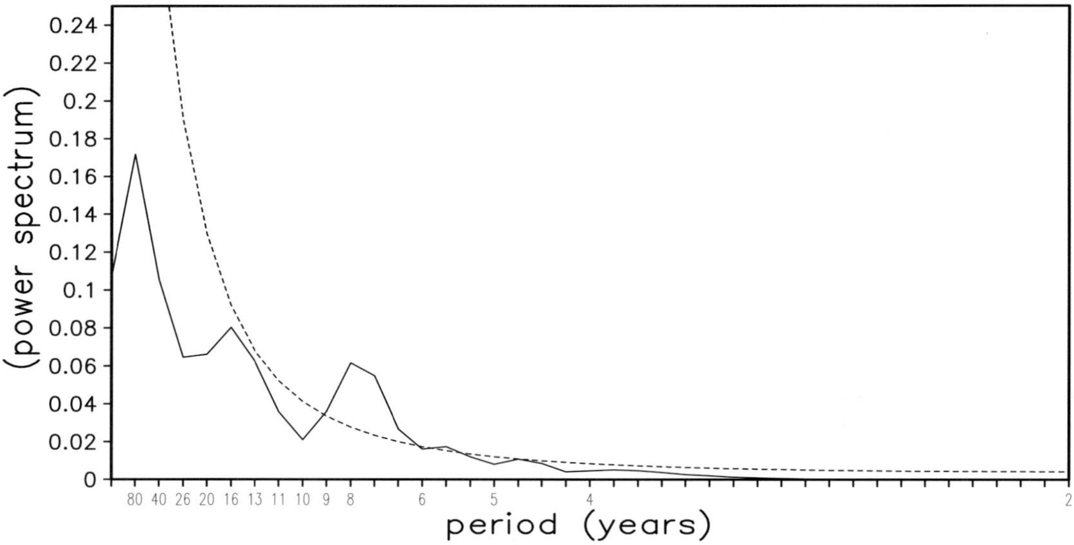

Figure 26. Power spectrum of 1000-hPa meridional winds averaged in the East Asia region (20°N – 30°N, 120°E – 150°E). The dashed curve shows the 95% confidence level.

It can be seen from this review that not many studies have been performed on these explosive cyclogenesis events probably because of the lack of data over the ocean. To understand more of such events requires extensive data coverage of the entire evolution of such systems so that the physical processes can be properly documented. Parameters of particular importance include the moisture transport (whether a critical value is needed for explosive cyclogenesis to occur), convective stability of the atmosphere, and vertical shear. These data are now available from various satellite observing platforms. More detailed studies of this phenomenon should now be possible.

5. Variations of the EAWM

5.1. *Temporal Variations of the EAWM*

The last two sections mainly focused on the synoptic-scale features associated with the EAWM. Since the EAWM is the main planetary-scale circulation of wintertime Asia, its variations should also be related to other circulation features on these spatial scales, the temporal variations of which can have time scales varying from intraseasonal to interannual and even interdecadal. This section therefore examines the variations of the EAWM on these longer time scales and attempts to establish relationships, and if possible causalities, between the EAWM and these other features. The focus here is on the factors that could modify the intensity of the EAWM or frequency of monsoon surges. Of course, the EAWM can also affect the other planetary-scale features such as convection over the maritime continent and ENSO (see Chapter 3 and 5).

5.1.1. Synoptic and Intraseasonal Variations of the Winter Monsoon

While the winter monsoon is always a dominant climate system in the East Asia/western North Pacific region every year, its activity in terms of frequency of occurrence and intensity varies during different months of the winter season. Since most of the work in this area do not distinguish between the NS and the ES (indeed, the surges studied are almost exclusively those belonging to the NS type), the synoptic and intraseasonal variations of the monsoon surges are reviewed here as a separate topic.

As discussed in section 3, the monsoon surges inevitably result from the passing of midlatitude troughs. Correspondingly, previous studies have identified a 6-8-day period associated with the cold-air outbreaks (Tao 1959; Murakami 1979). A spectral analysis of the 1000-hPa meridional winds over the East Asia/western North Pacific region (20°N – 30°N, 120°E – 150°E) indeed shows a peak of 5–8 days (Fig. 26). A similar analysis of the intensity of winter monsoon also shows a similar period (not shown).

Another peak with a 10–20 day period is also evident from Fig. 26. This quasi-biweekly oscillation has also been found in some previous studies (Pan and Zhou 1985; Chen and Xie, 1988). Variations of the EAWM on this time scale are mainly associated with the evolution of the blocking pattern of atmospheric circulation in the midlatitudes especially in the Ural mountain region, where the NS originates (see section 3.2).

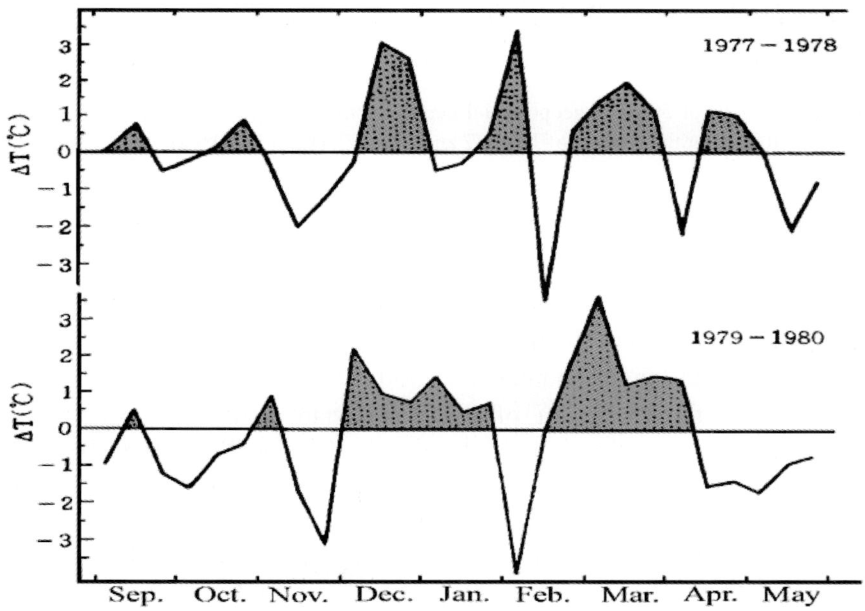

Figure 27. Temporal variation of the 10-day averaged surface temperature anomalies at Guangzhou station. (from Li 1989a)

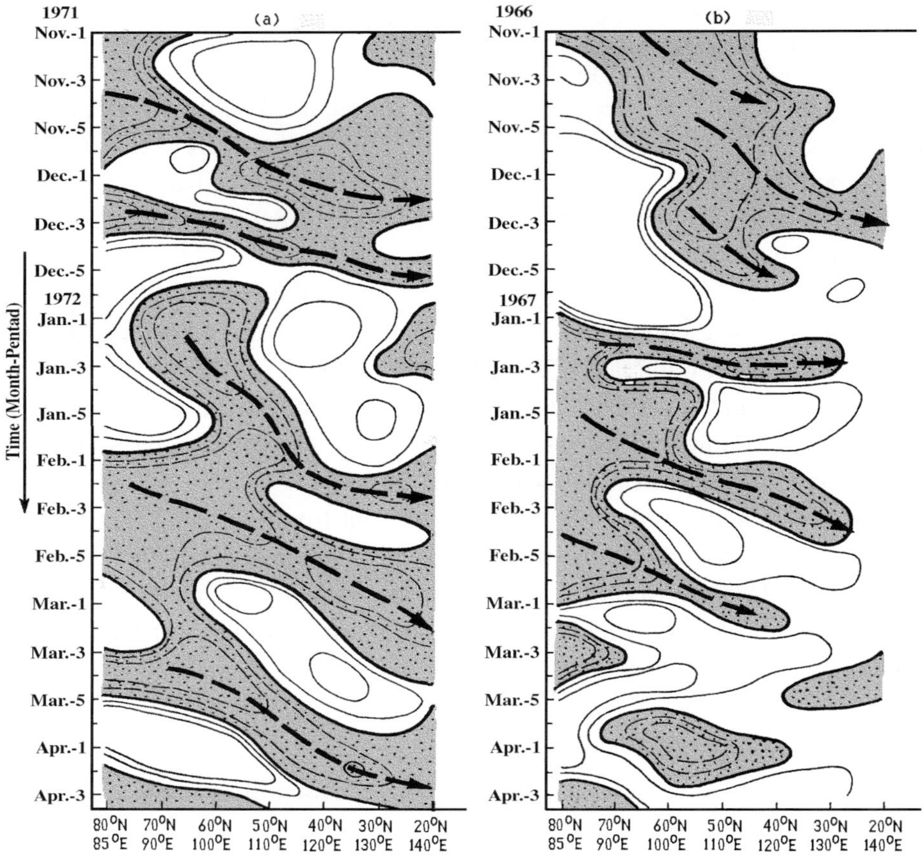

Figure 28. Time-space sections of 500-hPa geopotential height anomalies at along a northwest-southeast section [(80°N, 85°E), (20°N, 140°E)] for the winters of 1971/72 and 1966/67. (from Li 1989a)

On a slightly longer time scale, intraseasonal variations of the EAWM have also been identified (Li 1989a; Zhu *et al.* 1990). As an example, the temporal variation of surface temperature anomalies at Guangzhou in southern China shows a clear intraseasonal (30–60 day) oscillation (Fig. 27). Such oscillations are also apparent at the mid troposphere. A time-section of the geopotential height anomalies at 500 hPa along a northwest-southeast cross-section that represents the migration of the East-Asian troughs shows that the height anomalies propagate with a 30–60 day period (Fig. 28). In stronger EAWM years (e.g. 1971–72), the negative height anomalies can propagate very close to the equator (Li 1989a) than those in weaker EAWM years (e.g. 1966–67). It also means that the intraseasonal (30–60 day) variation is an important feature of the East-Asian winter monsoon activity.

5.1.2. Interannual and Interdecadal Variations of the EAWM

As discussed in section 3, the meridional wind in the lower troposphere in the East Asia/western North Pacific region is a good indicator to represent the activity of the EAWM.

By examining the temporal variation and power spectrum of the surface meridional wind anomalies averaged within the region (20°N – 30°N, 120°E – 150°E), Mu and Li (1999) found strong interannual variations with two major cycles: 2–year and 4–7-year (Fig. 29). The two-year cycle might suggest some type of biennial oscillation and the 4–7-year cycle is likely to be related to ENSO.

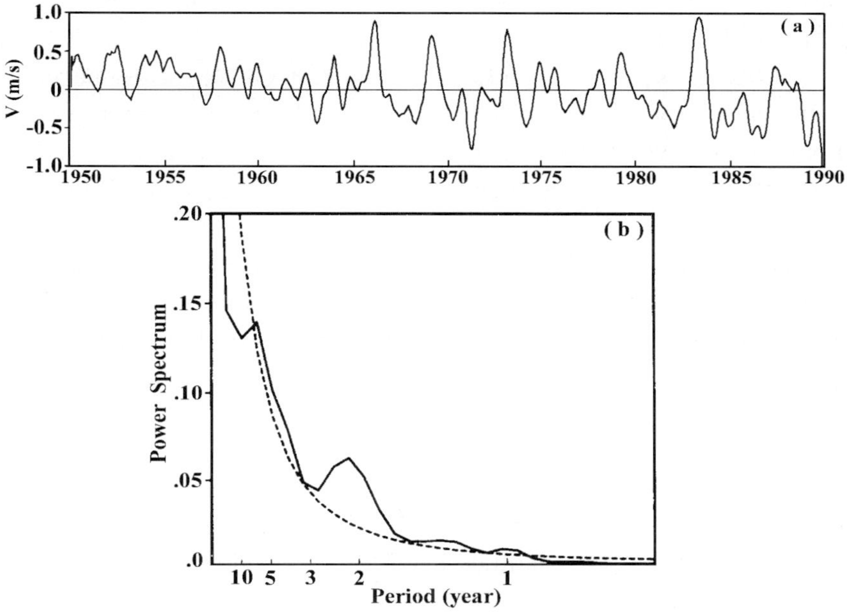

Figure 29. (a) Temporal variation and (b) power spectrum of the surface meridional wind anomalies in the (20°N – 30°N, 120°E – 150°E) region. The dashed curve in (b) shows the 95% confidence level. (from Mu and Li 1999)

Since the data for surface pressure are available much further back in time, they are better suited for studying the long-term variation of the EAWM. The MSLP difference between the region (100°E – 120°E, 35°N – 55°N) and the region (150°E – 170°E, 35°N – 55°N) (former minus latter) is used as a winter monsoon index. A large positive value of this index corresponds to strong northerly winds, and hence strong EAWM. A wavelet analysis of this index for the period 1895–1994 again shows a prominent 4–7-year period (Fig. 30). Although variations with 70–90-year period are difficult to be ascertained due to the relatively short data set, two other oscillations with 7–10-year and 20–30-year periods can be identified. In other words, in addition to interannual variations, the intensity of the EAWM also fluctuates on interdecadal time scales.

Variations in the intensity of the EAWM, particularly on interannual and interdecadal time scales, are usually affected by other planetary-scale systems in the atmosphere and their anomalies. In the following sub-sections (sections 5.2 and 5.3), the impact of some of these systems on the EAWM will be discussed.

5.2. *Relationship between EAWM and ENSO*

5.2.1. Influences of ENSO on the EAWM

The interaction between the EAWM and ENSO has been studied extensively. On the one hand, an anomalous East-Asian winter monsoon is a necessary (though not sufficient) condition for the occurrence of a warm ENSO event (Li *et al.* 1989; Li and Mu 1998; Xu and Chan 2001; see also chapter 5). On the other hand, a warm (cold) ENSO event can weaken (enhance) the EAWM through remote response and teleconnection. In this sub-section, only the latter is discussed. The influence of the EAWM on ENSO is considered in Chapter 5.

Li (1989b, 1995) showed that a warm winter (weak EAWM) usually occurs in East Asia during the following developing to mature phase of an El Niño year but the reverse occurs during a La Niña year. Two processes may be responsible for the influence of ENSO on the EAWM. Firstly, the Hadley and Ferrel cells are both enhanced (reduced) during an El Niño (a La Niña) event (Wu and Cubasch 1987; Li 1990b). As a result, the westerlies in the mid latitudes are strengthened (weakened), which then limits (enhances) and the development of the 500-hPa trough over East Asia is limited. Secondly, an anomalous anticyclonic (cyclonic) circulation usually exists over the western Pacific to the east of the Philippines during the El Niño (La Niña) winter (Wang *et al.* 2000). The anomalous southerlies (northerlies) will therefore weaken (enhance) the cold outbursts towards the south in East Asia and the western North Pacific region.

To demonstrate that the occurrence of an ENSO event indeed affects the subsequent winter over East Asia, the time-longitude sections of composite surface meridional wind anomalies within the 20°N – 30°N-latitude band for 8 El Niño and 6 La Niña cases during 1950–1996 are analyzed (Fig. 31). It is clear that not only do negative (positive) meridional wind anomalies exist prior to the occurrence of an El Niño (a La Niña) event, positive

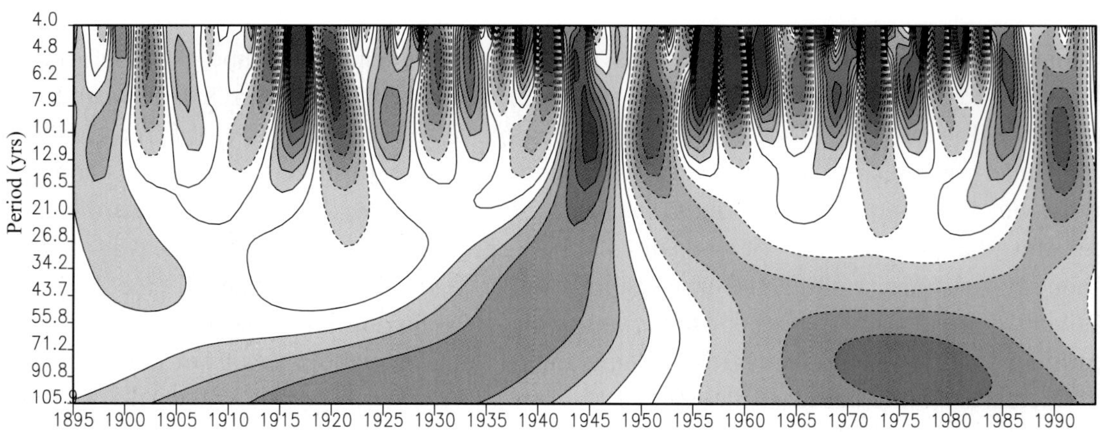

Figure 30. A wavelet analysis of East-Asian winter monsoon index for the period 1895–1994, which is defined by using the sea-level pressure difference between the regions (35°N – 55°N, 100°E – 120°E) and (35°N – 55°N, 150°E – 170°E). The ordinate is period (years).

(negative) meridional wind anomalies are also found during an El Niño (a La Niña) winter. Furthermore, the composite 500-hPa height anomalies during an El Niño (a La Niña) winter show positive (negative) anomalies over the East Asia region (Fig. 32). In other words, the planetary-scale circulations during an ENSO winter are modified, which in turn affects the intensity of the EAWM.

Wu and Chan (2000) compared the intensity of the winter monsoon over South China in ENSO and non-ENSO years, and found the results to be largely consistent with those for the whole of China. That is, in the winter during the mature phase of a warm (cold) event, the EAWM tends to be weak (strong) while no coherent signal can be detected in non-ENSO years, which suggests that factors other than ENSO also control the interannual variability of the EAWM.

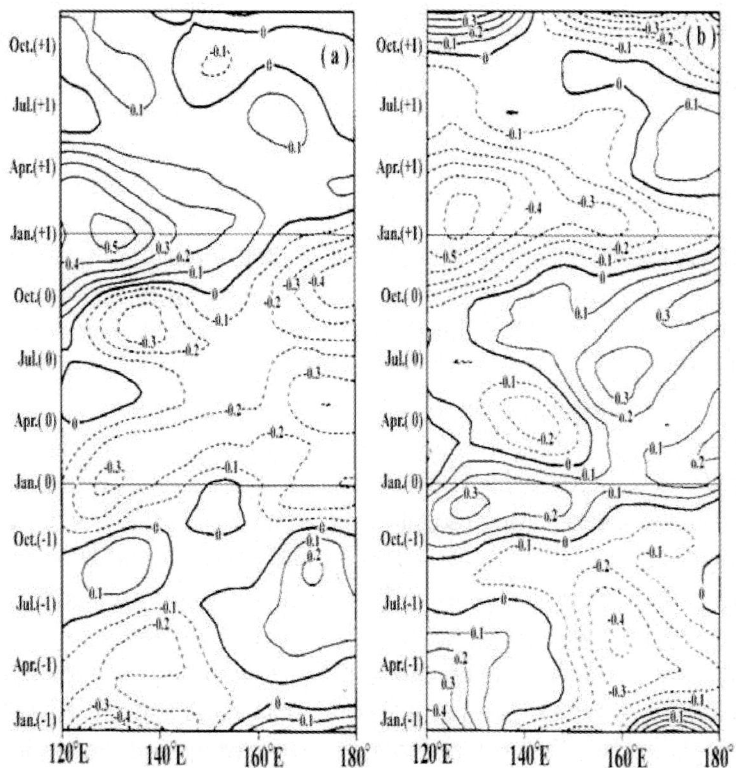

Figure 31. Time-longitude sections of composite surface meridional wind anomalies averaged between 20°N – 30°N latitudes in (a) El Niño years and (b) La Niña years. The labels (+1), (0) and (–1) represent next year, present year and before year of the ENSO occurrence, respectively. (from Mu 2001)

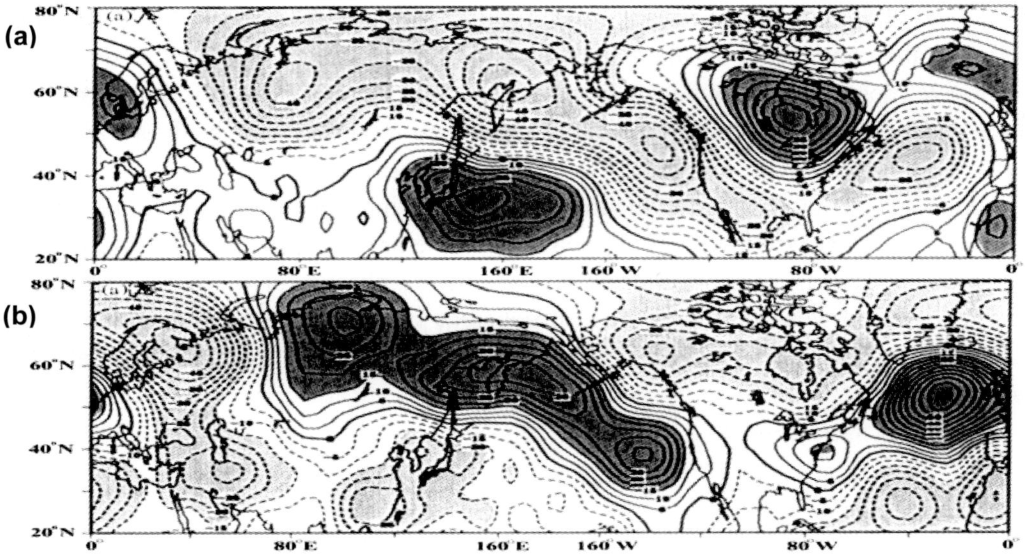

Figure 32. Composite patterns of 500-hPa geopotential height anomalies averaged between 30°N – 40°N in (a) El Niño and (b) La Niña winter. The shaded areas indicate that the statistical significance is > 95%. (from Mu 2001)

5.2.2. Impact of Anomalous EAWM on ENSO

Some studies have shown that the anomalous winter monsoon in East Asia plays an important role in the occurrence of El Niño through exciting westerly wind anomaly and stronger convection (intraseasonal oscillation) over the equatorial western Pacific (Li 1989a, 1990a; Li and Mu 1998).

The impact of the EAWM anomaly on El Niño can be seen from the composite temporal variations of 500-hPa height anomaly in the region (25°N – 35°N, 100°E – 130°E), sea level pressure anomaly (35°N – 50°N, 80°E – 100°E), surface air temperature anomaly (30°N – 40°N, 120°E – 140°E), surface meridional wind anomaly (25°N –35°N, 120°E – 140°E), zonal wind anomaly (6°S – 6°N, 140°E – 160°E) and SSTA in the Niño3 region (Fig. 33). Generally, strong (weak) winter monsoon in East Asia can be represented by a deep (shallow) 500-hPa trough over East Asia, a strong (weak) Siberia surface high, low (high) temperatures in Eastern China, and northerly (southerly) anomalies over the western North Pacific/East Asia region. Therefore, panels (a) to (d) in Fig. 33 can represent the EAWM situation and very clearly show its impact on the El Niño event. Westerly wind anomalies occur over the equatorial western Pacific prior to the occurrence of the El Niño (Fig. 33e), which is preceded by a strong winter monsoon in East Asia (Figs. 33a–d). This means a strong winter monsoon in East Asia could excite the occurrence of an El Niño. The same figures also show that in the winter of the El Niño, EAWM is weaker, which suggests the effect of the El Niño in weakening the EAWM.

The results for the composite pattern of La Niña (not shown) are opposite to those in Fig. 33, i.e., a weak winter monsoon in East Asia could excite the occurrence of a La Niña and the occurrence of a La Niña event would subsequently enhance the EAWM. Therefore, it can be suggested that an interactive relationship exists between ENSO and the EAWM: a strong (weak) winter monsoon in East Asia could excite the occurrence of an El Niño (a La Niña) and the subsequent winter monsoon will be weakened (enhanced) by an El Niño (a La Niña) event.

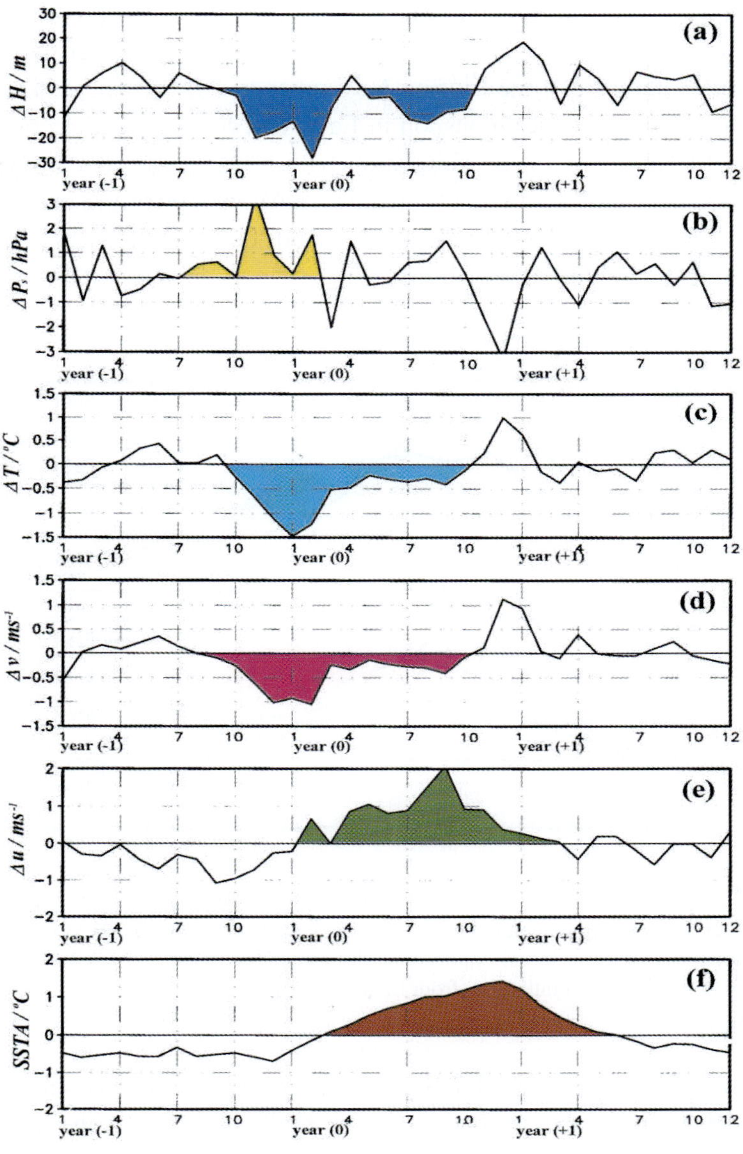

Figure 33. Composite temporal variations of all El Niño cases between 1950 and 1996 of (a) 500hPa height anomalies in the (30°N – 40°N, 100°E – 130°E) region, (b) sea surface pressure anomalies in the (33°N – 50°N, 80°E – 110°E) region, (c) surface air temperature anomalies in the (30°N – 40°N, 120°E – 140°E) region, (d) surface meridional wind anomalies in the (25°N – 35°N, 120°E – 140°E) region, (e), zonal wind anomalies in the (5°S – 5°N, 140°E – 160°E) region, and (f) the SSTA in the Nino3 region. The shading in each plot indicates the period during which a significant change of the parameter occurs. (from Mu and Li 1999)

Li and Mu (1998) have simulated the impact of anomalous winter monsoon in East Asia on the oceanic temperature in the tropical Pacific (or ENSO) using the coupled atmosphere-ocean general circulation model (CGCM) of Zhang and Endoh (1994). The difference between experiments with anomalies of the EAWM and the control run can be regarded as the response of the tropical Pacific to an EAWM forcing. The results (Fig. 34) clearly show that a strong winter monsoon can excite a positive SSTA response in the central and eastern equatorial Pacific (in the Nino1+2 and Nino3 regions) in the subsequent year although the SSTA does not last into the following winter because the strong winter monsoon forcing is applied for a shorter period of time (during November to April) than in the real case. This result suggests that a strong winter monsoon in East Asia could play an important role in the occurrence of an El Niño event. The simulation with a weak winter monsoon in East Asia also leads to the excitation of a La Niña (not shown). Anomalies of the low-level zonal wind over the equatorial western Pacific as well as anomalous oceanic Kelvin waves are also found in the model, in agreement with observations.

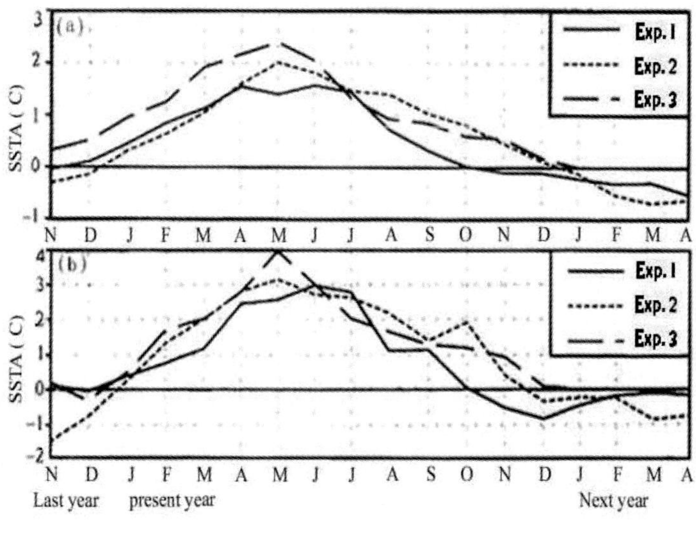

Figure 34. SSTA in (a) Nino1+2 region and (b) Nino3 region excited by a strong EAWM in the CGCM of Zhang and Endoh (1994) for three different initial conditions. The forcing of the strong EAWM was introduced only between November and April in the simulation. (from Li and Mu 1998)

The strongest El Niño event in the 20-century occurred in the early summer of 1997. Yu and Rienecker (1998) found that anomalies in the midlatitude atmospheric circulation over the East Asian region played an important role in the occurrence of this event, which again suggests that a strong EAWM is an important factor in exciting an El Niño event. Their analyses of data from the Special Sensor Microwave Imager showed the frequent occurrence of northerly surges in the 1996–1997 winter, originating north of 30°N over the East China Sea. These extratropical anomalies induced cyclone development, westerly surges and an eastward propagation of the MJO over the equatorial western Pacific. The resultant equatorial westerly wind bursts between 130°E – 150°E excited Kelvin wave packets as evidenced in the time-longitude sections of the SSTA and the sea-surface height anomalies (not shown). The evolution of anomalous surface winds in February-March 1997 shows clearly the equatorial westerly wind events and northerly surges from the midlatitudes (Fig. 35). The midlatitude surge over the East China Sea began on February 18 in the region 120°E – 140°E (Fig. 35a). By February 22, a weak cyclone formed to the east of Japan (Fig. 35b), and

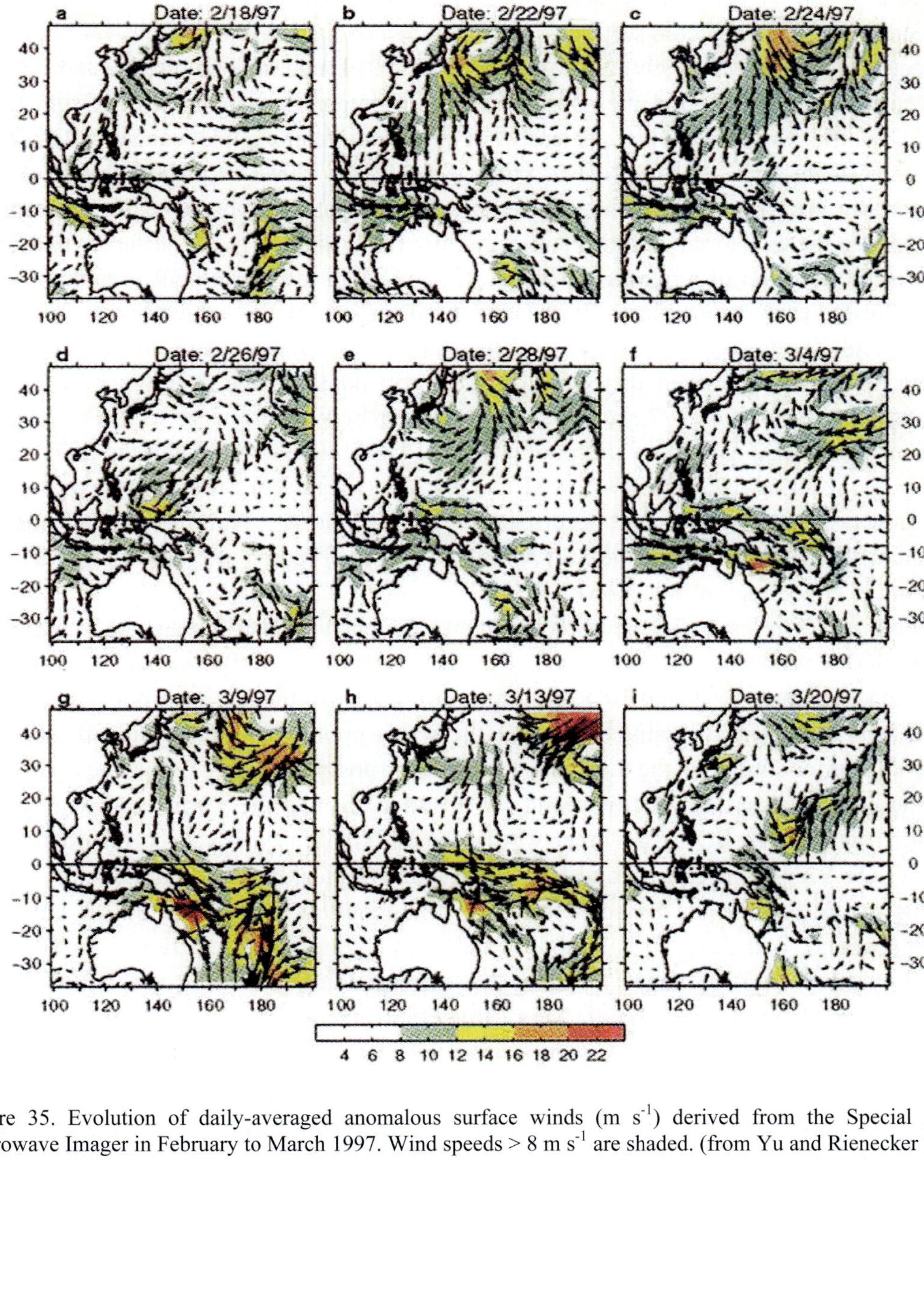

Figure 35. Evolution of daily-averaged anomalous surface winds (m s^{-1}) derived from the Special Sensor Microwave Imager in February to March 1997. Wind speeds > 8 m s^{-1} are shaded. (from Yu and Rienecker 1998)

intensified significantly during the next 4 days (Fig. 35c–e). Meanwhile, the northerly surge expanded eastward, faster in the midlatitudes than in the tropics. While the cyclone dissipated, the westerly winds along the equator strengthened and extended to 160°E (Fig. 35e). Another northerly surge struck the region between 120°E and 140°E in early March (Fig. 35f) and subsequently propagated through the region in a similar manner. The strongest influence of the northerly surges moved eastward with time. This, together with the formation of a cyclone on March 4, caused the region of anomalous equatorial westerlies to extend to near the dateline by March 20 (Fig. 35i). It is apparent that the onset and enhancement processes of the westerly wind events over the equatorial western Pacific were largely triggered by the northerly cold surges associated with a strong EAWM.

Recently, Li and Mu (1999) and Chao *et al.* (2002) found a close relationship between the occurrence of ENSO and the subsurface ocean temperature anomalies (SOTA) in the western Pacific warm pool and its eastward propagation along the equator. Before the occurrence of the 1997 El Niño event, positive SOTA had been in the equatorial western Pacific warm pool for a long time. The eastward propagation of positive SOTA from the warm pool to the equatorial eastern Pacific and upward extending to the surface was a direct reason of the El Niño occurrence. After that, the SOTA in the warm pool became negative and propagated eastward to surface in the equatorial eastern Pacific, which resulted in the La Niña of 1998. Zhou and Li (1999) also obtained similar results through a CGCM simulation.

The relationship between the SOTA in the warm pool and the occurrence of El Niño events is not only for the 1997 event, but also for other El Niño events as well. The temporal variations of the SOTA in the warm pool (10°S – 10°N, 140°E – 180°) and the SSTA in the Nino3 region during 1951–1994 showed that positive (negative) SOTA in the warm pool region always exists prior to the occurrence of an El Niño (La Niña) event, but after the outbreak of El Niño (La Niña), the SOTA in the warm pool region are out-of-phase with the SSTA in the Nino3 region (not shown).

As the SOTA propagates eastwards, an interesting phenomenon can also be found: when positive SOTA propagates eastwards along the equator, a westward propagation of negative SOTA appears along 10°N and 10°S, and the warm pool will then be controlled by negative SOTA. Similarly, when negative SOTA in the warm pool propagates eastward along the equatorial thermocline and the La Niña took shape, positive SOTA propagates westward from the equatorial eastern Pacific along 10°N and 10°S, and the warm pool will then be controlled by positive SOTA. In other words, the El Niño/La Niña can be regarded as the cycle of SOTA along the equator and 10°N/10°S in the tropical Pacific thermocline.

This cyclic feature can be seen from the time-longitude section of SOTA along the equator (Panel C) and along 10°N latitude (Panel A), the time-latitude section of SOTA in the tropical western Pacific (Panel B), and the corresponding SSTA in the Nino3 region (rightmost panel) during 1979–1993 (Fig. 36). For convenience, we begin the discussion from 1981. It is clear from Panel C that positive SOTA propagated eastward from the 140°E – 180° region, which upon reaching the equatorial eastern Pacific caused a strong positive SSTA there. Then, the positive SOTA propagated westwards along the 10°N latitude (Panel A) and reached the western Pacific by 1984, and the equatorial western Pacific through southward propagation (Panel B). After that, positive SOTA in the warm pool region propagated eastwards again in 1985 (left C) and formed a positive SOTA in the equatorial eastern Pacific in the summer 1986 when an El Niño event also occurred. Similarly, the negative SOTA went through the same cycle processes and formed the La Niña in 1984 and 1988.

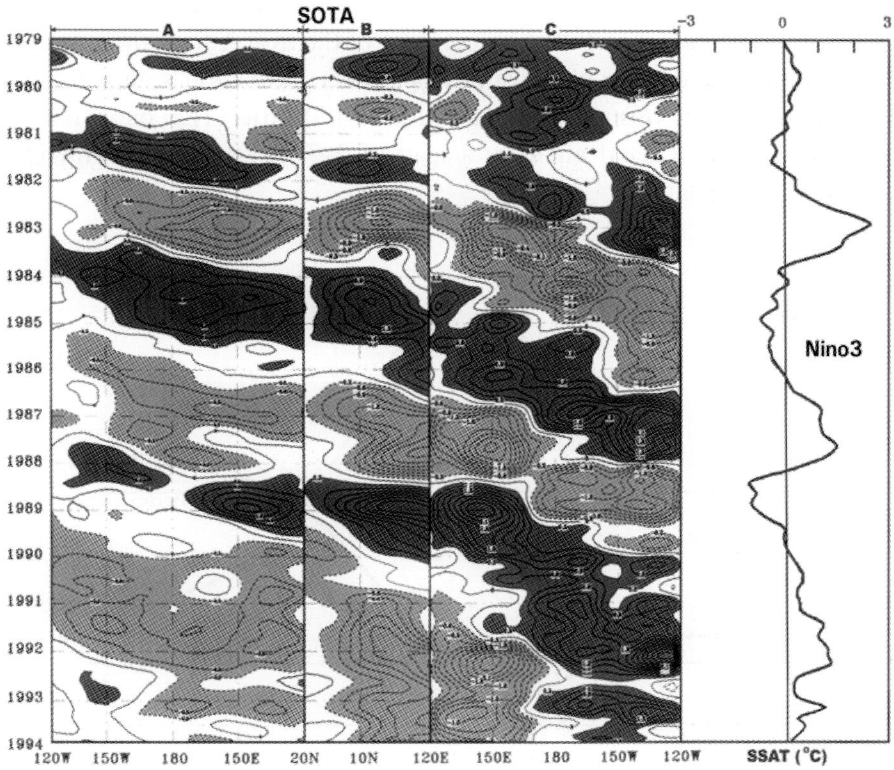

Figure 36. Time-longitude sections of SOTA in the tropical Pacific along the equator (6°S – 6°N, Panel C) and along 10°N (Panel A), time-latitude section of SOTA in the western Pacific (120°E – 160°E, Panel B) and the temporal variation of SSTA in the Nino3 region (rightmost panel). (from Li and Mu 2002)

Although only the SOTA propagations along the equator and 10°N latitude are shown in Fig. 36, the results along the equator and 10°S latitude are very similar (not shown). It may therefore be concluded that the ENSO cycle is actually a reflection in the equatorial eastern Pacific of the interannual cycle of the SOTA in the tropical Pacific along the equator and 10°N/10°S latitudes. The strongest SOTA is in the warm pool of the western North Pacific, but there the thermocline is deep (150–200m), so that SSTA in the equatorial western Pacific is not large. On the other hand, since the thermocline in the equatorial eastern Pacific is shallow, the SOTA has similar amplitudes as the SSTA. Thus the strongest SSTA is in the equatorial eastern Pacific, and it links directly with the ENSO so that it is natural to define the ENSO cycle by using SSTA in the equatorial eastern Pacific. But the essence of ENSO is a cycle of SOTA in the tropical Pacific thermocline in which the western Pacific warm pool plays an important role.

The important factor to drive this SOTA cycle is the zonal wind anomaly over the equatorial western Pacific caused mainly by the anomaly of the EAWM. To illustrate this, the eastward propagation of the SOTA in the equatorial Pacific is correlated with zonal wind anomalies over the equatorial western Pacific and with meridional wind anomalies over the western North Pacific (120°E – 140°E, 25°N – 35°N). Their time-lag correlation coefficients (Fig. 37) clearly show that the eastward propagation of the SOTA from the western Pacific to eastern Pacific along the equator has positive (negative) correlation with zonal (meridional) wind anomalies over the western equatorial (North) Pacific. The maximum correlation coefficient gives the zonal wind anomaly a lead of 2–3 months and the meridional wind anomaly a lead of 4–5 months.

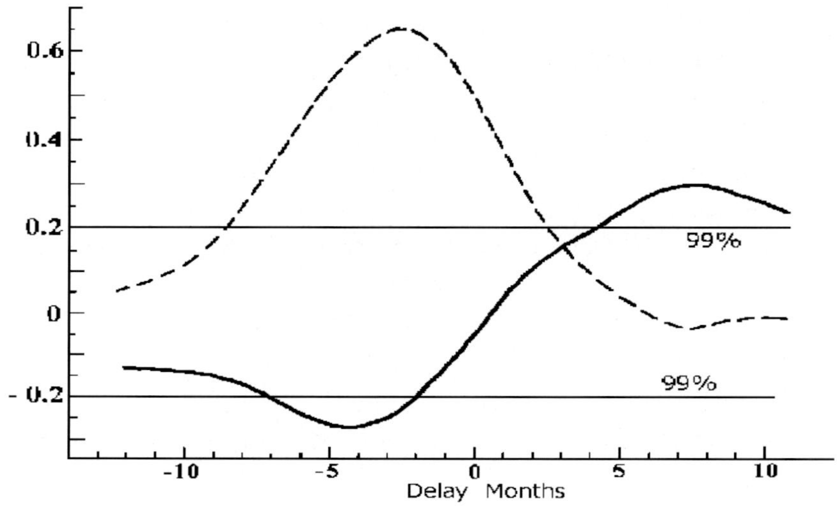

Figure 37. Temporal variations of the time-lag correlation of the eastward propagation of the SOTA along the equatorial with zonal wind anomalies over the equatorial western Pacific (dashed line) and with meridional wind anomalies over the western North Pacific (sold line). (from Mu and Li 2000)

Based on above analyses, it is not only clear that the ENSO is very closely related to interannual cycle of SOTA in the tropical Pacific, but also the cyclic feature of the SOTA in the tropical Pacific is driven by anomalous zonal winds caused mainly by the EAWM anomaly. For example, if the EAWM is continuously strong (persistent northerly wind anomalies), westerly wind anomalies will occur over the equatorial western Pacific and the positive SOTA in the warm pool will begin to propagate eastwards. The anomalous westerly winds will occur over the entire equatorial central and western Pacific due to the strong winter monsoon effect, and the positive SOTA will propagate eastward into the equatorial eastern Pacific and positive SSTA (or an El Niño event) will be excited. While the eastward propagation of positive SOTA occurs along the equator, negative SOTA propagates westwards along the 10°N/10°S latitudes so that the main negative SOTA will form in the equatorial western Pacific.

5.3. *Other Factors to Affect Variations of the EAWM*

5.3.1. The Polar Vortex

The polar region is the coldest in the Northern Hemisphere during winter, due to the absence of solar radiation. The "fountainhead" of cold air, which causes the cold waves in East Asia, is generally found in this region. To investigate the impact of the polar vortex on the EAWM, the distributions of composite 500-hPa height anomalies for eight strong and eight weak EAWM years during the period 1970–1997 are examined (Fig. 38). It is evident that the atmospheric circulation patterns in the Northern Hemisphere are different for these two types of years, with the 500-hPa trough over East Asia being obviously deeper in the strong EAWM years.

To focus on the polar vortex, the zonal mean 500-hPa height anomalies along 70°N – 80°N for these two types of years are examined (Fig. 39). For strong EAWM years, the 500-hPa heights in the polar region are higher than those in the weak years, particularly in the region 20°E – 70°E. It is interesting to note from these two results that while a strong (weak) EAWM is associated with deep (shallow) 500-hPa East-Asian trough at mid-high latitudes, the polar vortex is actually weaker (stronger), especially in the region 20°E – 70°E. The physical mechanisms for such correlations have not been identified and more in-depth studies are necessary to understand such relationships.

5.3.2. SSTA in the North Pacific

Recent research indicates that strong quasi-decadal and interdecadal signals exist in the SSTA over the northern Pacific Ocean. Since these anomalies can change the thermodynamic difference between the North Pacific and the Asian continent, the EAWM could be affected.

A recent analysis based on the dataset during the period 1900–1997 showed that the SSTA in this region possesses two major oscillation modes of 7–10 and 25–35-year periods. Similar anomalous fields of the planetary-scale atmospheric circulation can be found in these two modes. For example, corresponding to positive (averaged in 1954–1956 and 1962–1964) and negative (averaged in 1958–1960 and 1977–1979) phases of the 7–10-year mode in the SSTA patterns (Fig. 40), the anomalous 1000-hPa wind fields from the NCEP reanalyses show anomalous northerly and southerly wind anomalies respectively over East Asia (Fig. 41). Therefore, it appears that the SSTA in the northern Pacific has an impact on the EAWM, particularly in the quasi-decadal and interdecadal time scales. Corresponding to positive (negative) phase of the interdecadal mode of SSTA in the northern Pacific, the EAWM is likely to be stronger (weaker).

5.3.3. North Atlantic Oscillation (NAO)

As an important climate system, the NAO and its variations, have received much attention recently (Hurrell 1995; Jones *et al.* 1997). Li and Li (2000) also indicated that the interdecadal climate variations in China are related to the NAO, which suggests that the latter should have an influence on the EAWM. Wu and Huang (1999) have shown that the winter

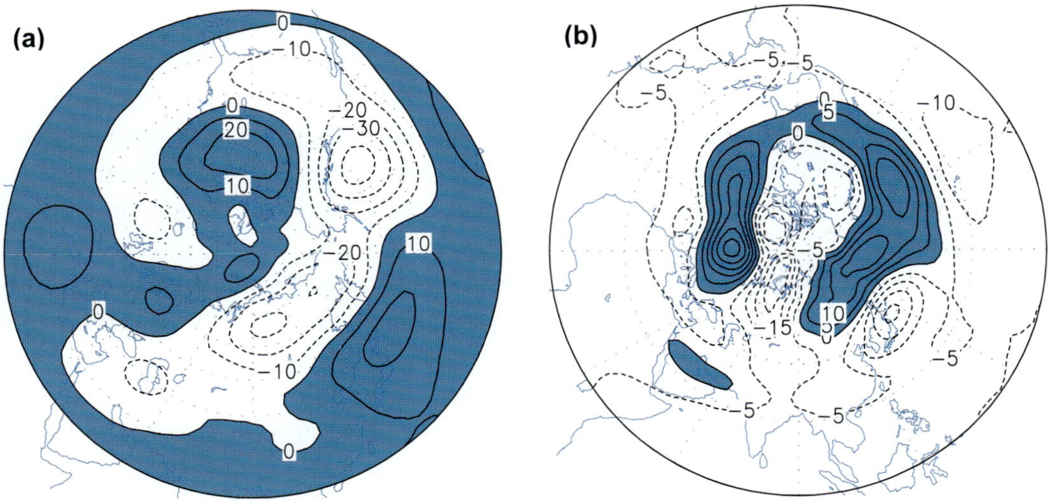

Figure 38. 500-hPa geopotential height anomalies [gpm, contour interval: 10 gpm in (a), 5 gpm in (b)] in the Northern Hemisphere in (a) strong and (b) weak East-Asian winter monsoon years.

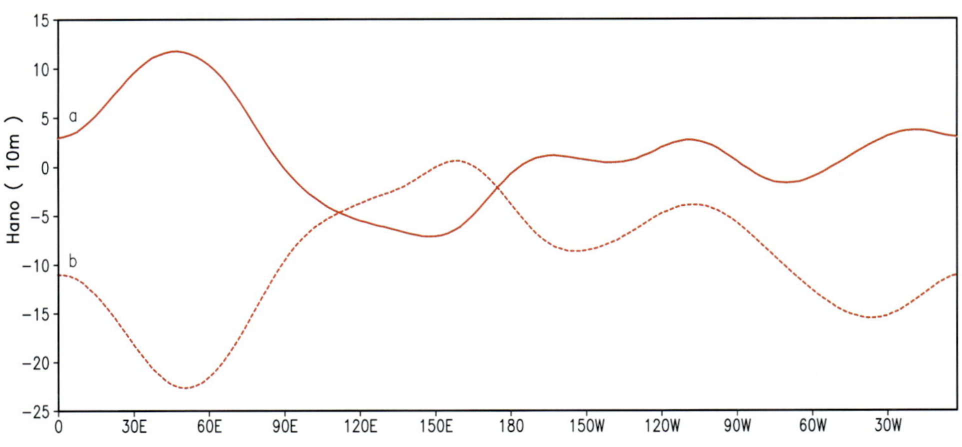

Figure 39. Zonal distributions of composite 500 hPa geopotential height anomalies at 70°N – 80°N in (a) strong and (b) weak East-Asian winter monsoon years.

NAO is negatively correlated with the intensity of Siberia high (Fig. 42). Since, as discussed in section 3, a strong (weak) EAWM corresponds to a strong (weak) Siberia high, such a correlation suggests that a high (low) winter NAO index will lead to a weakening (strengthening) of the EAWM.

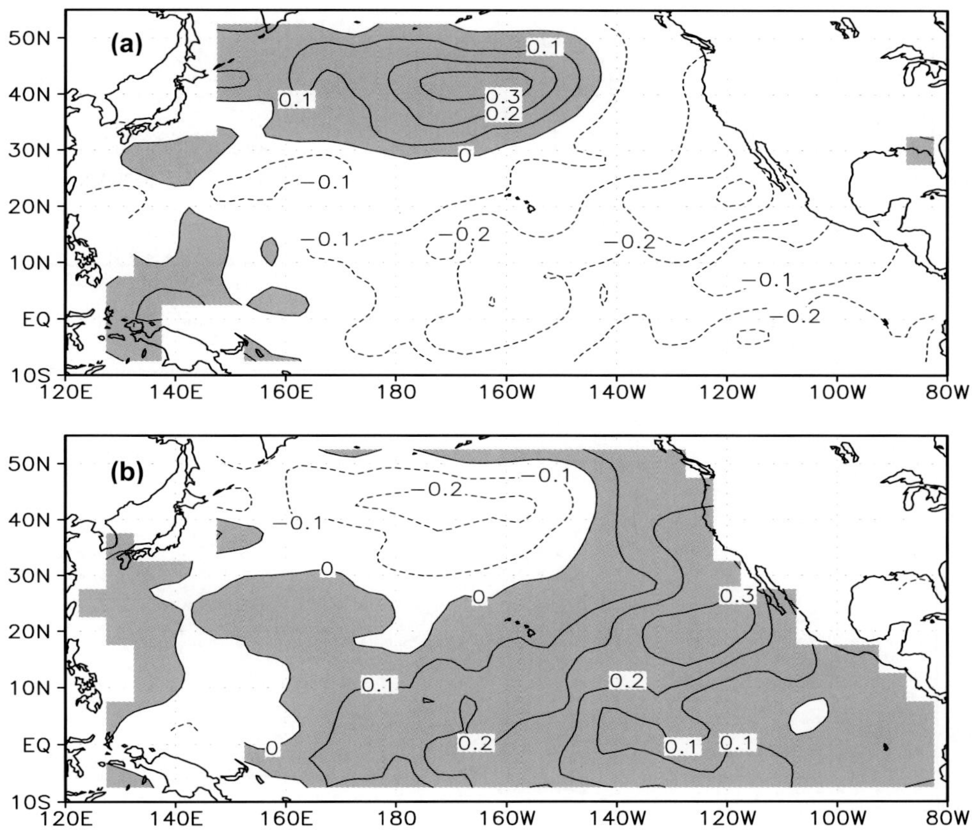

Figure 40. The pattern for the quasi 7–10 year mode of the North Pacific surface temperature (a) positive phase, and (b) negative phase of the 7-10 year mode. (from Xian and Li 2003)

5.3.4. The East Asia Summer Monsoon

Much of the research on the factors that might have an influence on the intensity of the EAWM focuses on the midlatitude or planetary-scale circulations. However, one feature that has not been extensively studied in this respect is the East Asia summer monsoon (EASM). Conceptually, an EASM that produces excessive amount of precipitation would leave the soil soaked for a long period of time so that radiational cooling in the fall and early winter might not be as efficient. This would then modify the north-south temperature contrast, and the subsequent baroclinity of the atmosphere could be altered. Such changes might have bearings

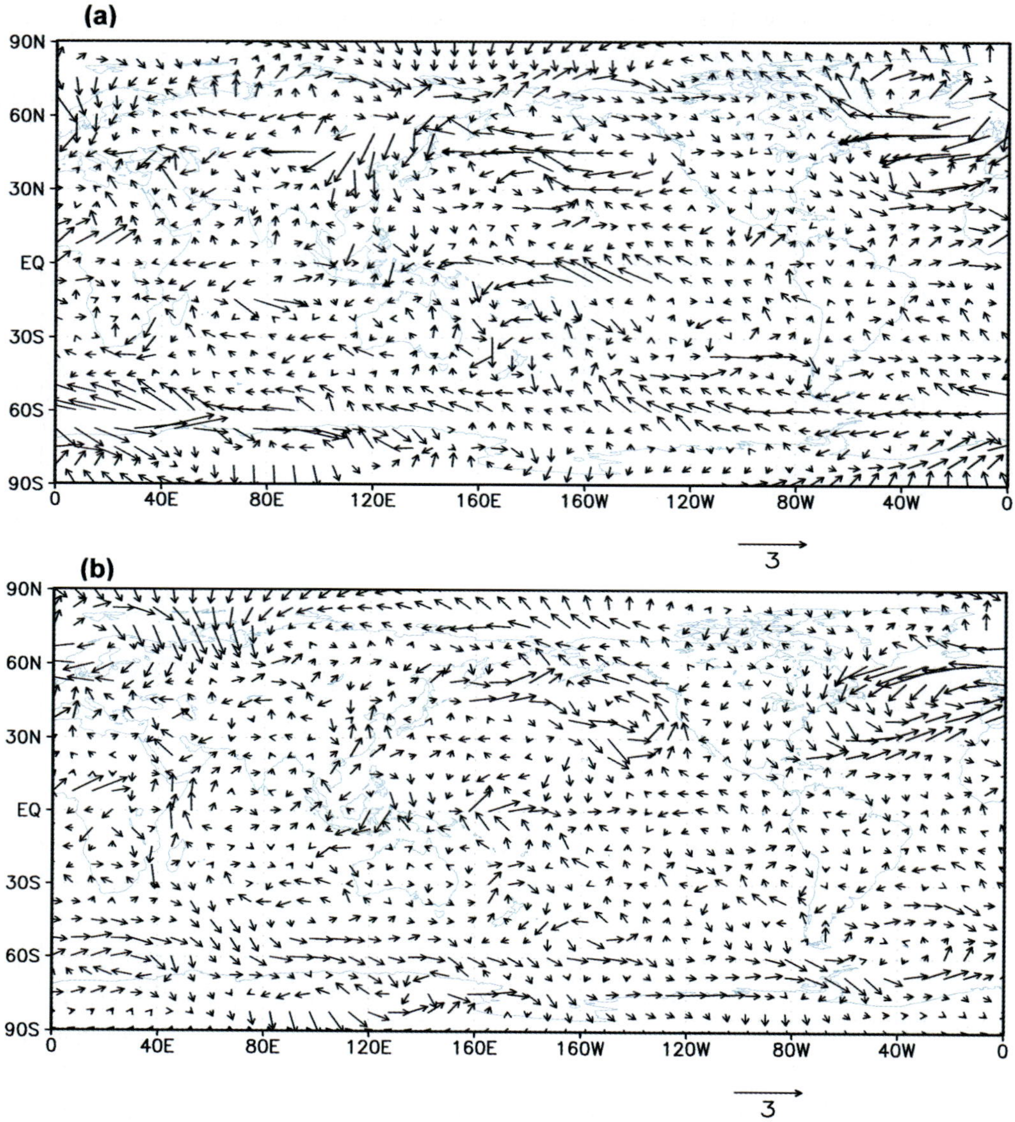

Figure 41. The anomalous wind fields at 1000 hPa corresponding to the (a) positive and (b) negative phase of the 7–10 year mode in the Northern Pacific surface temperature. (from Li and Xian 2003)

on the intensity of the EAWM since it is very much related to the baroclinity in the wintertime.

Wu and Chan (1997) and Wu (2003) have shown that indeed the intensity of the EASM can be related to that of the EAWM, and vice versa. While a basic two-year cycle appears to be the norm, such a cycle is broken up by the occurrence of an ENSO (warm or cold) event. The basic cycle is that a strong EAWM is followed by a weak EASM, then a weak EAWM and then a strong EASM. Although an ENSO event generally peaks in winter, the precursors of the occurrence of such an event often first appear in the spring. These precursors apparently also influence the EASM, which then couples with the ENSO event to influence the intensity of the subsequent EAWM (see also discussion in section 5.2). Thus, in studying the interannual variations of the EAWM, those of the EASM should also be examined.

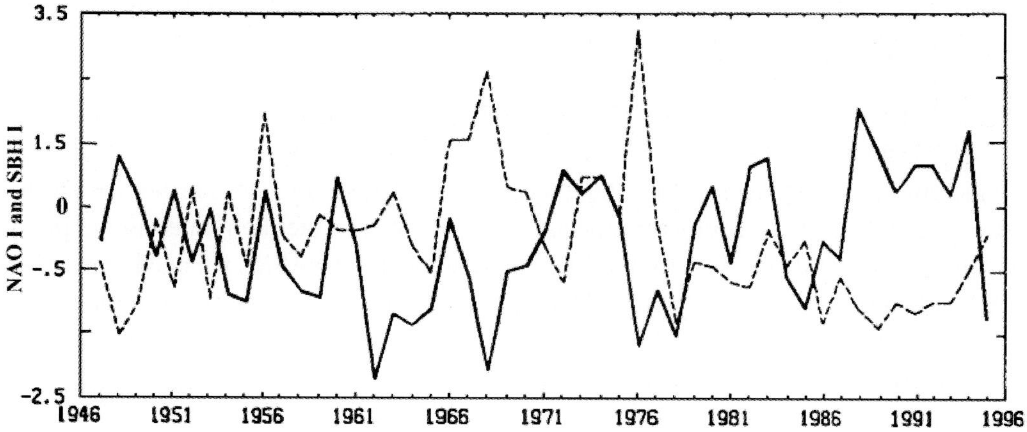

Figure 42. The relationship between standardized winter NAO index (solid) and standardized extent index of the winter Siberia high (dashed). (from Wu and Huang 1999)

5.4. *Summary*

Although the data length is not too long, these previous research results all suggest that the EAWM undergoes significant variations ranging from intraseasonal to interannual and interdecadal, which can be related to similar oscillations in other planetary-scale circulation features. In particular, the 10–20-day oscillation, the 30–60-day MJO, the polar vortex, ENSO, SSTA in the North Pacific, the NAO, and the EASM all have been shown to exhibit significant correlations with the EAWM on the various time scales. Unfortunately, our understanding of variations on the much longer temporal scales is far from adequate so that the underlying physical processes for any causality remain unknown. Much more research is obviously needed.

6. Overall Summary and Concluding Remarks

This chapter summarizes the main research results on the main features of the East Asia winter monsoon (EAWM), as well as factors that are likely to control its intensity. Specifically, it is shown that the EAWM is characterized not only by cold-air outbreaks (the northerly surges), but also the "coastal Kelvin waves" that propagate along the East China coast, as well as the explosive cyclogenesis off the East Asian continent. The origin of the EAWM comes from the midlatitudes and as such, much of the characteristics can be explained by baroclinic dynamics. Other wave-propagation features have also been shown to result from either gravity waves or Kelvin waves. The EAWM is also found to have significant temporal variations on intraseasonal, interannual as well as interdecadal time scales. The intraseasonal variations largely coincide with those synoptic- and low-frequency oscillations while the interannual variations are related to oscillations in the planetary-scale circulation features, including the ENSO. However, the physical processes associated with the interdecadal variations have yet to be identified.

Compared with its counterpart – the summer monsoon, the EAWM has not received as much attention because its impact on society is perceived to be not as large. Therefore, the amount of research in the study of the EAWM is much less. However, as evidenced by the discussion here, the effects of the EAWM could just be as hazardous. Further, the EAWM can actually influence other weather/climate phenomena and atmospheric circulations, e.g. its importance in the occurrence of El Niño events or in the ENSO cycle. Further, as discussed in section 5.3, the East Asia summer monsoon (EASM) can in fact be related to the previous EAWM. In other words, in order to improve predictions of the EASM, we need to understand the EAWM better. Li (1990) and Xu and Chan (2001) have also shown the importance of the EAWM in the development of an El Niño event. Therefore, it is essential that more research efforts be devoted to the study of the EAWM.

Acknowledgments

The first author (JCLC) benefited greatly from numerous discussions with Prof. Ding Yihui of the National Climate Center of the China Meteorological Administration and his former Ph.D. student, Dr. M. C. Wu, on the East Asia monsoons. Much of the literature search by JCLC was done during his visit to the Shanghai Typhoon Institute of the China Meteorological Administration, whose support, particularly from Mr. Yihong Duan, is much appreciated. Some of the figures were plotted using the NCEP/NCAR Reanalysis dataset. The continuous support from City University of Hong Kong and the Research Grants Council of the Hong Kong Special Administrative Region of China of JCLC's research on the East Asia monsoons is gratefully acknowledged.

References

Asai, T., Y. Kodama, and J.-C. Zhu, 1988: Long-term variations of cyclone activities in East Asia. *Adv. in Atmos. Sci.*, **5**, 149–158.

Bluestein, H. B., 1992: *Principles of Kinematics and Dynamics. Vol. I: Synoptic-Dynamic Meteorology in Midlatitudes*, Oxford Univ. Press, 429 pp.

Boyle, J. S., and T. J. Chen, 1987: Synoptic aspects of the wintertime East Asian monsoon. *Monsoon Meteorology*, C. P. Chang, and T. N. Krishnamurti (Eds.), Oxford Univ. Press, 125–160.

Chang, C.-B., D. J. Perkey, and W.-D. Chen, 1987: Observed dynamic structure of an intense oceanic cyclone. *Mon. Wea. Rev.*, **115**, 1127–1139.

Chang, C. P., and K. M. Lau, 1980: Northeasterly cold surges and near-equatorial disturbances over the winter MONEX area during December 1974. Part II: Planetary-scale aspects. *Mon. Wea. Rev.*, **108**, 298–312.

______, J. E. Millard, and G. T. J. Chen, 1983: Gravitational character of cold surges during winter MONEX. *Mon. Wea. Rev.*, **111**, 293–307.

Chao, J., S. Yuan, Q. Chao, and J. Tian, 2002: A data analysis study on the evolution of the El Nino / La Nina cycle. *Adv. in Atmos. Sci.,* **19**, 837–844.

Chen, L., and A. Xie, 1988: Westward propagating low-frequency oscillation and its teleconnection in the Eastern Hemisphere. *Acta Meteorologica Sinica*, **2**, 300–312.

Chen, Q., and A. Huang, 1989: Some features of the South China Sea cold surge during winter season. *Meteor. Monthly*, **15**, 49–53. (in Chinese with English abstract)

Chen, S.-J., and L. Dell'Osso, 1987: A numerical case study of East Asian coastal cyclogenesis. *Mon. Wea. Rev.*, **115**, 477–487.

Chen, T.-C., C.-B. Chang, and D. J. Perkey, 1983: Numerical study of an AMTEX oceanic cyclone. *Mon. Wea. Rev.*, **111**, 1818–1829.

______, ______, and ______, 1985: Synoptic study of a medium-scale ocean cyclone during AMTEX '75. *Mon. Wea. Rev.*, **113**, 349–361.

Chu, E. W. K., 1978: A method for forecasting the arrival of cold surges in Hong Kong. Tech. Note **No. 43**, Royal Observatory, Hong Kong, 31pp.

Compo, G. P., G. N. Kiladis, and P. J. Webster, 1999: The horizontal and vertical structure of East Asian winter monsoon pressure surges. *Quart. J. Roy. Meteor. Soc.*, **125**, 29–54.

Davidson, N. E., J. L. McBride, and B. J. McAvaney, 1983: The onset of the Australian monsoon during winter MONEX: Synoptic aspects. *Mon. Wea. Rev.*, **111**, 495–516.

Ding, Y., 1990a: Build-up, air mass transformation and propagation of Siberian high and its relation to cold surge in East Asia. *Meteor. Atmos. Phys.*, **44**, 281–292.

______, 1990b: A statistical study of winter monsoons in East Asia. *J. Tropical Meteor.*, **6**, 119–128. (in Chinese with English abstract)

______, 1994: *Monsoons over China.* Kluwer Academic Publishers, 419pp.

______, and T. N. Krishnamurti, 1987: Heat budget of the Siberian high and the winter monsoon. *Mon. Wea. Rev.*, **115**, 2428–2449.

______, S. Wen, and Y. Li, 1991: A study of dynamic structures of the Siberia high in winter. *Acta Meteorologica Sinica*, **49**, 428–439. (in Chinese with English abstract)

______, and M. Xiao, 1992: A case study of development and structure of a cold surge in East Asia. *Proc., 2nd Int'l Conf. East Asia and Western Pac. Meteor. and Climate*, W. J. Kyle and C. P. Chang, Eds., World Scientific, 311–328.

Dong, L., and D. Li, 1989: Coastal and oceanic explosive cyclones to the east of China. *Acta Meteorologica Sinica*, **47**, 371–375. (in Chinese with English abstract)

Gao, S., S. Tao, and Y. Ding, 1992: Upper wave-East Asia jet interaction during the period of cold wave outbreak. *Sci. Atmos. Sinica*, **16**, 718–724. (in Chinese with English abstract)

Hanson, H. P., and B. Long, 1985: Climatology of cyclogenesis over the East China Sea. *Mon. Wea. Rev.*, **113**, 697–707.

Holland, G. J., and L. M. Leslie, 1986: Ducted coastal ridging over S. E. Australia. *Quart. J. Roy. Meteor. Soc.*, **49**, 428–438.

Holton, J. R., 1992: *An Introduction to Dynamic Meteorology,* 3rd Ed.. Academic Press, 507pp.

Hurrell, J. W., 1995: Decadal trends in the North Atlantic Oscillation: regional temperature and precipitation. *Science,* **269**, 676–679.

Jones, P. M., T. Johnson, and D. Wheeler, 1997: Extension to the North Atlantic Oscillation using early instrumental pressure observations from Gibraltar and SW Iceland. *Intl. J. Climatol.*, **17**,

1433–1350.

Lau, K. M., C. P. Chang, and P. H. Chan, 1983: Short-term planetary-scale interactions over the tropics and midlatitudes. Part II: Winter MONEX period. *Mon. Wea. Rev.*, **111**, 1372–1388.

Lau, N. C., and K. M. Lau, 1984: The structure and energetics of midlatitude disturbances accompanying cold-air outbreaks over East Asia. *Mon. Wea. Rev.*, **112**, 1309–1327.

Li, C., 1989a: Frequent activities of stronger upper-level troughs in East Asia in wintertime and the occurrence of the El Nino event. *Science in China (B)*, **32**, 976–985.

______, 1989b: Warmer winter in Eastern China and El Nino. *Chinese Sci. Bull.*, **34**, 1801–1805.

______, 1990a: Interaction between anomalous winter monsoon in East Asia and El Niño events. *Adv. in Atmos. Sci.*, **7**, 36–46.

______, 1990b: On interaction between anomalous circulation/climate in East Asia and El Niño event. *Climate Change Dynamics and Modeling,* China Meteorological Press, Beijing. 101–126.

______, 1993: *Low-Frequency Oscillations in the Atmosphere.* China Meteorological Press, Beijing, 310pp. (in Chinese)

______, 1995: *Introduction to Climate Dynamics*. China Meteorological Press, Beijing, 461pp. (in Chinese)

______, Y. Chen, and Z. Yuan, 1989: Important cause of El Nino events – Frequent activities of stronger cold waves in East Asia. *Frontiers in Atmospheric Sciences*, New York: Allenton Press, 156–165.

______, and Y. Ding, 1989: A diagnostic study of an explosively deepening oceanic cyclone over the northwest Pacific Ocean. *Acta Meteorologica Sinica*, **47**, 180–190. (in Chinese with English abstract)

______, and G. Li, 2000: The NAO/NPO and interdecadal climate variation in China. *Adv. in Atmos. Sci.*, **17**, 555–561.

______, and M. Mu, 1998: Numerical simulations of anomalous winter monsoon in East Asia exciting ENSO. *Chinese J. Atmos. Sci.*, **22**, 393–403.

______, and ______, 1999 : ENSO occurrence and sub-surface ocean temperature anomalies in the equatorial warm pool, *Chinese J. Atmos. Sci.*, **23**, 217–225.

______, and ______, 2002: A further study of the essence of ENSO. *Chinese J. Atmos. Sci.,* **26**, 309–328.

______, and P. Xian, 2003: Atmospheric anomalies relation to interdecadal variability of SST in the North Pacific, *Adv. in Atmos. Sci.*, **20**, 859-874.

Lim, H., and C. P. Chang, 1981: A theory for midlatitude forcing of tropical motions during winter monsoons. *J. Atmos. Sci.*, **38**, 2377–2392.

Love, G., 1985: Cross-equatorial influence of winter hemisphere subtropical cold surges. *Mon. Wea. Rev.*, **113**, 1487–1498.

Lu, G., J. Yao, and Z. Tao, 1983: The dynamical cause of the cold dome's intensification in the cold outbreak process. *Acta Meteorologica Sinica*, **41**, 393–403. (in Chinese with English abstract)

Lu, W., and Q. Zhu, 1990: Theoretical study on the effect of Qinghai-Xizang Plateau on cold surges. *Acta Meteorologica Sinica,* **4**, 620–628.

Morrice, A. M., 1973: Quantitative forecasting of the winter monsoon in Hong Kong. Tech. Note **No. 35**, Royal Observatory Hong Kong, 41 pp.

Mu, M., 2001: A further research on the cyclic relationship between anomalous East-Asian winter monsoon and ENSO. *Climate Environ. Res.*, **6**, 273–285. (in Chinese)

______, and C. Li, 1999: ENSO signals in interannual variability of East Asian winter monsoon, Part

I: Observed data analyses. *Chinese J. Atmos. Sci.,* **23**, 134–143.

______, and C. Li, 2000: Interaction between subsurface ocean temperature anomalies in the western Pacific warm pool and ENSO cycle, *Chinese J. Atmos. Sci.*, **24**, 107–121.

Mullen, S. L., 1979: An investigation of small synoptic-scale cyclones in polar air streams. *Mon. Wea. Rev.*, **107**, 1636–1647.

Murakami, T., 1979: Winter monsoonal surge over East and Southeast Asia, *J. Meteor. Soc. Japan*, **57**, 133–158.

Nitta, T., and J.-I. Yamamoto, 1974: On the observational characteristics of intermediate scale disturbances generated near Japan and the vicinity. *J. Meteor. Soc. Japan*, **52**, 11–30.

Ouyang, Z., M. Lu, and Z. Hu, 1990: Some statistical characteristics of explosive cyclones in Asian and northwest Pacific. *J. Air Force Meteor.*, **11**, 23–30. (in Chinese)

Palmén, E., and C. W. Newton, 1969: *Atmospheric Circulation Systems – Their Structure and Physical Interpretation.* Int'l Geophys. Series **No. 13**, Academic Press, 603 pp.

Pan, H. L., and F. X. Zhou, 1985: The 10–20 day tropical-midlatitude interaction during winter monsoon season. *J. Meteor. Soc. Japan,* **63**, 829–843.

Qiu, Y., 1988: A case study on the intense cold wave accompanied by the polar front over East Asia. *Extratropical Cyclones: The Erik Palmén Memorial Volume*, C. Newton and E. Holopainen, Eds., Amer. Meteor. Soc., 150–156.

______, X. Li, and Y. Qiu, 1992: Statistical features of the cold waves invaded China and their relation to the snow cover area over the Eurasian continent. *Quart. J. Appl. Meteor.*, **3**, 235–241. (in Chinese with English abstract)

Ramage, C. S., 1971: *Monsoon Meteorology.* Academic Press, 296 pp.

Reed, R. J., 1979: Cyclogenesis in polar air streams. *Mon. Wea. Rev.*, **107**, 38–52.

Sanders, F., 1986: Explosive cyclogenesis in the west central North Atlantic Ocean, 1981–1984. *Mon. Wea. Rev.*, **114**, 1781–1794.

______, and J. R. Gyakum, 1980: Synoptic dynamic climatology of the "bomb". *Mon. Wea. Rev.*, **108**, 1589–1606.

Sun, S., and S. Gao, 1993: The effect of East Asia cold surge activity on the downstream genesis of explosive cyclones. *Acta Meteorologica Sinica*, **51**, 304–313. (in Chinese with English abstract)

Tao, S., 1959: East-Asian cold wave research in China during recent 10 years. *Acta Meteorologica Sinica*, **30**, 226–230. (in Chinese).

Wu, B., and R. Huang, 1999: Effects of the extremes in the North Atlantic oscillation on East Asia winter monsoon. *Chinese J. Atmos. Sci.*, **23**, 641–651.

Wu, G., and U. Cubasch, 1987: The impact of the El Niño anomaly on the mean meridional circulation and atmospheric transformations. *Scientia Sinica,* Series B, **30**, 533–545.

Wu, M. C., 2003: Relationships between summer and winter monsoons over East Asia. *Ph. D. Thesis*, City Univ. Hong Kong, 228pp.

______, and J. C. L. Chan, 1995: Surface features of winter monsoon surges over South China. *Mon. Wea. Rev.*, **123**, 662–680.

______, and ______, 1997a: Upper-level features associated with winter monsoon surges over South China. *Mon. Wea. Rev.*, **125**, 317–340.

______, and ______, 1997b: Relationship between summer and winter monsoons over South China. *Preprints, 22nd Conf. on Hurricanes and Trop. Meteor.*, Fort Collins, CO, Amer. Meteor. Soc., 19–23 May, 637–638.

______, and______, 2000: Winter monsoon over South China in ENSO and non-ENSO years.

Preprints, 24th Conf. Hurr. Trop. Meteor., Amer. Meteor. Soc., 29 May – 2 June, Ft. Lauderdale, Florida, USA, 110–111.

Xian, P., and C. Li, 2003: Interdecadal mode of sea surface temperature variation in the North Pacific Ocean and its evolution, *Chinese J. Atmos. Sci.*, **27**,118-126.

Xie, A., Y. Lu, and S. Chen, 1992: The evolution of Siberia high to the outbreak of cold air. *Sci. Atmos. Sinica*, **16**, 677–685. (in Chinese with English abstract)

Xu, J., and J. C. L. Chan, 2001: The role of the Asian/Australian monsoon system in the onset time of El Niño events. *J. Climate*, **14**, 418–433.

Xu, Y., and M. Zhou, 1999: Numerical simulations on the explosive cyclogenesis over the Kuroshio Current. *Adv. in Atmos. Sci.*, **16**, 64–76.

Yi, Q., and Y. Ding, 1992: A dynamic study of two explosively deepening cyclones over the East China Sea. *Acta Meteorologica Sinica*, **50**, 152–166. (in Chinese with English abstract)

______, and ______, 1993: Climatology of the explosive cyclogenesis over East Asia and the west Pacific. *Sci. Atmos. Sinica*, **17**, 302–309. (in Chinese with English abstract)

Yu, L., and M. Rienecker, 1998: Evidence of an extratropical atmospheric influence during the onset of the 1997–98 El Nino. *Geophys., Res. Lett.*, **25**, 3537–3540.

Zhang, P., and G. Chen, 1999: A statistical analysis of the cold wave high which influences China. *Acta Meteorologica Sinica*, **57**, 493–501. (in Chinese with English abstract)

Zhang, R. H., and M. Endoh, 1994: Simulation of 1986–87 El Niño and 1988 La Niña events with a free surface tropical Pacific Ocean general circulation model. *J. Geophys. Res.*, **99**, 7743–7759.

Zhang, Y., K. R. Sperber, and J. S. Boyle, 1997: Climatology and interannual variation of the East Asian winter monsoon: Results from the 1979–95 NCEP/NCAR Reanalysis. *Mon. Wea. Rev.*, **125**, 2605–2619.

Zhao, Q., and Y. Ding, 1991: A study of physical processes affecting the transformation of cold air over land after outbreak of cold waves in East Asia. *Acta Meteorologica Sinica*, **49**, 170–180. (in Chinese with English abstract)

Zhou, G., and C. Li, 1999: Simulation on the relation between the subsurface temperature anomaly in western Pacific and ENSO by using CGCM, *Climate and Environ. Res.*, **4**, 352–363. (in Chinese with English abstract).

Zhou, X., 1989: An analysis of two cold surges. *J. Trop. Meteor.*, **5**, 57–63. (in Chinese with English abstract)

Zhu, Q., X. Zhi, and Z. Lei, 1990: Low frequency summer monsoon in Indonesia-Northern Australia and its relation to circulation in both Hemispheres. *Acta Meteorologica Sinica,* **4**, 543–553.

3. MARITIME CONTINENT MONSOON: ANNUAL CYCLE AND BOREAL WINTER VARIABILITY

C.-P. CHANG AND PATRICK A. HARR
Department of Meteorology
Graduate School of Engineering and Applied Sciences
Naval Postgraduate School
Monterey, CA93943, USA
E-mail: cpchang@nps.edu; paharr@nps.edu

JOHN McBRIDE
Bureau Meteorology Research Centre, Australia
j.mcbride@bom.gov.au

HUANG-HSIUNG HSU
Department of Atmospheric Sciences
National Taiwan University
hsu@atmos1.as.ntu.edu.tw

The Maritime Continent and northern Australia is a region of strong seasonal variation in wind and rainfall regimes, which consist of a prevailing easterly wind and dry conditions during the boreal summer and prevailing westerly winds and wet conditions during the boreal winter. During the boreal winter, latent heat release associated with the wet season over the Maritime Continent and northern Australia contributes to a major component of the global circulation, which has been linked to significant tropical-extratropical interactions. Although substantial research has been conducted with respect to the northern Australia component of the boreal winter monsoon, the variations associated with the Maritime Continent component of the large-scale monsoon system have received less attention. In this Chapter, the annual cycle of rainfall and the interannual, sub-seasonal, and synoptic variabilities associated with the boreal winter monsoon over the Maritime Continent are described.

The equator serves as a general demarcation between the boreal summer (to its north) and boreal winter (to its south) monsoon rainfall regimes. However, locally the annual cycle is dominated by interactions between the complex terrain and an annual reversal of the surface winds. These interactions cause the summer and winter monsoon regimes to intertwine across the equator. In particular, the boreal winter regime extends far northward along the eastern flanks of the major island groups and landmasses. There is no complementary extension of the boreal summer regime into southern latitudes. The seasonal march is asymmetric during the transitional seasons, with the maximum convection following a gradual southeastward progression from the Asian summer monsoon to the Asian winter monsoon but a sudden transition in the reverse. This asymmetric march is explained by a hypothesis based on the redistribution of mass between land and ocean areas during spring and fall that results from different land-ocean thermal memories. This mass redistribution produces sea-level pressure patterns that lead to asymmetric wind-terrain interactions throughout the region, and a low-level divergence asymmetry in the region that promotes the southward march during boreal fall but opposes the northward march during boreal spring.

Interannual variability in boreal winter Maritime Continent monsoon rainfall is examined with respect to the El Nino-Southern Oscillation (ENSO). Significant variability in the ENSO-monsoon relationship exists across the Maritime Continent and its vicinity, particularly between the Sumatra-Malay Peninsula-western Borneo region and regions to its east and west. A significant part of this variability is linked to the influence of the Walker Circulation and its variations between warm and cold ENSO events. There is also evidence of interdecadal changes in the ENSO-monsoon relationship and its variability.

Both boreal summer and boreal winter rainfall in the Maritime Continent rainfall exhibit signals of the Tropical Biennial Oscillation. Atmosphere-ocean interaction processes appear to play an important role in the production of these signals.

Sub-seasonal and synoptic variability in boreal winter Maritime Continent rainfall is examined with respect to the Madden-Julian Oscillation (MJO), northeasterly cold surges, and the Borneo Vortex. Relationships among all three circulation systems are found to have significant impacts on large-scale atmospheric conditions that impact the spatial distribution of rainfall throughout the region. Convection over the southern South China Sea is strongest during the combination of a northeasterly surge and Borneo vortex. However, the frequency of surges is reduced when the MJO is present.

1. Introduction

The region "Maritime Continent" was defined by Ramage (1968) to consist of Malaysia, Indonesia, and the surrounding land and oceanic areas of the equatorial western Pacific between 10°S - 10°N. This is a region of strong seasonal reversal of winds that is often coupled with annual variation of rainfall, particularly near the equator where the prevailing wind is westerly during the wet season of boreal winter and easterly during the dry season of boreal summer (e.g., Ramage 1971; Masumoto 1992; McBride 1998; Hamada *et al.* 2002). The rainfall variation in this region is in phase with that of the northern Australia monsoon. Therefore, the terms northern winter monsoon or Australian (summer) monsoon are sometimes used to include the wet monsoon in both regions. During the boreal winter, the latent heat release over the combined Maritime Continent – northern Australia area is one of two major heat sources of the atmosphere, which is an important component of the global circulation. This heat source interacts directly with the strong East Asian baroclinic system and cold air outbreaks during the Asian winter monsoon, which may affect the East Asian jet (Chang and Lau 1982) and in turn influence the weather in North America (Yang *et al.* 2002). Recent analysis also suggest that flare-ups of this Maritime Continent heat source may lead to downstream wave-train propagation and may be related to storm development over Europe ten days later (Thorpe *et al.* 2002). In a general circulation model study, Neale and Slingo (2003) demonstrated that the tendency of a dry bias in the Maritime Continent region is a major source for systematic errors over both tropical Indian and Pacific oceans and extratropical North America and Northeast Europe. They concluded that the Maritime Continent plays a critical role in the global circulation and emphasized the need for better representation of convective organization over this region of complex land–sea terrains.

Substantial more research has been focused on the northern Australian sector of the northern winter monsoon than the Maritime Continent sector (e.g., McBride 1987; Manton and McBride 1992; Suppiah 1992; Power *et al.* 1999). In this chapter, the discussion will be focused on the Maritime Continent and will be based primarily on three recent research reports (Chang *et al.* 2004a,b,c). Section 2 will address the annual cycle and seasonal march of rainfall, which are dominated by the interaction between terrain and monsoon winds.

Section 3 will discuss interannual variations of monsoon rainfall, particularly the lower correlation between Indonesian rainfall and ENSO events during boreal winter. Section 4 will discuss synoptic to sub-seasonal variations of deep convection in the southern South China Sea and western Maritime Continent due to interactions among cold surges, quasi-stationary cyclonic disturbances near Borneo, and the Madden-Julian Oscillation (Madden and Julian 1972).

2. Annual Cycle and Seasonal March of Rainfall

2.1. *Annual Variation of Rainfall, Monsoon Onset and Withdrawal*

The Maritime Continent is often considered to be a part of the Asian winter monsoon regime because maximum rainfall over most of the region occurs during boreal winter (Ramage 1971). The western part of the Maritime Continent is adjacent to Indochina to the north, which is often considered as part of the Asian summer monsoon regime. Within the Maritime Continent, significant geographical variations of the seasonal march have been recognized

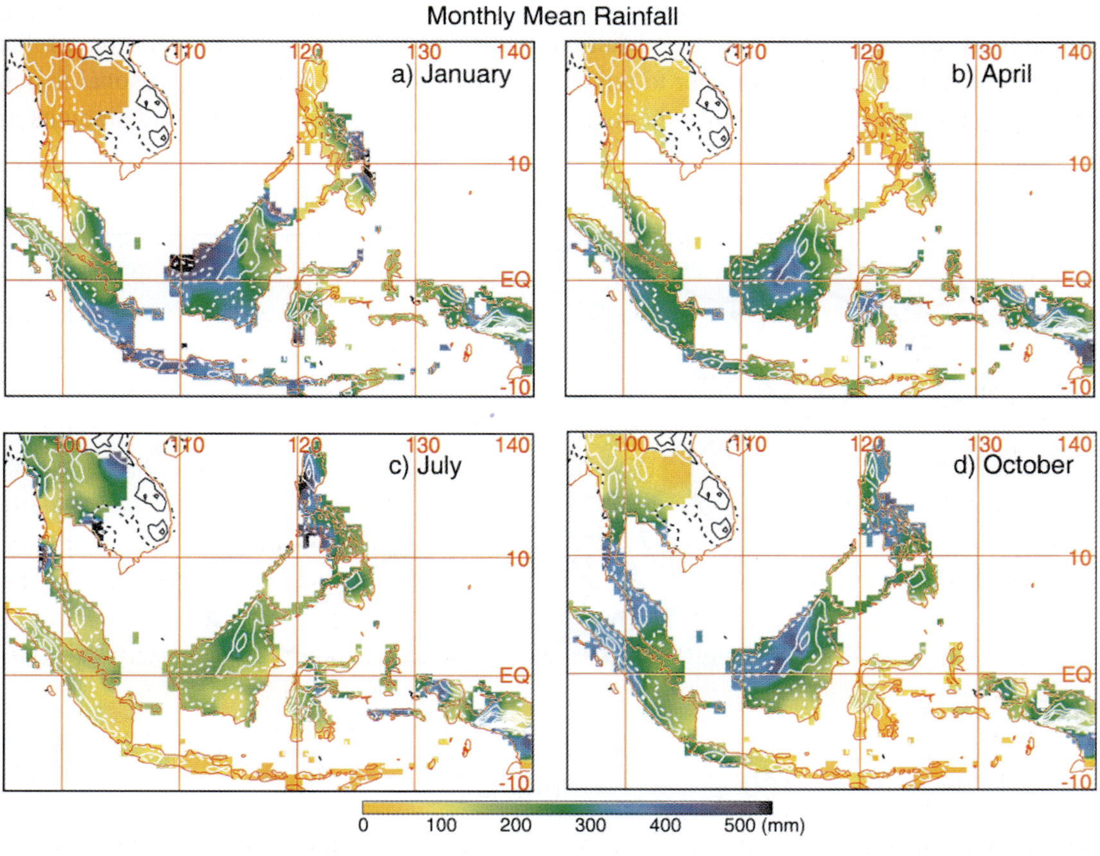

Figure 1. Monthly mean rainfall and topography for (a) January, (b) April, (c) July, and (d) October. Data are analyzed from long-term station rainfall reports of various periods with minimum length of 48 years.

since Braak (1921-1929, see Ramage 1971). A main reason for these variations is the complex terrain due to islands of different sizes interspersed among the surrounding seas. Figure 1 shows the monthly mean rainfall for January, April, July, and October based on an objective analysis of station rainfall data (Chang *et al.* 2004c). Rainfall amounts are high throughout the region with individual monthly totals on the order of 300 - 500 mm. Most of the region experiences a distinct dry season at some time of year, with the exception being parts of Borneo and New Guinea, which have high rainfall year round. (Place names are shown in Fig. 5). Figure 1 also shows strong rainfall gradients at all times of the year; for example, notice the east-west gradients of rainfall across the Philippines in January and across the Malay Peninsula in April. Chang *et al.* (2004c) pointed out that these patterns result largely from the interaction between the high topography in the region and the moisture-bearing low-level monsoon flow.

When considered together, the Southeast Asian landmasses, including Indochina, the Malay Peninsula, and the Maritime Continent, form the "land bridge" along which maximum convection marches gradually from the Asian (Indian) summer monsoon to the Asian winter (Australian summer) monsoon during boreal fall (e.g., Lau and Chan 1983; Meehl 1987; Yasunari 1991; Matsumoto 1992; Matsumoto and Murakami 2002; Hung and Yanai 2003; Hung *et al.* 2003). Figure 2, which was constructed by Lau and Chan (1983) using monthly climatology of outgoing longwave radiation (OLR) data, defines the progression of the maximum convection from summer to winter monsoon. Another view of this seasonal march can be seen in Fig. 3, which shows the monsoon onset dates based on high cloud amounts as

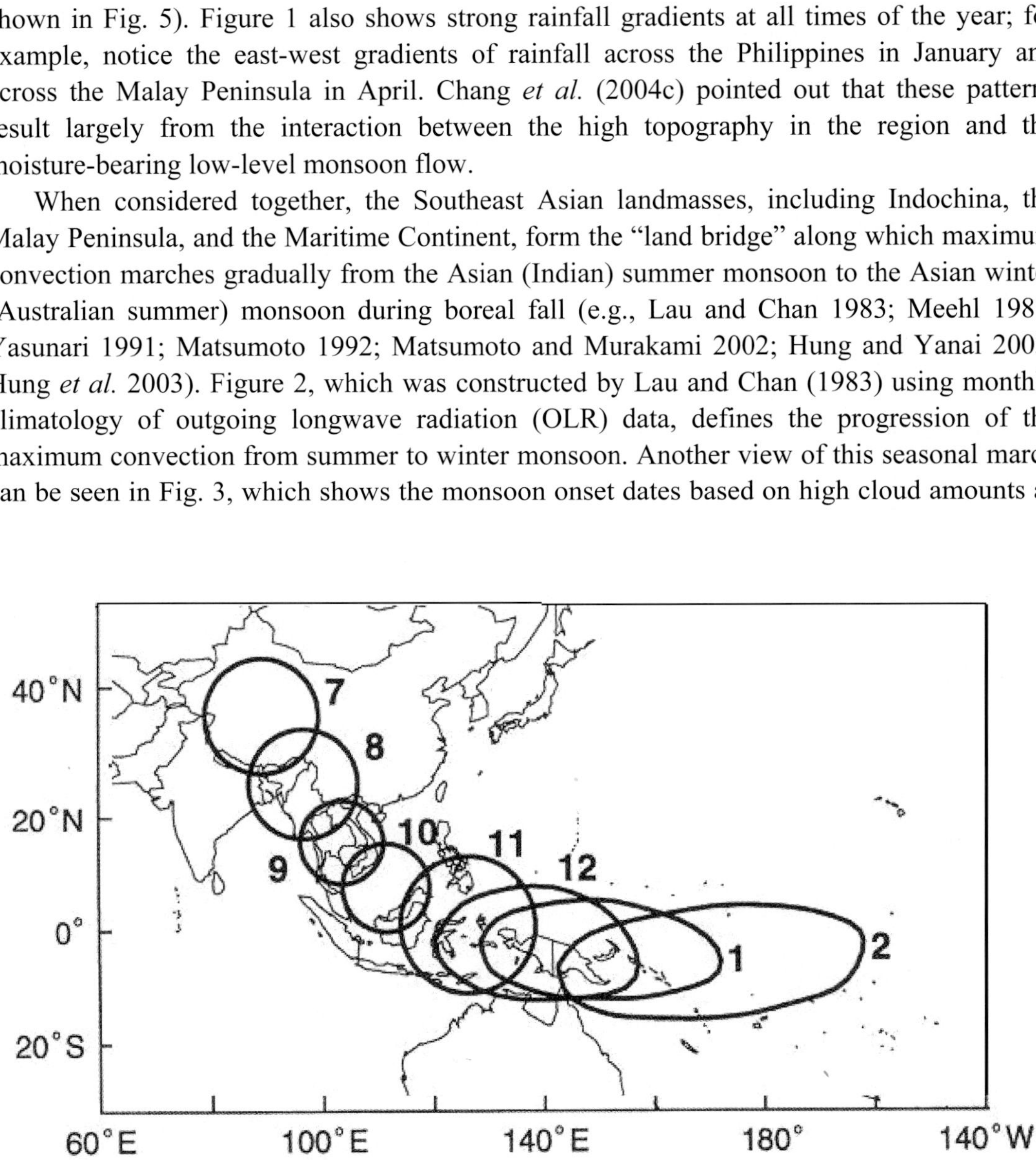

Figure 2. Seasonal migration of the monsoon diabatic heat sources during July-February (denoted by marching numerals). The extent of the diabatic heat sources is determined from the area with OLR < 225 W m^{-2} from monthly OLR climatology and is approximately proportional to the size and orientation of the schematic drawings. (Adapted from Lau and Chan 1983).

determined by Tanaka (1994). However, the seasonal march is not symmetric. During boreal spring the convection tends to stay near the equator as if most of it were blocked from moving northward of 5°N. The convection stays until the reversed meridional thermal gradient is established at which time the convection jumps over the northern equatorial belt to mark the onset of the Asian summer monsoon. This situation may be seen even in low-pass filtered OLR charts shown in Fig. 4, which shows the slow annual cycle of OLR (LinHo and Wang 2002) from April to August. In this figure, 95% of the fluctuations with periods less than 60 days have been removed. Still, it shows that geometric centers of equatorial convection areas mostly stay south of 5°N before pentad 30 (end of May), except a center at pentad 29 moved to ~ 6°N just north of Sumatra. This asymmetry of the seasonal march between boreal fall and spring will be discussed in Section 2.4.

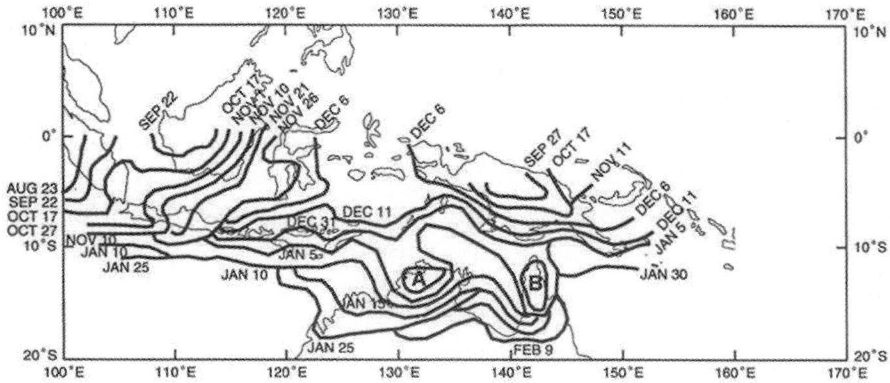

Figure 3. Monsoon onset dates defined by the threshold value of more than 30% of the mean high cloud amount for the monsoon season. Regions A and B had onset prior to 15 December and 26 December, respectively. (Adapted from Tanaka 1994).

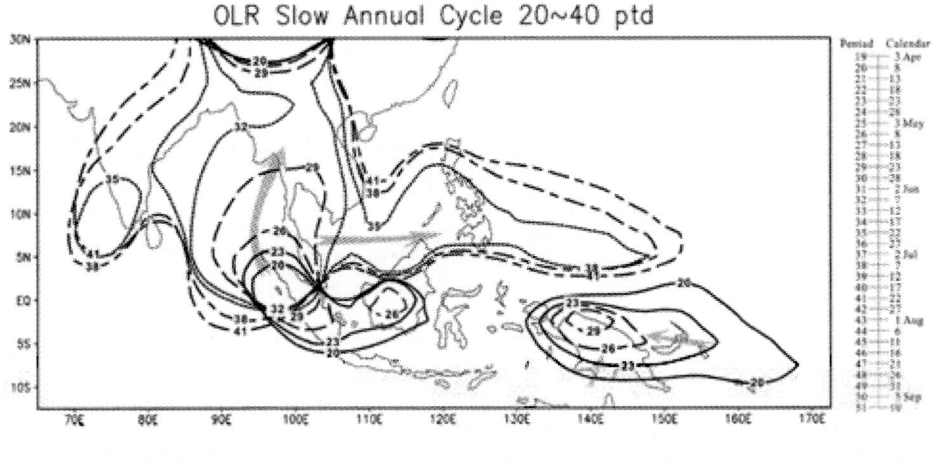

Figure 4. The slow annual cycle of areas enclosed by OLR ≤ 220 W m^{-2}, labeled by pentad. (From LinHo and Wang 2002).

2.2. *Annual Cycle and Semiannual Cycle Modes for Rainfall*

Hamada and Sribimawati (1998) and Hamada *et al.* (2002) constructed annual cycle and semiannual cycle modes by computing the first two harmonics of the climatologically averaged annual rainfall variation at Indonesian rainfall stations. Chang *et al.* (2004c) used longer periods of data and expanded the analysis to include reports from Malaysia, Singapore, Brunei, Thailand and Philippines. Fig. 5 shows their annual cycle mode at land stations, with the amplitude represented by the length of each arrow and the phase shown as a 12-month clock with a northward arrow indicating maximum rainfall in January. The arrow rotates clockwise with eastward, southward and westward arrows indicating April, July and October, respectively.

Over Indochina, the effect of the Asian summer monsoon is clearly indicated with most Thailand stations showing maximum rainfall around June. Over the Philippines, the Asian

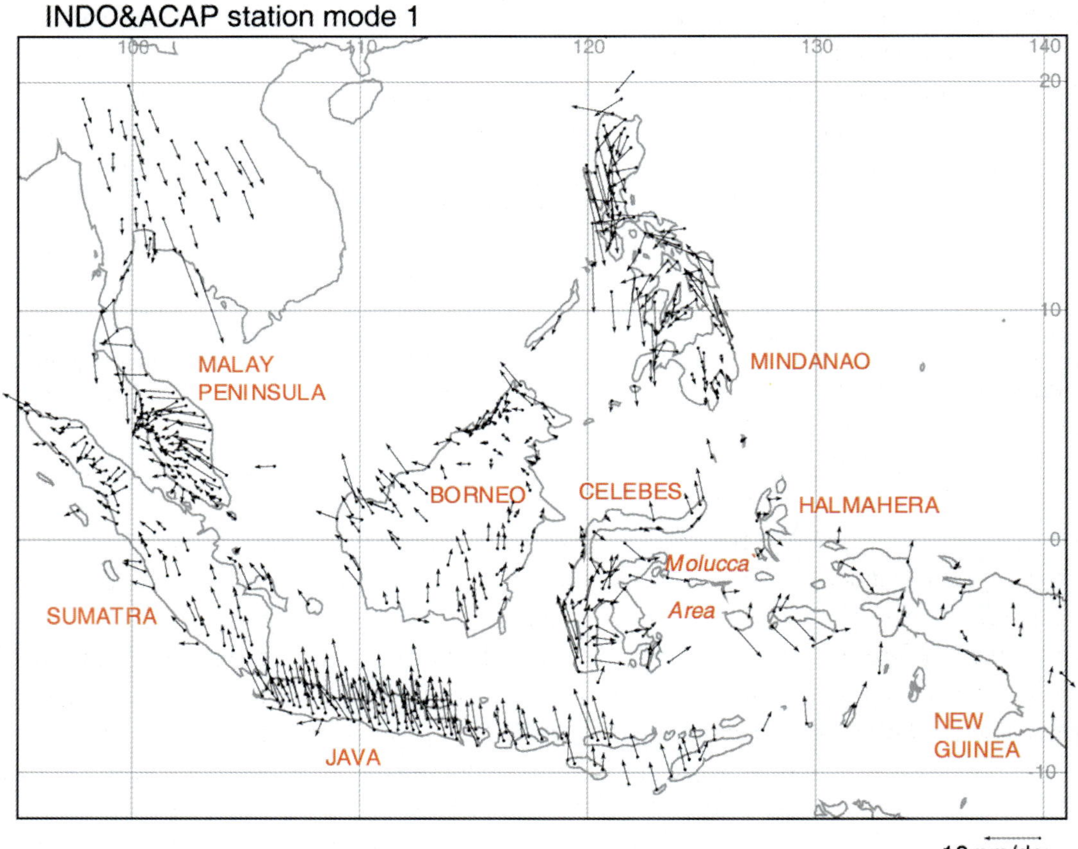

Figure 5. The annual cycle mode at rainfall stations. The phase of the cycle is shown as a 12-month clock with a northward arrow indicating maximum rainfall in January. The arrows rotate clockwise with eastward, southward and westward arrows indicating April, July and October, respectively. The length of the arrow defines the amplitude of the cycle. (From Chang *et al.* 2004c).

summer monsoon rainfall is defined at most stations in the south and west where the rainfall maximum occurs around July. These are stations on the windward side during the southwest monsoon. On the other hand, most northern and eastern stations show maximum rainfall in late summer or early fall. These are the stations on the leeward side of the high topography during the peak of the southwest monsoon in July and on the windward side during September through December when northeasterly flow occurs. In addition, the number of typhoons peaks in September therefore contributing to rainfall in the northeastern Philippines.

The southern Philippine island of Mindanao is a part of the Maritime Continent, based on Ramage's (1968) definition. Here, the seasonal cycle over most of the region is characterized by the summer monsoon rainfall. The exception is in its northeastern corner, where maximum rainfall occurs around November and December again as a result of the prevailing northeasterly onshore winds during the northern winter monsoon.

In Fig. 5 the distribution of the annual cycle over the Philippines appears as a counterclockwise pattern in the phase diagram. In contrast, over the western Maritime Continent from Indochina to the Malay Peninsula, the phase of the annual cycle moves in a clockwise manner reflecting the seasonal march of the deep convection that follows the sun (Lau and Chan 1983). Thus, the annual cycle maximum occurs mostly in northern fall in the Malay Peninsula and northern Sumatra and changes to around December in southern Sumatra. Over the rest of Indonesia, the maximum of annual cycle occurs during boreal winter in most places, especially in southern Borneo, Java and other islands near 10°S.

While the majority of the locations south of the equator have monsoonal rainfall with a wet boreal winter and a dry boreal summer, there are distinct exceptions. The most noteworthy exception is over the central Celebes and the Molucca area, where the maximum in the annual cycle of rainfall tends to occur during boreal spring and early boreal summer (Fig. 5). This has been noted by many previous investigators (e.g., Wyrtki 1956; Ramage 1971; Hamanda *et al.* 2002; Aldrian and Susanto 2003; Aldrian *et al.* 2003).

Chang *et al.* (2004c) showed that this "out of phase" effect is restricted to the southern coastline of the Molucca region where the annual cycle maximum occurs during boreal spring and early boreal summer while the northern coastline experiences a maximum of the annual cycle in January-February, which is closer to the phase of the larger scale region. The topography of the islands is such that a mountainous ridge lies along the islands in an east-west direction. The east and south sides of islands in the Molucca area are sheltered from the northeast monsoon winds during boreal winter, but they face the southeasterly monsoon winds that produce convergence against the terrain during boreal summer.

Chang *et al.* (2004c) also used objectively analyzed rainfall data to identify locations where the annual cycle is less dominant (Fig. 6), and where the semiannual cycle is important (Fig. 7). In Fig. 6 the annual cycle amplitude and phase are plotted against the maximum monthly rainfall. The locations where the annual cycle is less important are delineated by a black dot, which is plotted if the amplitude of the annual cycle is no more than 40% of the maximum rainfall and if the phase difference is at least two months. The locations where the

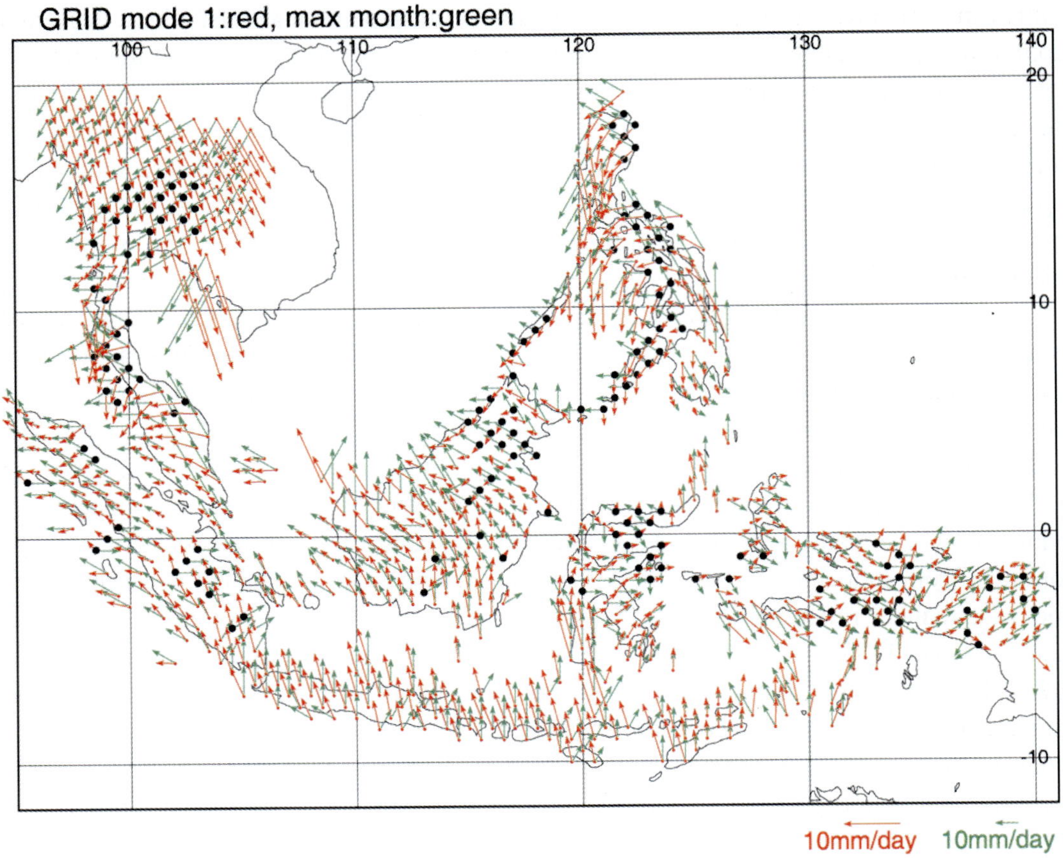

Figure 6. The annual cycle mode (red) and the month and amount of the maximum monthly rainfall (green) for the objectively-analyzed gridded rainfall. The maximum monthly rainfall arrows are scaled to 1:4 of the annual cycle arrows. Black dots indicate grid points where the amplitude of the annual cycle mode is no more than 40% of the maximum rainfall and the phase difference is at least two months. (From Chang *et al.* 2004c).

annual cycle plays a lesser role are the eastern and central Philippines, northeast Borneo, southern Thailand and parts of the northwest end of New Guinea.

In Fig. 7, the semiannual cycle is plotted at grid points where its amplitude is at least 80% of the annual cycle amplitude. In general, the semiannual cycle tends to be concentrated over three areas within the equatorial zone of 3°S - 7°N. The first is west of 110°E, which is the transitional zone between summer and winter monsoon with the annual mode peaking in boreal fall. North of the equator in western Malay Peninsula – northern Sumatra, the semiannual indicators orient mostly June/December versus the annual mode that peaks around September. Further southeast in the narrow equatorial sea region between Sumatra and Borneo, the semiannual indicators have a February/August orientation versus the annual mode peak of November. These semiannual mode indicators are normal to the annual mode vectors and suggest *the influence of both boreal summer and boreal winter rainfall.*

The second main area is in northeastern Borneo between 114°E - 120°E. The annual modes are small and the semiannual mode indicators again have a boreal summer/winter orientation, with a gradual clockwise shift from north to south. The orientation is January-February/July-August at about 3°N - 4°N, March/September near 1°N - 2°N, and April/October at the equator.

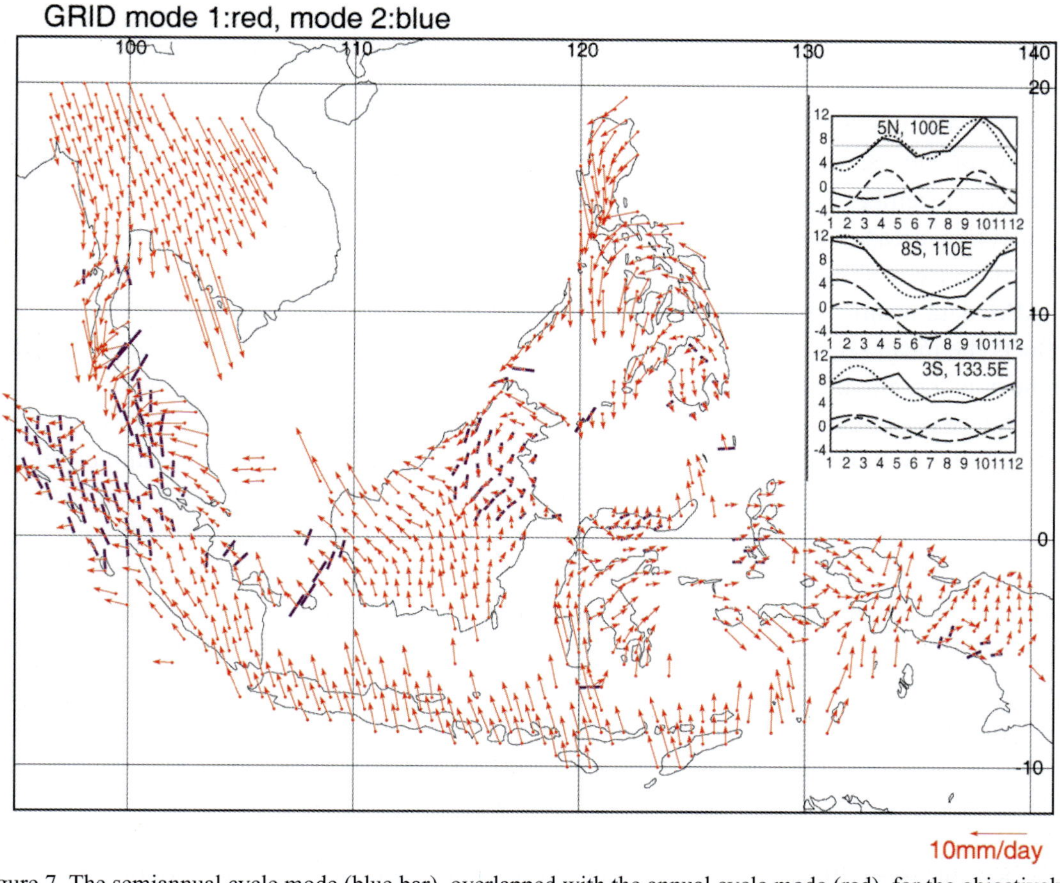

Figure 7. The semiannual cycle mode (blue bar), overlapped with the annual cycle mode (red), for the objectively-analyzed gridded rainfall. Data are plotted if the amplitude of the semiannual cycle is at least 80% of that of the annual cycle. Each bar is centered at the grid point and pointing in the two (opposite) directions of the semiannual cycle peaks. A vertical (north-south) bar indicates rainfall maximum in winter and summer and a horizontal (east-west) bar indicates rainfall maximum in spring and fall. The entire length of the bar is twice the length of an annual cycle vector with the same amplitude. Example of rainfall time series (mm day^{-1}) at grid points where the semiannual cycle is important (5°N, 100°E), the annual cycle is important (8°S, 110°E), and neither is important (3°S, 133.5°E) are shown in the upper-right insert. (From Chang *et al.* 2004c).

The few equatorial grid points define the third main area further to the east, which is the equatorial zone between 120°E - 130°E. This region covers the vicinity of the Molucca Sea and Molucca Passage, including northern Celebes and Halmahera. Here the semiannual mode indicators are oriented horizontally (March-April/September-October), suggesting the possible effect of the twice-a-year crossing of the sun over the equator. The double rainfall peak associated with the twice overhead crossing of the sun is sometimes considered as the classical mode for the annual cycle of tropical rainfall. One of the most significant aspects of Fig. 7 is how rarely the double rainfall peak occurs in the maritime continent, as it is effectively restricted to those few equatorial grid points where the semiannual mode indicators orient horizontally.

2.3. *Boreal Summer and Winter Monsoon Regimes*

Chang *et al.* (2004c) used Tropical Rainfall Measuring Mission (TRMM) precipitation radar data (Simpson *et al.* 1996) and the QuikSCAT scatterometer winds (Liu 2002) to define monsoon regimes for the four seasons. The boreal summer and winter monsoon regimes are defined in Fig. 8, which shows the difference in TRMM PR data and QuikSCAT winds between December-January-February (DJF) and June-July-August (JJA). Here the warm colors show areas of more rainfall in JJA and the cool colors show areas of more rainfall in DJF. Because of the limited duration (15 months for each season) and the narrow individual TRMM data swaths, some small scale features may be artificial, but the general patterns reflects the mean seasonal differences that the boreal summer rainfall regime dominates north of the equator, and the boreal winter rainfall regime dominates south of the equator.

The two regimes become mixed near the equator, but the extent of mixing is not symmetric. The boreal winter regime extends far northward into the boreal summer regime, whereas the southward extension of the boreal summer regime south of the equator is much more limited. Over Southeast Asia, the intrusions beyond 5°N occur in the following areas: east of the Philippines, northeast and northwest of Borneo in the South China Sea, east of Vietnam, eastern coast of Malay Peninsula, and north of Sumatra. In most of these areas the high boreal winter rainfall is due to the onshore northeasterly winter monsoon winds from the northwest Pacific and the South China Sea. These northeasterly monsoon winds are stronger than the southeasterly winds because of the intense baroclinicity over the cold Asian continent in boreal winter; a similar counterpart does not exist in the Southern Hemisphere in boreal summer. There are also very few coastal areas between 5°S - 10°S that face the prevailing seasonal wind as is the case in the northern tropics.

The intrusion of boreal winter rainfall regime into the southern South China Sea northwest of Borneo does not occur as a result of direct onshore winds, where the northeast monsoon is parallel to the coastline. This is the vicinity of the low-level quasi-stationary Borneo vortices that are associated with the heating of the island and can develop any time of the year. During boreal winter, the northeast cold surge winds increase periodically for periods of one to several weeks and enhance the low-level cyclonic shear vorticity off the northwest coast of Borneo. As a result, the Borneo vortices are particularly active during boreal winter (e.g., Johnson and Houze 1987; Chang *et al.* 2003). Due to the increased frequency of the Borneo vortex, the deep convection and heavy rainfall frequently extend offshore several hundred kilometers into the South China Sea during boreal winter, but much less so during boreal summer.

Over the Maritime Continent, the only significant intrusion of the boreal summer regime into the southern tropics is on the south side of western New Guinea between 135°E - 138°E.

During both winter and summer, this area faces some onshore winds. The southeasterly winds during boreal summer are stronger and have a larger angle with the southern coastline; the westerly winds during boreal winter are weaker and nearly parallel to the coastline.

Within the equatorial belt of 5°S - 5°N, the mixing of the boreal winter and boreal summer rain regimes is almost entirely the result of the sheltering by the mountains and the interaction of onshore winds with the complex island geography. Here, the TRMM data available over both land and water show this wind-terrain interaction within the Molucca region more completely than that can be revealed from the station rainfall data (Fig. 5).

Chang et al. (2004b) showed that the wind-terrain interaction also dominates the distribution of the summer monsoon rainfall around India and Indochina. They showed that this is a stronger effect in producing the monsoon rainfall than monsoon depressions.

2.4. *Regimes of the Transitional Seasons and the Asymmetric Seasonal March*

Chang *et al.* (2004c) also used the TRMM and QuikSCAT data to define the monsoon regimes during the transition seasons, boreal spring (March-April-May, or MAM) and boreal fall (September-October-November, or SON). Figure 9 is a summarized view of the monsoon rainfall in these seasons. Two steps are taken to create this chart. In the first step, at each grid point where the MAM rainfall (green) is the maximum among the four seasons, the difference between the MAM rainfall and the higher of the summer or winter rainfall is computed. The same procedure is followed at points where the SON rainfall (yellow) is the maximum of all four seasons. The resultant values are then plotted together in Fig. 9. The QuikSCAT winds for both seasons are also plotted, with the MAM wind in black and the SON wind in red.

In Fig. 9 the SON monsoon rainfall dominates areas north of the equator, particularly the western side of the domain. The MAM monsoon rainfall dominates areas south of the equator, particularly the eastern side of the domain. Within the equatorial belt of 10°S - 10°N, the area west of 114°E, or about the longitude of central Borneo, is mainly in the SON regime while the area to the east is mainly in the MAM regime. Away from the equator, significant Northern Hemisphere SON monsoon rainfall can also be found in the South China Sea and east of Philippines. Similarly significant monsoon rainfall does not exist in the Southern Hemisphere MAM regime.

As noted earlier, the advance of the annual cycle in space is not symmetric. During boreal fall, the maximum convection moves southeastward from South Asia around India and Bay of Bengal through the eastern Indian Ocean and the South China Sea in a track that roughly follows the Southeast Asia land bridge, to reach southern Indonesia and northern and eastern Australia during boreal winter. During boreal spring the maximum rainfall remains mostly south of the equator without a northwestward progression that would retrace the path of the boreal fall progression.

This asymmetric seasonal march of convection is consistent with the migration of the low-level wind field. Figure 10 shows the frequency of mesoscale-synoptic scale cyclonic circulation centers during boreal winter, spring, summer and fall, respectively. This frequency is compiled by identifying the centers of clockwise circulations between 5°S -

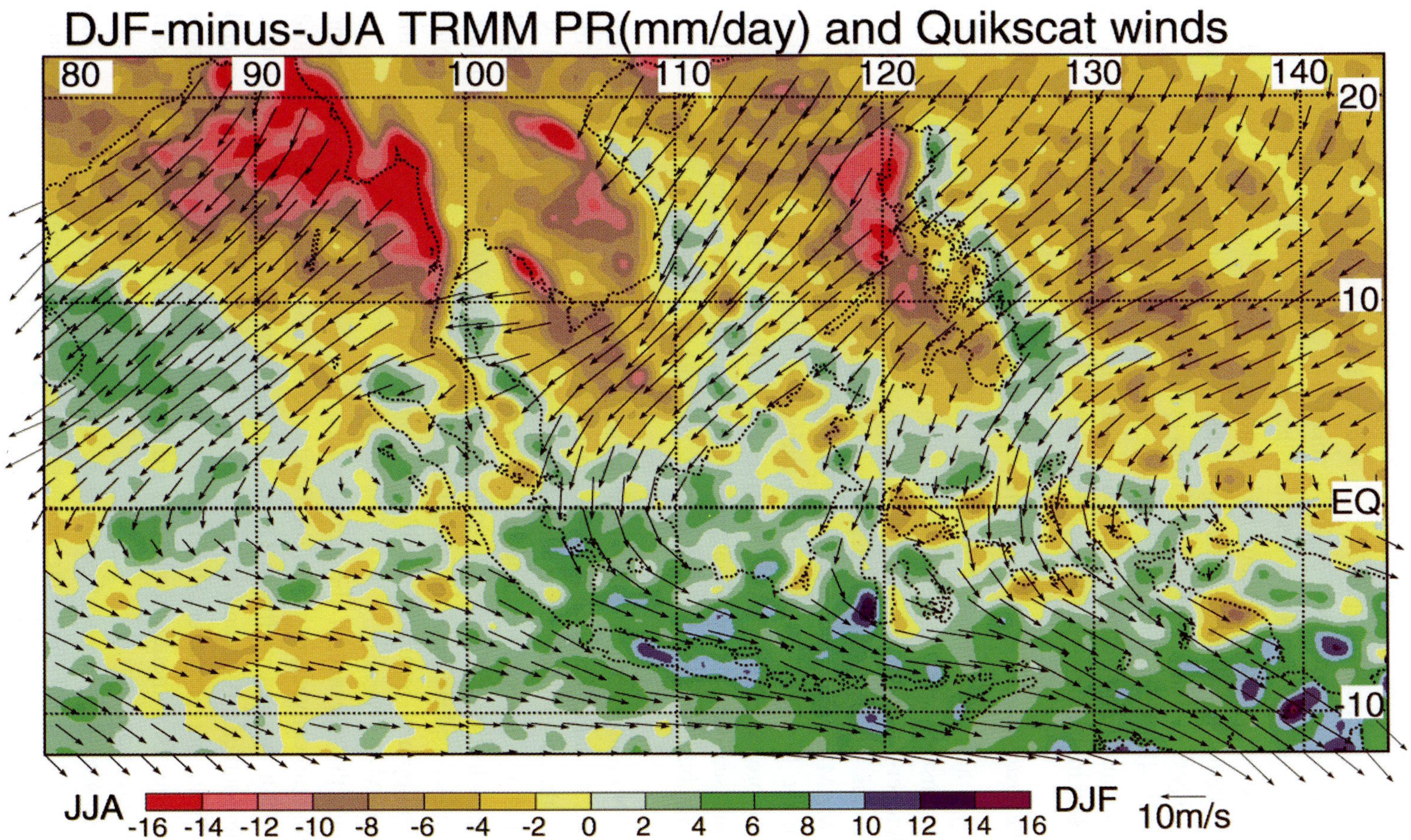

Figure 8. Differences of TRMM PR rainfall and QuikSCAT winds between boreal winter and boreal summer (DJF minus JJA). Warm colors are the boreal summer monsoon regime and cool colors are the boreal winter monsoon regime. (From Chang *et al.* 2004c).

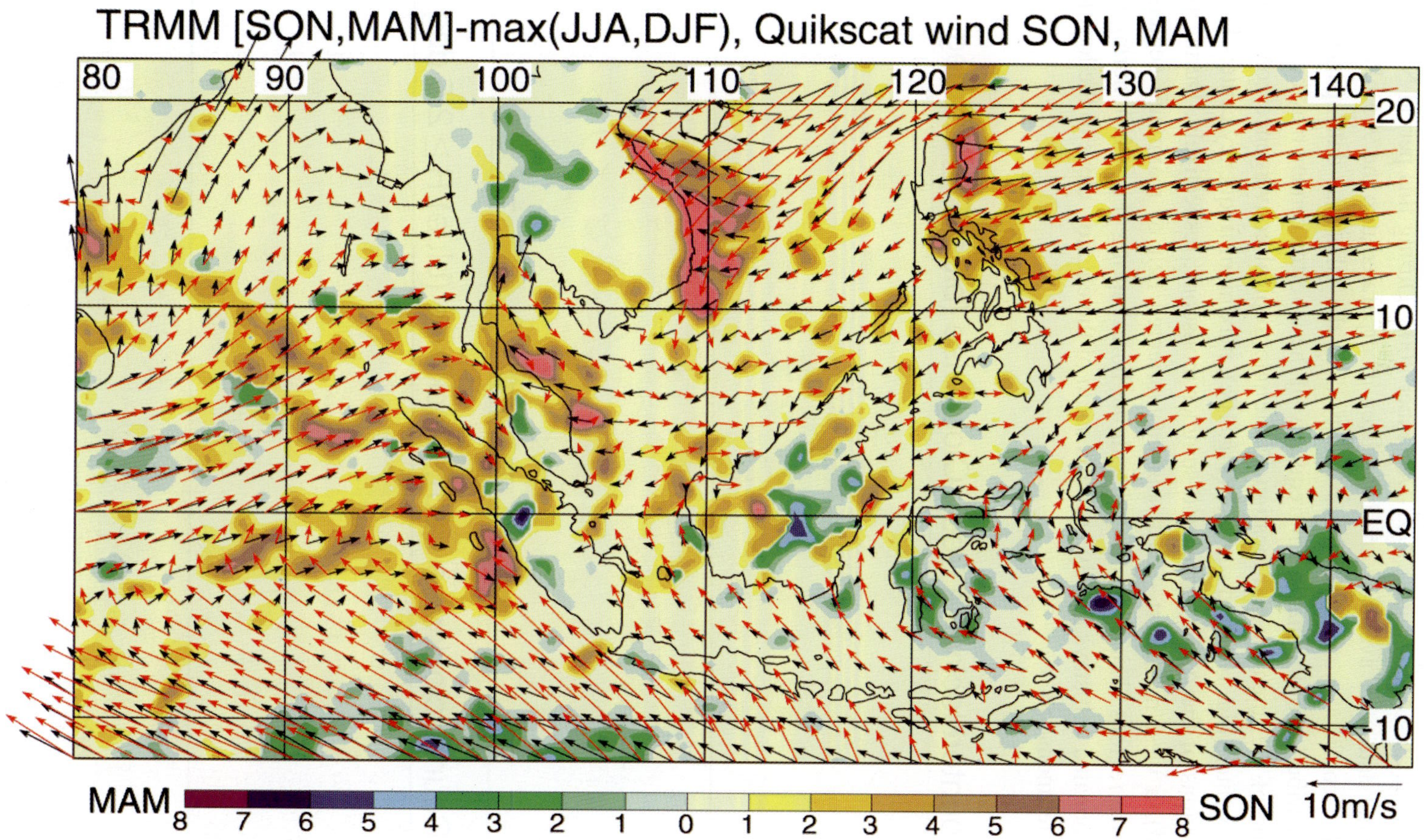

Figure 9. Monsoon regimes during transition seasons deduced form TRMM PR rainfall. A grid point is identified if the rainfall during one of the two transition seasons is the maximum in the annual cycle, and the value plotted is the difference between this transition-season rainfall and the boreal winter and or boreal summer whichever is highest. Warm colors are the boreal fall monsoon regime and cool colors are the boreal spring monsoon regime. The difference of QuikSCAT winds between the two transition seasons (SON minus MAM) is plotted for the entire domain. (From Chang *et al.* 2004c).

30°N and counterclockwise circulations between 30°S - 5°N from the 925 hPa winds of NCEP/NCAR Reanalysis for August 1997-December 2002. During this period NSCAT and QuikSCAT wind data were available and have been incorporated into the 925 hPa analysis, thus raising the confidence of the analysis quality over the otherwise data-sparse tropical oceans. The distribution of the maximum frequency of these vortex centers may be viewed as a manifestation of the intensity and location of the intertropical convergence zone (ITCZ). During boreal winter (Fig. 10a) and summer (Fig. 10c), maximum frequency centers lie close to the core regions of the Asian winter and summer monsoons, respectively. During boreal fall (Fig. 10d), a west northwest - east southeast oriented belt of maximum centers lies between the two regions. This is contrasted with the boreal spring (Fig. 10b), during which the zonally elongated belt stays mainly near and south of the equator, close to the location of the Asian winter monsoon.

a) DJF Counterclockwise (red), Clockwise (blue) R $\geq$ 2.5 deg

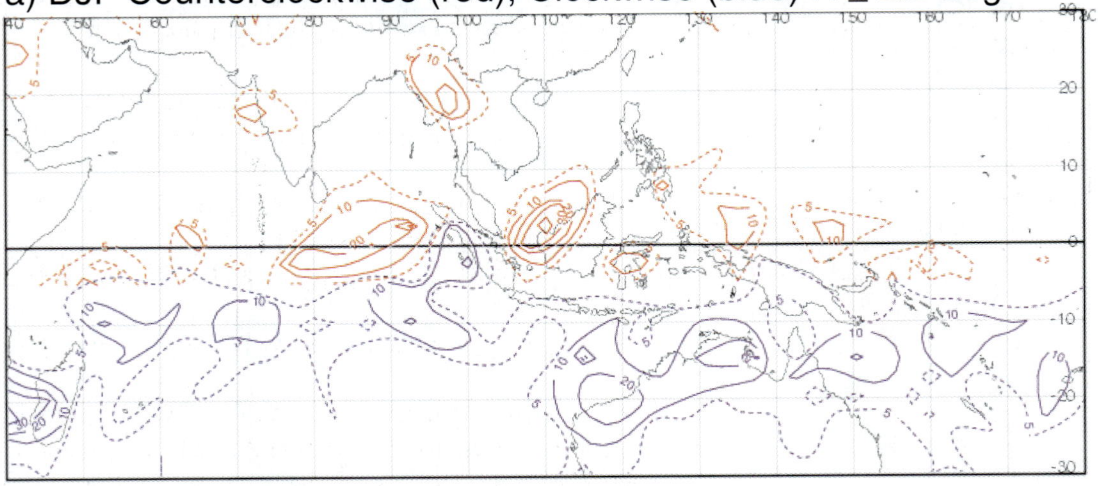

b) MAM Counterclockwise (red), Clockwise (blue) R $\geq$ 2.5 deg

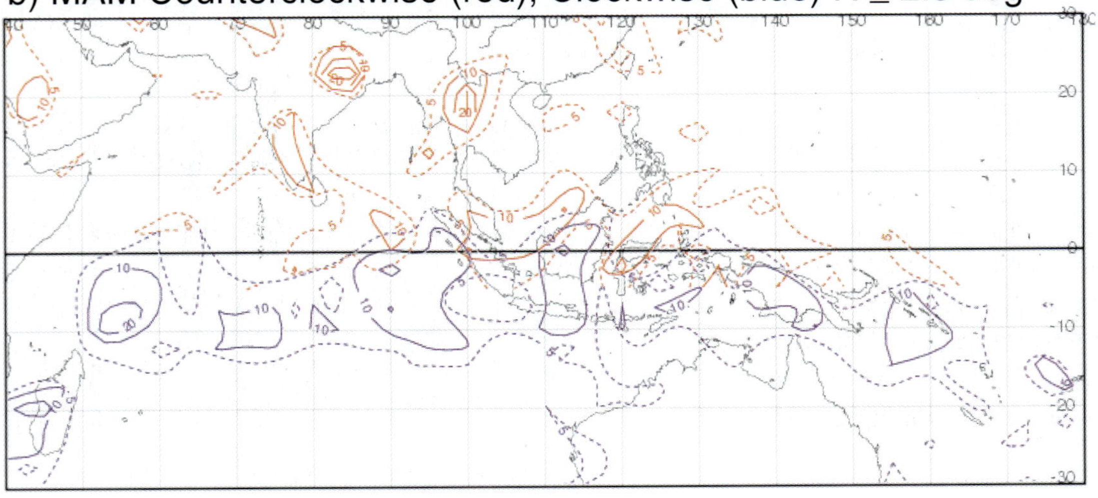

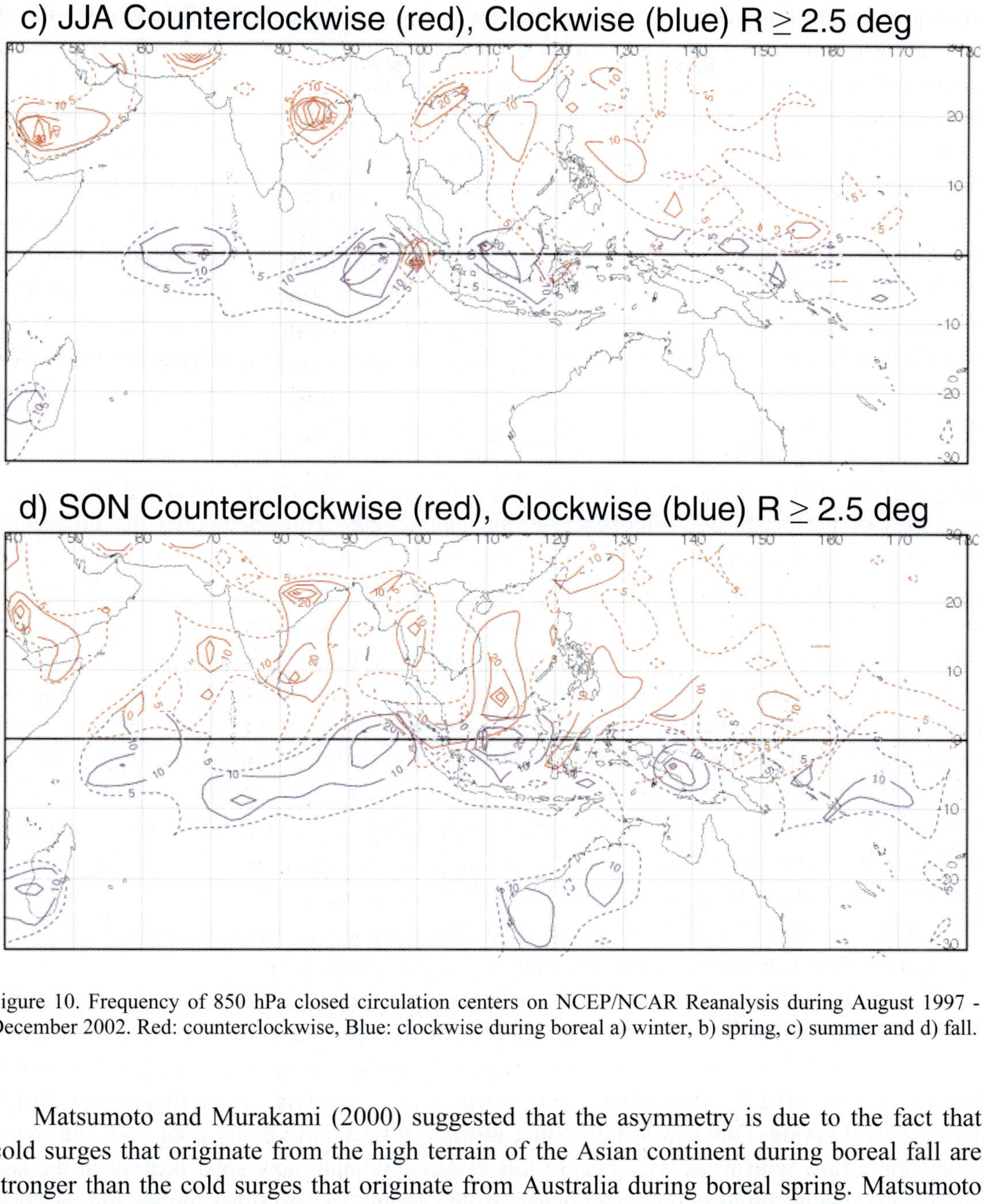

Figure 10. Frequency of 850 hPa closed circulation centers on NCEP/NCAR Reanalysis during August 1997 - December 2002. Red: counterclockwise, Blue: clockwise during boreal a) winter, b) spring, c) summer and d) fall.

Matsumoto and Murakami (2000) suggested that the asymmetry is due to the fact that cold surges that originate from the high terrain of the Asian continent during boreal fall are stronger than the cold surges that originate from Australia during boreal spring. Matsumoto and Murakami (2002) also proposed that this asymmetry is related to different equatorial basic flow characteristics over the western Pacific and Indian Ocean, and their respective annual variations. They suggested that the western equatorial Pacific with a strong Kelvin-type mean flow during boreal fall allows the convection to move southward; and that the flow weakens during boreal spring and inhibits the return journey during boreal fall. Over the equatorial Indian Ocean they suggested that the variation of a complex Rossby-type mean flow allows convection to move northward in boreal spring but inhibits the southward movement in boreal fall. Hung *et al.* (2003) proposed that the northward march of the ITCZ during boreal spring is blocked by subsidence in oceanic regions located to the west of heat sources (India, Indochina and the Philippines). They suggested that this subsidence, similar to that occurring to the west of the Tibetan Plateau during the summer monsoon (Yanai *et al.* 1992), is the result of the Rossby-wave response in the monsoon-desert mechanism suggested by Rodwell and Hoskins (1996). Tim Li (personal communication) proposed that the asymmetry is attributed to an internal atmospheric dynamics mechanism in which equatorial oceanic convection tends to propagate eastward due to the production of a boundary layer convergence to the east of the deep convection. Such a process will favor the southeastward seasonal march from Asian summer monsoon to Asian winter monsoon, but not the northwestward march from winter monsoon to summer monsoon.

Chang *et al.* (2004c) suggested that at least a part of this asymmetry is due to the sea level pressure (SLP) differences between land and ocean driven by the different thermal memories of the ocean and atmosphere-ocean interactions. They computed the difference between boreal spring and fall SLP (Fig. 11) and showed that in the Northern Hemisphere the land SLP is higher in boreal fall and the ocean SLP is higher in boreal spring. The reverse is true in the Southern Hemisphere. This pattern is seen throughout most of the global domain other than areas of the midlatitude storm tracks, and is consistent with a SLP difference that is driven by the different land-ocean thermal memories of the respective transition seasons. The largest SLP difference occurs over the Asian continent, leading to the stronger SON northeasterly winds in the northern South China Sea and northwestern Pacific (area A in Fig. 11), which causes the deep convection east of Vietnam and Philippines in boreal fall[a] (Fig. 9). This interpretation is consistent with Matsumoto and Murakami's (2000) observation that strong cold surges from Siberia leads to strong convection during boreal fall. In the southern South China Sea, the SLP gradient favors cyclonic flow and therefore deep convection in SON (area B in Fig. 11). The convection is further enhanced by the interaction of the wind with terrain on the east coast of Sumatra and west coast of Borneo.

The enhanced boreal fall convection in the eastern equatorial Indian Ocean may also be explained, at least partly, by the spring-fall SLP difference. The Bay of Bengal has higher SLP in boreal fall than in boreal spring, which suggests that in addition to the different thermal memory effect, atmosphere-ocean interaction is involved to warm the SST faster during boreal spring. One possibility is that in early spring the initially cool SST and more anticyclonic flow with weak winds cause less evaporation and more solar heating of the sea surface and downwelling in the upper ocean, so the spring SST becomes higher and the SLP becomes lower than in fall. But the land-sea redistribution of mass still contributes to lower

[a] The convection in these areas is much stronger than both boreal summer and winter, even though during boreal winter the northeasterly winds are stronger. This is because the colder and drier air and the colder sea-surface temperature (SST) make deep convection less likely to develop north of 10°N, so that strong winter cold surges actually produce drying conditions in northern and middle South China Sea (Chang *et al.* 2004a).

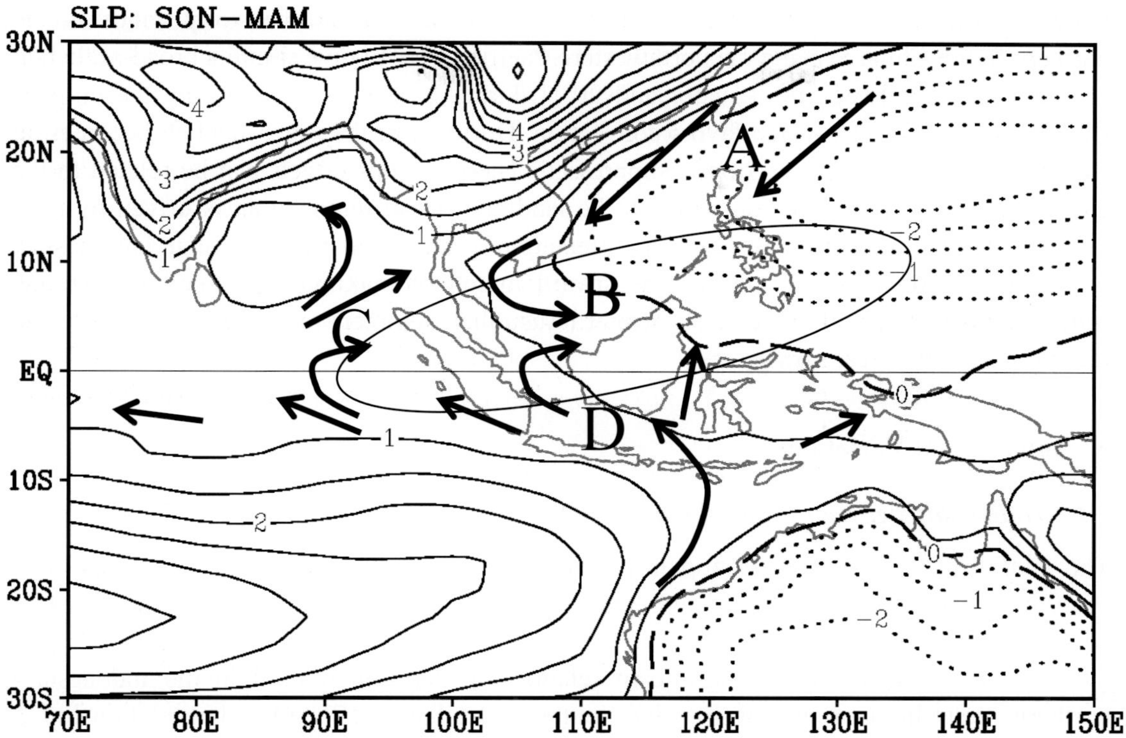

Figure 11. Differences of sea-level pressure between boreal fall and boreal spring (SON minus MAM), unit: hPa. Negative isobars are dotted and the zero line is dashed. Schematics of sea-level wind differences based on the differences in the sea-level pressure pattern are indicated. The elliptic–shaped area indicates preferred belt of convergence in fall and divergence in spring. See text for details. (From Chang *et al.* 2004c).

SLP in the Bay of Bengal during boreal fall compared to surrounding areas. The resulting difference in pressure gradient during boreal fall gives rise to cyclonic flow in the Bay of Bengal and favors increased cross-equatorial flow from the southern Indian Ocean.

The convection in and around the middle and southern South China Sea in boreal fall helps to induce southwesterly winds west of Sumatra (area C in Fig. 11). These southwesterly winds are enhanced by the tendency of cross-equatorial flow and the cyclonic flow in the Bay of Bengal. Other atmospheric and oceanic factors, such as the east-west pressure gradient across equatorial Indian Ocean, may also contribute to the development of equatorial westerly winds. These winds have two effects, both of which lead to more convection. The first is the onshore flow that causes convergence along the western coasts of northern Sumatra and Malay Peninsula. The second is the beta effect that produces convergence in the equatorial westerlies. The increased convection may further enhance the westerlies making a positive feedback possible.

South of the equator, the SLP difference between Australia and the South Indian Ocean favors counterclockwise flow towards the equator (area D in Fig. 11) in boreal fall. This also enhances the cross-equatorial flow that turns westerly north of the equator, which increases the wind-terrain interactions on the west coast of land areas, such as western Borneo (area B).

Furthermore, these winds from south of the equator and the northeasterly winds in the northern South China Sea and the northwestern Pacific (area A) give rise to a broad-scale belt of convergence between the equator and 20°N during boreal fall (marked by the elliptic-shaped area in Fig. 11), which favors convection. During boreal spring, this belt becomes an area of divergence, so convection is suppressed. These various effects can be related to the global-scale mass redistribution between the land and ocean regions that is driven by their different thermal memories. Because the orientation of the Asian and Australian landmasses, this redistribution facilitates the southeastward march of maximum convection from the Asian summer monsoon to the Asian winter (Australian summer) monsoon, but it deters the reverse march in boreal spring.

3. Interannual Variations

3.1. *Relationship between Rainfall and ENSO*

The rainfall in the Maritime Continent region has considerable interannual variations (e.g., Webster *et al.* 1998; McBride 1998; Ooi 1999). Many investigators (e.g., Braak 1919, Nicholls 1981, 1985; McBride and Nicholls 1983; Hastenrath 1987; Ropelewski and Halpert 1987; Allan 1991) noted the significant relationship between rainfall in the Indonesia - northern Australia area and the El Nino-Southern Oscillation (ENSO). This relationship is sometimes manifested in the rampant forest fires and the resulting haze in Indonesia during El Nino conditions (Nichol 1998). However, the ENSO – Indonesian rainfall relationship is strongest during the northern summer and fall, which are the dry and transitional seasons, respectively (e.g., McBride and Nicholls 1983; Ropelewski and Halpert 1987; Kiladis and Diaz 1989; Haylock and McBride 2001, Hendon 2003, McBride *et al.* 2003, Aldrian *et al.* 2003). Hamada *et al.* (2002) attributed this relationship to the different monsoon onset dates during ENSO. They reported that the onsets are earlier in La Nina years and later in El Nino years. During the wet season of northern winter, the negative correlation between Indonesian rainfall and eastern equatorial Pacific sea-surface temperature (SST) is the lowest in the annual cycle (Fig. 12). The correlation is low even though in northern winter the anomalous Walker circulation associated with ENSO events exhibits large-scale upper-level convergence over the Maritime Continent during warm events and divergence during cold events.

The change of correlation between ENSO and Indonesian rainfall during the different seasons motivated Hayden and McBride (2001) to examine the spatial coherence among Indonesian station rainfall. They reasoned that it is not possible for two or more stations to have uncorrelated interannual rainfall for a particular season, yet share the same predictors. Using a 63-station rainfall data over Indonesia for 1950-1998, selected according to the criteria of Kirono *et al.* (1999), they carried out a principal component analysis of the December-January-February (DJF) rainfall. Figure 13 shows the spatial patterns of loadings of the first two principal components. The first pattern is a horseshoe-shaped pattern around the island of Sulawesi; and the second pattern extends over an area from southern Sumatra to western Borneo. Together they explain only 21% of the total variance, and each has high loadings in a small region only. Since there is no high-variance principal component or any significant component with high loadings over a large part of the region, there is only limited coherence in the DJF rainfall. This implies that no single predictor is likely to explain a substantial proportion of rainfall variation over the entire region. Therefore, the DJF rainfall

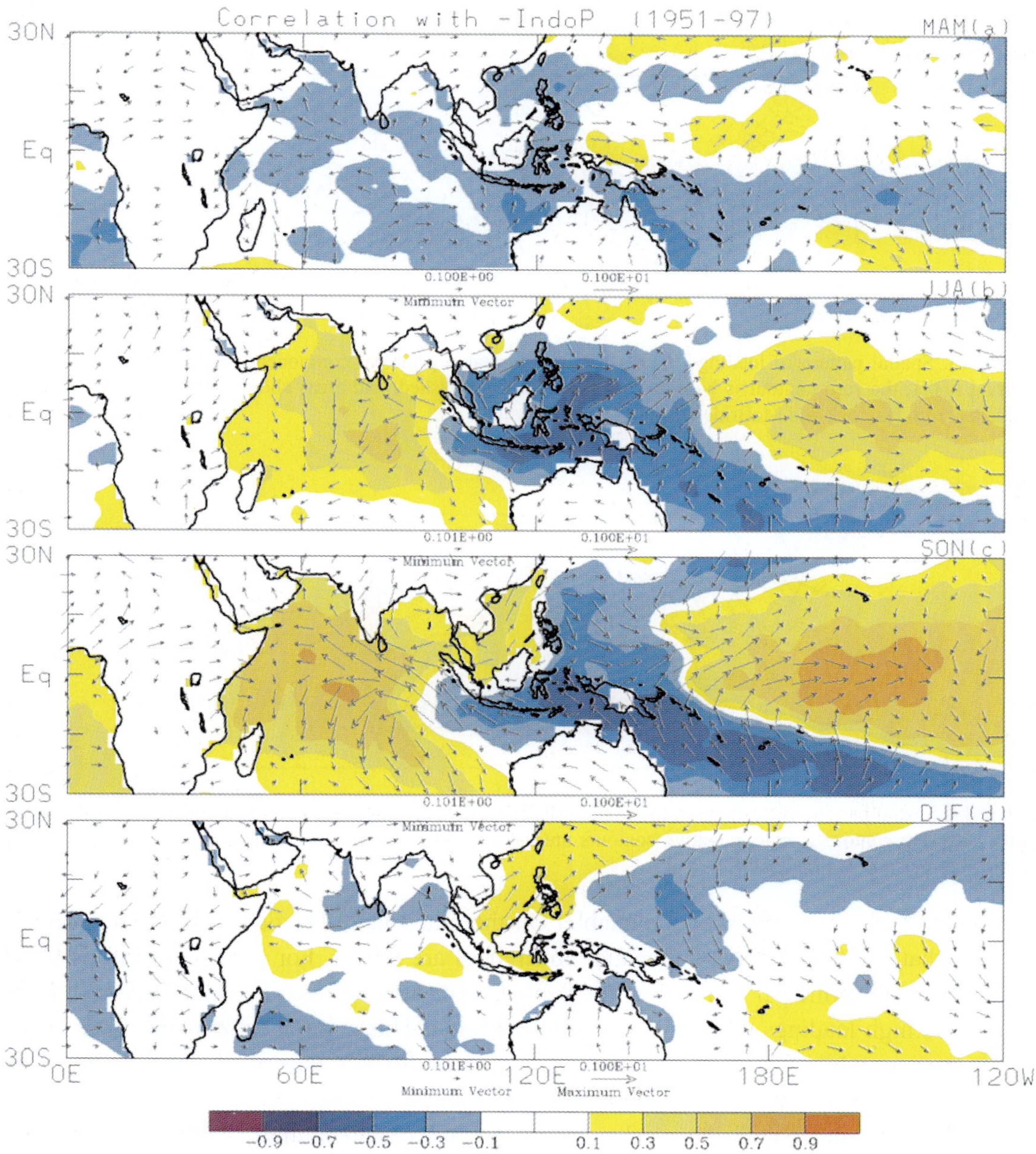

Figure 12. Correlation of monthly mean Indonesian rainfall with Nino34. The long dashed lines indicate significant correlation at the 95% level. (From Hendon 2003)

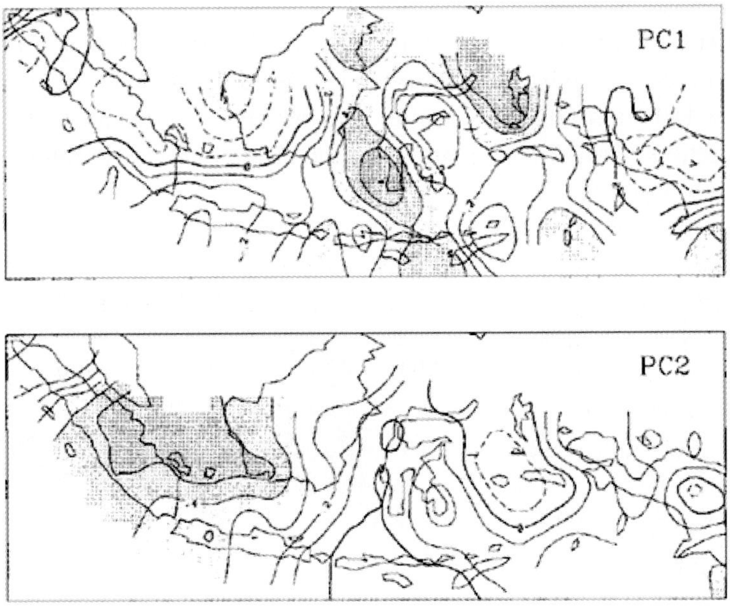

Figure 13. Spatial pattern of loadings of the first two unrotated principal components of DJF rainfall. Contour interval is 0.1, the zero contour is bold, negative contours are dashed, and areas above 10.3 and below 20.3 are shaded. (From Haylock and McBride 2001)

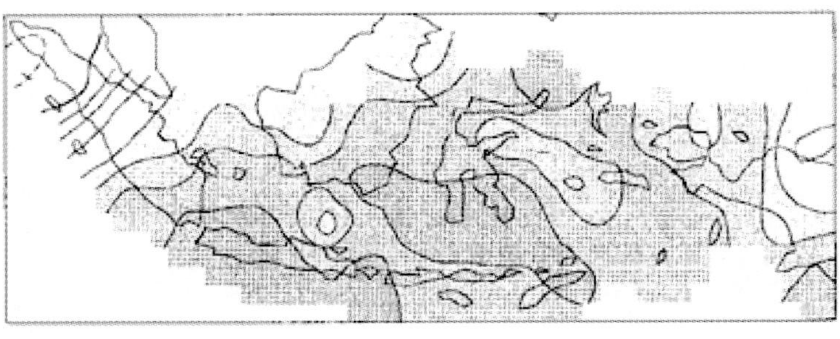

Figure 14. Spatial pattern of loadings of the first unrotated principal component of SON rainfall. Contour interval is 0.1, the zero contour is bold, negative contours are dashed, and areas above 10.3 and below 20.3 are shaded. (From Haylock and McBride 2001)

over Indonesia is inherently unpredictable. On the other hand, Hayden and McBride (2001) showed that in other seasons the spatial coherences are higher. For example, their first and only significant principal component for September-October-November (SON) rainfall (Fig. 14) shows high loadings over almost the entire region. This component explains 38% of the total variance, while the second components accounts for less than 8%. Thus, much of the variation in the SON rainfall over Indonesia is coherent.

In order to understand the weakening of the Indonesian rainfall – ENSO relationship from the dry season to the wet season, Hendon (2003) examined the relationships between

EAST ASIAN MONSOON

Errata

In Chapter 3, Figure 12 (p. 125) in color should be Figure 15 on page 127, and vice versa. (The original figure captions are correct.)

Indonesian rainfall, SSTs and atmospheric circulation over the entire tropical Indian Ocean and Pacific using 43 stations selected from the data set used by Hayden and McBride. Based on the result of the change of the correlation patterns (Fig. 15), he postulated that the weakening of the relationship results from seasonally varying feedback of ENSO on the local SST surrounding Indonesia. The local SSTA during ENSO tend to be of opposite sign to those in the eastern Pacific and western Indian Oceans during the dry season but of the same sign during the wet season (Rasmusson and Carpenter 1982). Coupled with the reversal of the prevailing surface winds (from easterlies during the dry season to westerlies during the wet season), during the dry season the local impact of SSTAs on rainfall and the Walker circulation tends to compound that due to the remote SSTAs (Rowell 2001) while during the wet season, the local impact of the SSTAs tends to oppose that produced by the remote SSTAs. As a result, there is an enhancement of the equatorial Pacific SST gradient and Walker circulation from the dry season to the transition season, and a rapid reduction in the SST gradient and Walker circulation during the wet season. This explains the lack of correlation between SSTs and rainfall during the wet season, and the reduction in spatial coherence of rainfall across Indonesia in going from the dry season to the wet season.

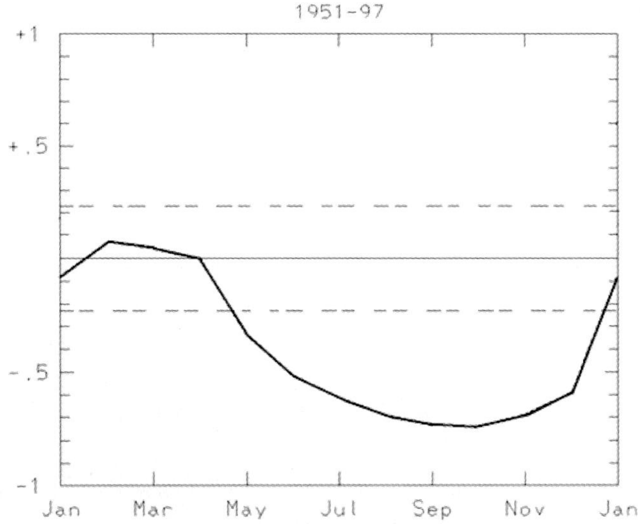

Figure 15. Correlation of seasonal mean Indonesian rainfall with SST (shading) and surface winds (maximum vector is length 1.) for the a) March-April-May, b) June-July-August, c) September-October-November, and d) December-January-February seasons. A correlation of approximately 0.24 is significantly different than zero at the 95% confidence level, assuming 46 degrees of freedom. (From Hendon 2003)

As discussed in Section 2, the interactions of the local terrain with the prevailing winds from the surrounding oceans have a dominant effect on the annual cycle of rainfall locally. Chang *et al.* (2004b) suggested that similar effects may give rise to local rainfall-ENSO relationships that vary among different sub-regions during northern winter, when the region is affected by cold surges off the Asian continent from the north, Indian Ocean zonal wind anomalies from the west, and cross-equatorial flow from the south. Therefore, certain parts of the domain may have localized relationships with ENSO, even though the correlation between ENSO and the rainfall averaged over the entire region is low. For example, Zhang *et al.* (1997) reported that a relationship between meridional wind in the South China Sea and ENSO exists only for the limited domain of 0°N - 20°N, 110°E - 130°E. Chang *et al.* (2004b) used the Climate Prediction Center Merged Analysis of Precipitation (CMAP) rainfall data (Xie and Arkin 1997) to show that during 1979-2002 the correlation between DJF rainfall and Nino3 SST (Fig. 16) is negative over a significant part of the Maritime Continent and vicinity. A notable exception is a low positive correlation area that is oriented northwest-southeast in the vicinity of Sumatra and the Malay Peninsula and includes the western parts of Java and Borneo. This area has been shown to have low correlation with ENSO in other seasons as well (Ropelewsk and Halpert 1987; Aldrian *et al.* 2003). It is interesting that this area also shows up in Lau *et al.*'s general circulation model (GCM) simulations (Chapter 7), although some higher resolution GCM simulations failed to distinguish this area from the rest of the Maritime Continent because the simulated boreal winter rainfall – Nino3 SST correlation was low across the entire domain (Aldrian *et al.* 2003). Chang *et al.* (2004b) then focused on the ENSO-rainfall relationship for three sub-regions where area-average rainfall indices are produced (Fig. 16):

SMP – "Sumatra and Malay Peninsula" area;

SWO – "Southwest Ocean" area, which is the oceanic area immediately adjacent to the southwest coast of Sumatra; and

CMC – "Central Maritime Continent" area.

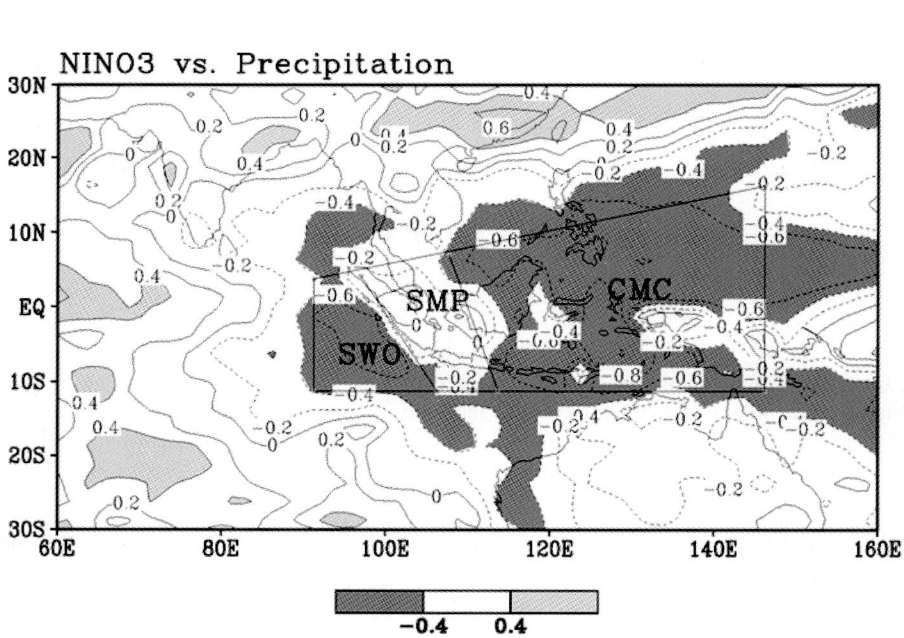

Figure 16. Correlations of 1979-2002 CMAP rainfall with Nino3 SST. Areas above the 5% significance level are shaded. The following rainfall index regions are delineated: SMP: Sumatra-Malay Peninsula; SWO: Southwest oceanic area southwest of Sumatra; and CMC: Central Maritime Continent. (From Chang *et al.* 2004b)

For northern winter, the correlation between the SMP index and the SWO and CMC indices are 0.41 and 0.34, respectively; while the correlation between the non-adjacent SWO and CMC indices is significantly higher at 0.68.

To study the reason for the different correlations, Chang *et al.* (2004b) constructed composites of the anomalous 850 hPa winds from the 1979-2002 NCEP Reanalysis with respect to the SMP, SWO and Nino3 SST indices, all based on an anomaly threshold of 0.75 standard deviation. Figures 16a,b show their Nino3 *cold* anomaly (*cold-minus-warm*) and SMP *wet* anomaly (*wet-minus-dry*) composites, respectively. The SWO wet anomaly composite (not shown) resembles closely the Nino3 cold anomaly composite (Fig. 17a). The main difference between Figs. 17a-b is the convergence of 850 hPa zonal wind near the equator. In the cold-anomaly composite (Fig. 17a) this convergence is a manifestation of the strong rising motion related to the Walker cells over the Pacific and Indian oceans that are typical of La Nina events, and appears to favor convection in the eastern part of the Maritime Continent. In addition, an anomalous cyclonic circulation is centered near the Philippines and extends southwestward into the equatorial South China Sea. In the SMP wet-anomaly (Fig. 17b), the general features are weaker and the strong westerlies between 100°E - 140°E in the cold anomaly composite are absent, resulting in the disappearance of both the strong equatorial zonal wind convergence and the vorticity anomaly centers east of Borneo. Thus, the effect of the eastern Walker cell and the Philippine Sea circulation associated with ENSO diminishes to the west and does not exert a strong influence on the SMP region.

Another difference between Figs. 17a-b is the anomalous westerlies from the equatorial Indian Ocean. In the SMP wet composite (Fig. 17b), the anomalous wind weakens near the longitudes of Sumatra-Malay Peninsula. In the Nino3 cold (and the SWO wet) composite, the anomalous wind, which is associated with the western anomalous Walker cell during cold events, is much stronger. It remains strong over the SMP area and does not attenuate until passing Borneo. Chang *et al.* (2004b) suggested that when strong anomalous westerlies prevail over both the SWO and SMP regions, the moist air from the equatorial Indian Ocean produces excess rainfall on the windward side (SWO) of the steep terrain along the west side of Sumatra, but the influence on the lee side (SMP) is much reduced. Therefore, the SMP region is also sheltered by the influence of the western Walker cell over the Indian Ocean.

Chang *et al.* (2004b) suggested that rainfall in the SMP area may also be affected by an anomalous cross-equatorial flow that is different from the typical circulation patterns associated with ENSO. In the cold anomaly composite (Fig. 17a), the zonal wind direction between the equator and northern Australia is mostly westerly between 125°E - 140°E while it is mostly easterly in the SMP wet anomaly composite (Fig. 17b). The mean 850 hPa wind field in this region during the northern winter monsoon (Fig. 17c) has prevailing westerly winds that are the result of the cross-equatorial flow from the Northern Hemisphere. Thus, for cold events the anomalous circulation enhances the northern-winter mean cross-equatorial flow, while for the SMP wet events it opposes the cross-equatorial flow. In the latter case, the cross-equatorial monsoon wind is opposed upstream (north) of the equator, which will favor an anomalous 850 hPa convergence and more rainfall in the SMP region.

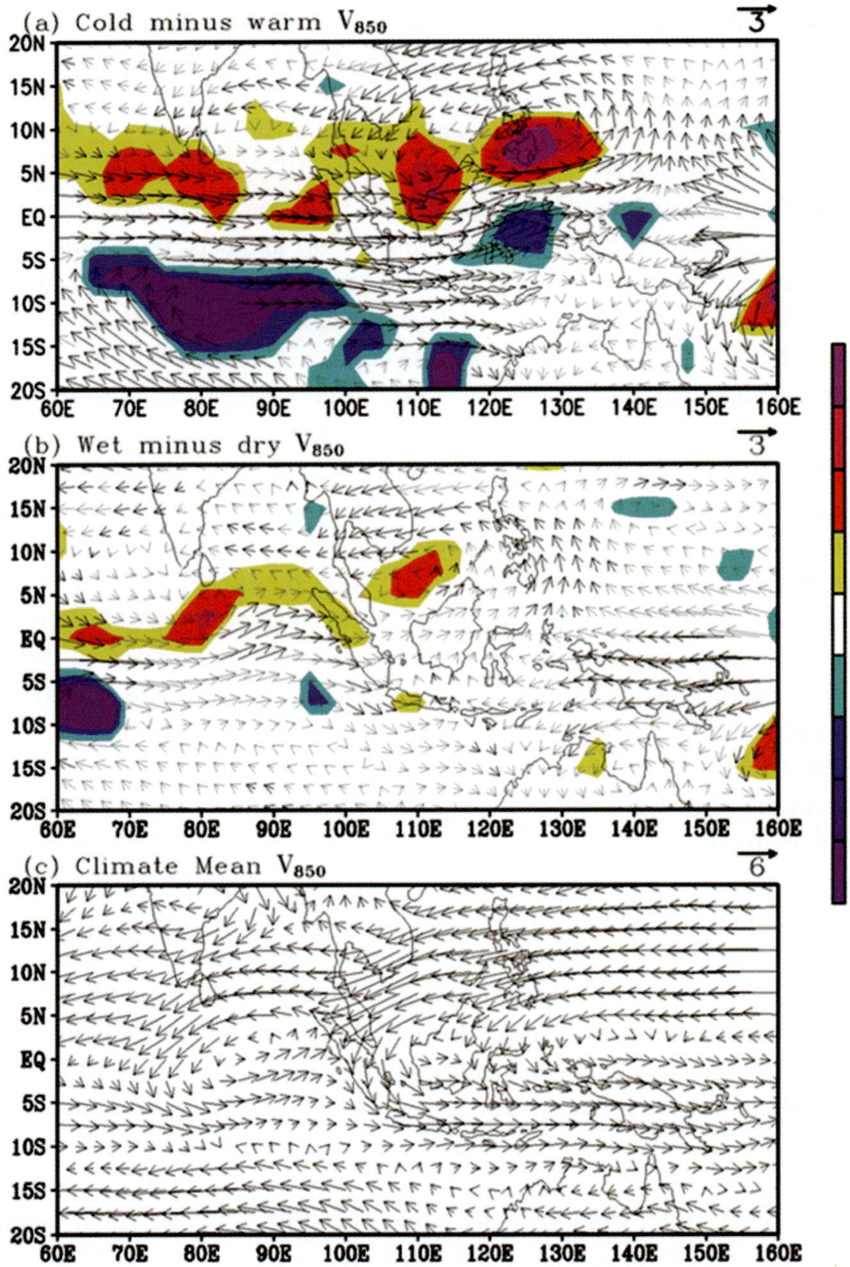

Figure 17. (a) Composite of Nino3 SST "cold minus warm" 850 hPa wind (heavy arrows indicate differences with at least a 95% confidence level in either the zonal or the meridional component) and vorticity (10^{-6} s^{-1}, only differences at 95% or higher confidence level are plotted) during 1979-2002, (b) Same as (a) except for "wet minus dry" in the SMP rainfall index. (c) Long-term seasonal (DJF) mean 850 hPa wind. (From Chang *et al.* 2004b)

3.2. *Interdecadal Variation of Indonesian Rainfall – ENSO Relationship*

The significant correlation between Nino3 SST and the CMAP rainfall amounts in the central Maritime Continent region during 1979-2002 contrasts with previous findings that the Indonesian wet season rainfall is uncorrelated with ENSO events. To reconcile this difference, Chang *et al.* (2004b) divided Haylock and McBride's (2001) Indonesian station data into two periods, 1950-1978 and 1979-1997, and computed the rainfall – Nino3 SST correlations for each period (Figs. 18a-b). In Fig. 18b, a mixture of positive (red) and

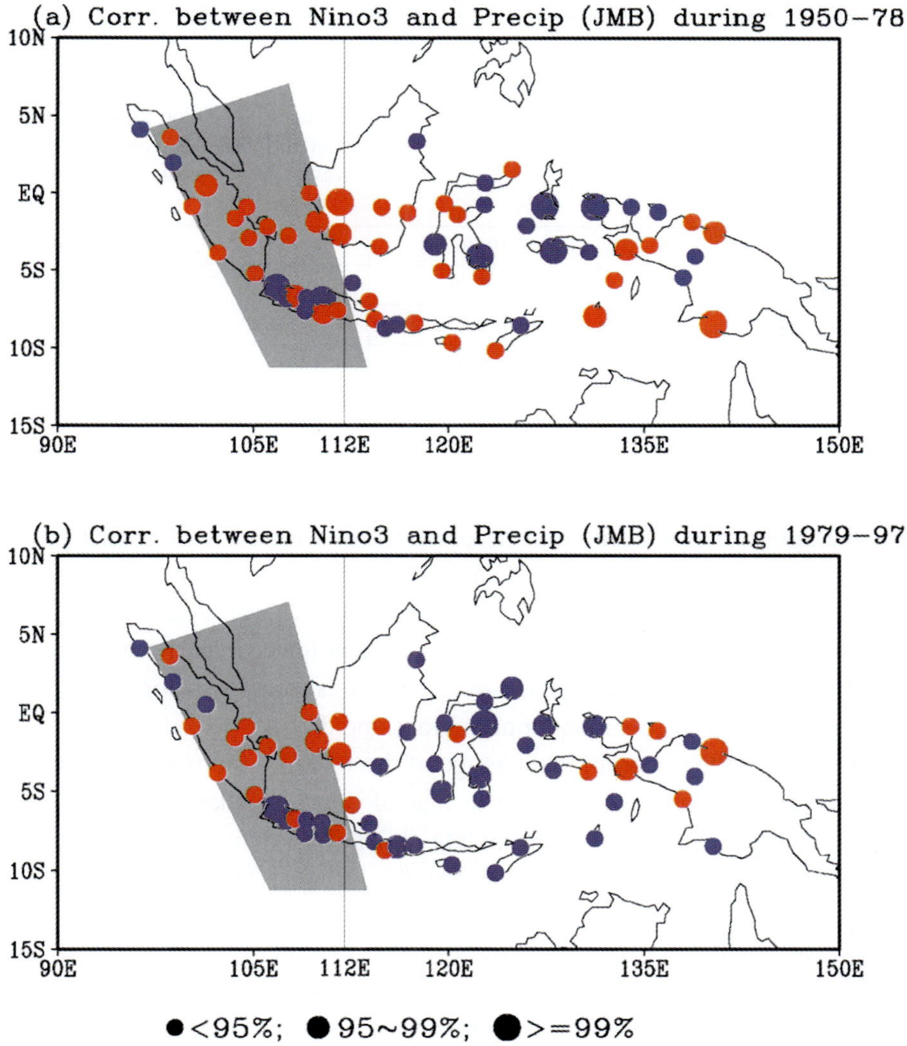

Figure 18. Correlations of Indonesian station rainfall data with Nino3 SST for (a) 1950-78, and (b) 1979-97. Positive (negative) correlations are indicated by red (blue) circles. The sizes of the circles indicate the significant levels: small: below 95%; medium: between 95% - 99%; large: above 99%. The SMP region used in the analysis of CMAP rainfall is shaded. (From Chang *et al.* 2004b)

negative (blue) correlations is seen in the SMP region, which indicates a weak correlation between Nino3 SST and the area-averaged rainfall. Outside of the SMP region, more negative (blue) correlations are generally observed for both periods and more so for 1979-1997. Thus, the Indonesia rainfall outside of the SMP region appears to become more correlated with Nino3 SST after the late 1970s. Table 1 compares the correlations between Nino3 SST and area-averaged station rainfall for western (west of 112°E), eastern (east of 112°E), and all Indonesia for 1950-78, 1979-97, and the entire period. The correlations are of opposite signs between western and eastern Indonesia, and the overall trend for the correlation is to be more negative in the recent decades. The negative correlation for eastern Indonesia is also significant for the entire 48-year period. By contrast, the correlation for all Indonesia remains insignificant regardless of the choice of data periods.

Table 1. Correlation between Nino3 SST and area-averaged Indonesian station rainfall for western Indonesia (west of 112°E), eastern Indonesia (east of 112°E), and all Indonesia. Values above the 1% significance are in italic bold. The significance of all other values is below the 7.5% level. (From Chang *et al.* 2004b)

Winter (DJF)	1950-78	1979-97	1950-97
Western Indonesia	***0.49***	0.26	0.22
Eastern Indonesia	-0.25	***-0.62***	***-0.44***
All Indonesia	0.16	-0.30	-0.17

These results suggest that the low correlation between All-Indonesian rainfall during the wet monsoon season (boreal winter) and ENSO reported by previous studies may be in part due to the averaging of rainfall across the eastern and western regions that have opposite characteristics. The results also suggest an interdecadal change in which the correlation between Indonesian monsoon rainfall and ENSO becomes more negative beginning in the late 1970s. However, the wet season correlations remain lower than those of the dry (boreal summer) and transitional (boreal fall) seasons. This is especially so for January, which has consistently insignificant correlations for all regions and decadal periods.

It is interesting that the interdecadal change of the ENSO – monsoon rainfall relationship for the Indonesian wet season is opposite to that of the Australian monsoon, in which the relationship has weakened beginning in the late 1970s (Power *et al.* 1999a, b). Power *et al.* (1999b) suggested that the change of the Pacific Decadal Oscillation from negative to positive phase around the mid-late 1970s is responsible for the decrease of the effect of ENSO in the western Pacific. Chang *et al.*'s (2004b) result suggests that the interdecadal change appears to concentrate the effect of ENSO during northern winter to a narrower equatorial region in the more recent period rather than a broad weakening of the effects.

3.3. *Biennial Oscillations*

Significant biennial variability in rainfall has been observed in the Maritime Continent, including Indonesia (e.g., Nicholls 1978; Yasunari 1991; Clarke *et al.* 1998) and Malaysia (e.g., Subramaniam 1993; Kwan 2002). These variations are similar to the biennial variations in many parts of the tropical atmosphere (e.g., Meehl 1987; Li et al 2001), and are called "Tropospheric Biennial Oscillation" (TBO) by Meehl (1987). In this oscillation, the rainfall and low-level wind often exhibit a tendency for alternating anomaly signs from one year to the next. Since Maritime Continent rainfall is correlated with ENSO and that ENSO often contains biennial signals (Rasmusson *et al.* 1990), some of the TBO signals in the Maritime Continent rainfall may be ENSO related. On the other hand, the biennial signals in the Maritime Continent and its surrounding equatorial ocean regions, as well as northern Australia, have a particularly interesting behavior that suggests that TBO in this region may also arise from local atmosphere-ocean interactions. In these regions the sign of the correlation between low-level wind and rainfall or surface pressure reverses seasonally. Brier (1978) proposed that if either the ocean-to-atmosphere forcing or the atmosphere-to-ocean forcing changes sign from one part to another part of the year, a biennial oscillation may result from the varying air-sea interactions. Thus, the seasonal sign reversals in the wind-pressure and wind-rainfall correlations may suggest the possibility of a biennially oscillating mechanism by itself.

South of the equator, the correlation between monthly 850 hPa wind speed and surface pressure changes from positive in boreal summer (southern winter) to negative in boreal winter (southern summer). Nicholls (1978) explained that during a positive pressure anomaly, easterly geostrophic wind south of the equator will enhance the prevailing easterly wind in boreal summer but retard the prevailing westerly wind in boreal winter. This wind changes lead to anomalously cool SST during boreal summer, which favors reduced surface pressure, and anomalously warm SST during boreal winter, which favors increased surface pressure. Thus, Brier's (1978) condition is satisfied and Nicholls (1978) demonstrated that a biennial oscillation could be simulated in a simple model.

In the entire Maritime Continent region, including the equatorial western Pacific both sides of the equator, a similar seasonal reversal exists in the correlation between low-level zonal wind and SST. Chang and Li (2000) theorized that this reversal is a part of the feedback mechanism that drives the TBO in the South Asia to Australia monsoon region (Meehl 1997, 2002). In their theory, the low-level zonal wind responds to the anomalous planetary scale east-west circulation (Krishnamurti 1971) that is forced by anomalous heating of the Asian monsoon during boreal summer, and the anomalous circulation of the Australian monsoon during boreal winter. Due to the seasonal reversal of the zonal wind in the region, the anomalous zonal wind enhances or reduces the surface wind speed depending on the season and the sign of the anomaly. This leads to a seasonal reversal of the zonal wind – SST correlation and a TBO. In this case the TBO is the result of the interactions between boreal summer and winter monsoon and the tropical Pacific and Indian Ocean, with the convective rainfall in the Maritime Continent – equatorial western Pacific region playing a critical role. Hendon (2003) used a similar mechanism to explain the TBO in the region, in which the anomalous zonal winds involved in the seasonal reversal of the local air-sea feedback are driven by ENSO instead of the broad-scale Asian-Australian monsoons.

4. Boreal Winter Convection and Synoptic and Intraseasonal Variations

4.1. *Boreal Winter Convection*

A major tropical feature of the northern winter monsoon is the area of deep convection over the large islands of the Maritime Continent. Figures 19 and 20 are reproduced from Chang *et al.* (2004a), who used NCEP/NCAR reanalysis 925 hPa winds for 21 boreal winters (December 1980-February 2001) and a convective index (CI) determined from the Geostationary Meteorological Satellite black-body temperature (T_{bb})[b] to describe the mean distribution of motion and convection fields. The CI is averaged over 00 and 12 UTC to remove the diurnal cycle. Figure 19a shows that maximum convection occurs over Java, which is connected to the convection maximum over Sumatra with an extension into the eastern Indian Ocean as part of the ITCZ south of the equator.

There is a lack of deep convection north of 5°N, which is due to the low-level divergence in the northeasterly monsoon flow (Fig. 19b). This is different from boreal fall, when strong convection occurs off the Vietnamese coast north of 10°N as a result of the low-level convergence produced by the northeast onshore winds (Chang *et al.* 2003). The winter mean northeast winds are stronger than fall, but the air is cooler and drier (Johnson and Houze 1987), and the SST in the northern and middle South China Sea is significantly lower. Thus, deep convection is less likely to develop until the air reaches the southern South China Sea after it is transformed by substantial surface sensible and latent heat fluxes (Johnson and Zimmerman 1986). As the low-level, northeast winds extend south of the South China Sea they are deflected to the west and south due to blocking and deflection by the terrain of the Malay Peninsula and Sumatra. Over the northern portion of the domain, the westerly deflection results in easterly winds that extend into the Bay of Bengal. To the south, the deflection results in northerly winds, which turn eastward after crossing the equator due to the conservation of potential vorticity (Lim and Chang 1981). The resulting westerlies support the Southern Hemisphere ITCZ where tropical cyclogenesis occurs frequently (Holland 1984; McBride 1995).

The two primary regions of low-level convergence (Fig. 19b) are associated with the maxima in deep convection (Fig. 19a). The convergence center near Sumatra is related to the interaction of the northeasterly monsoon flow with the terrain of the Malay Peninsula. However, the convergence center over Borneo is shifted east of the primary northeasterly wind belt and is in the region of the counterclockwise turning of the winds that cross the equator. Coincident with the convergence center over Borneo is a maximum in 925 hPa relative vorticity (Fig. 19c). In addition to the curvature contribution to the vorticity maximum, shear vorticity results due to the interaction between the northeasterly monsoon flow and the terrain of Borneo.

In general, the area of deep convection over the equatorial Maritime Continent started like a V-shape pattern and shrinks from December to January to February (Figs. 20a-c). The

---

[b] CI=250-T_{bb} if T_{bb} is less than 250 K and CI=0 otherwise.

region of minimum convection marches southward through the southern portion of the South China Sea and crosses into the Southern Hemisphere by February, which effectively separates the convective maximum over Borneo from the maximum over Sumatra. The southward extension of the convective minimum coincides with the intensification of the

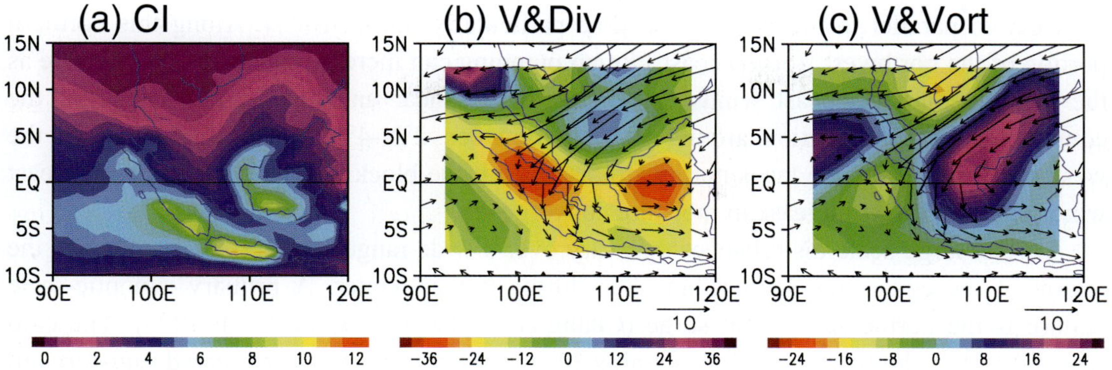

Figure 19. December-February (1979/80-2000/01) mean fields of (a) convective index, (b) 925 hPa winds (m s^{-1}) and divergence (shaded, 10^{-5} s^{-1}), and (c) 925 hPa winds (m s^{-1}) and vorticity (shaded, 10^{-5} s^{-1}). (From Chang *et al.* 2004a).

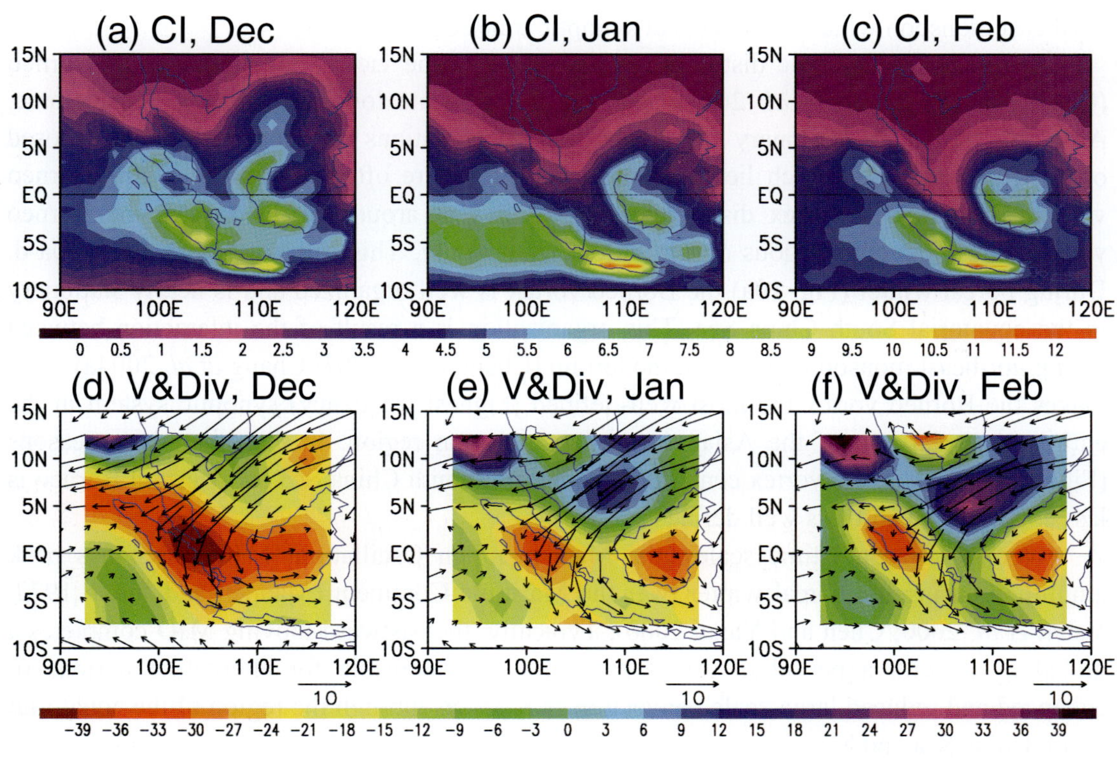

Figure 20. Individual monthly means (1979/80-2000/01) for (top row) convective index and (bottom row) 925 hPa winds (m s^{-1}) and divergence (shaded, 10^{-5} s^{-1}) for (left column) December, (middle column) January, and (right column) February. (From Chang *et al.* 2004a).

divergence over the South China Sea (Figs. 20d-f) associated with the strengthening of the northeast monsoon flow. Furthermore, the reduction in deep convection over Sumatra in January is related with the reduced 925 hPa convergence (Fig. 20e) and the increased drying influence of the strengthening northeast winds that flow toward Sumatra. The drying does not impact the convection center over Borneo as significantly since the northeast monsoon flow is nearly to parallel to the Borneo coastline particularly north of 4°N. Along this northern portion of the northwest Borneo coast Ekman pumping can increase moisture convergence as the strengthening northeast winds contribute to the shear and curvature vorticity of the counterclockwise circulation around Borneo. Between 2°N-4°N a substantial part of the western Borneo coastline faces northward, as a result the blocking of strengthened northeast winds causes a direct increase in moisture convergence.

Several large-scale disturbances that vary over a wide range of time scales influence the region of the equatorial South China Sea during boreal winter. A primary synoptic-scale feature is the northeasterly cold surge (Chang *et al.* 1983; Wu and Chan 1995). The cold surge signal is dominated by a freshening of northeasterly winds that spread equatorward around the eastern edge of low-level anticyclones located over eastern Asia. Topographic influences contribute to the channeling of the northeast winds toward the equator. Typically, one to two cold surges occur per month and may last from a few days to a week or longer (Ramage 1971). The cold surge frequency contributes to a regional oscillation in the northeast monsoon winds that occurs at synoptic time scales.

Synoptic-scale cyclonic disturbances also occur in the vicinity of the island of Borneo (Cheang 1977; Chang *et al.* 2003), where the basic-state low-level vorticity is cyclonic. Although these quasi-stationary low-level cyclonic circulations may not be completely closed on their eastern side, which lies over the island, they are often referred to as the Borneo vortex. Even though vortex disturbances exist all year around in the region, the Borneo vortices are most conspicuous during boreal winter only. This can be seen in Figs. 10a-d. During boreal winter (Fig. 10a) the Borneo vortex is well organized and is nearly stationary in the equatorial South China Sea. This organization is a result of the interaction between strong northeast monsoon winds and the terrain (Chang *et al.* 2003; Chang *et al.* 2004a). This makes the Borneo vortex the most well-defined (highest frequency) synoptic system in the entire equatorial belt of the Asian-Australian monsoon region. In the other three seasons (Figs. 10b, c and d) the vortex center near equatorial South China Sea and western Borneo is less concentrated and less well defined.

Over intraseasonal time scales, the Madden-Julian Oscillation (MJO) often has peak amplitude during the boreal winter over the Maritime Continent (Madden and Julian 1972; Yanai *et al.* 2000; Chen and Yanai 2000). Typically, the eastward-moving MJO completes a global cycle with a period of 30-60 days and is responsible for alternating periods of enhanced and reduced large-scale convection as it passes through the region of the equatorial South China Sea and Maritime Continent.

Taylor (1995) showed that interactions among cold surges, easterly waves and MJO led to tropical cyclone formation in both sides of the equator during the 1996-97 boreal winter. Simms (2000) also presented evidence that MJO may affect cold surge over the equatorial western North Pacific. Therefore, although the northeasterly cold surges, quasi-stationary Borneo vortex, and MJO differ greatly in their origin and do not vary over the same space and time scales. they may interact and influence the variability in deep convection over the Maritime Continent.

4.2. *Synoptic-Scale Circulations*

The impacts of the two synoptic-scale circulations on convection over the equatorial South China Sea and the western Maritime Continent have also been examined by Chang *et al.* (2004a). They used daily 00 UTC NCEP/NCAR reanalysis 925 hPa winds filtered over a period of 2-15 days for 21 boreal winters (December 1980-February 2001) to identify the synoptic motion systems. A cold surge index is defined as the average 925 hPa meridional wind along 15°N between 110°E - 117.5°E. The presence of a Borneo vortex is determined by a streamline analysis of unfiltered 925 hPa winds, where the centers of counterclockwise circulations during the 21 boreal winters are identified as shown in Fig. 21. In this figure the distribution of the circulation centers is summarized by isopleths produced by an analysis of the number of circulation centers in 2.5°x2.5° grid squares. The maximum occurrence of vortex centers is oriented in a northeast-southwest region along the west coast of Borneo with a maximum of more than 120 centers over the 21 seasons near 1.5°N, 111°E. Of the total 1895 days in the 21 boreal winter seasons, 628 days (nearly 1/3) have one or more vortex

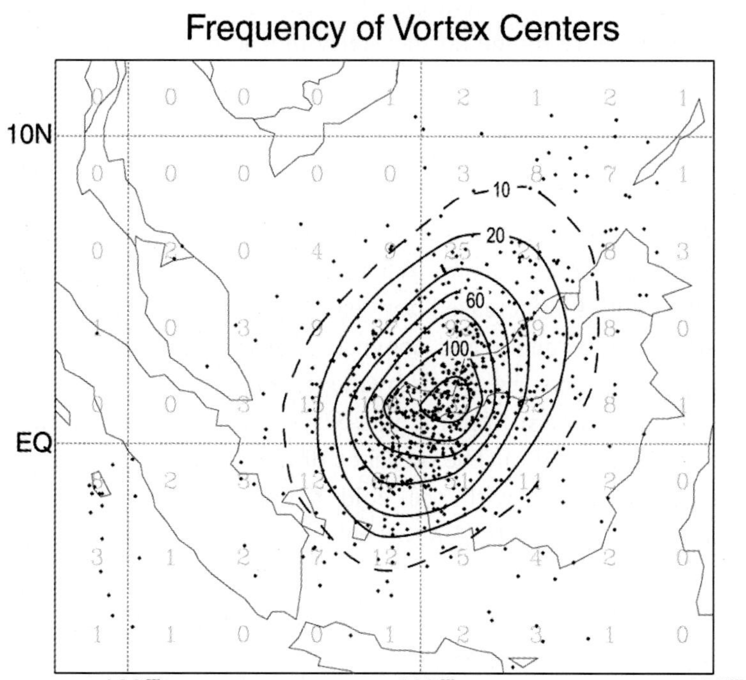

Figure 21. Analyzed Borneo vortex center locations based on streamlines of unfiltered 925 hPa winds. (From Chang *et al.* 2004a).

centers in the western Borneo-southern South China Sea region between 5°S - 10°N and 105°E - 115°E. From the surge index and Fig. 21, each of the 1895 days is classified as to whether a surge existed and to whether a Borneo vortex existed (Table 2).

Table 2. The distribution of Borneo vortex days with respect to the presence of a surge or no surge.

	No-Vortex Days	Vortex Days
No-Surge Days	1038	477
Surge Days	229	151

Composite maps (Fig. 22) of CI, 925 hPa winds, and 925 hPa divergence are used to examine the impact of the synoptic-scale surge and vortex circulations on deep convection over the equatorial South China Sea. When no surge and vortex occur (Figs. 22a,e), convection is reduced over the equatorial South China Sea and enhanced to the west, southwest, and south of the South China Sea. The area of reduced convection over the South China Sea is also an area of low-level divergence. During days when a Borneo vortex is present without a cold surge (Figs. 22b,f) the patterns of CI and low-level divergence are nearly opposite the patterns that occur when no surge and no vortex occur. Deep convection is enhanced over the equatorial South China Sea and along the western-most coast of Borneo near 2.5°N, 108°E, and reduced over regions surrounding the equatorial South China Sea. In accordance with the change in the pattern of deep convection, low-level convergence is found over the South China Sea. It appears that the presence of the Borneo vortex acts to intercept transport of low-level moisture by the northeasterly monsoon flow such that convection over the Malay Peninsula-Java region is reduced and there is also low-level divergence over these regions.

The presence of a surge without a vortex (Figs. 22c,g) acts to reduce deep convection over the equatorial South China Sea where there is also low-level divergence and increase convection and low-level convergence over the surrounding regions of the Malay Peninsula, Java, and surrounding equatorial regions. Whereas the low-level divergence over the South China Sea is associated with the increased northeasterly winds associated with the cold surge, the increased low-level convergence and convection downstream over the Malay Peninsula results from the blocking of the surge winds by the terrain.

Convection over the southern South China Sea is strongest during the combination of surge and vortex cases (Figs. 22d,h). The presence of the Borneo vortex restricts the region of low-level divergence associated with the cold surge winds to be over the Indochina Peninsula. Subsequently, a large region of enhanced low-level convergence covers the southern South China Sea in association with the Borneo vortex. Furthermore, convection is reduced over regions west and south of the South China Sea as the low-level moisture transport is intercepted by the Borneo vortex.

When a vortex is present without a surge (Fig. 22b), the cyclonic turning of the low-level winds occurs over the southern South China Sea. When a vortex occurs together with a surge (Fig. 22d), the axis of cyclonic turning is oriented along the west coast of Borneo. Therefore,

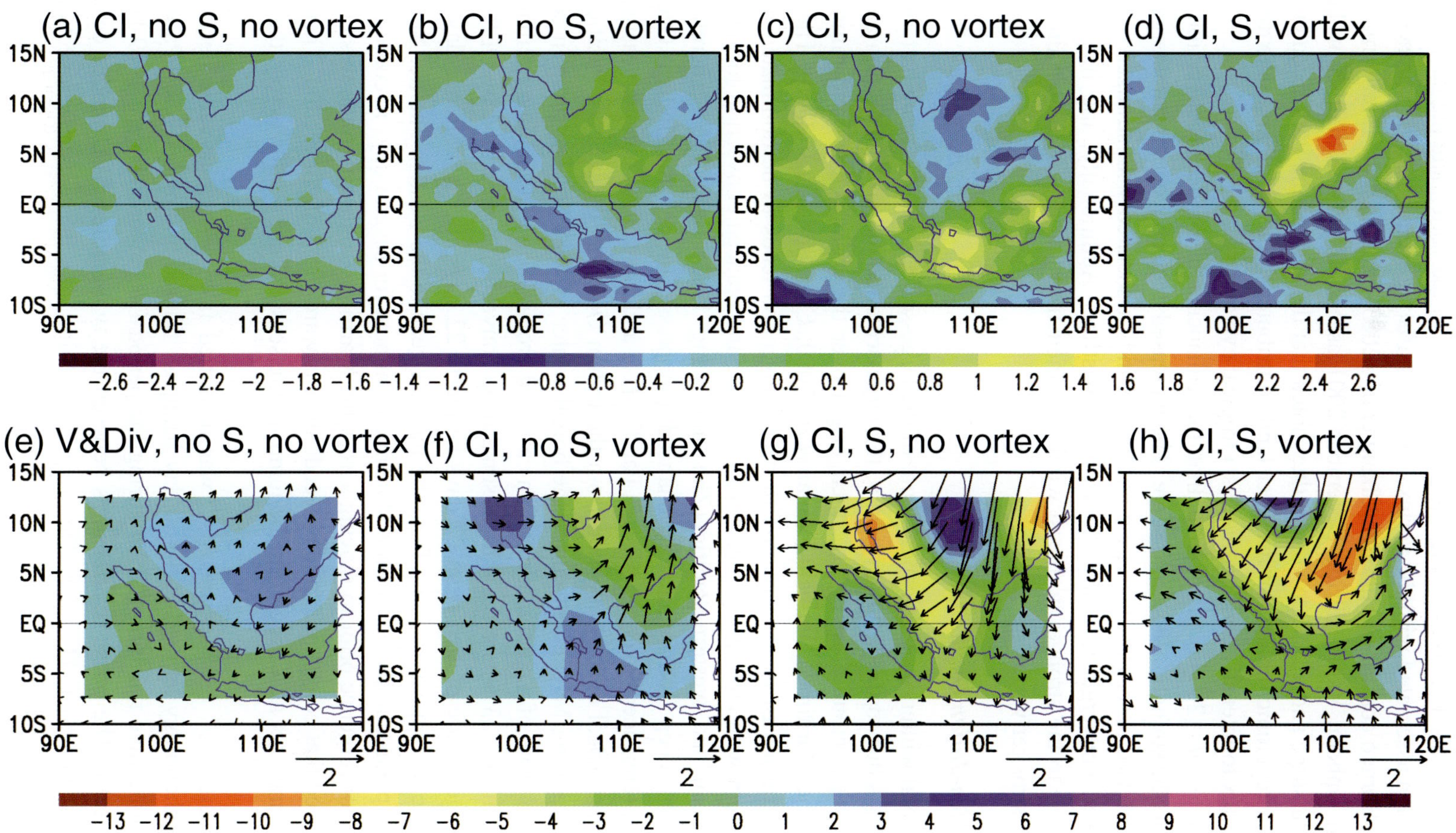

Figure 22. Composite maps of (top row) convective index and (bottom row) 925 hPa winds (m s^{-1}) and divergence (shaded 10^{-5} s^{-1}) for (a,e) no surge and no vortex cases, (b,f) no surge and vortex cases, (c,g) surge and no vortex cases, And (d,h) surge and vortex cases. (From Chang *et al.* 2004a).

the presence of a surge not only increases the strength of the vortex and increases the amount of convection associated with the vortex, but also shifts the location of the vortex to be more along the Borneo coastline. Although Chang *et al.* (2003) explained the rare formation of the equatorial typhoon Vamei during December 2001 as a result of the interaction between a strong cold surge and a Borneo vortex, they noted that the shift of the vortex center such that much of the cyclonic circulation lies over land contributes to the fact that it is extremely rare for the vortex to intensify and organize as a tropical cyclone.

Chang *et al.* (2004a) also noted that the enhancement of deep convection over the sensitive to the strength of the cold surge. The increased deep convection with surge intensity results from two processes. Increased northeast winds associated with stronger surges result in increased convergence along the Borneo coastline. Additionally, the increased shear vorticity due to the stronger northeast winds and their interaction with the terrain of Borneo result in a stronger vortex.

4.3. *Madden-Julian Oscillation*

During boreal winter, a significant amount of the large-scale, slowly-varying patterns of deep convection is due to the MJO. Typically, the large-scale motion and convection are coupled such that super cloud clusters have their origin over the Indian Ocean and migrate to the region of the Maritime Continent (Yanai *et al.* 2000; Chen and Yanai 2000). The eastward motion of the intraseasonal variation in deep convection tends to be dominated by a near-dipole pattern with alternating centers of enhanced and reduced convection located over the Indian Ocean and western Pacific (Zhu and Wang 1993; Hsu 1996). Therefore, when intraseasonal deep convection is enhanced (suppressed) over the Indian Ocean it is suppressed (enhanced) over the western Pacific. Hsu and Lee (2004) noted an eastward jump of the intraseasonal deep convection from the Indian Ocean to the western Pacific during a particularly strong MJO event during the boreal winter of 1985/86. The jump in convection often results from of a combination of an eastward propagation and standing oscillation. This is evident in Fig. 23 where the intraseasonal deep convection represented by OLR and low-level divergence are plotted from 22 days before to 22 days after a reference day when the deep convection appears to jump from the Indian Ocean to the western Pacific. Note that the convection persists to the east of the Maritime Continent for several days in conjunction with low-level convergence prior to moving across the Maritime Continent to the western Pacific. It is observed that cold surges often precede the eastward jump of convection. Therefore, the cold surge may act as a trigger for the jump in intraseasonal convection across the Maritime Continent.

In Chang *et al.* (2004a), the MJO was identified from a singular-value decomposition (SVD) methodology applied to 30-60 day bandpass filtered anomalous 850 hPa winds and outgoing longwave radiation (OLR) during the period December-February 1979-December-February 2001. The phase relationships between the two leading SVD modes defined four phases of MJO activity (Fig. 24). During phase 1 (Fig. 24a), the reduced convection regime of the MJO has passed through the Maritime Continent region where equatorial easterly anomalies exist between 80oE-150oE. During phase 2 (Fig. 24b), the Maritime Continent area is in a transition from the reduced convection portion of the MJO, which has moved eastward, to the approaching active convective portion of the MJO. Increased convection and low-level westerlies exist immediately west and south of the Malay Peninsula, Sumatra, and Java. During phase 3 (Fig. 24c), the enhanced convective portion of the MJO is centered over the eastern portion of the Maritime Continent. Low–level westerly anomalies exist

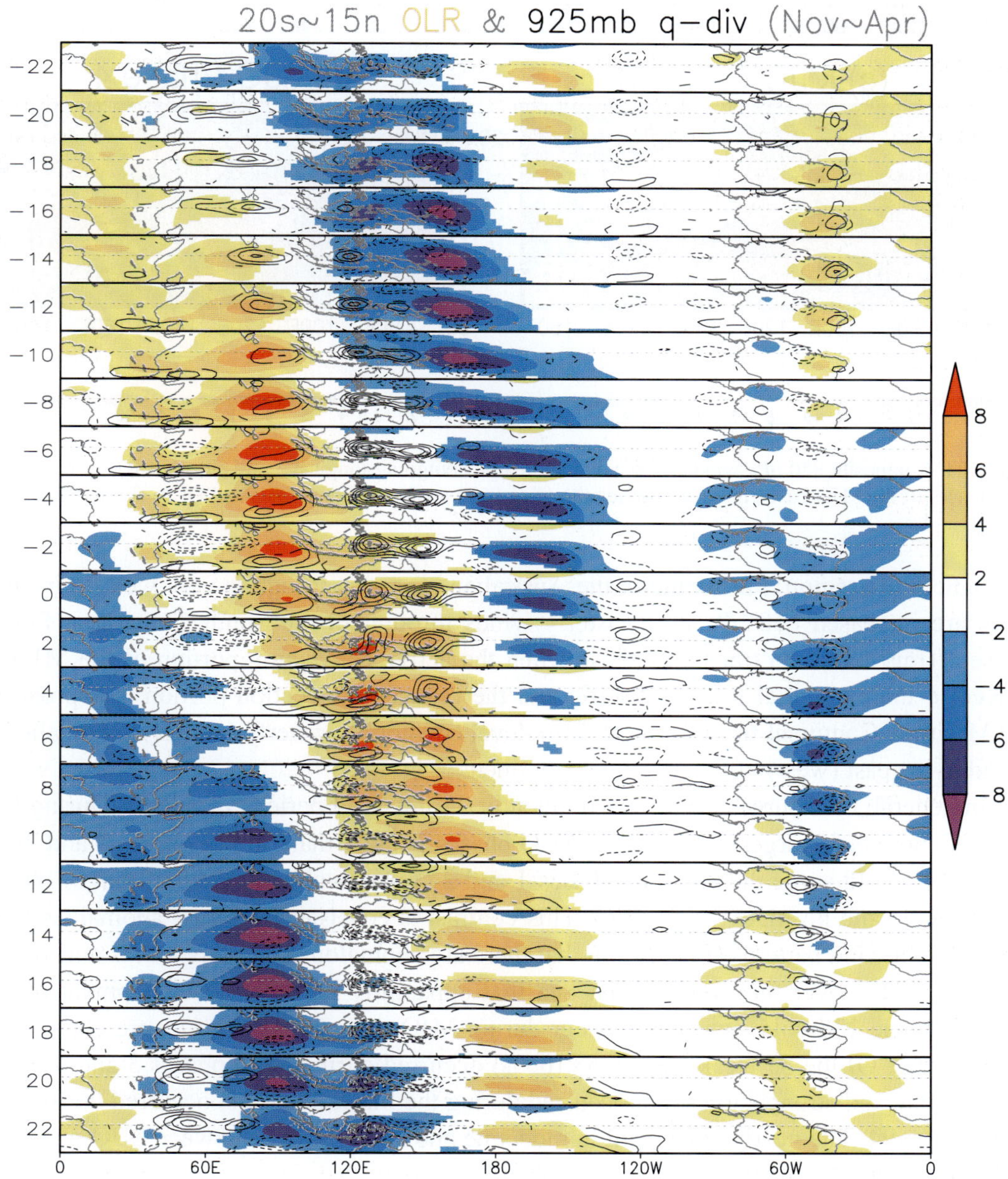

Figure 23. The spatial distribution of intraseasonal OLR and low-level divergence composite for 22 days before and 22 days after the time that the convection maximum proceeds across the Maritime Continent, which is defined as day 0. The horizontal dashed line indicates the equator. (From Hsu and Lee 2004)

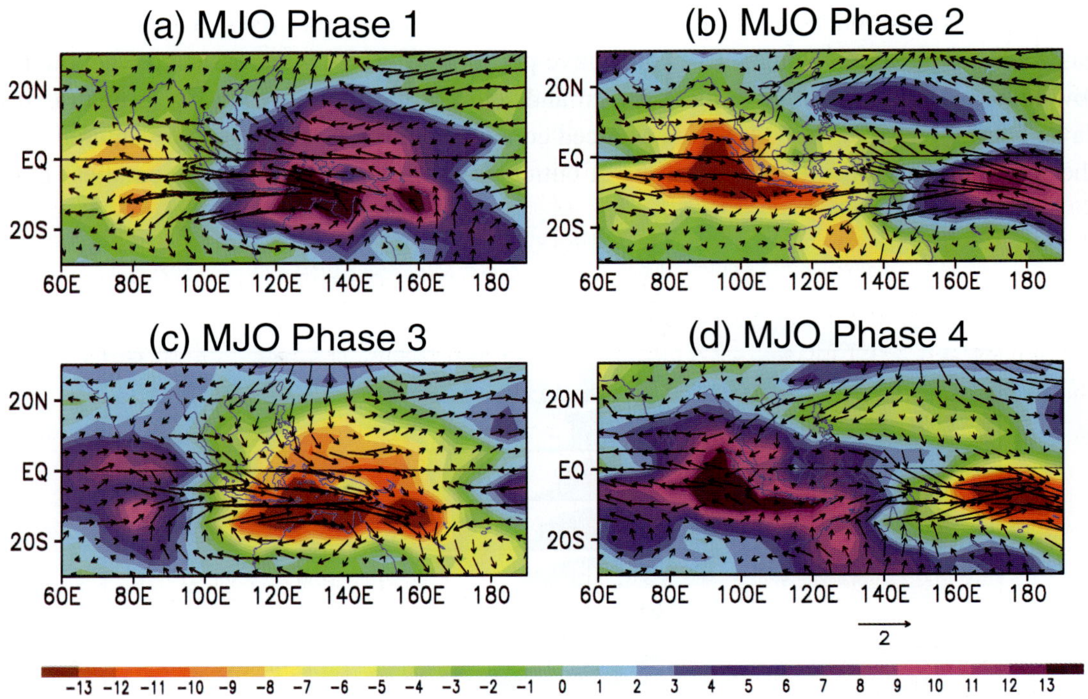

Figure 24. Composite 850 hPa winds (m s^{-1}) and anomalous outgoing longwave radiation (OLR), W m^{-2}) for the four phases of the MJO based on time coefficients of an SVD analysis of the winds and OLR. (From Chang *et al.* 2004a).

throughout the region. Furthermore, the Australian monsoon trough is very well defined and there are increased 850 hPa northeasterly winds throughout the South China Sea. Finally, phase 4 (Fig. 24d) is a transition phase when the active convection portion of the MJO has moved to the equatorial western Pacific and the reduced convection portion is approaching the Maritime Continent from the west. Low-level equatorial westerly (easterly) anomalies are found to the east (west) of the Maritime Continent.

In general, cold surges occur less frequently during MJO periods (Table 3). During MJO periods, more cold surges occur during MJO phases 3 and 4 than during phases 1 and 2. The chance of a cold surge during MJO phases 1 and 2 (15%) is almost one half that of phase 3 and 4 (28%) and no-MJO days (29%). This contrast is consistent with the anomalous 30-60 day 850 hPa wind patterns (Fig. 24), in that there are anomalous southerly winds during MJO phases 1 and 2 over the South China Sea that apparently inhibit the development of cold surges. Of the 138 cold surges that occur during MJO periods, 85 (62%) were classified as weak surges. Of the 85 weak surges, 53 (65%) occurred during MJO phases 3 and 4. Of the 53 moderate and strong surges that occurred during the MJO, 29 (55%) occurred during MJO phases 3 and 4. The nearly equal split in moderate and strong surges between MJO phases 1,2 and 3,4 compared to the larger number of weak surges that occur during phases 3 and 4 compared to phases 1 and 2 indicates that weak surges are inhibited during MJO phases 1 and 2.

Nearly twice as many vortex cases occur during no-MJO periods than MJO periods (Table 4). Although the largest number of vortex cases during MJO periods occur during MJO phase 2 (Table 4), the increase in the number of cases during phase 2 is only statistically larger than the number of cases during phase 4.

During the MJO phase 1, the composite Borneo vortex case (Figs. 25a,e) is similar to the composite of vortex cases with no surge (Figs. 22b,f). That is, there is a broad area of counter-clockwise turning of the low-level winds over the southern South China Sea with no closed circulation. Convection is increased over the southern South China Sea and reduced over the equatorial Southern Hemisphere, Malay Peninsula, and Sumatra. The similarity of the vortex and MJO phase 1 composite with the vortex and no surge pattern (Figs. 22b,f) indicates that during MJO phase 1 the Borneo vortex most likely occurs without a surge. This is consistent with the anomalous 30-60 day subtropical ridge over the western North Pacific with southerly anomalies over the South China Sea in Fig. 24a.

Table 3. The number of surge cases with respect to the MJO and phase of the MJO.

	No-MJO Days	MJO Days	MJO Phases 1-2	MJO Phases 3-4
Number of No-Surge Days	845	670	381	289
Number of Surge Days	242	138	56	82
Percentage of Surge Days	29%	21%	15%	28%

Table 4. The distribution of vortex cases (days) with respect to MJO periods and phases.

	No MJO	MJO			
No-Vortex	675	542			
		Phase 1	Phase 2	Phase 3	Phase 4
		148	163	157	124
Vortex	412	216			
		Phase 1	Phase 2	Phase 3	Phase 4
		57	69	50	40

During MJO phase 2, the deep convection over the equatorial South China Sea and the low-level winds associated with the Borneo vortex (Figs. 25b,f) are more organized than during MJO phase 1. Although there are northeast winds over the southern South China Sea, which might suggest the influence of a surge, their magnitude is much less than the vortex and surge composite (Figs. 22d,h).

Furthermore, the northeast winds over the southern South China Sea in the composite Borneo vortex based on 2-15 day bandpass filtered data (Figs. 25b,f) during periods of MJO phase 2 are opposite to the 30-60 day bandpass filtered southwest winds over the region (Fig. 24b). This may be attributed to two possibilities. One is that the MJO phase 2 is a transition phase between the inactive convection (phase 1) portion of the MJO and the active phase

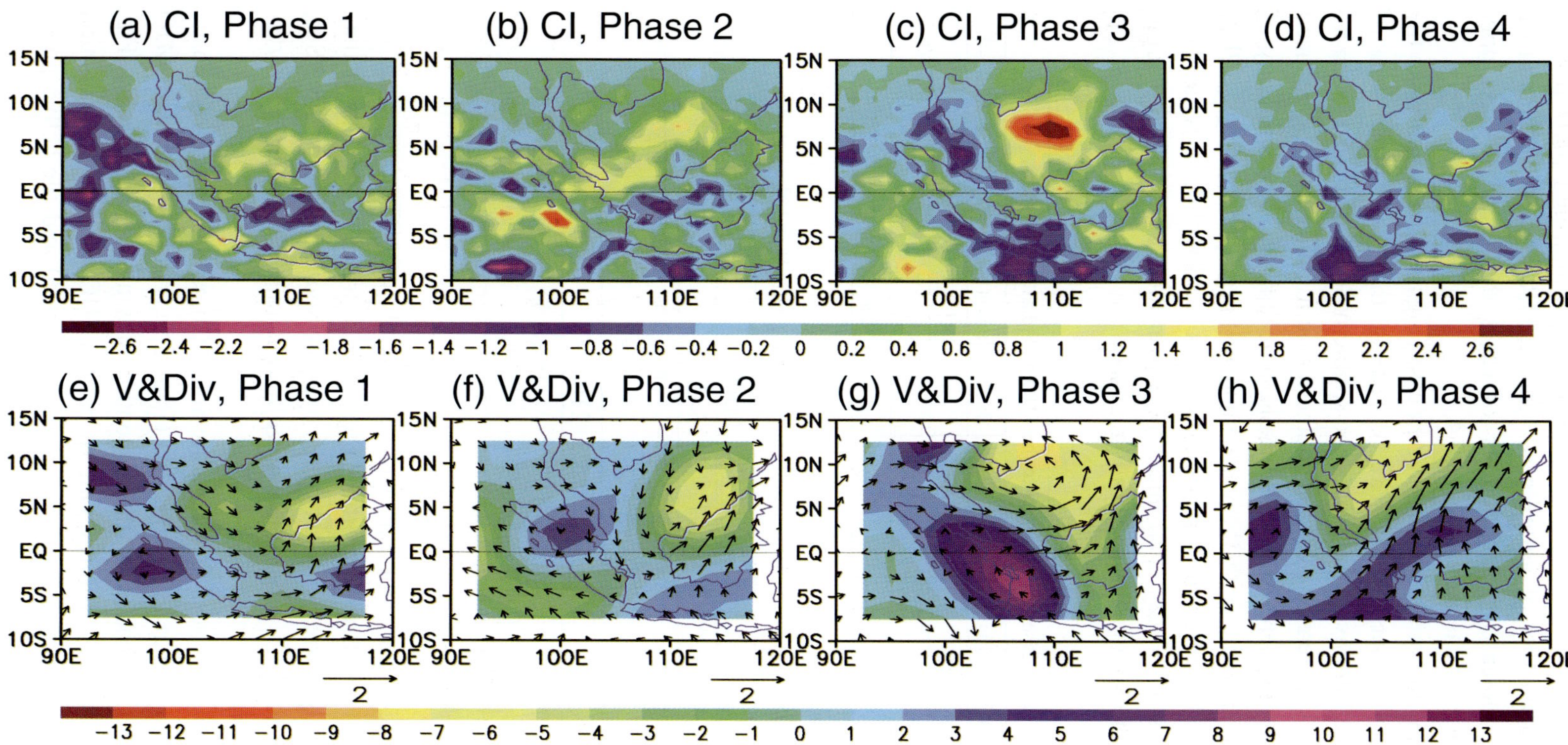

Figure 25. Composite maps of (top row) convective index and (bottom row) 925 hPa winds (m s^{-1}) and divergence (shaded, 10^{-5} s^{-1}) for MJO and vortex cases when the MJO is in (a,e) Phase 1, (b,f) Phase 2, (c,g) Phase 3, and (d,h) Phase 4. The MJO phases are as defined in Fig. 19. The number of cases in each composite is defined in the bottom row of Table 4. (From Chang *et al.* 2004a).

(phase 3). Therefore, the 30-60 day wind pattern may not be as stable throughout the period as the dominant active and inactive convection phases. Individual synoptic-scale cold-surge events may not be influenced as much by the transition MJO phase. Secondly, the anomalous strong 30-60 day subtropical ridge over the western North Pacific has moved eastward and northward (Fig. 24b) such that the MJO-scale opposition to potential synoptic-scale cold surges may be reduced. However, the reduced magnitude of the winds in the vortex-MJO phase 2 composite does indicate that the overall influence of the surge is reduced by the MJO phase 2 circulations, which again suggests that the MJO phase 2 influence may be to inhibit weak cold surges or reduce the intensity of strong cold surges, which is similar to the above discussion with respect to the distribution of cold surge cases during MJO periods (Table 3). Because vortex composite patterns associated with MJO phase 1 and 2 (Figs. 24a,b) are not very different from composites of vortex-only (Figs. 22b,f) and vortex-surge (Figs. 22d,h) cases, phases 1 and 2 of the MJO primarily influences the Borneo vortex via reduction of the influence of cold surges.

During MJO phase 3, the Borneo vortex (Figs. 25c,g) exhibits some fundamental differences from the MJO phase 1 and 2 composites and from all previous vortex-no surge (Figs 22b,f) and vortex-surge composites (Figs 22d,h). During the MJO phase 3, the cyclonic circulation of the vortex seems to be more linked to cyclonic horizontal shear associated with equatorial westerly winds rather than northeasterly winds that extend through the southern South China Sea. Consequently, the center of the vortex is located over the southern South China Sea (Fig. 25g) as is the maximum in deep convection (Fig. 25c). Although there is some indication that the MJO phase 3 may be related to increased cold surge frequency (Table 3), the composite vortex-MJO phase 3 pattern indicates that the Borneo vortex may be more influenced by the increased equatorial westerlies associated with the enhanced MJO-scale convection over the eastern portion of the Maritime Continent (Fig. 24c).

Again, the structure of the vortex and MJO phase 4 composite (Figs. 25d,h) is fundamentally different than composites constructed based on the presence of the vortex or the vortex and surge. During MJO phase 4, the vortex is very weak with cyclonic shear present only over the extreme western South China Sea and Malay Peninsula. Southwesterly anomalies exist over the primary region of the southern South China Sea and reduced deep convection spreads northeastward along the west coast of Borneo. The 30-60 day composite winds over the South China Sea during MJO phase 4 are primarily easterly (Fig 24d), which suggests that there is less penetration of northeasterly winds into the southern South China Sea on the MJO scale during phase 4. This may also be related to the reduced-convection branch of the MJO that is beginning to extend over the southern South China Sea. Enhanced MJO-scale downward motion and low-level diffluence (Fig 24d) may contribute to reduced vortex frequency and less influence of cold surges over the equatorial regions.

During periods of no-MJO activity, the composite vortex patterns of convection, winds, and divergence (not shown) are very similar to the composites of all vortex cases (Figs. 22b,d). This is consistent with the above results in that a majority of vortex cases occur during no-MJO periods (Table 4) and that the number of cold surges is reduced during MJO periods (Table 3)

The relationships among the MJO, cold surges, and the Borneo vortex are summarized by comparing the relative vortex frequency during periods of no surge, weak surges, moderate surges, and strong surges during MJO periods, no-MJO periods, and the total number of cases (Fig. 26). It is clear that the presence of the MJO is associated with fewer numbers of vortex cases, and the occurrence of vortex cases during periods of weak surges is most reduced during the MJO. In summary, while the presence of a surge acts to increase the strength of the Borneo vortex, the frequency of surges is reduced during periods when the MJO is present. Often the MJO-scale circulation pattern directly opposes the cold surge wind pattern. Therefore, weak surges may be more inhibited during periods when the MJO-scale circulations are strong. Primarily due to the impact of the MJO on cold surge intensity and frequency, 66% of the vortex cases occur during non-MJO periods. During periods when the active convection portion of the MJO is over the Maritime Continent, the occurrence of the Borneo vortex may be more related to the cyclonic shear to the north of anomalous equatorial westerlies rather than the northeasterly monsoon winds. During MJO periods, the Borneo vortex is least likely to be present when the inactive convective portion of the MJO extends to the Maritime Continent with large-scale low-level diffluence that acts to restrict the impact of cold surges on convection in the southern South China Sea.

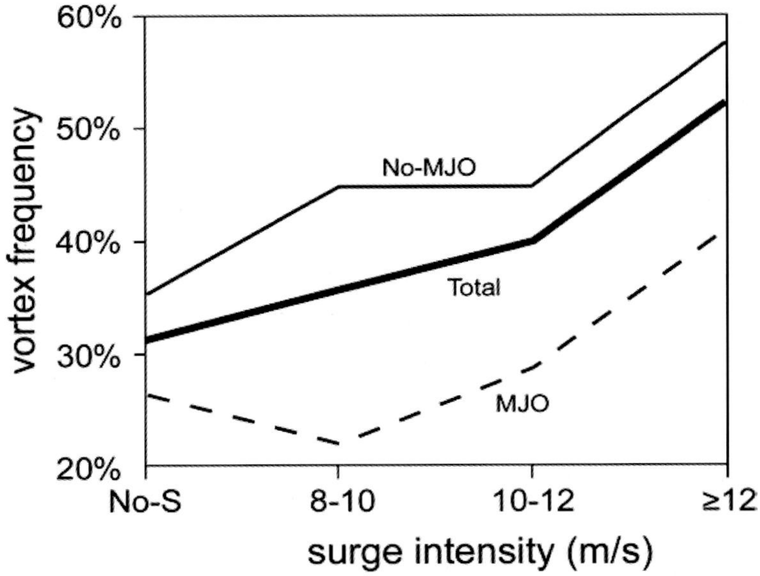

Figure 26. Summary of the percentage of days with no surge (No-S), weak surge (8-10), moderate surge (10-12), and strong surge (≥12) that also contain a vortex for all days, no-MJO days, and MJO days.

Acknowledgements

This work was supported in part by the National Oceanic and Atmospheric Administration under grant NA01AANRG0011, the National Science Foundation under grant ATM-0101135, and the Office of Naval Research Marine Meteorology Program.

References

Aldrian, E, and D. Susanto, 2003: Identification of three dominant rainfall regions within Indonesia and their relationship to sea surface temperature. *Intl. J. Climatol.*, **23**, 1435-1452.

______, L. Dumenil-Gates, and F. H. Widodo, 2003: Variability of Indonesian rainfall and the influence of ENSO and resolution in ECHAM4 simulations and in the reanalyses. Report No. 346, Max-Planck Institute, Hamburg, Germany.
(http://www.mpimet.mpg.de/de/web/download.php?src=max_scirep&file=pdfupload&id=350&filename=Rep346.pdf).

Allan, R. J., 1991: *Teleconnections linking worldwide climate anomalies*. M. Glantz *et al.*, Eds., Cambridge University Press. Cambridge, 73-120.

Braak, C, 1919: Atmospheric variations of short and long duration in the Malay Archipelago and neighbouring regions, and the possibility to forecast them. *Verhandelingen*, No. **5**, Koninklijk Magnetisch en Meteorologisch Obervatorium te Batavia, 57 pp.

______, 1921-29: Het Climaat van Nederlandsch Indie. *Magnet. Meteorol. Observ. Batavia, Verhand* No. **8**. (from Ramage, 1971)

Brier, G. W., 1978: The quasi-biennial oscillation and feedback processes in the atmosphere-ocean-earth system. *Mon. Wea. Rev.,* **106**, 938–946.

Chang, C.-P, P. Harr, and H. J. Chen, 2004a: Synoptic disturbances over the equatorial South China Sea and western Maritime Continent during boreal winter. *Mon. Wea. Rev.* (accepted).

______, and K. M. Lau, 1982: Short-term planetary-scale interactions over the tropics and midlatitude during northern winter. Part I: Contrasts between Active and Inactive Periods. *Mon. Wea. Rev.*, **110**, 933-946.

______, and T. Li, 2000: A theory for tropical tropospheric biennial oscillation. *J. Atmos. Sci.*, **57**, 2209-2224.

______, C. H. Liu, and H. C. Kuo, 2003: Typhoon Vamei: An equatorial tropical cyclone formation. *Geophys. Res. Lett.,* **30**, 50 1-4.

______, Z. Wang, J. Ju, and T. Li, 2004b: On the relationship between western Maritime Continent monsoon rainfall and ENSO during northern winter. *J. Climate*, **17**, 665-672.

______, ______, J. McBride, and C. H. Liu, 2004c: The annual cycle of Southeast Asia – Maritime Continent rainfall. *J. Climate* (accepted).

Cheang, B. K., 1977: Synoptic features and structures of some equatorial vortices over the South China Sea in the Malaysian region during the winter monsoon of December 1973. *Pure Appl. Geophys.*, **115**, 1303-1333.

Chen, B., and M. Yanai, 2000: Comparison of the Madden-Julian oscillation (MJO) during the TOGA COARE IOP with a 15-year climatology. *J. Geophys. Res.*, **105**, 2139-2149.

Chen, G. T. J., T. E. Gerish, and C.-P. Chang, 1986: Structure variations of the synoptic-scale cyclonic disturbances near Borneo during the WMONEX period. *Papers Meteor. Res.*, **9**, 117-135.

Clarke, A. J., X. Liu, and S. V. Gorder, 1998: Dynamics of the biennial oscillation in the equatorial Indian and far western Pacific Oceans., *J. Climate*, **11**, 987-1001.

Hamada, J.-I., and T. Sribimawati, 1998: Catalogue and sample plotting of daily rainfall data for 1961-90 at 157 stations in Indonesia. *Climatology of Indonesian Maritime Continent.* M. D. Yamanaka, ed., Research Report Topic 08041107, Kyoto University, 17-90.

______, M. D. Yamanaka, J. Matsumoto, S. Fukao, P. A. Winarso, and T. Sribimawati, 2002: Spatial and temporal variations of the rainy season over Indonesia and their link to ENSO. *J. Meteor. Soc. Japan*, **80**, 285-310.

Hastenrath, S., 1987: Predictability of Java monsoon rainfall anomalies: A case study. *J. Climate Appl. Meteor.*, **26**, 133-141.

Haylock, M., and J. McBride, 2001: Spatial and coherence and predictability of Indonesian wet season rainfall. *J. Climate*, **14**, 3882-3887.

Hendon, H. H., 2003: Indonesia rainfall variability: Impacts of ENSO and local air-sea interaction. *J. Climate*, **16**, 1775–1790.

Holland, G. J., 1984: On the climatology and structure of tropical cyclones in the Australian Southwest Pacific Region. *Aus. Meteor. Mag.*, **32**, 17-31.

Hsu, H.-H., 1996: Global view of the intraseasonal oscillation during northern winter. *J. Climate*, **9**, 2386-2406.

______, and M.-Y. Lee, 2004: Topographic effects on the eastward propagation and initiation of the Madden-Julian Oscillation. *J. Climate.* (submitted)

Hung, C.-W., and M. Yanai, 2003: Factors contributing to the onset of the Australian summer monsoon. *Quart. J. Roy. Meteor. Soc.* (in press)

______, X. Liu, and M. Yanai, 2003: Symmetry and asymmetry of the Asian and Australian summer monsoons. *J. Climate*. (submitted)

Johnson, R. H., and R. A. Houze, Jr., 1987: Precipitating cloud systems of the Asian monsoon. *Monsoon Meteorology*, C.-P. Chang and T. N. Krishnamurti, Eds., Oxford University Press, 298-353.

Kiladis, G. N., and H. F. Diaz, 1989: Global climate anomalies associated with extremes in the Southern Oscillation. *J. Climate*, **2**, 1069-1090.

Kirono, D. G. C., N. J. Tapper, and J. L. McBride, 1999: Documenting Indonesian rainfall in the 1997/1998 El Nino event. *Phys. Geogr.*, **20**, 422-435.

Kwan, K. F., 2002: A conceptual model relating the quasi-biennial oscillation and the tropospheric biennial oscillation. *M.S. Thesis*, University of Malaya, Kuala Lumpur.

Krishnamurti, T. N., 1971: Tropical east-west circulations during northern summer. *J. Atmos. Sci.*, **28**, 1342-1347.

Lau, K. M., and P. H. Chan, 1983: Short-term climate variability and atmospheric teleconnections from satellite-observed outgoing longwave radiation. Part II: Lagged correlations. *J. Atmos. Sci.*, **40**, 2751-2767.

______, and C.-P. Chang, 1987: Planetary scale aspects of winter monsoon and teleconnections. *Monsoon Meteorology*, C.-P. Chang and T. N. Krishnamurti, Eds., Oxford University Press, 161-202.

Li, T., B. Wang, and C.-P. Chang, 2001: Theories on the tropospheric biennial oscillation: A review. *Dynamics of Atmos. and Oceanic Circulations and Climate*, M. Wang *et al.*, Eds., Chinese Academy of Sciences, China Meteor. Press, Beijing, 872pp.

LinHo and B. Wang, 2002: The time-space structure of the Asian-Pacific summer monsoon: A fast annual cycle view. *J. Climate,* **15**, 2001-2019.

Liu, W. T., 2002: Progress in scatterometer application. *J. Oceanogr.*, **58**, 121-136.

Maddan, R. A., and P. R. Julian, 1972: Description of global-scale circulation cells in the tropics with a 40-50 day period. *J. Atmos. Sci.*, **29**, 827-836.

Manton, M. J., and J. L. McBride, 1992: Recent research on the Australian monsoon. *J. Meteor. Soc. Japan*, **70**, 275-285.

Matsumoto, J., 1992: The seasonal changes in Asian and Australian monsoon regions. *J. Meteor. Soc. Japan*, **70**, 257-273.

______, and T. Murakami, 2000: Annual changes of tropical convective activities as revealed from equatorial symmetric OLR Data. *J. Meteor. Soc. Japan*, **78**, 543-561.

______, and ______, 2002: Seasonal migration of monsoons between the Northern and Southern Hemisphere as revealed from equatorially symmetric and asymmetric OLR data. *J. Meteor. Soc. Japan*, **80**, 419-437.

McBride, J., 1987: The Australian summer monsoon. *Monsoon Meteorology*, C.-P. Chang and T. N. Krishnamurti, Eds., Oxford University Press, 203-231.

______, 1995: Tropical cyclone formation. *Global Perspective on Tropical Cyclones*, R. L. Elsberry, Ed., Tech. Docu. 693, World Meteorological Organization, 63-105.

______, and N. Nicholls, 1983: Seasonal relationships between Australian rainfall and the southern oscillation. *Mon. Wea. Rev.*, **111**, 1998-2004.

______, 1998: Indonesia, Papua New Guinea, and tropical Australia: The southern hemisphere summer monsoon. *Meteorology of the Southern Hemisphere*, Meteor. Monogr., No. **49**, Amer. Meteor. Soc., 89–99.

______, M. Haylock and N. Nicholls, 2003: Relationships between the Maritime Continent heat source and the El Niño–Southern Oscillation Phenomenon. *J. Climate*, **16**, 2905-2914.

Meehl, G. A., 1987: The annual cycle and interannual variability in the tropical Pacific and Indian Ocean region. *Mon. Wea. Rev.*, **115**, 27-50.

______, 1997: The South Asian monsoon and the tropospheric biennial oscillation (TBO). *J. Climate*, **10**, 1921-1943.

______, and J. M. Arblaster, 2002: The tropospheric biennial oscillation and Asian–Australian monsoon rainfall. *J. Climate,* **15**, 722-744.

Neale, R., and J. Slingo, 2003: The Maritime Continent and its role in the global climate: A GCM Study. *J. Climate,* **16**, 834-848.

Nichol, J. 1998: Smoke haze in Southeast Asia: a predictable recurrence. *Atmos. Environ.* **31**, 1209-1219.

Nicholls, N., 1978: Air-sea interaction and the quasi-biennial oscillation. *Mon. Wea. Rev.,* **106**, 1505-1508.

______, 1981: Air-sea interaction and the possibility of long-range weather prediction in the Indonesian Archipelago. *Mon. Wea. Rev.*, **109**, 2435-2443.

______, 1985: Towards the prediction of major Australian droughts. *Intl. J. Climatol.*, **5**, 553-560.

Ooi, S. H. 1999: Impacts of ENSO on monsoons over Malaysia. ESCAP/WMO Typhoon Committee Annual Review 1998, World Meteorological Organization, Geneva, 153 - 173.

Power, S., F. Tseitkin, V. Mehta, B. Lavery, S. Torok, and N. Holbrook, 1999a: Decadal climate variability in Australia during the twentieth century. *Intl. J. Climatol.*, **19**, 169-184.

______, T. Casey, C. Folland, A. Colman, and V. Mehta, 1999b: Interdecadal modulation of the impact of ENSO on Australia. *Climate. Dyn.*, **15**, 319-324.

Ramage, C. S., 1968: Role of a tropical "Maritime Continent" in the atmospheric circulation. *Mon. Wea. Rev.*, **96**, 365-369.

Ramage, C. S., 1971: *Monsoon Meteorology*. Acad. Press, 296pp.

Rasmusson, E. M., and T. H. Carpenter, 1982: Variations in tropical sea surface temperature and surface wind fields associated with the Southern Oscillation/El Niño. *Mon. Wea. Rev.*, **110**, 354-384.

______, X. Wang, and C. F. Ropelewski, 1990: The biennial component of ENSO variability. *J. Mar. Syst.*, **1**, 71-96.

Ropelewski, C. F., and M. S. Halpert, 1987: Global and regional scale precipitation patterns associated with the El Nino/Southern Oscillation. *Mon. Wea. Rev.*, **115**, 1606-1626.

Rowell, D. P., 2001: Teleconnections between the tropical Pacific and the Sahel. *Quart. J. Roy. Meteor. Soc.*, **127**, 1683-1706.

Rodwell, M. J., and B. J. Hoskins, 1996: Monsoons and the dynamics of deserts. *Quart. J. Roy. Meteor. Soc.*, **122**, 1385-1404.

Simpson, J., C. Kummerow, W. - K. Tao, and R. F. Adler, 1996: On the Tropical Rainfall Measuring Mission (TRMM). *Meteor. Atmos. Phys.*, **60**, 19-36.

Simms, J. W. IV., 2000: A composite study of the Madden-Julian Oscillation and northeasterly cold surges during the northern winter monsoon. *M. S. thesis*, Naval Postgraduate School, Monterey, California, 93pp.

Sukanto, M. 1969: Climate of Indonesia. *World Survey of Climatology*, H. Arakawa, Ed., **Vol. 8**, Climates of Northern and Eastern Asia, Elsevier, 215-229.

Subramaniam, M., 1993: Multiple time scales in rainfall variability. *Proc. Indian Acad. Sci.*, **102**, 249-263.

Suppiah, R., 1992: The Australian summer monsoon: a review. *Progress Phys. Geogr.*, **16**, 283-318.

Thorpe, A., M. Shapiro, and R. Langland, 2003: Presentation of THORPex Project to WMO/CAS, February 13, 2002, Oslo.
(http://www.mmm.ucar.edu/uswrp/powerpoint/thorpex/v3_document.htm).

Tanaka, M., 1994: The onset and retreat dates of the Austral summer monsoon over Indonesia, Australia and New Guinea. *J. Meteor. Soc. Japan*, **72**, 255-267.

Taylor, S. C., 1998: Interactions of large-scale tropical motions systems during the 1996-1997 Australian monsoon. *M. S. thesis*, Naval Postgraduate School, Monterey, California, 46pp.

Webster, P. J., V. O. Magana, T. N. Palmer, J. Shukla, R. A. Tomas, M. Yanai, and T. Yasunari, 1998: Monsoons: Processes, predictability, and the prospects for prediction. *J. Geophys. Res.,* **103**, **C7**, 14,451-14,510.

Wyrtki, K., 1956: The rainfall over the Indonesian waters. Kementerian Perhubungan Lembaga Meteorologi dan Geofisik Verhand, **No. 49**. (from Ramage, 1971).

Yanai, M., B. Chen, and W.-W. Tung, 2000: The Madden-Julian oscillation observed during the TOGA-COARE IOP: Global view. *J. Atmos. Sci.*, **57**, 2374-2396.

Yanai, M., C. Li, and Z. Song, 1992: Seasonal heating of the Tibetan Plateau and its effects on the evolution of the Asian summer monsoon. *J. Meteor. Soc. Japan,* **70**, 319-351.

Yang, S., K.-M. Lau, and K.-M. Kim, 2002: Variations of the East Asian jet stream and Asian–Pacific–American winter climate anomalies, *J. Climate*, **15**, 306-325.

Yasunari, T., 1991: The monsoon year - a new concept of the climatic year in the tropics. *Bull. Amer. Meteor. Soc.*, **72**, 1331-1338.

Xie, P., and P. A. Arkin, 1997: Global precipitation: A 17-year monthly analysis based on gauge observations, satellite estimates, and numerical model outputs. *Bull. Amer. Meteor. Soc.,* **78**, 2539-2558.

Zhang, Y., K. R. Sperber, and J. S. Boyle, 1997: Climatology and interannual variation of the East Asian winter monsoon: Results from the 1979-95 NCEP–NCAR reanalysis. *Mon. Wea. Rev.,* **125**, 2605–2619.

Zhu, B., and B. Wang, 1993: The 30-60 day convection seesaw between the tropical Indian and western Pacific Oceans. *J. Atmos. Sci.*, **50**, 184-199.

Part II

Interannual Variations

4. INTERANNUAL VARIABILITY, GLOBAL TELECONNECTION, AND POTENTIAL PREDICTABILITY ASSOCIATED WITH THE ASIAN SUMMER MONSOON

WILLIAM K.-M. LAU AND KYU-MYONG KIM
Laboratory for Atmospheres
NASA/Goddard Space Flight Center
Greenbelt, MD 20771, USA
E-mail: William.K.Lau@nasa.gov

JUNE-YI LEE
Department of Atmospheric Sciences
Seoul National University
Seoul, Korea
E-Mail: jylee@climate.snu.ac.kr

In this Chapter, aspects of global teleconnection associated with the interannual variability of the Asian summer monsoon (ASM) are discussed. The differences in the basic dynamics of the South Asian Monsoon and the East Asian monsoon, and their implications on global linkages are discussed. Two teleconnection patterns linking ASM variability to summertime precipitation over the continental North America were identified. These patterns link regional circulation and precipitation anomalies over East Asia to continental North America, via coupled atmosphere-ocean variations over the North Pacific. The first pattern, nicknamed as the "Tokyo-Chicago Express", is associated with the fluctuations of the East Asian jet stream, and downstream amplification linking precipitation anomalies of regions of northern China, Japan, and Korea (the Baiu rain belt) to those in the Pacific Northwest and northern Great Plains of the US. The second pattern, dubbed the "Shanghai-Kansas Express" reflects features of Rossby wave dispersion linking precipitation anomalies in central East Asia to the US Midwest. Both patterns possess strong sea surface temperature (SST) expressions in the North Pacific. Results suggest that the two teleconnection patterns are associated with intrinsic modes of sea surface temperature variability in the extratropical oceans, which are forced in part by atmospheric variability and sustained by air-sea interaction. The first pattern may also be related to, while the second is independent of, tropical SST forcing. The potential predictability of the ASM associated with SST variability in different ocean basins is explored using a new canonical ensemble correlation prediction scheme. It is found that SST anomalies in tropical Pacific, i.e., El Niño, is the most dominant forcing for the ASM, especially over the maritime continent and eastern Australia. SST anomalies in the Indian Ocean may overshadow the influence from El Niño in western Australia and western maritime continent. Both El Niño and North Pacific SSTs contribute to monsoon precipitation anomalies over Japan, southern Korea, and northern and central China. By optimizing SST variability signals from different world ocean basins, the overall predictability of ASM can be substantially enhanced.

1. Introduction

Floods and droughts caused by variability of the Asian summer monsoon (ASM) are among the most devastating natural disasters which impact a large segment of the world population. ASM floods and droughts are extremely variable in space and time. During the summer of 1998, the great flood over the Yangtze River caused a record 3700 deaths, 223 million people displaced, and up to 30 billion US dollars in crop losses and property damages. In the same year, excessive monsoon rainfall also caused severe flooding over Bangladesh displacing 30 million people and wrecking up economic losses to over 3 billion US dollars. Yet, one year before, in the summer of 1997, the above regions had nearly normal summer rainfall, while northern China had been stricken with a record drought which began since the early 1990's. During the same year, southern China was gripped by wide spread flooding, but the averaged rainfall over the entire India subcontinent was nearly normal.

Because of the need for mitigating the adverse socio-economic impacts caused by ASM anomalies, understanding and predicting ASM monsoon variability are paramount. Recent studies have shown that the large variability within the ASM may be related to the intrinsic variability, such as intraseasonal oscillations (ISO), tropospheric biennial oscillation (TBO), sea surface temperature anomalies in the Indo-Pacific region, and their interactions with the El Niño Southern Oscillation (ENSO) (Lau and Wu 2001; Lau and Nath 2000; Meehl and Arblaster 1998; Shen and Kimoto 1999; Meehl 1997; Ju and Slingo 1995 and many others). Lau and Wu (1999, 2001) found that both regional and basin-scale processes contributed significantly to the observed ASM rainfall anomalies in 1997 and 1998. Lau *et al.* (2000) demonstrated that, based on the regional characteristics of dynamics and lower boundary forcings, at least two major sub-components of the ASM, i.e., the South Asia monsoon and the East-Asia monsoon need to be distinguished. These two components, while connected to each other, may interact differently with ENSO and other global climate phenomena, e.g., the North Atlantic Oscillation, decadal-to-interdecadal variability, and global change, to yield very different responses to global scale forcings (Kumar *et al.* 1999; Chang *et al.* 2001).

There is now growing evidence that not only is the ASM affected by global scale forcings, the ASM itself can influence global climate (Yasunari and Seki 1992). Kirtman and Shukla (2000) and Chung and Nigam (2000) found that momentum and heat fluxes associated with anomalies of the ASM may increase the variability of ENSO cycles. Kim and Lau (2000) demonstrated that the 6-month time lag between ASM surface wind forcings and peak SST in the tropical eastern Pacific may induce biennial tendency in ENSO cycles. Hoerling *et al.* (2001) showed that decadal climate signals in the North Atlantic may have their roots in tropical convection in the Indo-Pacific regions. Teleconnection arising from monsoon processes on seasonal to interannual time scales have been explored in previous studies. Nitta (1987) and Huang and Sun (1992) showed that summertime convection in the vicinity of the northern South China Sea and the Philippines may excite Rossby wavetrain that span the North Pacific to North America. Lau (1992) and Lau and Peng (1992) suggested that similar wavetrain signal may stem from instability of the summertime large-scale basic flow, and hence may not be sensitive to the exact location of the forcing. More recently, Lau and Weng (2000, 2002) have shown evidence that summertime rainfall anomalies over the East Asian continent and North America may be linked by teleconnection patterns, with pronounced SST signals in the extratropical Pacific. Ting and Wang (1997) noted that North

America summertime rainfall anomalies may be correlated with extratropical Pacific SST. However, it is not clear whether the extratropical SST is by the atmosphere, by tropical SST, or by air-sea interaction in the extratropics. If extratropical air-sea interaction is important, it is possible that extratropical SST, in addition to tropical SST, may have influence on predictability of the ASM.

In this Chapter, we focus on the discussion of the climate teleconnection associated with the ASM and implications on its potential predictability. Aspects of the interannual variability of the ASM including effects of intraseasonal oscillations, tropospheric biennial oscillations, and ENSO are covered in other Chapters of this book. In Section 2, we review the basic structures of the two basic components of the ASM, i.e., the South Asian Monsoon (SAM) and the East Asian Monsoon (EAM). In Section 3, we discuss the boreal summer teleconnection patterns associated with monsoon variability and with North American summertime precipitation anomalies. In Section 4, we show results demonstrating that the teleconnection patterns may be manifestations of intrinsic, recurrent summertime SST modes in the extratropical oceans. In Section 5, we assess the potential predictability of the ASM, and its sources associated with SST variability in different ocean basins. Concluding remarks are presented in Section 6.

2. Interannual Variability of the ASM

The interannual variability of the ASM encompasses a complex mix of regional and sub-regional scale characteristics. To better define monsoon regionality and ENSO impacts, a number of ASM regional indices have been proposed by different authors (Goswami *et al.* 1999; Wang and Fan 1999; Lau *et al.* 2000; Wang *et al.* 2001). Lau *et al.* (2000) defined two indices RM1 and RM2, respectively for the South Asian Monsoon (SAM) and the East Asian Monsoon (EAM). RM1 is defined by the vertical shear of the meridional wind over the longitude sector of the Indian subcontinent including the Bay of Bengal. The meridional wind shear is a measure of the balance of the vorticity generated by vertical stretching effect due to a deep heat source, and the advection of planetary vorticity by the meridional wind (see Section 3 for further discussion). RM2 is defined by the zonal wind gradient between the mid-latitude and the subtropics, and is a measure of the lateral displacement of the East Asian subtropical jet stream, due, in part, to strengthening or weakening of the local Hadley circulation associated with variability of monsoon heat sources in East Asia and Southeast Asia. Specifically, the two indices are defined by:

$$\mathrm{RM1} = (v_{850\,\mathrm{hPa}} - v_{200\,\mathrm{hPa}}) \text{ over the region } [10°\mathrm{N} - 35°\mathrm{N}, 70°\mathrm{E} - 110°\mathrm{E}] \quad (1)$$

$$\mathrm{RM2} = (u_{200\,\mathrm{hPa}} [40°\mathrm{N} - 50°\mathrm{N}, 110°\mathrm{E} - 150°\mathrm{E}] - u_{200\,\mathrm{hPa}} [25°\mathrm{N} - 35°\mathrm{N}, 110°\mathrm{E} - 150°\mathrm{E}]) \quad (2)$$

where the zonal and meridional winds are averaged over the geographic regions indicated. It should be pointed out that while these two indices bring out salient features separating the

SAM and EAM especially with respect to teleconnection, the results are largely independent of the indices used, as far as separating the two components of the ASM. Similar patterns can be obtained from EOF or combined EOF analyses (not shown) or by using other indices (e.g. Wang *et al.* 2001).

2.1. *SAM Variability*

The patterns shown in Figs. 1 and 2 are regressions of RM1 and RM2 based on CPC Merged Analysis of Precipitation (CMAP) monthly rainfall and National Center for Environmental

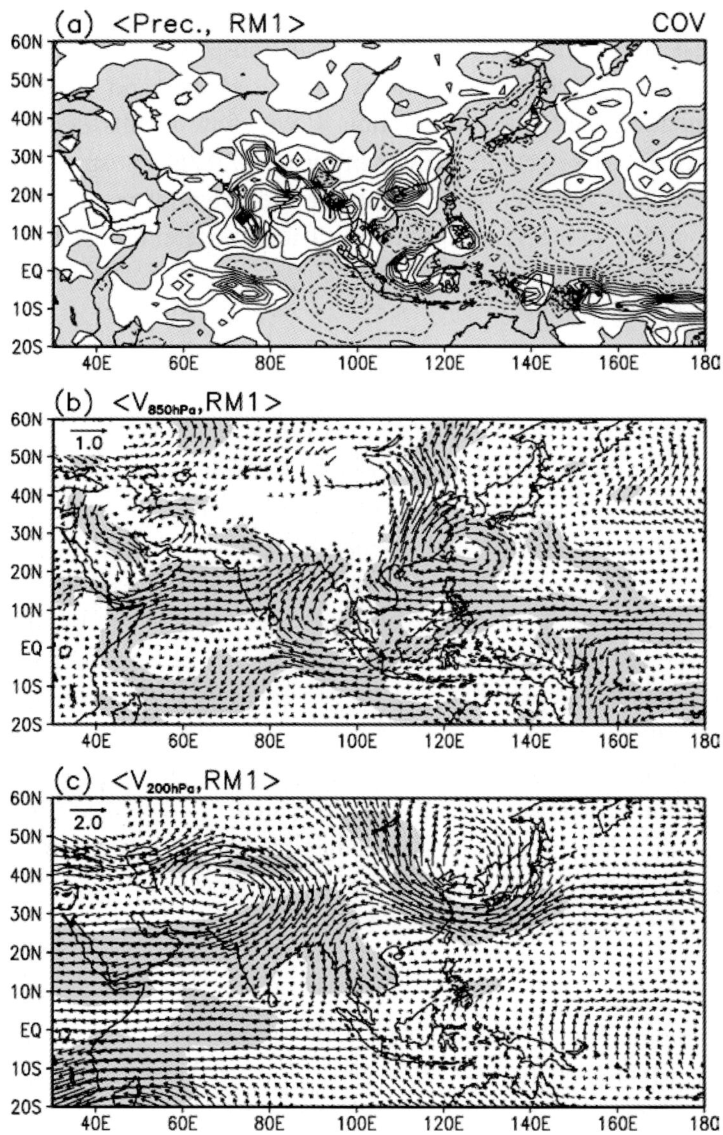

Figure 1. Linear regression of South Asia Monsoon index, RM1, against a) rainfall anomalies (mm/day), b) 850 hPa wind (m s^{-1}), and c) 200 hPa wind (m s^{-1}). Areas with correlation exceeding the 95% confidence level are shaded in b) and c).

Prediction (NCEP) reanalysis wind for the period 1979-1996 and for June through August (JJA). Fig. 1a depicts a classic monsoonal inverse relationship in rainfall between land and ocean, associated with RM1, reflecting the interannual variability of SAM. Large positive rainfall anomalies are found over land and coastal regions of South Asia, including the Indian subcontinent, the Bay of Bengal, northern Indo-China and southern East Asia. Negative rainfall anomalies are found in the surrounding oceanic areas. Pronounced negative anomalies are found over the southeastern Indian Ocean, the South China Sea, and

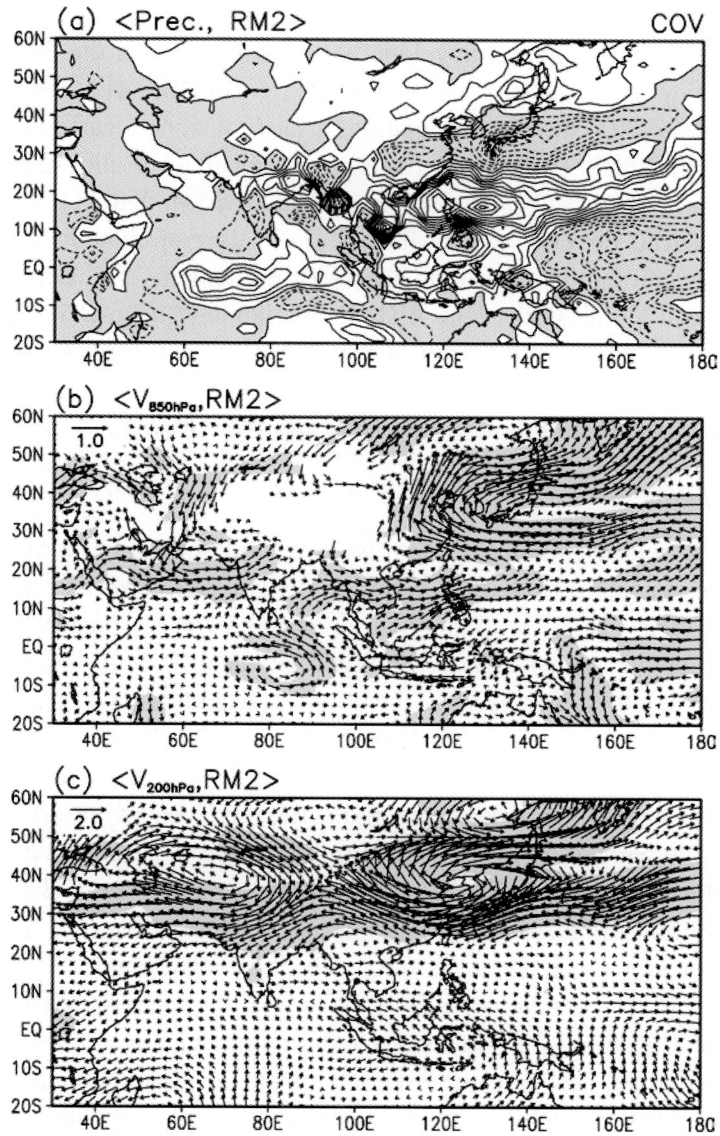

Figure 2. Same as in Fig. 1, except for the East Asian Monsoon index, RM2.

the subtropical and tropical western Pacific. Associated with the positive rainfall anomalies are strong low-level westerly and southwesterly flows over the Arabian Sea and the Bay of Bengal (Fig. 1b). This circulation pattern suggests an enhancement of the climatological ASM circulation. Two branches of cross-equatorial flow can be identified: one located along the east coast of Africa and a second one emerging from southeastern Indian Ocean. The latter appears to be the main supply of the low-level inflow into the Bay of Bengal. Another regional of strong meridional flow is found around 110°E, from southern to northern East Asia, associated with a low-level anticyclone over the East China Sea. This anticyclone represents an enhancement and westward extension of the West Pacific Subtropical High, which is associated with surface easterlies and subsidence over the South China Sea region. At the upper level, a prominent anticyclone center, identified as the South Asia High, is found to the northwest of the positive heating (Fig. 1c). The circulation pattern is also consistent with induced Rossby-type circulation by latent heating located off the equator (Gill 1980, Lau and Lim 1982). Further downstream, an upper level anticyclone is found over northern Korea (near 40°N, 120°E). Recent results have shown that interannual variability of the SAM is influenced directly by the shift of the Walker circulation during an El Niño. Rainfall variability in the land regions in SAM particularly are sensitive to the local overturnings induced by the shift of the Walker circulation (Annamalai and Slingo 2001; Ju and Slingo 1995; Lau and Nath 2000). Because the local overturnings are themselves affected by strong intraseasonal variability in SAM, the regional responses of SAM to large scale SST forcing are highly variable from year to year (Lau and Wu 2001).

2.2. *EAM Variability*

In contrast to RM1, the precipitation pattern for RM2 in Fig. 2 shows a well-defined zonally oriented structure, indicating alternating dry and wet belts over the subtropical western Pacific and East Asia sector, with only weak signal over the Indian subcontinent. The meridional extent of the monsoon-related rainfall anomalies is very large, covering regions from the equator to 50°N. In Fig. 2b, the low-level circulation field indicates a strong influence from the West Pacific Subtropical High (WPSH). When convection is strongly developed over Southeast Asia/South China Sea region, low-level westerly flow prevails from the southern Bay of Bengal, across Indochina and central South China Sea to the western Pacific. The circulation pattern signals a northward advance of the WPSH over East Asia. The easterly inflow near 25°N - 30°N is associated with local sinking motion and the returning westerly flow is linked to rising motion near 35°N - 45°N (not shown). At 200 hPa (Fig. 2c), the most prominent feature is the presence of an intensive anticyclone over East Asia, associated with the meridional shift of the axis of the climatological subtropical jet in response to tropical heating in the Southeast Asia region (Lau *et al.* 2000). The East Asian anticyclone is connected upstream to an anomalous anticyclone associated with the South Asian High. Note that the 850 hPa patterns are quite different from, while the 200 hPa wind pattern is somewhat similar to, those of RM1 (Figs.1b and c). The difference in the 850 hPa wind is due to the fact that low level moisture convergence associated with latent heating from precipitation is quite different between SAM and EAM. The similarity in the upper level large scale flow suggests the excitation of intrinsic teleconnection mode that is governed by the structure of the summertime basic flow, rather than by the detailed distribution of the heat sources and sinks (Lau and Peng 1992). Because the East Asian jet stream is controlled by extratropical dynamics as much as by tropical heating, the EAM has been called a "hybrid" monsoon system (Lau *et al.* 2000). As shown in Lau *et al.* (2000), the interannual variability of EAM is closely tied to SST variations in the subtropical and extratropical North Pacific, more so than to those in the tropical eastern Pacific. Given the large variability of responses of EAM to El Niño from event to event, it is possible that the EAM anomalies may be strongly influenced by additional forcings derived from intrinsic climate modes governing the West Pacific Subtropical High, the subtropical jet stream and related air-sea interaction in the extratropical oceans. This possibility is further explored in Sections 3 and 4.

3. Teleconnection Dynamics

Analysis of the vorticity balance in the RM1 region shows an approximate cancellation between the meridional advection of planetary scale vorticity and generation by the divergence of the anomalous flow, suggesting that the vorticity balance of the SAM is of the classic Sverdrup-type i.e.,

$$\beta v' + (\overline{\zeta} + f)\nabla \bullet \mathbf{V}' \approx 0 \qquad (3)$$

A similar analysis shows that the dominant balance for the EAM in the RM2 region, (30°N - 40°N) is between the vorticity generation induced by divergence and the advection of the anomalous vorticity by the mean circulation, i.e.,

$$(\overline{\zeta} + f)\nabla \bullet \mathbf{V}' + \overline{\mathbf{V}} \bullet \nabla \zeta' = 0 \qquad (4)$$

As shown in the following discussions, the fundamental difference in the vorticity balance between SAM and EAM has important implication on the dynamics of teleconnection patterns generated by SAM and EAM, respectively.

3.1. *ASM Teleconnection*

Figure 3a and 3b show the correlation maps of the northern hemisphere 300 hPa geopotential field with RM1 and RM2 for JJA respectively. Areas with correlation coefficients equal to or exceeding the 95% significance level are shaded. For RM1 (Fig. 3a), the anticyclone over northwestern India due to heating over South Asia appears as a part of an elongated ridge system extending from northern Africa. The South Asia anticyclone is linked to a high over northeastern Asia. As discussed previously, this high is associated with the fluctuation of the WPSH. The teleconnection pattern associated with RM2 is broadly similar to that of RM1, but much stronger and more well-defined with a strong zonally oriented trough/ridge system spanning the coastal region of East Asia, Japan, and the north central Pacific (Fig. 3b). This system is connected to the north-south migration of the subtropical jet stream associated with anomalous heating over the South China Sea and western Pacific region (See Fig. 2a). The similarities between the RM1 and RM2 teleconnection patterns in the Asian and western Pacific sectors indicate that the northward shift of the jet stream and the fluctuation of the West Pacific High may be tied to the development of the South Asia High, with enhanced heating over the South Asian land regions (see Fig. 1a). Note that the wavetrain appears to extend over North America, downstream of the north Pacific anomalies. The RM2 teleconnection pattern is similar to the Pacific-Japan pattern associated with tropical heating near the Philippines and zonally banded convective pattern over the subtropical western Pacific reported by Nitta (1987). As we show in the next subsection, teleconnection pattern similar to those shown in Fig. 3 can also be recovered based on North America precipitation anomalies. The results are consistent with the vorticity balance of Eq. (4) which suggests that the basic state vorticity gradient associated with the East Asia jet stream is a major source of vorticity and that the teleconnection pattern may be more strongly controlled by the media in which the planetary waves propagate or grow, than energy dispersion from the specific location of the heat source.

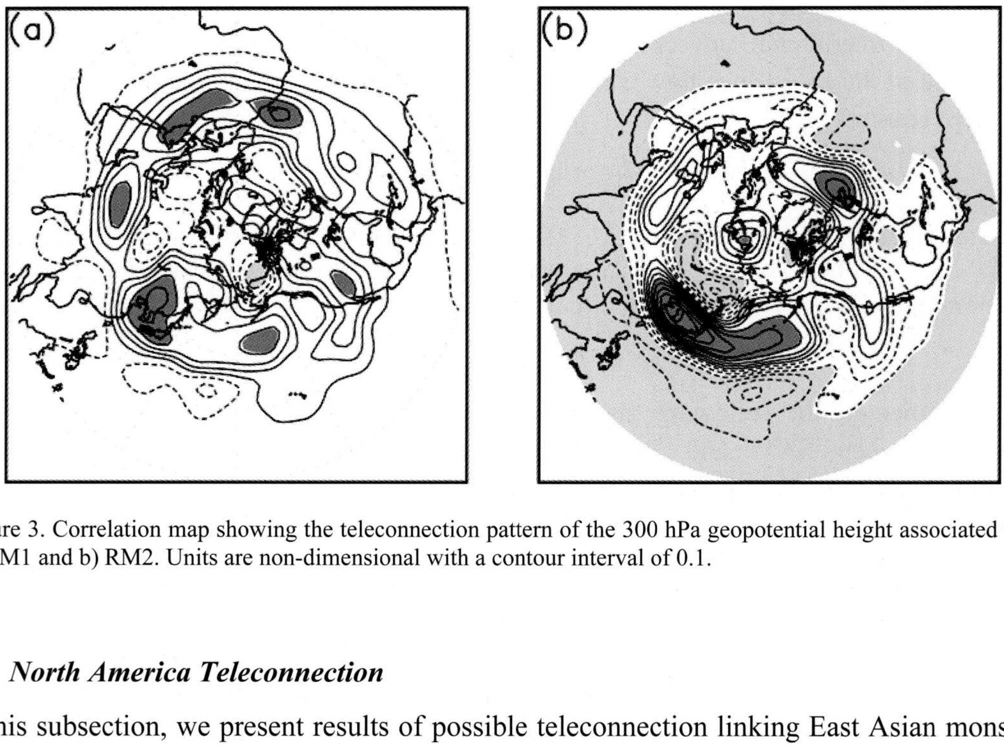

Figure 3. Correlation map showing the teleconnection pattern of the 300 hPa geopotential height associated with a) RM1 and b) RM2. Units are non-dimensional with a contour interval of 0.1.

3.2. *North America Teleconnection*

In this subsection, we present results of possible teleconnection linking East Asian monsoon variability to North America summertime rainfall anomalies. Teleconnection between North America summertime climate and ASM variability may arise not only from direct ASM influence but also from the fact that both systems are simultaneously under the control of the same global scale forcings, e.g. El Niño, the Pacific Decadal Oscillation or the Arctic Oscillation. Hence, understanding of the teleconnection between climate anomalies in Asia and in North America may shed new light on the mechanism and predictability of summertime climate fluctuations for both continents.

Two characteristic patterns linking US summer precipitation to large-scale circulation and SST anomalies have been identified (Lau and Weng 2000, 2002). Figures 4 shows the patterns of 850-hPa wind, rainfall and SST associated with the most dominant mode (Mode-1) based on a Singular Value Decomposition of the US rainfall and 500hPa geopotential height. Mode-1 explains 35% of the co-variability between US rainfall and 500 hPa geopotential height. Its principal component (not shown) projects strongly on the severe flood over the US Midwest in 1993. It depicts a zonally oriented Pan-Pacific pattern rainfall/circulation stretching from East Asia/Japan region to North America (Fig.4a). Excessive rainfall is found over the northern and northwestern North America, and deficient rainfall over the eastern and southeastern U. S. The rainfall pattern is coupled to an

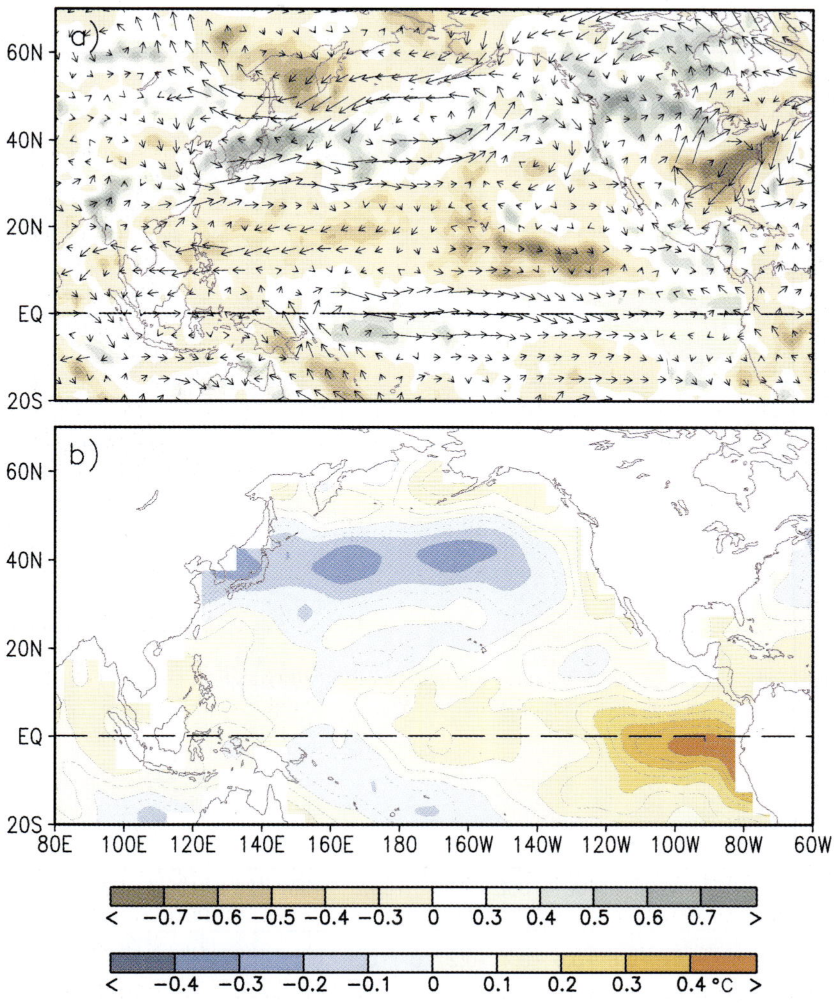

Figure 4. Linear regressions of principal component of North America summer-time rainfall for Mode-1, the Tokyo-Chicago Express (TCE) against a) global precipitation (mm day^{-1}) and 850 hPa winds (m s^{-1}), and b) sea surface temperature (^{o}C).

anomalous low-level anticyclonic flow over the eastern US, which favors the transport of warm moist air from the Gulf coast to the US Midwest. Along the equator, there is a weaker signal indicating generally enhanced rainfall in a large fetch of enhanced westerlies in the central and eastern equatorial Pacific. The zonally oriented rainfall and circulation patterns over the Asian-West Pacific sector are quite similar to that associated with EAM as shown in Fig. 2a. The pattern links rainfall anomalies of the same sign over Japan, western Canada, and the US Midwest. We shall refer to this pattern as the "Tokyo-Chicago Express" (TCE).

The regressed SST anomaly pattern for TCE suggests possible El Niño influence, as evidenced in the positive SST over the equatorial eastern and central Pacific (Fig 4b). However, the dominant feature here is an extensive cold tongue in the extratropical Pacific (near 40°N), stretching from the Sea of Japan to the west coast of North America, but not

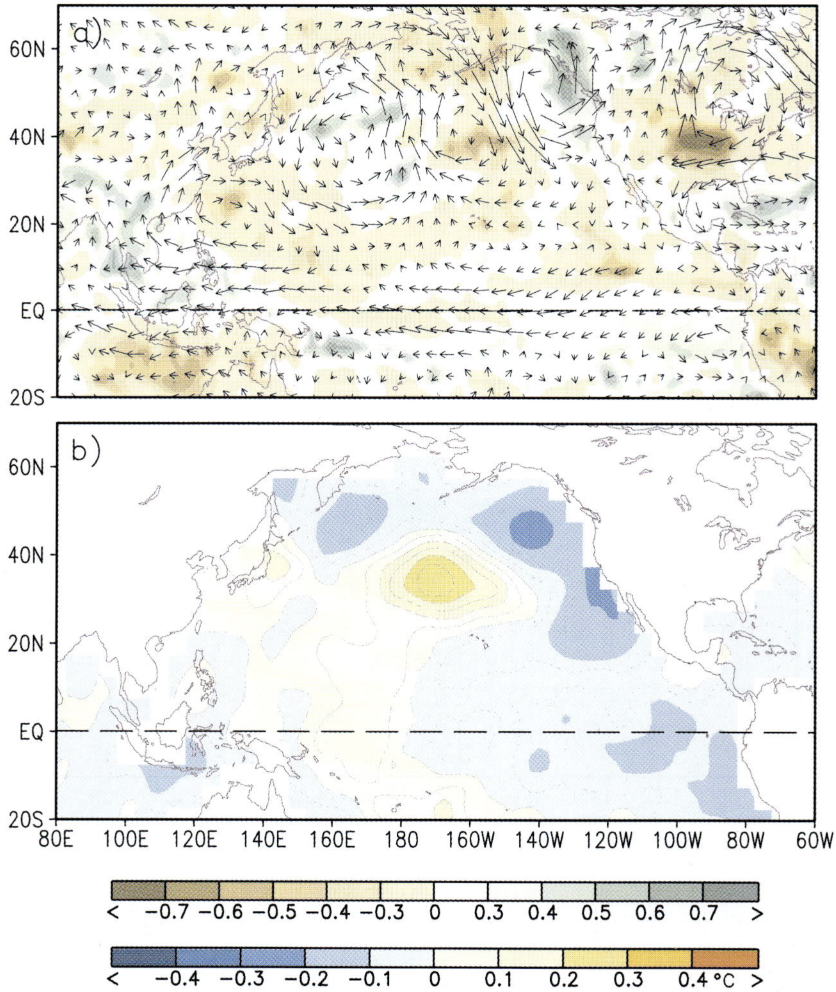

Figure 5. Same as in Fig. 4, except for Mode-2: the Shanghai-Kansas Express.

SST in the tropical eastern Pacific. Also found is anomalous warm SST in the extreme northern part of the North Pacific. The cold SST anomaly appears to underlie the anomalous low-level cyclonic circulation and rainfall anomalies over the North Pacific (Fig.4a). Increased cyclonic activities imply stronger surface wind, and cloudy and rainy conditions, leading to the cooling of the upper ocean by increased evaporation and by shortwave cloud shielding. Hence, the result suggests that the SST anomalies in the extratropics may arise from forcings of the extratropical ocean by atmospheric wind that has been shown to operate during boreal winter (Lau and Nath 1996, 2001). In a more detailed analysis, Lau *et al.* (2004) has shown that the cyclonic development is associated with a southward migration of the storm track and the subtropical jet stream, which may lead to further SST cooling (through surface evaporation) to the south and warming (through increased surface shortwave) irradiance to the north. Their results suggest that air-sea interaction may play an important role in sustaining and/or amplifying TCE.

Mode-2 explains 30% of the covariability between US rainfall and global geopotential height. The associated 850-hPa wind and rainfall patterns suggest that US summer time rainfall variability may be associated with monsoon rainfall variability over Indo-China and the tropical western Pacific region (Figs. 5a). The principal component (not shown) of this mode shows a strong projection on the severe drought over the US in the summer of 1988. Excessive rainfall is found over the west coast of Canada and below normal rainfall over central US. The associated low-level flow indicates a large anticyclone over northeastern North America coupled to a cyclone over the Gulf region. This anticyclone/cyclone couplet induces anomalous low-level easterlies in southern U.S., effectively cutting off moisture supply from the Gulf of Mexico, leading to reduced rainfall over the central US. The wave pattern over North America appears to be a part of a much larger and well-organized wavetrain emanating from the subtropical western Pacific, in an arc across the north Pacific to North America. Regions of enhanced (reduced) rainfall appear to align along the direction of the wavetrain, coinciding with low-level cyclonic (anticyclonic) circulation that can be traced back to enhanced convection over Indo-China. The anticyclone over the subtropical western Pacific near the Philippine has been identified as one of the key features of the Asian summer monsoon variability affecting droughts and floods in China, Japan, and Korea (Lau and Wu 2001; Wang *et al.* 2001). To distinguish from the TCE, we shall refer to the Mode-2 teleconnection as the "Shanghai-Kansas Express" (SKE) as they affect rainfall variability linking these two regions simultaneously.

Similar to TCE, SKE is also associated with well-defined extratropical SST anomalies, with positive (negative) SST anomalies underlying the anticyclones (cyclones) (Fig. 5b). Again, the displacement of the circulation pattern with respect to the SST suggests that the SST anomalies may be forced by local atmospheric circulation. However, the large-scale SST gradient so generated is likely to provide a strong feedback to the overlying atmosphere. There is no SST signal in the tropical eastern Pacific. Hence, the SKE appears to be independent of El Niño/La Niña, but more closely related to variability of the Asian/West Pacific monsoon convection.

3.3. *Global Linkages*

To confirm that the teleconnection patterns shown in Figs 4 and 5 are truly global, Fig. 6 shows the regression of the Principal Component (PC) for TCE against the global 500 hPa geopotential and 850 hPa wind and 200 hPa velocity potential. From Fig. 6a, the TCE emerges as a distinct feature in the 500hPa field, embedded in a global pattern that includes signals over western Europe possibly connecting with the Atlantic region via the polar regions. The 850 hPa wind anomaly is consistent with the 500 hPa height anomaly, indicating quasi-geostrophic flow, with deep vertical structure at least up to the mid-troposphere. Figure 6a strongly suggests that the regional circulation over the US is a part of a global pattern associated with extratropical stationary waves and jet stream variability along latitude belts between 30° - 60°N. TCE appears also to be linked to strong low-level wind signal in the equatorial central Pacific and the maritime continent. The 200 hPa velocity potential regression (Fig. 6b) suggests that TCE may be related to the fluctuation of the Walker

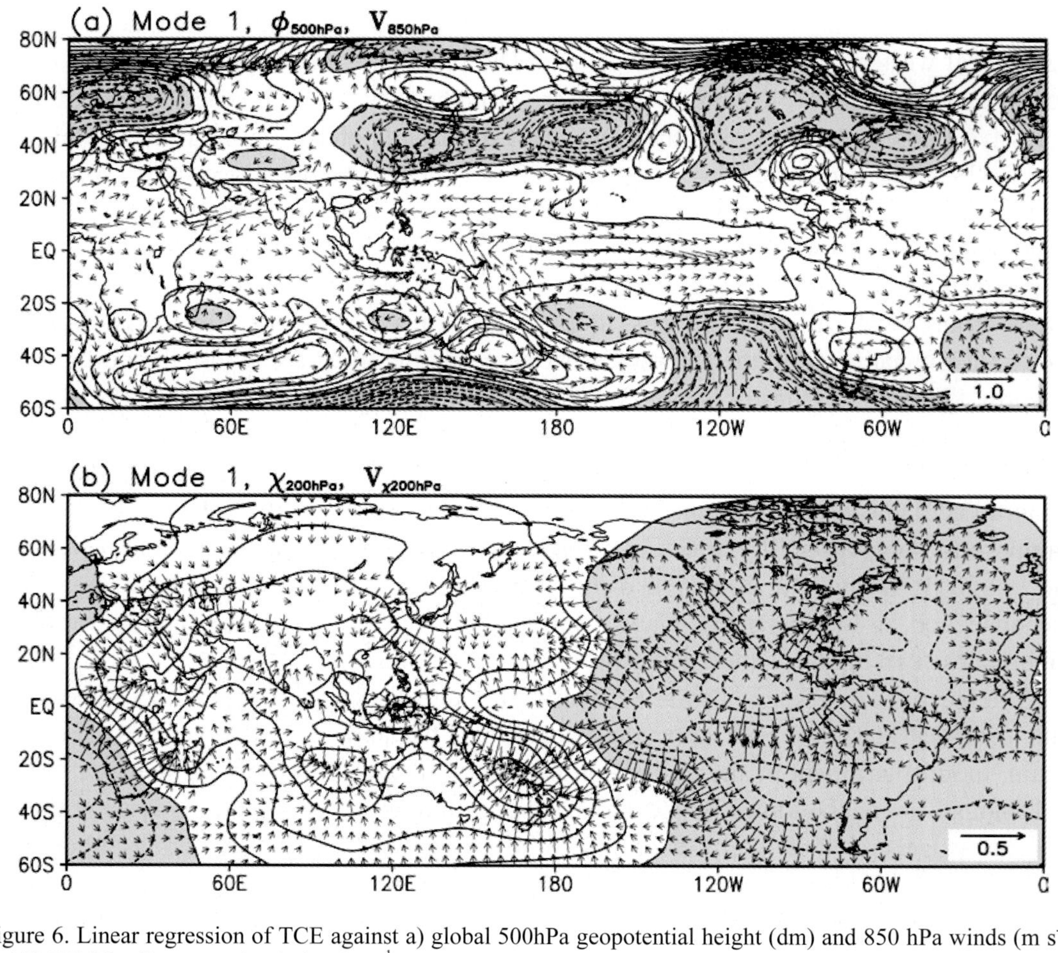

Figure 6. Linear regression of TCE against a) global 500hPa geopotential height (dm) and 850 hPa winds (m s^{-1}), and b) 200 hPa divergent circulation (m s^{-1}).

circulations, driven by heat sources and sinks in the western Pacific/maritime continent, as well as the Mexican monsoons. The divergent circulation associated with these overturning cells may cause through vorticity advection, the midlatitude jet stream to accelerate or decelerate, effectively providing a source of vorticity for downstream wave development (Sardesmukh and Hoskins 1988; Lau and Peng 1992). Large-scale coherent signals are also found over the Southern Hemisphere extratropics, which experiences the austral winter during JJA.

The regression pattern associated with SKE (Fig. 7a) clearly suggests that it is a part of a circum-global Rossby wave pattern, spanning the northern extratropics. A major apparent heat source is evidence in the divergent center in the Indian Ocean in Fig. 7b. Elsewhere, the upper level divergent and convergent patterns are consistent with the wavetrain signal shown in Fig. 7a. The overall velocity potential pattern suggests that in addition to heat sources of the Indo-Pacific monsoon, those associated with the South Pacific Convergence Zone, the northwestern Pacific and the eastern equatorial Pacific may be important in forcing the SKE wavetrain.

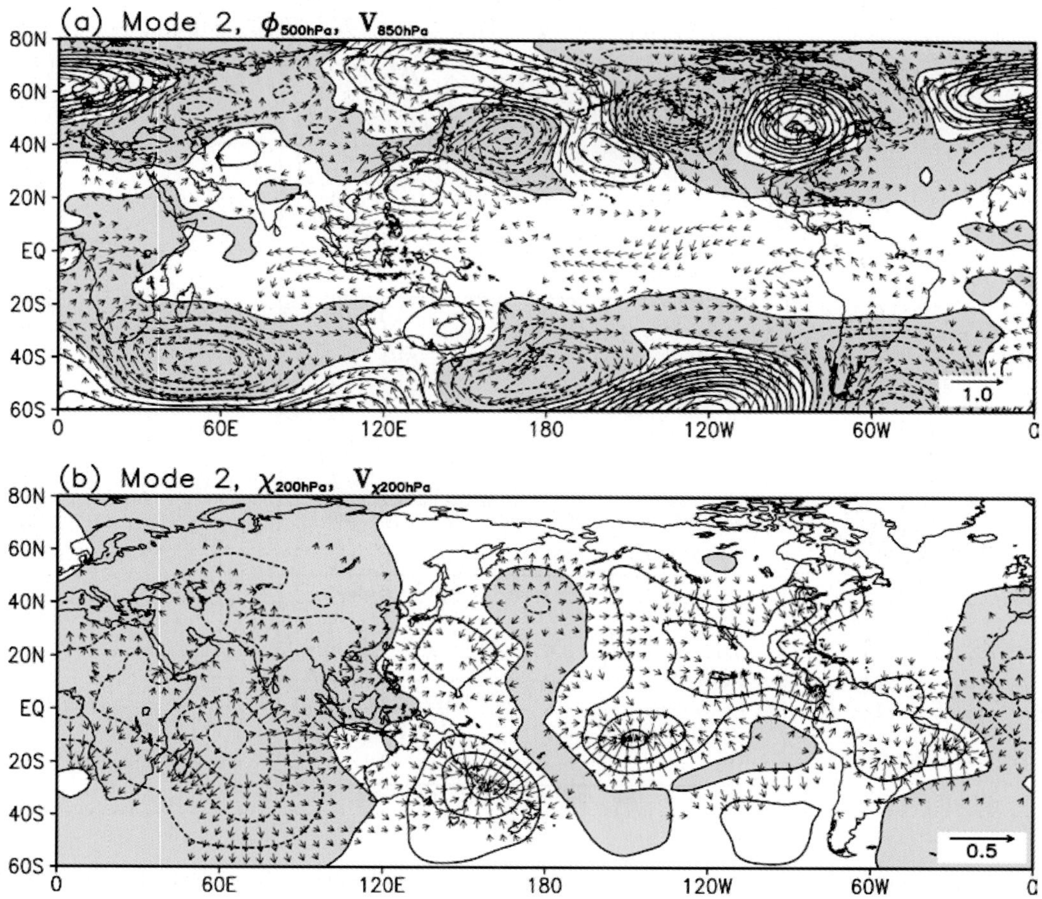

Figure 7. Same as in Fig. 6, except for SKE.

4. Extratropical Climate Modes

The strong coherent SST signals in the North Pacific, and global linkages associated with the teleconnection patterns suggest that they may stem from summertime intrinsic SST modes in the subtropical and extratropical oceans, distinct from ENSO. This notion is confirmed in Figs. 8a and b, which show the first two EOF modes of SST interannual variability of the North Pacific and North Atlantic (>20°N) during June-August. Analyses using only North Pacific SST or only North Atlantic SST yield similar results. It is clear that the SST patterns associated with the first and second SST modes have strong similarities in the North Pacific to that associated with TCE and SKE, respectively (See Figs. 4b and 5b). The first (ESST1), and the second (ESST2) modes explain respectively 17% and 13% of the variance of the North Pacific SST during boreal summer. In ESST1, the North Pacific SST varies coherently with that over the North Atlantic, having anomalies of the same sign, centered at 40°N.

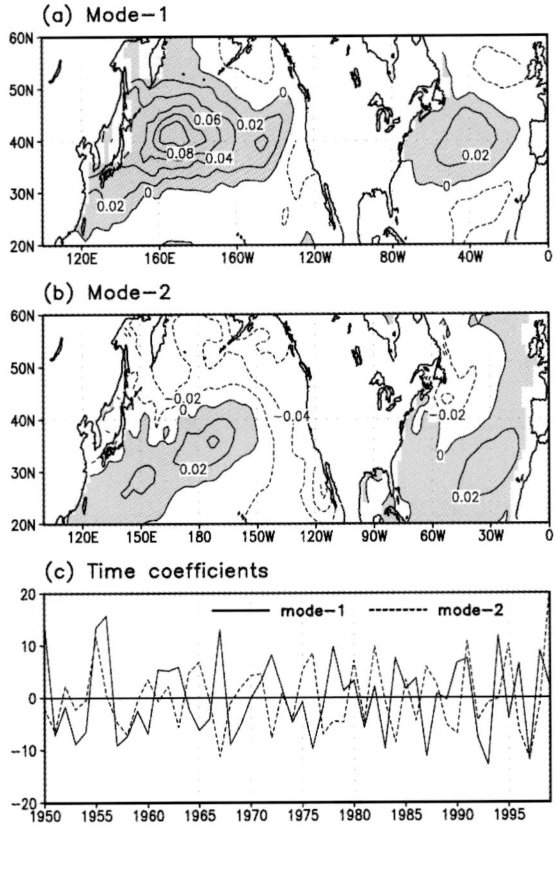

Figure 8. Eigenvector of JJA seasonal SST in the North Pacific and North Atlantic for a) the first mode and b) the second mode, and c) corresponding principal components. Multiplication of contour values and time coefficients yield units in °C.

Similarly, in ESST2, the two oceans have similar patterns with the same polarity (Fig.8b). This coherence in oceanic signals suggests the underpinning of an "atmospheric bridge" linking the two ocean basins (cf. Lau and Nath 1996).

The lagged correlation analysis of the principal components of ESST1 and ESST2 with Niño 3.4 SST is shown in Fig. 9. ESST1 is correlated with Niño 3.4 SST (>95 % confidence level) for the current and previous seasons. Yet, it is not significantly correlated with Niño 3.4 SST in the previous or following winters, when the El Niño signals are strongest. The lagged correlation SST spatial patterns associated with ESST1 (not shown) are distinctly different from the evolution of El Niño. The results suggest that while the ESST1 pattern may be linked to or even triggered by tropical SST forcing, i.e., El Niño, in the spring, it becomes increasingly independent of El Niño during the boreal summer. ESST2 shows no significant correlation with Niño 3.4 SST exists at any lag, and is therefore independent of El Niño.

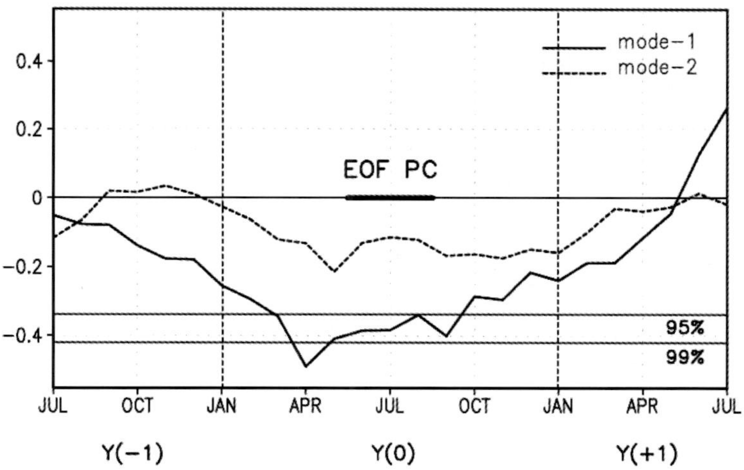

Figure 9. Lagged correlation of ESST1 and ESST2 principal components with monthly SST in Niño-3.4 region for 12 months before and after current summer season. Confidence levels of 1% and 5% are marked.

Composite analyses with respect to the positive and negative phases of the ESST1 and ESST2 yield teleconnection patterns in 850 hPa wind and 500hPa geopotential as shown in Figs. 10 and 11. The zonally oriented structure in the North Pacific in the ESST1 composite is quite similar to TCE shown in Figs 4 and 6. The simultaneous occurrence of low-level anticyclones in the northwestern Pacific and the Atlantic over regions of warm SST is the most distinct feature (Fig. 10a). Here, one gets the impression that the Pacific signal is a part of global pattern including large anomalies not only over the extratropical ocean but also over the polar land regions. The teleconnection pattern reverses sign with respect to the phases of SST in the extratropics (Fig. 10b). Overall, the patterns suggest the presence of zonally symmetric structures in the global atmosphere-ocean, interrupted by land-ocean thermal contrasts in the northern hemisphere.

Likewise, the ESST2 composite (Fig. 11) shows a circum-global wavetrain signal, that bear strong similarity, in the North Pacific region, to SKE pattern shown previously. This wavetrain appears to have shorter wavelength (approximately wave number 6) compared to its wintertime counterpart, i.e., the Pacific North American pattern (Wallace and Gutzler 1981). The signs of the wavetrain are reversed during the positive and negative phases of ESST2 (Fig. 11b). The wavetrain seems to be associated with the fluctuations of the subtropical jets. Its linkage to the tropics is through an anomalous anticyclone in the subtropical western Pacific northeast of the Philippines. As suggested in the discussion of Fig.5a, this mode may be associated with the fluctuation of large-scale heat sources and sinks in the equatorial Indo-Pacific region.

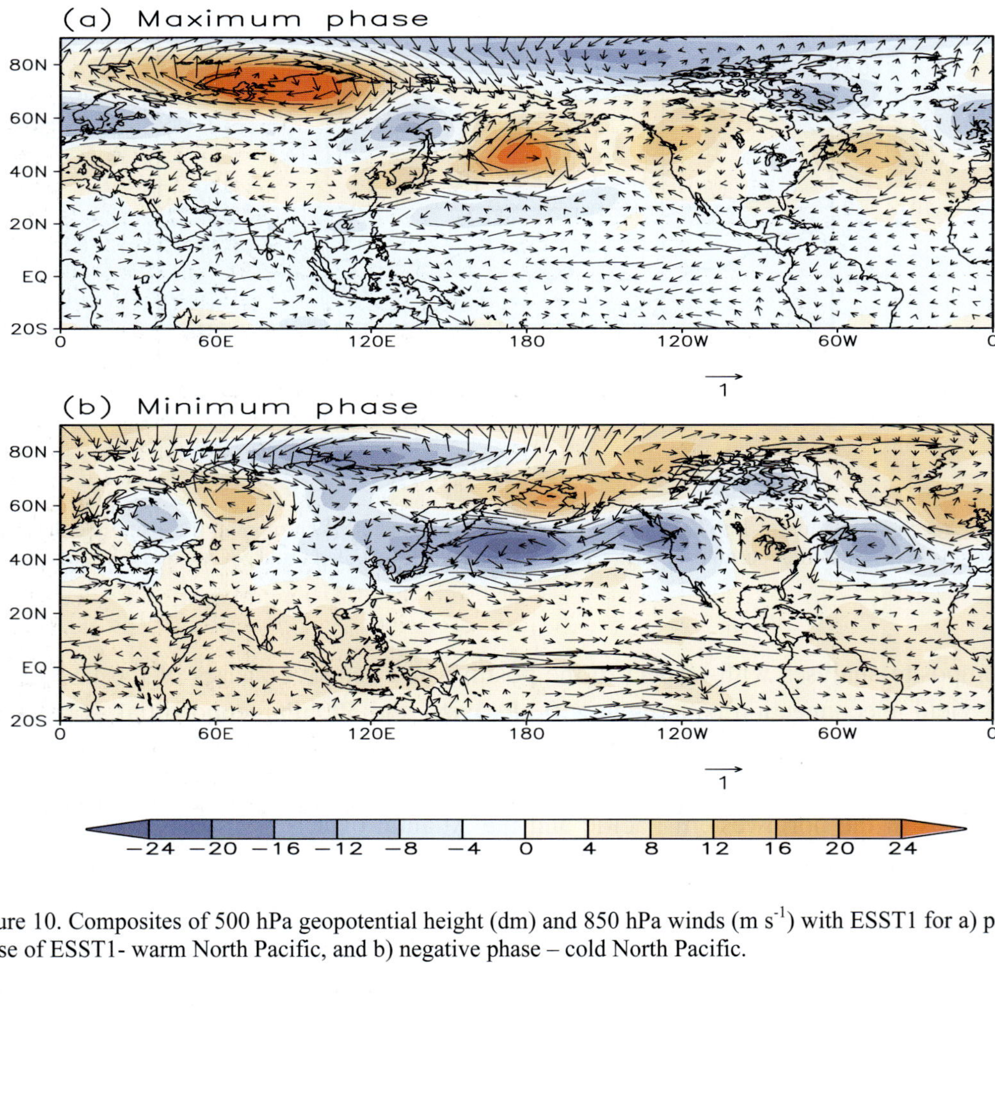

Figure 10. Composites of 500 hPa geopotential height (dm) and 850 hPa winds (m s^{-1}) with ESST1 for a) positive phase of ESST1- warm North Pacific, and b) negative phase – cold North Pacific.

5. ASM Rainfall Predictability

Interannual variability of the ASM can arise from internal dynamics and from changes in boundary conditions such as sea surface temperature (SST) and land surface conditions (Shukla 1984, Webster *et al.* 1998). For seasonal-to-interannual predictions, the former constitutes the chaotic or unpredictable component, while the latter, the potentially predictable component of the ASM. The degree to which the ASM is controlled by each component determines the overall predictability of the ASM (Lau *et al.* 2000).

Up to now, seasonal-to-interannual predictions of the ASM have depended on the existence of a physical relationship between ASM and ENSO (Rasmusson and Carpenter 1983; Webster *et al.* 1998; Ju and Slingo 1995; Lau and Bua 1998; and others). However, this relationship has undergone significant interdecadal variations. For example, the linear correlation between Indian summer monsoon rainfall and the Niño 3.4 SST has decreased dramatically in the last two decades (Kumar *et al.* 1999; Krishnamurthy and Goswami 1999; Chang *et al.* 2001). Previous studies and forecast experience have also shown that, while

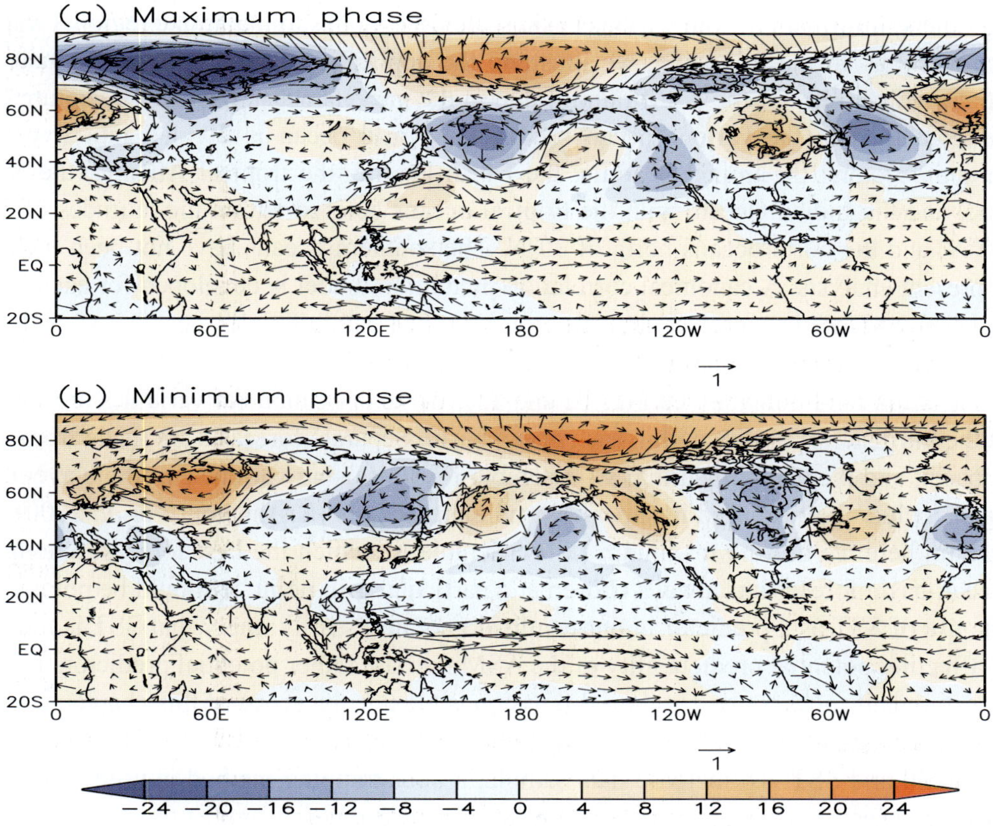

Figure 11. Same as in Fig. 10, except for ESST2.

there may be some useful skill in forecasting ASM using ENSO-related indices, many extreme ASM droughts and floods occur without ENSO. By some estimate, more than half of the major monsoon rainfall extremes in the ASM occurred in non-ENSO years, suggesting that other factors must also contribute to the predictability (or lack thereof) of the ASM (Webster *et al.* 1998). Other predictors of ASM rainfall, such as regional SST, wintertime snow cover, springtime soil moisture, and large-scale circulation parameters have been used in conjunction with ENSO. However, some of these predictors may be a function of the SST itself. Furthermore, many previous predictability studies tend to focus on the impact of the ASM by ENSO and paid relatively little attention to the possible contribution by possible intrinsic variability of SST outside the tropical Pacific.

Discussions in the previous sections suggest there may be natural variability in the extratropical oceans that are distinct from ENSO, but can profoundly affect climate of East Asia and North America during the boreal summer. This raises the possibility that regional SST variability outside the tropical Pacific, even though relatively smaller in variance compared to ENSO, may provide additional information that will improve regional ASM predictions. Recently, Lau *et al.* (2002) and Shen *et al.* (2001) have developed a new Canonical Ensemble Correlation (CEC) prediction and diagnostic procedure that utilizes, among others, information from various basin scale SST modes to maximize regional rainfall prediction skills. In the following, we illustrate the application of the CEC to evaluate the potential predictability of the ASM rainfall due to the various ocean basins. To capture SST signals contained in the various ocean basins, the world ocean is divided into five non-overlapping sectors: the tropical Pacific, the North Pacific, the tropical Atlantic, the North Atlantic, and the Indian Ocean, with the tropical oceans defined within the latitudes of 30°S and 30°N, and the north oceans north of 30°N. The southern oceans have not been included in the present analysis. A canonical correlation prediction model is developed for each of the ocean basin and used to predict the same target ASM rainfall field, i.e., 2.5 x2.5 degree rainfall from the optimal interpolation from land observation (Chen *et al.* 2002) over the Asian-Australia land domain (see Fig. 12 and 13). Up to six canonical variables are used to depict the intrinsic variability within each ocean basin. Separate forecasts are made from each ocean basin, and then combined into an ensemble forecast using various ensemble averaging strategies including those used in super-ensemble predictions (Krishnamurti *et al.* 2000). For illustration, here we use the maximum skill ensemble, in which the best forecast among the five ocean-basin forecasts is always chosen at a particular grid point. It should be pointed out that the use of SST as a predictor in our results do not necessarily imply that SST is the only factor affecting potential predictability of the ASM rainfall. Other potential predictors include soil moisture and snow cover, but their discussion is outside the scope of this chapter.

The results shown in the following are based on 49 years (1951-1999) of seasonal mean ASM rainfall and SST. To increase data sample, a bootstrapping method is used in which a forecast for a particular year is carried out by taking out that year from the training period (48 years), and all the statistics are recomputed based on the new 48 years. To quantify prediction skill, we chose a three category hit score (HS). The observed rainfall is ranked in terms of three equal-size categories, above normal, normal, and below normal at each grid point, according to the magnitude of the rainfall anomaly. The HS is defined as the percentage of the number of hits in any one category to the total number of forecasts. Hence, a hit score of 33.3% will be considered no skill. For the sample we used, the 5% and 1% confidence level corresponds respectively to a HS of 45% and 50% respectively.

The CEC is a prediction scheme that can be used for empirical forecasting ASM seasonal rainfall anomalies with different lead times, $\Delta t > 0$. Here, we define the potential predictability as the HS obtained when $\Delta t = 0$, i.e., when there is perfect knowledge of the predictor (SST) field. Using the above procedure, the potential predictability is obtained as the climatological HS computed based on 49 forecasts. To put the ASM rainfall variability and predictability in

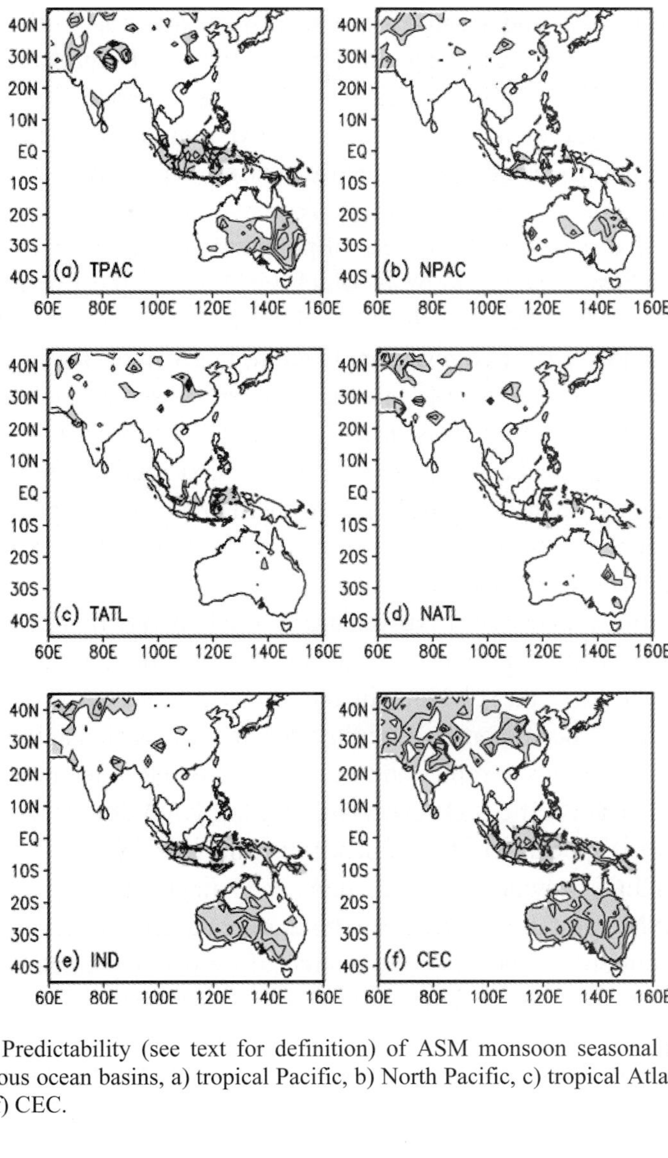

Figure 12. Potential Predictability (see text for definition) of ASM monsoon seasonal rainfall based on SST variability in the various ocean basins, a) tropical Pacific, b) North Pacific, c) tropical Atlantic, d) North Atlantic, e) Indian Ocean and f) CEC.

a broader context, we show results for the entire Asian-Australian monsoon (AAM) region. Fig. 12 shows the potential predictability, or climatological HS from each model forecasts (with zero lag) based on SST variability in each ocean basin and the overall potential predictability based on the CEC. For the individual ocean basins, it can be seen that the tropical Pacific and the Indian Ocean provide the largest contribution to potential predictability (>45%) to the ASM, mostly over Indonesia and Australia. The North Pacific, the tropical Atlantic, and the North Atlantic also contribute to potential predictability in various regions, but in decreasing order of importance. Obviously, there is a large degree of spatial variability of potential predictability attributed to SST variability in the different ocean basins. For example, the tropical Pacific affects most strongly eastern Australia, while the Indian Ocean has the dominant impact on western Australia and southwestern Indonesia. Potential predictability over India and East Asia is very low based on SST variability of individual ocean basins only, including the tropical Pacific. Remarkably, using the CEC technique, in which SST variability associated with intrinsic modes in each ocean basin is optimized, the potential predictability over the AAM region as a whole is substantially enhanced. The potential predictability is much increased over Australia and Indonesia, and also over central East Asia and central Asia, and regions in southern and eastern India.

Fig. 13 shows the distribution of sources of potential predictability for the AAM region. The color shading indicates the ocean basin which yields the highest mean HS (based on 49 forecasts) for that location. Although the following discussion is focused on JJA, all four seasons are shown in Fig. 13 for comparison. Overall, the SST in the tropical Pacific stands out as the most important source of potential predictability, as evidenced in the large proportion of areas shaded red for all seasons. This result is not surprising, confirming the notion that El Niño is the primary contributor to monsoon rainfall variability in the AAM region. In some regions, such as central and northern India, the impact from the tropical Pacific (El Niño) alone, while ranked the best, may still be sub-critical, in the sense of yielding a significant HS score based on tropical Pacific SST alone. During JJA, the source of potential predictability in western Australia and western rim of Indonesia appear to be from the Indian Ocean (colored purple). This means that the impact of the Indian Ocean SST anomalies trumps that from the El Niño in the above regions. Over southern Japan and coastal region of China, the importance of the SST in the North Pacific (colored blue) is noted. Indeed, in the northern subtropics and extratropics, much of the potential predictability can be attributed to North Pacific SST, especially in MAM and SON. During DJF, the North Atlantic (colored green) appears to gain influence over the extratropical latitudes.

In summary, CEC provides a prediction and diagnostic tool whereby potential predictability of ASM rainfall arising from both regional and remote influences can be estimated. Most importantly, via CEC, the source of potential predictability can be easily identified and traced to specific canonical climate modes (not shown here). Therefore CEC can be very useful in unraveling the physical mechanisms and the sources for ASM predictability. It also provides a common framework in which ASM predictions from general circulation models and from empirical schemes can be quantitatively compared.

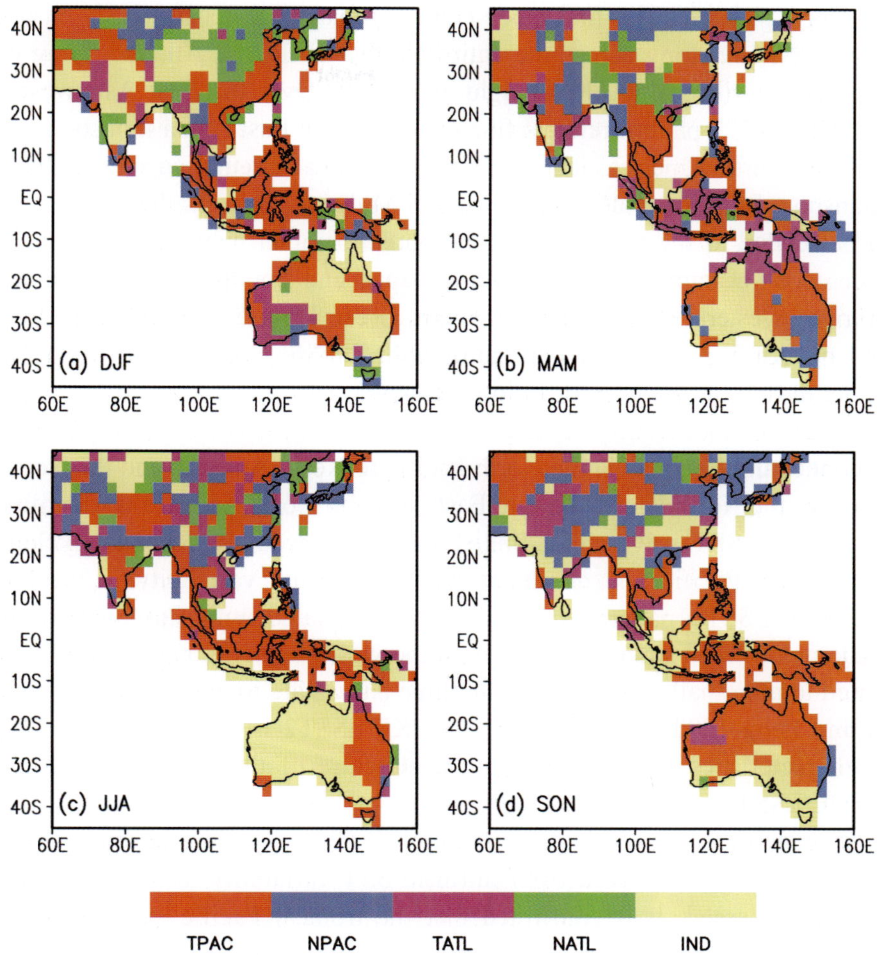

Figure 13. Source function showing distribution of most influential (based on anomaly correlations) ocean basin on predictability in ASM region for a) DJF, b) MAM, c) JJA, and d) SON.

6. Conclusions

In this Chapter, we have presented results of the latest research on the global teleconnection associated with interannual variability of the ASM and its possible implications on potential predictability of ASM. We have pointed out differences in basic dynamics in interannual variability between SAM and EAM. One of the main consequences in the fundamental difference in dynamics between SAM and EAM is that the interannual variability in the EAM is underpinned by more pronounced teleconnection patterns. Two summertime wavetrains signals named the Tokyo-Chicago Express (TCE) and the Shanghai-Kansas Express (SKE) have been identified linking precipitation variability over East Asia and North America. The wavetrain patterns project strongly, respectively on two intrinsic SST modes in the North

Pacific and North Atlantic. The first mode (ESST1) features extensive body of cold (warm) water along 40°N, stretching across the entire North Pacific and Atlantic, lying underneath zonally oriented low (high) 500 hPa height anomalies. Over the East Asia/western Pacific region, the TCE/ESST1 pattern reflects fluctuations of the East Asian monsoon heat sources and sinks. Over North America, it projects on a regional circulation pattern that regulates moisture transport from the Gulf coast to the US Midwest. Overall, the TCE/ESST1 pattern appears to be related to the variability of the boreal summer subtropical jet stream linking warm/dry or cool/wet summertime climate of Japan and South Korea to anomalies of the same sign over western Canada, the northern Great Plains, and the US Midwest. The TCE/ESST1 pattern has significant correlation with El Niño SST only during the antecedent spring, but not during the previous or following winter.

The second SST mode (ESST2) depicts a circum-global wavetrain (approximate wave number 6) connecting Eurasia via the North Pacific to North America and the North Atlantic. The portion over the North Pacific linking East Asia and North America is identified with the SKE wavetrain. The SKE wavetrain is linked to a pronounced low-level anticyclone over the subtropical western Pacific and dominant heat source/sink variability in the Indo-Pacific region. The SKE/ESST2 pattern is likely to be associated with Rossby wave dispersion forced by anomalous heat sources and sinks in the western Pacific and Indian Ocean. It may also be related to marginally unstable summertime basic state found in previous studies (Nitta 1987; Lau and Peng 1992). Furthermore, the SKE/ESST2 is unrelated to El Niño, but its dispersion characteristics may depend on the structure of the basic flow, which may be altered by TCE/ESST1.

Using a canonical ensemble correlation (CEC) prediction scheme, the potential predictability of the ASM derived from intrinsic SST variability in the major world ocean basins has been assessed. It is confirmed that the tropical Pacific is the major source of predictability in the Asian-Australia monsoon region, especially over Indonesia and eastern Australia. However, in many monsoon sub-regions over land, the major source of predictability is derived from ocean basins outside the tropical Pacific. Predictability is low over the India subcontinent from all individual ocean basins, including the tropical Pacific. In JJA, the Indian Ocean contribution to rainfall variability in the western Australia, and the southwestern Indonesia is stronger than those from the tropical Pacific. The North Pacific contributes more to rainfall variability over southern Japan and northern China compared to the tropical Pacific, in boreal summer and fall. Most important, by optimizing contributions from the different ocean basins, the CEC prediction yields overall improvement in potential predictability in all ASM regions, including the Indian subcontinent.

References

Annamalai, H., and J. M. Slingo, 2001: Active/break cycles: Diagnosis of the intraseasonal variability of the Asian summer monsoon. *Climate Dyn.*, **18**, 85-102.

Chang, C. P., P. Harr, and J. Ju, 2001: Possible role of Atlantic circulations on the weakening Indian monsoon rainfall-ENSO relationship. *J. Climate.,* **14**, 2376-2380.

______, and T. Li, 2000: A theory of the tropospheric biennial oscillation. *J. Atmos. Sci.,* **57**, 2209-2224.

Chen, M., P. Xie, J. E. Janowiak, and P. A. Arkin, 2002: Global land precipitation: A 50-year monthly analysis based on gauge observations. *J. Hydrometeorology,* **3(3)**, 249-266.

Chung, C., and S. Nigam, 2000: Asian summer monsoon-ENSO feedback on the Cane-Zebiak model ENSO. *J. Climate*, **12**, 2787-2807.

Gill, A. E., 1980: Some simple solutions of heat-induced tropical circulation. *Quart. J. Roy. Meteor. Soc.*, **106**, 447-462.

Goswami, B. N., V. Krishnamurthy, and H. Annamalai, 1999: A broad scale circulation index for the interannual variability of the Indian summer monsoon. *Quart. J. Roy. Meteor. Soc.*, **125**, 611-634.

Hoerling M. P., J. W. Hurrell, and T. Xu, 2001: Tropical origin for recent North Atlantic climate change. *Science*, **292**, 90-92.

Huang, R.-H., and F. Y. Sun, 1992: Impacts of the tropical western Pacific on the East Asian summer monsoon. *J. Meteor. Soc. Japan*, **70**, 243-256.

Ju, J., and J. Slingo, 1995: The Asian summer monsoon and ENSO. *Quart. J. Roy. Meteor. Soc*.**,** **121**, 1133-1168.

Kim, K. M., and K. M., Lau, 2000: Dynamics of monsoon-induced biennial variability in ENSO. *Geophys. Res. Lett*., **28**, 315-318.

Kirtman, B. P., and J. Shukla, 2000: Influence of the Indian summer monsoon on ENSO. *Quart. J. Roy. Meteor. Soc.*, **126**, 213-239.

Krishnamurthy, V., and B. N. Goswami, 2000: Indian Monsoon-ENSO relationship on interdecadal timescale. *J. Climate*, **13**, 579-595.

Krishnamurti, T. N., C. M. Kishtawal, D. W. Shin, and C. E. Williford, 2000: Improving tropical precipitation forecasts from a multianalysis superensemble. *J. Climate*, **13**, 4217-4227.

Kumar, K. K., B. Rajagopalan, M. A. Cane, 1999: On the weakening relationship between the Indian Monsoon and ENSO. *Science*, **284**, 2156-2159.

Lau, K. M, 1992: East Asian summer monsoon variability and climate teleconnection. *J. Meteor. Soc. Japan*, **70**, 211-242.

______, and W. Bua, 1998: Mechanism of monsoon-southern oscillation coupling: insights from GCM experiments. *Climate Dyn.,* **14**, 759-779.

______, K.-M. Kim, and S. Shen, 2002: Potential predictability of seasonal precipitation over the United States from canonical ensemble correlation predictions. *Geophys. Res. Lett.* **29(7)**, 10.1029/2001GL014263.

______, ______, and S. Yang, 2000: Dynamical and boundary forcing characteristics of regional components of the Asian summer monsoon. *J. Climate,* **13**, 2461-2482.

______, J.-Y. Lee, K.-M. Kim, and I.-S. Kang, 2004: The North Pacific as a regulator of summertime climate over Eurasia and North America. *J. Climate,* **17(4)**, 819-833.

______, and H. Lim, 1982: Thermally induced motions in an equatorial b-plane: Hadley and Walker circulation during the winter monsoon. *Mon. Wea. Rev*., **110**, 336-353.

______, and L. Peng, 1992: Dynamics of atmospheric teleconnections during the northern summer. *J. Climate*, **5**, 140-158.

______, and ______, 2001: Coherent modes of global SST and summer rainfall over China: an assessment of the regional impacts of the 1997-98 El Niño. *J. Climate*, **14**, 1294-1308.

______, and ______, 2000: Teleconnection linking summertime rainfall variability over North America and East Asia. *CLIVAR Exchanges*, **5**, 18-20.

______, and H. T. Wu, 1999: An assessment of the impact of the 1997-98 El Niño on the Asian-Australian monsoon. *Geophys. Res. Lett.*, **26**, 1747-1750.

______, and ______, 2001: Intrinsic modes of coupled rainfall/SST variability for the Asian summer monsoon: a re-assessment of monsoon-ENSO relationship. *J. Climate*, **14**, 2880-2895.

Lau, N. C., and M. J. Nath, 1996: The role of "atmospheric bridge" in linking tropical Pacific ENSO events to extratropical SST anomalies. *J. Climate*, **9**, 2036-2057.

______, and ______, 2000: Impact of ENSO on the variability of the Asian-Austraalian monsoons as simulated in GCM experiments. *J. Climate*, **13,** 4287-4309.

______, and ______, 2001: Impact of ENSO on SST variability in the North Pacific and North Atlantic: seasonal dependence and role of extratropical air-sea interaction. *J. Climate.,* **14**, 2846-2866.

Meehl, G. A., 1997: The South Asian monsoon and the tropospheric biennial oscillation. *J. Climate*, **10**, 1921-1943.

______, and J. Arblaster, 1998: The Asian-Australian monsoon and El Niño-Southern Oscillation in the NCAR climate system model. *J. Climate*, **11**, 1356-1385.

Nitta, T., 1987: Convective activities in the tropical western Pacific and their impact on the northern hemisphere summer circulation. *J. Meteor. Soc. Japan*, **41**, 373-390.

Rasmusson, E. M, and T. H. Carpenter, 1983: The relationship between eastern equatorial Pacific sea surface temperature and rainfall over India and Sri Lanka. *Mon. Wea. Rev.,* **111**, 517-528.

Sardesmukh, P., and B. Hoskins, 1988: The generation of global rotational flow by steady idealized tropical divergence, *J. Atmos. Sci.*, **45**, 1228-1251.

Shen S. S. P., K. M. Lau, K. M. Kim, and G. Li, 2001: A canonical ensemble correlation prediction model for seasonal precipitation anomaly. *NASA/TM-2001-209989*, 53pp.

Shen, X., and M. Kimoto, 1999: Influence of El Niño on the 1997 Indian summer monsoon. *J. Meteor. Soc. Japan,* **77**, 1023-1037.

Shukla. J., 1984: Predictability of time averages, Part II: The influence of boundary forcing. *Problems and prospects in long and medium range weather forecasting*. D. M. Burridge and E. Kallen, Eds, Springer-Verlag, 115-206.

Ting, M., and H. Wang, 1997: Summertime U. S. precipitation variability and its relation to Pacific sea surface temperature. *J. Climate*, **10,** 1853-1873.

Wallace, J. M., and D. Gutzler, 1981: Teleconnections in the geopotential height field during the northern hemisphere winter. *Mon. Wea. Rev.*, **109**, 784-812.

Wang, B., and Z. Fan, 1999: Choice of South Asian summer monsoon indices. *Bull. Amer. Meteor. Soc.*, **80**, 629-638.

______, R. Wu, and K. M. Lau, 2001: Interannual variability of the Asian summer monsoon: Contrasts between the Indian and the western North Pacific- East Asian monsoon. *J. Climate*, **14**, 4073-4090.

Webster, P. J., V. O. Magana, T. N. Palmer, J. Shukla, R. A. Tomas, M. Yanai, and T. Yasunari, 1998: Monsoon: Processes, predictability, and the prospects for prediction. *J. Geophys. Res.*, **103**, 14,451-14,510.

Yasunari, T., and Y. Seki, 1992: Role of the Asian monsoon on the interannual variability of the global climate system. *J. Meteor. Soc. Japan,* **70**, 177-189.

5. EAST ASIAN MONSOON-ENSO INTERACTIONS

BIN WANG AND TIM LI
International Pacific Research Center and Department of Meteorology
University of Hawaii, Honolulu, Hawaii
1680 East West Road, POST 401
Honolulu, HI 96816
E-mail: wangbin@hawaii.edu

In this chapter, we describe characteristic spatial-temporal structures of the East Asian monsoon anomalies associated with the ENSO. A Pacific-East Asian teleconnection hypothesis is put forth to explain how the El Niño affects the "upstream" climate in East Asia. The key circulation system that conveys the impact of El Niño to East Asia is the anomalous anticyclone in the western North Pacific (WNP). This anomalous anticyclone is maintained through a positive thermodynamic air-sea feedback in the WNP. The wind fluctuations associated with the monsoon anomaly may further remotely impact the El Niño evolution through exciting consecutive equatorial oceanic Kelvin waves. The rainfall variability over the Meiyu region exhibits a strong biennial signal in the correlation with the equatorial eastern Pacific and Indian Ocean SST. This interannual monsoon-ENSO relationship is subject to an interdecadal variation. For 1951-77 an El Niño-like condition precedes both enhanced Meiyu and southeast China monsoon seasons, whereas for 1978-96 an El Niño warming precedes an enhanced Meiyu but a deficient rainfall season in southeast China. It is hypothesized that this results from the interdecadal change of the basic state. The possible impacts of East Asian winter monsoon on the onset of El Niño is also reviewed and discussed.

1. Introduction

The East Asian monsoon (EAM) and El Niño-Southern Oscillation (ENSO) interact with each other on interannual time scales. In this chapter, we will review recent progress and major issues that need to be addressed (section 1), describe the spatial-temporal structure of the anomalous EAM as background information (section 2), and discuss how El Niño (La Niña) events affect East Asian climate (section 3), how the EAM influences evolution of ENSO (section 4), and how the EAM-ENSO relationship might change in interdecadal time scale (section 5). Specific emphases are placed on discussion of the physical processes that are responsible for the described phenomena.

1.1. *EAM as a Distinctive Component of the Asian-Australian Monsoon System*

The EAM is normally referred to as a subtropical monsoon encompassing eastern China, Japan, Korea, and adjacent marginal seas and monsoon oceans, i.e., the area between 20°N and 45°N and from 100°E to 140°E (Zhang *et al.*1996). The major rain-producing system is the monsoon subtropical front, which is also known as Meiyu or Baiu front during early summer when it is located along the Yangtze River Valley extending to southern Japan. The

Meiyu contains mixed tropical and baroclinic properties (Chen and Chang 1980); thus it is distinguished from the tropical monsoon in South Asia (the Indian monsoon) and the western North Pacific (WNP) (0° - 20°N, 110°E - 160°E). The seasonal cycle and interannual variations of the EAM, however, couple with the WNP monsoon trough and Australian High (Tao and Chen 1987; Wang *et al.* 2001a).

The Meiyu is preceded by the onset of the South China Sea summer monsoon, which, on average, occurs in mid-May when heavy convective rainfall suddenly develops over the northern South China Sea (10°N - 20°N, 110°E - 120°E) and westerly flows control the central South China Sea (5°N - 15°N, 110°E - 120°E) (e.g., Wang and Wu 1997; Hsu and Chen 1997; Xie *et al.* 1998; Chan *et al.* 2000). The onset of the summer monsoon over the South China Sea marks a transition of the EAM front from a mid-latitude-controlled rainfall system (the pre-monsoon rainy period in southern China) to a subtropical rainy system that is strongly influenced by the southwest monsoon from the tropics and characterized by a successive northward progression toward higher latitudes (Lau and Li 1984; Lau *et al.* 1988; Liang and Wang 1998; Huang *et al.* 2003). The South China Sea onset is first followed by an establishment of a rain band over the northwest flank of the WNP subtropical high extending from the southern coast of China, Taiwan, to east of Okinawa (Johnson *et al.* 1993; Tanaka 1997; Wang and LinHo 2002). Around June 10, the EAM front and associated rain band move rapidly northward to the Yangtze River Valley and southern Japan, where continuous rain and cloudiness last for about a month (a.k.a Meiyu in China, Baiu in Japan, and Changma in South Korea). In mid-July the rain belt further advances northward to northern and northeast China (Tao and Chen 1987; Ding 1992, 1994). In late August, the EAM begins to withdraw southward in northern China. The sub-seasonal stepwise progression of the EAM rain band (Lau *et al.* 1988; Kang *et al.* 1999) is a feature distinctive from the Indian monsoon. The sub-seasonal abrupt changes were found to be intimately associated with the climatological intraseasonal oscillation and monsoon singularities rooted in the WNP (Nakazawa 1992; Ueda *et al.* 1995; Wang and Xu 1997).

1.2. *EAM Variability Associated with ENSO*

It is not intended to give a comprehensive review here. Chapters 2, 3 and 6 of this book provide additional review materials. Here we focus on areas where rapid progress has been made and issues remain.

The summer rainfall anomaly patterns in East Asia have notable spatial complexity. A conventional EOF analysis of the East Asian summer monsoon (EASM) rainfall normally reveals a pronounced band structure. Yet, the results are sensitive to the choice of the spatial domain and the data period (Weng *et al.* 1999). For this reason, the meridional rain band structures of the interannual variations presented by different authors using different domains and data show somewhat different patterns with respect to the central and southeastern China (e.g., Huang and Wu 1989; Liu and Ding 1992; Huang and Sun 1992; Tian and Yasunari 1992; and Shen and Lau 1995).

The year-to-year variability of the EAM is determined by complex air-sea-land interaction and tropical-extratropical interaction (Huang *et al.* 2003). However, the ENSO, which exhibits the strongest influence on the interannual variability of the global climate (Webster *et al.* 1998), has been generally recognized as the most prominent factor affecting the EAM among others, such as changes in Tibetan Plateau heating, Eurasian snow cover, mid-latitude circulation, Pacific decadal oscillation (PDO) and polar ice coverage. While the ENSO is the principle factor causing the EAM variation, it nevertheless can only explain a small fraction of the variability. The EAM was found to have strong relationship with Eurasian circulation and north Pacific SST variability, which may or may not be related to ENSO (Lau 1992; Lau and Weng 2002; Samel *et al.* 1999; Yang *et al.* 2002).

While our major concern is with summer monsoon rainfall, the ENSO often reaches maximum strength toward the end of the calendar year (Rasmusson and Carpenter 1982); thus ENSO has stronger impacts on East Asia (EA) winter monsoon. ENSO normally weakens winter monsoon along the east coast of Asia and over the South China Sea (e.g., Zhang *et al.* 1996; Tomita and Yasunari 1996; Ji *et al.* 1997). Consequently, the climate in southeast China and Korea is warmer and wetter than normal during ENSO winter and the ensuing spring (e.g., Hwu and Wang 1994; Kang and Jeong 1996; Tao and Zhang 1998). On the other hand, the winter monsoon variability may exert influence on tropics (Chang and Lau 1980; Lau *et al.* 1982).

During boreal summer of the El Niño developing year, the anomalous ENSO forcing is relatively weak. Huang and Wu (1989) found that the developing stage of a warm ENSO event tends to coincide with drought in the southern and northern China and flood in between, i.e., the Yangtze-Hui River Valley. In the summer after an El Niño the Meiyu/Baiu rainfall tends to be abundant (Huang and Wu 1989; Wang and Li 1990; Zhang *et al.* 1996; Ye and Huang 1996; Soman and Slingo 1997; Kawamura 1998). The physical mechanisms behind this delayed or prolonged impact of ENSO on EA summer monsoon have been unexplained until recently when some works have been specifically devoted to address these issues (Wang *et al.* 2000; Zhang and Sumi 2002; see also Chapter 7).

Previous studies have found that summer precipitation or winter temperature in various regions of the EA has a quasi-biennial variation. This has been loosely referred to as the tropospheric biennial oscillation (TBO). Unlike other phenomena (such as ENSO), to the authors' view, the TBO has not been well defined (Li *et al.* 2001a). The biennial variation holds if El Niño-type and La Niña-type conditions alternate from one year to the next. Shen and Lau (1995) argued that the biennial relationship in the SST-EAM interaction is more robust than the one on the ENSO time scale (three to five years). Tomita *et al.* (2003) found that the biennial mode of rainfall over EA is represented by the second EOF mode, which displays a meridional seesaw like fluctuation with node at 28°N. This biennial variation is closely related to the Indian and western Pacific Ocean SST anomalies (SSTA). If Asian monsoon and ENSO are coupled, it remains unclear why so many monsoon regions (such as India, South China Sea, Indonesia, India, Australia, and WNP) all have a major biennial component, while the ENSO has a dominant period of 3-5 years.

Besides the remote ENSO impacts, local SST anomalies in the warm pool oceans were speculated to be important factors for EAM rainfall. Nitta (1987) documented an inverse relationship between Japan summer rainfall and the strength of convection in the WNP around (25°N, 140°E). With atmospheric general circulation model simulations, Huang and Sun (1992) showed that the convection activity over the Philippines is largely driven by the western Pacific warm pool SST anomalies, which causes a northward shift of the WNP subtropical high and drought in the Yangtze River Valley, as well as flood in southern and northern China. Other regions of warm ocean anomalies were also contemplated to be important to EAM, including Indian Ocean (e.g., Huang and Lu 1989; Shen and Lau 1995; Nitta and Hu 1996), the South China Sea, and Kuroshio region. The relationship between the warm ocean SST anomalies and summer rainfall in East Asia remains controversial. More importantly, we should note that it may be inadequate to attribute the warm pool SST anomalies as causes to EAM anomalies, because these warm ocean SST anomalies themselves result from anomalous monsoon conditions and are, to some degrees, related to remote ENSO forcing (Wang *et al.* 2003b). It may be more meaningful to consider that the local monsoon–warm ocean interaction, rather than the SST anomalies in adjacent monsoon regions, is one of the causes of the year-to-year EAM variability.

The powerful Asian monsoon has been hypothesized to be influential to ENSO (Yasunari 1990; Webster and Yang 1992; Kirtman and Shukla 2000). It has been proposed that the East Asian winter monsoon could trigger the onset of an El Niño (Li and Hu 1987; Li 1990; Li 1996; Li *et al.* 2001). Evidence has been shown that the WNP wind anomalies, which are induced in part by the remote El Niño in northern fall, could feed back to the equatorial Pacific Ocean and lock the turnabout of the ENSO to the boreal winter (Wang *et al.* 1999).

It is notable that, in the last 50 years, the Pacific SST and circulation regimes have undergone noticeable changes since the middle 1970 (e.g., Nitta and Yamata 1989; Trenberth 1990; Trenberth and Hurrell 1994). A number of properties of ENSO (i.e., the onset and propagation, amplitude, frequency, coupled mode structure) changed coherently and concurred with the Pacific climate shift (Wang 1995; Gu and Philander 1995; Wang and Wang 1996; An and Wang 2000). There have been apparent increases of warm events compared to cold events and decreases of the sea-level pressure in the tropical Pacific. This change may be due to a general warming of mean SSTs (Trenberth and Hoar 1996; Meehl and Washington 1996) or the occurrence of an in-phase relationship between the decadal, long-term trend and interannual variations (Lau and Weng 1999). In any case, this variation has likely affected the relationship between the ENSO and the EAM. For example, Nitta and Hu (1996) and Weng *et al.* (1999) found that certain EOF components of rainfall over China underwent a significant shift in the late 1970s; Tanaka (1997) found that the maximum negative correlation between Niño3 SST and July rainfall over eastern China and Japan as a whole increased after 1978. Thus, how these interdecadal changes may affect the interannual EA summer monsoon and tropical SST relationship becomes an important issue.

1.3. *Major Scientific Issues to be Discussed Concerning EAM-ENSO Interactions*

The review of the previous studies has raised some important questions about the relationship between the EAM and ENSO. In this chapter we intend to specifically focus our discussion on the following issues.

1) What is the characteristic large-scale spatial structure of the anomalous EA summer monsoon? What determine the meridional structure of the monsoon rainfall over the southeast and central China? What is the nature of the EAM biennial tendency? Addressing these questions provides a necessary background to understand the ENSO-EAM relationship.

2) How does the El Niño affect the East Asian monsoon (in its developing, mature, and decaying phases)? Why does the influence of ENSO differ in its developing and decaying years? In particular, why does the ENSO have a delayed impact in the summer after the ENSO peak when the El Niño disappears?

3) How does the EAM affect the ENSO evolution? Especially, what is the impact of the East Asian winter monsoon on the El Niño onset? How do WNP monsoon anomalies (e.g., anomalous WNP anticyclone), which are in part maintained by local monsoon-ocean interactions, feed back to El Niño turnabout?

4) Is the EAM-ENSO relationship steady over the interdecadal time scale? If not, what causes the change of the EAM-ENSO relationship?

These issues will be discussed sequentially in section 2 through section 5.

2. Spatial-Temporal Structures of the EASM Variability

2.1. *Composite Anomalous East Asia-Western North Pacific Monsoon*

The interannual variations of the EA and WNP summer monsoons are closely linked (Nitta 1987; Huang and Lu 1989). The circulation index for the entire Asian summer monsoon (Webster and Yang 1992) cannot adequately depict the variability of individual subsystems such as the Indian or East Asian summer monsoon (Goswami 1998; Wang and Fan 1999). To quantify the interannual variability of the summer monsoon variability in the EA-WNP sector (EA-WNP monsoon hereafter), Wang and Fan (1999) proposed a monsoon circulation index, which is defined by the meridional difference between 850 hPa zonal winds averaged over a southern rectangular region (5°N - 15°N, 100°E - 130°E) and a northern rectangular region (20°N - 30°N, 110°E - 140°E). The EA-WNP monsoon index reflects the vorticity variations in the low-level monsoon flows, thus directly measuring the variability of the WNP monsoon trough and the subtropical ridge, as well as the Philippine Sea convection. It is highly correlated with the principal component of the dominant EOF pattern of the 850 hPa wind anomalies in the EA-WNP monsoon domain (0° - 40°N, 110°E - 140°E) (Wang *et al.* 2001a). Because of the coupling between the EA and WNP summer monsoons, this index, to some extents, reflects also the variability of the East Asian summer monsoon rainfall.

Figure 2.1 shows the structure of a composite of 10 strongest minus 10 weakest EA-WNP summer monsoon years, which are selected using the EA-WNP monsoon index from 1950 to 2000. During a strong EA-WNP monsoon, rainfall is above normal in a zonal band along 10°N - 20°N extending from the South China Sea to the central North Pacific (Fig. 2.1c), whereas negative rainfall anomalies are found over the Meiyu front extending from the lower reach of the Yangtze River Valley to east of Japan, indicating an out-of-phase variation between the WNP rainfall and Meiyu/Baiu, confirming previous results (Nitta 1987; Huang and Wu 1989). In correspondence to the rainfall anomalies, pronounced 200-hPa

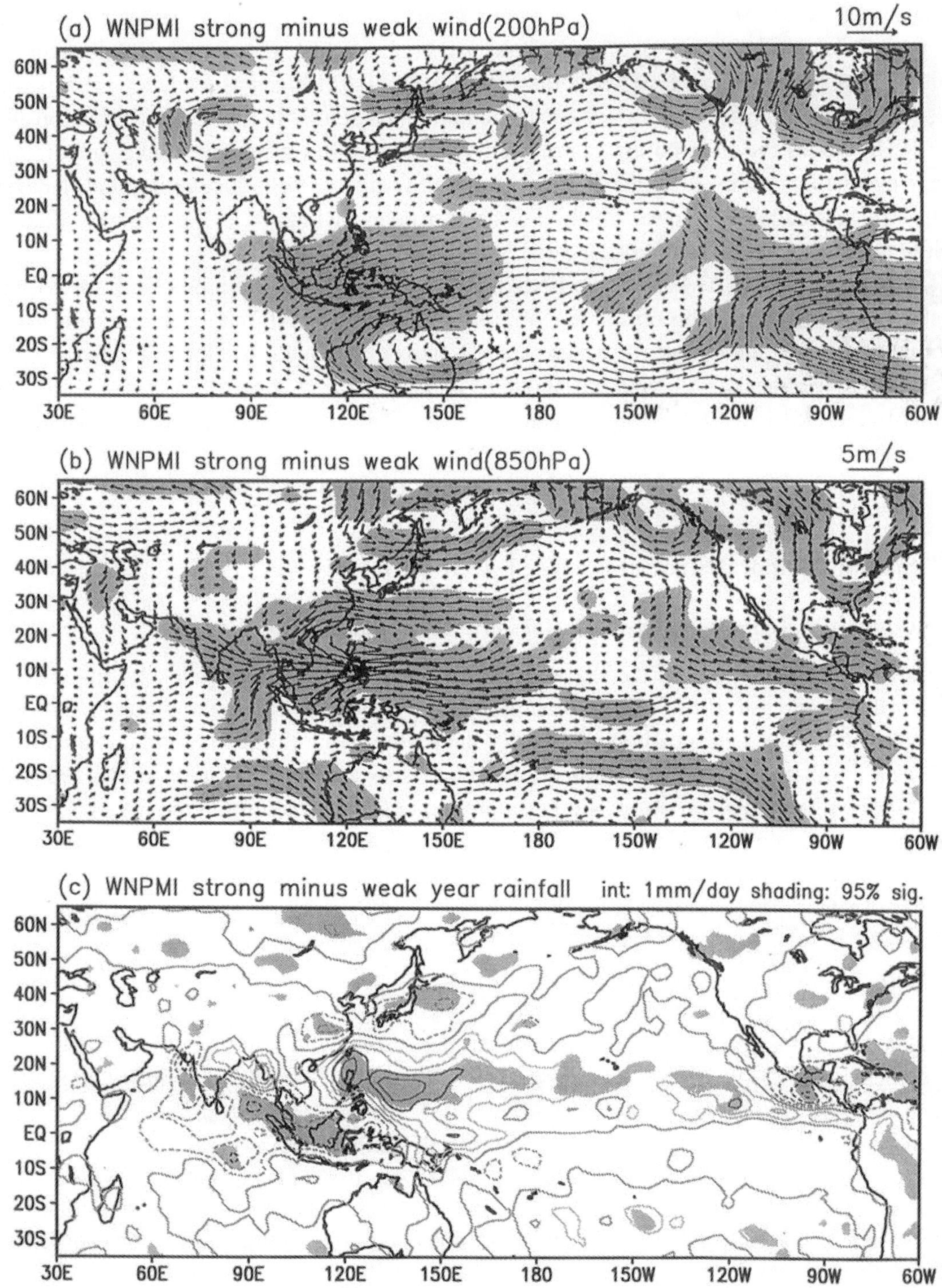

Figure 2.1. Composite differences in (a) 850 hPa winds, (b) 200 hPa winds, and (c) precipitation between the strong and weak monsoon years with respect to the EA-WNP monsoon index defined by Wang and Fan (1999). The wind scale is displayed at the upper-right of the panels. The contour interval for precipitation rate is 2 mm/day. Shading denotes regions of difference at 95% confidence level. (Wang *et al.* 2001a)

divergent flows are found in the WNP along 20°N, which produce the northeasterly cross-equatorial flows over the Maritime Continent, connecting with the anomalous Australian High (Fig. 2.1b). At the 850-hPa level, westerly anomalies prevail over the Philippine Sea, linking with cross-equatorial southerly anomalies west of Sumatra (Fig. 2.1a, also see Li *et al.* 2002). The surface temperature, precipitation, and the lower-middle tropospheric circulation anomalies in the EA-WNP region all display prominent east-west band structures with opposing centers located along 20°N and 35°N, respectively. These anomalies are caused by northward displacement of the WNP monsoon trough, subtropical high, and the EA westerly jet during a strong monsoon. This northward shift causes a strong EASM with the subtropical rain band being pushed to the north of the normal position and Meiyu/Baiu being reduced.

The summertime teleconnection patterns linking EA-WNP and North American summertime rainfall variability have been recently investigated (Wang *et al.* 2001a; Lau and Weng 2002). As shown in Fig. 2.1, this teleconnection pattern was found in association with anomalous WNP summer monsoon. During a strong monsoon year, a pronounced wave train pattern emanates from the WNP, arching around the North Pacific and extending to North America (Fig. 2.1). The wave train consists of five circulation cells. They are, respectively, the east-west elongated cyclone along 20°N in the subtropical WNP off the coast of East Asia, the elongated anticyclone along 35°N extending from the Yellow Sea to the dateline, the elongated cyclone extending from the Okhotsk Sea to the Bering Sea, the anticyclone over the Gulf of Alaska and northwestern Canada, and the cyclone over the Great Lakes of North America (Fig. 2.1a, b). The wave train exhibits a dominant barotropic structure north of 30°N. But, in the tropics between 10°N and 30°N where the wave train is originated, the circulation exhibits a baroclinic structure with a cyclone at 850 and 500 hPa capped by 200 hPa divergent flows (Fig. 2.1 a, b). A strong WNP summer monsoon tends to correlate with an anomalous low pressure over the Great Lakes and above normal precipitation in the Great Plains (105°W - 85°W, 32°N - 45°N) (Fig. 2.1c). For the 50-year (1948-1997) period, the seasonal mean (JJA) WNP monsoon index shows a significant correlation coefficient (0.42) with the US Great Plains precipitation index used by Ting and Wang (1997). In particular, there were eight weakest WNP monsoon years (1954, 1955, 1959, 1966, 1983, 1988, 1995, and 1996), during which the summer rainfalls over the Great Plains were *all* deficient.

Possible mechanisms responsible for this WNP-North America teleconnection have been speculated by Wang *et al.* (2001a) and discussed in Chapter 4 of this book. Many questions, however, remain open concerning its mechanisms.

2.2. *Circulation and SST Anomalies Associated with Chinese Rainfall Variability*

The composite pattern above is based on a large-scale circulation index for the entire EA-WNP monsoon. Now let us focus on regional characteristics of rainfall variability over China. Using Chinese rainfall station data during the period of 1956-1985, Shen and Lau (1995) identified two major rainfall variability patterns over China, one with a maximum center over Yangtze River Valley (YRV) and the other with two opposite sign centers in the northern and southern China. They found a strong biennial signal in the correlations between the YRV rainfall and the tropical SST. Their lag correlations show that a wet YRV season is preceded by warm eastern and central equatorial Pacific and Indian Ocean SST anomalies in the previous winter. These SSTA decay and switch sign in northern spring, leading to cold anomalies after the monsoon rain starts. The cooling continues through northern fall and peaks in the following winter.

Chang *et al.* (2000a, b) focused on summer (MJJA) rainfall over the Yangtze River Valley (27°N - 34°N, 110°E - 125°E), where maximum rainfall during the Meiyu occurs, and the southeastern coastal area of China (SEC, 20°N - 27°N, 110°E - 125°E), which is the favored location of the maximum rainfall belt during the early season (pre-Meiyu). The summer rainfalls over both regions are not significantly correlated with Indian monsoon rainfall for the period of 1946-1996. The northern China summer rainfall is much less than either the YRV or the SEC rainfall, but shows a significant positive correlation with the all-Indian rainfall, in agreement with the results of Yatagai and Yasunari (1995).

Figure 2.2 shows the monthly SST anomaly composite derived from the reconstructed Reynolds SST data (Smith *et al.* 1996) based on the YRV index, starting from December prior to the summer rainy season (DEC-1) and ending with the November following the rainy season (NOV0). A wet summer season is preceded by warm SSTA in the equatorial eastern Pacific and the Indian Ocean that can be traced back to the preceding winter. A conspicuous feature in May is the development of prolonged warm SSTA in the South China Sea that expand to the western Pacific east of the Philippines. This warm pattern persists through October, while the eastern Pacific cools. The South China Sea SST anomaly reverses its sign in November. Beginning in June (0) the northeastern tropical Pacific shows a cooling trend. The cooling area extends southwestward toward the equator near the dateline and by September it forms a northeast-southwest oriented cool tongue extending from Baja California to the equatorial central Pacific. In October and November the cool area expands eastward along the equator, resulting in a pattern that seems to be the opposite of the December distribution one year before.

The changes of the Pacific and Indian Ocean SST are consistent with the lag-correlation results between the monsoon and equatorial SST indices reported by Shen and Lau (1995), and also resemble correlation results for the Indian monsoon (Yasunari 1990; Lau and Yang 1996; Webster *et al.* 1998). *In the above results, a strong YRV rainfall is preceded by warm equatorial eastern Pacific and Indian ocean SSTA in the preceding northern winter and followed by cool anomalies in the next winter.* Thus, it appears that the YRV and Indian monsoon rainfall shares gross characteristics of a TBO (Meehl 1987, 1994; Chang and Li 2001; Li *et al.* 2001b; Li and Zhang 2002) in terms of their relationship with the equatorial eastern Pacific and Indian Ocean SST. The biennial tendency with regard to the growth and decay phase of the 1997-99 ENSO and the enhanced 1998 EASM (i.e., heavy rain over YRV) is demonstrated in Lau and Wu (2001).

Figure 2.3 shows the monthly evolution of composite (wet-minus-dry) 850 hPa wind field from March to October in association with the MJJA YRV rainfall variation for a 46-year (1951-96) period. In general, the equatorial wind characteristics are consistent with the tropical SST-YRV rainfall relationship shown in Fig. 2.2. A conspicuous feature during

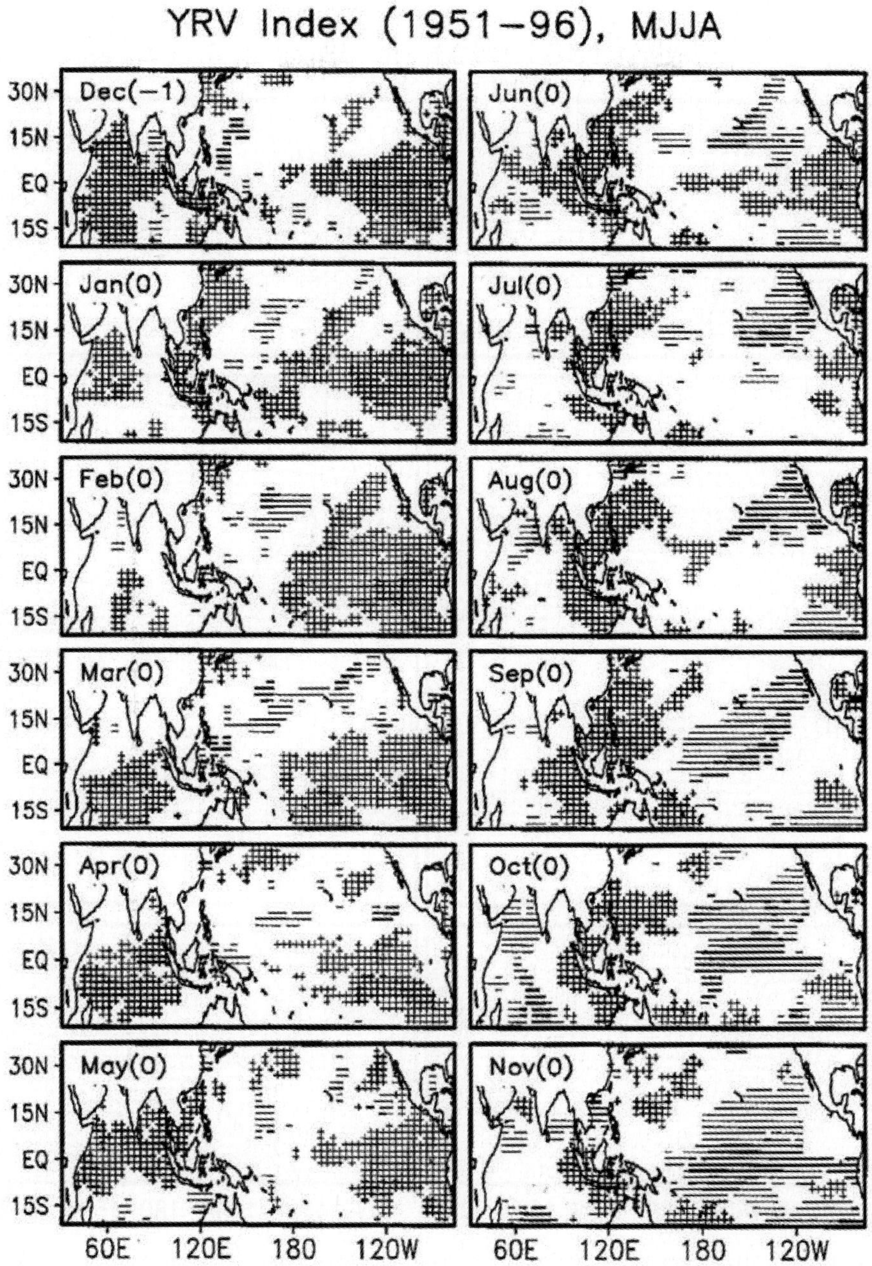

Figure 2.2. The composite relationship between monthly SSTA, from December before the monsoon season to November after the monsoon, and the all-summer YRV monsoon rainfall anomalies, for the 46-year data set. Bold +: SSTA positive for both very wet and wet, and negative for both very dry and dry; light +: SSTA positive for very wet and negative for very dry, and either very wet in-phase with wet or very dry in-phase with dry but not both; bold -: SSTA negative for both very wet and wet, positive for both very dry and dry; light -: SSTA negative for very wet and positive for very dry, and either very wet in-phase with wet or very dry in-phase with dry but not both. (Chang *et al.* 2000a)

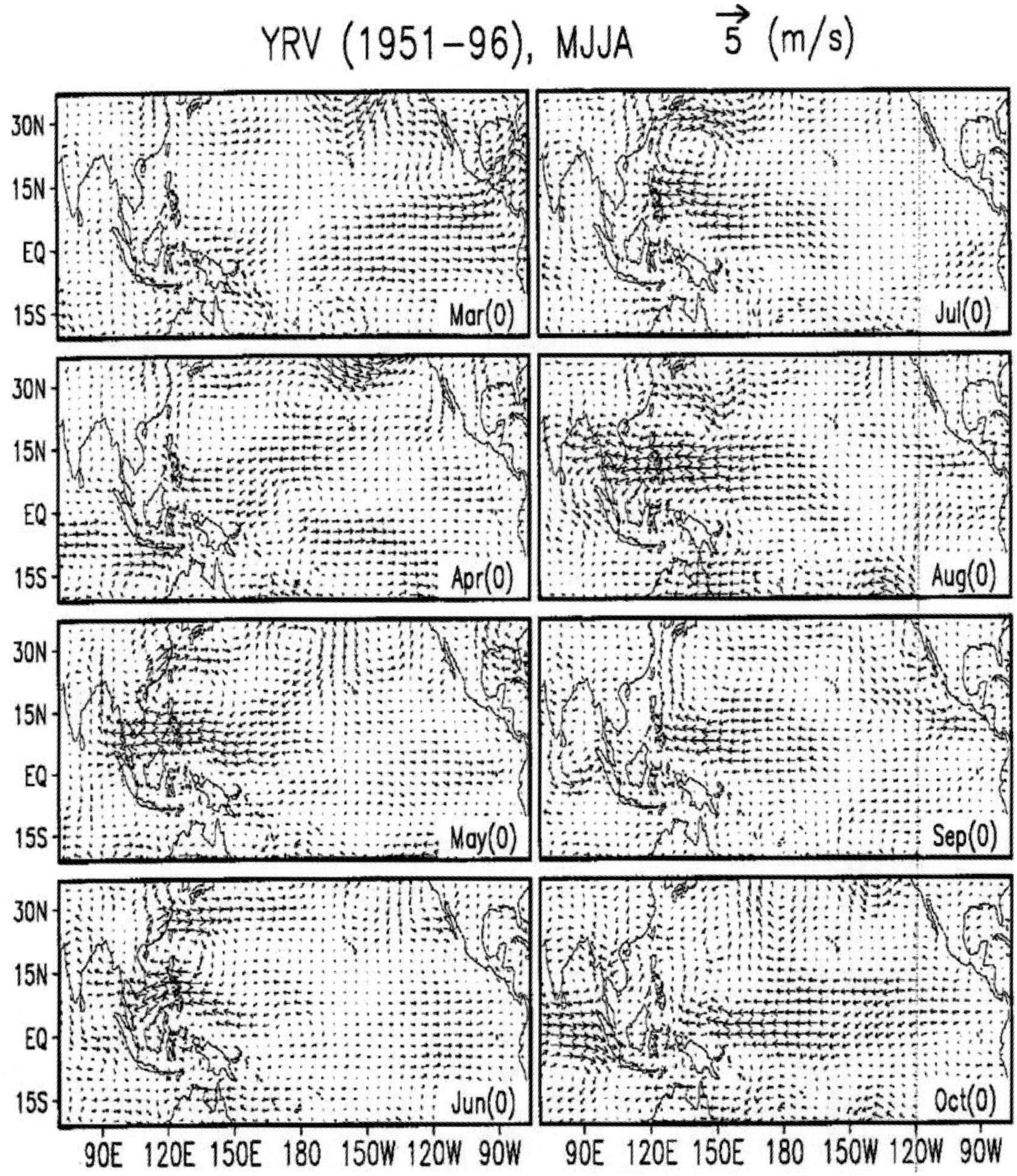

Figure 2.3. Wet-minus-dry composite of the 850 hPa wind differences (m s^{-1}), from March before the monsoon season to October after the monsoon, as computed from the all-summer YRV monsoon rainfall anomalies for the 46-year data set. (Chang *et al.* 2000a)

summer is an anomalous anticyclone cell near the southeast coast of China, which can be seen every month from May to August. The cell is slightly east-west elongated and can be identified by westerly anomalies in the East China Sea to its north and easterly anomalies in the South China Sea and Philippine Sea to its south. This anomalous cell indicates that the western end of the western Pacific subtropical ridge is stronger than normal. Since the Meiyu rainfall is mostly produced from Meiyu fronts that developed after midlatitude baroclinic systems have moved into southeastern China (Chang and Chen 1995; Chang and Chen 1998), the immediate effect of the anomalous anticyclone cell is a tendency to block the front from moving into the SEC region. This increases the time the Meiyu front stays in YRV. Furthermore, the strengthened subtropical ridge may lead to stronger pressure gradients to its northwest, resulting in a more intense Meiyu front. Both of these effects favor a wet season.

Similar to the YRV composites, an anomalous anticyclonic cell is observed off the southeastern coast of China during the SEC rainy (MJ) season (Chang *et al.* 2000b). As far as the relationship composed from the entire 46-year data set is concerned, El Niño-like equatorial eastern Pacific conditions in northern winter lead to a wet YRV summer monsoon but a dry SEC monsoon. However, this SST-rainfall relationship is subject to an interdecadal variation, which will be discussed in section 5.

3. Impact of El Niño (La Niña) on the EA-WNP Monsoon

3.1. *Precipitation and Circulation Anomalies Associated with El Niño/La Niña Events*

The composite East Asian-western Pacific monsoon anomalies (925 hPa wind and 500 hPa vertical p-velocity) for the ten El Niño events in the last 50 years are presented in Fig. 3.1. This figure is essentially the same as the leading extended singular value decomposition mode between Pacific SST and Asian monsoon anomalies (Wang *et al.* 2003b). The ascent (descent) regions correspond well to the enhanced (suppressed) precipitation anomalies. In general, the La Niña tends to have an opposite impact as an El Niño event (the figure not shown here).

During El Niño developing summer, the WNP monsoon trough extends eastward, which enhances the rainfall in the southeast quadrant (0° - 17°N, 140°E - 180°E) (Fig. 3.1a). Over East Asia the rainfall anomalies are weak, implying a weak EA monsoon-ENSO relationship during El Niño developing year. The impacts of ENSO on tropical storm activity over the WNP are enormous. Wang and Chan (2002) found that in five strong El Niño years, the number of tropical storm formation in the southeast quadrant is, on average, 6.4 during July to September, while during six strong La Niña years, only 0.3 tropical storms form in the same period. During strong El Niño (La Niña), the mean tropical storm life-span is about 7 (4) days, and the mean number of days of their occurrence is 159 (84) days; in addition, during strong El Niño years 2.5 times more tropical storms in the fall recurve northward across 35°N than during strong La Niña years.

The most remarkable low-level circulation anomalies associated with ENSO turnaround is the anticyclonic anomalies over the WNP. This anomalous anticyclone originates near the northern Philippine during the fall of an El Niño developing year (Fig. 3.1b), rapidly develops and shifts eastward over the Philippine Sea during the boreal winter when El Niño matures (Fig. 3.1c), and persists through the ensuing spring (Fig. 3.1d) and summer (Fig. 3.1e) with the intensity decreasing and the center moving eastward toward the central North Pacific subtropics.

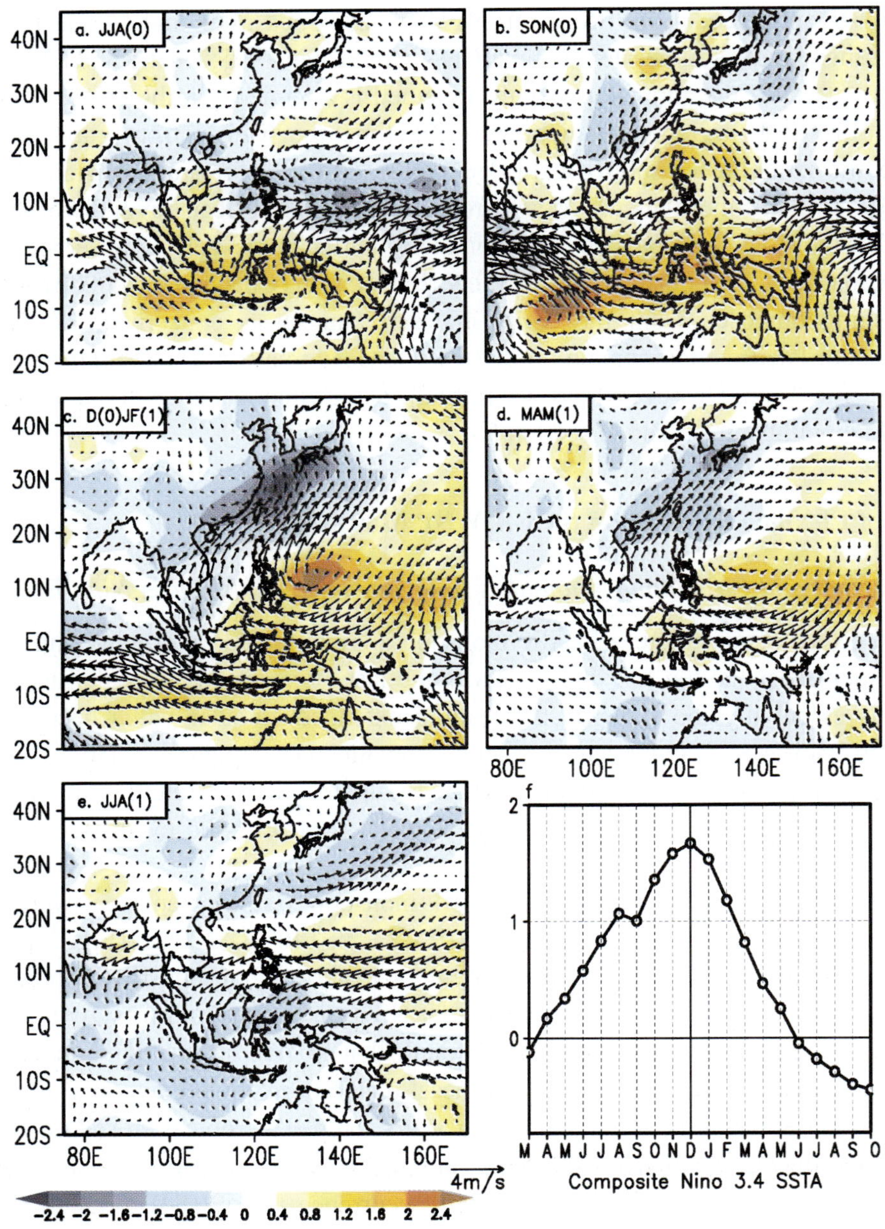

Figure 3.1. The Composite East Asian-western Pacific monsoon anomalies (925 hPa wind and 500 hPa vertical p-velocity (shading unit: 10^{-5} hPa/s)) for the ten El Niño events in the last 50 years (1951, 1957/8, 1963, 1965, 1972, 1976/7, 1982/3, 1991/2, 1994, 1997/8): (a) JJA(0), (b) SON(0), (c) D(0)/JF(1), (d) MAM(1), and (e) JJA(1). Year 0 and year 1 denote, respectively, the year during which El Niño develops and decays. The composite ten-year time series of Niño3.4 SSTA is shown in (f). (Wang *et al.* 2003)

Precipitation anomalies over the EA and WNP are closely linked to the evolution of the WNP anomalous anticyclone. During the fall of the El Niño developing year, the anticyclone over northern Philippines (Fig. 3.1b) induces drought and warmer than normal climate over the Philippines and the South China Sea. During an El Niño winter, the anticyclone severely suppresses the convection over the Philippine Sea, whereas it enhances the precipitation along the East Asian polar front extending from southern China to south of Japan (Fig. 3.1c). The southwesterly anomalies to the west of the anticyclone imply a weakening of the East Asian winter monsoon. This warmer and wetter climate continues into the ensuing spring. Noticeable enhanced ascending motion and precipitation remain along the East Asian polar front (Fig. 3.1d). This concurs with and extends the results of the previous studies of ENSO impacts on regional climates in southern China (Zhang *et al.* 1996; Tao and Zhang 1998), Taiwan (Wang and Hwu 1994), and Korea (Kang and Jeong 1996). It is important to note that the increased rainfall from March to May coincides with the local rainy seasons in those regions. Thus, ENSO exerts a significant impact on the annual rainfall in the East Asian spring polar front.

Due to the persistence of the anticyclone, the Western Pacific Subtropical Ridge extends farther to the west from the previous winter to the following summer, resulting in an 850 hPa anomalous anticyclone near the southeast coast of China during early summer, which then causes the abundant precipitation in the lower reach of the YRV, as discussed in section 2. That is why most of the YRV flood events occur in the year after peak El Niño (e.g., Fu and Teng 1993; Ye and Huang 1996), or they are preceded by warm SSTA in the eastern Pacific (Lau and Sheu 1988; Chang *et al.* 2001). Since during the El Niño developing summer, the rainfall anomalies in the East Asia are hardly significant (Fig. 3.1a). If one does not distinguish the developing and decaying phase of El Niño (La Niña), the simultaneous correlation between summer rainfall in East Asia and the SST anomalies in the eastern equatorial Pacific would not be significant (Chen *et al.* 1992). Over the WNP, after an El Niño, significantly fewer storms form before July; the WNP experiences a dry summer.

Note that the precipitation and circulation anomalies flip-flop their signs from the developing to the decaying year of El Niño or La Niña (Figs. 3.1a and e), suggesting that strong biennial tendency of the EA-WNP monsoon is associated with ENSO turnabout.

3.2. *Mechanism for Establishment and Maintenance of ENSO-EAM Teleconnection*

How can eastern Pacific warming affect the "upstream" (with respect to the direction of the midlatitude westerlies) climate in the East Asia? Wang *et al.* (2000) found that the circulation system that conveys the impact of El Niño to East Asia is the *WNP anomalous low-level anticyclone* (Figs. 3.1c, d, and e), which is not a direct response to the enhanced central Pacific warming. The WNP anomalous anticyclone has a baroclinic structure and is primarily confined to the lower troposphere. The ENSO-East Asian teleconnection differs fundamentally from the downstream Pacific-North America teleconnection (Bjerknes 1969; Wallace and Gutzler 1981; Horel and Wallace 1981).

This anomalous anticyclone forms in the fall of El Niño development (Fig. 3.1b) and persists for three seasons, providing a prolonged influence on EA-WNP monsoon. Figure 3.2

shows monthly mean SLP and SST anomalies averaged along a latitudinal band between 10°N and 20°N. The positive pressure anomalies with magnitude of one hPa occur near the Philippines (120°E) in October during El Niño development. The center of WNP anticyclone shifts eastward from 120°E in October to 152°E in the following February with a speed about 8 longitudes per month. The eastward extension of the WNP anticyclone is also evident.

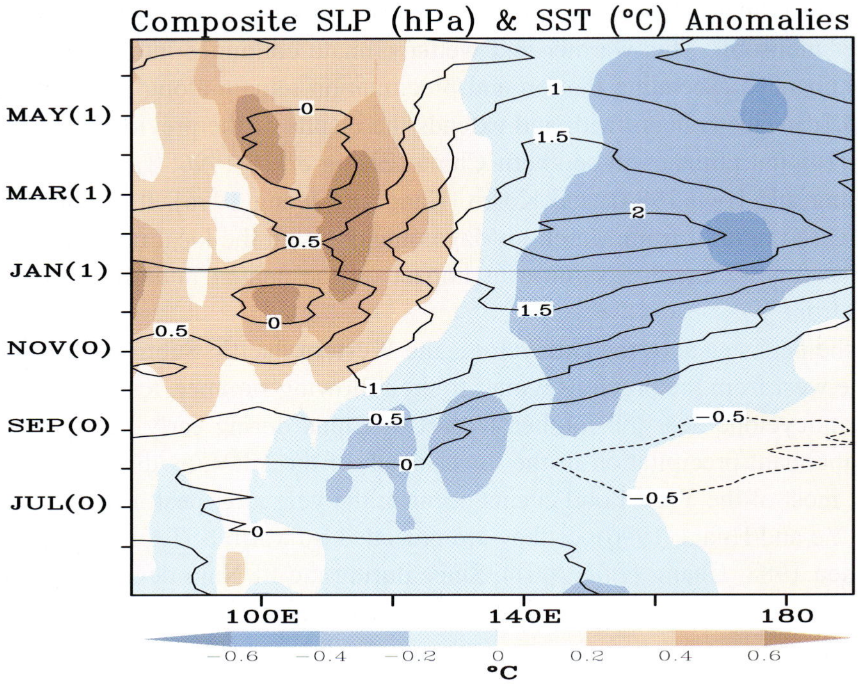

Figure 3.2. The longitude-time diagram of monthly mean sea-level pressure (contours in units of hPa) and SST (color shading in units of °C) anomalies averaged for the latitude belt between 10°N and 20°N. The anomaly fields are composed of El Niño events. The time ordinate runs from July of year 0 to June of year 1, where the year 0 and year 1 denote the year during which ENSO warming develops and decays, respectively. (Wang and Zhang 2002)

3.2.1. Formation Mechanisms for the WNP Anomalous Anticyclone

Why is the anticyclone formed first near the Philippines in October (not in other place or not in the summer when the eastern Pacific is already warming and affecting the Indian monsoon)? Wang and Zhang (2002) specifically addressed these questions. They showed that the anomalous WNP anticyclone forms concurrently with abnormal deepening of the mid-upper troposphere East Asian trough and accompanied by early retreat of the East Asian summer monsoon and by increasing number of northward recurvature of tropical storms in the western Pacific. These anomalous large-scale conditions are attributed to the combined effects of the remote El Niño forcing, tropical-extratropical interaction, and monsoon-ocean interaction. The schematic diagram shown in Fig. 3.3 summarizes this complex processes (Wang and Zhang 2002). The enhanced central Pacific heating due to developing El Niño induces a pair of upper tropospheric anticyclones and land surface cooling in northeast Asia;

both favor deepening of the East Asian trough in fall, which enhances the cold air outbreak and initiation of the anticyclone. The central Pacific heating also enhances low-level WNP monsoon trough and increases the tropical storm formation and northward recurvature (Wang and Chan 2002), which along with the frequent cold air outbreak from continental midlatitude induces vigorous tropical-extratropical exchange of air mass and heat. Through exciting descending Rossby waves, the El Niño induces Indonesian subsidence, which generates low-level anticyclonic vorticity over South Asia that is advected by mean monsoon westerly, providing anticyclonic vorticity source for the formation of WNP anticyclone. The WNP anticyclonic normally establishes rapidly (within 3 to 5 pentads), following a large-amplitude intraseasonal oscillation (ISO), suggesting the importance of the fast atmospheric processes, especially the intraseasonal oscillation. They have shown that the *in situ* air-sea interaction may strongly regulate ISO activity. For more details, the reader is referred to Wang and Zhang (2002).

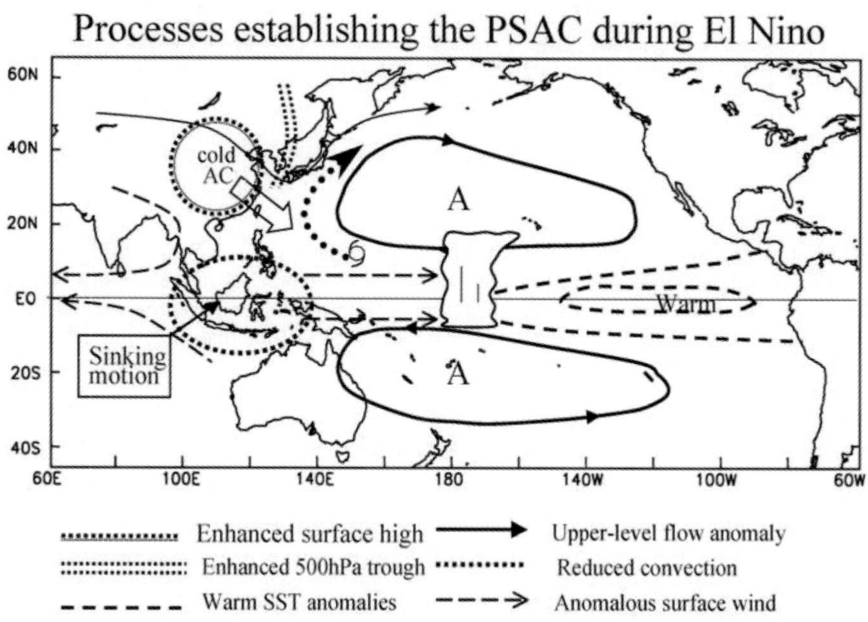

Figure 3.3. A schematic diagram illustrating the processes by which remote El Niño forcing-induced tropical-extratropical interaction initiates the anomalous Philippine Sea anticyclone. (Wang and Zhang 2002)

3.2.2. Maintenance Mechanism of the WNP Anomalous Anticyclone

In the summer following an El Niño, the SSTA in the equatorial eastern Pacific nearly vanishes (Fig. 3.1f), how can El Niño have a delayed impact to EA climate? This is a fundamental question concerning the physical basis of the climate prediction in the EA. Because the warming in the equatorial central Pacific disappears by the next summer, the WNP anticyclonic anomalies cannot be regarded as forced by the remote eastern Pacific El

Niño. Given the chaotic nature of atmospheric motion and the decaying remote forcing by El Niño, what mechanisms sustain the WNP anticyclone?

Figure 3.2 shows that the eastward displacement of the anticyclone is accompanied by significant development of negative SST anomalies to the east and positive SST anomalies to the west of the anticyclonic center. Both the SLP and negative SST anomalies extend eastward in a coherent manner, suggesting a coupling between the Philippine Sea Anticyclone and underlying mixed layer ocean temperature.

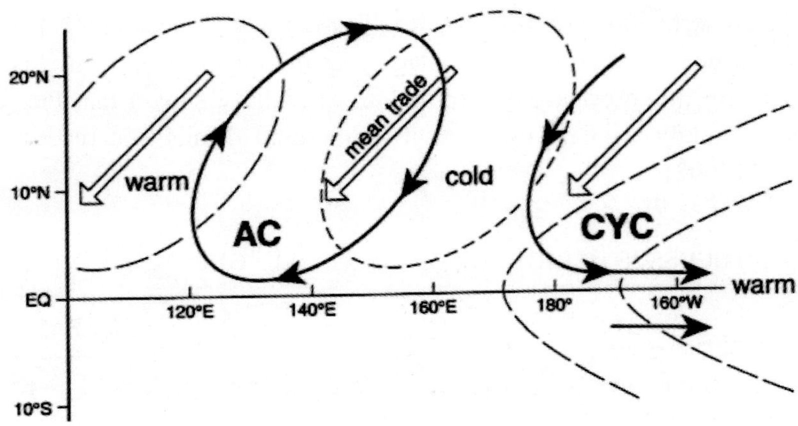

Figure 3.4. Schematic diagram showing the air-sea interaction in the western North Pacific which maintains the Philippine Sea anticyclonic anomalies and associated negative SST anomalies in the western North Pacific. The double arrows denote the mean trade winds. The heavy lines with black arrows represent the anomalous winds. The long (short) dashed lines indicate contours of positive (negative) SST anomalies. (Wang *et al.* 2000)

Wang *et al.* (2000) have put forth a hypothesis that attributes the persistence of the WNP anomalous anticyclone to positive thermodynamic feedback between the anticyclone and the underlying warm-pool ocean. The positive feedback idea is illustrated by the schematic diagram in Fig. 3.4 and explained as follows. Over the Philippine Sea, the mean winds (basic states) are dominated by the northeasterly trade winds in the cold season from November to May. To the east of an anomalous anticyclone, the total wind speed and associated evaporation and entrainment cooling are enhanced. The wind-induced SST variation would favor mixed layer cooling to the east of the anticyclone center. This is confirmed by Fig. 3.2, which shows the cooling region being located about 20° longitudes to the east of the anticyclone. The negative SST anomaly located east of the anticyclone, in turn, favors amplification of the anomalous anticyclone by exciting descending Rossby waves in their westward journeys. Therefore, the positive feedback between the anticyclonic wind and SST through evaporation and entrainment processes in the western Pacific can play a critical role in the development and maintenance of the western Pacific surface wind anomalies against thermal dissipation. The cooling in the western Pacific and the warming in the East Asian marginal seas also result from this interaction.

The results presented here suggest that, in addition to remote ENSO forcing and land surface processes, the monsoon and ocean interaction may play an important role in Asian summer monsoon variations. This hypothesis has been corroborated by the numerical experimental results with a coupled general circulation model at Geophysical Fluid Dynamics Laboratory (GFDL). Lau *et al.*, in their recent numerical experiments using the GFDL atmospheric general circulation model coupled to a mixed layer ocean model, have demonstrated that the local air-sea interaction in the WNP warm pool can indeed generate and maintain the WNP anomalous anticyclone. Detailed discussion is referred to the Chapter 4 of this book.

4. Impacts of the East Asian Monsoon on ENSO Evolution

4.1. *Impacts on El Niño Turnabout*

The possible role of WNP winds in ocean thermocline adjustment and ENSO evolution were discussed by Weisberg and Wang (1997) and Wang *et al.* (1999). Weisberg and Wang (1997) speculated that the oceanic upwelling Rossby waves excited by the central Pacific westerly wind anomalies associated with eastern Pacific warming might induce atmospheric anticyclones in the western Pacific. Of course, such anticyclonic anomalies must have a narrow meridional scale corresponding oceanic Rossby wave radius, which does not resemble the anomalous anticyclone shown in Fig. 4.1 that links ENSO and East Asian climate anomalies. However, they did postulate that easterly anomalies associated with these hypothetical anticyclones might help to terminate the warm events.

4.1.1. WNP-Anticyclone-Induced Equatorial Easterlies Preceding El Niño Peak Warming

The establishment of the anomalous anticyclone over the WNP during the developing phase of a major El Niño is about two months earlier than the peak warming in the eastern-central equatorial Pacific (Wang *et al.* 2001b). After the establishment of the WNP anticyclone, the easterly anomalies to the south of the anticyclone dominate the Maritime Continent and far western equatorial Pacific (Fig. 3.1b and Fig. 3.1c). This has important implications for its possible role in ENSO turnabout.

To show the linkage between the WNP anticyclone and equatorial easterly anomalies, we display in Fig. 4.1 the tendency of the monthly mean 1000 hPa winds at one month prior to the warm peaks for the composite warm and cold episodes. There is a notable feature common to all major warm episodes, i.e., sudden enhancement of the anticyclone in the Philippine Sea and the easterlies north of New Guinea.

4.1.2. Sudden Change of the Western Pacific Winds Generates Oceanic Kelvin Waves

Using the data derived from the NCEP Ocean Data Assimilation System (Ji *et al.* 1995), Wang *et al.* (2001b) have shown that the sudden emergence of the equatorial easterly (westerly) anomalies over the western Pacific generated oceanic upwelling (downwelling) Kelvin waves that propagated along the equator into the eastern Pacific. The forced equatorial Kelvin waves have been well documented in the previous observations (e.g., Knox and Halpern 1982; Eriksen *et al.* 1983; Lukas *et al.* 1984). Figure 4.2 shows local rates of change of the subsurface temperature (referred to as temperature tendency hereafter) in a vertical section along the equator. The tendencies of the subsurface temperature result

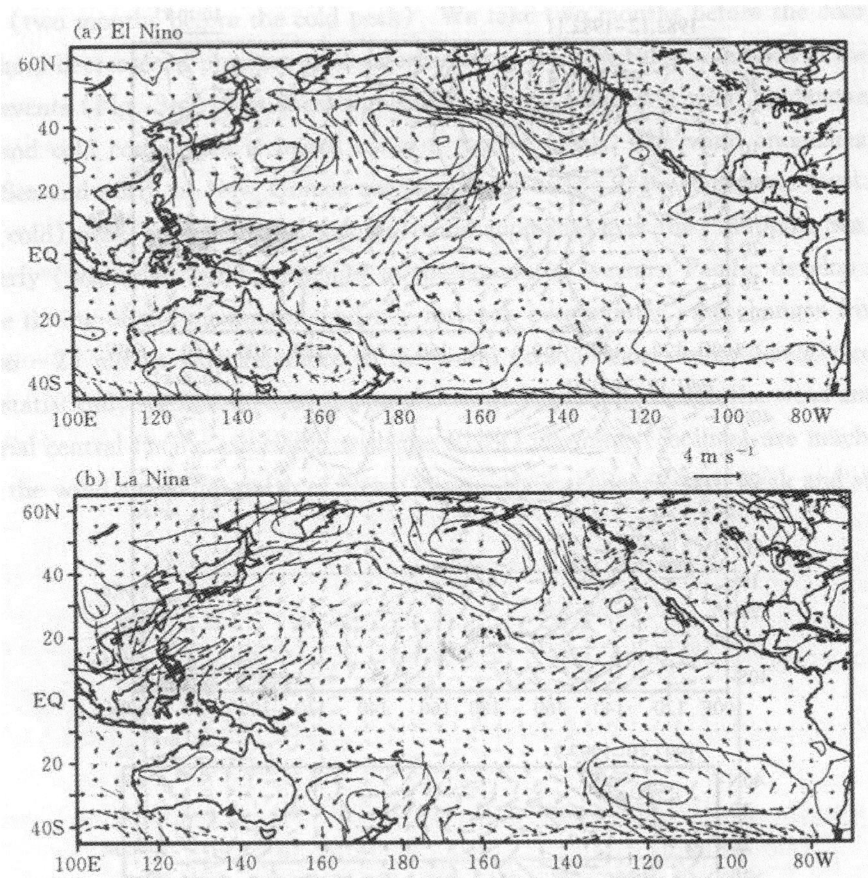

Figure 4.1. Monthly tendency of 1000 hPa winds at one month prior to the peaks of the composite (a) six strongest warm and (b) five strongest cold episodes during 1957-2000. The data are from NCEP reanalysis except for 1957 which is from COADS. (Wang *et al.* 2001b)

primarily from vertical displacement of the thermocline that is mainly caused by the passage of the equatorial Kelvin waves. In December 1991, one month before the peak of the 1991/92 warm event, a sudden increase of easterlies over the equatorial western Pacific (the figure not shown here) induced upwelling and locally elevated thermocline, causing temperature decreases at its mean depth (Fig. 4.2b). The region of decreased temperature (signals of upwelling or cold Kelvin waves) then migrated eastward at a speed of about 40-50 longitudes per month along the thermocline. The enhancement of anomalous easterlies in February 1992 triggered a second set of cold Kelvin waves that propagated all the way to the equatorial eastern Pacific (Fig. 4.2b). The consecutive eastward passage of upwelling Kelvin waves contributed to the recovery of the thermocline slope in the eastern Pacific and to the decay of the warming. Similar sequences in the 1982-83 and 1997-98 events are illustrated in Wang *et al.* (2001b). Conversely, prior to the peak of the 1988-89 cold event in October 1988, the abrupt enhancement of westerlies in the equatorial western Pacific (associated with the WNP anomalous cyclone) deepened the thermocline, triggering a downwelling or warm Kelvin wave that propagated eastward along the tilted thermocline (Fig. 4.2a). The enhanced westerlies in December 1988 sent another warm Kelvin wave to the eastern Pacific. A similar sequence of downwelling Kelvin wave episodes occurred in the 1984-85 cold event.

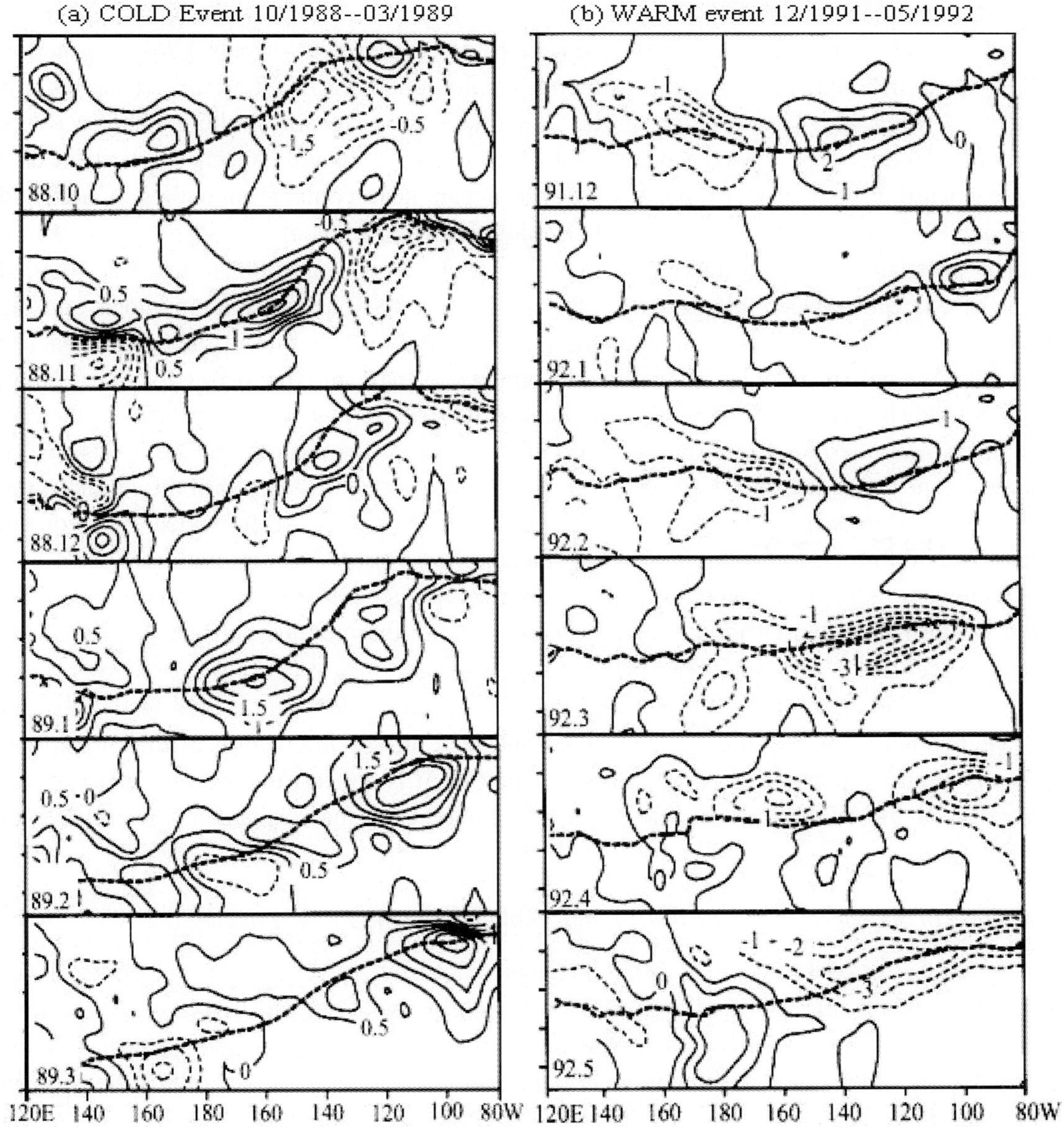

Figure 4.2. Monthly mean local rates of change of subsurface ocean temperature in a vertical section along the equator during the turnabouts of (a) the 1988/89 cold event (October 1988-March 1989), and (b) the 1991/92 warm event (December 1991-May 1992). The ordinate denotes the depth (m) below the surface. The thick dashed lines are the 20°C isotherm, showing the mean position of the thermocline during that month. The contour interval is 1°C/month in (a) and 0.5°C/month in (b). (adapted from Wang *et al.* 2001b)

4.1.3. Accumulative Effects of Forced Equatorial Kelvin Waves on ENSO Turnabout

The consecutive generation and passage of upwelling (downwelling) Kelvin waves into the eastern Pacific can effectively erode equatorial surface cooling (warming) as they journey to the east. Because the sudden change of zonal winds in the equatorial western Pacific leads the warm (cold) peak of the ENSO cycle, it is plausible that these wind anomalies play an active role in initial reversal of the warming (cooling) trend.

The equatorial thermocline adjustment that is critical to ENSO turnabouts appears to involve multiple time scales. The fast time scale is associated with the eastward propagation of Kelvin waves forced by the month-to-month wind fluctuations or intraseasonal oscillations in the western Pacific. The variation on the slow time scale is seen from the gradual recovery of the thermocline slope after the warm peak (Figs. 4.2). The latter also manifests itself as a slow eastward propagation of negative upper ocean heat content anomalies (Chao and Philander 1993), which is in tandem with the slow eastward movement of the central Pacific easterly anomalies and suppressed convection. This slow variation reflects a coupled atmosphere-ocean mode (Philander *et al.* 1984; Hirst 1986). The fast Kelvin waves and the slowly coupled mode are likely interactive. A rise of the thermocline in the eastern Pacific due to the arrival of a forced upwelling Kelvin wave may initially offset local warming, which would in turn intensify SST gradients and restore equatorial easterlies, thereby inducing a slow thermocline shoaling and SST decrease in the eastern Pacific. Consecutive generation and passage of forced Kelvin waves may stimulate slow eastward migration of the heat content through the "fetch extension" mechanism (Kessler and McPhaden 1995) or the zonal advection process (Lukas *et al.* 1984; Picaut *et al.* 1996). The simulation of the 1997-98 warm event indicates that the upwelling Kelvin waves generated by the easterly anomalies in the western Pacific contributed to the shoaling of the thermocline and the demise of the warming (McPhaden and Yu 1999).

McBride and Nicholls (1983) showed that SST in the Indonesian region leads those in the eastern Pacific by 4-6 months and he emphasized air-sea interaction in the Indonesian region as being responsible for turnaround for ENSO. It is conceivable that both the air-sea interaction in the WNP and over the Indonesian region can be relevant to the turnaround ENSO.

4.1.4. Role of Season-Dependent WNP Wind Anomalies in the Phase Locking of ENSO

The positive feedback between the atmospheric Rossby wave and ocean mixed layer thermodynamics occurs only when mean northeasterly trades prevail in the western North Pacific. This favorable mean state happens only from late fall through the next spring. Thus, the atmosphere-ocean thermodynamic interaction, which generates and maintains the western North Pacific wind anomalies, depends critically on the presence of the boreal cold season (from October to May) mean surface winds. This helps to explain why most El Niños (La Niñas) prefer to reach peak toward the end of the calendar year. It also helps to explain why strong El Niño episodes tend to decay quickly after their mature phase.

The above hypothesis was tested using a modified Cane-Zebiak coupled model in which the original atmosphere model was replaced by an empirical atmosphere model derived from the singular value decomposition of the observed SST and surface wind anomalies (Wang *et al.* 2001b). The empirical atmospheric model contains significant seasonally dependent wind anomalies in the western Pacific to mimic the Philippine Sea wind anomalies occurring in association with the El Niño and La Niña. The model results indicate that the annual variation of the anomalous western Pacific winds favors reversal of the warming and cooling trends toward the end of the calendar year. Without this process, the coupled model fails to reproduce the preferred occurrence of the cold peaks in boreal winter. The roles of the wind anomalies in the equatorial western Pacific have been discussed in other studies. Kim and Lau (2001) have shown that the wind anomalies in the western Pacific, lagged six months to the eastern Pacific SST maximum, can lead to a strong biennial tendency in the ENSO cycle.

There are three moderate warm events (1968-69, 1976-77, and 1986-87) that failed to reverse the warming trend by the end of the ENSO developing year. Inspection of the western Pacific wind anomalies indicates that the easterly anomalies in the equatorial western Pacific were not well established and did not persist in the boreal winter of these El Niño developing years (1968, 1976, and 1986). An important feature common to all three prolonged events is the insufficient strength of the central Pacific warming (Niño3.4 SSTA less than 1.5 standard deviation) toward the end of the ENSO development year, whereas during the other six strong events, the Niño3.4 SSTA were above 1.5 standard deviations. This suggests that the strong warming in the equatorial central Pacific is necessary for the rapid establishment of the western Pacific wind anomalies. Wang and Zhang (2002) have shown that the strength and persistence of the WNP anticyclonic anomalies increase with increasing strength of the El Niño forcing.

4.2. *Does the East Asian Winter Monsoon Initiate an El Niño Event?*

During the January through April preceding the six major El Niño developments, the equatorial westerly anomalies are found over the equatorial western Pacific. Li (1990) proposed that anomalous winter monsoon in East Asia contribute to the generation of the equatorial westerly and thus be responsible for the initiation of El Niño events. He noted that during the El Niño events in the period of 1950-1979, the Mongolia cold highs were stronger and surface temperature in eastern China was lower than normal during wintertime prior to each El Niño event except 1965. This prompted him to hypothesize that the strong winter monsoon in East Asia result in the development of the warm episode in the equatorial eastern Pacific. The physical argument given for this linkage is that strong cold surges associated with a strengthened Asian winter monsoon penetrates into South China Sea and equatorial western Pacific, triggering strong convection over the warm tropical ocean (Chang and Lau 1980; Lau *et al.* 1982, 1983). The anomalous convection induces westerly wind anomalies to its west as a Rossby wave response, which excite downwelling oceanic Kelvin waves and lead to the anomalous warming in the eastern Pacific.

While this observational study pointed out a potentially important linkage between the East Asian winter monsoon and El Niño, caution is needed with regard to its application to real forecast. The period of the data used for this analysis covers only from 1950 to 1979. What happened to El Niño episodes after 1979 (such as a strong warming episode in 1982-

83)? Even for the 30-year period examined, a strengthened winter monsoon did not always precede an El Niño episode. Sometimes it preceded a La Niña or normal condition. Thus, it is necessary to test the statistical significance of this winter monsoon-El Niño relationship.

In the following we show results from an EOF analysis with use of the NCAR/NCEP reanalysis data. We specifically analyzed the winter mean air temperature at 925 hPa, surface pressure, and meridional and zonal wind components in the East Asian domain (100°E - 140°E and 0° - 50°N). The first EOF mode of the winter mean air temperature field accounts for 42% of total variance. It represents a strong East Asian winter monsoon scenario with a maximum cold temperature anomaly center appearing in northern China (Fig. 4.3a). The second EOF mode represents a meridional dipole over East Asia continent (figure not shown), which accounts for approximately 19% of total variance.

The time coefficient of the first EOF mode shows pronounced interannual variability (Fig. 4.3b). Note that in some years (e.g., in 1950, 52, 56) a strong winter monsoon indeed preceded an El Niño event as pointed out by Li (1990), but in other years (e.g., 1964, 71, 81, 86, 96) the relationship is not clear or even reversed.

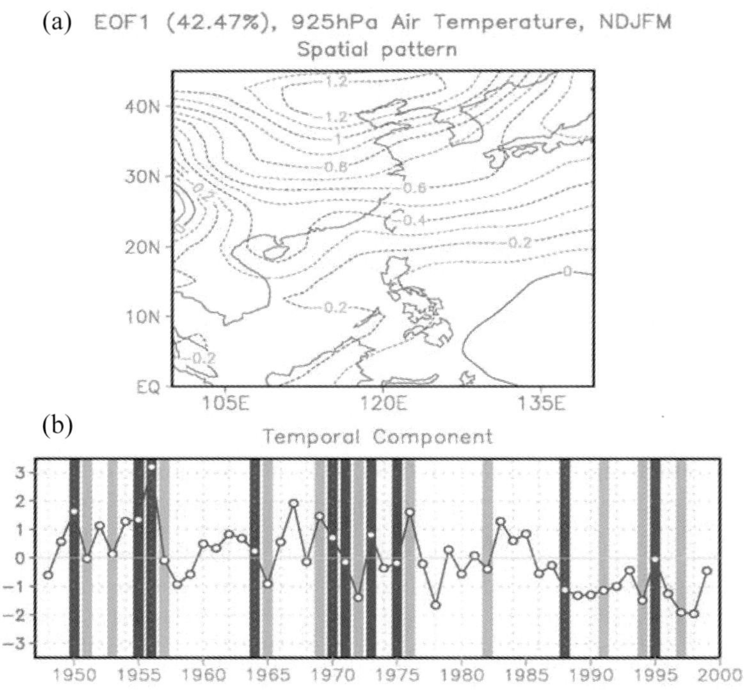

Figure 4.3. (a) Horizontal pattern and (b) time coefficient of the first EOF mode of the winter (DJFM) mean air temperatures at 925 hPa derived for the period of 1948-1999. Light (dark) bars in (b) denote peak El Niño (La Niña) winters.

By calculating the lagged correlation of the EOF1 time series with tropical SSTs, one can readily see that maximum correlation with the eastern Pacific SST anomalies occurs either simultaneous or prior to the winter monsoon (Fig. 4.4). The shaded areas of Fig. 4.4 represent the lagged correlation exceeding -0.3 (over 95% significance level). This negative, simultaneous correlation implies that the Asia winter monsoon weakens (strengthens) during the mature phase of an El Niño (La Niña), suggesting the significant impact of ENSO on the Asia winter monsoon. A relatively weak, positive correlation appears a year later (DJF+1), implying that the Asian winter monsoon might lead the change of the eastern Pacific SSTA and thus on the initiation of El Niño. However, this positive lagged correlation is not statistically significant for the entire 50-year period. The similar conclusions can be derived from the analyses of other meteorological variables such as the winter mean surface pressure or mean meridional wind field.

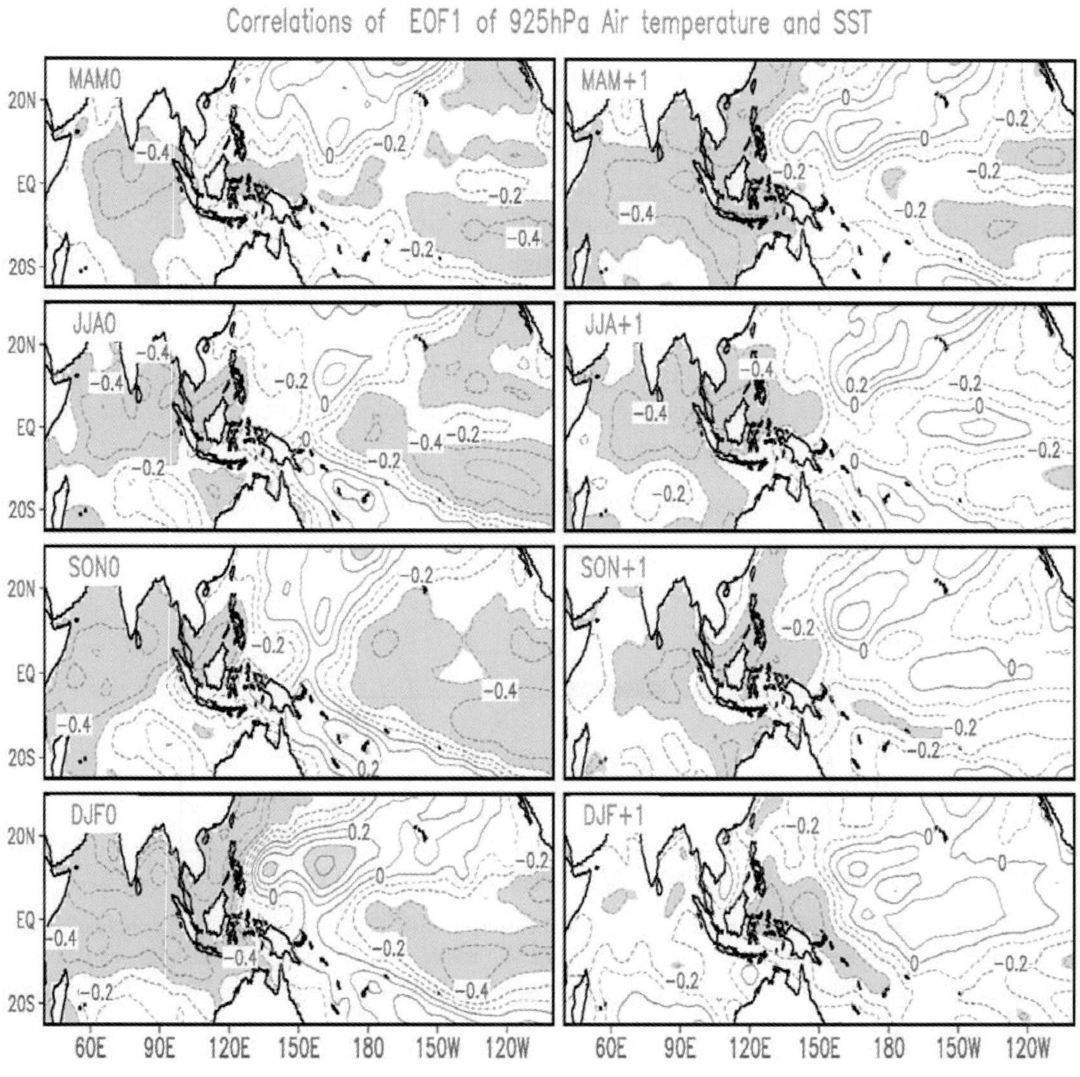

Figure 4.4. Lagged correlations of tropical SSTs with the time coefficient of the first EOF mode of winter air temperature shown in Fig. 4.3. DJF0 denotes a simultaneous correlation. DJF+1 denotes a lagged SST correlation with the previous-year winter air temperature in Asia.

5. Interdecadal Changes of the EAM-ENSO Relationship

5.1. *Rainfall Anomalies over Central and Southeast China*

Chang *et al.* (2000a, b) separated the YRV and SEC rainfall data into the first interdecadal period (1951-1977, ID1) and the second interdecadal period (1978-1996, ID2) and noted that the rainfall-SST relationships in both regions were subject to strong interdecadal variations. In the following we describe the observed aspect of these interdecadal variations and discuss possible mechanisms.

5.1.1. Rainfall Variability in Yangtze River Valley and Associated ENSO Transition

Figure 5.1 shows the evolution difference of the equatorial eastern Pacific SST anomalies between ID1 and ID2 in reference to the YRV MJJA rainfall anomaly. Here the normalized wet-minus-dry area-averaged eastern Pacific SST anomalies are plotted relative to the YRV MJJA rain season, from October before the rain season (OCT-1) to December after the rain season (DEC+1). For both epochs, the equatorial eastern Pacific SST anomalies are positive in the fall and winter prior to a wet season and negative in the fall and winter immediately following a wet season. However, the change of sign occurs in early spring before the summer rain season for ID1, similar to Shen and Lau (1995), whereas in ID2 the change of sign occurs in early fall after the rain season. Furthermore, in ID1 the SSTA turns to warm in the ensuing spring, indicating a TBO signal. However, for ID2 the negative SSTA, although decreased in magnitude in the ensuing spring, become colder in the summer and appear to reach coldest value in the following fall.

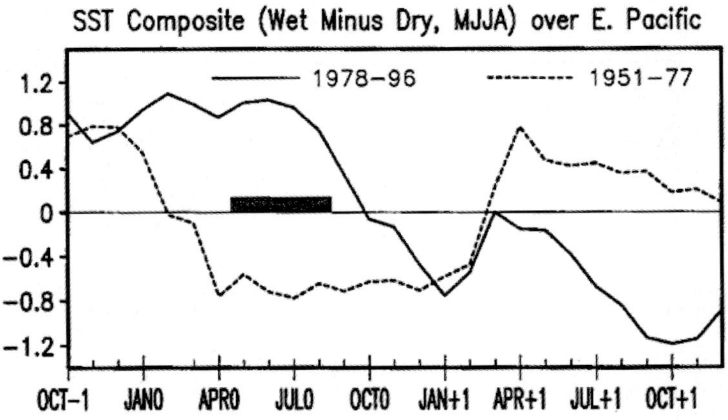

Figure 5.1. The wet-minus-dry composite of equatorial eastern Pacific SST evolution from October before a wet YRV monsoon to October after the monsoon in the (105°W - 85°W, 5°S - 5°N) for ID1 (dashed) and ID2 (solid). The May-August monsoon rainfall season is marked by a horizontal bar. (Chang *et al.* 2000a)

The interdecadal difference in the equatorial eastern Pacific SST evolution can be clearly related to anomalous winds at the equator (Fig. 5.2). For ID1, anomalous easterlies appear east of the dateline starting in early spring. They soon strengthen and expand so that a wide longitudinal span covering both sides of the dateline is occupied by anomalous easterlies through summer and fall. This sequence of events is consistent with the cooling of the equatorial eastern Pacific in the northern spring before the wet YRV monsoon starts during ID1. In contrast, for ID2 a belt of anomalous easterly winds first appears in June in the western Pacific and South China Sea. It then propagates slowly eastward and reaches the eastern Pacific in October.

The mechanisms that cause the interdecadal difference for the SSTA phase reversal are hypothesized as follows. The long-term mean equatorial eastern Pacific SST is several degrees below the threshold of that required for convective conditional instability (Chang and Li 2000). In addition, the low-level easterly wind produces a mean divergence. Therefore,

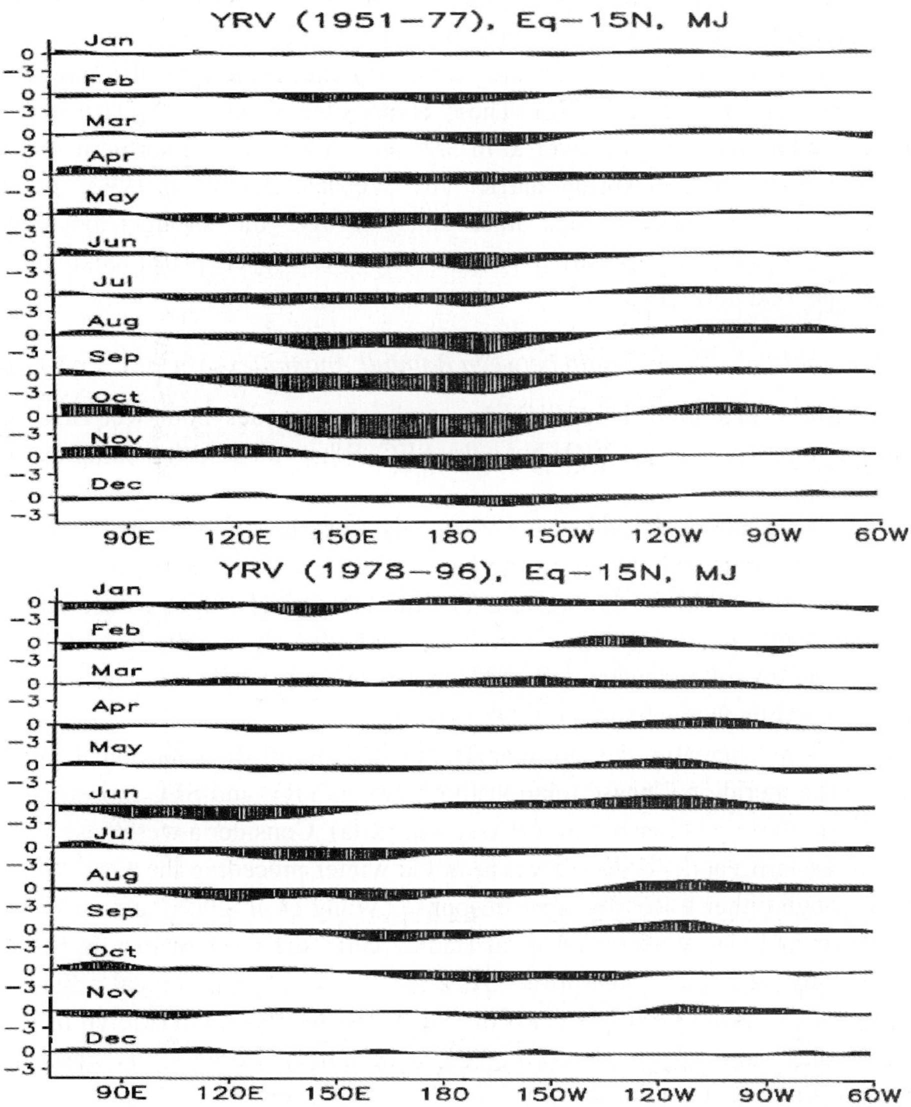

Figure 5.2. Time-longitude section of the wet-minus-dry composite (based on MJ YRV rainfall anomalies) of the 850 hPa wind differences (m s^{-1}), averaged between equator and 15°N, for the ID1 (a) and ID2 (b) period. (Chang *et al.* 2000a)

even under warm anomalies, it is difficult to sustain a positive feedback between zonal wind stress and east-west asymmetric thermocline variations through the anomalous Walker cell-SST interaction[a]. In ID2 the mean equatorial eastern Pacific SST is nearly 1K higher than in ID1. Meanwhile, the warmer mean SST also produces a stronger mean low-level convergence, due to zonal convergences along the equator and beta-induced meridional convergences. During a warm phase within ID2 the combination of these two mean effects and the warm SSTA may be sufficient to elevate the environment to a favorable condition for convection. Therefore, the positive feedback between zonal wind stress and thermocline variations will be more likely to occur.

This interdecadal difference implies that a warm SST phase in ID1 is more likely to reverse its sign after the SST peak in northern winter, such as those observed in the TBO composites (e.g., Yasunari 1990; Shen and Lau 1995) and in the typical ENSO composites (e.g., Rasmusson and Carpenter 1982). These composites all show a phase lock with the seasonal cycle such that a sign of reversal in SST anomaly occurs in northern spring (which has been termed the "boreal spring barrier" by Webster and Yang 1992; Torrence and Webster 1998; and others). On the other hand, in ID2 the stronger positive Walker circulation-thermocline feedback may overcome the seasonal cycle phase locking and sustain a warm phase beyond northern spring.

5.1.2. Meridional Phase Relationship between Rainfall Anomalies over SEC and YRV

Observations show that in ID1 an El Niño condition precedes both wet SEC and YRV whereas in ID2 it precedes a wet YRV but a dry SEC. What cause such a complex meridional phase relationship?

Chang *et al.* (2000b) hypothesized that this complex rainfall-SST relationship is attributed to the interdecadal change of the basic state. A key element that links the eastern Pacific SST and the EAM is the western Pacific subtropical ridge, whose strengths and spatial extents vary greatly and are correlated with the equatorial eastern Pacific SST anomalies. A westward extension of the subtropical ridge may project an anomalous anticyclone at 850 hPa that enhances rainfall to its north.

Figure 5.3 is a schematic diagram that depicts the role of the subtropical ridge and the mean SST on the meridional phase relationship between YRV and SEC monsoons. First, let us discuss the in-phase relationship in ID1 (see Fig. 5.3a). Consider a wet-phase scenario that the equatorial eastern Pacific SSTA is warm in the winter preceding the East Asian summer monsoon. Through either a Rossby wave response (Wang *et al.* 2000) and/or an anomalous overturning that includes Walker and local Hadley cells, the western Pacific SSTA is cold and an anomalous subtropical ridge forms. As a result, 500 hPa subtropical ridge (solid line) extends westward near the southeast coast of China, and the monsoon (Meiyu or pre-Meiyu) front tends to stay stationary to its north. The South China Sea SSTA is warm due to the downwelling effects of the subtropical ridge (Chu and Chang 1997; Chu *et al.* 1997). This

---

[a] In this interaction warmer eastern Pacific SSTA enhance an anomalous Walker cell with increased surface westerlies, which further enhance the warm SSTA (Li 1997).

enhances the surface moisture flux into both YRV and SEC. This configuration helps to promote excess rainfall in both the SEC and the YRV regions.

In the ID2 period, the stronger mean subtropical ridge due to the higher mean eastern Pacific SSTs compounds the interannual variation. As a result, during the YRV wet phase the 500 hPa subtropical ridge near Southeast China moves northward by about 5° from its position in 1951-1977. This northern position enhances the YRV rainfall but suppresses the SEC rainfall; therefore it corresponds to the SEC dry phase. Therefore, rainfall anomalies in the two regions are *out-of-phase* (Fig. 5.3b).

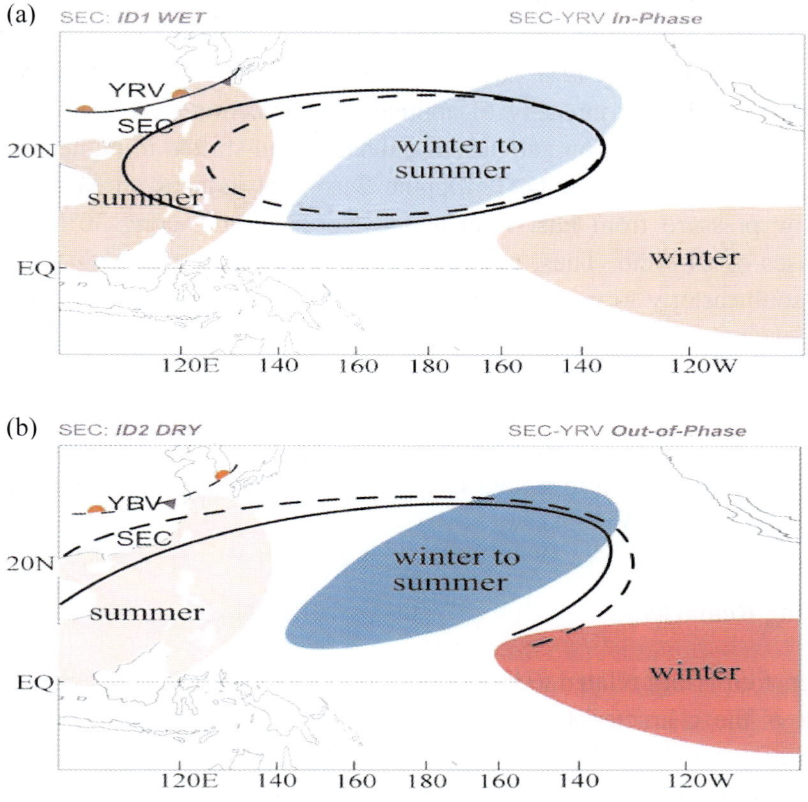

Figure 5.3. Schematic diagram illustrating (a) the *in-phase* relationship between SEC and YRV monsoon rainfall during ID1 and (b) the out-of-phase relationship between SEC and YRV monsoon rainfall during ID2. (adapted from Chang *et al.* 2000b)

5.2. *Rainfall Anomalies over the Northeast Asia*

The rainfall in the East Asia region normally exhibits sandwiched distribution: the anomalous rainfall in Meiyu/Baiu region often shows an opposite polarity to those over the Philippine Sea and over the northern China (Fig. 3.1). According to Chang *et al.* (2000a), above normal May-June rainfall in the Yangtze River Valley follows a warm equatorial eastern Pacific in the preceding winter for both 1951-1977 and 1978-1996 periods. However, using station rainfall data and the NCEP-NCAR reanalysis, Wu and Wang (2001) show that, different from the Yangtze River region and the WNP, the ENSO-monsoon relationship in northern

China and Japan has experienced a significant change since 1978, which delineates the pre- and post shift period. This change is mainly seen in the decaying phases of ENSO while the changes in developing phases of ENSO are less significant. The following description is focused on the decaying phase of ENSO only. For brevity, figures are not shown but interested readers are referred to Wu and Wang (2002). From the pre- to post-shift period, the summer rainfall anomaly in eastern North China during decaying phases of El Niño changed from above to below normal, whereas that in central Japan changed from negative to normal. Consistent with this, a barotropic anticyclonic anomaly over the sea of Japan changed to cyclonic; the associated anomalous winds changed from southerly to northerly over the Yellow Sea-eastern North China and from northeasterly to northwesterly over central Japan. The change in ENSO-related East Asian summer circulation anomaly owes its origin to changes in the location and intensity of anomalous convection over the WNP and India . After 1977, the WNP convection anomaly is enhanced and shifted to higher latitudes due to increased summer mean SST in the Philippine Sea. This induces an eastward shift of an anomalous low pressure from East Asia to the North Pacific along 30°N - 45°N during decaying phases of El Niño. Thus, anomalous winds over eastern North China and Korea switch from southeasterly to northeasterly. Before the late 1970s, an anomalous barotropic anticyclone develops over East Asia and anomalous southerlies prevail over eastern North China during decaying phases of El Niño. This relates to anomalous Indian convection through a zonal wave train pattern along 30° - 50°N (Enomoto and Hoskins 2003). After the late 1970s, the amplitude of the anomalous Indian convection decreases due to a decreased influence of ENSO in relation to ENSO frequency change, which reduces the impact of the Indian convection on the EAM.

6. Concluding Remarks

In this chapter, four issues related to the EAM-ENSO relationship have been addressed. They are: 1) What is the characteristic spatial structure of the anomalous East Asian summer monsoon and associated circulation and SST patterns? 2) How does the El Niño affect the East Asian monsoon? 3) How does the East Asian monsoon impact the El Niño evolution? and 4) How and why does the EAM–ENSO relationship vary on the interdecadal time scale?

As a subtropical monsoon system, the East Asian monsoon exhibits complex spatial and temporal structures. Its annual cycle consists of staged progression of zonally oriented rain belts from the pre-Meiyu in the southeast coast of China starting in mid-May to the Meiyu front in the Yangtze River Valley and southern Japan starting in mid-June. In late June Changma started in Korean peninsula and in early July the rain belt advances to Yellow River Valley and northern Japan (Wang and LinHo 2002). The rainfall variability over the Meiyu region exhibits a strong biennial signal in association with the equatorial eastern Pacific and Indian Ocean SST. A conspicuous feature associated with the Meiyu rainfall-SST relationship is an anomalous anticyclone cell near the southeast coast of China, which can be seen from May to August. The cell is slightly east-west elongated and can be readily identified by westerly anomalies in the East China Sea to its north and easterly anomalies in the South China Sea and Philippine Sea to its south. Although the summer mean rainfall over northern China is much less than that over the southern and central China, it shows a significant positive correlation with the all-Indian rainfall.

As far as the interannual EAM-SST relationship is concerned, El Niño-like equatorial eastern Pacific conditions in northern winter lead to a wet Meiyu/Baiu/Changma but a dry condition in southeast China. However, such an interannual relationship is subject to an interdecadal variation. For the first interdecadal period (ID1, 1951-77), an El Niño-like condition in the eastern Pacific precedes both enhanced Meiyu and southeast Chinese monsoon seasons, but for the second interdecadal period (ID2, 1978-96), the El Niño warming precedes an enhanced Meiyu but a deficient monsoon in southeast China. A pronounced interdecadal change of the rainfall-SST relationship is also found in northern China and northern Japan. In northern China, the correlation between local rainfall and SSTA of the eastern Pacific was significantly positive before 1977 but switched to negative in the latter period. In central Japan and South Korea the significant negative correlation disappeared after the late 1970s. For both the ID1 and ID2 epochs, the equatorial eastern Pacific SSTA is warm in the fall and winter prior to an enhanced Meiyu and cool in the fall and winter immediately following the enhanced Meiyu. However, the change of sign occurs in early spring before the summer rainy season for ID1, whereas in ID2 the change of sign occurs in early fall after a wet YRV season.

A significant teleconnection pattern was found in association with anomalous monsoon rainfall over the WNP. During a strong WNP monsoon, a pronounced wave train pattern is seen in the lower, middle, and upper troposphere, emanates from the WNP, arching around the North Pacific, and extends to North America. This western Pacific-North America teleconnection has an important implication for short-term climate prediction over North America.

An important question with regard to the East Asian monsoon-ENSO relationship is how the eastern-central Pacific warming affects the "upstream" climate in East Asia. In particular, in the summer following an El Niño, the SSTA in the equatorial eastern Pacific nearly vanishes, how could El Niño have a delayed impact on the East Asian climate? A Pacific-East Asian teleconnection hypothesis is put forth to explain this observed phenomenon (Wang *et al.* 2000). The key circulation system that conveys the impact of El Niño to East Asia is an anomalous WNP anticyclone. The anticyclonic anomaly originates during the fall of an El Niño developing year, rapidly develops over the Philippine Sea during the boreal winter when El Niño matures, and persists until the ensuing summer. It has a baroclinic structure and is primarily confined to the lower troposphere. The maintenance of the anomalous anticyclone is attributed to a positive thermodynamic air-sea feedback between the anomalous anticyclone and the underlying warm-pool ocean in the WNP (Fig. 3.4 and text in section 3).

The anomalous easterly to the south of the WNP anticyclone may impact the El Niño evolution through induced oceanic upwelling Kelvin waves. Given that the establishment of the WNP anticyclone is about two months earlier than the peak warming of El Niño, it is conceivable that the WNP anticyclone plays a role in the El Niño turnaround. Physically, a rise of the thermocline in the eastern Pacific due to the arrival of forced upwelling Kelvin waves may initially offset local warming, which would in turn intensify SST gradients and restore equatorial easterlies, thereby inducing a slow thermocline shoaling and SST decrease in the eastern Pacific. Consecutive generation and passage of forced Kelvin waves may stimulate slow eastward migration of the heat content through the "fetch extension" mechanism or the zonal advection process.

Another question with regard to the EAM-ENSO relationship is why an El Niño condition precedes a wet southeast China rainy season in ID1 but a dry season in ID2. It was hypothesized that this results from the interdecadal change of the basic state (Chang *et al.* 2000a, b). A key element that links the eastern Pacific SST and the EAM is the western Pacific subtropical ridge, whose strength and spatial extent vary greatly and are correlated with the equatorial eastern Pacific SSTA. In both the Meiyu and southeast China rainfall composites, the WNP subtropical ridge is, in general, stronger in ID2 than in ID1. This stronger ridge is likely a result of the higher mean SST in eastern Pacific in ID2, which produces stronger subsidence over the WNP. The stronger mean subtropical ridge in ID2 tends to increase Meiyu rainfall, as it prevents the pre-Meiyu/Meiyu fronts from moving southward. However, because it is associated with stronger mean subsidence south of the Meiyu front, it tends to decrease the mean rainfall in southeast China. As a result, rainfall anomalies over the central and southeast China tend to be out-of-phase in ID2 but in-phase in ID1.

To sum up, the relationship between the East Asian monsoon rainfall and eastern Pacific SST anomalies is complex. It displays seasonal dependence (Zhang *et al.* 1999), subseasonal change (Chang *et al.* 2000a, b), ENSO phase dependence, and interdecadal modulation (Chang *et al.* 2000a, b; Wu and Wang 2002). On one hand, the El Niño remotely influences the EA climate, and on the other hand the EAM impacts the El Niño evolution. More work is needed to document and understand the complexity of the EAM-ENSO relationship. It is also interesting to explore the possible role of midlatitude mean circulation changes in modulating this relationship since the EAM is a subtropical monsoon that involves both tropical and mid-latitude systems (Tao and Chen 1987; Zeng *et al.* 1994). Another topic for future study is the interdecadal change in the Eurasian land surface condition and its possible impact on the EAM-ENSO relationship.

Acknowledgements

The discussions with Drs. C.-P. Chang, Randy Wu, Chongyin Li, and Yongsheng Zhang are greatly appreciated. Dr. Yongsheng Zhang helps the drawing of figures.

References

An, S.-I., and B. Wang, 2000: Interdecadal change of the structure of the ENSO mode and its impact on the ENSO frequency. *J. Climate*, **13**, 2044-2055.

Bjerknes, J., 1969: Atmospheric teleconnections from the equatorial Pacific. *Mon. Wea. Rev.*, **97**, 163–

172.

Chan, J. C. L, Y. Wang, and J. Xu, 2000: Dynamic and thermodynamic characteristics associated with the onset of the 1998 South China Sea summer monsoon. *J. Meteor. Soc. Japan*, **78**, 367-380.

Chang, C.-P., and G. T.-J. Chen, 1995: Tropical circulations associated with southwest monsoon onset and westerly surges over the South China Sea. *Mon. Wea. Rev.*, **123**, 3254-3267.

______, and G. T. Chen, 1998: The development of an intense East Asian summer monsoon disturbance with strong vertical coupling. *Mon. Wea. Rev.*, **126**, 2692-2712.

______, P. Harr, and J. Ju, 2001: Possible roles of Atlantic circulations on the weakening Indian monsoon rainfall-ENSO relationship. *J. Climate*, **14**, 2376-2380.

______, and K. M. Lau, 1980: Northeasterly cold surges and near-equatorial disturbances over the winter MONEX area during December. Part II: Large-scale aspects. *Mon. Wea. Rev.*, **108**, 298-312.

Chang, C.-P., and T. Li, 2000: A theory of the tropical tropospheric biennial oscillation. *J. Atmos. Sci.*, **57**, 2209-2224.

______, and______, 2001: Tropical tropospheric biennial oscillation and ENSO. *East Asian and Western Pacific Meteorology and Climate*, C.P. Chang *et al.* , Eds., Book Series on East Asian Meteorology, **Vol. 1**, World Scientific Publishing Company, Singapore, 167-179.

______, Y. Zhang, and T. Li, 2000a: Interannual and interdecadal variations of the East Asian summer monsoon and tropical Pacific SSTs. Part I: Roles of the subtropical ridge. *J. Climate*, **13**, 4310-4325.

______, ______, and ______, 2000b: Interannual and interdecadal variations of the East Asian summer monsoon and tropical Pacific SSTs. Part II: The meridional structure of the monsoon. *J. Climate*, **13**, 4326-4340.

Chao, Y., and S. G. H. Philander, 1993: On the structure of the southern oscillation. *J. Climate*, **6**, 450-469.

Chen, G. T., and C.-P. Chang, 1980: Structure and vorticity budget of early summer monsoon trough (Mei-Yu) over southeastern China and Japan. *Mon. Wea. Rev.*, **108**, 942-953.

Chen, L., M. Dong, and Y. Shao, 1992: The characteristics of interannual variations on the East Asian monsoon. *J. Meteor. Soc. Japan*, **70 (1B)**, 397-421.

Chen, T. -C., S. -P. Weng, N. Yamazaki, and S. Kiehne, 1998: Interannual variation in the tropical cyclone formation over the western North Pacific. *Mon. Wea. Rev.*, **126**, 1080-1090.

Chu, P., and C.-P. Chang, 1997: South China Sea warm pool in boreal spring. *Adv. in Atmos. Sci.*, **14**, 195-206.

______, H. C. Tseng, C.-P. Chang, and J. M. Chen, 1997: South China Sea warm pool detected in spring from the Navy's Master Oceanographic Observational Data Set (MOODS). *J. Geophs. Res.*, **102, (C7)**, 15761-15771.

Ding, Y. H., 1992: Summer monsoon rainfalls in China. *J. Meteor. Soc. Japan*, **70**, 373-396.

______, 1994: *Monsoon over China*, Kluwer Academic Publishers, 420pp.

Eriksen, C. C., M. B. Blumenthal, S. P. Hayes, and P. Ripa, 1983: Wind generated equatorial Kelvin waves observed across the Pacific Ocean. *J. Phys. Oceanogr.*, **13**, 1622-1640.

Enomoto T., B. J. Hoskins, and Y. Matsuda, 2003: The formation mechanism of the Bonin high in August. *Quart. J. Roy. Meteor. Soc.*, **129**, 587, 157-178.

Fu, C., and X. Teng, 1993: Relationship between summer climate in China and the El Niño/Southern Oscillation phenomenon. *Institute of Atmospheric Physics, Frontiers in atmospheric sciences.*, New York, Allerton Press, Inc. 166-178.

Goswami, B. N., 1998: Interannual variations of Indian summer monsoon in a GCM: External conditions versus internal feedbacks. *J. Climate*, **11**, 501-522.

Gu D, and S. G. H. Philander, 1995: Secular changes of annual and interannual variability in the tropics during the past century. *J. Climate*, **8**, 864-876.

Hirst, A. C., 1986: Unstable and damped equatorial modes in simple coupled ocean-atmosphere models. *J. Atmos. Sci*., **43**, 606-630.

Horel, J. D., and J. M. Wallace, 1981: Planetary-scale atmospheric phenomena associated with the Southern Oscillation. *Mon. Wea. Rev*., **109**, 813-829.

Hsu, H. H., and C. T. Chen, 1997: Evolution of large-scale circulation and heating during the first transition of the Asian summer monsoon. *J. Climate*., **12**, 793-810.

Huang, R.-H., and L. Lu, 1989: Numerical simulation of the relationship between the anomaly of subtropical high in East Asia and the convective activity in the tropical western Pacific. *Adv. in Atmos. Sci*., **6**, 202-204.

______, and F. Sun, 1992. Impacts of the tropical western Pacific on the East Asian summer monsoon. *J. Meteor. Soc. Japan*, **70**, 243-256.

______, and Y. Wu, 1989: The influence of ENSO on the summer climate change in China and its mechanism. *Adv. Atmos. Sci*., **6**, 21-32.

______, L. T. Zhou, and W. Chen, 2003: The progress of recent studies on the variabilities of the East Asian monsoon and their causes. *Adv. in Atmos. Sci*., 20, 55-69.

Hwu, J.-W., and J.-T. Wang, 1994: Springtime East Asia large scale characteristics post ENSO. Part II: The circulation pattern of different levels. *Atmospheric Sciences*, Taiwan, Republic of China, **22**, 361-385.

Ji, L., S. Sun, and K. Arpe, 1997: Model study on the interannual variability of Asian winter monsoon and its influence. *Adv. in Atmos. Sci*., **14**, 1–22.

Ji, M., A. Leetmaa, and J. Derber, 1995: An ocean analysis system for seasonal to interannual climate studies. *Mon. Wea. Rev*., **123**, 460-481.

Johnson, R. H., Z. Wang, and J. F. Bresch, 1993: Heat and moisture budgets over China during the early summer monsoon. *J. Meteor. Soc. Japan*, **71**, 137-152.

Kang, I.-S., C.-H. Ho, Y.-K. Lim, and K. M. Lau, 1999: Principal modes of climatological seasonal and intraseasonal variations of the Asian summer monsoon. *Mon. Wea. Rev*., **127**, 322-340.

______, and Y. Jeong, 1996: Association of interannual variations of temperature and precipitation in Seoul with principal modes of Pacific SST. *J. of Korean Meteor. Soc*., **32**, 339-345.

Kawamura, R., 1998: A possible mechanism of the Asian summer monsoon-ENSO coupling. *J. Meteor. Soc. Japan*, **76**, 1009-1027.

Kessler, W. S., and M. J. McPhaden, 1995: Oceanic equatorial waves and the 1991-1993 El Niño. *J. Climate*, **8**, 1757-1774.

Kim, K. M., and K. M. Lau (2001). Dynamics of monsoon-induced biennial variability in ENSO. *Geophys. Res. Lett*., **28**(2), 315-318.

Kirtman, B. P., and J. Shukla, 2000: Influence of the Indian summer monsoon on ENSO. *Quart. J. Roy. Meteor. Soc*., **126**, 213-239.

Knox, R. A., and D. Halpern, 1982: Long range Kelvin wave propagation of transport variations in Pacific Ocean equatorial currents. *J. Mar. Res*., **40** (suppl.), 329-339.

Lau, K. M., 1992: East Asian summer monsoon variability and climate teleconnection. *J. Meteor. Soc. Japan*, **70**, 211-242.

______, C. P. Chang, and P. H. Chan, 1982: Short-term planetary scale interactions over the tropics and midlatitudes, Part I: Contrast between active and inactive periods. *Mon. Wea. Rev.*, **110**, 933-946.

Short-term planetary scale interactions over the tropics and midlatitudes, Part II: Winter-MONEX period. *Mon. Wea. Rev.*, **111**, 1372-1388.

______, K. M. Kim, and S. Yang, 2000: Dynamical and boundary forcing characteristics of regional components of the Asian summer monsoon. *J. Climate*, **13**, 2461-2482.

______, and M. T. Li , 1984: The monsoon of East Asia and its global associations - a survey, *Bull. Amer. Meteor. Soc.*, **65**, 114-125.

______, and P. J. Sheu, 1988: Annual cycle, quasi-biennial oscillation, and Southern Oscillation in global precipitation. *J. Geophys. Res.*, **93**, 10975-10988.

______, and H. Weng, 1999: Interannual, decadal-to-interdecadal and global warming signals in sea surface temperature during 1955-1997. *J. Climate*, **12**, 1257-1267.

______, and______, 2002: Recurrent teleconnnection patterns linking summertime precipitation variability over East Asia and North America. *J. Meteor. Soc. Japan*, **80**, 1309-1324.

______, and S. Yang, 1996: The Asian monsoon and predictability of the tropical ocean-atmosphere system. *Quart. J. Roy. Meteor. Soc.*, **122**, 945-957.

______, G. J., Yang, and S. Shen 1988: Seasonal and intraseasonal climatology of summer monsoon rainfall over East Asia. *Mon. Wea. Rev.*, **116**, 18-37.

______, and H. T. Wu, 2001: Intrinsic modes of coupled rainfall/SST variability for the Asian summer monsoon: a re-assessment of monsoon-ENSO relationship. *J. Climate*, **14**, 2880-2895.

______, and______, 2001: Principal modes of rainfall-SST variability of the Asian summer monsoon: A reassessment of the Monsoon-ENSO relationship. *J. Climate*, **14**, 2880-2895.

Li, C., 1990: Interaction between anomalous winter monsoon in East Asia and El Niño Events. *Adv. in Atmos. Sci.*, **7**, 36-46.

______, 1996: A further study on interaction between anomalous winter monsoon in East Asia and El Niño. *Acta Meteorologica Sinica*, Beijing, China, **10**, 309-320.

______, and J. Hu, 1987: Study on interaction between the East Asia atmospheric circulation and El Niño. *Scientia Atmospheric Sinica*, Beijing, **11**, 359-364.

______, S. Shuqing, and M. Mingquan, 2001: Origin of the TBO-interaction between anomalous East-Asian winter monsoon and ENSO cycle. *Adv. in Atmos. Sci.*, **18**, 554-566.

Li, T., 1997: Phase transition of the El Niño-Southern Oscillation: A stationary SST mode. *J. Atmos. Sci.*, **54**, 2872-2887.

______, B. Wang, and C.-P. Chang, 2001a: Theories on the tropospheric biennial oscillation: a review. *Dynamics of Atmospheric and Oceanic Circulations and Climate*, M. Wang *et al.* , Eds., Chinese Academy of Sciences, China Meteorological Press, Beijing, China, 252-276.

______, Y. Zhang, C. P. Chang, and B. Wang, 2001b: On the relationship between Indian Ocean SST and Asian summer monsoon. *Geophy. Res. Lett.*, **28**, 2843-2846.

______, and Y. Zhang, 2002: Processes that determine the quasi-biennial and lower-frequency variability of the South Asian monsoon. *J. Meteor. Soc. Japan*, **80**, 1149-1163.

______, ______, E. Lu, and D. Wang, 2002: Relative role of dynamic and thermodynamic processes in the development of the Indian Ocean dipole. *Geophy. Res. Lett.*, **29**, 2110-2113.

Liang, X.-Z., and W. C. Wang, 1998: Associations between China monsoon rainfall and tropospheric jets. *Quart. J. Royal Meteor. Soc.*, **124**, 2597-2623.

Liu, Y., and Ding, Y. H., 1992; Influence of El Niño on weather and Climate in China. *Acta meteorological Sinica*, **6(1)**, 117-131.

Lukas, R., S. P. Hayes, and K. Wyrtki, 1984: Equatorial sea level response during the 1981-1983 El Niño. *J. Geophys. Res.*, **89**, 10425-10430.

McBride, J. L., and N. Nicholls, 1983: Seasonal relationships between Australian rainfall and the Southern Oscillation. *Mon. Wea. Rev.*, **111**, 1998-2004.

McPhaden, M. J., and X. Yu, 1999: Genesis and evolution of the 1997-98 El Niño. *Science*, **283**, 950-954.

Meehl, G. A., 1987: The annual cycle and interannual variability in the tropical Pacific and Indian Ocean region. *Mon. Wea. Rev.*, **115**, 27-50.

______, 1994: Coupled land-ocean-atmosphere processes and South Asian monsoon variability. *Science*, **266**, 263-267.

______, and W. M. Washington, 1996: El Niño-like climate change in a model with increased atmospheric CO2 concentrations. *Nature*, **382**, 56-60.

Nakazawa, T., 1992: Seasonal phase lock of intraseasonal variation during the Asian summer monsoon. *J. Meteor. Soc. Japan*, **70**, 597-611.

Nitta, T., 1987: Convective activities in the tropical western Pacific and their impacts on the Northern Hemisphere summer circulation. *J. Meteor. Soc. Japan*, **65**, 373-390.

______, and Z. Z. Hu, 1996: Summer climate variability in China and its association with 500 hPa height and tropical convection. *J. Meteor. Soc. Japan*, **74**, 425-445.

______, and S. Yamada, 1989: Recent warming of tropical sea surface temperature and atmospheric flow patterns. *J. Meteor. Soc. Japan*, **67**, 375-383.

Philander, S. G. H., T. Yamagata, and R. C. Pacanowski, 1984: Unstable air-sea interactions in the tropics. *J. Atmos. Sci.*, **41**, 604-613.

Picaut, J., M. Loualalen, C. Menkes, T. Delcroix, and M. J. McPhaden, 1996: Mechanism of the zonal displacements of the Pacific warm pool: Implications for ENSO. *Science*, **274**, 1486-1489.

Rasmusson, E. M., and T. H. Carpenter, 1982: Variations in the tropical sea surface temperature and surface wind fields associated with the Southern Oscillation/El Niño. *Mon. Wea. Rev.*, **110**, 354-384.

Samel, A. N., W. C. Wang, and X. Liang, 1999: The monsoon rainband over China and relationship with the Eurasian circulation. *J. Climate*, **12**, 115 -131.

Shen, S., and K. M. Lau, 1995: Biennial oscillation associated with the East Asian monsoon and tropical sea surface temperatures. *J. Meteor. Soc. Japan*, **73**, 105-124.

Simmons, A. J., J. M. Wallace, and G. W. Branstator, 1983: Barotropic wave propagation and instability, and atmospheric teleconnection patterns. *J. Atmos. Sci.*, **40**, 1363-1392.

Smith, T. M., R. W. Reynolds, R. E. Livezey, and D. C. Stokes, 1996: Reconstruction of historical sea surface temperatures using empirical orthogonal functions. *J. Climate*, **9**, 1403–1420.

Soman, M. K., and J. Slingo, 1997: Sensitivity of the Asian summer monsoon to aspects of sea-surface-temperature anomalies in the tropical Pacific Ocean. *Quart. J. Roy. Meteor. Soc.*, **123**, 309-336.

Tanaka, M., 1997: Interannual and interdecadal variations of the western North Pacific monsoon and Baiu rainfall and their relationship to the ENSO cycle. *J. Meteor. Soc. Japan*, **75**, 1109-1123.

Tao, S. Y., and L. X. Chen, 1987: A review of recent research of the East Asian summer monsoon in China. *Monsoon Meteorology*, C.-P. Chang and T. N. Krishnamurti, Eds., Oxford University Press, 60-92.

______, and Q. Zhang, 1998: Response of the East Asian summer monsoon to ENSO events (in Chinese). *Sci. Atmos. Sinica*, **22**, 399-407.

Tian, S. F., and T. Yasunari, 1992: Time and space structure of interannual variations in summer rainfall over China. *J. Meteor. Soc. Japan*, **70**, 585-596.

Ting, M., and H. Wang, 1997: Summertime U.S. precipitation variability and its relation to Pacific sea surface temperature. *J. Climate*, **10**, 1853-1873.

Tomita, T., and T. Yasunari, 1996: Role of the northeast winter monsoon on the biennial oscillation of the ENSO/monsoon system. *J. Meteor. Soc. Japan*, **74**, 399–413.

______, T. Yoshikane, and T. Yasunari, 2003: Biennial and low-frequency variability of early summer climate in East Asia. *J. Climate*. (Submitted)

Trenberth, K. E., 1990: Recent observed interdecadal climate changes in the Northern Hemisphere. *Bull. Amer. Meteor. Soc.*, **71**, 988-993.

______, and T. J. Hoar, 1996: The 1990-1995 El Niño-Southern Oscillation event: Longest on record. *Geophys. Res. Lett.*, **23**, 57-60.

______, and J. W. Hurrell, 1994: Decadal atmosphere-ocean variations in the Pacific. *Climate Dyn.*, **9**, 303-319.

Torrence, C., and P. J. Webster, 1998. The annual cycle of persistence in the El Niño/Southern Oscillation. *Quart. J. Roy. Meteor. Soc.*, **124**, 1985-2004.

Ueda, H., T. Yasunari, and R. Kawamura, 1995: Abrupt seasonal change of large-scale convection activity over the western Pacific in northern summer. *J. Meteor. Soc. Japan*, **73**, 795-809.

Wallace, J. M., and D. S. Gutzler, 1981: Teleconnections in the geo-potential field during the Northern Hemisphere winter. *Mon. Wea. Rev.*, **109**, 784–812.

Wang, B., 1995: Interdecadal changes in El Niño onset in the last four decades. *J. Climate*, **8**, 267-285.

______, and S.-I. An, 2001: Why the properties of El Niño change in the late 1970s? *Geophys. Res. Lett.*, **28**, 3709-3712.

______, and J. C. L. Chan, 2002: How strong ENSO events affect tropical storm activity over the Western North Pacific. *J. Climate*, **15**, 1643-1658.

______, S. C. Clemons, and P. Liu, 2003: Contrasting the Indian and East Asian monsoons: implications on geologic timescales. Marine Geology, 201 (1-3), 5-21.

______, and Z. Fan, 1999: Choice of South Asia summer monsoon indices. *Bull. Amer. Meteo. Soc.*, **80**, 629-638.

______, I.-S. Kang, and J.-Y. Lee, 2003a: Ensemble simulation of Asian-Australian monsoon variability by 11 AGCMs. *J. Climate.* (In press.)

______, and LinHo, 2002: Rainy seasons of the Asian-Pacific monsoon. *J. Climate*, **15**, 386-398.

______, and Y. Wang, 1996: Temporal structure of the Southern Oscillation as revealed by waveform and wavelet analysis. *J. Climate*, **9**, 1586-1598.

______, and R. Wu, 1997: Peculiar Temporal Structure of the South China Sea Summer Monsoon. *Adv. in Atmos. Sci.*, **14**, 177-194.

______, ______, and X. Fu, 2000: *J. Climate*, **13**, 1517-1536.

______, ______, and K.-M. Lau, 2001a: Interannual variability of the Asian summer monsoon: Contrasts between the Indian and the western North Pacific-East Asian monsoons. *J. Climate*, **14**, 4073-4090.

______, ______, R. Lukas, and S. I. An, 2001b: A possible mechanism for ENSO turnabout. *Dynamics of Atmospheric General circulation and Climate*, Ed., IAP/Academia Sinica, China Meteor. Press, 552-578.

______, ______, and T. Li, 2003b: Atmosphere-warm ocean interaction and its impact on Asian-

Australian monsoon variation. *J. Climate*, **16**, 1195-1211.

______, ______, and R. Lukas, 1999: Roles of the western North Pacific wind variation in thermocline adjustment and ENSO phase transition. *J. Meteor. Soc. Japan*, **77**, 1-16.

______, and X. Xu, 1997: northern Hemisphere summer monsoon singularities and climatological intraseasonal oscillation. *J. Climate.*, **10**, 1071-1084.

______, and Q. Zhang, 2002: Pacific-East Asian teleconnection, part II: How the Philippine Sea anticyclone established during development of El Niño. *J. Climate*, **15**, 3252-3265.

Wang, J. T., and J. W. Hwu, 1994: Springtime East Asia large scale characteristics post ENSO. Part I: The spatial distribution of SST and OLR. *Atmos. Sci.*, **22**, 339-360.

Wang, W. C., and K. Li, 1990: Precipitation fluctuation over semiarid region in northern China and the relationship with El Niño/southern oscillation. *J. Climate*, **3**, 769–783.

Webster, P. J., and S. Yang, 1992: Monsoon and ENSO: Selectively interactive systems. *Quart. J. Roy. Meteor. Soc.*, **118**, 877-926.

______, P. J., V. O. Magana, T. N. Palmer, J. Shukla, R. A. Tomas, M. Yanai, and T. Yasunari, 1998: Monsoons: Processes, predictability, and the prospects for prediction. *J. Geophys. Res.*, **103-C7**, 14451-14510.

Weisberg, R. H., and C. Wang, 1997: Slow variability in the equatorial west-central Pacific in relation to ENSO. *J. Climate*, **10**, 1998-2017.

Weng, H., K-M. Lau, and Y. Xue, 1999: Multi-scale summer rainfall variability over China and its long-term link to global sea surface temperature variability. *J. Meteor. Soc. Japan*, **77**, 845-857.

Wu, R., and B. Wang, 2001: Multi-stage onset of summer monsoon over western North Pacific. *Climate Dyn.*, **17**, 277-289.

______, and ______, 2002: A contrast of the East Asian summer monsoon and ENSO relationship between 1962-1977 and 1978-1993. *J. Climate*, **15**, 3266-3279.

Xie, A., Y.-S. Chung, X. Liu, and Q. Ye, 1998: The interannual variations of the summer monsoon onset over the South China Sea. *Theor. and Appl. Climatol.*, **59**(4), 201-213.

Yang, S., K. M. Lau, and K. M. Kim, 2002: Variations of the East Asian jet stream and Asian-Pacific-American winter climate anomalies. *J. Climate*, **15**, 306-325.

Yasunari, T., 1990: Impact of Indian monsoon on the coupled atmosphere/ocean system in the tropical Pacific. *Meteor. Atmos. Phys.*, **44**, 29-41.

Yatagai A., and T. Yasunari 1995: Interannual variations of summer precipitation in the arid/semi-arid regions in China and Mongolia: their regionality and relationship to the Asian monsoon. *J. Meteor. Soc. Japan*, **73**, 909-923.

Ye, D.-Z., and R.-H. Huang, 1996: Study on the regularity and formation reason of drought and flood in the Yangtze and Huaihe River regions (in Chinese). *Shandon Sci. and Tech. Press*, 387 pp.

Zeng, Q.-C., B.-L. Zhang, Y.-L. Liang, and S.-X. Zhao, 1994: The Asian summer monsoon-A case study. *Proc. Indian Natn. Sci. Acad.*, **60 A**, 81-96.

Zhang, Y., T. Li, B. Wang, and G. Wu, 2002: Onset of Asian summer monsoon over Indo-China and its interannual variability. *J. Climate*, **15**, 3206-3221.

Zhang, R., and A. Sumi, 2002: Moisture circulation over East Asia during El Niño episode in Northern winter, spring and autumn. *J. Meteor. Soc. Japan*, **80**(2), 213-227.

______, ______, and M. Kimoto, 1996: Impact of El Niño on the East Asian monsoon: a diagnostic study of the '86/87 and '91/92 events. *J. Meteor. Soc. Japan*, **74**, 49-62.

______, ______, and ______, 1999: A diagnostic study of the impact of El Niño on the precipitation in China. *Adv. in Atmos. Sci.*, **16**, 229-241.

6. CLIMATE VARIATIONS OF THE SUMMER MONSOON OVER CHINA

RONGHUI HUANG AND GANG. HUANG
Institute of Atmospheric Physics
Chinese Academy of Sciences, P. O. Box 2718
Beijing 100080, China
E-mail: hrh@lasg.iap.ac.cn

ZHIGANG WEI
Cold and Arid Regions Environment and Engineering Institute
Chinese Academy of Sciences
Lanzhou 730000, China
E-mail: wzg@ns.lzb.ac.cn

The progress in recent studies on climate variations of the summer monsoon in China is reviewed in this paper. Especially, the characteristics of the intraseasonal, interannual and interdecadal variations of the summer monsoon in China and its surrounding regions, which can cause severe drought and flooding disasters there, are discussed from the analyses of observed data. Moreover, some causes and physical processes of these variations are also sought from the components of the East Asian climate system. Besides, the scientific problems which need to be studied in the near future are suggested.

1. Introduction

China is in East Asia, which is one of the strong monsoon regions in the world. There are many characteristic weather systems in different seasons, such as the Meiyu in China, the Baiu in Japan, the Changma in Korea, tropical cyclones in summer, and persisting northwesterly and northeasterly winds and cold surges in winter. Climate variations including interdecadal, interannual, and intraseasonal variabilities of the summer monsoon over China and its surrounding regions have been found to be significant. And due to these variations, serious climatic disasters such as extensive droughts and floods frequently occur there. Since the 1980s, these climatic disasters have become more severe and caused great damage to agricultural and industrial production and to the daily life of people in China. In each year of the 1980s and the 1990s, the climatic and synoptic disasters caused grain loss of about 20 billion kg and economic loss of about 200 billion RMB (Chinese yuan), approximately 3% - 6% of the GNP of China. The extremely severe flood that occurred in the Yangtze River valley in the summer of 1998 caused an economic loss of over 260 billion RMB (e.g., Huang *et al.* 1998; Huang and Zhou 2002). Thus, it is an important issue for Chinese meteorologists to study the regularity and causes of climate variations of the summer monsoon over China, so that seasonal prediction of climate variations in China can be improved and the predictability of interannual variations of climate in China can be understood further.

Since the East Asian monsoon has a great impact on climate in China and its surrounding regions, more than sixty years ago, Zhu (1934) first proposed the possible relationship between summer rainfall in China and the East Asian summer monsoon. Later on, Tu and Huang (1944) investigated the advance and retreat of the Asian summer monsoon in East Asia. These studies opened the way to search for the regularity of the East Asian monsoon variability and its impact on climate variations in China. Following Tu and Huang's (1944) investigation, many Chinese meteorologists (e. g., Tao and Chen 1957; Yeh *et al.* 1959; Gao and Xu 1962) studied the characteristics of monsoon climate in China and the influence of the onset and retreat of the East Asian summer monsoon on climate in China. Tao and Chen (1987) made a systematic review on these studies. After their review, studies on the regularity and causes of different time scale climate variations of the summer monsoon over China have been greatly made using observed data for a longer period. For example, many scholars (e. g., Huang and Li 1987; Chen *et al.* 1991; Ding 1992; Huang *et al.* 1993; Huang and Sun 1994b) investigated the intraseasonal variations of the summer monsoon over China. Recently, many scholars also studied the regularity and causes of interannual variations of the summer monsoon over China (e.g., Miao and Lau 1990; Lau and Shen 1992; Huang and Sun 1992; Huang and Sun 1994a; Chang *et al.* 2000; Huang *et al.* 2003). In addition, the decadal and interdecadal variations of the summer monsoon over China and its surrounding regions have attracted many researchers' interest (e.g., Wang 1993; Nitta 1993; Huang *et al.* 1999; Chang *et al.* 2000; Huang 2001).

Since the late 1970s, many investigations emphasized the thermal effect of the Tibetan Plateau on the interannual variations of the summer monsoon over China and its surrounding regions (e.g., Ye and Gao 1979; Nitta 1983; Luo and Yanai 1984; Huang 1984, 1985). The thermal effect of the West Pacific warm pool on the interannual variations of the summer monsoon over East Asia have also been investigated by many scholars (e.g., Nitta 1987; Huang and Li 1987; Kurihara 1989; Huang and Sun 1992).

Since observed data for relatively short periods were used in previous studies of climate variations of the summer monsoon over China and its surrounding regions, it is necessary to re-examine the summer monsoon climate variabilities using longer datasets. Thus, the observed datasets of monthly precipitation and surface air temperatures at 160 observational stations of China from 1951 to 2000 and daily precipitation at 598 stations of China from 1961 to 2000, daily snow cover at 72 stations in the Tibetan Plateau, provided by the National Climate Center (NCC) of the China Meteorological Administration (CMA), and the NCEP/NCAR reanalysis data of height, moisture, and wind fields, the observed high cloud amount published by the Monthly Report on Climate System, JMA, and the dataset of GISST, provided by the Hadley Center, UK are used to analyze the intraseasonal, interannual, and interdecadal variabilities of the summer monsoon over China in this chapter. Moreover, recent studies of climate variations of the summer monsoon over China and the relationship between these variabilities and the East Asian climate system including the East Asian monsoon, the western Pacific subtropical high and the disturbances in the middle latitudes in the atmosphere, the thermal effect of the West Pacific warm pool, the ENSO cycle in the tropical Pacific, the land surface processes of the Eurasian continent, and the thermal and dynamical effects of the Tibetan Plateau are also reviewed in this chapter.

2. Intraseasonal Variability of the Summer Monsoon over China

2.1. *The Summer Monsoon Climate over China*

2.1.1. Precipitation

In order to study climate variations of the summer monsoon over China, the climatological mean monthly precipitation and surface air temperature over China averaged for 30 years from 1961 to 1990 are analyzed by using observed monthly precipitation and surface air temperature at 160 stations of China, whose locations are shown in Figure 2.1. Figures 2.2a-f are the distributions of climatological monthly precipitation in April, May, June, July, August, and September, respectively. These figures feature the zonal characteristics in rainfall distributions, which are known as the rainband in China, Japan and Korea. The rainband exhibits a quasi-stationary state during a period and moves from South China to North China from April to August. In April and May, large precipitations of about 250 mm are mainly concentrated in South China and to the south of the Yangtze River. This period is called the

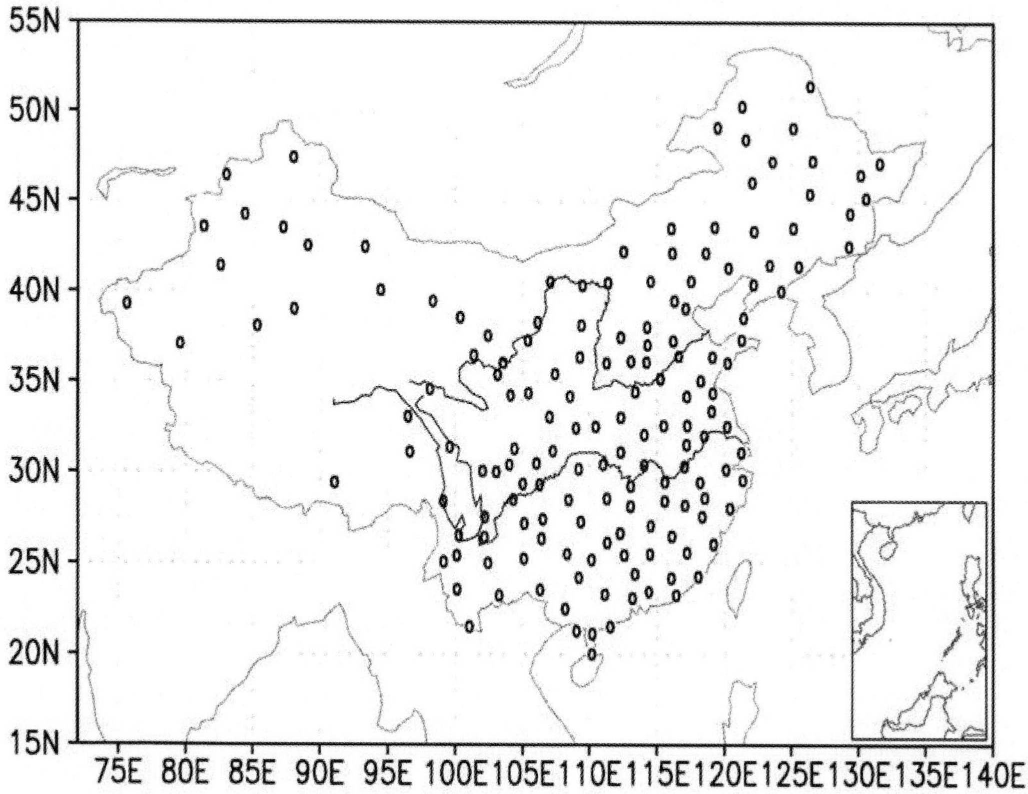

Figure 2.1. Distribution of 160 meteorological observation stations in China.

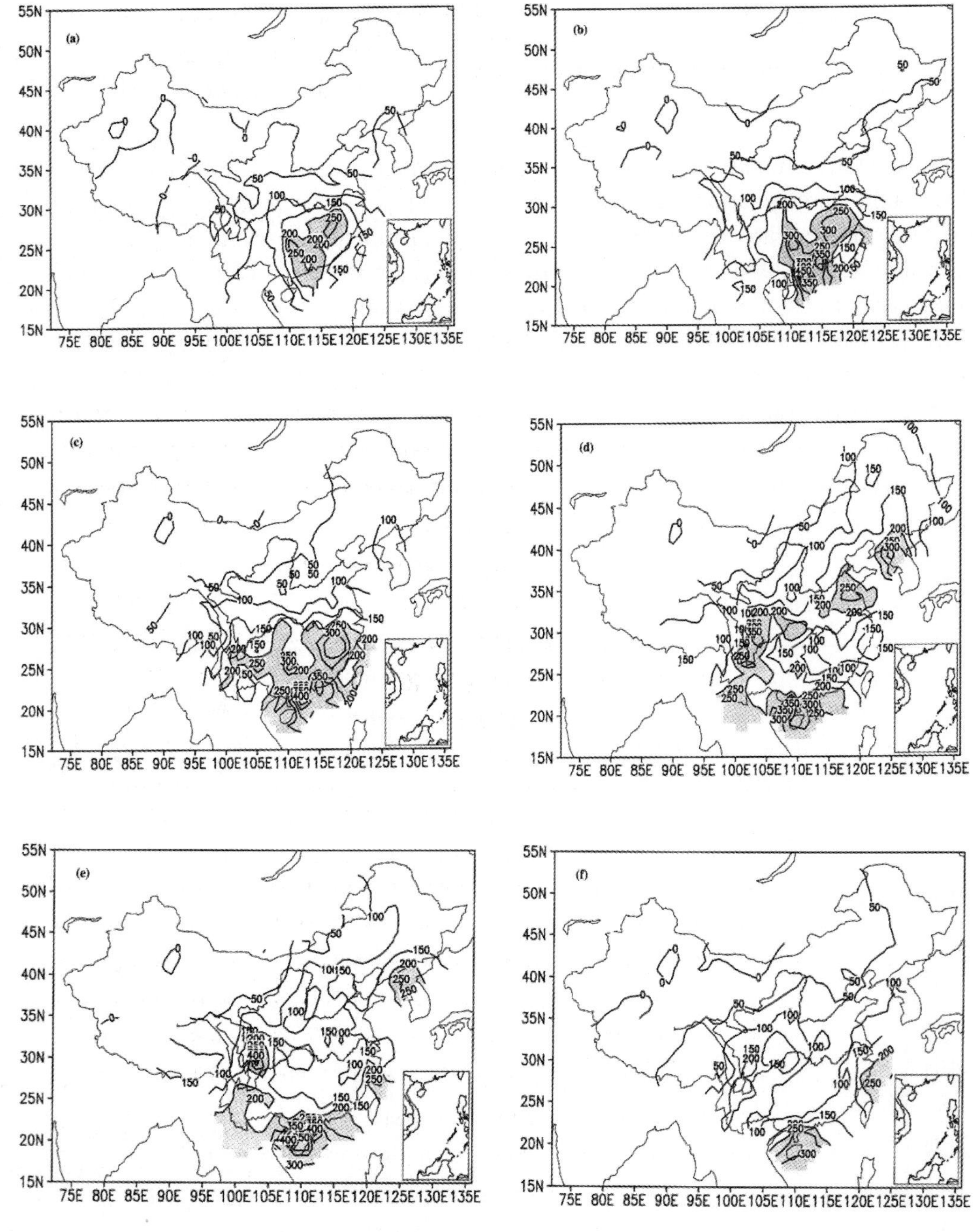

Figure 2.2. Distributions of climatological monthly precipitation averaged for 30 years from 1961 to 1990. Units: mm. (a) April; (b) May; (c) June; (d) July; (e) August; (f) September. The areas of monthly precipitation amount over 200 mm are shaded, and contour interval is 50 mm in these figures.

pre-rainy season of South China, but the precipitation is small in North China and Northeast China in the period. From June, precipitation begins to increase in the middle and lower reaches of the Yangtze River (i.e., the area denoted by "3" in Fig. 3.1) and the Yangtze River and Huaihe River valley (see Fig. 3.1, i.e., the Jianghuai valley in Chinese. This valley is the area between the Yangtze and Huaihe Rivers and the area denoted by "2" in Fig. 3.1), and the monthly precipitation can reach about 250 mm in these regions in June and July, respectively. This period is called the Meiyu season in China. From July to August, precipitation begins to increase in North China and Northeast China, and large amounts of precipitation totally about 150-200 mm are mainly concentrated in the upper reaches of the Yangtze River, North China and Northeast China. From September, the areas of large precipitation retreat southward and are located in the upper and lower reaches of the Yangtze River and South China.

From these distributions of climatological monthly precipitation, it is clearly seen that the belt of large precipitation moves northward from April to August, which is schematically depicted in Fig. 2.3, but the rainband moves southward from September. This may be closely associated with the advance and retreat of the East Asian summer monsoon (EASM) (e.g., Tu and Huang 1944).

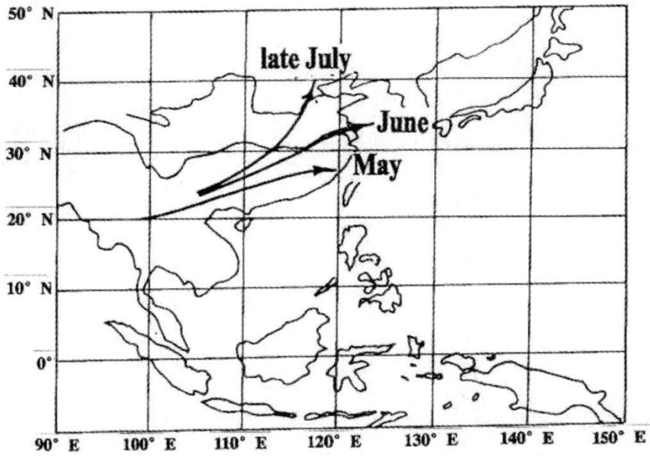

Figure 2.3. Schematic diagram of the locations of the summer monsoon rainband over East Asia.

2.1.2. Surface Air Temperature

Figures 2.4a-f are the distributions of climatological monthly mean surface air temperature in April, May, June, July, August and September, respectively. These figures have the same characteristics as the distributions of monthly precipitation shown in Figs. 2.2a-f. In April and May, the area of higher surface air temperature is mainly over South China and to the south of the Yangtze River. This area abruptly extends northward to North China and Northeast China from June and is maintained in the eastern part of China in July and August. However, the area of higher surface air temperature retreats southward to the Yangtze River valley and South China from September. This may be associated with not only the solar radiation, but also the retreat of the EASM.

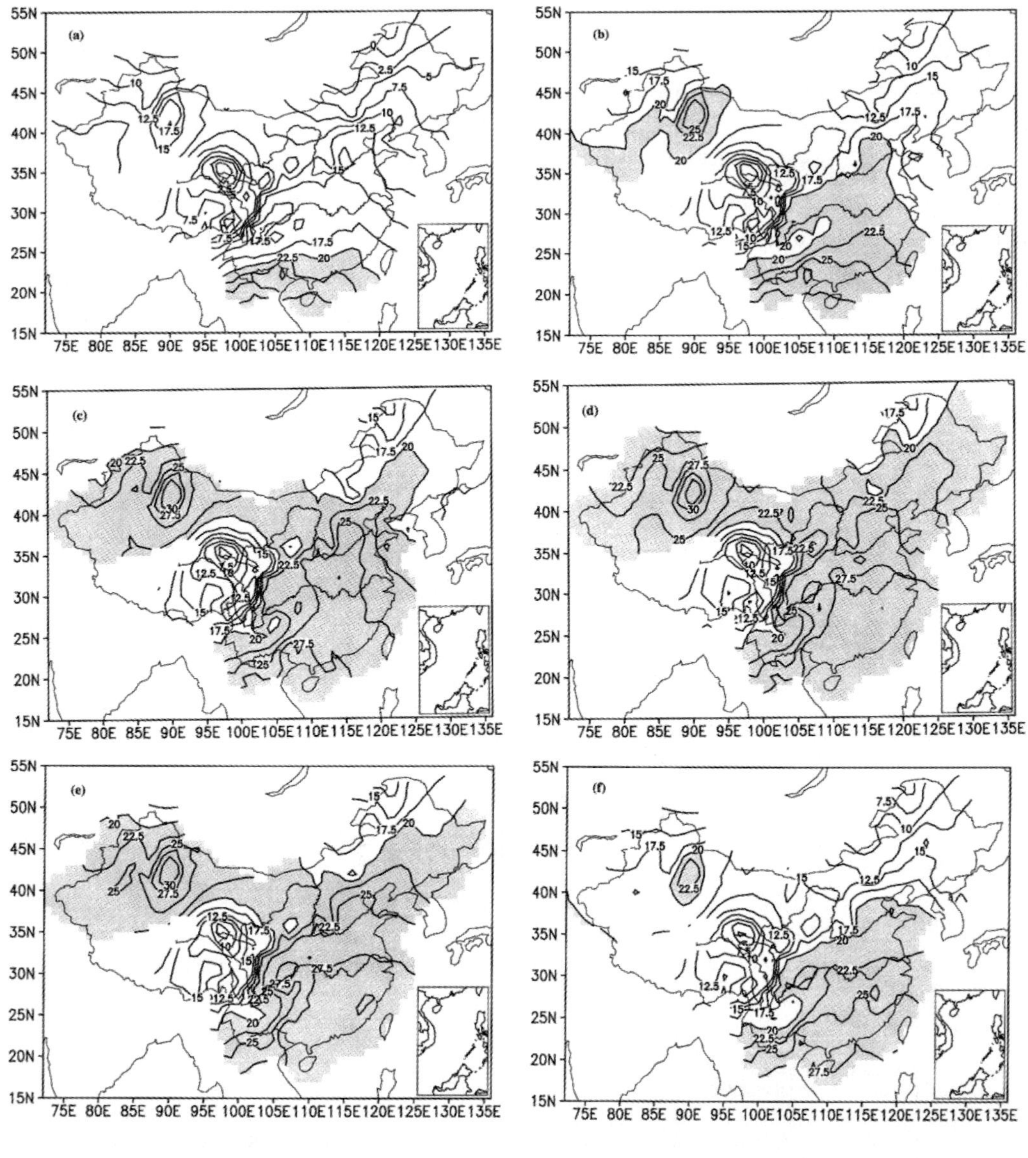

Figure 2.4. Distributions of climatological monthly mean surface air temperature for 30 years from 1961 to 1990. Units: °C. (a) April; (b) May; (c) June; (d) July; (e) August; (f) September. The areas of monthly mean surface air temperature over 20°C are shaded, and contour interval is 2.5°C in these figures.

2.2. *Intraseasonal Variability of the Summer Monsoon Rainband over East Asia*

Perhaps the variation of the monsoon rainband over East Asia from May to August is the most important intraseasonal variability of the EASM. In order to analyze well the intraseasonal variations of the monsoon rainband over China, a detailed dataset of precipitation is needed. Therefore, the observed dataset of daily precipitation at 598 stations in China, provided by NCC/CMA, is used to analyze the intraseasonal variability of the summer monsoon rainband over East Asia. However, considering the spatial distribution and the time length of the establishment of these observational stations, only the observed data at 255 stations are selected from the dataset of daily precipitation at 598 stations to be used in this study. Figure 2.5 is the latitude-time cross section of 5-day precipitation along 115°E (averaged between 110°E - 120°E) averaged for 30 years from 1961 to 1990 by using the observed daily precipitations at 255 observational stations of China. The intraseasonal variability of the summer monsoon rainband can be clearly seen from Fig. 2.5. As shown in the figure, the monsoon rainband is located over the area to the south of the Yangtze River during the period from May to early June, and then it moves abruptly northward and is located over the Yangtze River and Huaihe River valleys of China in mid-June. This is the

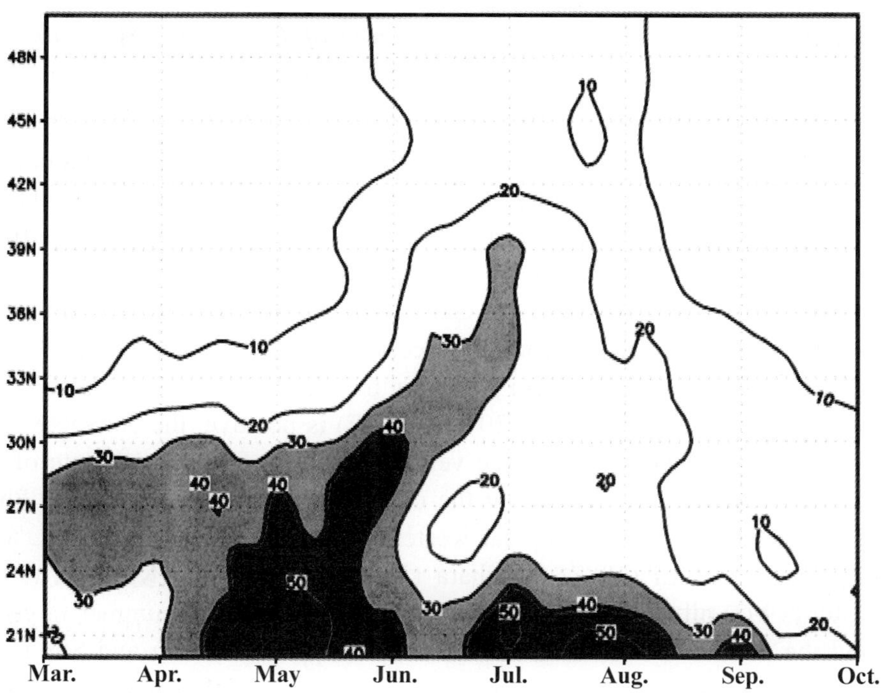

Figure 2.5. Latitude-time cross section of 5-day precipitation along 115°E (averaged between 110°E - 120°E) averaged for 30 years from 1961 to 1990. Units: mm. The precipitation amounts over 30 mm are shaded, and contour interval is 10 mm in the figure.

beginning of the Meiyu season in China (i.e., the Changma season in South Korea and the Baiu season in Japan). Moreover, the rainband again moves abruptly northward to North China in mid-July. Therefore, the northward movement of the maximum rainfall area shown in Fig. 2.5 is in agreement with the intraseasonal variations of the summer monsoon rainband shown schematically in Fig. 2.3.

The abrupt movement of the rainband from the south to the north is closely associated with the abrupt northward shift of the western Pacific subtropical high (e. g., Huang and Sun 1992; Ding 1992). Yeh *et al.* (1959) first discovered that the abrupt movement of the summer rainband is closely associated with the abrupt change of the large scale circulation over East Asia during early and mid-June. This abrupt change of planetary-scale circulation will bring the onset of the EASM in the Yangtze River and Huaihe River valley. Later on, Krishnamurti and Ramanatahan (1982), and McBride (1987) also pointed out the abrupt change of the Indian and Australian summer monsoon circulations.

The transition from the winter to the summer monsoon circulation and the northward movement of the monsoon rainband over East Asia are abrupt in the climatological mean sense (see Yeh *et al.* 1959). However, it is different in different years and is dependent on the thermal state of the tropical western Pacific, i.e., dependent on the SST anomaly in the surface and subsurface of the tropical western Pacific. Thus, this is discussed in the following two cases.

2.3. *Relationship between the Intraseasonal Variations of the Summer Monsoon Rainband and Convective Activities around the Philippines*

As well known, the tropical western Pacific is a region of the highest SST in the global sea surface and is known as "the warm pool". Due to the warm state of this region, the air-sea interaction is very strong, and the ascending branch of the Walker circulation is in the region. Thus, the strong convergence of the air and moisture leads to strong convective activities and heavy rainfall there.

Nitta's (1986) study showed that the convective activities over the tropical western Pacific are closely associated with the sea surface temperature anomaly of the West Pacific warm pool. When the West Pacific warm pool is in the warming state, i.e., the SST anomaly in the area of NINO-west (EQ - 14°N, 130°E - 150°E) is positive, the convective activities are strong around the Philippines, and vice versa. Therefore, the thermal state of the West Pacific warm pool can be described using the convective activities around the Philippines because the SST anomalies in the tropical western Pacific are relatively small. Moreover, Nitta (1987), Huang and Li (1987), Kurihara (1989), Tsuyuki and Kurihara (1989), and Huang and Sun (1992) all showed a close relationship between the summer monsoon over East Asia and the convective activities around the Philippines. Huang and Sun (1992) and Huang *et al.* (1993) showed that the intraseasonal variations of the summer monsoon over China during the summers with strong convective activities around the Philippines are different from those during the summers with weak convective activities in this region. In a summer with strong convective activities around the Philippines, the abrupt northward shift of the summer monsoon rainband from the Yangtze River valley to the Yellow River valley is obvious in early or mid-July, and the summer monsoon rainfall may be weak in the Yangtze River and Huaihe River valley of China, South Korea and Japan. On the other hand, in a summer with weak convective activities around the Philippines, the abrupt northward shift of the summer monsoon rainband from the Yangtze River valley to the Yellow River valley may be not obvious, and the summer monsoon rainfall may be heavy in the Yangtze River and Huaihe River valley of China, South Korea and Japan. Therefore, in the following, the intraseasonal variability of the summer monsoon over China is discussed by using the observed data for a longer period.

2.3.1. In the Summers with Strong Convective Activities around the Philippines

Figure 2.6 is the interannual variations of normalized high cloud amount anomaly averaged for spring (March-May), summer (June-August), autumn (September-November) and winter (December-February) around the Philippines (i.e., 10°N - 20°N, 110°E - 140°E). If the normalized high cloud amount anomaly averaged for a summer season (June-August) is positive around the Philippines, the convective activities may be considered as a case of strong convective activities in the summer (e.g., Monthly Report on Climate System, JMA). By this way, it can be seen from Fig. 2.6 that in the summers of 1978, 1981, 1985, 1988, 1994 and 1999, the convective activities were strong around the Philippines. On the other hand, if the normalized high cloud amount anomaly averaged for a summer is negative around the Philippines, the convective activities may be considered as a case of weak

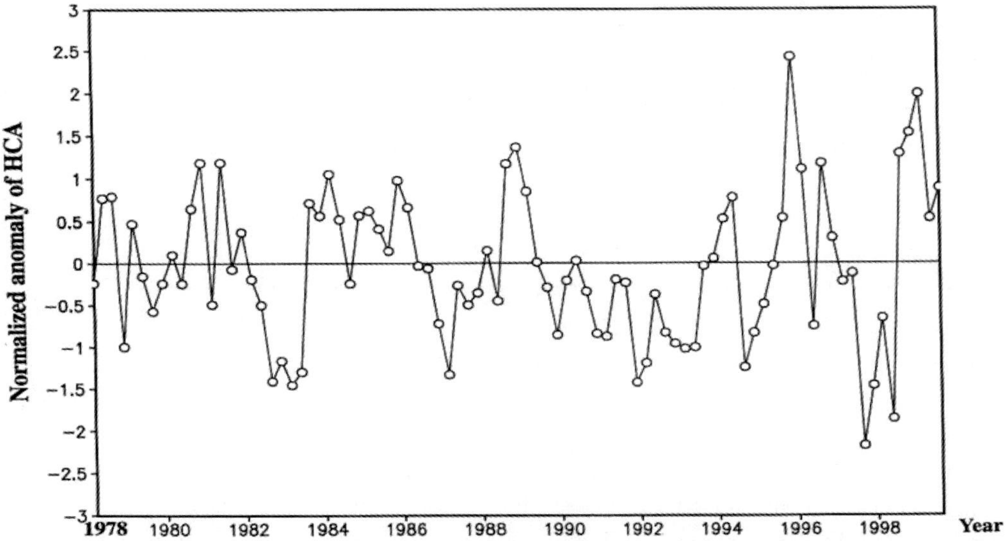

Figure 2.6. Interannual variations of the normalized seasonal mean high cloud amount (HCA) around the Philippines (i.e., 10°N - 20°N, 110°E - 140°E). The seasons mentioned in this analysis include spring (March-May), summer (June-August), autumn (September-November) and winter (December-February). Data are obtained from Monthly Report on Climate System, JMA.

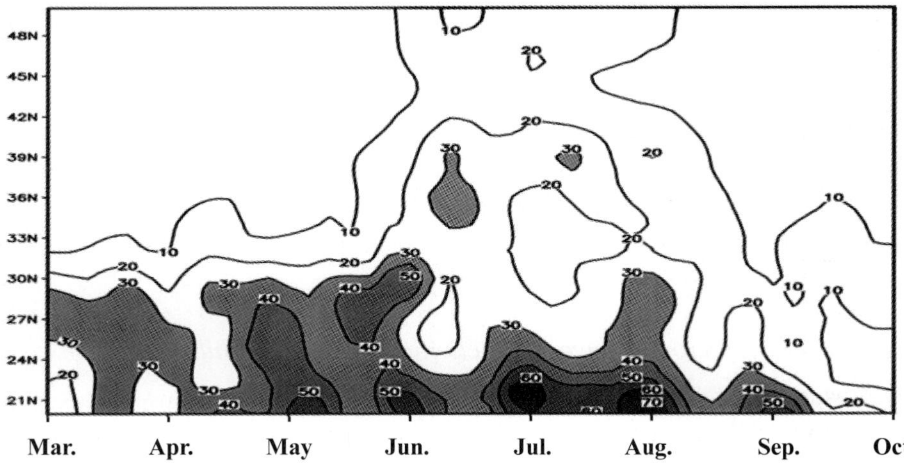

Figure 2.7. The composite latitude-time cross section of 5-day precipitation along 115°E (averaged between 110°E - 120°E) for the summers with strong convective activities around the Philippines (i.e., 10°N - 20°N, 110°E - 140°E). Units: mm. The precipitation amounts over 30 mm are shaded, and contour interval is 10 mm in the figure.

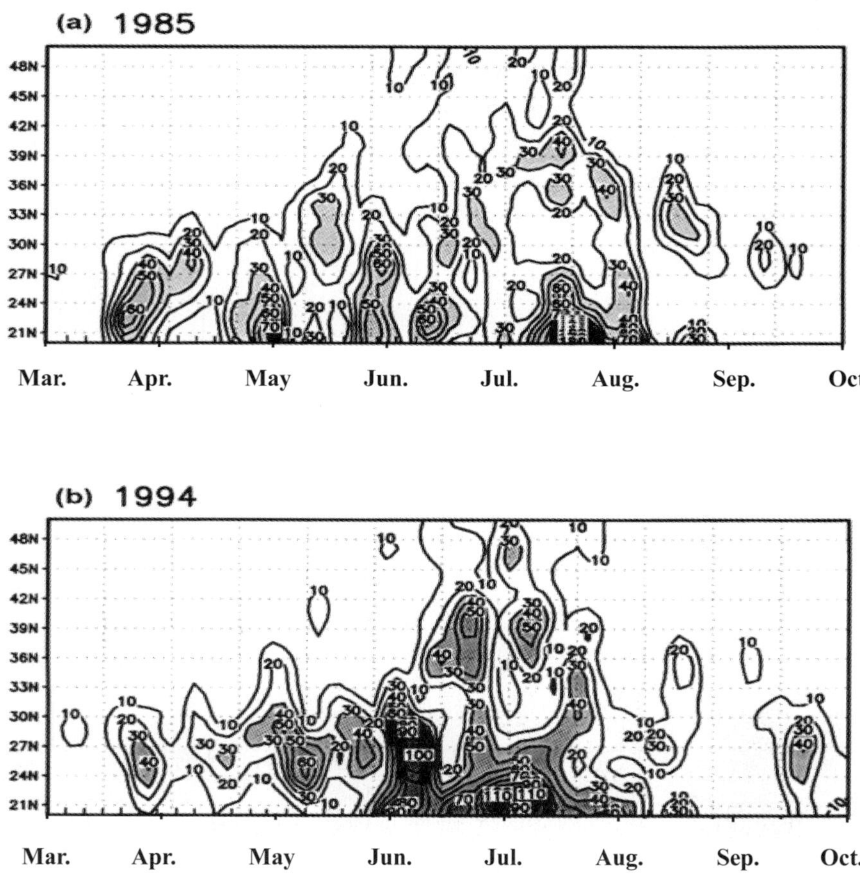

Figure 2.8. The latitude-time cross section of 5-day precipitation along 115°E (averaged between 110°E - 120°E) in the summers of (a) 1985 and (b) 1994. Units: mm. The precipitation amounts over 30 mm are shaded, and contour interval is 10 mm in the figure.

convective activities in the summer. It may be also seen from Fig. 2.6 that in the summers of 1980, 1982, 1983, 1991, 1992, 1993, 1996 and 1998, the convective activities were weak around the Philippines, and in this case, the convective activities were intensified over the equatorial central Pacific (e.g., Huang *et al.* 1998).

In order to discuss the different characteristics of the intraseasonal variability of the summer monsoon over China, the composite latitude-time cross sections of monsoon rainfall for the strong and weak convective activities around the Philippines are analyzed respectively.

Figure 2.7 is the composite latitude-time cross section of 5-day precipitation along 115°E (averaged between 110°E - 120°E) for the summers with the strong convective activities around the Philippines. Fig. 2.7 features the abrupt northward-shift of the summer monsoon rainband to the Yangtze River and Huaihe River valley in mid-June and then to the Yellow River valley in mid-July. To illustrate further this feature, the intraseasonal variations of the summer monsoon rainfall in the summers of 1985 and 1994 are taken as examples and shown in Figures 2.8a and 2.8b, respectively. It may be seen from Figs. 2.8a and 2.8b that during May and early June, the monsoon rainband was maintained over the area to the south of the Yangtze River, and the rainband abruptly moved northward to the Yellow River valley in mid-July. In these two summers, the monsoon rainfalls were very weak and were 30-50% below normal in the Yangtze River and Huaihe River valley. Due to the anomalous northward movement of the rainband, hot and dry summers occurred in the Yangtze River and Huaihe River valley in 1985 and 1994.

2.3.2. In the Summers with Weak Convective Activities around the Philippines

As shown in Fig. 2.6, in the summers of 1980, 1982, 1983, 1987, 1991, 1992, 1993, 1996 and 1998, the convective activities were weak around the Philippines, and the convections were

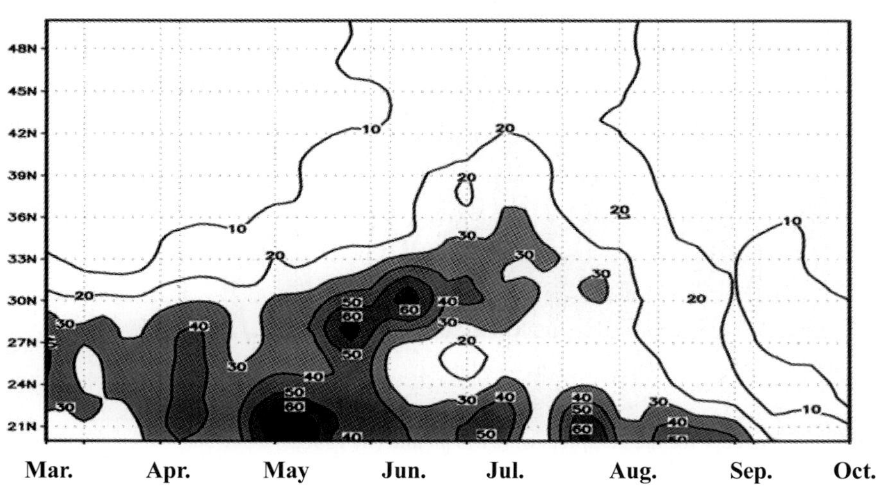

Figure 2.9. As in Fig. 2.7 except for the summers with weak convective activities around the Philippines (i.e., 10°N - 20°N, 110°E - 140°E). Units: mm.

intensified over the equatorial central Pacific in this case (e.g., Huang *et al.* 1998). Figure 2.9 is the composite latitude-time cross section of 5-day precipitation along 115°E (averaged between 110°E - 120°E) for the summers with weak convective activities around the Philippines. The figure shows that the northward shift of the summer monsoon rainband was not abrupt, and the summer monsoon rainband was maintained for a longer period over the Yangtze River and Huaihe River valley. In this case, the intraseasonal variability of summer rainfall is not remarkable in the eastern part of China.

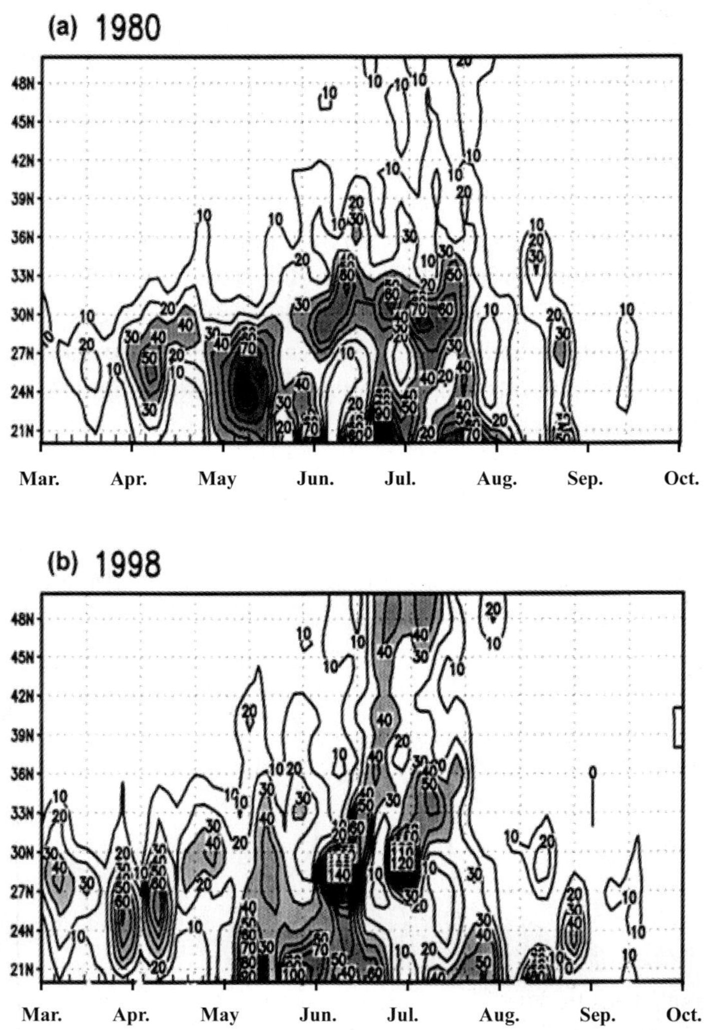

Figure 2.10. As in Fig. 2.8 except for the summers of (a) 1980 and (b) 1998.

Similarly, the summers of 1980 and 1998 are taken as examples, and the intraseasonal variations of rainfall in these two summers are shown in Figs. 2.10a and 2.10b. It can be seen that in these two summers, the abrupt northward shift of the monsoon rainband from the

Yangtze River valley to North China was not obvious and the monsoon rainband was maintained over the Yangtze River and Huaihe River valley from late May to late July. In the summer of 1980, the precipitation anomaly percentage was 30% above normal in the Yangtze River and Huaihe River valley and severe flooding occurred in this region. And in the summer of 1998, the precipitation was double normal amount and particularly severe flooding also occurred in the Yangtze River valley, as shown in Figure 2.11.

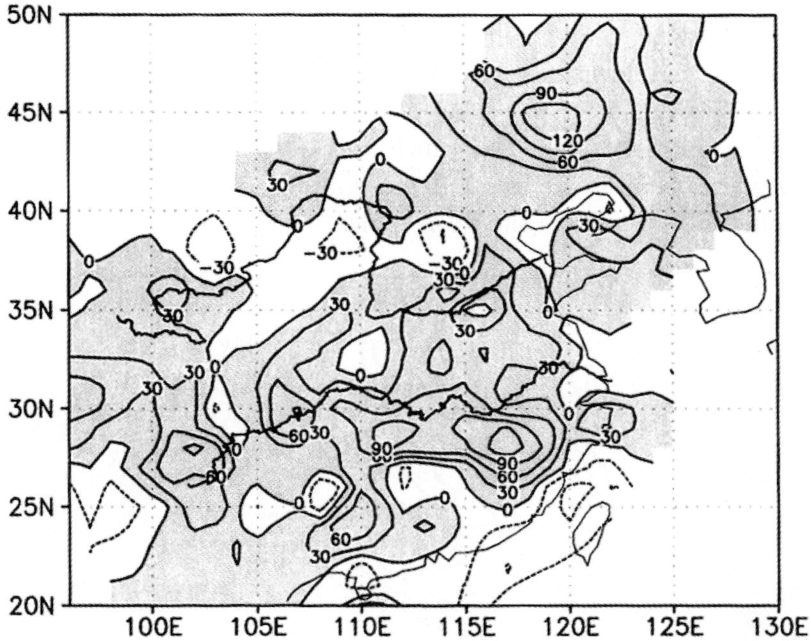

Figure 2.11. Distribution of monsoon rainfall anomaly in percentage in China in the summer (June-August) of 1998. The solid and dashed contours denote positive and negative anomalies of rainfall, respectively. The areas of positive rainfall anomalies are shaded, and contour interval is 30% in the figure. The distributions of climatological monthly precipitation in June, July and August shown in Figs. 2.2 c-e are taken as the normals, respectively.

2.4. *Relationship between the Intraseasonal Variations of the Western Pacific Subtropical High and Convective Activities around the Philippines*

Since the intraseasonal variations of summer monsoon rainfall over China are closely associated with the intraseasonal evolution of the western Pacific subtropical high, it is necessary for the further understanding of the causes of intraseasonal variations of summer monsoon rainfall over China to discuss the characteristics of the intraseasonal evolution of the western Pacific subtropical high in different cases of convective activities around the Philippines.

2.4.1. In the Summers with Strong Convective Activities around the Philippines

As described above, in the summers with strong convective activities around the Philippines, i.e., the summers of 1978, 1981, 1985, 1988, 1994 and 1999, the abrupt northward movements of the summer monsoon rainband in either mid-June or mid-July were obvious over China, as shown in Fig. 2.7. This is closely associated with the abrupt variation of the summer monsoon circulation system, especially the abrupt northward shift of the western Pacific subtropical high in mid-June over East Asia. Figure 2.12 is the composite distribution of latitude-time cross section of the 500 hPa height along 135°E for the summers with strong convective activities around the Philippines. It can be seen from the figure that the ridge line of the western Pacific subtropical high shifts abruptly from 18°N to 25°N in mid-June, which will cause the abrupt movement of the summer monsoon rainband to the Yangtze River and Huaihe River valley. Moreover, the subtropical high abruptly shifts northward again in mid-July and its ridge moves to 33°N, which will cause the abrupt shift of the summer monsoon rainband to the Yellow River valley and North China and the close of the Meiyu season in the Yangtze River and Huaihe River valley.

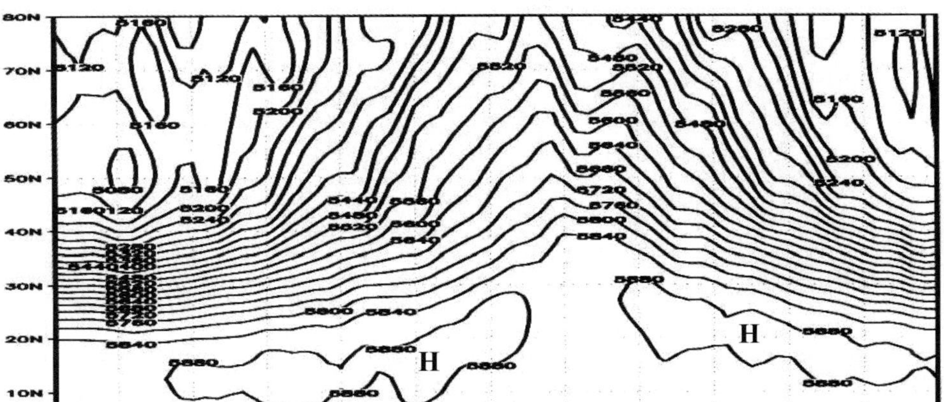

Figure 2.12. The composite distribution of latitude-time cross section of the 500 hPa height along 135°E for the summers with strong convective activities around the Philippines (i.e., 10°N - 20°N, 110°E - 140°E). Units: gpm. Contour interval is 20 gpm in the figure.

2.4.2. In the Summers with Weak Convective Activities around the Philippines

In the summers with weak convective activities around the Philippines, i.e., the summers of 1980, 1982, 1983, 1987, 1991, 1992, 1993, 1996 and 1998, the abrupt northward movements of the summer monsoon rainband were not remarkable over China either in mid-June or in mid-July, and the monsoon rainband stayed easily in the Yangtze River and Huaihe River valley, as shown in Fig. 2.9. This mainly is due to the fact that the western Pacific subtropical high remained over the region to the south of the Yangtze River. Figure 2.13 is the composite distribution of latitude-time cross section of the 500 hPa height along 135°E for the summers with weak convective activities around the Philippines. As shown in the figure, the abrupt northward shifts of the subtropical high are not obvious in either mid-June or mid-July, as it slowly moves northward and its ridge line remains at about 20°N from early June to mid-July.

From the above-mentioned analyses, it may be shown that the character of the intraseasonal variability of the western Pacific subtropical high in a summer with strong convective activities around the Philippines is different from that in a summer with weak convective activities around the Philippines. What causes this difference? Recently, Cao *et al.* (2002) applied the theory of multiple equilibrium proposed by Charney and Devore (1979) to the study of the nonlinear evolution of the western Pacific subtropical high. Their result showed that whether the northward shift of the subtropical high in mid-June is abrupt or not, it may depend on the thermal forcing due to convective activities around the Philippines. If the thermal forcing by convective activities around the Philippines is strong and exceeds a critical value, due to the strong wave-mean flow interaction and wave-wave interactions among the waves responding to the thermal forcing, the abrupt transition from a winter circulation pattern to a summer pattern can occur over East Asia in mid-June. In this case, the western Pacific subtropical high can abruptly shift northward. In contrast, if the thermal forcing by convective activities around the Philippines is weaker, since the wave-mean flow interaction and wave-wave interactions among the waves responding the thermal forcing are very weak, the transition from a winter circulation pattern to a summer pattern slowly occurs over East Asia, and the subtropical high oscillates with the thermal forcing oscillation.

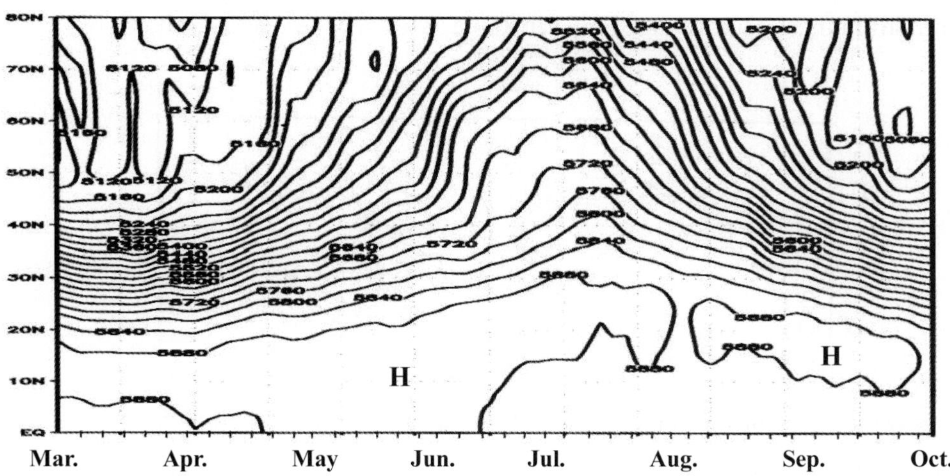

Figure 2.13. As in Fig. 2.12 except for the summers with weak convective activities around the Philippines (i.e., 10°N - 20°N, 110°E - 140°E).

Besides, the influence of convective activities around the Philippines on the intraseasonal variability of the summer monsoon over China and its surrounding regions may be associated with the Madden-Julian Oscillation. Huang (1994), Sun and Huang (1994) showed the intraseasonal variability of the summer monsoon rainfall in East Asia is closely associated with the 30-60 day oscillation propagated from the tropical eastern Indian Ocean and the Indo-China Peninsula to the East Asian summer monsoon region.

3. Interannual Variability of the Summer Monsoon over China

The summer climate in China, Korea and Japan is mainly influenced by the EASM, and summer rainfall anomaly in East Asia is also an important criterion measuring the interannual variability of the EASM (e. g., Tao and Chen 1987; Huang *et al.* 2003). Therefore, the interannual variability of the summer monsoon rainfall in China is emphasized in this section. However, the EASM is influenced by not only the Indian monsoon, but also the western Pacific subtropical high (e. g., Tao and Chen 1987; Huang and Sun 1992), thus, the interannual variations of the summer monsoon over China are complex (e. g., Huang *et al.* 2003).

3.1. *Interannual Variability of Summer Monsoon Rainfall in China*

Because China is located over the regions from the tropics to the temperature zone, and the eastern part of China is faced with the Pacific and the Tibetan Plateau is in the western part of China, the interannual variations of summer rainfall and surface air temperature in various regions of China are very different. Considering the characteristics of climate variability, China is generally divided into about 7 climate regions as shown in Figure 3.1 in the study of climate variability over China. The division of these climate regions of China and the number of observational stations used in the study of the intraseasonal and interannual variations of summer rainfall in these regions are shown in Table 3.1.

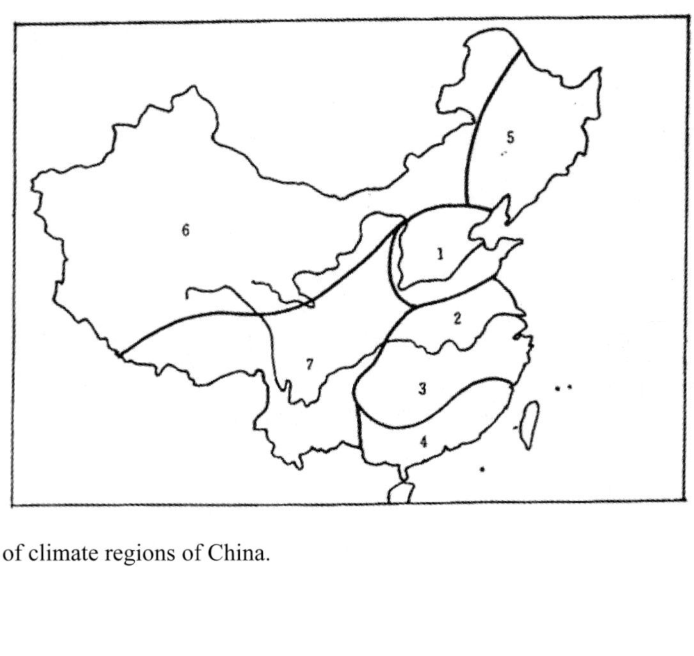

Figure 3.1. A division of climate regions of China.

Table 3.1 The division of climate regions of China and the number of observational stations in each region used in this study.

Region number	1	2	3	4	5	6	7
Name of Region	North China	The Yangtze River and Huaihe River valley	The middle and lower reaches of the Yangtze River	South China	Northeast China	Northwest China	Southwest China
Number of stations	24	20	33	28	33	68	49

Moreover, being the same as the analysis of intraseasonal variability of the summer monsoon over China described in the previous section, the observed data at 255 observational stations selected from the observed dataset of daily precipitation at 598 stations in China are used to study the interannual variations of summer rainfall in various climate regions of China, especially in the Yangtze River and Huaihe River valley, the middle and lower reaches of the Yangtze River, South China, and North China, whose locations are shown in Fig. 3.1. And the interannual variations of summer rainfall in these regions are discussed in this section.

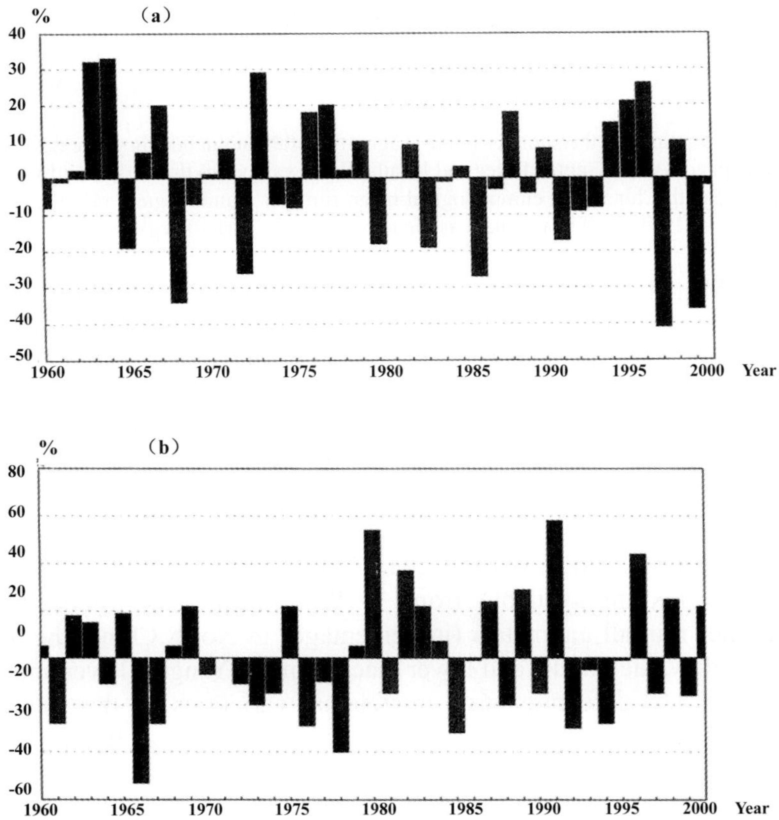

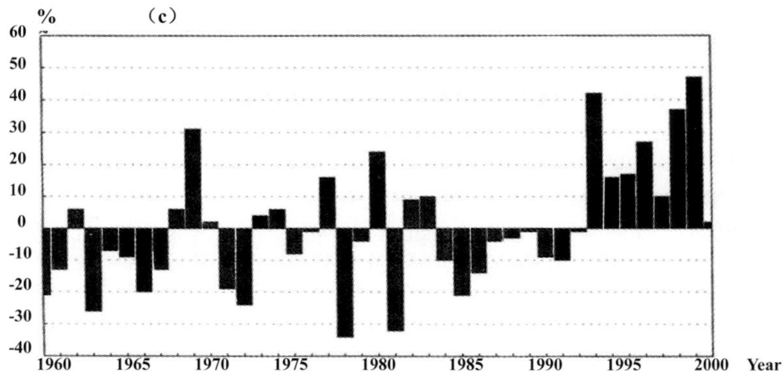

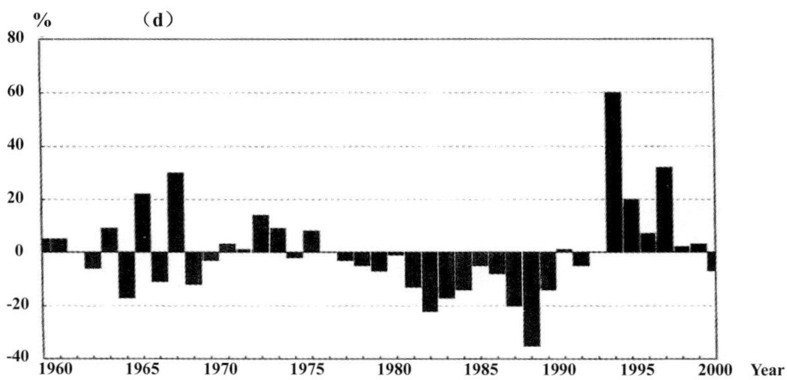

Figure 3.2. Interannual variations of the summer (June-August) rainfall anomalies (in percentage) averaged for the regions of (a) North China, (b) the Yangtze River and Huaihe River valley, (c) the middle and lower reaches of the Yangtze River, and (d) South China. The climatological mean summer (June-August) rainfalls in various regions averaged for 30 years from 1961 to 1990 are taken as the normals in the various regions.

Since the EASM mainly influences the interannual variability of summer rainfall in the southern and eastern parts of China, the observed data of daily precipitation at the selected observational stations shown in Table 3.1 in North China, the middle and lower reaches of the Yangtze River, the Yangtze River and Huaihe River valley and South China are used to analyze the interannual variations of summer (June-August) monsoon rainfall in these regions. Moreover, the climatological mean of precipitation in the various regions averaged for 30 summers from 1961 to 1990 is taken as the normal of precipitation in each region. In this section, the interannual variations of summer monsoon rainfall in these regions from 1978 are emphasized because the observed data of high cloud amount over East Asia and the tropical western Pacific are available from 1978. Figures 3.2 a-d are the interannual variations of summer rainfall anomalies (in percentage) in North China, the Yangtze River and Huaihe River valley, the middle and lower reaches of the Yangtze River and South China, respectively. Figures 3.2a-d feature large interannual variations of the summer monsoon rainfall in these regions of the eastern and southern parts of China. It may be clearly seen from Fig. 3.2b that in the summers of 1978, 1981, 1985, 1988, 1990, 1992, 1994, 1997 and 1999, the monsoon rainfall was 10% - 20% below normal in the Yangtze River and Huaihe River valley and droughts occurred in this region in the summers of 1978, 1985, 1988, 1992 and 1994, respectively. On the other hand, in the summers of 1980, 1982, 1983, 1987, 1989, 1991, 1996, 1998 and 2000, the summer rainfall was over 20% above normal, and severe floods occurred there in the summers of 1980, 1982, 1991 and 1996 respectively, and the rainfall was at least 30% greater than the mean in this region in these summers.

It may be also seen from Fig. 3.2a that the interannual variations of summer rainfall are also significant in North China. After 1978, the rainfall was more than 10% below normal in the summers of 1980, 1983, 1986, 1991, 1997, 1999 and 2000, respectively, and in particular, severe droughts occurred in the summers of 1997 and 1999, the rainfall was at least 30% less than the mean in this region in these two summers.

Comparing Fig. 3.2a with Fig. 3.2b, it is clearly seen that after 1978, flood summers frequently appeared in the Yangtze River and Huaihe River valley, but drought summers frequently appeared in North China. Thus, the interannual variations of summer rainfall in these two regions seem to be opposite.

Similarly, from 1978 to 1993, the summer monsoon rainfall was below normal, and drought summers frequently occurred in the middle and lower reaches of the Yangtze River and South China. However, flood summers continuously appeared in these two regions from 1994 to 1999, as shown in Fig. 3.2c and Fig. 3.2d, respectively.

From the above-mentioned results, it may be seen that there are obvious interannual oscillations of 2-6 years in the variations of the summer monsoon rainfall in North China, East China and South China. However, the interannual variability of summer monsoon rainfall is different in these regions although these regions all are located in the East Asian monsoon region.

3.2. *The Quasi-Biennial Oscillation of Summer Monsoon Rainfall in China and its Link with that in Its Surrounding Regions*

In order to reveal the regularity of the interannual variations of summer monsoon rainfall in China, the Empirical Orthogonal Function (EOF) analysis method is applied. Figures 3.3a and 3.3b are the spatial distribution and corresponding time-coefficient series of the first component of EOF analysis (EOF1) (it explains 16.28% of the variance) of summer rainfall in China. As shown in Fig. 3.3a, the spatial distribution of EOF1 of summer monsoon rainfall shows a meridional tripole pattern in China, and the large-scale strong negative signal is in the middle and lower reaches of the Yangtze River and in the Yangtze River and Huaihe River valley, and the large-scale positive signals are in South China and North China. This also shows that the monsoon rainfall anomaly in North China is opposite to that in the Yangtze River and Huaihe River valley in summer. The corresponding time-coefficient series shown in Fig. 3.3b obviously exhibits a prevailing quasi-biennial oscillation from the mid-1970s to the late 1990s, but the quasi-biennial oscillation was not obvious from the 1950s to the mid-1970s. By using entropy spectrum analysis, the dominant oscillation period of the EOF1 of summer rainfall in China is about 2.2 years, as shown in Fig. 3.4.

From the above-mentioned result, it is clearly seen that there is an obvious quasi-biennial oscillation of the summer monsoon rainfall in East China. This result is in agreement with that studied by Miao and Lau (1990), Yasunari (1991), Lau and Shen (1992).

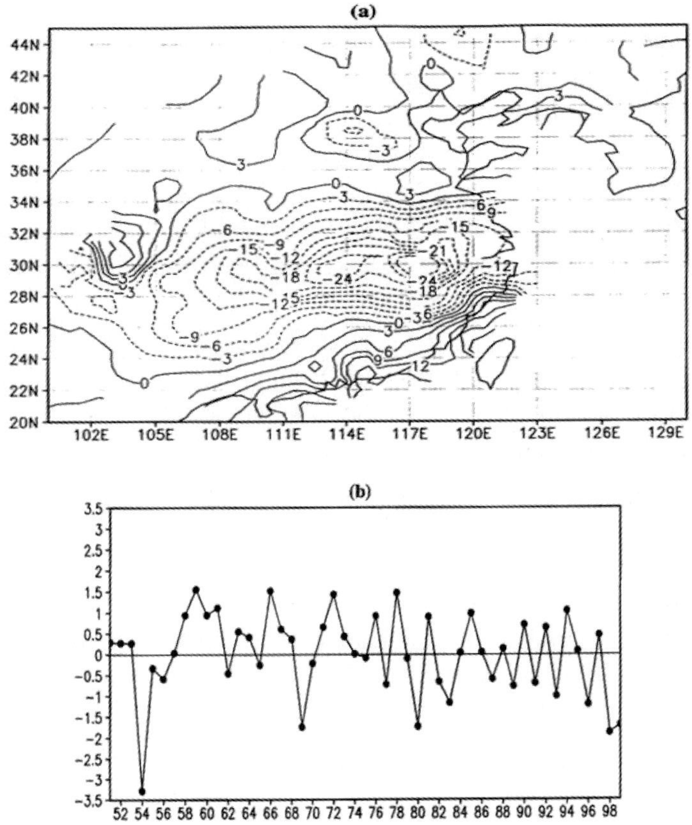

Figure 3.3. (a) The spatial distribution and (b) corresponding time-coefficient series of the first component of EOF analysis (EOF1) of summer rainfall in China from 1951 to 1999, analyzed using the dataset of monthly precipitation in 160 observational stations of China. The solid and dashed contours in Figure 3.3a indicate positive and negative signals, respectively, and the first EOF explains 16.28% of the variance.

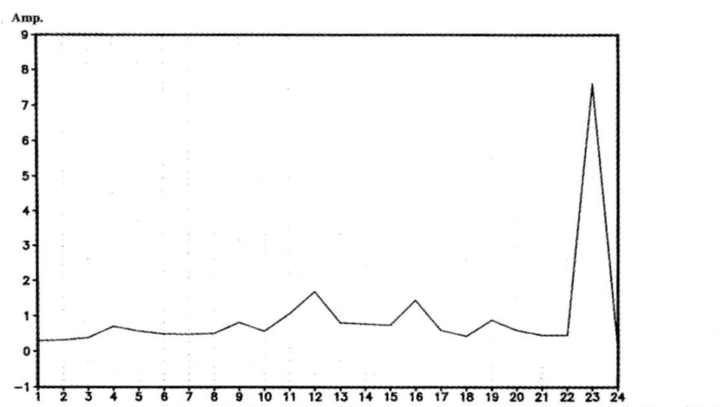

Figure 3.4. The entropy spectrum of the corresponding time-coefficients of EOF1 of summer rainfall in China

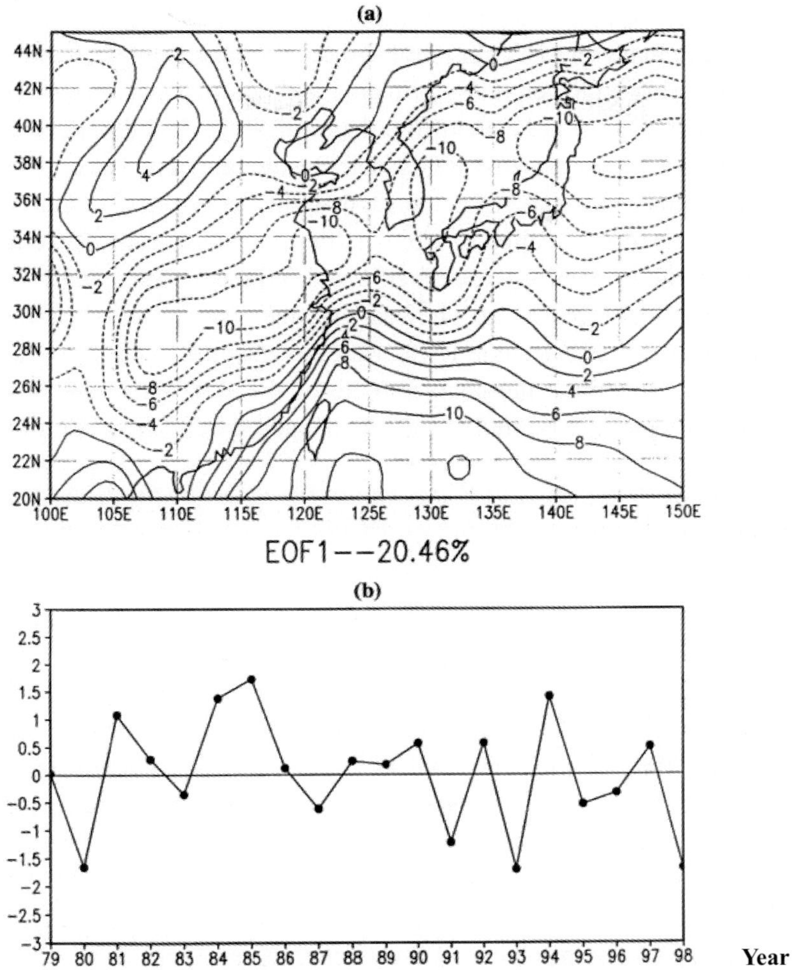

Figure 3.5. As in Fig. 3.3, except for using the dataset of precipitations analyzed by Xie and Arkin (1997) for 20 summers from 1979 to 1998. Note that the first EOF explains 20.46% of the variance.

The interannual variability of summer monsoon rainfall in Korea and Japan is also similar to that in the Yangtze River and Huaihe River valley of China. In order to investigate the linkage between the interannual variation of summer monsoon rainfall in China and that in its surrounding regions, the dataset of precipitation analyzed by Xie and Arkin (1997) for 20 summers from 1979 to 1998 is used to analyze the spatial distribution and the corresponding time-coefficient series of EOF analysis of summer monsoon rainfall in China and its surrounding regions. Figures 3.5a and 3.5b are the spatial distribution and the corresponding time-coefficient series of the EOF1 (it explains 20.46% of the variance) of summer rainfall in China and its surrounding regions, respectively. It may be found from Fig. 3.5a that the spatial distribution of the summer rainfall signal also exhibits a meridional tripole pattern. The large-scale negative signal is in the area from Japan, Korea to the middle and lower reaches of the Yangtze River, and the Yangtze River and Huaihe River valley of China. This can explain that there is a close linkage between the summer monsoon rainfall in the Yangtze River and Huaihe River valley of China and that in Japan and Korea. Moreover, the large-scale positive signal is in the area from the tropical western Pacific, the South China Sea and South China to the Indo-China Peninsula. Beside, another weaker positive rainfall signal is in North China. This pattern features that the large-scale rainfall signal in the Yangtze River and Huaihe River valley of China is the same as that in Korea and Japan and the opposite to that in the tropical western Pacific and Southeast Asia, and North China and Northeast China. Lu *et al.* (1995) also showed that the interannual variation of summer rainfall in the Yangtze River and Huaihe River valley is generally similar to those in Korea. Moreover, Fig. 3.5b also exhibits the characteristic of the quasi-biennial oscillation in the interannual variations of summer rainfall in East Asia and the tropical western Pacific.

3.3. *The Quasi-Biennial Oscillation of Water Vapor Transport by the EASM and its Impact on Summer Monsoon Rainfall over China and its Surrounding Regions*

As mentioned above, the interannual variability of the East Asian climate system has a great impact on the summer monsoon over China and its surrounding regions. Furthermore, the interannual variability of the summer monsoon over China greatly influences summer rainfall in China through water vapor transport. According to the result investigated by Huang *et al.* (1998), the characteristics of water vapor transport over the East Asian monsoon region including South China, East China, Korea and Japan are greatly different from those over the Indian monsoon region. Over the latter, the zonal component of water vapor transport is dominant. However, over the East Asian monsoon region, the meridional component of water vapor transport is large and has the same order as the zonal component. Moreover, the convergence of water vapor, which can form the monsoon rainfall, is mainly caused by the moisture advection over South China, East China, Korea and Japan. Therefore, the interannual variability of water vapor transport caused by monsoon flow is very important for the interannual variability of summer monsoon rainfall in China and its surrounding regions. Therefore, discussion of the impact of the water vapor transport by the EASM on summer rainfall in China and its surrounding regions is needed.

Assuming no water vapor above 100 hPa, the transport flux vector of water vapor $\vec{Q} = (\vec{Q}_\lambda, \vec{Q}_\varphi)$ can be described as follows:

$$Q_\lambda = \frac{1}{g}\int_{100}^{P_0} q \cdot u dp \tag{3.1}$$

and

$$Q_\varphi = \frac{1}{g}\int_{100}^{P_0} q \cdot v dp \tag{3.2}$$

where Q_λ and Q_φ are the zonal and meridional components of water vapor transport flux, respectively, q is the specific humidity, u and v are the zonal and meridional components of wind field, respectively, and P_0 =1000 hPa for the sake of simplicity.

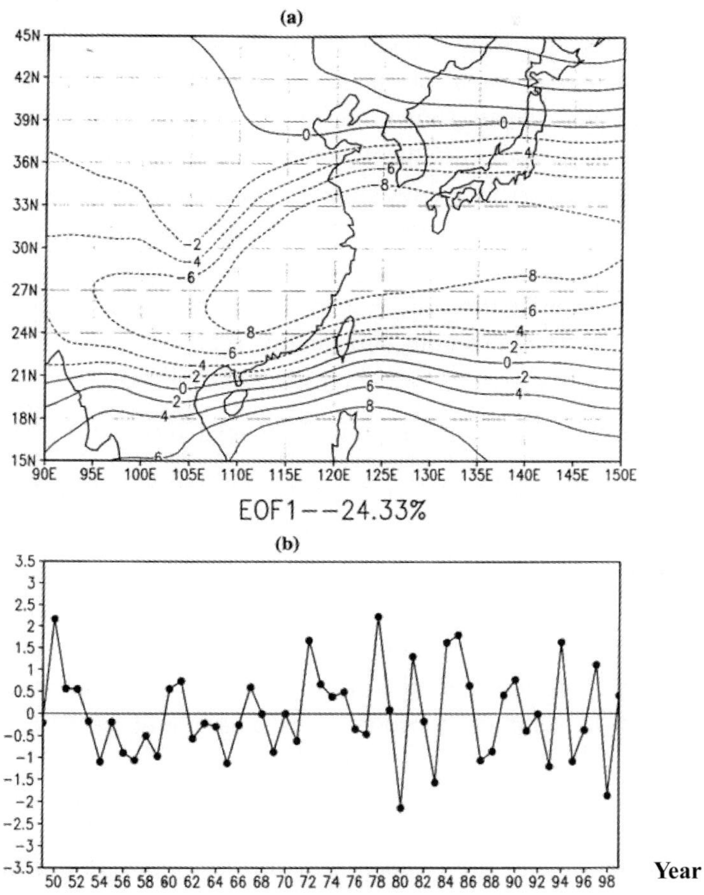

Figure 3.6. (a) The spatial distribution and (b) corresponding time-coefficient series of the first EOF component of zonal water vapor transports during summer, analyzed using the NCEP/NCAR reanalysis dataset of moisture and wind fields from 1951 to 1999. The solid and dashed contours in Figure 3.6a indicate positive and negative anomaly signals of zonal water transports, respectively, and the first EOF explains 24.33% of the variance.

The water vapor transport fluxes over East Asia and the tropical western Pacific in the summers (June-August) of 1951-1999 are calculated using Formulas (3.1) and (3.2) and the NCEP/NCAR reanalysis data of moisture. Moreover, the EOF analysis is applied to study the characteristics of spatial-temporal variability of the zonal and meridional components of water vapor transport over the regions. Figures 3.6a and 3.6b are the spatial distribution and the corresponding time-coefficient series of the first EOF component (EOF1) of zonal water vapor transport in summer, respectively. The spatial distribution of the EOF1 of zonal water vapor transport shown in Fig. 3.6a exhibits a meridional tripole pattern, which is the same as the spatial pattern of the EOF1 of summer rainfall in China and its surrounding regions shown in Fig. 3.3a and Fig. 3.5a. Moreover, it may be seen from the corresponding time-coefficient series of the first component of EOF shown in Fig. 3.6b that there is an obvious oscillation of two-three years from the mid-1970s. That is to say, there is a characteristic of quasi-biennial oscillation in the interannual variation of zonal water vapor transport over China and its surrounding regions from the mid-1970s. This is in agreement with the characteristic of quasi-biennial oscillation in the interannual variations of summer rainfall in China, Korea and Japan shown in Fig. 3.3b and Fig. 3.5b. Moreover, the corresponding time-coefficient series of the second EOF component of meridional water vapor transport also exhibits the characteristic of quasi-biennial oscillation (figure omitted).

The above-mentioned analysis shows that there is an obvious quasi-biennial oscillation in the interannual variations of water vapor transport by the summer monsoon flow over East Asia and the tropical western Pacific, which may cause the quasi-biennial oscillation of summer monsoon rainfall over China and its surrounding regions.

4. Interdecadal Variation of the Summer Monsoon over China

As in other monsoon regions, the decadal and interdecadal variations of the summer monsoon over China are also significant (e.g., Chang *et al.* 2000; Huang 2001). Thus, the discussion on decadal and interdecadal variations of the summer monsoon over China will be emphasized in this section. Moreover, being different from the interannual variability of summer surface air temperature, the interdecadal variability of summer surface air temperature is as significant as the summer monsoon rainfall in East Asia. Therefore, in order to explain the interdecadal variability of the summer monsoon over China, the observed dataset of monthly precipitation and surface air temperature at 160 observational stations of China, whose distribution is shown in Fig. 2.1, provided by NCC/CMA, are analyzed in this section.

4.1. *Decadal Variations of Surface Air Temperature and Precipitation in China*

In order to analyze the decadal variability of summer (June-August) surface air temperature and precipitation in China, the climatological mean monthly precipitation and surface air temperature averaged from 1961 to 1990 as shown in Figs. 2.2b-d and Figs. 2.4b-d are taken as the normals of summer (June-August) precipitation and surface air temperature, respectively. By this way, the decadal-mean anomaly distributions of summer surface air temperature and precipitation can be obtained by averaging for various decades from the 1950s to the 1990s.

Figures 4.1a-e indicate the decadal-mean anomaly distributions of summer (June-August) surface air temperature over China in the 1950s, 1960s, 1970s, 1980s and 1990s, respectively. From these figures, it can be found that a warming trend was obvious in North China, Northwest China, Northeast China and South China. From the 1980s to the 1990s, especially in the 1990s, the surface air temperature anomalies reached 0.4-0.8°C in North China and

Northwest China. However, there was a cooling trend in the middle and upper reaches of the Yangtze River from the 1980s to the 1990s, and the maximum anomaly of surface air temperature reached –0.6°C.

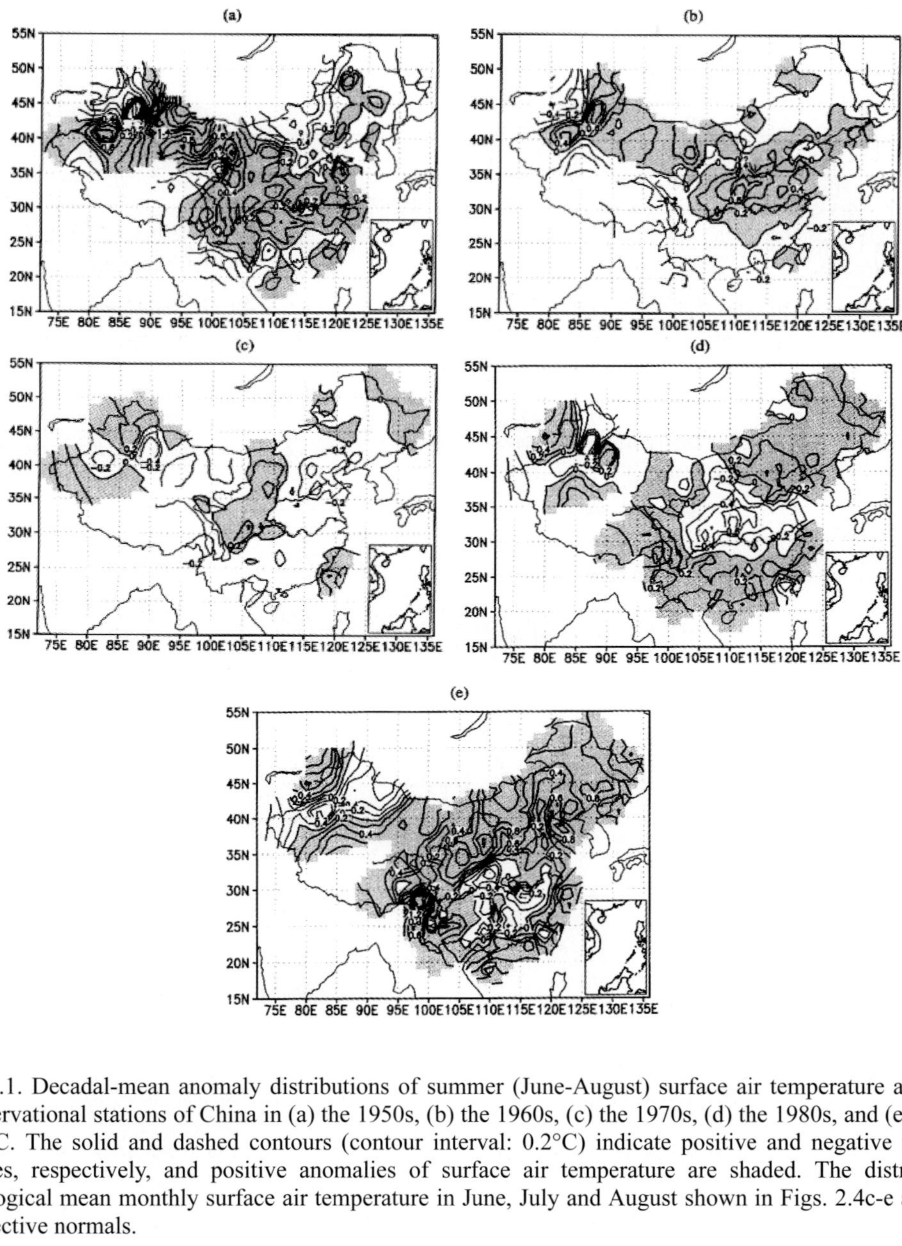

Figure 4.1. Decadal-mean anomaly distributions of summer (June-August) surface air temperature anomalies at 160 observational stations of China in (a) the 1950s, (b) the 1960s, (c) the 1970s, (d) the 1980s, and (e) the 1990s. Units: °C. The solid and dashed contours (contour interval: 0.2°C) indicate positive and negative temperature anomalies, respectively, and positive anomalies of surface air temperature are shaded. The distributions of climatological mean monthly surface air temperature in June, July and August shown in Figs. 2.4c-e are taken as the respective normals.

Compared with the decadal variations of summer surface air temperature shown in Figs. 4.1a-e, the interdecadal fluctuation of summer precipitation is more obvious in China. Figures 4.2a-e are the decadal mean anomaly distributions of summer (June-August) precipitation in the 1950s, 1960s, 1970s, 1980s and 1990s, respectively. It may be seen from these figures that the anomaly distributions of summer precipitation in the 1980s and 1990s were obviously different from those in the 1950s, 1960s, 1970s. The summer monsoon rainfall obviously increased in the 1980s and 1990s in the Yangtze River valley, especially in the middle and lower reaches of the Yangtze River where the summer precipitation anomalies were 30%-40% greater than the climatological mean. Thus, flood disasters frequently

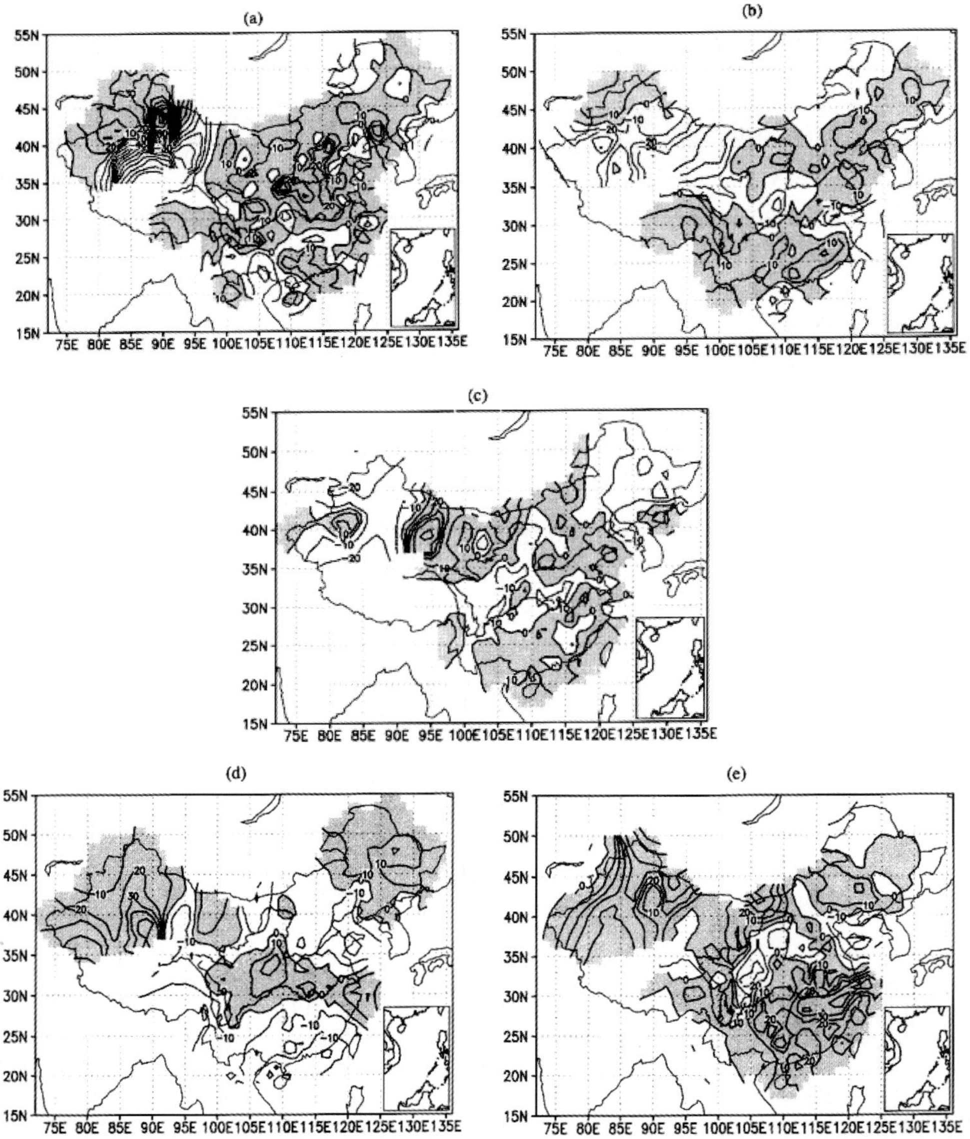

Figure 4.2. As in Fig. 4.1, except for the summer rainfall anomalies (in percentage). The solid and dashed contours (contour interval: 10%) indicate positive and negative rainfall anomalies, respectively, and positive rainfall anomalies are shaded. The distribution of climatological mean monthly rainfall in June, July and August shown in Figs. 2.2c-e are taken as the respective normal.

occurred in this region in the 1990s, as shown in Fig. 3.2c. Moreover, the monsoon rainfall obviously increased in South China in the 1990s although it decreased in the 1980s. However, the opposite phenomenon appeared in North China and the Yellow River valley. In the 1980s and 1990s, the summer precipitation was 10%-20% less than the climatological mean and prolonged droughts occurred in North China. Besides, Figures 4.2a-e also show that the summer rainfall obviously increased in Northwest China in the period of the 1980s and 1990s.

4.2. *The Climate Jump in the Late 1970s*

The interdecadal variations of summer monsoon rainfall in China may be also seen from the interannual variations of summer precipitation in various regions of China. Figures 3.2a-d show that the interannual variations of summer precipitation in North China and the middle and lower reaches of the Yangtze River before 1976 were different from those after 1976. Thus, the difference between the summer precipitation anomalies (in percentage) averaged for 1977-2000 and those averaged for 1967-1976 over China is analyzed and is shown in Fig. 4.3. The figure shows that there was a large difference between the summer precipitation anomalies averaged for 1977-2000 and those averaged for 1967-1976 in North China, Northwest China and the Yangtze River valley. After the late 1970s, the summer monsoon rainfall obviously decreased in North China and South China, but the summer precipitation obviously increased in the Yangtze River valley and Northwest China. Thus, a climate jump of summer rainfall in China occurred in the late 1970s. Due to this climate jump, North China has become drier, while the Yangtze River valley has become wetter, and summer precipitation has obviously increased in Northwest China.

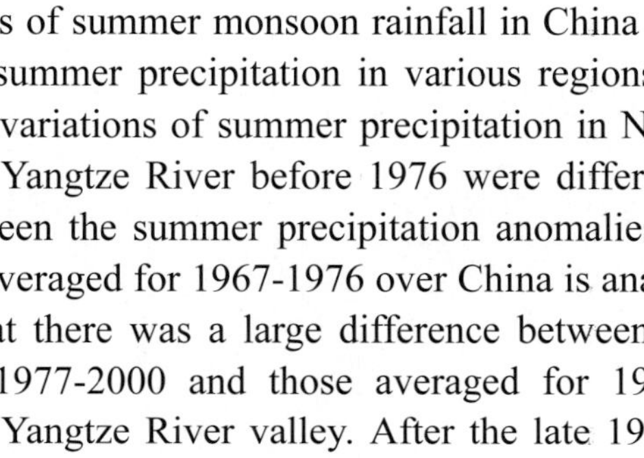

Figure 4.3. Differences between the summer precipitation anomalies (in percentage) averaged for 1977-2000 and those averaged for 1967-1976 at 160 observational stations in China. The solid and dashed contours (contour interval: 10%) indicate positive and negative values of rainfall differences, and positive rainfall differences are shaded. The distributions of climatological mean monthly precipitation in June, July and August shown in Figs. 2.2c-e are taken as the respective normal.

The climate jump can be also seen from the interdecadal variations of summer (June-August) surface air temperature. The difference between the summer surface air temperature anomalies averaged for 1977-2000 and those averaged for 1967-1976 in China is also analyzed. As shown in Fig. 4.4, there was also a large difference between the summer air temperature anomalies in China averaged for 1977-2000 and those averaged for 1967-1976. The figure shows the positive surface air temperature differences in North China, Northeast China, Northwest China and South China, respectively. Moreover, it may be seen that the negative surface air temperature difference appeared in the Yangtze River valley, with the maximum of about –0.6 in the upper reaches of the Yangtze River. This shows the decrease of surface air temperature in the upper reaches of the Yangtze River from the late 1970s, and it may be associated with the increase of summer rainfall there.

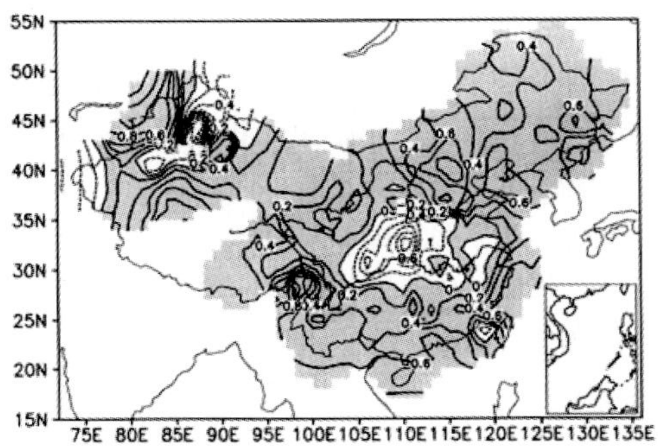

Figure 4.4. As in Fig. 4.3, except for summer surface air temperature anomalies in China. Units: °C. The solid and dashed contours (contour interval: 0.2°C) indicate positive and negative values of surface air temperature differences, respectively, and positive surface air temperature differences are shaded.

From Figs. 4.3 and 4.4, it may be found that the regional characteristics of the interdecadal variations of summer climate in China are significant. In the eastern and southern parts of China which are in the East Asian monsoon region, the increase (decrease) of summer monsoon rainfall is accompanied by a decrease (increase) of surface air temperature. For example, from the late 1970s to the 1990s, the summer monsoon rainfall obviously increased in the Yangtze River valley, which was accompanied by a decrease of surface air temperature in summer. Oppositely, the summer monsoon rainfall obviously decreased in North China, which was accompanied by an increase of surface air temperature in summer. However, in Northwest China which is not in the East Asian monsoon region, the increase of summer rainfall was accompanied by an increase of surface air temperature from the late 1970s to the 1990s.

Furthermore, the analysis of observed winter (December-February) surface air temperature also shows that the climate jump also occurred in the late 1970s. Since the late 1970s, the winter surface air temperature has obviously increased in Northeast China, North China and Northwest China. Because climate variations of the winter monsoon over China are described in another chapter, the detailed results will not be repeated in this section.

4.3. *The Interdecadal Variability of Water Vapor Transport by the EASM and its Impact on Summer Monsoon Rainfall over China*

Huang and Yan (1999) and Huang *et al.* (1999) investigated the cause of the interdecadal variations of summer climate in North China from the interdecadal variability of the southerly wind at 700 hPa over East Asia and pointed out that the southerly winds in the lower troposphere over North China began to become weak from 1965 and became more weak from the late 1970s. This led to the weakening of the meridional water vapor transport in North China and caused the beginning of drought in North China from 1965 and the prolonged droughts in this region from the late 1970s.

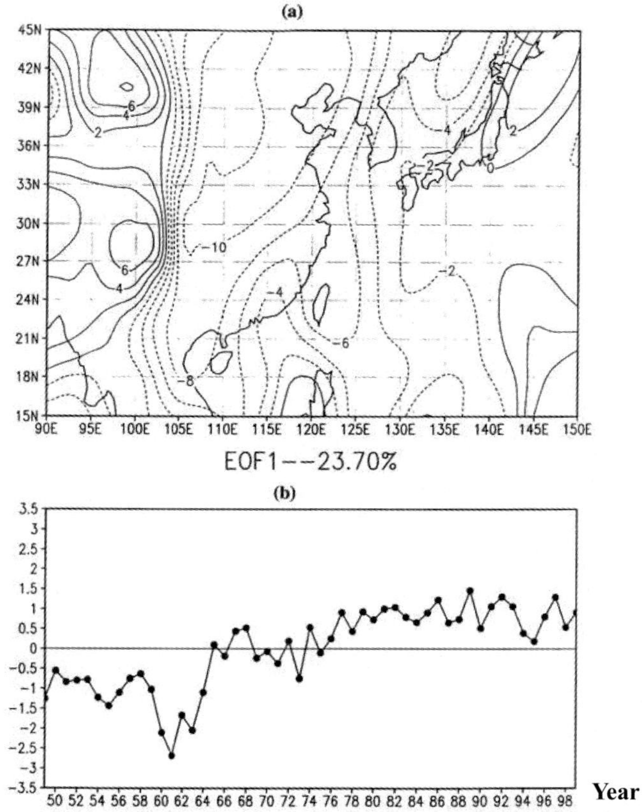

Figure 4.5. As in Fig. 3.6, except for meridional water vapor transport. The solid and dashed contours in Figure 4.5a indicate positive and negative anomaly signals of meridional water vapor transport, respectively, and the first EOF explains 23.70% of the variance.

The interdecadal variability of water vapor transport over East Asia is also analyzed using the water vapor transport fluxes over East Asia and the tropical western Pacific during summer (June-August) calculated in Section 3. As mentioned above, the meridional transport of water vapor is large over the eastern part of China, thus, the interdecadal variability of water vapor transport may be clearly shown in the meridional component of water vapor transport in China and its surrounding regions. Figures 4.5a and 4.5b are the spatial distribution and the corresponding time-coefficient series of the first EOF component of meridional water vapor transport in summer, respectively. It may be seen from Fig. 4.5a that the spatial distribution of meridional water vapor transport signals exhibits a negative single pattern over the eastern part of China, the Korean Peninsula and the western part of Japan, which is very different from the spatial distribution of the first EOF component of zonal water vapor transport shown in Fig. 3.6a. Moreover, it may be seen from the corresponding time-coefficient series of the first component of EOF shown in Fig. 4.5b that there is an obvious interdecadal variability in the interannual variations of meridional water vapor transport. As shown in this figure, in the summer of 1965, the time-coefficient of the EOF 1, which explains 23.7% of the variance, abruptly became a positive value from a negative value, and then oscillated about the normal up to 1976. Moreover, from 1977 to 1999, the time-coefficients of the EOF1 became larger positive values. Combined with the spatial pattern shown in Fig. 4.5a, it may be clearly seen that the meridional water vapor transport remarkably decreased over the eastern part of China, the Korean Peninsula and the western part of Japan from the late 1970s. This led to a serious decrease of water vapor transported into North China and North Korea. The serious decrease of water vapor transported into North China is closely associated with the weakening of southerly winds in the lower troposphere over North China (e.g., G. Huang 1999). This may be an important cause of the prolonged drought disaster which has occurred in North China and North Korea from the late 1970s.

5. The East Asian Climate System and its Impact on the Interannual Variations of the Summer Monsoon over China

As mentioned above, the interannual and interdecadal variabilities of the summer monsoon over China are significant, which have caused severe droughts and floods in China. Recently, the study of the East Asian climate system and its impact on the climate variation of the summer monsoon over China has been greatly promoted. Thus, the East Asian climate system influencing the interannual variations of the summer monsoon over China and its surrounding regions will be simply discussed in this section.

5.1. *The East Asian Climate System*

The summer monsoon climate variabilities in East Asia are mainly influenced by the interactions among the components of the East Asian climate system (e.g., Huang *et al.* 2003). As shown schematically in Fig. 5.1, the East Asian climate system includes the East Asian monsoon, the western Pacific subtropical high and the disturbances in middle latitudes in the atmosphere, the thermal effect of the West Pacific warm pool, the ENSO cycle in the tropical Pacific, the land surface processes in the Eurasian continent and the thermal and dynamical effects of the Tibetan Plateau. Due to the interactions among these components of the East Asian climate system, the climate variabilities of the summer monsoon with different timescales are caused in China and its surrounding regions. Therefore, the relationships between the interannual variations of the summer monsoon over China and these components of the East Asian climate system are analyzed using the observed data in the following subsections.

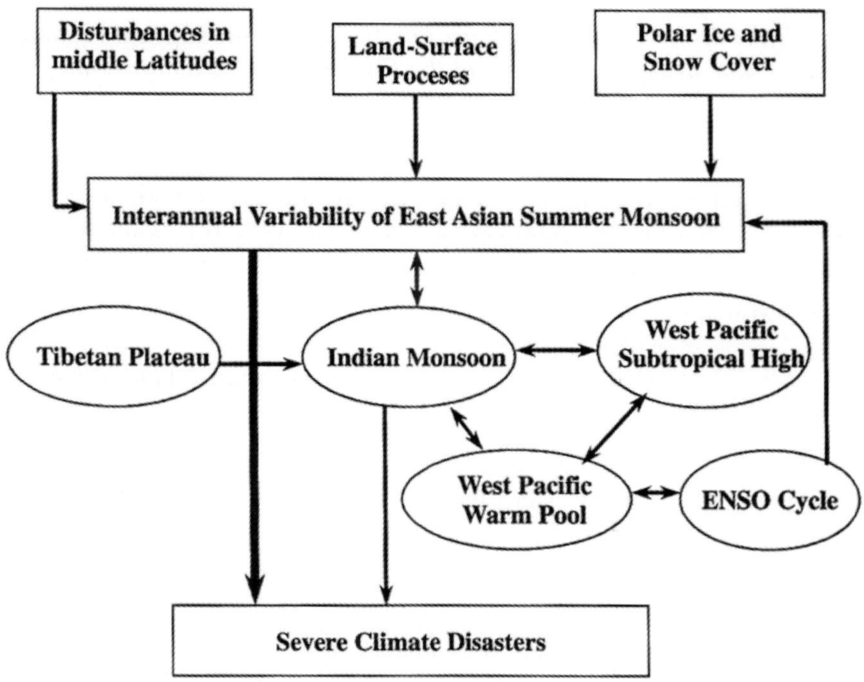

Figure 5.1. Conceptual diagram of the East Asian climate system.

5.2. *Dynamic and Thermal Effects of the Tibetan Plateau on the Interannual Variability of the Summer Monsoon over China*

The Tibetan Plateau has an important dynamical effect on the interannual variability of the EASM. By using the GFDL 9-layer GCM simulations, Hahn and Manabe (1975) showed that due to the dynamic effect of the Tibetan Plateau, the strong southwesterly (SW) flow can extend to the eastern part of China including South China, the Yangtze River valley, the Yangtze River and Huaihe River valley and North China from the Bay of Bengal. In the late 1970s, Ye and Gao (1979) first pointed out the thermal effect of the Tibetan Plateau on the Asian monsoon. Later on, many scholars also emphasized the thermal effect of the Tibetan Plateau on the Asian summer monsoon (e.g., Nitta 1983; Luo and Yanai 1984; Huang, 1984, 1985), and pointed out that the heating anomaly over the Tibetan Plateau has a large impact on the Asian summer monsoon circulation anomalies. Wu and Zhang (1997) explained that the Tibetan Plateau may play an air-pumping effect on the Asian summer monsoon onset through the sensible heating. Recently, Zhang *et al.* (2002) pointed out that the heating oscillation over the Tibetan Plateau can trigger the east–west oscillation of the South Asian high, which has a significant influence on the summer monsoon anomalies over the Yangtze River valley. As for the dynamic and thermal effects of the Tibetan Plateau on the interannual variability of the summer monsoon over China, these have been described in another chapter and will be omitted here.

5.3. *Thermal Effect of the Tropical Western Pacific on the Interannual Variability of the Summer Monsoon over China and the EAP Pattern Teleconnection*

The thermal state of the tropical western Pacific and the convective activities around the Philippines has a large impact on the interannual variability of the EASM. This impact may be through the western Pacific subtropical high and the onset of the South China Sea summer monsoon (SCSM). Thus, in this subsection and the next subsection, the impacts of the thermal state of the tropical western Pacific on the western Pacific subtropical high and the onset of the SCSM will be discussed.

5.3.1. Impact of the Thermal State of the Tropical Western Pacific on the Western Pacific Subtropical High

Since the summer monsoon rainband over the eastern part of China, Korea and Japan is located on the north side of the western Pacific subtropical high, the location of the western Pacific subtropical high greatly influences the distribution of the summer monsoon rainband over East Asia. Thus, the western Pacific subtropical high is an important component of the East Asian summer monsoon system (e.g., Tao and Chen 1987). If its location is shifted northward, a hot and drought summer may appear in the Yangtze River and Huaihe River valley. On the other hand, if its location is shifted southward, a flood summer may occur there.

The studies by many scholars (e.g., Nitta 1987; Huang and Li 1987; Kurihara 1989; Tsuyuki and Kurihara 1989; Huang and Sun 1992) showed that the thermal states of the tropical western Pacific and the convective activities around the Philippines play an important role in the interannual variations of the EASM. Moreover, Nitta (1987), Huang and Li (1987) and Kurihara (1989) all pointed out from the analyses of observed data that the thermal states of the tropical western Pacific and the convective activities around the Philippines can greatly influence the interannual variability of the western Pacific subtropical high.

The interannual variations of the western Pacific subtropical high are closely related to the thermal state of the tropical western Pacific or the convective activities around the Philippines. Figures 5.2a and 5.2b are the composite distributions of monthly mean 500 hPa height fields in July over East Asia for the summers with strong convective activities and for the summers with weak convective activities around the Philippines, respectively. From Fig. 5.2a, it can be seen that when the tropical western Pacific is in a warming state, i.e., the summer (June-August) SST anomaly in the area of NINO-west (EQ - 14°N, 130°E - 150°E) is above normal, the convective activities are strong around the Philippines, then the 500 hPa height field in July over East Asia will display a similar distribution to that shown in Fig. 5.2a.

In this case, the western Pacific subtropical high will shift unusually northward and will be located over the Yangtze River and Huaihe River valley. This will cause a hot and drought summer in the Yangtze River and Huaihe River valley. On the other hand, when the tropical western Pacific is in a cooling state, i.e., the summer SST anomaly in NINO-west is below normal, the convective activities are weak around the Philippines. In this case, the 500 hPa height field in July over East Asia will display a distribution as shown in Fig. 5.2b, which is very different from Fig. 5.2a. Under this situation, the western Pacific subtropical high will shift unusually southward and will be located to the south of the Yangtze River. This is favorable for the long-term maintenance of the Meiyu front in the Yangtze River and Huaihe River valley, thus, a flood summer will occur there.

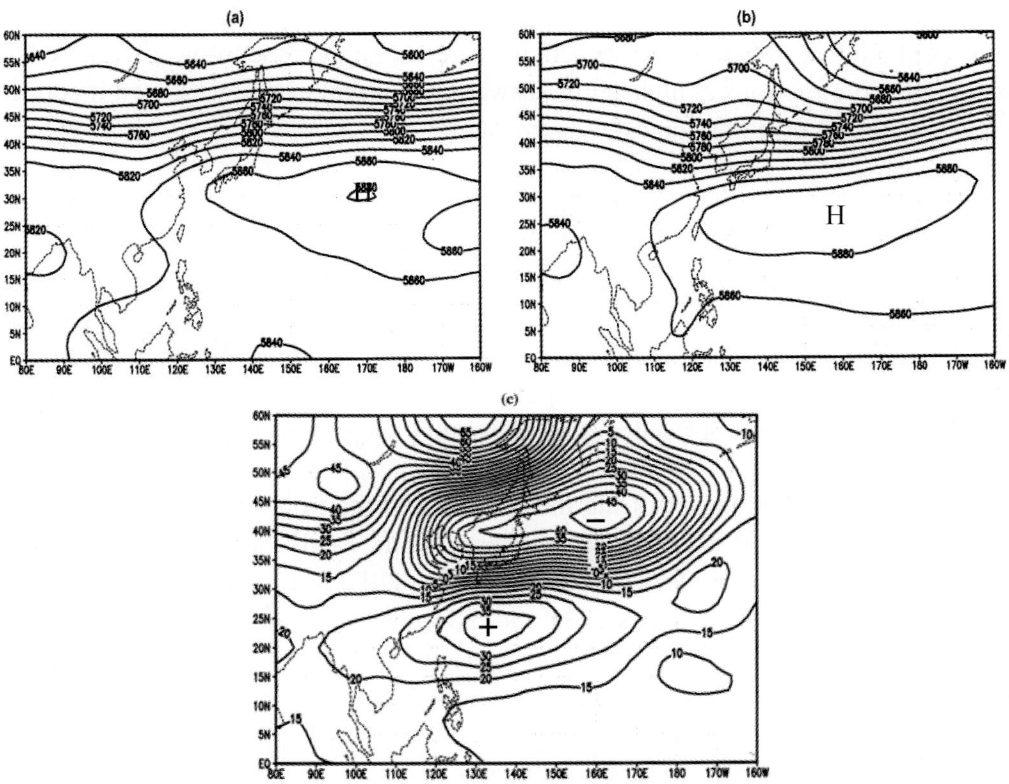

Figure 5.2. The composite distributions of monthly mean 500 hPa height field in July over East Asia for (a) summers with strong convective activities and (b) summers with weak convective activities around the Philippines (i.e., 10°N - 20°N, 110°E - 140°E) and (c) the difference between the distribution shown in Fig. 5.2b and that shown in Fig. 5.2a. Units: gpm. The NCEP/NCAR reanalysis data are used in this analysis. The contour interval is 20 gpm in Figs. 5.2a and 5.2b, and the solid and dashed contours (contour interval; 5 gpm) indicate positive and negative values of the 500 hPa height differences in Fig. 5.2c, respectively.

In order to show clearly the impact of the thermal state of the tropical western Pacific or the convective activities around the Philippines on the western Pacific subtropical high, the difference between the composite 500 hPa height field in July for the summers with weak convective activities around the Philippines and that for the summers with strong convective activities around the Philippines is shown in Fig. 5.2c. As shown in the figure, the positive differences are located in South China and eastern Siberia, respectively, and the negative differences appear in the Yangtze River and Huaihe River valley and North China. This features similar characteristics to the EAP pattern circulation anomalies over East Asia. It may explain that the location of the western Pacific subtropical high in a summer with weak convective activities around the Philippines is obviously different from that in a summer with strong convective activities around the Philippines.

In the above-mentioned analysis, the thermal state of the tropical western Pacific greatly influences the anomalous location of the western Pacific subtropical high. Recently, Lu (2001) and Lu and Dong (2001) also showed that the convective activities over the tropical western Pacific have a significant impact on the zonal shifts of the western Pacific subtropical high. They pointed out that if the convective activities are strong over the tropical western Pacific, the western Pacific subtropical high shifts eastward, but it will extend westward if convective activities are weaker over the tropical western Pacific.

5.3.2. Impact of the Thermal State of the Tropical Western Pacific on Convective Activities over the Tropical Western Pacific and East Asia

Nitta (1986) systematically studied the long-term variability of convective activities over the tropical western Pacific and showed that the interannual variations of high cloud amount in this region are closely associated with the thermal state of the tropical western Pacific. Nitta (1987) and Huang and Li (1987, 1988) pointed out that the thermal state of the tropical western Pacific can also influence the convective activities over East Asia through teleconnection.

In order to show the impact of the thermal state of the tropical western Pacific on the convective activities over the tropical western Pacific and East Asia, GMS-observed $1^\circ \times 1^\circ$ data of monthly mean black body temperature (TBB) from 1980 to 1998, provided by JMA, are used in this study. Figures 5.3a and 5.3b are the composite distribution of TBB anomalies in East Asia and the tropical western Pacific averaged for the summers (June-August) with the warming state and for the cooling state of the tropical western Pacific, respectively. Since TBB indicates the temperature of a black body at the surface in a cloud free area, the value of TBB is higher in the area. Thus, a high value of TBB or positive anomaly of TBB can denote weak convective activities. Oppositely, in a cloudy area, TBB indicates the temperature of a black body at the cloud top, and the value of TBB may be lower. Thus, a low value of TBB or negative anomaly of TBB can indicate strong convective activities in the area. By this way, it may be shown from Fig. 5.3a that in a summer with a warming state of the tropical western Pacific, i.e., the summer SST anomaly in the area of NINO-west (EQ - 14°N, 130°E - 150°E) is above normal, convective activities are weak from the Yangtze River and Huaihe River valley to South Korea and Japan, and strong from the Indo-China Peninsula and the South China Sea to the tropical western Pacific around the Philippines. In contrast, Fig. 5.3b shows that in a summer with a cooling state of the tropical western Pacific, i.e., the summer SST anomaly in the area of NINO-west is below normal, distributions of strong and weak areas of convective activities are just opposite to those shown in Fig. 5.3a. Moreover, Fig. 5.3b shows that convective activities are strong over the equatorial central Pacific in this case. Beside, from Figs. 5.3a and 5.3b, it may be seen that the distributions of convective activities over the tropical Pacific and East Asia also exhibit a meridional tripole pattern, which corresponds to that shown in Figs. 3.5a.

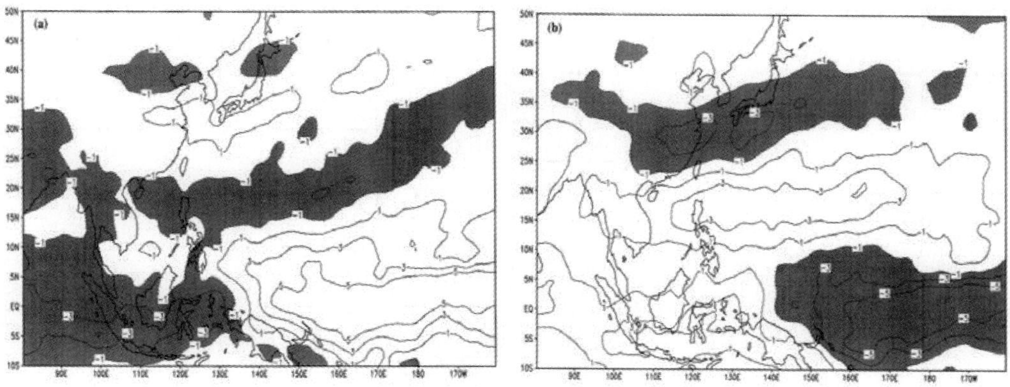

Figure 5.3. The composite distributions of TBB (black body temperature) anomalies in East Asia and the tropical western Pacific for (a) summers with the warming state and (b) summers with the cooling state of the tropical western Pacific. Units: K. The solid and dashed contours denote positive and negative TBB anomalies, respectively, and the TBB anomalies lower than -1K (e.g., strong convective activities) are shaded. The climatological mean summer (June-August) TBB is obtained by averaging the mean values of daily TBB from 1 June to 31 August. Data are obtained from JMA.

5.3.3. The East Asia/Pacific Pattern Teleconnection

The analyses of observed data showed that there is a teleconnection pattern of the summer circulation pattern over the Northern Hemisphere, i.e., the so-called Pacific-Japan (PJ) Oscillation (e.g., Nitta 1987) or the so-called East Asia/Pacific pattern teleconnection (e.g., Huang and Li 1987, 1988) and Lau (1992) also suggested a similar teleconnection pattern. The above-mentioned analyses of the detailed data can demonstrate further the existence of the EAP teleconnection pattern in the summer circulation anomalies over the Northern Hemisphere shown in Fig. 5.4, suggested by Nitta (1987) and Huang and Li (1987, 1988) from the observed OLR and circulation anomalies, respectively. It is seen from this teleconnection pattern that the quasi-stationary planetary wavetrain can propagate from Southeast Asia to the western coast of North America through East Asia during the Northern Hemisphere summer. What cause the formation of the EAP teleconnection pattern of the summer circulation anomalies over the Northern Hemisphere? This may be explained from the theory of planetary wave propagation (e.g., Huang and Li 1987, 1988; Huang and Sun 1992). As shown in Section 2, since the tropical western Pacific is a region of the highest SST in the global sea and the strong ascending branch of the Walker circulation is also over this region, the strong convergence of air and moisture leads to strong convective activities over this region. Thus, there is a strong heat source caused by strong convective activities. Huang and Li (1987, 1988) showed that due to the forcing by the heat source caused by strong convections, the quasi-stationary planetary waves responding to this forcing can propagate over the Northern Hemisphere. They calculated the propagating ray of planetary waves forced by an idealized heat source around the Philippines in a realistic summer basic flow with the formula of propagating ray path of planetary waves on the sphere, proposed by Hoskins and Karoly (1981). It may be seen from Fig. 5.5 that the planetary waves forced by a

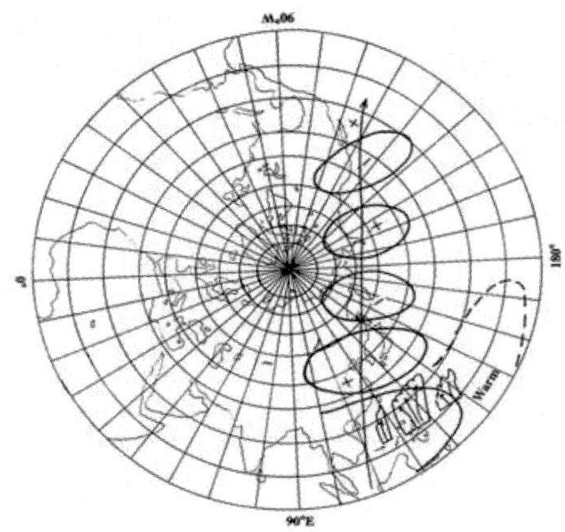

Figure 5.4. Schematic diagram of the East Asia/ Pacific (EAP) teleconnection pattern of summer circulation anomalies over the Northern Hemisphere (from Nitta 1987; Huang and Li 1987). + and − in the figure indicate positive and negative height anomalies at 500 hPa, respectively.

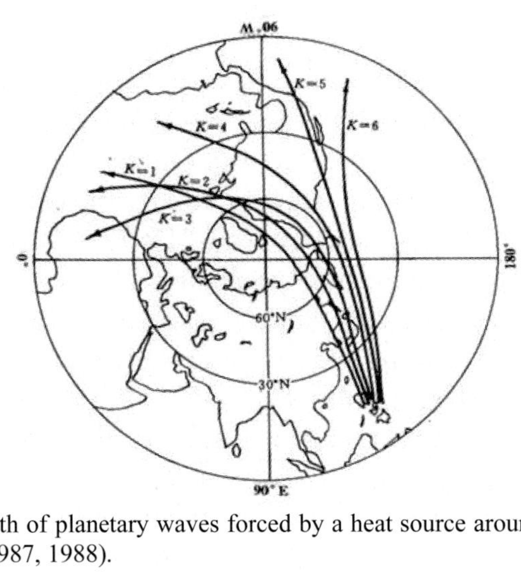

Figure 5.5. Propagating ray path of planetary waves forced by a heat source around the Philippines during boreal summer (from Huang and Li 1987, 1988).

heat source around the Philippines can propagate from the area around the Philippines toward the western coast of North America through East Asia and the North Pacific, which is in agreement with the teleconnection pattern shown in Fig. 5.4, obtained from the observed data (e.g., Nitta 1987; Huang and Li 1987; 1988; Lau 1992). Moreover, this teleconnection pattern has been demonstrated by Huang and Lu (1989), Nikaido (1989) and Huang and Sun (1992) using numerical simulations with general circulation models.

The above-mentioned analyses show that the impacts of the thermal state of the tropical western Pacific and convective activities around the Philippines on the interannual variations of the location of the western Pacific subtropical high, monsoon circulation and convective activities over East Asia during summer may be through the EAP pattern teleconnection, i.e., through the propagation of the quasi-stationary planetary wavetrain forced by the heat source around the Philippines.

5.4. *Impact of the Thermal State of the Tropical Western Pacific on the Onset of the South China Sea Monsoon (SCSM)*

The onset of the EASM is earlier in the area over the South China Sea, which generally is in May (e.g., Tao and Chen 1987), and this summer monsoon is called the South China Sea summer monsoon (hereafter SCSM). The onset of the SCSM is not only an important phenomenon of the intraseasonal variations of the EASM, but also has a large influence on the interannual variability of the Meiyu in the Yantze River and Huaihe River valley, the Baiu in Japan and the Changma in Korea. Figure 5.6 is the correlation between the summer (June-August) precipitation in China and the onset date of the SCSM defined by Lian and Wu

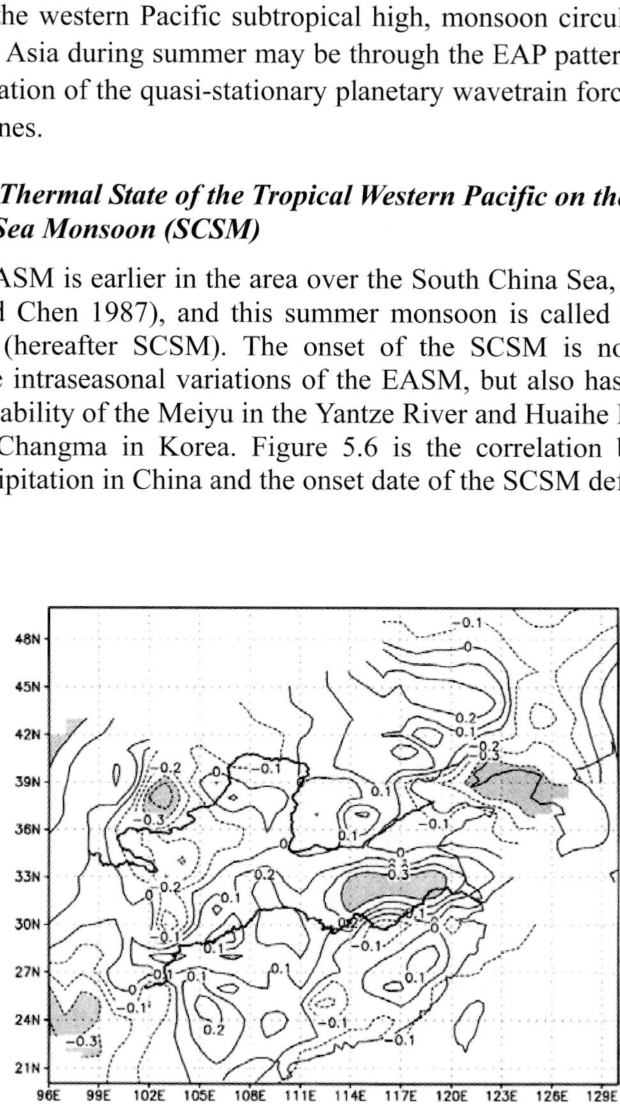

Figure 5.6. Correlation between the summer (June-August) rainfall in China and the onset date of the South China Sea monsoon (SCSM) during the period of 1951-1999. The shaded areas denote the correlations over the 95% confidence level, and the dashed and solid contours indicate negative and positive correlations, respectively. The onset date of the SCSM is provided by Lian and Wu (2002).

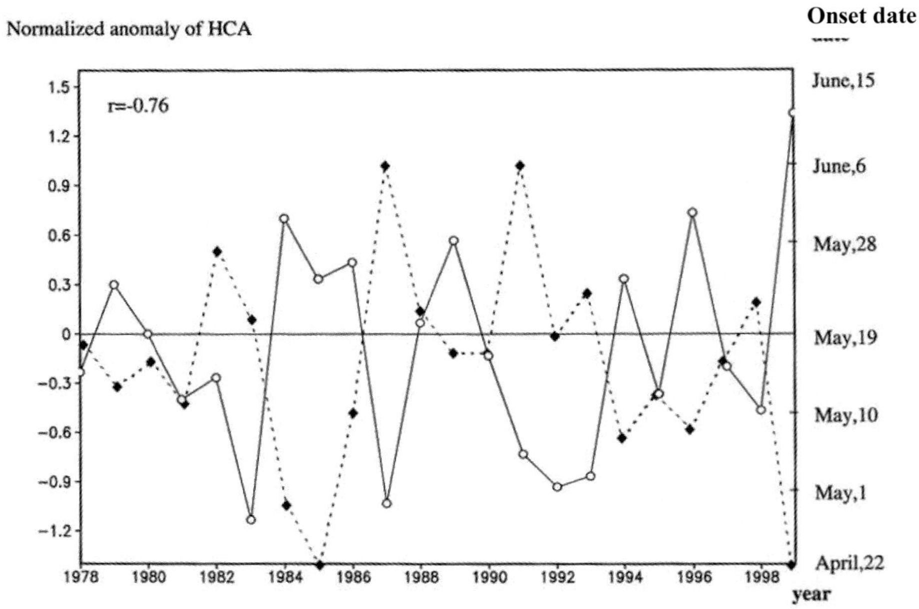

Figure 5.7. The interannual variations of onset date of the South China Sea summer monsoon (dashed line) (from Lian and Wu 2002) and the normalized high cloud amount anomaly (HCA) around the Philippines (i.e., 10°N - 20°N, 110°E - 140°E) in spring (March-May) (solid line) during the period of 1951-1999. Data of high cloud amount are obtained from Monthly Report on Climate System, JMA.

(2002) for 49 summers from 1951 to 1999. It may be clearly seen from the figure that there is an obvious positive correlation between the summer rainfall in the Yangtze River and Huaihe River valley and the onset date of the SCSM, the correlation coefficients between them are larger than 0.3 and exceed the 95% significance level. This is to say, if the onset of the SCSM is early, then the summer rainfall will be below normal in the Yangtze River and Huaihe River valley, but if the onset of the SCSM is late, such as in 1998, the summer rainfall will be above normal in this region. This has been used in prediction of summer monsoon rainfall anomalies in the Yangtze River and the Huaihe River valley.

The onset of the SCSM is closely associated with the thermal state of the tropical western Pacific in the leading winter and spring, especially with the convective activities around the Philippines in spring. As shown in Fig. 5.7, the correlation coefficient between the onset date of the SCSM and the normalized high cloud amount anomaly around the Philippines (i.e., 10°N - 20°N, 110°E - 140°E) in spring (March-May) can reach –0.76, which greatly exceeds the 99% significant level. Therefore, as shown in Fig. 5.7, the onset of the SCSM is early in a spring with a warming state of the tropical western Pacific, while it is late in a spring with a cooling state of the tropical western Pacific.

The impact of the convective activities around the Philippines on the onset of the SCSM may be interpreted from the Walker circulation. Cornejo-Garrido and Stone (1977) and Hartmann *et al.* (1984) showed that the heating caused by strong convective activities over the tropical western Pacific supplies the energy to drive the strong Walker circulation over the

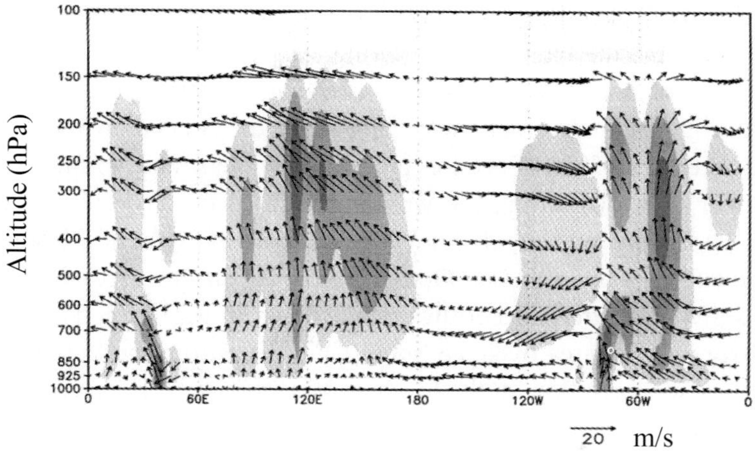

Figure 5.8. The composite distribution of zonal-altitude circulation along the equator (averaged between 5°S - 5°N) for the periods of the preceding November-April during the warming state of the tropical western Pacific (e.g., the positive SST anomaly in the area of NINO-west, i.e., EQ - 14°N, 130°E - 150°E). The vertical velocities in the figure are multiplied by 100. The NCEP/NCAR reanalyzed data are used in this analysis.

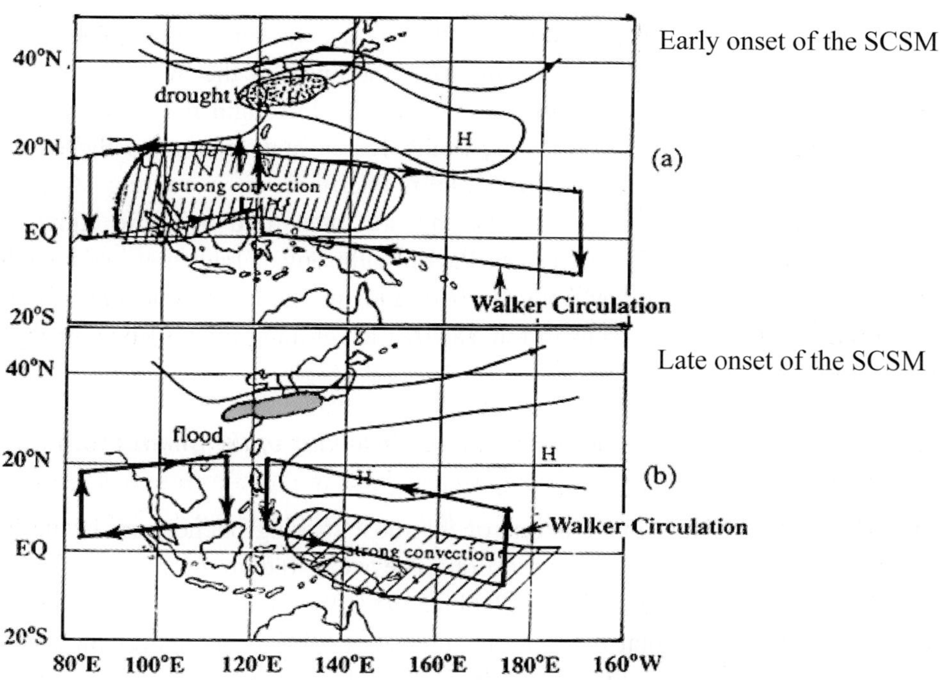

Figure 5.9. Schematic map of the relationships among the SST in the tropical western Pacific (TWP), the convective activities around the Philippines, the western Pacific subtropical high, the onset of the South China Sea summer monsoon (SCSM) and the summer rainfall in China and its surrounding regions. (a) in the warming state of the TWP; (b) in the cooling state of the TWP.

tropical western Pacific. When the tropical western Pacific is in a warming state, convective activities are strong around the Philippines in spring, as shown in Fig. 5.7. Thus, in this case, the ascending branch of the Walker circulation is strong over the tropical western Pacific. Figure 5.8 is the composite distribution of zonal-altitude circulation over the equatorial region (averaged between 5°S - 5°N) for preceding winters and springs (i.e., from the preceding November to April) during the warming state of the tropical western Pacific. It features a strong ascending branch of the Walker circulation over the equatorial western Pacific and the equatorial eastern Indian Ocean. Moreover, in this case, the western Pacific subtropical high shifts eastward and northward, as shown in Fig. 5.2a. Thus, the southwest flow can enter early into the area over the South China Sea from the Indo-China Peninsula because of the strong westerly flow in the lower troposphere over the tropical eastern Indian Ocean, as shown in Fig. 5.8. As a consequence, the onset of the SCSM may be early in a warming case of the tropical western Pacific. Oppositely, when the tropical western Pacific is in a cooling state, the onset of the SCSM may be late.

From the above-mentioned analyses, the impacts of the thermal states of the tropical western Pacific on the western Pacific subtropical high, convective activities over the tropical western Pacific and East Asia, summer monsoon rainfall in East Asia and onset date of the SCSM may be schematically summarized in Fig. 5.9. As shown in Fig. 5.9a, when tropical western Pacific is in a warming state, convective activities are intensified from the Indo-China Peninsula to the area east of the Philippines, and the western Pacific subtropical high may shift unusually northward. In this case, the onset of the SCSM may be early, and the summer monsoon rainfall may be below normal and drought may occur in the Yangtze River and Huaihe River valley of China, South Korea and Japan. On the other hand, as shown in Fig. 5.9b, when the tropical western Pacific is in a cooling state, convective activities are weak around the Philippines and are intensified over the equatorial central Pacific near the dateline, and the western Pacific subtropical high may shift southward. In this case, the onset of the SCSM may be late, and the summer monsoon rainfall may be above normal and flood may occur in the Yangtze River and Huaihe River valley of China, South Korea and Japan.

5.5. *Impact of ENSO Cycle on the Interannual Variability of the Summer Monsoon over China*

It is well known that ENSO cycle is one of the most striking phenomena in the tropics and has a great influence on the Asian monsoon. The weak Indian summer monsoon tends to occur in El Niño years (e.g., Shukla and Paolina 1983; Webster *et al.* 1998). Huang and Wu's (1989) study first showed that the summer monsoon rainfall anomalies in East Asia may depend on the stages of ENSO cycle. Recently, Huang and Zhou (2002) analyzed the composite distributions of the summer monsoon rainfall anomalies during different stages of 14 El Niño events that occurred in the period of 1951-2000. During the 50 summers of 1951-2000, the summers of 1951, 1957, 1963, 1965, 1969, 1972, 1976, 1982, 1987, 1991, 1993 and 1997 were just in the developing stage of the El Niño event, respectively, while the summers of 1953, 1958, 1977, 1983, 1992 and 1998 were just in the decaying state of the El

Niño event, respectively. Figures 5.10a and 5.10b are the composite distributions of the summer monsoon rainfall anomalies for 12 summers in the developing stage of El Niño events and for 6 summers in the decaying stage of El Niño events, respectively. The composite distributions of monsoon rainfall anomalies for the summers in the different stages of ENSO cycles show that during a summer in the developing stage of an El Niño event, the summer monsoon rainfall is strong and flood tends to occur in the Yangtze River and Huaihe River valley of China, but the summer monsoon rainfall is weak and drought may be caused in North China, as shown in Fig. 5.10a. In contrast, during a summer in the decaying stage of an El Niño event, as shown in Fig. 5.10b, the summer monsoon rainfall is weak and drought tends to occur in the Yangtze River and Huaihe River valley of China, but the summer monsoon rainfall may be normal or above normal in North China, and large positive monsoon rainfall anomalies may appear to the south of the Yangtze River and severe flood tends to occur there. For example, in the summer of 1998 when the 1997/98 El Niño event was in its developing stage, a particularly severe flood disaster occurred to the south of the Yangtze River, as shown in Fig. 2.11.

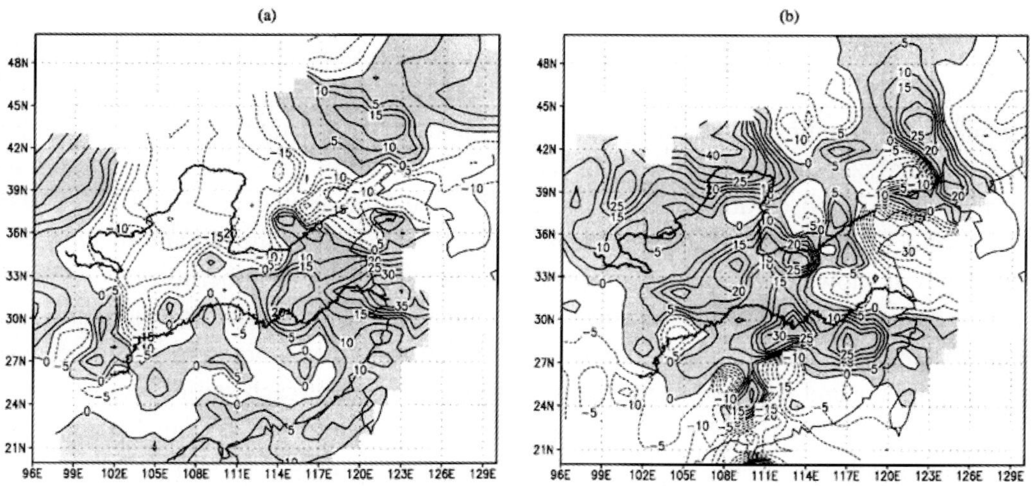

Figure 5.10. The composite distributions of summer rainfall anomalies (in percentage) in China (a) for 12 summers (June-August) when ENSO events were in their developing stage and (b) for 6 summers when ENSO events were in their decaying stage. The solid and dashed contours (contour interval: 5%) in these figures indicate positive and negative anomalies of rainfall, respectively, and the shaded areas denote positive rainfall anomaly regions. The climatological mean monthly precipitation in June, July and August shown in Figs. 2.2c-e are taken as the respective normals.

What causes severe flood disasters to occur in a summer in the decaying stage of an El Niño event in the Yangtze River valley? This may be associated with the water vapor transport from the tropical western Pacific and the Bay of Bengal. Zhang *et al.* (1996) and Zhang (2001) pointed out that since the western Pacific subtropical high shifts southward during and after the mature phase of an El Niño event, the southerly wind anomalies can appear in the lower troposphere along the southeast coast of China. The intensified southerly winds will be favorable for the transport of water vapor from the Bay of Bengal and the tropical western Pacific to the eastern part of China, which can provide the condition of sufficient water vapor for monsoon rainfall in this region.

However, it should be pointed out that there is an interaction between the Asian monsoon and ENSO cycle. Diagnostic and modeling studies have revealed that the variability of the Asian monsoon has a significant effect on the atmosphere/ocean coupled system in the equatorial Pacific (e.g., Yamagata and Matsumoto 1989; Li 1990; Yasunari 1990; Yasunari and Seki 1992; Webster and Yang 1992; Li *et al.* 2001). Because of the interaction between the Asian monsoon and ENSO cycle, the influence of ENSO cycle on the interannual variation of the summer monsoon over China is very complex and needs to be studied further.

5.6. *Impact of the Snow Cover in the Tibetan Plateau on the Interannual Variability of the Summer Monsoon over China*

The interannual variability of the Asian summer monsoon is also influenced by the Eurasian snow cover, especially the snow cover in the Tibetan Plateau. Chen and Yan (1981) and Wei and Luo (1996) pointed out that there is a positive correlation between the snow cover in the Tibetan Plateau and the summer monsoon rainfall in the upper and middle reaches of the Yangtze River.

Recently, the interannual variations of the days and depth of snow cover in the Tibetan Plateau were analyzed using the observed dataset of daily snow cover from the preceding October to May at 72 observational stations located in the Tibetan Plateau during 1960-1999 (Figs. 5.11a and 5.11b). Considering the quality of the observed data, the climatological mean days and depth of snow cover during the period from the preceding October to May averaged for the period of 1965-1999 at the various observational stations are taken as the respective normals. Moreover, both the days and depth anomalies of snow cover are normalized because the variance of days and depth of snow cover are very different at each station. As shown in Figs. 5.11a and 5.11b, both the days and depth of snow cover in the Tibetan Plateau have obvious interannual variations, and their interannual variations are similar. Thus, the normalized depths of snow cover over 2 and below –2 are considered as the criteria measuring a strong snow year and a weak snow year in this study, respectively. By this way, it can be seen from Fig. 5.11a that 1962, 1968, 1978, 1982, 1983, 1989, 1990, 1995 and 1998 were strong snow years, while 1960, 1965, 1967, 1971, 1976, 1984, 1995, 1991 and 1999 were weak snow years in the Tibetan Plateau.

The interannual variations of snow cover in the Tibetan Plateau have an important impact on the summer monsoon rainfall in the Yangtze River valley. Figure 5.12 is the correlation between the summer (June-August) rainfall in China and the normalized depth of snow cover in the period from the preceding October to May averaged for 72 observational stations in the Tibetan Plateau. It may be seen in the figure that there are larger positive correlations in the middle and upper reaches of the Yangtze River and negative correlations in South China and Northeast China. This may explain that if the snowfall in the Tibetan Plateau is heavy in the preceding winter and spring of a year, the following summer rainfall may be strong in the middle and upper reaches of the Yangtze River. For example, in the winter of 1997 and the spring of 1998, a particularly heavy snowfall appeared in the Tibetan Plateau as shown in Figs. 5.11a and 5.11b, and a heavy monsoon rainfall occurred in the middle and upper reaches of the Yangtze River and a particularly severe flood was caused in the Yangtze River valley in the summer of 1998, as shown in Fig. 2.11.

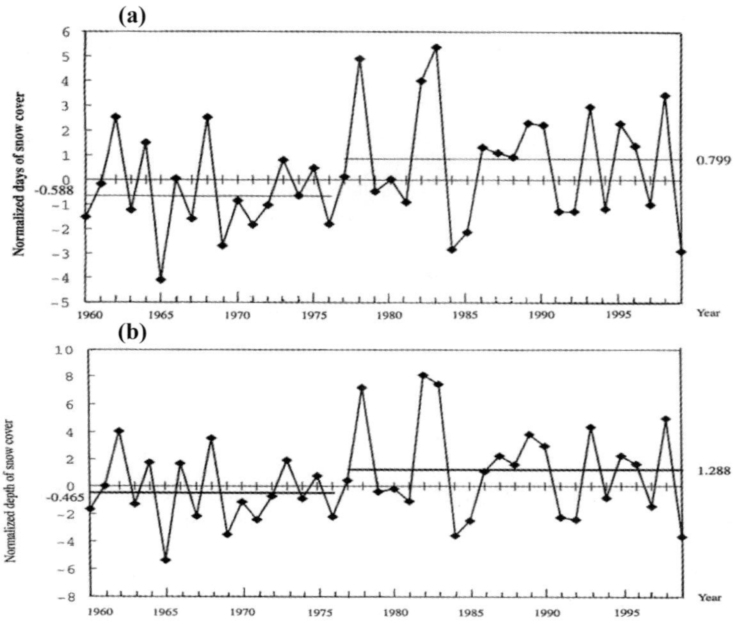

Figure 5.11. Interannual variations of the normalized anomalies of (a) days and (b) depth of snow cover over 0.5 cm during the period of the last October-May averaged for 72 observational stations in the Tibetan Plateau. The climatological mean values of days and depth of snow cover over 0.5cm during the period of the last October-May averaged for 1965-1999 are taken as the respective normals.

5.7. *Impact of the East Asian Winter Monsoon (EAWM) on the Summer Monsoon over China*

East Asia is also a region of strong winter monsoon. The winter monsoon features strong northwesterlies over North China and Northeast China, Korea and Japan and strong northeasterlies along the coast of China (e.g., Staff members of Academia Sinica 1957; Chen *et al.* 1991; Ding 1994). The strong winter monsoon can bring disasters such as low temperature, severe snow storms in Northwest China and Northeast China, North Korea and North Japan in winter, and severe sand dust-storms in North China, Northwest China and Korea in spring. Moreover, the Asian winter monsoon can cause strong convective activities over the maritime continent of Borneo and Indonesia (e.g., Chang *et al.* 1979; Lau and Chang 1987). Besides, strong and frequent activities of cold waves caused by a strong EAWM may trigger the occurrence of an El Niño event (see Li 1988). Tomita and Yasunari (1996) also pointed out that the EAWM might play a key role in the biennial oscillation of the ENSO/monsoon system. Thus, the EAWM may have an important impact on the EASM, and the interannual variability of the EAWM and its impact on the EASM are also an interesting scientific issue.

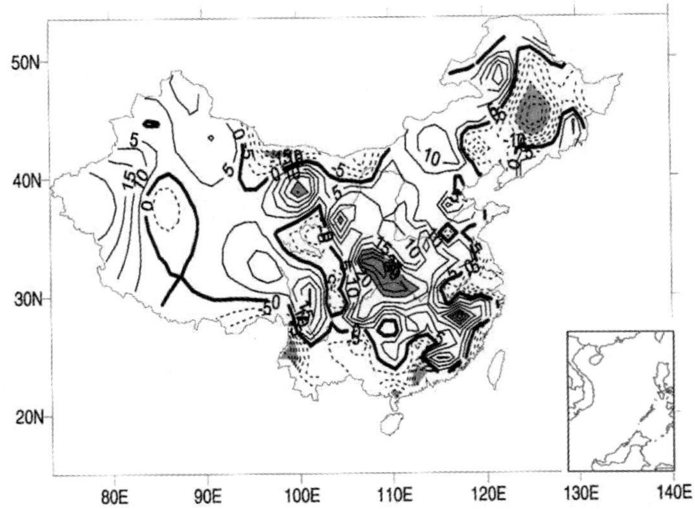

Figure 5.12. The correlation between the summer (June-August) rainfall in China and the normalized depth of snow cover averaged for 72 observational stations in the Tibetan Plateau. The thick black contour (contour interval: 5×10^{-2}) indicates zero correlation, and the solid and dashed contours denote positive and negative correlations, respectively, and the correlations over the 95% confidence level are shaded.

The East Asian summer monsoon and winter monsoon are a phenomenon of the annual cycle in both wind field in the lower troposphere and rainfall over East Asia (e.g., Chen *et al.* 2000). The anomalous EAWM can influence the following EASM (e.g., Sun and Sun 1994; Chen *et al.* 2000).

Chen and Graf (1998) and Chen *et al.* (2000) systematically investigated the interannual variability of the EAWM and its relation to the EASM with a new definition of the EAWM index. Chen *et al.* (2000) pointed out that after a strong EAWM, since the western Pacific subtropical high will shift northward in the following summer, generally a drought summer could occur in the middle and lower reaches of the Yangtze River. On the other hand, the western Pacific subtropical high will shift southward and a flood summer will occur in the Yangtze River and Huaihe River valley following a weak EAWM. For example, a severe flood occurred in the Yangtze River valley in the summer of 1998 following the anomalously weak EAWM in the winter of 1997 and the spring of 1998. This confirms the connection between the EAWM and the following EASM. As for the interannual variability of the EAWM and its impact on the EASM variability, these have been described in another chapter and will be omitted here.

6. Interdecadal Variability of the East Asian Climate System and its Impact on the Summer Monsoon over China

As described in Section 4, the interdecadal variability of the summer monsoon over China is significant, and has caused severe droughts in North China. In order to investigate the causes of the interdecadal variability of the summer monsoon over China and its surrounding regions, the interdecadal variability of the East Asian climate system will be analyzed by using the observed data.

6.1. *Interdecadal Variability of the SST Anomaly in the Tropical Pacific and its Impact on the Summer Monsoon over China*

The SST anomalies in the tropical Pacific can influence not only the interannual variability of the summer monsoon over East Asia, but also its interdecadal variability. In order to study the interdecadal variability of the SST anomaly in the tropical Pacific and its impact on the summer monsoon over China, the nine-year running mean SST anomalies in the equatorial Pacific (averaged between 5°S - 5°N) are analyzed using the database of QISST, provided by the Hadley Center, UK. As shown in Fig. 6.1, the SST anomalies display an obvious interdecadal variability in the equatorial central and eastern Pacific, and the SST in the equatorial central and eastern Pacific were below normal from the mid 1960s to the mid 1970s and remarkably increased during the early and middle 1980s and the 1990s in these regions. In order to show this variability clearly, the difference between the summer SST anomalies averaged for 1977-2000 and those averaged for 1967-1976 in the Pacific is also analyzed (Fig. 6.2). As shown in Fig. 6.2, an obvious El Niño -like SST anomaly pattern appeared in the tropical central and eastern Pacific in the period from 1977 to 2000 (e.g., Huang 2001). This may explain that a "decadal El Niño event" seems to have occurred from the late 1970s to now, while a "decadal La Niña event" occurred in the 1970s. Therefore, there seems to be a "decadal ENSO-like cycle" in the interdecadal variability of SST anomalies of the tropical Pacific.

The interdecadal variability of the summer monsoon over China and its surrounding regions may be in close association with the interdecadal variations of SST anomalies in the tropical Pacific. However, because it is difficult to reproduce the relationship between the monsoon and ENSO obtained from the observed data in climate models (e.g., Webster *et al.* 1998), this close relationship cannot be explained well with numerical simulations so far. Thus, it can only be simply discussed with correlation analysis. Figure 6.3 is the correlation between the summer (June-August) rainfall anomalies over China and the difference of the SST anomalies in the area of NINO.3 between summer and last autumn during 1951-1996 (e.g., Huang *et al.* 1998). From Fig. 6.3, it is seen that the positive correlation appears in the Yangtze River and Huaihe River valley and the lower reaches of the Yangtze River, and the correlation coefficients between them exceed the 95% significance level in these regions. Moreover, the negative correlations are located in North China and the area to the South of the Yangtze River, and these correlations also exceed the 95% significance level in these regions. Although Fig. 6.3 is the correlation between the summer rainfall in China and the

SST anomaly in the equatorial eastern Pacific on the interannual timescale, it may be used to estimate the relationship between them on the interdecadal timescale. The figure also shows that when the equatorial central and eastern Pacific is in a decadal warming episode, the summer monsoon rainfall may be strong in the Yangtze River and Huaihe River valley and the lower reaches of the Yangtze River and weak in North China. From the late 1970s to the

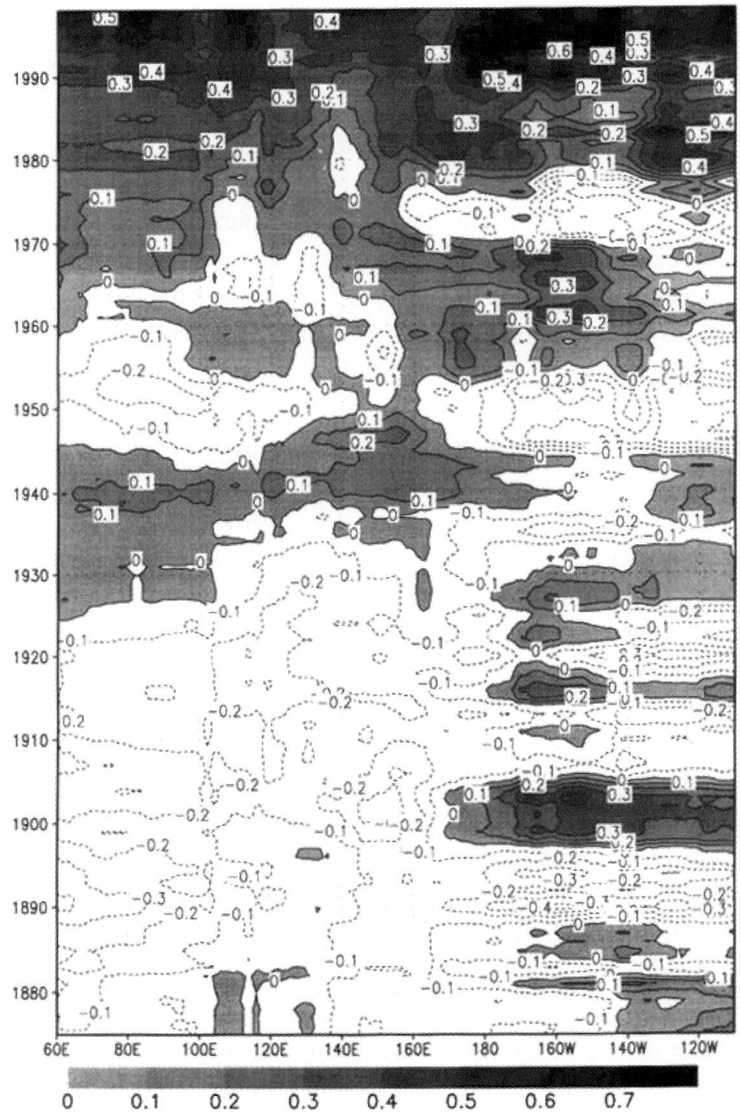

Figure 6.1. The longitude-time cross section of the 9-years running-mean SST anomalies along the equatorial Pacific (averaged between 5°S - 5°N). Units: °C. The solid and dashed contours in the figure denote positive and negative SST anomalies, respectively, and the shaded areas indicate the periods of positive SST anomalies. The dataset of QISST, Hadley Center, U.K. is used in this analysis.

1990s, the SST anomalies in the area of NINO.3 were positive, and according this correlation, the summer monsoon rainfall might be strong in the Yangtze River and Huaihe River valley and weak in North China during this period. Thus, the prolonged droughts that occurred in North China from the late 1970s to the 1990s may be closely associated with the obvious warming trend in the equatorial central and eastern Pacific from the late 1970s to the 1990s. In fact, the prolonged droughts still continue to appear in North China.

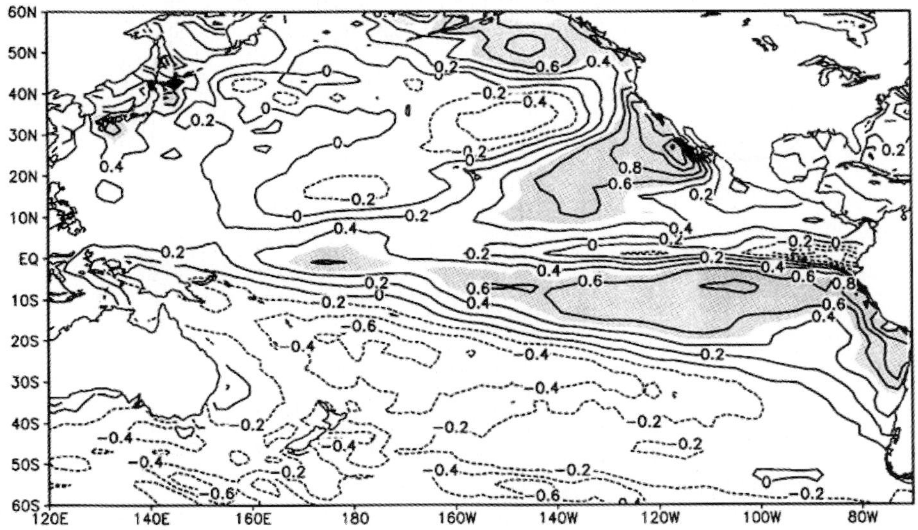

Figure 6.2. Distributions of the difference between the summer (June-August) SST anomalies in the Pacific averaged for 1977-2000 and those averaged for 1967-1976. Units: °C. The solid and dashed contours denote positive and negative SST anomalies, respectively, and the positive SST anomalies over +0.5°C are shaded.

In Section 4, it was discussed from the EOF analysis of the meridional component of water vapor transport that there was a serious decrease of water vapor transport by the EASM in North China from the late 1970s. What caused the decrease of water vapor transport by the summer monsoon in North China from the late 1970s? This may be closely related to the interdecadal variations of SST in the tropical Pacific. Influenced by the interdecadal variations of SST in the tropical Pacific, the interdecadal anomaly distribution of water vapor transport flux over East Asia and the tropical western Pacific after the late 1970s is different from that before the late 1970s. Figures 6.4a and 6.4b are the anomaly distributions of water vapor transport flux averaged for the period of 1967-1976 and for the period of 1977-1999, respectively. In the figures, the climatological mean distribution of water vapor transport averaged for 1961-1990 is taken as the normal. From Fig. 6.4a, it may be seen that during the period of 1967-1976, the westward transport flux anomalies of water vapor by the trade winds along the equator were strong, and the northward transport flux anomalies of water vapor appeared over the South China Sea and around the Philippines. Moreover, the water vapor anomaly fluxes were transported from the Bay of Bengal toward the Indo-China Peninsula and South China. Therefore, a large amount of water vapor was transported into North China from the tropical western Pacific, the South China Sea and the Bay of Bengal in the period of 1967-1976. This provided the condition of sufficient water vapor transport for the strong summer rainfall in North China during 1967-1976.

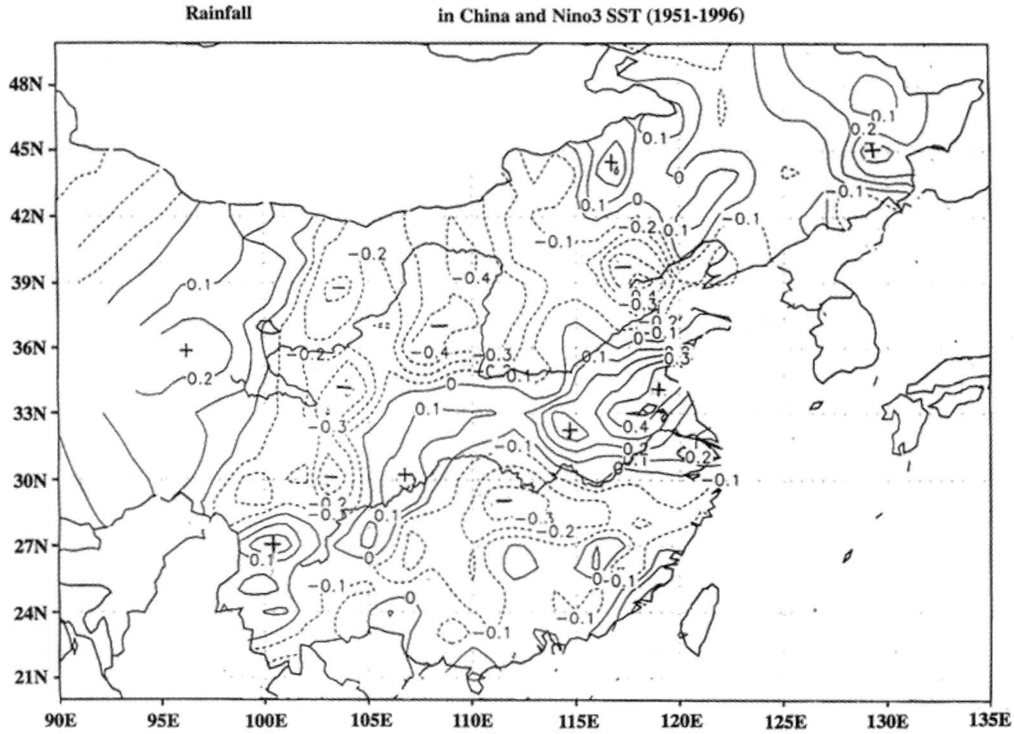

Figure 6.3. Correlation between the summer (June-August) rainfall anomalies in China and the differences of SST anomaly in the area of NINO.3 (i.e., 150°W - 90°W, 4°S - 4°N) between summer and the preceding autumn from 1951 to 1996. The areas of solid and dashed contours in the figure denote the positive and negative correlations, respectively.

However, in the period of 1977-1999, influenced by the warming of the tropical Pacific as shown in Fig. 6.2, the trade winds became weak over the tropical Pacific, and the distribution of water vapor transport anomalies shown in Fig. 6.4b is opposite to that during 1967-1976 as shown in Fig. 6.4a. Since the trade winds became weak over the tropical Pacific from the late 1970s, the eastward transport flux anomalies of water vapor appeared over the tropical Pacific, and the southward transport flux anomalies of water vapor appeared over the South China Sea and around the Philippines. Moreover, the water vapor anomaly fluxes as shown in Fig. 6.4b show the weak water vapor transport from the Bay of Bengal to the Indo-China Peninsula and South China. Because the southward transport flux anomalies of water vapor appeared over North China and East China, as shown in Fig. 6.4b, the water vapor transported into the eastern part of China from the tropical western Pacific, the South China Sea and the Bay of Bengal became weak from the late 1970s. As a consequence, the water vapor transport by the summer monsoon flow became weak in North China, and a large amount of water vapor converged in the Yangtze River valley. This caused the remarkable decrease of summer precipitation in North China and the significant increase of summer precipitation in the Yangtze River valley from the late 1970s to the 1990s.

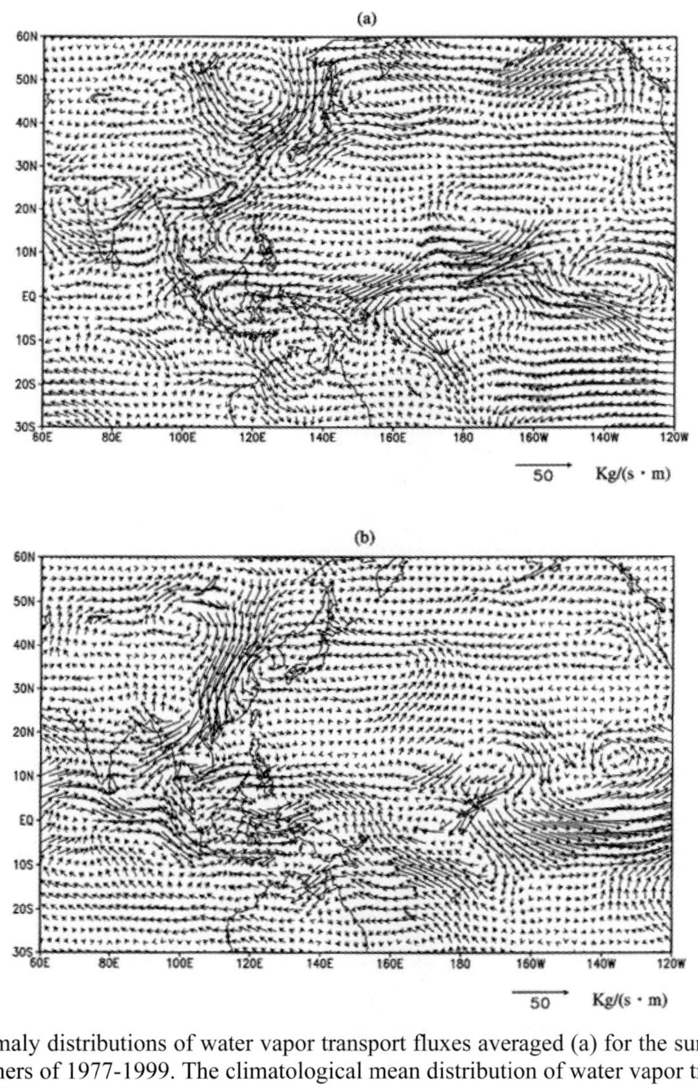

Figure 6.4. The anomaly distributions of water vapor transport fluxes averaged (a) for the summers of 1967-1976, and (b) for the summers of 1977-1999. The climatological mean distribution of water vapor transport fluxes for 30 years from 1961 to 1990 are taken as the normal.

6.2. *Interdecadal Variability of Snow Cover in the Tibetan Plateau and its Impact on the Summer Monsoon over China*

As shown by Chen *et al.* (2000), the EAWM became weak from the late 1980s. According to Tao and Zhang's (1998) study, if the EAWM is weak in a winter or spring, the disturbances over the south side of the Tibetan Plateau tend to be more active, which cause stronger rainfall or snowfall in the Tibetan Plateau. In fact, in comparing the time and depth of snow cover in the Tibetan Plateau in the period from the late 1970s to the late 1990s with those in the period from the 1960s to the mid 1970s as shown in Figs. 5.11a and 5.11b, the time of snow cover has become longer and the snow depth has become deeper in the Tibetan Plateau from the late 1970s.

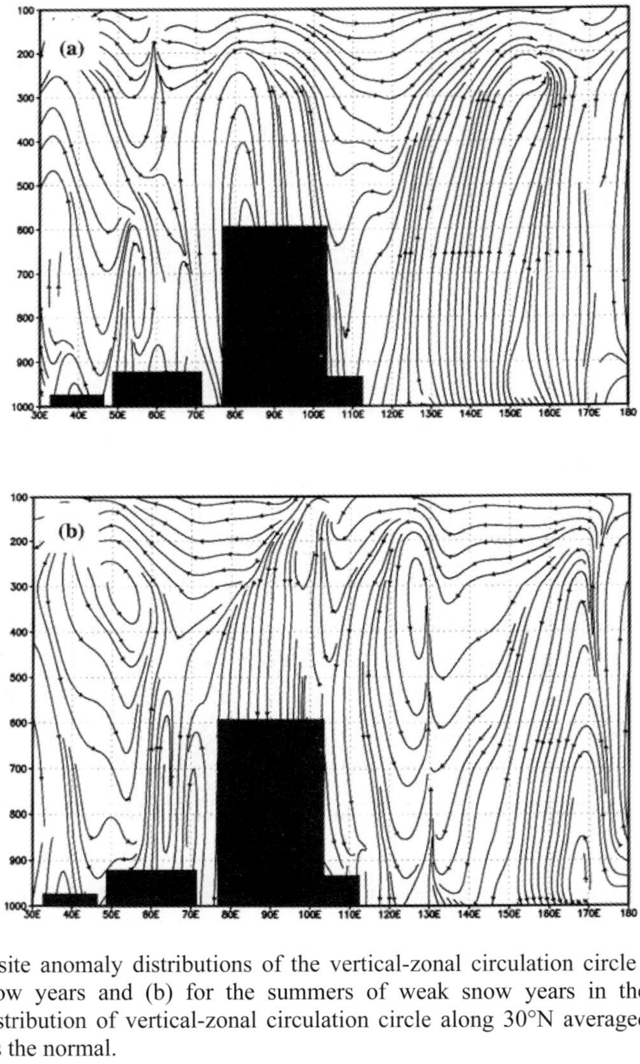

Figure 6.5. The composite anomaly distributions of the vertical-zonal circulation circle along 30°N for (a) the summers of strong snow years and (b) for the summers of weak snow years in the Tibetan Plateau. The climatological mean distribution of vertical-zonal circulation circle along 30°N averaged for 35 summers from 1965 to 1999 is taken as the normal.

The interdecadal variability of snow cover in the Tibetan Plateau in winter and spring may have a significant effect on the interdecadal variations of summer monsoon rainfall in the central and eastern parts of China, especially in the middle and upper reaches of the Yangtze River. As shown in Fig.5.12, the correlations between the summer rainfall and the normalized depth of snow cover in the Tibetan Plateau in winter and spring exceeded the 95% confidence level in these regions. Since the time of snow cover became longer and the depth of snow cover became deeper in the Tibetan Plateau from the late 1970s to the late 1990s, it may be also explained from Fig. 5.12 that the summer rainfall became stronger in the middle and upper reaches of the Yangtze River.

Why can the snow cover in the Tibetan Plateau in winter and spring influence the following summer monsoon rainfall in the Yangtze River valley on the interdecadal timescale? This problem has not been well studied up to now, especially using numerical simulations. Thus, it can only be simply explained from the observed data. Based on a preliminary investigation, the snow cover in the Tibetan Plateau has an important influence on the vertical-zonal circulation circle in the east and west sides of the Tibetan Plateau. Figures 6.5a and 6.5b are the composite anomaly distributions of the vertical-zonal circulation circle along 30°N for the summers of strong snow years and for the summers of weak snow years as defined in Section 5, respectively. In the figures, the vertical-zonal circulation along 30°N averaged for the period of 1965-1999 is taken as the climatological mean. From Fig. 6.5a, it may be seen that when the depth of snow cover in the Tibetan Plateau is deeper from the preceding winters to springs during a period, a stronger ascending flow may appear in the area from 110°E to 160°E in the following summers during the period. This may show that when the snowfall in the Tibetan Plateau is heavy in winters or springs during a period, a stronger ascending flow will appear in the east side of the Tibetan Plateau during the following summers. This may cause strong summer rainfall in the middle and upper reaches of the Yangtze River during the period. Since the time of snow cover became longer and the depth of snow cover became deeper in the Tibetan Plateau from the late 1970s, according to the composite distribution of vertical-zonal circulation anomalies shown in Fig. 6.5a, the stronger ascending flow might appear in the area from the middle and upper reaches of the Yangtze River to the subtropical western Pacific from the late 1970s. This may cause strong convective activities and strong monsoon rainfall in these regions. Therefore, the interdecadal variability of snow cover in the Tibetan Plateau also has a significant influence on the interdecadal variability of summer monsoon rainfall in the Yangtze River valley.

7. Summary and Discussion

It is seen from the above review that significant progress in recent research on the intraseasonal, interannual and interdecadal variations of the summer monsoon over China and its surrounding regions has been achieved. Moreover, the causes and physical processes of these variabilities have been preliminarily sought. Through these studies, the seasonal prediction of the summer monsoon over China has been improved to some extent and the predictability of the interannual and interdecadal variabilities of the summer monsoon over

East Asia has been understood further. However, since climate variations of the summer monsoon over East Asia and their physical mechanism are very complex, there are many unknown phenomena in the intraseasonal, interannual and interdecadal variabilities of the summer monsoon over China and its surrounding regions, and the causes and physical processes of these variabilities are still unclear so far. Besides, the summer monsoon rainband can not be simulated well in GCMs or coupled models. Therefore, the following problems need to be studied further:

(1) Interaction between the monsoon activity and the low-frequency oscillation in East Asia.
(2) Internal-dynamic process of monsoon variability in East Asia.
(3) Physical mechanism of the interannual variability of the EASM and its linkage with the EAWM as a phenomenon of the annual cycle.
(4) Processes of the interaction between the Asian monsoon and ENSO cycle and its reproduction in coupled models.
(5) Modeling of the EASM, especially the summer monsoon rainband.
(6) Causes of the interdecadal variability of the EASM and its impact on interannual variability.

The authors believe that with the implementation of the CLIVAR Program, the above-mentioned problems related to climate variations of the summer monsoon over China will be well investigated from observational, theoretical and modeling aspects in the early part of the 21st century.

Acknowledgments

The authors are much indebted to the detailed reviews of the manuscript by Dr. S. Yang and related referees and wish to thank Drs. B.-H. Ren and W. Chen for providing new observed data of TBB over East Asia and the tropical western Pacific and new results associated with the East Asian winter monsoon. Moreover, the authors are grateful to Mr. Lian and Mr. Wu for providing the onset date of the SCSM and to Mr. D. P. Griffith for his valuable comments on the style of presentation of the manuscript in English. This paper was written under the supports of Project 40231005, one of the projects of the Major Research Program for Global Change and Regional Response, National Natural Science Foundation of China, and Project KZCX3-SW-218, one of the Projects of the program for Knowledge Innovation, Chinese Academy of Sciences.

References

Cao, J., R. H. Huang, Y. Q. Xie, and Y. Tao, 2002: Research on the evolution mechanism of the western Pacific subtropical high. *Science in China (Series D)*, **45**, 659-666.

Chang, C. P., J. Erickson, and K. M. Lau, 1979: Northeasterly cold surges and near-equatorial disturbances over the winter-MONEX area during 1972. Part I: Synoptic aspects. *Mon. Wea. Rev.*, **107**, 812-829.

______, Y. S. Zhang, and T. Li, 2000: Interannual and interdecadal variations of the East Asian

summer monsoon and tropical Pacific SSTs. I, II. *J. Climate*, **13**, 4310-4340.
Charney, J. G., and J. G. Devore, 1979: Multiple flow equilibrium in the atmosphere and blocking. *J. Atmos. Sci.*, **36**, 1205-1216.
Chen, L. T., and Z. X. Yan, 1981: The statistical analysis of the impact of the anomalous snow cover over the Tibetan Plateau on the early-summer monsoon. *Collected Papers of Medium and Long-Range Hydrological-Meteorological Forecast, II.* Water Resources and Electric Power press, Beijing, 133-141. (in Chinese)
Chen, L. X., Q. G. Zhu, and H. B. Luo, 1991: *East Asian Monsoon.* China Meteorological Press, Beijing, 362pp. (in Chinese)
Chen, W., and H. F. Graf, 1998: The interannual variability of East Asian winter monsoon and its relationship to global circulation. Max-Planck-Institute fur Meteorologie, Report **No. 250**.
______, ______, and R. H. Huang, 2000: Interannual variability of East Asian winter monsoon and its relation to the summer monsoon. *Adv. in Atmos. Sci.*, **17**, 48-60.
Cornejo-Garrido, A. G., and P. H. Stone, 1977: On the heat source of the Walker circulation. *J. Atmos. Sci.*, **34**, 1155-1162.
Ding, Y. H., 1992: Summer monsoon rainfall in China. *J. Meteor. Soc. Japan*, **70**, 373-396.
______, 1994: *Monsoon over China.* Kluwer Academic Publishers, 420pp.
Gao, Y. X., and S. Y. Xu, 1962: Problem on monsoon over East Asia. Collected Papers of the Institute of Geophysics & Meteorology. Chinese Academy of Sciences, Beijing, Science Press, **No. 5**, 1-106. (in Chinese)
Hahn, D. C., and S. Manabe, 1975: The role of mountains in the South Asian monsoon circulation. *J. Atmos. Sci.*, **32**, 1515-1541.
Hoskins, B. J., and D. J. Karoly, 1981: The steady linear response of a spherical atmosphere to thermal and topographic forcing. *J. Atmos. Sci.*, **38**, 1179-1196.
Hartmann, D., H. Hendon, and R. A. Houze, 1984: Some implications of the mesoscale circulations in tropical cloud clusters for large-scale dynamics and climate. *J. Atmos. Sci.*, **41**, 113-121.
Huang, G., and Z. W. Yan, 1999: The East Asian summer monsoon circulation anomaly index and its interannual variation. *Chinese Sci. Bull.*, **44**, 1325-1328.
______, 1999: Study on the relationship between summer monsoon circulation anomaly index and the climatic variations in East Asia. *Quart. J. Appl. Meteor.*, **10**, 61-69. (in Chinese)
Huang, R. H., 1984: The characteristics of the forced planetary wave propagations in the summer Northern Hemisphere. *Adv. in Atmos. Sci.*, **1**, 85-94.
______, 1985: Numerical simulation of the three-dimensional teleconnections in the summer circulation over the Northern Hemisphere. *Adv. in Atmos. Sci.*, **2**, 81-92.
______, and W. J. Li, 1987: Influence of the heat source anomaly over the tropical western Pacific on the subtropical high over East Asia. *Proceedings of the International Conference on the General Circulation of East Asia*, Chengdu, April 10-15, 1987, 40-51.
______, and ______, 1988: Influence of the heat source anomaly over the tropical western Pacific on the subtropical high over East Asia and its physical mechanism. *Chinese J. Atmos. Sci.*, **14**, Special Issue, 95-107. (in Chinese)
______, and L. Lu, 1989: Numerical simulation of the relationship between the anomaly of the subtropical high over East Asia and the convective activities in the western tropical Pacific. *Adv. in Atmos. Sci.*, **6**, 202-214.
______, and Y. F. Wu, 1989: The influence of ENSO on the summer climate change in China and its mechanisms. *Adv. in Atmos. Sci.*, **6**, 21-32.

______, and F. Y. Sun, 1992: Impact of the tropical western Pacific on the East Asian summer monsoon. *J. Meteor. Soc. Japan.* **70** (1B), 243-256.

______, B. Y. Yin, and A. D. Liu 1993: Intraseasonal variability of the East Asian summer monsoon and its association with the convective activities in the tropical western Pacific. *Climate Variability*, D. Z. Ye *et al.*, Eds., China Meteorological Press, Beijing, 134-155.

______, and F. Y. Sun, 1994a: Impact of the thermal state and convective activities over the western Pacific warm pool on summer climate anomalies in East Asia. *Chinese J. Atmos. Sci.*, **18**, 262-272. (in Chinese)

______, and ______, 1994b: Impact of the convective activities over the West Pacific warm pool on the intraseasonal variability of summer climate in East Asia. *Chinese J. Atmos., Sci.*, **18**, 456-465. (in Chinese)

______, 1994: Interaction between the 30-60 day oscillation, the Walker circulation and the convective activities over the tropical western Pacific and their relations to the interannual oscillation. *Adv. in Atmos. Sci.*, **11**, 367-384.

______, Z. Z. Zhang, G. Huang, and B. H. Ren, 1998: Characteristics of the water vapor transport in East Asian monsoon region and its difference from that in South Asian monsoon region in summer. *Chinese J. Atmos. Sci.*, **22**, 460-469. (in Chinese)

______, B. H. Ren, and G. Huang, 1998: Further investigation on the impact of the tropical western Pacific on the East Asian summer monsoon. *Unraveling the Mystery of Climate System Variability*. Series of Books of Climate System Research, A. Sumi, Ed., CCSR Tokyo University, **No. 3** 107-117. (in Japanese)

______, Y. H. Xu, P. F. Wang, and L. T. Zhou, 1998: The features of the particularly severe flood over the Changjiang (Yangtze River) basin during the summer of 1998 and exploration of its cause. *Clim. & Envir. Res.*, **3**, 300-313. (in Chinese)

______, Y. H. Xu, and L. T. Zhou, 1999: The interdecadal variation of summer precipitations in China and the drought trend in North China. *Plateau Meteor.*, **18**, 465-476. (in Chinese)

______, 2001: Decadal variability of the summer monsoon rainfall in East Asia and its association with the SST anomalies in the tropical Pacific. *CLIVAR Exchange*, **2**, 7-8.

______, and L. T. Zhou, 2002: Research on the characteristics, formation mechanism and prediction of severe climate disasters in China. *J. Natural Disasters*, **11**, 1-9. (in Chinese)

______, ______, and W. Chen, 2003: The progresses of recent studies on the variabilities of the East Asian monsoon and their causes. *Adv. in Atmos. Sci.*, **20**, 55-69.

Japan Meteorological Agency: Monthly Report on Climate System 1978-2000.

Krishnamurti, T. N., and Y. Ramanathan, 1982: Sensitivity of monsoon onset to differential beating. *J. Atmos., Sci.*, **39**, 1290-1306.

Kurihara, K., 1989: A climatological study on the relationship between the Japanese summer weather and the subtropical high in the western northern Pacific. *Geophys. Mag.*, **43**, 45-104.

Lau, K. M., and C. P. Chang, 1987: Planetary scale aspects of the winter monsoon and atmospheric teleconnections. *Monsoon Meteorology*, C. P. Chang and T. N. Krishnamurti, Eds, Oxford Univ. Press, 161-201.

______, 1992: East Asian summer rainfall variability and climate teleconnection. *J. Meteor. Soc. Japan.* **70**, 211-242.

______, and S. H. Shen, 1992: Biennial oscillation associated with the East Asian summer monsoon and tropical sea surface temperature. *Climate Variability*, D. Z. Ye *et al.*, Eds., China Meteorological Press, Beijing, 53-58.

Li, C. Y., 1990: Interaction between anomalous winter monsoon in East Asia and El Niño events. *Adv. in Atmos. Sci.*, **7**, 36-46.

______, S. Q. Sun, and M. Q. Mu, 2001: Origin of the QBO-Interaction between anomalous East Asian winter monsoon and ENSO cycle. *Adv. in Atmos. Sci.*, **18**, 554-566.

Lian, J. Y., and S. S. Wu, 2002: A Study of southwest monsoon onset date over the South China Sea and its impact factors. *Chinese J. Atmos. Sci.*, **26**, 844-855. (in Chinese)

Lu, R. Y., Y. S. Chang, and R. H. Huang, 1995: Interannual variations of the precipitation in Korea and the comparison with those in China and Japan. *J. Korean Environ. Sci. Soc.*, **4**, 345-356.

______, 2001: Interannual variability of the summertime North Pacific subtropical high and its relation to atmospheric convection over the warm pool. *J. Meteor. Soc. Japan*, **79**, 771-783,

______, and B. W. Dong, 2001: Westward extension of North Pacific subtropical high in summer. *J. Meteor. Soc. Japan*, **79**, 1229-1241.

Luo, H. B., and M. Yanai, 1984: The large-scale circulation and heat sources over the Tibetan Plateau and surrounding areas during the early summer of 1979. *Mon. Wea. Rev.*, **108**, 1849-1853.

McBride, J. J., 1987: The Australian summer monsoon. *Monsoon Meteorology*, C. P. Chang and T. N. Krishnamurti, Eds., Oxford University Press, 203-232.

Miao, J. H., and K. M. Lau, 1990: Interannual variability of the East Asian monsoon rainfall. *Quart. J. Appl. Meteor.*, **1**, 377-382. (in Chinese)

Nikaido, Y., 1989: The P-J like north-south oscillation found in 4-month integration of the global spectral model T42. *J. Meteor. Soc. Japan*, **67**, 587-604.

Nitta, T., 1983: Observational study of heat sources over the eastern Tibetan Plateau during the summer monsoon. *J. Meteor. Soc. Japan*, **61**, 590-605.

______, 1986: Long-term variations of cloud amount in the western Pacific region. *J. Meteor. Soc. Japan*, **64**, 373-300.

______, 1987: Convective activities in the tropical western Pacific and their impact on the Northern Hemisphere summer circulation. *J. Meteor. Soc. Japan*, **64**, 373-390.

______, 1993: Interannual and decadal scale variations of atmospheric temperature and circulations. *Climate Variability*, D. Z. Ye *et al.*, Eds., China Meteorological Press, Beijing, 15-22.

Shukla, J., and D. A. Paolina, 1983: The southern oscillation and long range forecasting of the summer monsoon rainfall over India. *Mon. Wea. Rev.*, **111**, 1830-1837.

Staff members of Academia Sinica, 1957: On the general circulation over Eastern Asia (I). *Tellus*, **9**, 432-446.

Sun, A. J., and R. H. Huang, 1994: Low frequency oscillation characteristics of 500 hPa geopotential height fields in summer of 1983 and 1985 over the Northern Hemisphere. *Chinese J. Atmos. Sci.*, **18**, 365-375. (in Chinese)

Sun, B. M., and S. Q. Sun, 1994: The analysis on the features of the atmospheric circulation in preceding winter for the summer drought and flooding in the Yangtze and Huaihe River valley. *Adv. in Atmos. Sci.*, **11**, 79-90.

Tao, S. Y., and L. X. Chen, 1957: The structure of the summer circulation over Asia, *Acta Meteorologica Sinica*, **28**, 237-247. (in Chinese)

______, and ______, 1987: A review of recent research on the East Asian summer monsoon in China. *Monsoon Meteorology*, C. P. Chang and T. N. Krishnamurti, Eds., Oxford University Press, 60-92.

______, and Q. Y. Zhang, 1998: Response of the Asian winter and summer monsoon to ENSO events. *Chinese J. Atmos. Sci.*, **22**, 399-407. (in Chinese)

Tomita, T., and T. Yasunari, 1996: Role of the northeast winter monsoon on the biennial oscillation of the ENSO/Monsoon system. *J. Meteor. Soc. Japan*, **74**, 399-413.

Tsuyuki, T., and K. Kurihara, 1989: Impact of convective activity in the western tropical Pacific on the East Asian summer circulation. *J. Meteor. Soc. Japan*, **67**, 231-247.

Tu, C. W., and S. S. Huang, 1944: The advance and retreat of the summer monsoon. *Meteor. Mag.*, **18**, 1-20.

Wang, S. W., and D. Z. Ye, 1993: An analysis of global warming during the last one hundred years. *Climate Variability*, D. Z. Ye *et al.*, Eds., China Meteorological Press, Beijing, 23-32.

Webster, P. J., and S. Yang, 1992: Monsoon and ENSO: Selectively interactive systems. *Quart. J. Roy. Meteor. Soc.*, **118**, 877-926.

______, V. O. Magana, T. B. Palmer, J. Shukla, R. A. Tomas, M. Yanai, and T. Yasunari 1998: Monsoons: Processes, predictability, and the prospects for prediction. *J. Geophys. Res.*, **103**, 14451-14510.

Wei, Z. G., and S. W. Luo, 1996: Impact of the snow cover in western China on the precipitation in China during the rainy season. Collected Papers of the Project "Disastrous Climate Prediction and Its Impact on Agriculture and Water Resources", II. *Processes and Diagnosis of Disastrous Climate*, R. H. Huang *et al.*, Eds., China Meteorological Press, Beijing, 40-45. (in Chinese)

Wu, G. X., and Y. S. Zhang, 1997: Tibetan Plateau forcing and the timing of the monsoon onset over South Asia and the South China Sea. *Mon. Wea. Rev.*, **126**, 917-927.

Xie, P., and P. A. Arkin, 1997: Analyses of global monthly precipitation using gauge observations, satellite estimates and numerical model prediction. *J. Climate.*, **9**, 804-858.

Yamagata, T., and Y. Matsumoto, 1989: A simple ocean-atmosphere coupled model for the origin of a warm El Niño/Southern Oscillation event. *Phil. Trans. Roy. Soc. London*, A, **329**, 225-236.

Yasunari, T., 1990: Impact of Indian monsoon on the coupled atmosphere/ocean systems in the tropical Pacific. *Meteor. & Atmos. Phys.*, **44**, 29-41.

______, 1991: The monsoon year-A new concept of the climatic year in the tropics. *Bull. Amer. Meteor. Soc.*, **72**, 133-138.

______, and Y. Seki, 1992: Role of the Asian monsoon on the interannual variability of the global climate system. *J. Meteor. Soc. Japan*, **70**, 179-189.

Ye, D. Z., and Y. X. Gao, 1979: *Tibetan Plateau Meteorology*. Science Press, Beijing, 279pp. (in Chinese)

Yeh, T. C. (Ye, D. Z.), S. Y. Tao and M. C. Li, 1959: The abrupt change of circulation over the Northern Hemisphere during June and October. *Atmosphere and the Sea in Motion*, Rockefeller Institute, New York, 249-267.

Zhang, Q., G. X. Wu, and Y. F. Qian, 2002: The bimodality of the 100hPa South Asian high and its relationship to the climate anomaly over East Asia in summer. *J. Meteor. Soc. Japan*, **80**, 733-744.

Zhang, R. H., A. Sumi, and M. Kimoto, 1996: Impact of El Niño on the East Asian monsoon: A diagnostic study of the 86/87 and 91/92 events. *J. Meteor. Soc. Japan*, **74**, 49-62.

______, 2001: Relations of water vapor transport from Indian monsoon with those over East Asia and the summer rainfall in China. *Adv. in Atmos. Sci.*, **18**, 1005-1017.

Zhu, K. Z., 1934: Southeast monsoon and rainfall in China. *J. Chinese Geogr. Soc.*, **1**, 1-27. (in Chinese)

Part III

General Circulation Modeling

7. SIMULATIONS BY A GFDL GCM OF ENSO-RELATED VARIABILITY OF THE COUPLED ATMOSPHERE-OCEAN SYSTEM IN THE EAST ASIAN MONSOON REGION

NGAR-CHEUNG LAU AND MARY JO NATH
Geophysical Fluid Dynamics Laboratory / NOAA
Princeton University
P.O. Box 308
Princeton, NJ 08542, USA
E-mail: Gabriel.Lau@noaa.gov

HAILAN WANG
Program in Atmospheric and Oceanic Sciences
Princeton University
Princeton, NJ 08544, USA

The impact of El Niño-Southern Oscillation (ENSO) on the East Asian Monsoon (EAM) has been examined using a general circulation model (GCM). The observed monthly changes in sea surface temperature (SST) in the equatorial Pacific east of 172°E during 1950-99 were inserted as the lower boundary condition of the model. For all oceanic grid points lying outside of the region of SST prescription, the atmosphere was coupled to an oceanic mixed layer model.

The typical evolution of the atmosphere-ocean system during ENSO was analyzed using composite charts. These patterns show that the key changes in the EAM sector are related to a prominent sea level pressure anomaly simulated over the South China Sea and subtropical northwestern Pacific. During warm ENSO events, the circulation anomalies associated with this anomalous anticyclone correspond to weaker winter monsoon flows along the East Asian coast, as well as above-normal precipitation over southern China. These signals move systematically eastward during the following spring and summer. Stationary wave modeling indicates that the atmospheric anomaly in the EAM region is essentially a Rossby-wave response to the ENSO-related diabatic heating pattern over the equatorial western Pacific.

The wintertime atmospheric circulation anomalies over the western Pacific generate strong SST anomalies in the following spring. Further model diagnoses indicate that the feedback of these SST changes on the atmosphere leads to eastward propagation of the pressure anomaly in the EAM region, and to amplification of rainfall anomalies along the Meiyu-Baiu front.

The atmospheric and oceanic changes in the EAM sector described in this chapter are discussed in the broader context of ENSO influences on the entire Asian-Australian monsoon system.

1. Introduction

The El Niño-Southern Oscillation (ENSO) is recognized as the prominent mode of interannual variability of the atmosphere-ocean system in the tropics. In considering the mechanisms responsible for fluctuations of the East Asian Monsoon (EAM) in individual years, an obvious avenue of research is the assessment of the impact of ENSO on the atmospheric circulation over East Asia and the conditions of the nearby maritime regions. During the past decades, considerable efforts have been devoted to discerning empirical relationships between ENSO and EAM by performing statistical analyses or case studies based on observational records gathered at mostly ground-based stations. The composite precipitation patterns compiled by Ropelewski and Halpert (1987, 1989) for multiple ENSO events are characterized by weak spatial coherence in the East Asian sector. In view of the disorganized signals in that region, the ENSO-related rainfall variations in the EAM region have received limited attention in the series of global surveys undertaken by these authors. The comprehensive review by Chen *et al.* (1992) on EAM variability also reported mostly weak correspondence between El Niño events and rainfall anomalies in different parts of China. However, they noted that the precipitation at a few Chinese sites do exhibit some temporal relationships with ENSO at time scales of 2-3 years. Generally speaking, many studies in the existing literature suggest that the nature of the covariability between ENSO and changes in EAM appears to depend on the specific region and particular episode under investigation. Due to these complex factors, it is often difficult to make broad and robust statements on the impact of ENSO on the overall EAM region.

In more recent years, further progress in our understanding of the influences of ENSO processes on EAM variability has been made possible by the following developments:

- Availability of data products (e.g., satellite-derived precipitation estimates and gridded reanalyses of other meteorological fields) with near-global coverage, which facilitates the extension of previous land-based investigations to include pertinent phenomena occurring over the oceanic regions off the East Asian coast (e.g., Zhang *et al.* 1996).
- More definitive identification of the teleconnection between ENSO forcing and large-scale circulation features in the East Asian sector; as well as more systematic documentation of the seasonal dependence and interdecadal variations of these relationships (e.g., Wang *et al.* 2000; Chang *et al.* 2000a, 2000b; Wang and Zhang 2002).
- More incisive physical interpretation of the observational results on the basis of dynamical considerations of atmospheric responses to heat sources and sinks, and of the nature of in situ coupling between the atmospheric and oceanic anomalies over the subtropical western Pacific (e.g., Wang *et al.* 2000; Lau and Wu 2001; Wang and Zhang 2002).
- Efforts to conduct and diagnose general circulation model (GCM) experiments aimed at simulating the influences of ENSO on the Asian-Australian monsoon (e.g., Lau and Nath 2000, hereafter referred to as LN00; and the studies cited therein). These studies typically entail the prescription of observed or idealized ENSO-related sea surface temperature (SST) anomalies as the lower boundary condition of an atmospheric GCM. The impact of ENSO on the monsoon system of interest could then be inferred from the model-generated anomalies in the presence of this SST forcing.

There are several advantages in adopting the modeling approach mentioned in the last bullet of the above list. Within the controlled environment of a GCM, hypotheses on the relative importance of various processes (e.g., atmospheric response to remote SST anomalies, local air-sea interaction, etc.) can be tested by comparing model runs in which each process is incorporated separately. It is much more difficult to isolate the effects of individual mechanisms using observations. Insofar as a GCM is capable of mimicking the actual climate system, the model output is also a comprehensive resource for diagnosing the key processes in considerably greater detail than is feasible with observational datasets, since the latter still do not provide many crucial quantities (e.g., diabatic heating, surface energy fluxes, vertical motion, etc.) with adequate accuracy and spatial/temporal resolution. Due to the limited duration of the available observational records (mostly less than a half-century), it is often not possible to establish empirical relationships between ENSO and EAM with a high level of confidence. This situation can be remedied by extending the length of GCM experiments that are subjected to a temporally fixed SST anomaly, or by performing multiple runs that are forced by temporally evolving SST forcing.

During the past decade, a concerted effort has been made at the Geophysical Fluid Dynamics Laboratory (GFDL) to use GCM tools to investigate the linkages between ENSO and the variability of the global atmosphere-ocean system on a wide range of time scales. Much of the attention in the earlier phase of this project has been focused on extratropical changes accompanying ENSO (Lau and Nath 1994, 1996, 2001). More recently, the scope of this endeavor has been extended to include tropical phenomena, with specific emphasis on features in the Asian-Australian monsoon region. The results obtained thus far (see LN00) illustrate that the GCM used in our studies provides a satisfactory simulation of the climatological monsoonal characteristics. It has been demonstrated in LN00 that the warm (cold) ENSO events are accompanied in the model atmosphere by weakened (strengthened) summer monsoons over India and northern Australia, as well as a weakened (strengthened) winter monsoon over East Asia. The findings reported in LN00 also indicate that ENSO-induced atmospheric anomalies over the monsoon regions could initiate air-sea feedback processes, thus affecting the subsequent development of the local atmospheric circulation and SST conditions.

Building upon the experience gained from the LN00 study, and taking into account our expanding observational and theoretical knowledge of the ENSO-monsoon relationships, a new suite of GCM experiments have recently been completed at GFDL. One of the primary goals of these new model runs is to further delineate the contributions of remote ENSO forcing and in situ air-sea coupling to monsoon variability. As will be described in greater detail in Section 2, the novelty of these experiments mainly lies in the strategic choices of maritime domains where SST forcing is prescribed and where two-way atmosphere-ocean interactions are incorporated, the use of a more realistic ocean mixed layer model to simulate such interactions, and the much larger number of independent integrations performed for each SST forcing scenario.

Results from the new series of GCM experiments will be presented in this chapter, with a specific focus on the influences of ENSO on EAM in different phases of the seasonal cycle. Whenever feasible, the model findings will be compared with available observations. The simulations will be interpreted in light of empirical findings on this subject as reported in the recent literature, some of which are reviewed in other chapters here. In view of the main theme of this volume, we shall devote much of our discussion to phenomena in the East Asian sector. However, it must be borne in mind that ENSO also exerts strong influences on monsoon systems throughout the western Pacific and Indian Ocean basins. Model results on such larger-scale effects will only be summarized in brief in the concluding section of this chapter. Details of these findings will be described in another article (Lau and Nath 2003; hereafter referred to as LN03).

The design for the GCM experiments analyzed in this study is outlined in Section 2. The model climatology in the East Asian sector for all seasons of the year is compared with the observations in Section 3. The evolution of the simulated and observed atmospheric and oceanic anomalies related to the EAM during selected ENSO events is examined in Section 4. The role of the energy fluxes at the air-sea interface in driving the SST changes off the East Asian coast is documented in Section 5. The nature of the two-way feedbacks between ENSO-related perturbations in the atmospheric circulation and SST conditions over the western Pacific is further diagnosed in Section 6 using supplementary model integrations. In the concluding section, the results for East Asia are discussed in the broader context of ENSO modulation of the atmospheric and marine environments in the entire Asian-Australian monsoon region.

2. Experimental Design

The atmospheric GCM used in the current study is identical to that analyzed in LN00. It is a global spectral model with rhomboidal truncation at 30 wavenumbers and 14 vertical layers. The numerical formulation and physics packages used in this model have been described by Gordon and Stern (1982) and Broccoli and Manabe (1992). The essential climatological statistics of this model have been documented by Alexander and Scott (1995).

Two basic sets of experiments have been conducted with this GCM. In both of these sets, observed month-to-month SST variations during the 50-year period of 1950-99 were imposed as the evolving lower boundary condition in the deep tropical eastern/central Pacific (DTEP), which extends from 15°S to 15°N, and from 172°E to the South American coast. The oceanic conditions prescribed in this region represent the primary warm (cold) SST forcing during El Niño (La Niña) events. The SST observations were based on the routine analyses provided by the National Centers for Environmental Prediction (NCEP) using the reconstructed empirical orthogonal function approach (Smith *et al.* 1996). The distinction between the two sets of experiments lies in the treatment of SST conditions outside of the DTEP region, as described below:

- In the first set, the atmospheric GCM was coupled to a mixed layer model of the ocean at all ice-free maritime grid points beyond the DTEP region. The incorporation of the mixed layer model in the framework of the GFDL GCM has been described by Alexander *et al.* (2000). The basic formulation of the mixed layer model has been outlined in Gaspar (1988). At each grid point, this columnar 31-layer model computes the variations of temperature, salinity and thickness of the mixed layer according to air-sea energy fluxes, precipitation and wind stress forcing directly above the same grid point.

There is no lateral communication between the mixed layer models at adjacent sites. A flux correction procedure (see Manabe *et al.* 1991) has been used to ensure that the seasonal SST climatology of the mixed layer model is close to the corresponding observations. Model runs that incorporate this mixed layer model (hereafter referred to as the "MLM" experiment) would include the impact of ENSO forcing originating from the DTEP area, as well as the effects due to air-sea interactions taking place in the other parts of the world's oceans. Altogether 16 integrations have been performed, with each 50-year run being initiated from an independent set of atmospheric conditions. The initial conditions for the oceanic mixed layer have been obtained by averaging the data for the last 20 years of a 40-year uncoupled integration of the mixed-layer model, which was subjected to surface fluxes and flux correction values generated by the atmosphere-only GCM.

- In the second set, the seasonal SST climatology has been prescribed at all grid points outside of DTEP, so that there is no interannual variability in the oceanic conditions at those locations. The SST climatology used in this experiment was obtained by averaging the output from the MLM runs described above[a]. This setup explicitly eliminates the air-sea coupling in areas outside of DTEP, and hence serves as a control experiment for the MLM runs. We shall refer to this set as the "CTRL" runs. An ensemble of 8 integrations has been made under this scenario.

The MLM and CTRL experiments examined in the present study are analogous to the TOGA-ML and TOGA experiments, respectively, analyzed in LN00. In the TOGA series, the region of variable SST prescription has a relatively broader meridional width, and extends across the entire tropical Pacific. The corresponding forcing region in MLM and CTRL (i.e., DTEP) is confined to the central and eastern equatorial Pacific, so that air-sea coupling can be incorporated over the western part of that basin in the MLM runs. Another noteworthy distinction between the two suits of experiments is that the mixed layer model previously used in TOGA-ML consists of a single slab with a fixed depth of 50 m; whereas the multi-layered mixed layer model for the current work predicts the variations of the mixed layer thickness, and thereby provides a more realistic depiction of the seasonal evolution as well as anomalous changes of the thermal capacity of the surface ocean.

3. Model and Observed Climatology

Since previous studies (e.g., Sperber and Palmer 1996) indicate that realistic model climatology is an important factor for determining the skill of simulating interannual monsoon variability, we begin our presentation with the mean state of the EAM region. The climatological distributions of the simulated near-surface atmospheric circulation

[a] The rationale for applying the MLM climatology (as opposed to the observed climatology) here is that the mean SST conditions in both sets of experiments would be identical to each other, so that differences between the two experiments cannot be attributed to differences in the SST climatology. As a result of the flux adjustment procedure (see discussion in the first bullet), the SST climatology of the MLM experiment is actually very similar to the observed climatology.

(streamlines) and precipitation (shading) for the four seasons are displayed in the left panels of Fig. 1. These patterns have been obtained by averaging the 50-year output from all of the 16 MLM runs. The corresponding observations, as deduced from long-term means of the NCEP reanalyses (Kalnay *et al.* 1996) for the surface wind and the Climate Prediction Center Merged Analysis of Precipitation (CMAP, see Xie and Arkin 1997) for the precipitation rates, are shown in the right panels of the same figure. In the following discussion, we shall first devote our attention to salient features in the simulated patterns, and then point out some noteworthy differences between model and observations.

During the boreal spring (Fig. 1a), the mean surface easterly wind belt in the model atmosphere extends all the way from the subtropical northwestern Pacific to the southern Chinese coast. Anticyclonic flow prevails over northeastern China, Korea and part of Japan. The zone extending northeastward from southern China to southern Japan is characterized by confluence of the streamlines and enhanced precipitation. The rainbelt along the East Asian seaboard is reminiscent of the observed Meiyu-Baiu front in this season. The wind field in the 30°N - 40°N zone just west of the dateline is dominated by an anticyclonic pattern accompanying the subtropical pressure ridge in that region.

In the summer (Fig. 1c), the simulated southwesterly monsoon flow originating from the Indian Ocean basin protrudes into the South China Sea and Philippine Sea. At about 25°N, this air current turns cyclonically towards the East Asian seaboard. The formation of this prominent monsoon trough is accompanied by large precipitation amounts on its southern flank. The subtropical ridge centered near the dateline is seen to migrate poleward from spring to summer, and extend farther westward to northeastern China.

The autumn conditions (Fig. 1e) are characterized by the prevalence of an anticyclonic circulation over the East Asian landmass, so that much of the coastal region is under the influence of northeasterly flows. The rainfall over Indochina decreases noticeably from the previous season, and the main precipitation centers over the subtropical Pacific now retreats equatorward to the Indonesian Archipelago.

Noticeable intensification and eastward extension of the East Asian anticyclone occur in winter (Fig. 1g). Comparison between the flow patterns for winter and those for summer (Fig. 1c) reveals a complete reversal of the wind direction over much of East Asia and the surrounding waters during these two solstitial seasons. The principal wintertime precipitation centers reside near the Equator and along a belt extending northeastward from eastern China to the extratropical Pacific, whereas dry conditions prevails within the 10°N - 20°N zone west of the dateline.

A good match exists between the simulated seasonal features mentioned in the above paragraphs and the observed climatological fields (see right panels of Fig. 1), with the following exceptions. The model monsoon trough during summer is too strong and extends too far east to the subtropical Pacific, so that the maritime region east of the Philippines is too wet in the model atmosphere (compare Figs. 1c and 1d). The wind speed of the southeasterly flow over the South China and Philippine Seas in the simulated atmosphere is higher than the observations (see Figs. 1c and 1d in LN00). The GCM produces too much rainfall over southwestern China during winter (Figs. 1g and 1h). Throughout the year, there is little

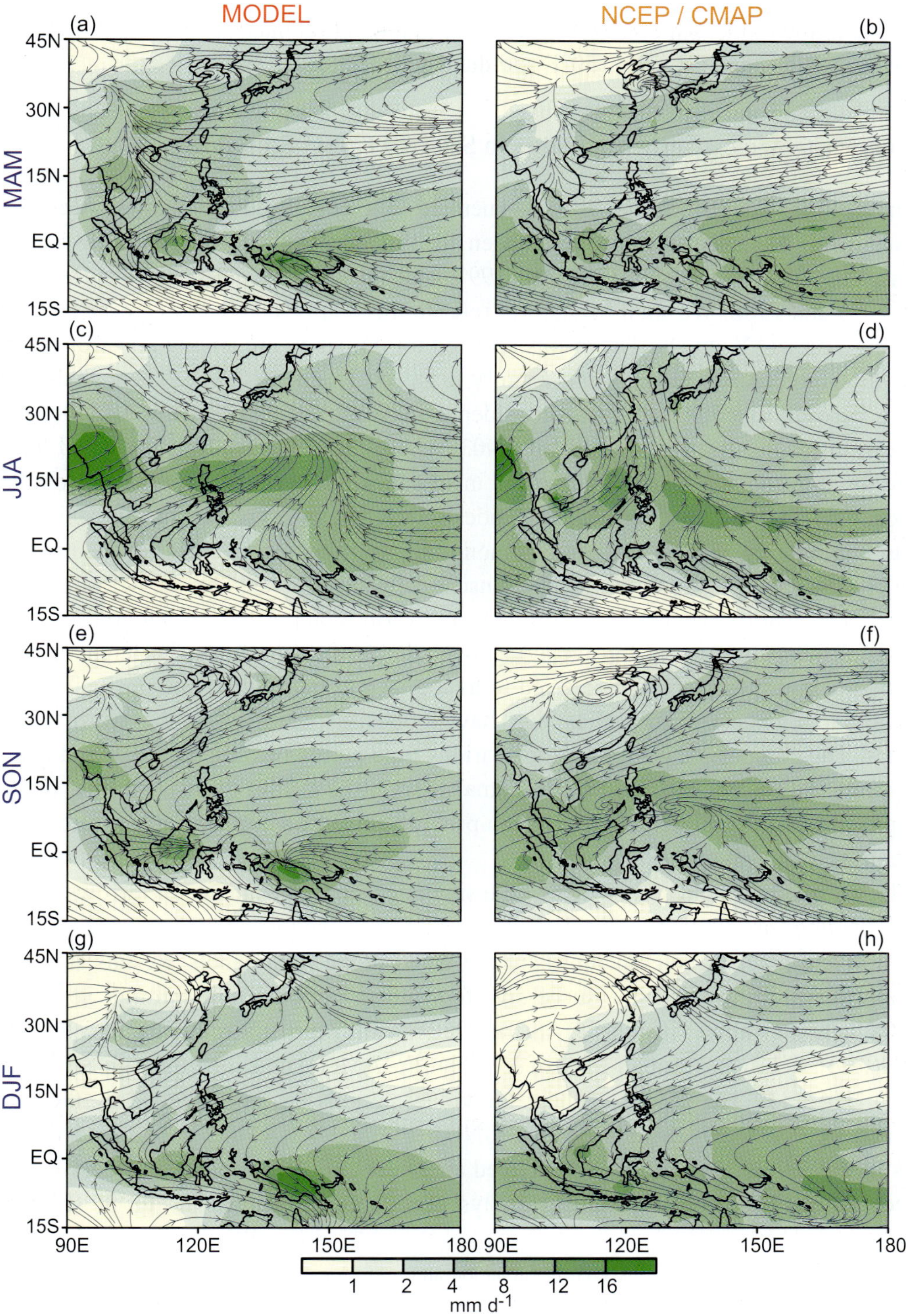

Figure 1. Climatological distributions of the near-surface circulation (streamlines) and precipitation (shading, see scale bar at bottom) for the four conventional seasons, as computed using output from the MLM experiment (left panels) and the NCEP/CMAP datasets (right panels).

indication in the model patterns of the observed precipitation maximum along a narrow band situated near 10°N and between 150°E and the dateline.

4. Composite Atmospheric Anomalies in Selected ENSO Episodes

We proceed to describe the model-generated and observed anomalies occurring in outstanding ENSO events. The cases chosen for this purpose include the five warm events starting in 1957, 1972, 1982, 1991 and 1997, and the cold events starting in 1955, 1970, 1973, 1975 and 1988. We shall henceforth refer to the years listed here as "Year(0)", and the following year as "Year(1)". The events selected for our study are among the strongest in the observational record for the past half-century (e.g., see Trenberth 1997). There is also a large overlap between our list and those considered in Wang *et al.* (2000) and LN00. Further justification for our list is provided by LN03, whose statistical analysis indicated that both remote forcing from the eastern Pacific and in situ air-sea interaction over the western Pacific are most evident during the particular episodes selected here. A composite approach has been used to highlight the signals that are common to these prominent events. For the model data, averages were taken over the 16-member ensemble mean generated by the MLM experiment for the five warm events, thus forming a "warm composite". Analogously, a "cold composite" was constructed by averaging over the five cold events. Differences between these two composites (hereafter referred to as the "warm minus cold composite") were then computed to emphasize anomalies that have opposite polarities in the warm and cold episodes. The composites were taken for various time periods within Years(0) and (1), so as to delineate the typical evolution of the signals during the ENSO cycle. A similar procedure was applied to the NCEP reanalyses (which provide only one sample for each selected event) to obtain the observational composites.

The warm minus cold composites of the simulated surface wind vector (arrows) and wind speed (shading) are plotted in the left panels of Fig. 2. Patterns for the sea level pressure (contours) and precipitation (shading) are presented in the right panels. Results are arranged from top to bottom rows for the successive four-month periods: June through September of Year(0) [hereafter abbreviated as JJAS(0)], October of Year(0) through January of Year(1) [ONDJ(0/1)], February through May of Year(1) [FMAM(1)] and June through September of Year(1) [JJAS(1)]. The choice of these individual periods has been guided by the findings presented in LN03, which indicate that ONDJ(0/1) and FMAM(1) correspond to distinct stages in the development of ENSO-induced air-sea coupling in the Asian-Australian sector. The composites based on the NCEP reanalysis products[b] are shown in Fig. 3. The features

[b] Note that the composite for the observed precipitation anomalies has been estimated using NCEP reanalyses; whereas the observed precipitation climatology (right panels in Fig. 1) has been constructed using CMAP. The latter dataset covers the 1979-1996 period only, and hence cannot be used in our composite procedure, since many of the selected ENSO events did not occur within this period. Considering that the reanalysis products for precipitation are dependent on the model tools used to generate such data, the accuracy of these precipitation estimates should be regarded with caution. It is, however, noteworthy that precipitation anomaly charts based on the CMAP dataset for the 1982, 1991 and 1988 events exhibit many of the characteristics seen in the corresponding composite patterns constructed using NCEP reanalysis data (right panels of Fig. 3).

appearing in Figs. 2 and 3 may be viewed as anomalies that typically occur in the course of warm ENSO events, whereas cold events would be associated with changes of the opposite sign. As in Section 3, we shall devote much of the following discussion to the model simulations, and then provide some comments on how these results compare with the observational estimates.

The simulated composite pattern for the summer period of Year(0) (Figs. 2a-2b) is dominated by below-normal sea level pressure over the western Pacific north of 10°N, and by above-normal pressure in the Indonesian sector. These pressure changes are associated with anomalous westerlies over much of the maritime region south of 20°N, and by northerlies off the east China coast. Reduction in the strength of the onshore flow accompanying the climatological monsoon trough over southern and eastern China (see Fig. 1c) leads to dry conditions within the 20°N - 25°N zone extending from Indochina to the waters south of Japan. Near the Equator, the below-normal rainfall over Indonesia and above-normal rainfall near the dateline are indicative of the eastward displacement of the Walker Circulation during warm ENSO events.

During the ONDJ period following the onset of ENSO episodes (Figs. 2c-2d), a prominent high pressure anomaly is simulated over the subtropical western Pacific and the Philippines. We shall henceforth refer to this western Pacific high feature as the WPH. The southwesterly surface wind anomalies north and west of WPH oppose the northeasterly climatological winter monsoon flow over eastern and southern China (see Fig. 1g), thus resulting in reduced surface wind speeds over the East and South China Seas (see blue shading in Fig. 2c). Conversely, the anomalous flow south of WPH is parallel to the time mean circulation, leading to enhanced local wind speeds. Below-normal precipitation prevails near the center of WPH. Weakening of the mean dry northeasterly current over south China results in positive rainfall anomalies in that region. The dipole-like precipitation pattern in the equatorial region between 120°E and the dateline, as noted in the JJAS(0) period, persists through the cold season. Over western Indonesia, a confluence zone is discernible at the boundary between the northeasterly anomalies over the southern portion of the South China Sea and the southeasterly anomalies over the eastern edge of the Indian Ocean south of the Equator, so that above-normal precipitation occurs over Sumatra and the nearby waters.

In the spring of Year(1), the center of the WPH in the previous season is seen to maintain its intensity and migrate eastward by about 30° of longitude (Figs. 2e-2f). The organization of surface wind and rainfall changes about this feature is analogous to that described in the previous paragraph, i.e., southeasterly anomalies, lower wind speeds and wetter conditions in the north and west; and northeasterly anomalies, higher wind speeds and dryness in the south. The wet anomaly stretching northeastward from south China to southern Japan is collocated with the springtime climatological rainbelt in that region (see Fig. 1a).

During the JJAS(1) period (Figs. 2g-2h), the WPH center and the attendant wind circulation over the subtropical western Pacific continue to move northeastward, and undergo considerable weakening. Comparison between the composite patterns for this period with

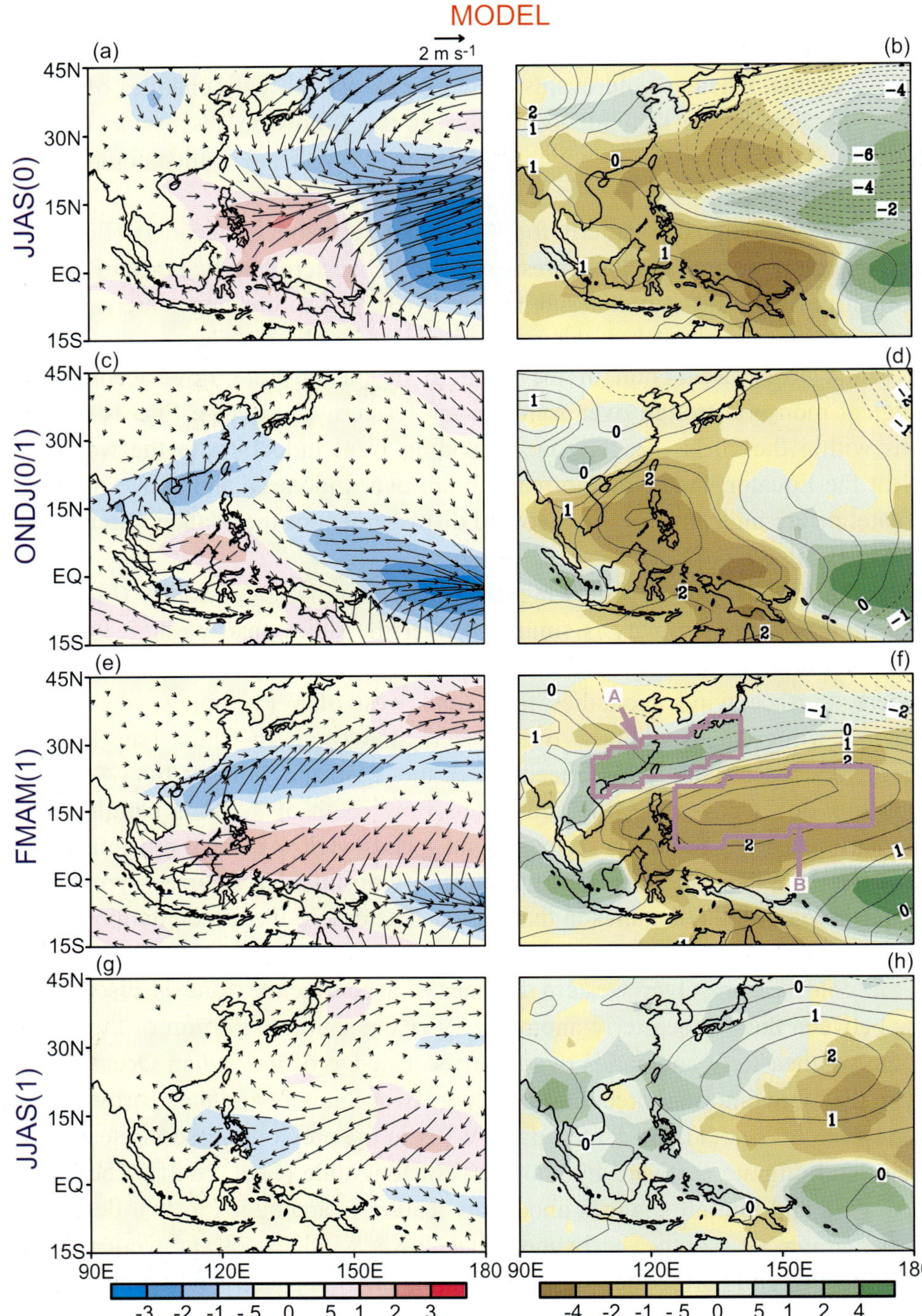

Figure 2. Warm minus cold composites of (left panels) near-surface wind vector (arrows, see scale at top) and wind speed (shading, see scale bar at bottom), and of (right panels) sea level pressure (contours; interval: 0.5 mb) and precipitation (shading, see scale at bottom), for different stages of the selected ENSO events. All patterns are based on output from the MLM experiment. Shown in panel (f) are the boundaries of Regions A and B for computing the areal averages analyzed in Fig. 4.

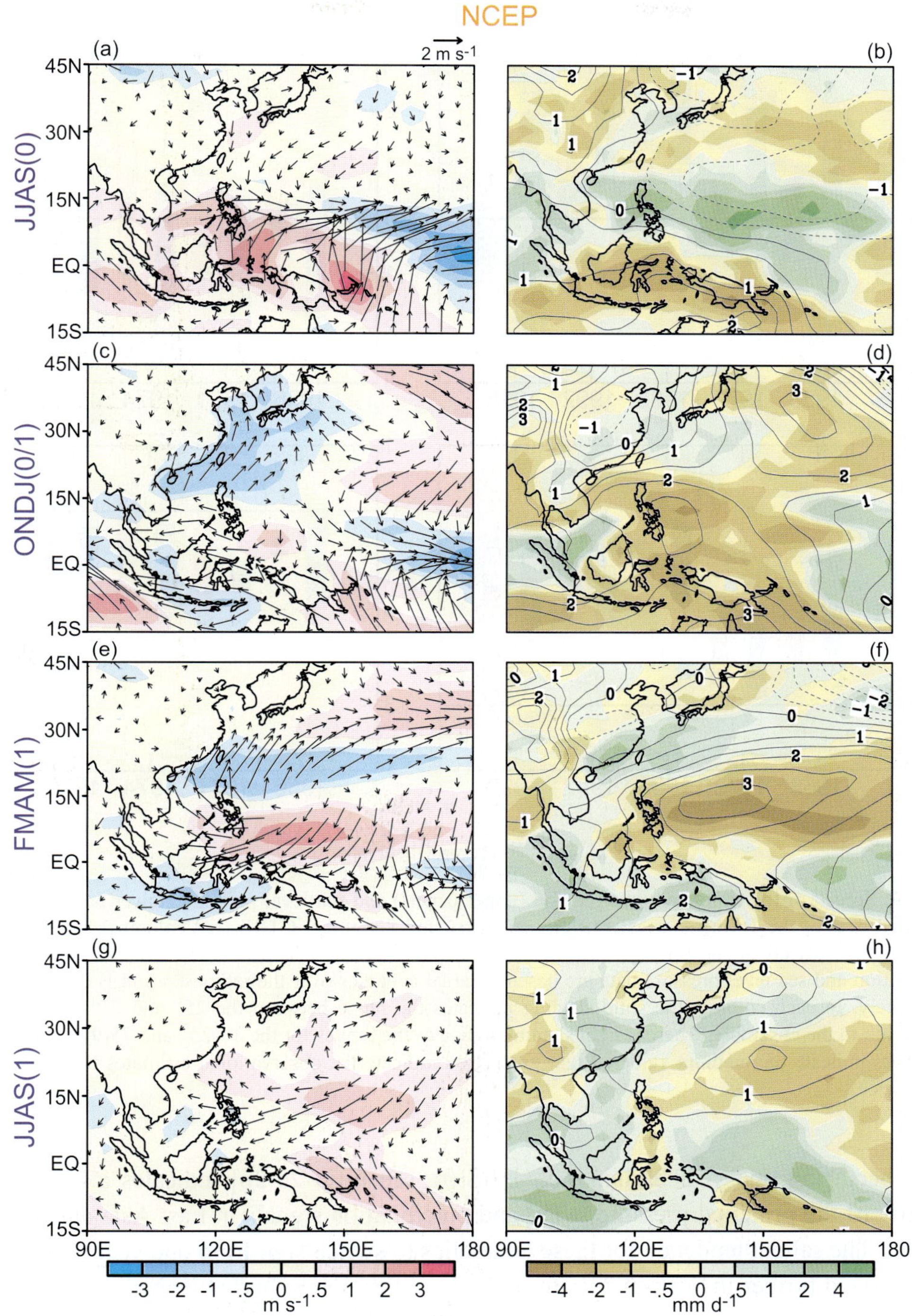

Figure 3. As in Fig. 2, except for composites based on the NCEP reanalyses.

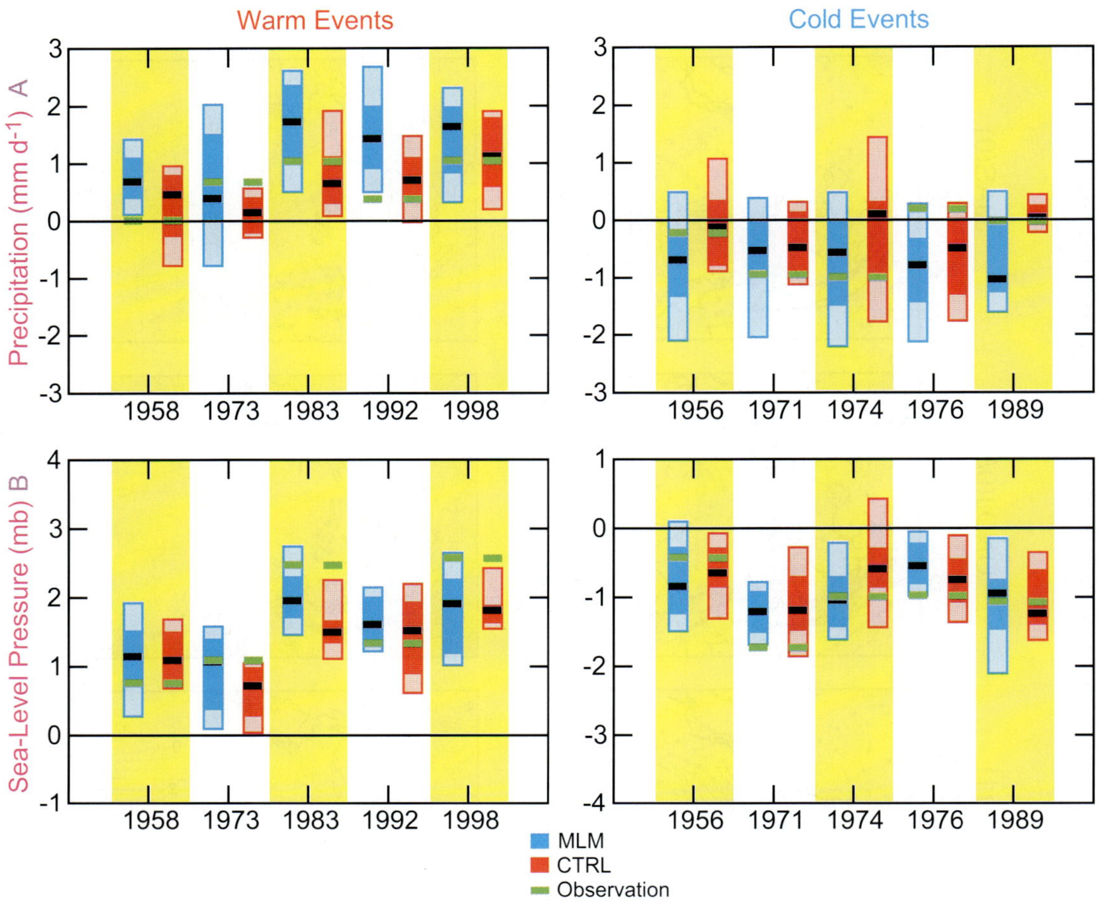

Figure 4. Spread of the simulated precipitation (upper panels) and sea level pressure (lower panels) anomalies in individual runs of the 16-member MLM (blue columns) and 8-member CTRL (red columns) experiments during the FMAM(1) period of the selected warm (left panels) and cold (right panels) ENSO events. Results for the precipitation and sea level pressure data are based on areal averages over the regions A and B depicted in Fig. 2f, respectively. In each column, the solid shading indicates the range between the 25th and 75th percentiles of the distribution in the ensemble, whereas light shading indicates the ranges for the 10-25th and 75-90th percentiles. The black bars indicate the medium values. Light green bars indicate the observational estimates based on the NCEP reanalyses.

those occurring one year earlier [i.e., JJAS(0), see Figs. 2a-2b] indicates a reversal in the polarity of the principal pressure, wind and precipitation signals in East Asia and subtropical Pacific. The salient tendency for these anomalies to switch sign from one year to the next will be discussed further in Section 7, in the context of the tropospheric biennial oscillation.

The corresponding composites based on NCEP reanalyses (Fig. 3) provide considerable observational support for the simulated pressure, surface circulation and rainfall anomalies associated with WPH during ONDJ(0/1) and FMAM(1). It is seen that the GCM reproduces the observed propagation and amplitude characteristics of WPH through much of the 12-

month period from November(0) through October(1). The observational studies of Wang *et al.* (2000), Chang *et al.* (2000a, 2000b) and Wang and Zhang (2002) have also demonstrated that many of the atmospheric changes occurring during ENSO are related to fluctuations of WPH. The correspondence between model and observational results is much less evident during the JJAS(0) period, in which the NCEP analyses indicate comparatively weaker negative pressure and cyclonic wind anomaly over the subtropical western Pacific. This discrepancy may be linked to an overestimation of the subtropical atmospheric response to the diabatic heating field in the central equatorial Pacific, as will be pointed out in Section 6.1.

We proceed to examine the variability of the precipitation and sea level pressure signals in different ENSO events and in individual runs of the 16-member ensemble. The simulated rainfall and pressure anomalies in the FMAM(1) period have been averaged over the regions denoted as “A” and “B” in Fig. 2f, respectively. These selected regions correspond to sites of extrema in the model fields being considered. For a given ENSO event, the distribution of the areal averaged anomalies in the 16 realizations of the MLM experiment is illustrated using a vertical column with markers indicating the 10th, 25th, 50th, 75th and 90th percentiles of the population. These columns are plotted in Fig. 4 separately for the five warm events (left panels) and five cold events (right panels). Precipitation data averaged over “A” are shown in the top row, and pressure data averaged over “B” are shown in the bottom row. In each column, the 50th percentile (or median) is marked by a black bar, the range between the 25th and 75th percentiles are depicted using solid blue shading, and the ranges within the 10th-25th and 75th-90th percentiles are indicated using lighter blue shading. The corresponding observational estimates as obtained from the NCEP reanalyses are also shown using light green bars. The above procedure is repeated for the eight realizations of the CTRL experiment, and the results are displayed in Fig. 4 using red columns. We shall focus on the simulations of the MLM experiment in the following discussion. Comparison between the MLM and CTRL statistics presented in Fig. 4 will be made later in Section 6.2, where the effects of air-sea coupling on the model responses is considered.

For the MLM experiment (see blue columns in Fig. 4), the medians of simulated rainfall and pressure anomalies at the chosen sites (black bars) are positive in all five warm events, and negative in all cold events. The observations (light green bars) exhibit a similar dependence. Of the 20 blue columns displayed in Fig. 4, the observational estimates lie within the 10th-90th percentiles of the simulated values in 17 columns. The notable spread of the 16 individual samples indicates that inferences drawn for a single model realization or a small ensemble of runs could be subject to considerable uncertainties. However, it is seen that a large fraction (usually more than 75%) of the individual sample runs share the same polarity as the median for a given event. The evidence presented here illustrates the degree of the reproducibility of the relationships between ENSO and East Asian-western Pacific anomalies in different episodes, and in individual MLM runs for the same episode.

5. Evolution of SST and Surface Fluxes over the Western Pacific during ENSO

It is anticipated that the ENSO-related atmospheric anomalies described in the previous section would alter various forms of energy transfer across the local air-sea interface. Fluctuations in the surface fluxes would in turn lead to changes in the temperature of the ocean mixed layer. The nature of such interactions between the atmosphere and ocean in the EAM region is delineated here by examining the typical space-time evolution of the anomalies in surface fluxes and SST during ENSO. It should be noted at the outset that, within the context of the mixed layer model used in this study, surface flux variation is not the only mechanism that could lead to SST changes. Several other processes, including deepening or shoaling of the mixed layer, convective mixing, diffusion, as well as temperature entrainment and penetration of shortwave radiation at the base of the mixed layer, could also modify the SST field. Hence the following discussion on the relationships between the SST and surface flux anomalies is concerned with a particular facet of the heat budget of the surface ocean. A more comprehensive description of the contributions to SST tendency by the other processes mentioned above is given in LN03.

The warm minus cold composite of the simulated SST (shading) and total surface flux at the ocean surface (contours, hereafter referred to as SF) are shown in Fig. 5 for the (a) ONDJ(0/1) and (c) FMAM(1) periods. The composite charts of the observed SST field for the same periods are displayed in panels (b) and (d). The composite procedure is the same as that used in constructing Figs. 2-3. The sign convention used for SF is that positive values correspond to energy transfer from the atmosphere to the mixed layer, thus leading to oceanic warming. Note that the SST at most grid points in the plotting domain used here has been computed using the mixed layer model (see Section 2). Observed SST variations have been prescribed only near the lower right edge of this domain (see blank region in Figs. 5a and 5c) and at near-equatorial sites east of the dateline.

During ONDJ(0/1), a positive anomaly in SF is discernible over the South and East China Seas (Fig. 5a). Analysis of the contributions of various flux components in this region (not shown) indicates that latent heat flux accounts for about half of the total anomaly, whereas sensible heat and shortwave radiative fluxes each contribute to about a quarter of the sum. The effect of longwave radiation is weak and acts to dissipate the warm anomaly. The reduced latent and sensible heat loss (which corresponds to positive anomalies in these components according to the sign convention adopted here) are evidently linked by the bulk aerodynamic law to the decreased wind speeds to the north and west of the WPH (Fig. 2c). The positive shortwave radiative flux anomaly in this region is attributable to the reduction in cloud cover accompanying the suppressed precipitation near the WPH (Fig. 2d). Negative SF anomalies prevail in ONDJ(0/1) over the subtropical Pacific east of about 150°E. Decomposition of the SF anomaly at this site into its individual components (not shown) indicates that the local energy transfer is dominated by latent heat flux, with secondary contribution from sensible heat flux. The radiative fluxes are comparatively much weaker.

The composite pattern of the simulated SST field in ONDJ(0/1) is characterized by warm anomalies in the waters off the coasts of China and Indochina, and cold anomalies farther

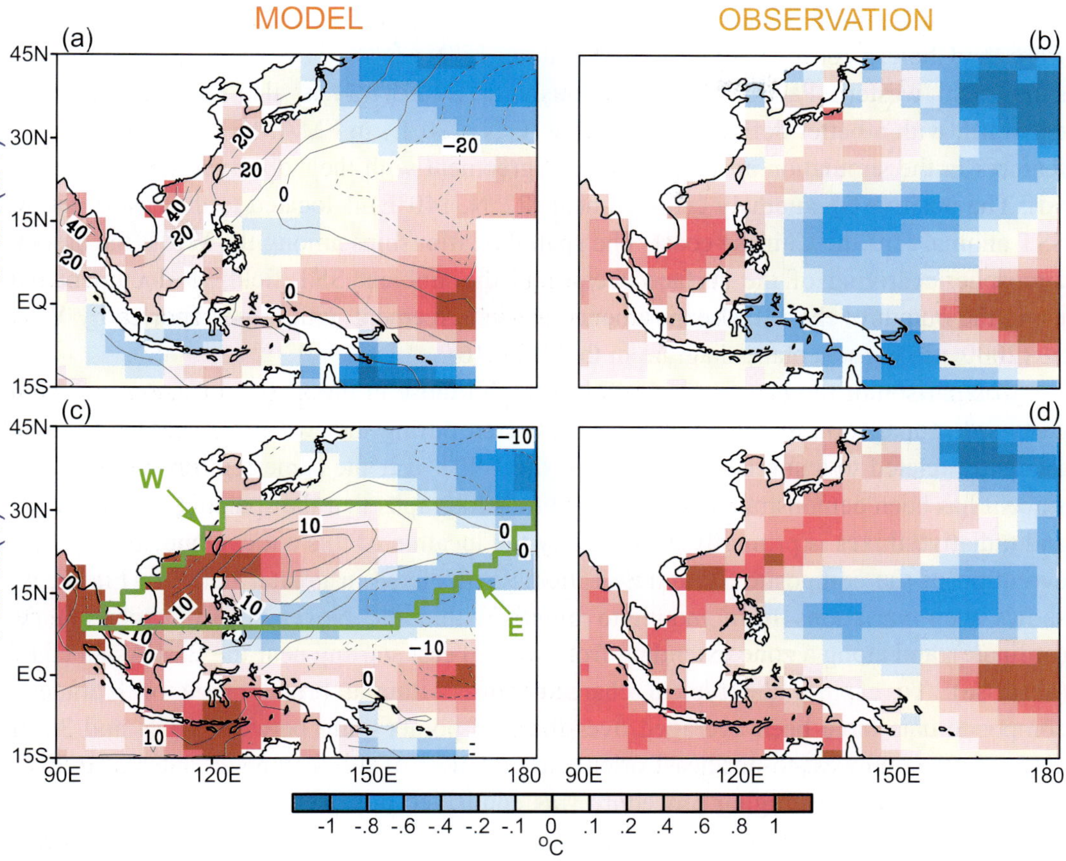

Figure 5. Warm minus cold composites of atmosphere-to-ocean heat flux (contours; interval: (a) 10 Wm^{-2}, (c) 5 Wm^{-2}) and SST (shading, see scale bar at bottom), for the (top panels) ONDJ(0/1) and (bottom panels) FMAM(1) time periods. The left panels are based on output from the MLM experiment. The right panels (showing SST patterns only) are based on observations. No model result is displayed near the lower right edge of the plotting domain of the left panels, where observed SST values have been prescribed in MLM. Shown in panel (c) are the boundaries used in analyzing the time-space development of the flux and SST fields in Fig. 6.

east (shading in Fig. 5a). The broad-scale features of the simulated SST pattern are comparable to those observed (see Fig. 5b). The observational chart indicates a relatively stronger cold anomaly east of the Philippines, and a more eastward extension of the warm anomaly to the waters south of Japan. Such discrepancies between the model and observed SST patterns could partially be attributed to absence in the mixed layer model of dynamical processes in the ocean, such as Ekman flows, propagation of various wave modes and advection by ocean currents.

In the spring of Year(1), a positive SF anomaly is discernible to the northeast of the Philippines (contours in Fig. 5c), on the northwestern flank of the WPH (Fig. 2f). Analysis of the individual flux components contributing to this anomaly (not shown) indicates that increased shortwave radiation plays a primary role. Weak, negative SF anomalies are simulated along the eastern and southern seaboards of China during FMAM(1). The SST pattern for this period (shading in Fig. 5c) indicates eastward spreading of the warm anomaly from the Chinese coast to western Pacific along 15°N -25°N. The weak cold anomaly in the subtropical western Pacific noted previously in ONDJ(0/1) is also seen to be displaced eastward, with a better defined center near (10°N, 170°E) in FMAM(1). The prominent SST features in the simulation are in reasonable agreement with their observed counterparts (Fig. 5d). The positive SF anomaly centered near (22°N, 140°E) is located to the east of the warm SST anomaly off the Chinese coast. The spatial relationships among the WPH, the associated atmospheric driving of the surface ocean and the ambient SST field, as documented here using the output of the MLM experiments, are in agreement with those reported by Wang *et al.* (2000, see the schematic diagram in their Fig. 16).

Comparison of the patterns in ONDJ(0/1) with those in FMAM(1) indicates a systematic eastward propagation of the SF and SST extrema during the two periods. This behavior is evidently related to a similar movement of the WPH over the same region (see Figs. 2c-2f), since it has been demonstrated in the preceding paragraphs that the spatial distribution of heat and radiative fluxes is primarily governed by the location of this anticyclone. The space-time evolution of the composite SF and SST fields is depicted in greater detail in Fig. 6, which shows their variations from month to month within the November(0)-October(1) period (ordinate), and along a zone extending from west to east starting near the East Asian coastline (abscissa). To capture the typical characteristics within a broad range of latitudes, the composite data have been averaged over the grid points lying between 10°N and 30°N. In view of the southwest-to-northeast orientation of most of the features of interest in Figs. 5a and 5c, this spatial averaging procedure was conducted not over north-south meridians, but instead along slanted paths that are approximately parallel to the isolines of the SF and SST fields shown in those figures. The path selected for computing the average for the last entry along the abscissa (right end) in Fig. 6 is indicated by the eastern border near the dateline (marked as "E") in Fig. 5c. The set of grid points that are located just to the west of this green line was then used to form the average for the second-to-last entry along the abscissa, and so forth. By translating the entire path to the west by one grid point in each step, we could hence calculate various entries along the abscissa in Fig. 6. This procedure was followed until we reached the western border near the Asian coast (shown as another line marked as "W" in Fig. 5c), which was used to compute the first entry along the abscissa (left end). The same set of paths was used for evaluating the averages in all months considered here. Only data at maritime grid points lying along a given path were used in the averaging process; all land points (as encountered near the Asian coast) on the path, as well as several ocean points in the Gulf of Thailand, were ignored.

The pattern for SF (Fig. 6a) illustrates the tendency for the positive anomaly just off the Asian seaboard to move progressively eastward with time during the November(0)-March(1) period. This anomaly attains peak intensity in January (1) near the coast, and is replaced locally by a negative feature with lower amplitude in April-May(1). A negative anomaly is also evident during December(0)-February(1) at about 40-50° of longitude to the east of the concurrent positive anomaly. The eastward traveling signals are even more evident in the SST chart (Fig. 6b), with peak amplitudes occurring in March-April(1), i.e., about 2-3

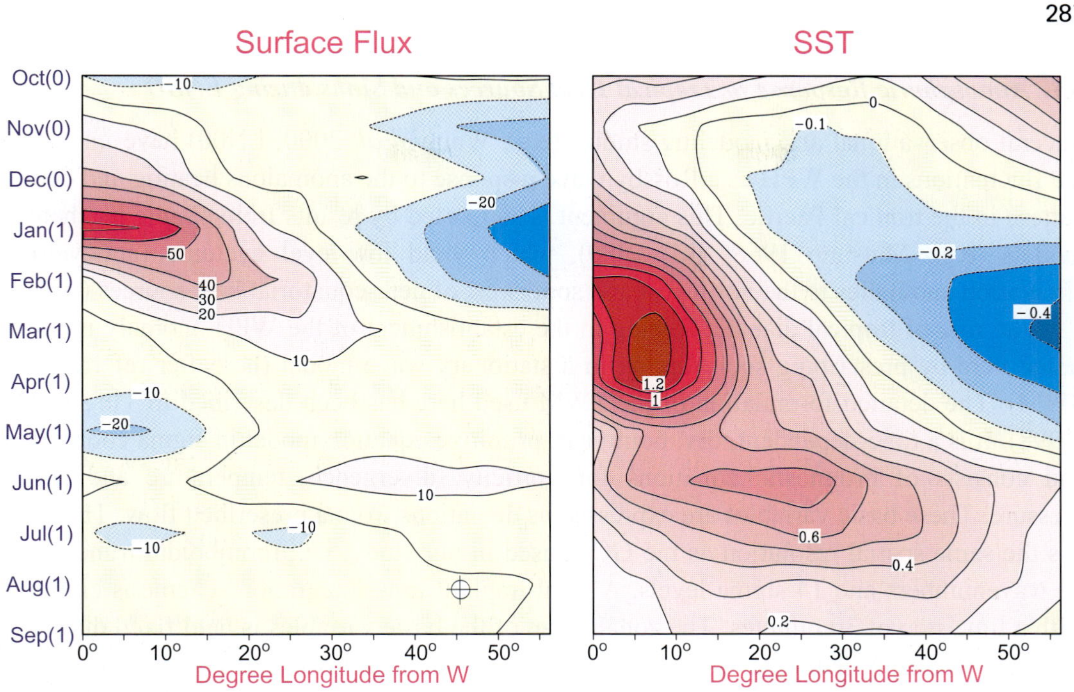

Figure 6. Distributions of the warm minus cold composites of atmosphere-to-ocean heat flux (left panel) and SST (right panel) with respect to time (ordinate) and longitude (abscissa). The left and right ends of the abscissa axis correspond to the western (W) and eastern (E) edges of the domain depicted in Fig. 5c. The patterns are based on averages from 10°N to 30°N, and along southwest-to-northeast oriented paths that are parallel to the edges shown in Fig. 5c. See text for further details.

months following the corresponding extrema in the surface flux forcing. Prior to May(1), the SST distribution within the 10°N - 30°N zone is dominated by a dipole-like pattern, with warm anomalies near the Asian coast, and cold anomalies just to the west of the border "E" in Fig. 5c. The negative SF anomaly near the western border "W" in April-May(1) is seen to be accompanied by attenuation of the warm SST anomaly in that region. The eastward spreading of the warm anomaly continued during the summer of Year(1), leading to above-normal SST throughout the plotting domain in that period.

6. Nature of ENSO-Induced Atmosphere-Ocean Coupling over the Western Pacific

The results reported in Sections 4-5 highlight the central role of the WPH in organizing various atmospheric and oceanic anomalies in the EAM region during ENSO. We proceed to investigate in greater detail the air-sea feedback processes associated with this anticyclone by conducting and analyzing additional experiments. Such supplementary model diagnoses have been designed to delineate individual links in the chain of interactions operating over the western Pacific, such as the formation of the WPH, and the effects of atmosphere-ocean coupling on the behavior of the WPH.

6.1. *Atmospheric Response to Tropical Heat Sources and Sinks during ENSO*

Several observational and modeling studies (e.g., Wang *et al.* 2000; LN00) have interpreted the fluctuations in the WPH as a Rossby wave response to the anomalous heating and cooling pattern in the tropical Pacific. This argument is supported by results from simple mechanistic models (e.g., Matsuno 1966; Gill 1980), which yield low-level cyclonic (anticyclonic) circulation anomalies to the northwest and southwest of near-equatorial heat sources (sinks).

The role of tropical diabatic forcing in the establishment of the WPH anomaly is further tested here by prescribing such forcing in a stationary wave model (hereafter referred to as SWM). The detailed formulation of the SWM used here has been described in Ting and Yu (1998). It is a time-dependent, dry, nonlinear, primitive equation model in sigma coordinates, and consists of prognostic equations for vorticity, divergence, temperature and surface pressure. These basic variables are expressed as deviations from a prescribed flow. The SWM has the same spatial resolution as the GCM used in our study, i.e., rhomboidal truncation at 30 wavenumbers and 14 sigma levels. A semi-implicit time integration scheme is employed with a time step of 30 minutes. The zonal mean of the basic variables is held fixed during the integration. Strong damping is applied to suppress model-generated transient disturbances. The SWM has been used by Ting *et al.* (2001) to diagnose the mechanisms for maintaining the seasonal evolution of the climatological stationary waves simulated by the GFDL GCM. For the purpose of our study, the warm minus cold composite of the three-dimensional diabatic heating field, as generated in the ensemble of MLM runs for the ONDJ(0/1) period, has been inserted as a forcing term in the SWM. The zonally averaged ONDJ climatological flow from the MLM runs was used as the basic state. For this diabatic forcing scenario, the steady state was reached after the SWM was integrated for approximately 20 days. The SWM solutions presented below were obtained by averaging the solutions over days 21-40.

The response of the SWM to the imposed global diabatic heating field is portrayed in Fig. 7a, which shows the departure from zonal mean of the streamfunction at the 0.935 sigma level[c] (contours). The degree to which this SWM solution reproduces the full GCM solution may be assessed by comparing this pattern with that in Fig. 7b, which shows the warm minus cold composite of the streamfunction field as simulated in the MLM experiment. The warm minus cold composite of the vertically averaged diabatic heating field, which has been extracted from the MLM runs for use in forcing the SWM, is depicted in both Figs. 7a and 7b with shading. As noted in the discussion of Fig. 2, the diabatic forcing field in the deep tropics is dominated by heating over the central Pacific, and cooling between 120°E and 160°E.

The streamfunction patterns based on both SWM and GCM data exhibit well-defined spatial relationships with the diabatic heating field. Anticyclonic circulations [i.e., positive (negative) streamfunction anomaly in the Northern (Southern) Hemisphere] are generated to the northwest and southwest of the negative heating center over eastern Indonesia, whereas cyclonic features emanate in an analogous fashion from the positive heating center over

[c] The particular sigma level chosen for display here is located in the lower troposphere, so that the streamfunction pattern can be examined in relation to other near-surface charts presented in the previous sections.

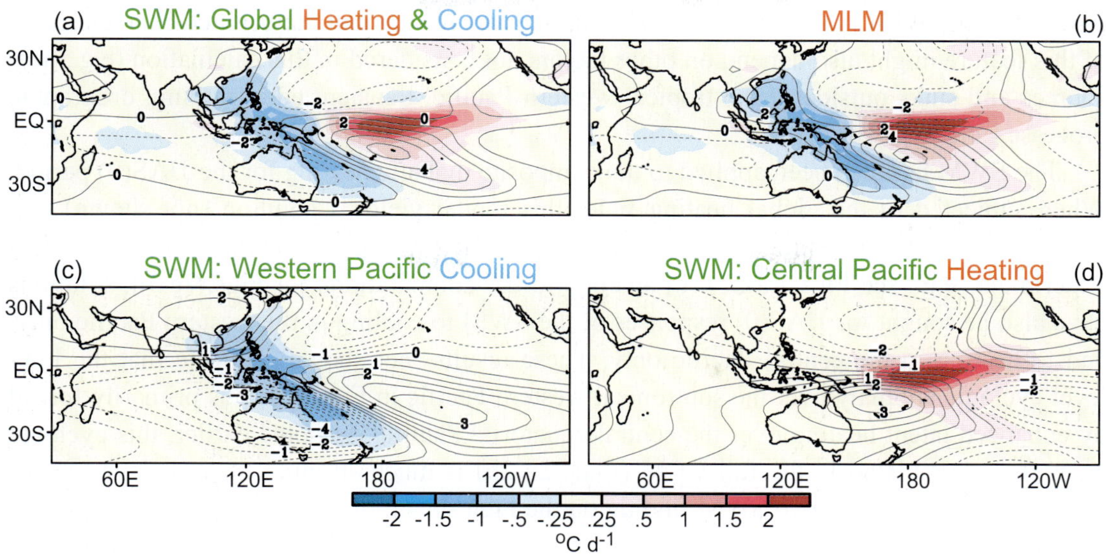

Figure 7. Distributions of the departure from zonal mean of the streamfunction at the 0.935 sigma level (contours; interval: 1 x 10^6 m^2s^{-1} for upper panels; 5 x 10^5 m^2s^{-1} for lower panels), as deduced from the response of the nonlinear stationary wave model to (a) diabatic heating and cooling over the entire globe, (c) cooling over the western Pacific only and (d) heating over the central and eastern Pacific only. The diabatic heating and cooling rates are based on warm minus cold composites of the output from the MLM experiment for the ONDJ(0/1) period, and their vertical means are displayed in the corresponding panels using shading (see scale bar at bottom). The warm minus cold composite of the corresponding streamfunction field generated in the MLM experiment is shown in panel (b).

central Pacific. Of particular interest to the current study is the appearance of the positive streamfunction anomaly center over the South China Sea-Philippines region, which corresponds closely to the WPH. Considering that diabatic heating is the only explicit anomalous forcing mechanism prescribed in the SWM integration, the general agreement between its solution (Fig. 7a) and the full GCM result (Fig. 7b) implies that the principal features in that latter pattern, including the WPH, owe their existence to the heat sources and sinks associated with ENSO.

In order to separate the effects of the two centers in the dipole-like heating pattern in Figs. 7a-7b, two more integrations of the SWM have been performed: one subjected only to the negative heating anomaly over the western Pacific; and the other only to the positive heating anomaly over the central Pacific. The distributions of the diabatic forcing used in these computations are depicted using shading in Figs. 7c-7d. The resulting SWM streamfunction solutions for these monopolar diabatic forcings are displayed as contours in the respective panels. The pattern in Fig. 7c indicates that the anticyclonic anomalies in the EAM and Indian Ocean-Australian sectors are primarily responses to the cooling center over the western Pacific; whereas the heating center (Fig. 7d) exerts the strongest influences on the circulation over the central and eastern Pacific. The East Asian anticyclonic center in the response pattern to the western Pacific cooling (Fig. 7c) does not coincide exactly with the position of the WPH (see Figs. 2d and 7a-7b), thus suggesting that the detailed characteristics of this feature might also depend on other factors not considered in this calculation (e.g., heat sources and sinks outside of the tropical western Pacific, transient eddy forcing, orographic processes, etc.).

The SWM has also been applied to the composite diabatic heating for the JJAS(0) period. When subjected to the global heating field, this model yields a solution (not shown) that bears a strong resemblance to the composite patterns based on output from the MLM experiment (e.g., see Figs. 2a-2b). In analogy to the computations for Figs. 7c-7d, solutions have also be sought for JJAS(0) responses of the SWM to cooling in the western Pacific only, and to heating in the central Pacific only. These results (not shown) indicate that the low-level cyclonic anomaly over the subtropical western Pacific in Figs. 2a-2b is primarily forced by enhanced latent heating over the central equatorial Pacific. As noted earlier, this cyclonic anomaly as simulated in the full GCM (Figs. 2a-2b) is much stronger than observed (Figs. 3a-3b). In view of the above SWM finding on the high sensitivity of the flow pattern over the subtropical western Pacific to central Pacific heating, the excessive strength of the subtropical anomalous cyclone in the GCM in JJAS(0) could be the result of incorrect representation of the equatorial Pacific heat source in the model atmosphere, or inadequate simulation of other processes that might interfere with the link between tropical heating and subtropical circulation.

6.2. *Feedbacks between Simulated SST Anomalies in the Western Pacific and the Ambient Atmospheric Circulation*

The diagnoses in Section 5 indicate that the western Pacific SST anomalies appearing in Year(1) result from anomalous atmospheric driving associated with the formation of the WPH in that region. In March-April(1), the amplitude of the SST changes exceeds 1°C near the East Asian coast (Figs. 5c and 6b). It is of interest to ascertain the feedback effects of these SST perturbations on the atmosphere, particularly with regards to their influences on the subsequent evolution of the WPH. We shall address these issues here by contrasting the atmospheric anomalies simulated in GCM experiments with and without air-sea coupling in the western Pacific, as well as by subjecting the GCM to anomalous SST forcing in the western Pacific and then analyzing the atmospheric response.

6.2.1. Differences between the MLM and CTRL Experiments

Recalling the experimental setup as described in Section 2, the MLM experiment has been designed to incorporate the effects of both the ENSO-related SST forcing in the deep tropical eastern/central Pacific (DTEP), and air-sea interaction in all ocean areas lying outside of DTEP; whereas the CTRL experiment considers the SST forcing in DTEP only. The contribution of atmosphere-ocean coupling outside of DTEP to the model signals may hence be estimated by taking the differences between the output from the MLM and CTRL experiments. Warm minus cold composites of selected fields have been constructed using the ensemble means of both experiments. Differences between the two sets of composites thus obtained (MLM minus CTRL) were then taken. The distributions of such difference charts for the FMAM(1) period are shown in Fig. 8 for (a) surface wind vectors (arrows) and SST[d] (shading), and (b) sea level pressure (contours) and precipitation (shading).

The patterns in Fig. 8b indicate that the presence in the MLM experiment of air-sea interaction over the western Pacific leads to a stronger WPH relative to the CTRL simulation. Also apparent in this panel is the enhancement in MLM of a low pressure zone which extends northeastward from Indochina to Japan. The intensified WPH in MLM is seen to be accompanied locally by a cold SST anomaly generated in that experiment (shading in Fig. 8a), and by a strengthened anticyclonic circulation and suppressed precipitation. Conversely, the trough feature off the Asian coast is situated over a warm SST anomaly, and is characterized by stronger surface flow convergence and larger rainfall amounts in MLM as compared to CTRL. The occurrence of these coastal anomalies at the same location as the climatological Meiyu-Baiu front (see Fig. 1a) suggests that the intensity of the latter phenomenon may be modulated by atmosphere-ocean coupling.

Some appreciation of the differences between the MLM and CTRL experiments for individuals ENSO events and individual members of the ensembles may be gained by comparing the statistical spread of the signals appearing in MLM (blue columns) and CTRL (red columns), as shown in Fig. 4. In a large majority of the selected warm and cold events, the median amplitude (see black bars) of the precipitation anomaly along the East Asian seaboard and the pressure anomaly near the WPH is higher in MLM than in CTRL. For a given event, there exists considerable overlap between the ranges of the anomalies generated by individual members of MLM and the corresponding ranges produced in CTRL, thus indicating that the differences between the MLM and CTRL signals are subject to large sampling fluctuations.

6.2.2. GCM Response to Prescribed SST Anomalies in the Western Pacific

As a more direct test of the impact of western Pacific SST anomalies on the atmospheric circulation, another suite of model runs has been conducted with the same GCM as that used for the MLM and CTRL experiments. Altogether three integrations were performed with the solar forcing set perpetually at the mid-April condition:

- The climatological ("CLIMO") run, with the long-term averaged SST field[e] in April being prescribed at all ocean grid points.
- The 1xSST run, with the warm minus cold composite of the SST field in the western Pacific region (approximately, 5°N - 40°N, 100°E - 170°E) for April(1) being added to the climatological SST data.

---

[d] Since climatological SST data have been prescribed outside of DTEP in CTRL, the anomalous SST field for that experiment is zero everywhere outside of DTEP. The shading pattern in Fig. 8a is therefore identical to that of the anomalous SST field for the MLM experiment (see Fig. 5c).

[e] For the sake of consistency, the climatological SST data used here is identical to those used in the CTRL experiment, i.e., they have been obtained by averaging the global SST field in the ensemble of MLM runs. As pointed out in footnote a, the SST climatology computed in this manner is very similar to the observed climatology.

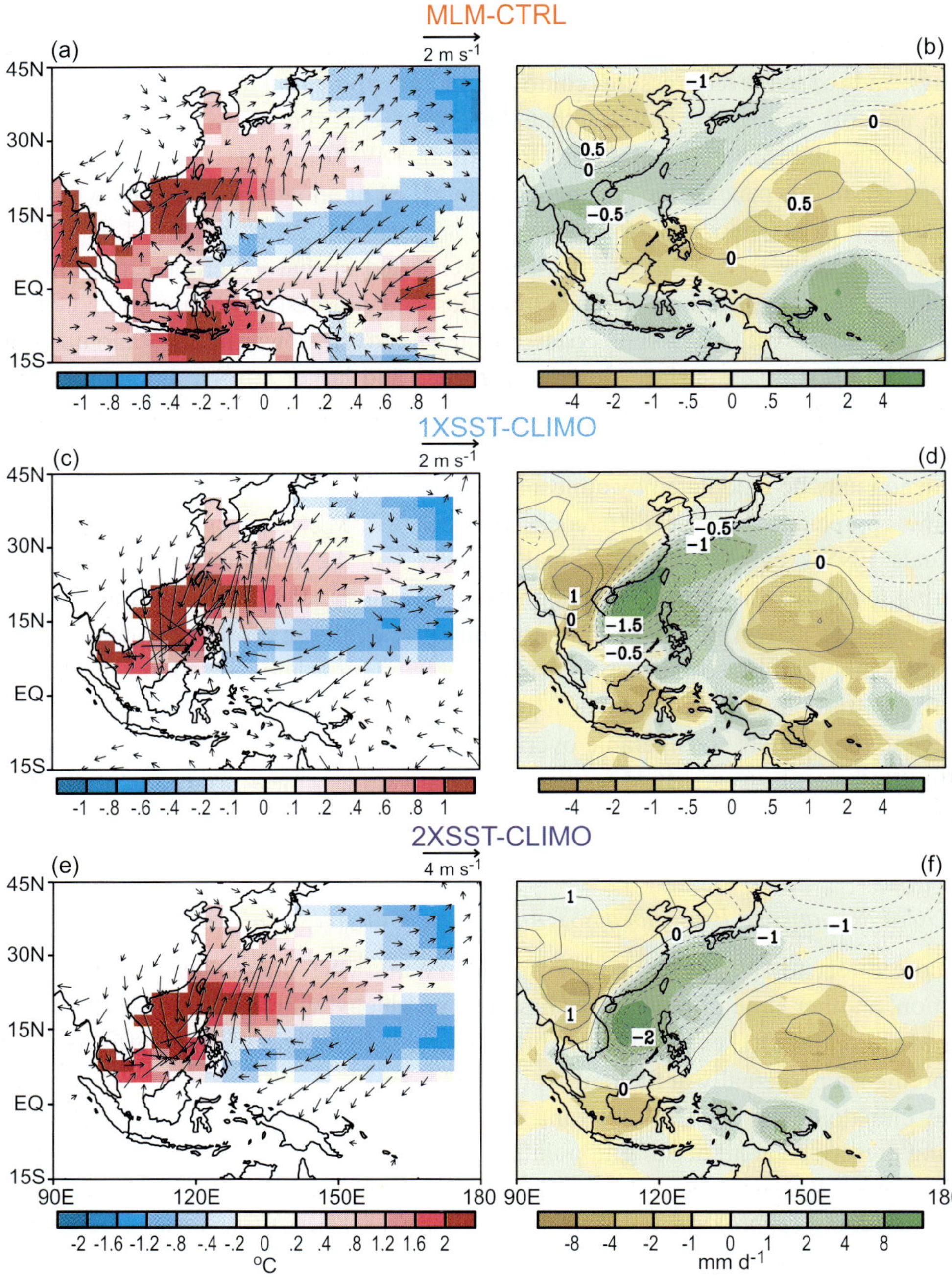

Figure 8. Distributions of (left panels) near surface wind vector (arrows, see scale at upper right) and SST (shading, see scale bar at bottom), and (right panels) sea level pressure [contours; interval: 0.25 mb for panels (b), (d); 0.5 mb for panel (f)] and precipitation (shading, see scale bar at bottom), as obtained by taking the differences between (a,b) warm minus cold composites of the output from the MLM and CTRL experiments for the FMAM(1) period, (c,d) long-term averages of the output from the 1xSST and CLIMO experiments, and (e,f) long-term averages of the output from the 2xSST and CLIMO experiments.

- The 2xSST run, similar to the 1xSST run, except that the amplitude of the SST composite data for the western Pacific has been doubled prior to summation with the climatological field.

The CLIMO run has been conducted for a duration of 5000 days, whereas the length of each of the 1xSST and 2xSST experiments is 1000 days. The following results are based on averages over the entire course of each of these integrations.

The differences between the 1xSST and CLIMO experiments (1xSST minus CLIMO) are shown in Fig. 8 for (c) surface wind vector (arrows) and SST (shading) and (d) sea level pressure (contours) and precipitation (shading). The corresponding difference charts obtained by subtracting the 2xSST output from CLIMO are displayed in Figs. 8e-8f. The response patterns in Figs. 8c-8f reveal most of the features noted in the differences between the MLM and CTRL experiments (Figs. 8a-8b)[f]. Of particular interest are the anticyclonic winds and suppressed precipitation associated with the high pressure center over the prescribed cold SST anomaly situated at 5°N - 15°N and 130°E - 170°E; as well as the convergent flows and enhanced rainfall along the elongated trough lying above the warm SST forcing off the East Asian coast. The simulation of these atmospheric features in the 1xSST and 2xSST experiments offers strong support for our earlier interpretation (based on MLM and CTRL) that such phenomena are partially attributable to forcing by the western Pacific SST anomalies. The results from the 1xSST experiment are qualitatively similar to those from the 2xSST run, with the response amplitudes in the latter run being higher by about a factor of two, thus implying quasi-linearity of the model responses to the imposed SST forcing.

The statistical significance of the precipitation and sea level pressure signals displayed in Figs. 8d and 8f has been assessed using Student's t-test (e.g., see Chervin and Schneider 1976), after making conservative estimates of the number of degrees of freedom in the time series generated by the CLIMO, 1xSST and 2xSST runs. The results from this testing procedure indicate that the precipitation and sea level pressure responses near the East Asian coast in both the 1xSST and 2xSST experiments are significantly different from the corresponding fields in CLIMO at the 99% level. Over the subtropical western Pacific, the departure of the sea level pressure field in the 2xSST experiment from CLIMO is also significant at the 99% level.

In summary, the MLM-CTRL comparison as well as the 1xSST and 2xSST experiments indicate that the western Pacific SST anomalies in the spring and summer of Year(1), which were generated by atmospheric forcing in the previous winter (see Fig. 6), can exert noticeable feedbacks on the atmosphere. The resulting atmospheric responses could then influence the evolution of the atmosphere-ocean system in the following ways:

- The positive sea level pressure tendencies on the eastern flank of the WPH (note the near-quadrature spatial relationship between the anomaly pattern of sea level pressure in

---

[f] The prominent positive precipitation anomaly located to the east of New Guinea in Fig. 8b is not discernible in Figs. 8d and 8f. This precipitation feature is apparently associated with the warm SST anomaly in the same region (see Fig. 8a), which is situated outside the forcing domain in the 1xSST and 2xSST experiments (see Figs. 8c and 8e).

Fig. 2f and the response of this field to SST forcing in the right panels of Fig. 8) are conducive to eastward migration of this feature.

- The strengthening of the northeasterly trade winds within the 0° - 15°N belt over the western Pacific (left panels of Fig. 8) supports the transition from El Niño conditions in Year(0) to La Niña conditions in Year(1).
- The intensification of the subtropical frontal zone off the East Asian seaboard favors the reversal from the dry conditions in this region during the summer of Year(0) to wet conditions in the spring of the following year (see Figs. 2b, 2f and right panels of Fig. 8).

In the following section, we shall discuss further the above-mentioned reversals in the trade wind and precipitation anomalies during successive years of the ENSO cycle.

7. Discussion

7.1. *Important Role of the WPH Anomaly in EAM Variability*

The model experimentation and diagnosis performed in this study have identified the WPH anomaly as the crucial phenomenon about which much of ENSO-related variability in the East Asian sector is organized. Solutions from the SWM (Fig. 7) indicate that the formation of this anticyclonic feature in ONDJ(0/1) can primarily be attributed to the east-west oriented dipole-like diabatic forcing in the tropical Pacific, with the effects of the cooling center over the western Pacific being relatively stronger. The wind anomaly in the northwestern quadrant of the wintertime WPH is directed against the climatological flow, resulting in weaker wind speeds (Fig. 2c), which then lead to reduced heat loss and above normal SST in the marginal seas of East Asia (Fig. 5a). Conversely, enhancement of the northeasterly winter monsoon in the eastern quadrant of the WPH is conducive to cooling of the subtropical Pacific waters east of about 135°E. Both the warm and cold SST anomalies attain highest amplitudes in FMAM(1). Results from the 1xSST and 2xSST experiments (Figs. 8c-8f), as well as comparison of the MLM and CTRL runs (Figs. 8a-8b), indicate that these SST changes could in turn affect the atmospheric circulation by imparting an eastward tendency to the WPH. This chain of cooperative feedbacks between the atmospheric flow and the underlying oceanic mixed layer supports the continuous eastward propagation of both the WPH and the accompanying SST signals throughout the ONDJ(0/1) and FMAM(1) periods (Figs. 2 and 6).

The WPH also exerts a strong influence on the precipitation pattern in the EAM region. During the winter of Year(0/1), the anomalous circulation on the northwestern flank of the WPH opposes the climatological southwestward intrusion of dry continental air masses from the Asian landmass, thus resulting in wetter than normal condition over southern China (Fig. 2d). During the spring and summer of Year(1), the WPH feature steers moisture-laden air currents from the tropics towards the East Asian sea board, so that above normal precipitation prevails along the climatological Meiyu-Baiu front (Fig. 2f). Comparison of the summertime patterns for Year(0) and Year(1) reveals a quasi-biennial tendency for the sea level pressure and precipitation fields in the EAM region (Figs. 2b, 2f and 2h), with the WPH in Year(1) substituting the low pressure center that appeared near the dateline in Year(0); and with the dryness in Year(0) along the 20°N zone from Indochina to the Philippine Sea being replaced by wetter conditions in Year(1).

The above-mentioned reversals in polarity of the atmospheric anomalies from Year(0) to Year(1) suggest that the WPH is a key element of the family of monsoonal phenomena associated with the "tropospheric biennial oscillation" (TBO). The contributions of coupling within the land-atmosphere-ocean system, near-equatorial zonal circulation cells, tropical convection and tropical-midlatitude interactions to the evolution of the TBO have been reviewed by Meehl (1997). Different facets of the TBO in the Asian-Australian monsoon region have been examined using a variety of research tools, including observational datasets (Meehl and Arblaster 2002a), simplified mechanistic modeling (Chang and Li 2000), atmospheric GCM (Meehl and Arblaster 2002b), and atmosphere-ocean coupled GCM (Ogasawara *et al.* 1999). Furthermore, the results from an intermediate coupled model (Kim and Lau 2001) indicate that the quasi-biennial tendency in ENSO development may be linked to TBO-related monsoon wind forcing in the western Pacific.

7.2. *ENSO Signals beyond the EAM Region*

In concert with the theme of this monograph, our attention has been focused on ENSO-related atmospheric and oceanic changes in the EAM region only. However, it is important to stress that the impacts of ENSO are felt not just in the East Asian sector, but in other parts of the Indo-Pacific Basin as well. A broader perspective of the linkages between various components of the Asian-Australian monsoon complex and ENSO events may be gained by surveying the patterns in Fig. 9, which cover a much wider spatial domain than that considered thus far in this chapter. The charts in Fig. 9 show the warm minus cold composites of the data generated in the MLM experiment during ONDJ(0/1), for (a) surface wind direction and speed, (b) sea level pressure and precipitation, and (c) SST. Note that features of the corresponding fields within the EAM region, as shown earlier in Figs. 2c, 2d and 5a, are embedded in the more expanded patterns in Fig. 9. A brief discussion of the salient phenomena simulated outside of the EAM region is given here. Interested readers are referred to LN03 for further details of the interrelationships and temporal evolution of these signals, as well as their roles in atmosphere-ocean feedbacks.

It is evident from Fig. 9 that the composite signals in the western Pacific exhibit some degree of symmetry about the Equator. Particular noteworthy is the simulation in both hemispheres of prominent anomalies of above-normal pressure and below-normal rainfall (Fig. 9b). The northern center (i.e., the WPH) is situated over the Philippines, and its counterpart in the Southern Hemisphere is located off the eastern Australian coast. The anticyclonic circulation cells associated with this pair of features are accompanied by decreased wind speeds off the southeastern coasts of China and Australia, and increased speeds to the south of the WPH and to the northeast of the Australian anticyclone (Fig. 9a). The reduction and intensification of the surface flows over the subtropical southwestern Pacific are in turn collocated with warm and cold SST anomalies, respectively (Fig. 9c). Solutions of the stationary wave model (Fig. 7) indicate that the subtropical near-surface anticyclones over eastern Asia and eastern Australia are primarily Rossby-wave responses to

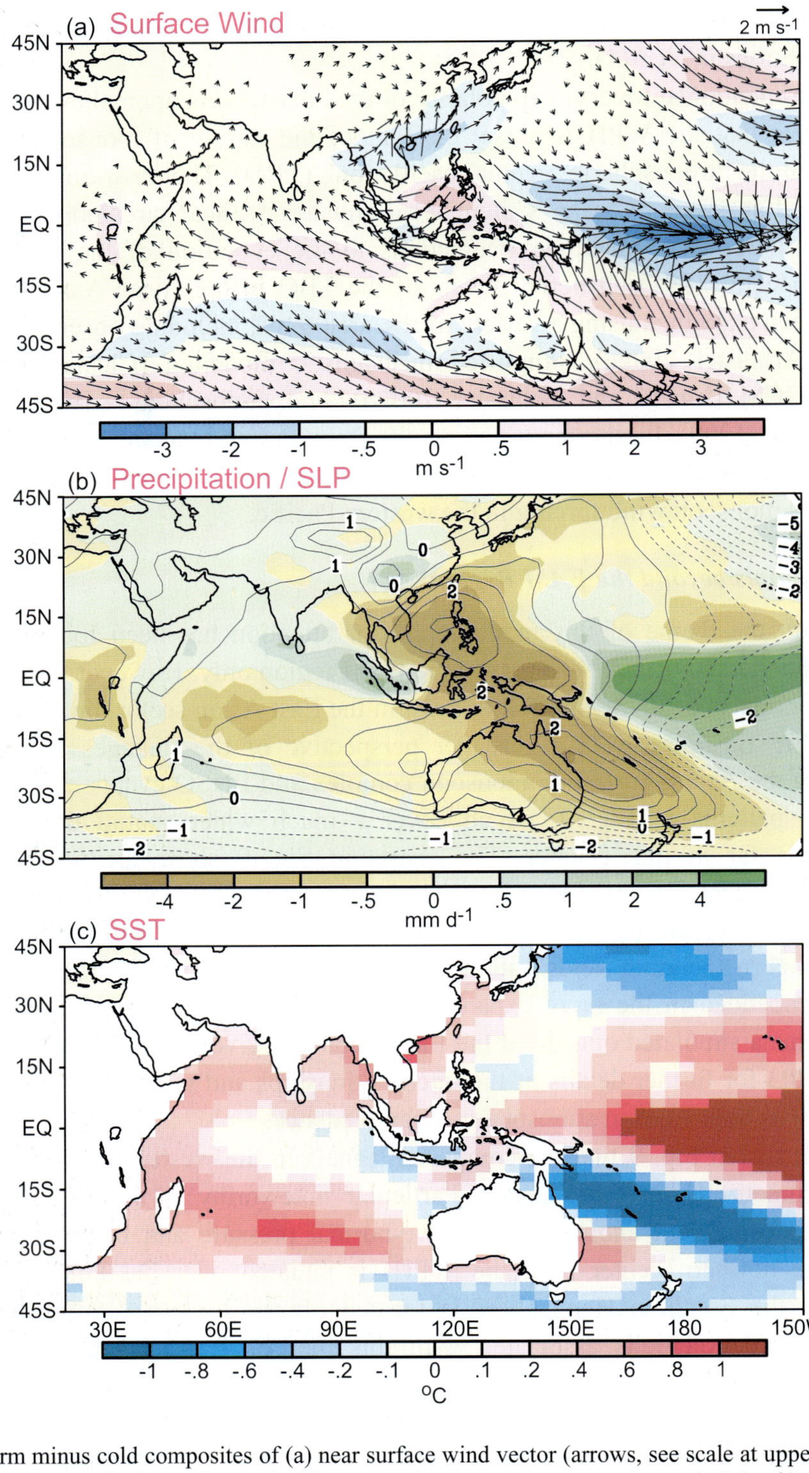

Figure 9. Warm minus cold composites of (a) near surface wind vector (arrows, see scale at upper right) and wind speed (shading, see scale bar at bottom), (b) sea level pressure (contours; interval: 0.5 mb) and precipitation (shading, see scale bar at bottom), and (c) SST (shading, see scale bar at bottom), for the entire Asian-Australian monsoon region during the ONDJ(0/1) period. All patterns are based on output from the MLM experiment.

reduced condensational heating over the tropical western Pacific during warm ENSO events.

In the Indian Ocean sector, anomalous surface easterly outflow from the divergent center over the Indonesian Archipelago is prevalent in the 0° - 15°S belt (Fig. 9a). The anomalous high pressure center extending westward from the northwestern coasts of Australia (Fig. 9b) may be viewed as part of the Rossby-wave response to the heat sink over the western Pacific (see Fig. 7). Northwesterly wind anomalies prevail on the southwestern flank of this anticyclone, thus leading to weakening of the climatological southeasterly trades (Fig. 9a) and SST warming (Fig. 9c) in that region. Warm SST anomalies are also simulated in the MLM experiment in parts of the Arabian Sea and Bay of Bengal during ONDJ(0/1), partly as a result of air-sea feedbacks initiated by the weakened summer monsoon in Year(0) of warm events (see LN00 for details). The surface wind anomalies within the 0°-10°N belt in the Indian Ocean are mostly directed northward, i.e., towards the warm SST anomalies in the northern portion of that ocean basin.

The effects of SST anomalies in the Indian Ocean Basin on the atmospheric circulation have been investigated in LN03 by prescribing such SST anomalies in perpetual GCM integrations analogous to the 2xSST experiments described in Section 6.2. The atmospheric responses thus obtained (not shown) act to further reduce the intensity of the time-averaged southeasterly trades over the southern Indian Ocean as well as the northeasterly winter monsoon over parts of the Arabian Sea and Bay of Bengal, and thereby reinforce the existing warm SST anomalies in these regions through alteration of the surface fluxes. These results hence reveal interesting air-sea feedbacks in the Indian Ocean sector. Such processes tend to maintain local atmospheric and oceanic changes initiated by ENSO events.

The cumulative evidence presented in this chapter indicates that the influences of ENSO are mainly transmitted to the remote maritime sites through the atmosphere. Specifically, the SST anomalies in the central and eastern tropical Pacific affect the distribution of heating sources, thus modifying the equator-wide atmospheric circulation through excitation of various modal (e.g., Rossby-type) patterns as well as displacements of the local Walker and Hadley Cells. These atmospheric anomalies alter the surface fluxes at oceanic areas lying outside of the DTEP region, which in turn lead to SST changes in the western Pacific and Indian Oceans. This chain of processes, in which the "atmospheric bridge" plays a critical role, is known to explain many aspects of the teleconnections between ENSO events and SST variability in the tropical and extratropical oceans (Lau and Nath 1996, 2001; Klein *et al.* 1999; Alexander *et al.* 2002). The MLM experiment, which accommodates the prescription of ENSO forcing in DTEP and two-way air-sea coupling elsewhere, has been specifically designed to study the atmospheric bridge mechanism. On the other hand, model integrations with prescription of SST forcing throughout the World Oceans would be much less suited for investigating the impact of this mechanism on interannual climate variability.

7.3. *Outstanding Issues*

The findings presented in this study demonstrate the utility of GCM experiments with appropriate designs for studying various linkages between ENSO-related processes and EAM variability. It is seen that the broad-scale impacts of ENSO on monsoon features in the East

Asian sector are simulated in such model experiments with a reasonable degree of fidelity. However, further improvements in both model resolution and model physics would be needed for investigating the detailed regional characteristics of monsoon changes on interannual and interdecadal time scales, as noted in the observational studies of Chang *et al.* (2000a, 2000b) and Lau and Weng (2001). In addition, a full account of the implications of the atmospheric circulation anomalies associated with the WPH on the subsequent development of the ENSO episodes in the latter half of Year(1) can only be made using coupled models with full ocean dynamics, so as to incorporate the effects of crucial processes such as advection by ocean currents, and propagation of characteristic modal structures. By examining the typical atmospheric and oceanic changes in the EAM region during different stages of the ENSO cycles simulated in these fully coupled GCM experiments, it would be feasible to explore in greater detail the myriad processes accounting for the evolution of the WPH. Such diagnoses would also yield fresh insights into the feedback of the atmospheric anomalies associated with WPH to the underlying ocean. These feedback effects could contribute to phase transitions of the ENSO cycle.

The results shown in Fig. 4 indicate notable differences between the model responses in individual ENSO events. For a given event, the responses simulated in individual members of the ensemble experiments also exhibit considerable variability. Further study is required to identify the contributions to these inter-event and inter-sample variations by factors such as distinctive space-time evolution of the SST forcing in individual ENSO episodes, dependencies of the model responses on the ambient flow pattern, and the stochastic nature of the atmospheric circulation. Resolution of these complex issues would advance our understanding of the predictability of the fluctuations in the EAM system.

The SST anomalies in the western Pacific and Indian Oceans accompanying ENSO in the MLM experiment (see Fig. 9c), as produced by the tropical atmospheric bridge mechanism discussion in the previous subsection, could in turn generate extratropical teleconnection patterns and thereby affect the atmospheric circulation over the North Pacific and North America. The rudimentary evidence presented by Alexander *et al.* (2002) suggests that such effects may compete with the direct extratropical response to ENSO-related SST forcing in the tropical Pacific. Experiments analogous to the 1xSST and 2xSST runs, with prescription of perpetual SST forcing in a variety of oceanic sites, could be used to quantify the relative contributions of SST anomalies in different parts of the tropical oceans to midlatitude atmospheric variability.

Acknowledgments

We would like to thank John Lanzante for conducting the suite of MLM and CTRL experiments examined in this study. We are also grateful to Anthony Broccoli, In-Sik Kang, Brian Soden and an anonymous official reviewer for offering constructive comments on initial drafts of this chapter. The first author has benefited from many stimulating discussions with Bin Wang throughout the course of this work.

References

Alexander, M. A., and J. D. Scott, 1995: *Atlas of climatology and variability in the GFDL R30S14 GCM.* CIRES, University of Colorado, 121 pp. [Available from the authors at CDC/NOAA, R/CDC1, 325 Broadway, Boulder, CO 80305-3328].

______, C. Deser, and J. D. Scott, 2000: Processes that influence sea surface temperature and ocean mixed layer depth variability in a coupled model. *J. Geophys. Res.*, **105**, 16823-16842.

______, I. Blade, M. Newman, J. R. Lanzante, N.-C. Lau, and J. D. Scott, 2002: The atmospheric bridge: the influence of ENSO teleconnections on air-sea interaction over the global oceans. *J. Climate*, **15**, 2205-2231.

Broccoli, A. J., and S. Manabe, 1992: The effects of orography on midlatitude Northern Hemisphere dry climates. *J. Climate*, **5**, 1181-1201.

Chang, C.-P., and T. Li, 2000: A theory for the tropical tropospheric biennial oscillation. *J. Atmos. Sci.*, **57**, 2209-2224.

______, Y. Zhang, and T. Li, 2000a: Interannual and interdecadal variations of the East Asian summer monsoon and tropical Pacific SSTs. Part I: Roles of the subtropical ridge. *J. Climate*, **14**, 4310-4325.

______, ______, and ______, 2000b: Interannual and interdecadal variations of the East Asian summer monsoon and tropical Pacific SSTs. Part II: Meridional structure of the monsoon. *J. Climate*, **14**, 4326-4340.

Chen, L.-X., M. Dong, and Y.-N. Shao, 1992: The characteristics of interannual variations on the East Asian monsoon. *J. Meteor. Soc. Japan*, **70**, 397-421.

Chervin, R. M., and S. H. Schneider, 1976: On determining the statistical significance of climate experiments with general circulation models. *J. Atmos. Sci.*, **33**, 405-412.

Gaspar, P., 1988: Modeling the seasonal cycle of the upper ocean. *J. Phys. Oceanogr.*, **18,** 161-180.

Gill, A. E., 1980: Some simple solutions for heat-induced tropical circulation. *Quart. J. Roy. Meteor. Soc.*, **106**, 447-462.

Gordon, C. T., and W. F. Stern, 1982: A description of the GFDL global spectral model. *Mon. Wea. Rev.*, **110**, 625-644.

Kalnay, E., and Coauthors, 1996: The NCEP/NCAR 40-year Reanalysis Project. *Bull. Amer. Meteor. Soc.*, **77**, 437-472.

Kim, K.-M., and K.-M. Lau, 2001: Dynamics of monsoon-induced biennial variability in ENSO. *Geophys. Res. Lett.*, **28**, 315-318.

Klein, S. A., B. J. Soden, and N.-C. Lau, 1999: Remote sea surface temperature variations during ENSO: Evidence for a tropical atmospheric bridge. *J. Climate*, **12**, 917-932.

Lau, K.-M., and H. Weng, 2001: Coherent modes of global SST and summer rainfall over China: An assessment of the regional impacts of the 1997-98 El Niño. *J. Climate*, **14**, 1294-1308.

______, and H. T. Wu, 2001: Principal modes of rainfall-SST variability of the Asian summer monsoon: A reassessment of the monsoon-ENSO relationship. *J. Climate*, **14**, 2880-2895.

Lau, N-C., and M. J. Nath, 1994: A modeling study of the relative roles of tropical and extratropical SST anomalies in the variability of the global atmosphere-ocean system. *J. Climate*, **7**, 1184-1207.

______, and ______, 1996: The role of the "atmospheric bridge" in linking tropical Pacific ENSO events to extratropical SST anomalies. *J. Climate*, **9**, 2036-2057.

______, and ______, 2000: Impact of ENSO on the variability of the Asian-Australian monsoons as simulated in GCM experiments. *J. Climate*, **13**, 4287-4309.

______, and ______, 2001: Impact of ENSO on SST variability in the North Pacific and North Atlantic: Seasonal dependence and role of extratropical sea-air coupling. *J. Climate*, **14**, 2846-2866.

______, and ______, 2003: Atmosphere-ocean variations in the Indo-Pacific sector during ENSO episodes. *J. Climate*, **16**, 3-20.

Manabe, S., R. J. Stouffer, M. J. Spelman, and K. Bryan, 1991: Transient responses of coupled ocean-atmosphere model to gradual changes of atmospheric CO2. Part I: Annual mean responses. *J. Climate*, **4**, 785-818.

Matsuno, T., 1966: Quasi-geostrophic motions in the equatorial area. *J. Meteor. Soc. Japan*, **44**, 25-43.

Meehl, G. A., 1997: The South Asian monsoon and the tropospheric biennial oscillation. *J. Climate*, **10**, 1921-1943.

______, and J. Arblaster, 2002a: The tropospheric biennial oscillation and Asian-Australian monsoon rainfall. *J. Climate*, **15**, 722-744.

______, and ______, 2002b: Indian monsoon GCM sensitivity experiments testing tropospheric biennial oscillation transition conditions. *J. Climate*, **15**, 923-944.

Ogasawara, N., A. Kitoh, T. Yasunari, and A. Noda, 1999: Tropospheric biennial oscillation of the ENSO-monsoon system in the MRI coupled GCM. *J. Meteor. Soc. Japan*, **77**, 1247-1270.

Ropelewski, C. F., and M. S. Halpert, 1987: Global and regional scale precipitation patterns associated with the El Niño/Southern Oscillation. *Mon. Wea. Rev.*, **115**, 1606-1626.

______, and ______, 1989: Precipitation patterns associated with the high index phase of the Southern Oscillation. *J. Climate*, **2**, 268-284.

Smith, T. M., R. W. Reynolds, R. E. Livezey, and D. C. Stokes, 1996: Reconstruction of historical sea surface temperatures using empirical orthogonal functions. *J. Climate*, **9**, 1403-1420.

Sperber, K. R., and T. N. Palmer, 1996: Interannual tropical rainfall variability in general circulation model simulations associated with the Atmospheric Model Intercomparison Project. *J. Climate*, **9**, 2727-2750.

Ting, M., and L. Yu, 1998: Steady response to tropical heating in wavy linear and nonlinear baroclinic models. *J. Atmos. Sci.*, **55**, 3565-3582.

______, H. Wang, and L. Yu, 2001: Nonlinear stationary wave maintenance and seasonal cycle in the GFDL R30 GCM. *J. Atmos. Sci.*, **58**, 2331-2354.

Trenberth, K. E., 1997: The definition of El Niño. *Bull. Amer. Meteor. Soc.*, **78**, 2771-2777.

Wang, B., and Q. Zhang, 2002: Pacific-East Asian teleconnection. Part II: How the Philippine Sea anomalous anticyclone is established during El Niño development. *J. Climate*, **15**, 3252-3265.

______, R. Wu, and X. Fu, 2000: Pacific-East Asian teleconnection: How does ENSO affect East Asian climate? *J. Climate*, **13**, 1517-1536.

Xie, P., and P. A. Arkin, 1997: Global precipitation: A 17-year monthly analysis based on gauge observations, satellite estimates, and numerical model outputs. *Bull. Amer. Meteor. Soc.*, **78**, 2539-2558.

Zhang, R., A. Sumi, and M. Kimoto, 1996: Impact of El Niño on the east Asian Monsoon: A diagnostic study of the '86/87 and '91/92 events. *J. Meteor. Soc. Japan*, **74**, 49-62.

8. CURRENT STATUS OF AGCM MONSOON SIMULATIONS

IN-SIK KANG
School of Earth and Environment Sciences
Seoul National University
Seoul 151-742, Korea
E-mail: kang@climate.snu.ac.kr

This chapter assesses the overall performance of current GCMs in simulating the summer monsoon rainfall, particularly its seasonal evolution and the anomalies during the 1997-98 El Niño period. Eleven GCM data used in the present study are from the CLIVAR/Monsoon GCM intercomparison project.

1. Introduction

In the Asian monsoon region, a major portion of water resources is provided by the summer monsoon (Lau and Li 1984; Lau *et al.* 1988). The prediction of summer rainfall is thus very important in Asian countries. General circulation models (GCMs) have been developed and used for monsoon diagnostic and prediction studies (Shukla and Fennessy 1994; Zhang *et al.* 1997; Webster *et al.* 1998; Sperber *et al.* 2000; Lau and Nath 2000). Studies have shown that GCM results may be sensitive to physical parameterization (Zachary and Randall 1999) and to model resolution (Sperber *et al.* 1994; Lal *et al.* 1997; Martin 1999). For GCMs to be useful for monsoon studies, it is essential that main features of the summer monsoon should be simulated with reasonable accuracy. In this chapter, based on the results of Kang *et al.* (2002a and 2002b) and Wang *et al.* (2003), we assess the overall performance of recent GCMs in simulating the summer monsoon rainfall, particularly its seasonal evolution and the anomalies during the 1997-98 El Niño period. The GCM data used by Kang *et al.* and Wang *et al.* are from the CLIVAR/Monsoon GCM intercomparison project.

Previously, a number of GCM intercomparison studies have been conducted using the data from the Atmospheric Model Intercomparison Project (AMIP). Sperber and Palmer (1996) found that AMIP GCMs which have a better rainfall climatology generally simulate better the interannual variability of rainfall. Gadgil and Sajani (1998) and Soman and Slingo (1997) showed that realistic simulation of variability in the western Pacific may be a key to simulate strong or weak monsoons associated with ENSO. However, the performance of GCMs with regard to the impacts on the monsoon by the exceptional ENSO of 1997-98 has not been evaluated. Under this situation, the Monsoon GCM intercomparison project described in this paper was initiated by the CLIVAR/Asian Australian Monsoon Panel to evaluate a number of current atmospheric GCMs in simulating the ENSO and monsoon anomalies associated with the 1997-98 El Niño. The present intercomparison is similar to the AMIP except that the simulation period is September 1996 – August 1998. The present study thus examines the atmospheric GCM responses to a given SST forcing over the globe. Given the primacy of tropical SST anomalies in forcing the global circulation anomalies (Lau and Nath 1994; Graham *et al.* 1994), it is anticipated that the present GCM results are dominated by the models' responses to tropical SST anomalies, particularly for the 1997-98 El Niño period.

The monsoon GCM intercomparison project has also requested each participating group to carry out an AMIP type simulation for the 20-year period from 1979 – 98 to obtain the climatological cycle. These long-term simulations provided a unique opportunity for us to assess the performance of the GCMs in simulating the climatological seasonal variations of summer rainfall over the Asian monsoon region. The detail description of the GCM's climatological mean performance can be found in Kang *et al.* (2002b). The climatological seasonal variations are the most distinctive phenomena among many complex time scales of variations in the Asian monsoon region (Lau *et al.* 1988). This climatological cycle is characterized not by a simple smooth variation but by rather complicated variations with diverse regional characteristics. In particular, the climatological onset dates of regional rainy season are different for different regions of the Asian monsoon and appear to propagate northward (Tao and Chen 1987; Tanaka 1992; Lau and Yang 1996). It is also well known that the monsoon rainbands associated with the regional features undergo abrupt changes and appear at different phases of the monsoon cycle (Tao and Ding 1981; Lau and Li 1984). Kang *et al.* (1999) demonstrated that the regional rainbands are linked to large-scale monsoon rainfall patterns by showing the principal modes of climatological variations of the Asian summer monsoon. Recently, several investigators have shown that the climatological variations of the Asian summer monsoon include distinctive intraseasonal components on top of the smoother seasonal cycle (Krishnamurti 1985; Wang and Xu 1997; Kang *et al.* 1999). The phase-locked intraseasonal variations appear to control the climatological onset dates of the rainy season over much of the Asian monsoon region (Kang *et al.* 1989; Nakazawa 1992; Wang and Xu 1997; Kang *et al.* 1999). All of these previous studies have demonstrated that the climatological variations of Asian summer monsoon rainfall exhibit diverse temporal and spatial characteristics, which should be represented reasonably well by the models if dynamical forecasts of the Asian monsoon are to be skillful.

Section 2 describes the experimental design of the present intercomparison project and the data utilized. Section 3 examines the climatological summer mean precipitation simulated by various GCMs. Section 4 examines the climatological seasonal and intraseasonal variations during boreal summer from May to September. Section 5 describes the ENSO precipitation anomalies simulated by various GCMs. Section 6 intercompares various Monsoon indices of the models for 1996-98 and examines the simulated ENSO-Monsoon relationships. Summary and concluding remarks are given in Section 7.

2. Experimental Set Up and Data

For the CLIVAR/Monsoon AGCM Intercomparison, all participating groups have performed a set of 10-member ensemble experiments for the period from 1 September 1996 to 31

August 1998 with the same SST boundary condition but with different initial conditions. In addition, each GCM group has carried out an AMIP-type simulation, a 20-year run with the observed SST from 1 January 1979 to 31 December 1998, to determine the model's climatological cycle. The only exception is the GEOS model climatology, which is for the shorter period of 1980 – 1992. The anomalies of each model for the 1996-98 period are obtained by subtracting the climatological cycle from the two-year simulation of the corresponding model. The SST data used in these experiments are the observed pentad means from the Global Sea Ice and Sea Surface Temperature (GISST) data set (Parker *et al.* 1995) created by the Hadley Centre for Climate Prediction for the period up to January 1979 – October 1981 and the Optimum Interpolation Sea Surface Temperature (OISST) analyses data set (Reynolds and Smith, 1994) created by the Climate Prediction Center of the National

Table 1. Description of the atmospheric GCMs participating in the CLIVAR/Asian-Australian Monsoon GCM intercomparison project.

Model	Institution	Resolution	Radiation	Convection	Land Surface Process
COLA	Center for Ocean–Land–Atmosphere Studies (U.S.A.)	R40L18	Lacis and Hansen (74) Harshvardhan *et al.* (87)	Relaxed Arakawa-Schubert (Moorthi and Suarez 92)	Sib model, Xue *et al.* (91)
DNM	Institute of Numerical Mathematics (Russia)	4°×5°, L21	Slingo (89) Chou *et al.* (93)	Betts (86)	Volodin and Lykossov (98)
GEOS	NASA/GSFC (U.S.A.)	2°×2.5°, L43	Chou and Suarez (94)	RAS (Moorthi and Suarez 92)	Schemm *et al.* (92)
GFDL	Geophysical Fluid Dynamics Laboratory (U.S.A.)	T42L18	Lacis and Hansen (74) Schwarzkopf and Fels (91)	RAS (Moorthi and Suarez 92)	Deardorff (78)
IAP	Institute of Atmospheric Physics (China)	R15L9	ESFT Shi (81)	MCA (Manabe *et al.* 65), No shallow convection	Sib model, Xue *et al.* (91)
IITM	Indian Institute of Tropical Meteorology (India)	2.5°×3.75°, L19	Ingram (96) Slingo & Wilderspin (86)	Mass flux scheme (Gregory and Rowntree 90)	Smith (90)
MRI	Meteorological Research Institute (Japan)	4°×5°, L15	Lacis and Hansen (74) Shibata and Aoki (89)	Arakawa-Schubert scheme	Katayama (78), Kitoh *et al.* (88)
NCAR	National Center for Atmospheric Research (U.S.A.)	T42L18	Kiehl *et al.*(98)	Mass flux scheme (Zhang and McFarlane 95)	Land surface model (Bonan 98)
NCEP	National Centers for Environmental Prediction (U.S.A.)	T42L28	Chou *et al.* (93) Fels and Schwarzkopf (75)	RAS (Moorthi and Suarez 92)	Pan and Mahrt (87)
SNU	Seoul National University (Korea)	T31L20	Nakajima and Tanaka (86)	RAS (Moorthi and Suarez 92)	Land surface model (Bonan 98),
SUNY	State University of New York (U.S.A.)	4°x 5°, L17	Harshvardhan *et al.* (87)	Modified Arakawa-Schubert	Deardorff (78)

Centers for Environmental Prediction (NCEP) for the period November 1981 – December 1998. Because of the unavailability of the data for recent years, the sea ice is prescribed by the climatological monthly mean data utilized by the AMIP.

Eleven modeling groups (COLA, DNM, GEOS, GFDL, IAP, IITM, MRI, NCAR, NCEP, SNU, and SUNY/GLA) have provided experimental data to Climate Environment System Research Center (CES) in Seoul National University, Korea. The names of the participating institutes and their model descriptions are given in Table 1. Although the spatial resolution of the models varies from rhomboidal truncation at wavenumber 15 to triangular truncation at wavenumber 42, all GCM data were converted to a spatial resolution of 2.5°×2.5° using a linear interpolation method. Note that the regridding algorithm conserves area mean quantities and does not affect the main results of the paper.

To compare the GCM results with observations, we used the CPC Merged Analysis of Precipitation [CMAP; see documentation by Xie and Arkin (1997)] and atmospheric circulation statistics from the National Centers for Environmental Prediction (NCEP) reanalysis database (Kalnay *et al.* 1996). The CMAP data utilized are the pentad data for the 20 year period from 1979 to 1998. It is noted that the CMAP precipitation is an estimate based on satellite data and in situ observations over the continents (Xie and Arkin 1997). But, there are some differences between the CMAP and another estimate such as the Global Precipitation Climatology Project (GPCP) precipitation, particularly over the ocean regions, because of different algorithms used to retrieve the rainfall from the satellite data. Some of distinctive differences occur in the western Pacific (Gruber *et al.* 2000). However, most of the conclusions of this paper are not sensitive to the choice of the precipitation data. The observational issue of the circulation statistics may be less sensitive to the choice of the NCEP reanalysis, since there should be little difference between the NCEP and European Center for Medium-Range Weather Forecast (ECMWF) reanalyses in the rotational flows.

3. Climatological Summer Mean Precipitation

The most basic feature, which should be examined before describing the monsoon variations, may be the distribution of climatological summer mean precipitation over the region of interest. In this section, we examine how well the models simulate the climatological mean precipitation for the northern summer. The climatological means are defined here as the averages of the 20 year data for 1979-98. Firstly, all-model composite of climatological summer-mean (June, July and August) precipitation is made by averaging the 10 model results, and it is compared with the corresponding observations. All GCMs in Table 1 except the NCEP model are used for the climatological mean and variation studies, because the NCEP model did not provide the climatological mean data for the 20 years. The model composite of the summer-mean precipitation shown in Fig. 1b has a spatial pattern similar to that of the CMAP precipitation shown in Fig. 1a, although there are certain differences, especially in the western Pacific. Over the South China Sea and the subtropical western Pacific, where maximum precipitations appear in the observations, the model precipitation is rather weak compared to the observed. Over the Bay of Bengal and off the west coast of

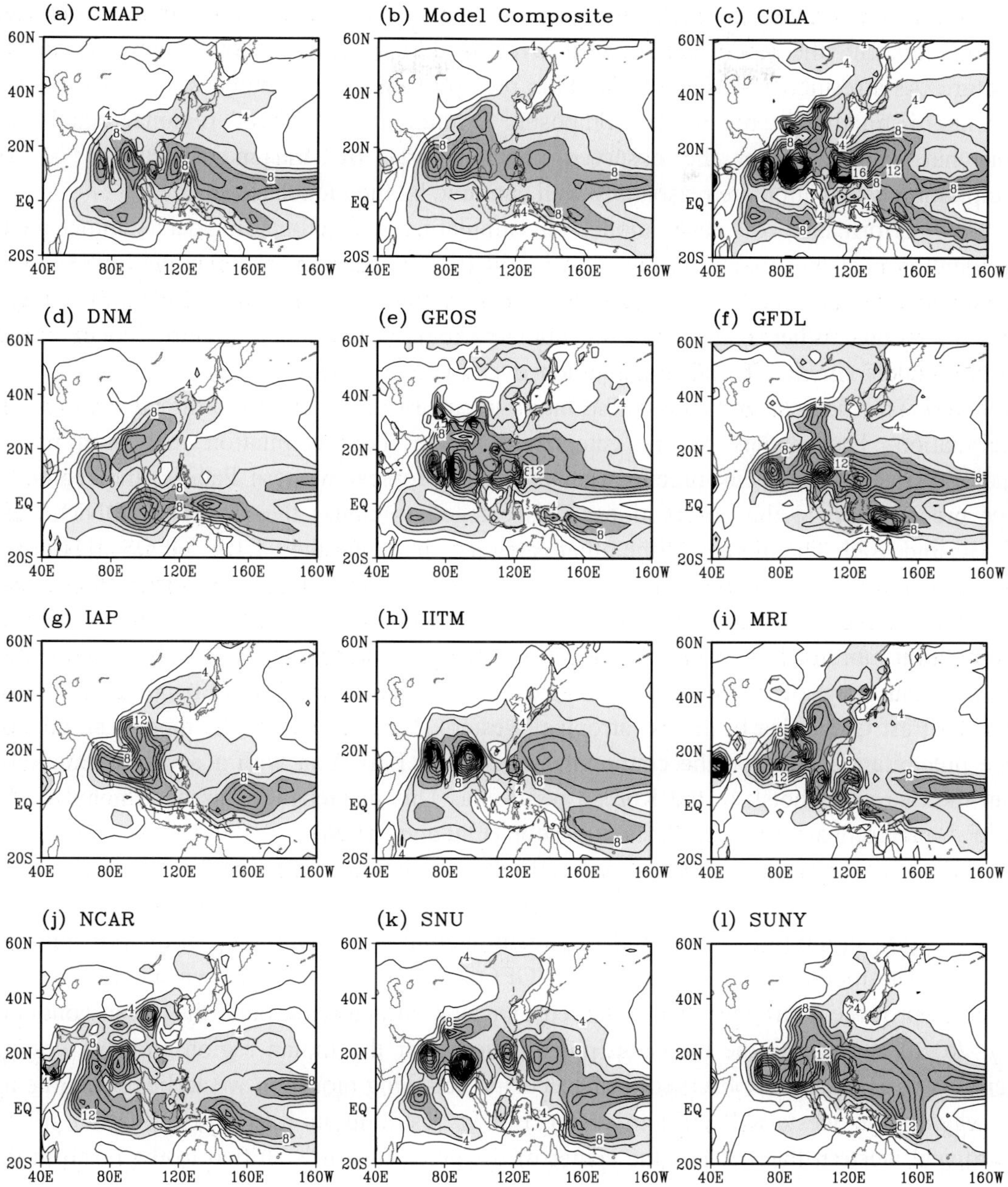

Figure 1. Distribution of climatological summer-mean precipitation for June, July, and August. (a) is for the observation from CMAP data, (b) is for the model composite and (c) – (l) are for various GCMs. Contour interval is 2 mm day^{-1}. Light and Dark shadings indicate the rainfall rate more than 4 mm day^{-1} and 8 mm day^{-1}, respectively.

India, on the other hand, the model composite reproduces reasonably well the observed large precipitation. But the model precipitation over the Indian regions shows a broader spatial structure. As a result, the observed minimum precipitation off the southeast coast of India is not very distinctive in the model composite. Overall, the simulated regional details are somewhat different from the observed counter parts, but the major features of the precipitation distribution are reasonably well captured by the model composite.

The performance of each model in simulating the climatological summer-mean precipitation is also examined in Fig. 1. All models show precipitation maxima in the Indian Ocean and western Pacific monsoon regions. However, regional details differ from one model to another. Whereas most models simulate excessive precipitation in the Indian region, several models simulate weak precipitation over the subtropical western Pacific (e.g., IAP and MRI). The models can be classified into two categories according to the western Pacific precipitation. The models of one category simulate the precipitation maximum in the equatorial Pacific but weak precipitation in the subtropical western Pacific, the region of large precipitation in the observations. This category includes the models of DNM, IAP, NCAR, and MRI. The models of the other category, including the COLA, GEOS, IITM, and SNU models, simulate excessive precipitation in the subtropical western Pacific compared to the observations. The GFDL and SUNY/GLA models simulate the western Pacific precipitation more or less in the right location, although their intensity is rather strong. It is also note that none of the models reproduces the observed precipitation band in the region from the East China Sea to the extratropical western Pacific. Most of the models simulate a large precipitation band on the continental side along East Asia with a dry area extending from the Sea of Okhotsk to the Korean peninsula. This discrepancy has been noted in the AMIP-I intercomparison of the Asian monsoon (Lau *et al.* 1996).

The two categories classified according to the simulation of western Pacific precipitation are more clearly shown in Fig. 2. The figure shows the composites of the four models for each category. In the figure, the 850 hPa streamfunction is plotted on top of the precipitation (shaded). The model composite of COLA, GEOS, IITM, and SNU shown in Fig. 2b (first category) produce a stronger than observed monsoon circulation around the Asian continent (Fig. 2a). Accordingly, the monsoon precipitation over the Indian region and the western Pacific is larger than the observed. In particular, the strong monsoon westerlies penetrate all the way to the western Pacific to about 150°E, resulting in heavy precipitation in the subtropical western Pacific. But the center of the precipitation is shifted to the north in the western Pacific by about 5 degrees compared to the observations. Also note that the model composite in Fig. 2b shows a ridge line penetrating to Japan and Korea from the center of the Pacific high. This relates to the models to be dry over the region of the observed rainband near Japan and pushes the rain band to the continental side. For the models of DNM, IAP, MRI, and NCAR (second category), on the other hand, the simulated Asian monsoon circulation is rather weak, resulting in weak precipitation in both in India and the western Pacific (Fig. 2c). Clearly, the weak precipitation in the subtropical western Pacific is related to the abnormal strength of the Pacific high extending to the South China Sea, which prevents the monsoon westerlies in the western Pacific. For those models, the subtropical western

Pacific is occupied by the ridge, which pushes the rainband to the equatorial region. The comparison of the two categories of model simulations suggests that the strength of Asian monsoon circulation relative to the Pacific high strongly affects the monsoon westerlies and precipitation in the subtropical western Pacific. This finding is consistent with previous results which showed that better simulation of the Asian monsoon in AMIP GCMs is dependent upon how well the models simulate the northward migration of the convective zone over the tropical western Pacific (Gadgil and Sajani 1998).

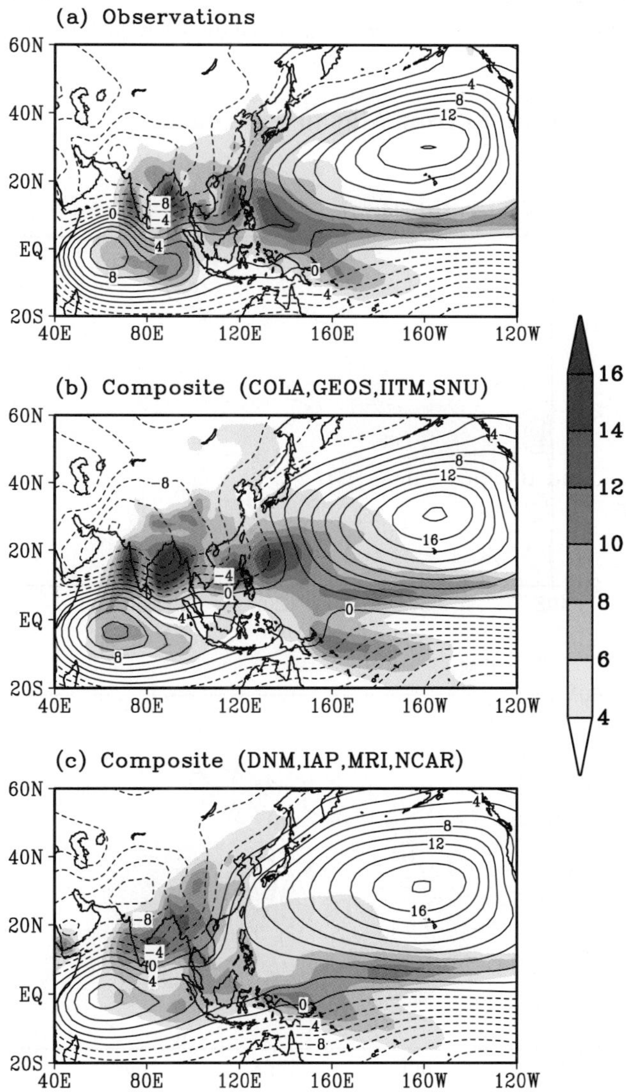

Figure 2. (a) Climatological summer mean precipitation and NCEP 850hPa streamfunction for the observations. (b) As in (a) except for the four model composites of COLA, GEOS, IITM, and SNU, and (c) for the composites of DNM, IAP, MRI, NCAR. Precipitation is shaded. The precipitation scale (mm day^{-1}) is shown in the bar at the right side of the figure. Unit of the streamfunction is 10^6 m^2 sec^{-1}.

The different performance of the two model groups should be related to the physical parameterizations, particularly for the convection, adapted in the models. Most models of the first category use the Arakawa-Schubert scheme for the convection, except the IITM model which uses the mass flux scheme. The GFDL and SUNY/GLA models also use the Arakawa-Schubert scheme, which appears to simulate the precipitation distribution in the western Pacific more reasonably than other schemes do. On the other hand, the models of the second category use different convection schemes (Table 1). It is pointed out that the MRI model uses the Arakawa-Schubert scheme, although the model belongs to the second category. This indicates that physical parameterizations other than the convection scheme also play important roles in the simulation of precipitation over the western Pacific.

4. Climatological Variations from May to September

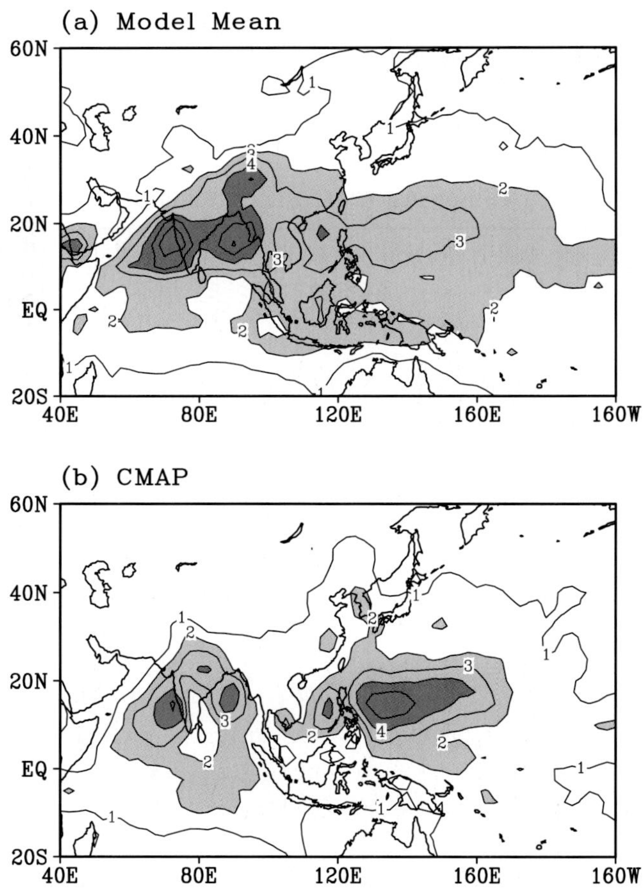

Figure 3. Standard deviation of climatological variations of pentad-mean precipitation for the five months from May to September. (a) and (b) are for the average of the 10 model standard deviations and for the CMAP observations, respectively. Contour interval is 1 mm day^{-1}. Light and dark shadings indicate the standard deviations larger than 2 mm day^{-1} and 4 mm day^{-1}, respectively.

In this section, the climatological variations of precipitation during boreal summer simulated by various GCMs are examined by analyzing the pentad data of the 20-year mean climatological cycle from May to September. In particular, the amplitude of the variation from the five-month mean is examined in terms of the standard deviation. Fig. 3a shows the square root of the average of the 10 model variances. This is compared to the corresponding standard deviation of CMAP precipitation shown in Fig. 3b. The models generally capture the large climatological variations over the Asian monsoon region. But, the amplitude of the simulated variations in the western Pacific is weaker than that of the observed. It is also noted that both the observed and model climatological cycles show large variations along the subtropical latitudes between 10°N and 20°N.

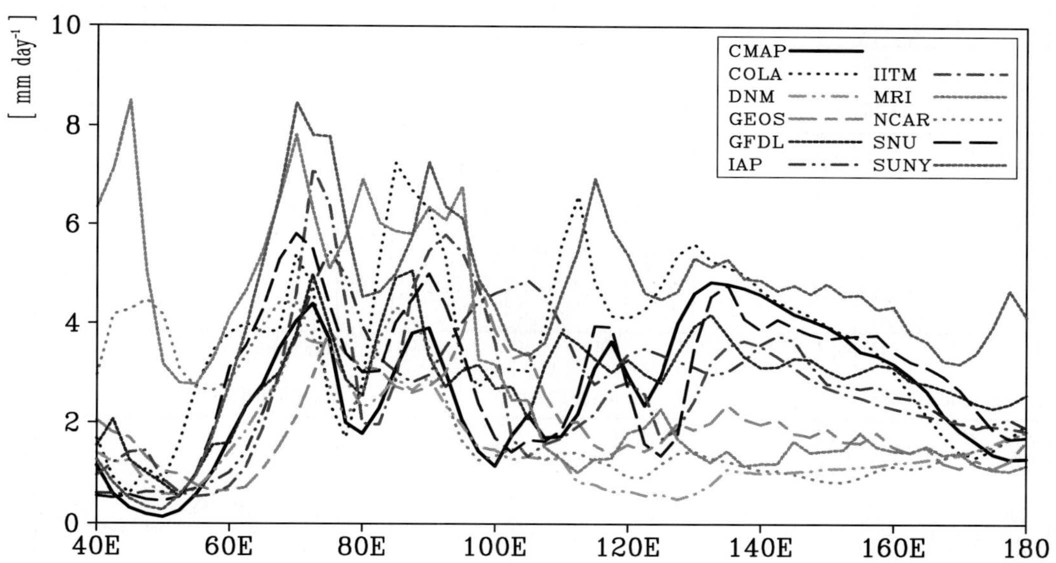

Figure 4. Longitudinal distribution of the standard deviation of climatological pentad-mean precipitation from May to September, for the latitudinal mean of 10°N - 20°N. The lines, illustrated at the right upper corner of the figure, indicate the standard deviation of each model.

We next examine how the variability of each model distributes along the subtropical latitude belt. Fig. 4 shows the standard deviation of each model for the latitudinal means between 10°N and 20°N. The thick black line indicates the observed CMAP standard deviations along the subtropical belt between 40°E and 160°W. The figure clearly shows that in the Indian region (70°E - 100°E), most of models overestimate the amplitude of the climatological variations. In the western Pacific, on the other hand, most of the models simulate weaker variations. In particular, the model variations of DNM, NCAR, MRI, and GEOS are significantly weaker than those of the observed. As shown in the previous section, these models, except GEOS, simulate weak summer-mean precipitation in the western Pacific. The other groups of models, such as SUNY/GLA, COLA, SNU, and IITM, which simulate relatively large summer-mean precipitation, produce reasonable amplitudes of the climatological variations in the western Pacific. These results indicate that the simulated climatological variability depends on the model ability of simulating the climatological mean, consistent with previous results (Sperber and Palmer 1996).

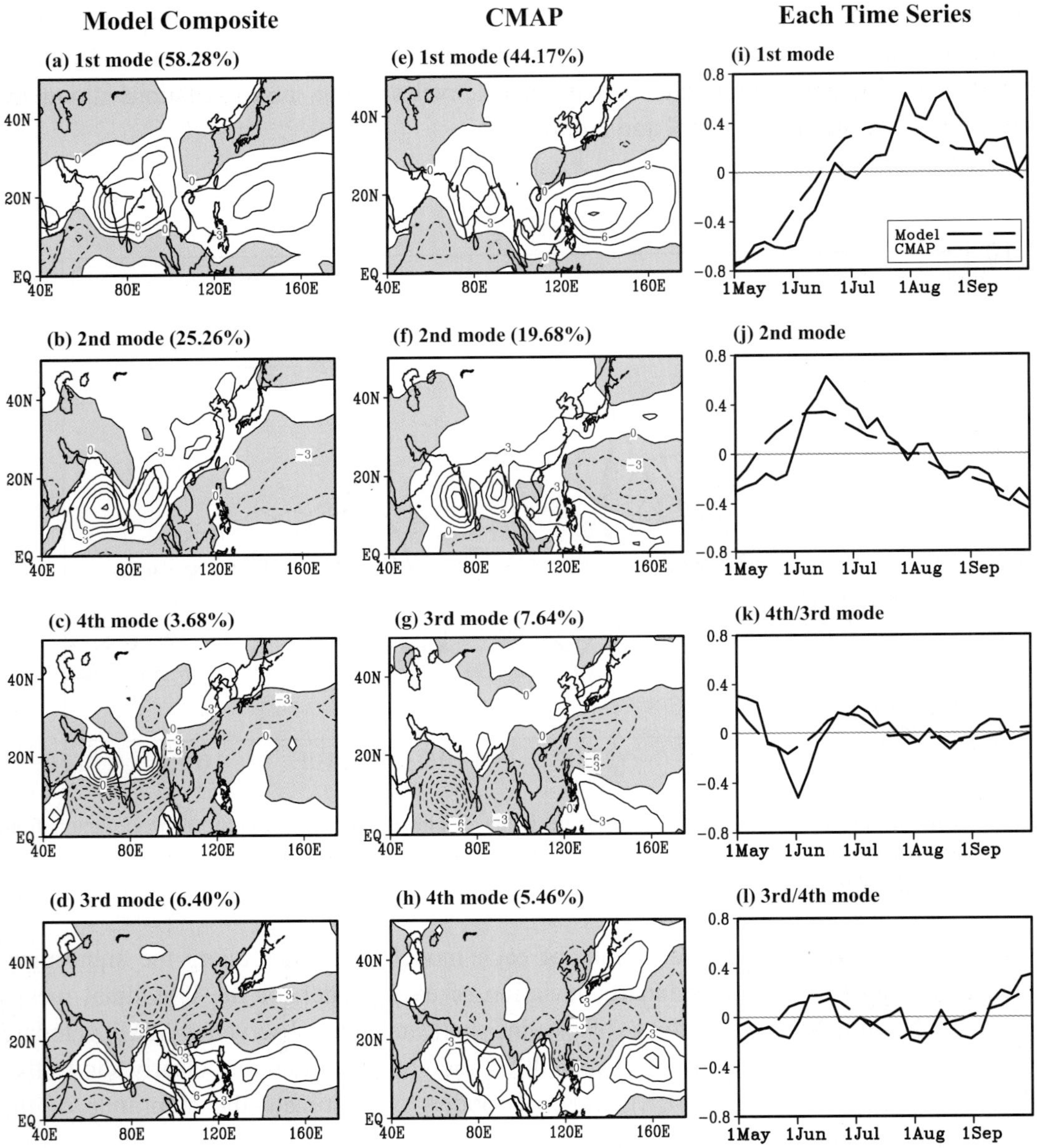

Figure 5. Four leading EOF eigenvectors and associated time series of the climatological pentad-mean precipitation for May – September. (a) – (d) are for the model composite, and (e) – (h) for the CMAP observations. The associated time series of CMAP and the model composite are shown in (i) – (l) with a line and a dashed line, respectively. Units are arbitrary.

The principal modes of the climatological variations of precipitation over the Asian-Pacific region are obtained by applying the empirical orthogonal function (EOF) analysis to the climatological mean pentad data from May to September. Fig. 5 shows the four leading EOF eigenvectors of the model composite (Figs. 5a – 5d) and the observations (Figs. 5e – 5h). The model composite reproduces the observed four leading eigenvectors with reasonable accuracy. The first eigenvector shows simultaneous development and decay of the Indian and western Pacific monsoons representing the seasonal march of broad-scale monsoon. Consistent with results in the previous section, the magnitudes of the model eigenvector in the western Pacific are smaller than those of the observed. The time series associated with the eigenvector (Fig. 5i) shows earlier development of the model monsoon. The peak phase of the model composite appears about one month earlier than that of the observed. The second eigenmode represents the onset of the Indian monsoon, with a linkage to the East Asian rainband. Kang *et al.* (1999) obtained an eigenvector similar to the present second one using high cloud data, and they related the mode to the onset of the Baiu-Meiyu rainband. The time series of the second mode of the model composite is much smoother than the observed, and the onset timing and peak phase are somewhat earlier in the model than those in the observations. The observed third (Fig. 5g) and fourth (Fig. 5h) modes appear in the model as the fourth (Fig. 5c) and third (Fig. 5d) modes, respectively. Although the order of the modes is interchanged, the model composite reproduces the observed third and fourth modes reasonably well. As noted in Kang *et al.* (1999), the observed third mode may be related to the climatological intraseasonal oscillation (CISO) of Asian summer monsoon. Although the associated time series of the model composite shown in Fig. 5k and 5l are much smoother than the observed counterparts, the major fluctuations appear to agree well.

We examine how each model produces the leading eigenmodes. For this purpose, the pattern correlations between the eigenvectors of individual models and the observed counter parts over the domain of Fig. 5 are shown in Fig. 6. The correlations for the first eigenvector are shown with a black bar. The spatial patterns of the first eigenvector simulated by individual models can be found in Kang *et al.* (2002b). Many models reproduce the observed first eigenvector reasonably well. However, several models including COLA, DNM, MRI, and NCAR produce the first eigenvector somewhat different from the observed. The percent variance explained by the first eigenvector ranges from about 40 to 55 % of the total variance of the individual model. The pattern correlations of the second, third, and forth eigenvectors are also shown in Fig. 6 with different bars. Whereas all correlation values of the model composite are quite high, most individual models have a large correlation only for the first eigenvector but not for the higher modes. One exception, the SNU model, appears to reproduce all of the observed leading eigenmodes reasonably well. The SUNY/GLA model reproduces the first and second modes most realistically. It is also noted that the MRI and NCAR models poorly simulate the first eigenvector but produce the second and third eigenvectors better than those of the other models.

It is well known that the Asian monsoon is dominated by the two components, namely the Indian monsoon and the western Pacific-East Asian monsoon (Lau and Li 1984; Wang *et al.* 2001). Over these regions, the climatological rainfall variations undergo abrupt changes,

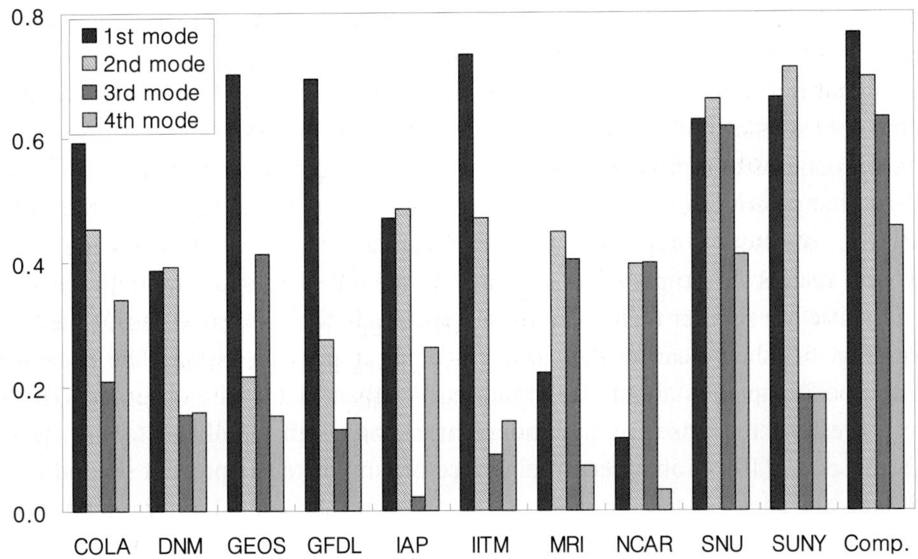

Figure 6. Pattern correlations between the EOF eigenvectors of CMAP observation and the corresponding eigenvectors of individual model over the domain of 40°E - 160°E and EQ - 40°N. The correlation values for the first, second, third, and fourth eigenvectors are shown by the bars with different shadings. See details in text.

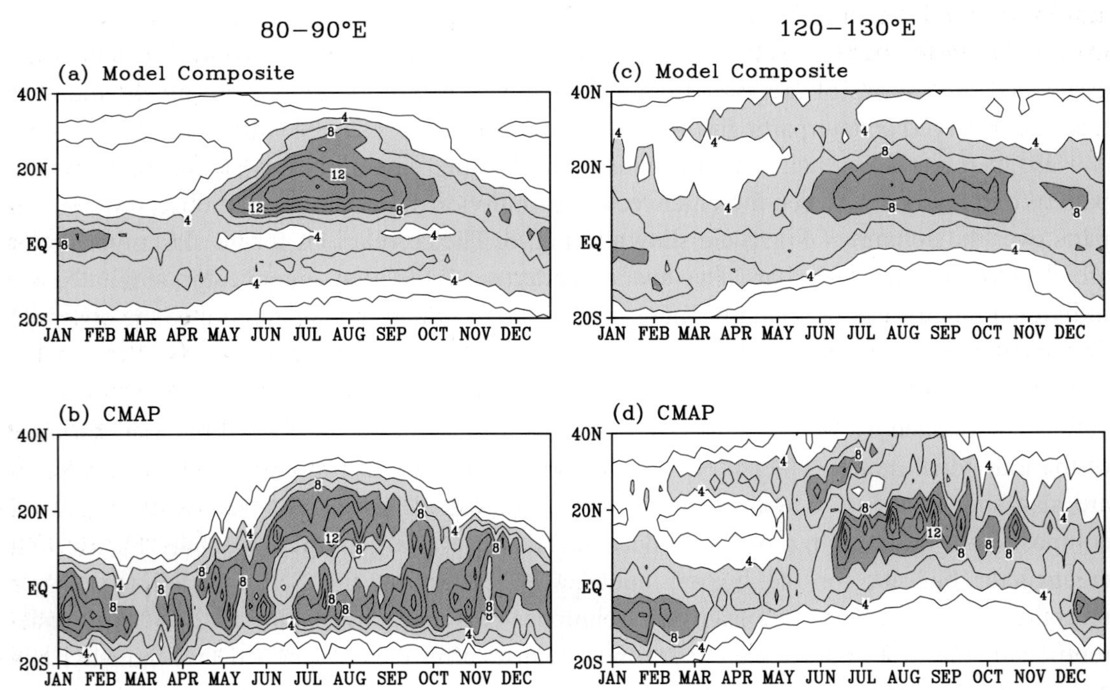

Figure 7. Time-latitude cross section of climatological pentad-mean precipitation along 80°E - 90°E and 120°E - 130°E. (a) and (b) are, respectively, for the model composite and the CMAP observations along the latitude of 80°E - 90°E. (c) and (d) are as in (a) and (b) except for 120°E - 130°E.

which are the characteristic feature of monsoon onset, and have significant intraseasonal components (Wang and Xu 1997). These climatological intraseasonal variations control the climatological onset date of regional rainy seasons over India and East Asia (Nakazawa 1992; Kang *et al.* 1999). Here, we show how the intraseasonal components appear in the climatological variations, particularly along the longitude belts covering India (80°E - 90°E) and East Asia (120°E - 130°E). Figs. 7a and 7b show the time-latitude cross sections along 80°E - 90°E for the model composite and the CMAP observations, respectively. The model composite produces the sudden onset of Indian monsoon in early May, which is similar to the observed. The mature phase of the model lasts until early July and the rainfall rate decays afterward. In the observations, on the other hand, maximum precipitation appears intermittently in June, July and August, indicating the existence of climatological intraseasonal oscillation (CISO). The CISO is more evident near the equator and tropical Southern Hemisphere in the observations, but the CISO is missing in the model composite.

Along 120°E - 130°E longitude, the model composite (Fig. 7c) is characterized by the maximum precipitation located at the equator during spring until late May, with a sharp transition to the latitude 10°N - 15°N in early June, where it stays until the end of summer. Although this model behavior is also seen in the observations (Fig. 7d), the jump of rain belt in the subtropical western Pacific appearing in early June is less sudden in the observations, where enhanced precipitation appears off the equator during early June and then propagates northward until the end of August. This northward progression does not appear in the model composite. In the observations, alternative northward propagations of rainband and dry zone are also evident in the extratropics between 15°N and 40°N during the summer. This extratropical character is largely missing in the model composite. Kang *et al.* (1999) showed that the northward movement of the extratropical rainband strongly affects the rainy seasons in East Asia. They ascribed this alternative appearance of rainband and dryness to the climatological intraseasonal oscillation with a time scale of about two months. The absence of the extratropical CISO in the model composite can be due to the cancellation of the CISO's of individual models having slightly different phases from each other. To check this possibility, we examine the CISO's of individual models along the 130°E longitude line.

Fig. 8 shows the time-latitude cross sections of the CISO component of precipitation along 120°E - 130°E for the CMAP observations (Fig. 8a), the model composite (Fig. 8b), and individual models (Figs. 8c-8l). As in Kang *et al.* (1999), the CISO component is obtained by removing the smooth seasonal cycle, the sum of first four harmonics of the climatological cycle, from the original climatological cycle. In the CMAP, the large positive anomalies near 20°N appearing in mid-May propagate northward up to 40°N. The dry anomalies following the rainband locate at 20°N in early June and at 40°N in early August. The positive CISO anomalies again follow the negative anomalies between 20°N and 40°N. This northward propagating character of the CISO component can be found in the model composite, but it is generally very weak and the phase is shifted so that the model CISO occurs 20-30 days earlier than the observations. Several models such as COLA, GEOS, IITM, and SNU show the northward moving CISO components, which have equivalent amplitudes to those of the CMAP. But the CISOs simulated by several models are very weak.

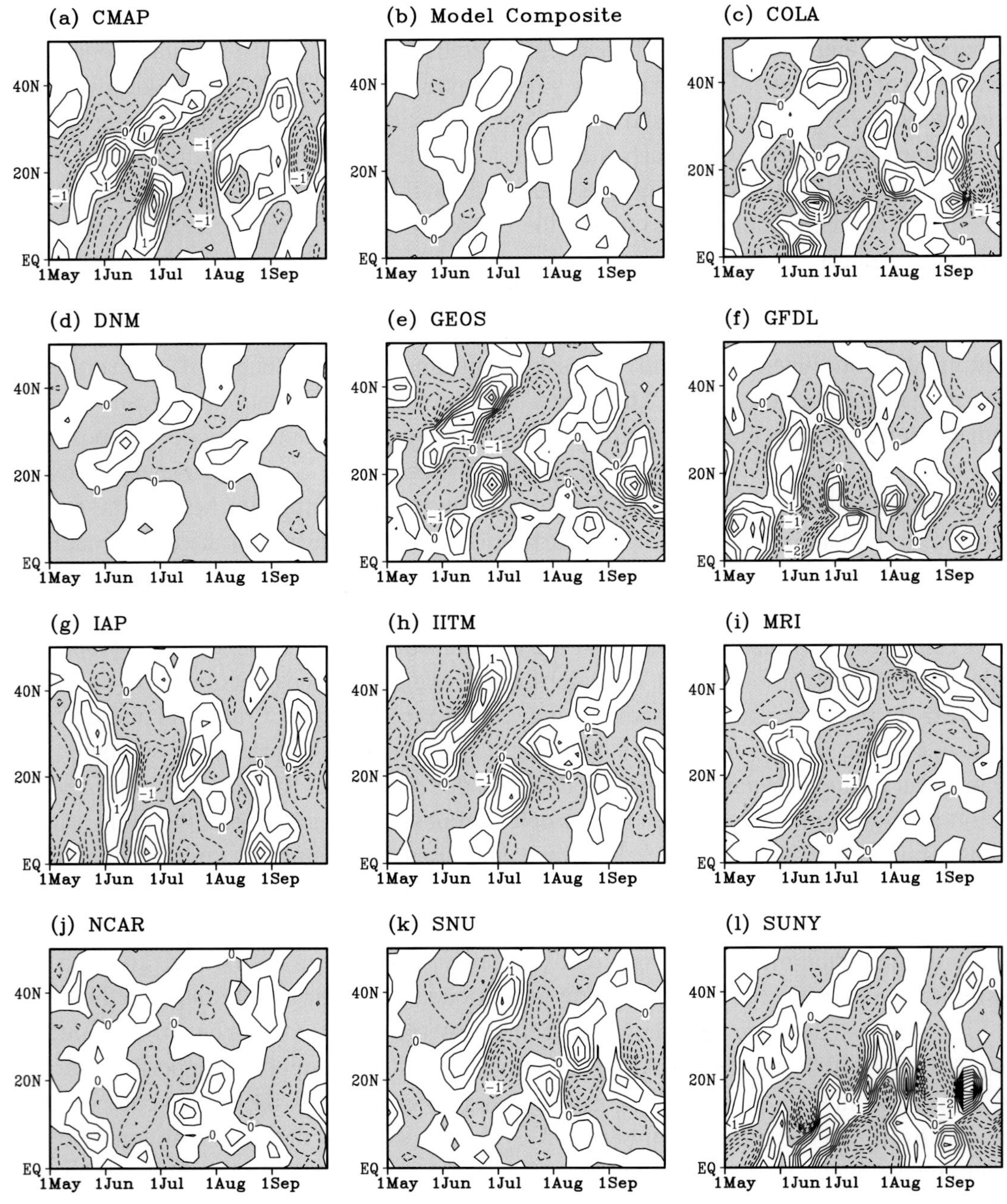

Figure 8. Time-latitude cross section of the climatological intraseasonal component of precipitation along 120°E - 130°E. (a) and (b) are for the CMAP and the model composite, respectively, and (c) – (l) for each model. Contour interval is 0.5 mm day^{-1} and negative values are shaded.

The SUNY model simulates the tropical CISO reasonably well but poorly reproduces the extratropical CISO. The figures show that some of the individual models contain strong CISO components, whose phases are somewhat different from one model to another. As a result, the model composite shows unrealistically weak CISO components.

5. ENSO Anomalies during 1997-98

Before presenting the monsoon anomalies during the 1997-98 El-Niño episode, we examined the model performance in simulating the tropical precipitation anomalies for 1997-98. The SST anomalies prescribed in the GCM experiment for the two-year period from September 1996 to August 1998 can be represented by the NINO3.4 index, defined as the average of SST anomalies over the region of 5°S - 5°N and 120°W - 170°W (Trenberth 1997). The time evolution of the NINO3.4 index, shown in Fig. 9a, consists of a normal winter (December

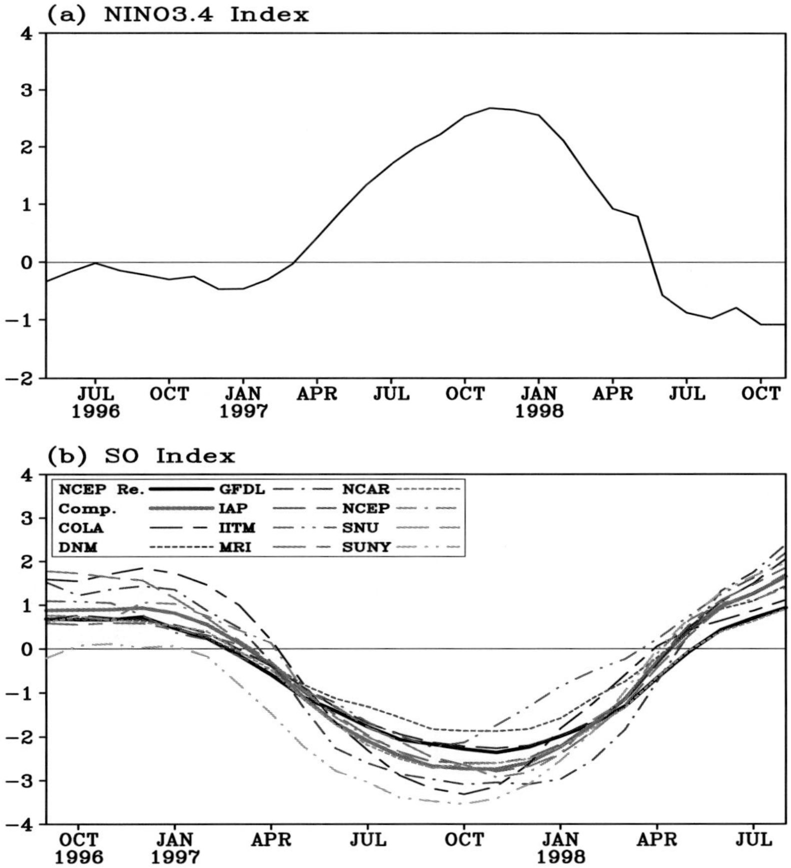

Figure 9. (a) Time evolution of the NINO3.4 index for June 1996 - October 1998. (b) Southern Oscillation index (SOI) for September 1996 - August 1998. Observed SOI is shown by the thick black solid line, and the model composite index is shown by the thick gray solid line. The line patterns for the SOI of the individual models are indicated in the upper left corner box.

1996 – February 1997), a summer season (June – August 1998), a rather strong El Niño summer (June – August 1997) and a strong El Niño winter season (December 1997 – February 1998). The spatial distribution of the SST anomalies for the El Niño winter is shown in Kang *et al.* (2002a).

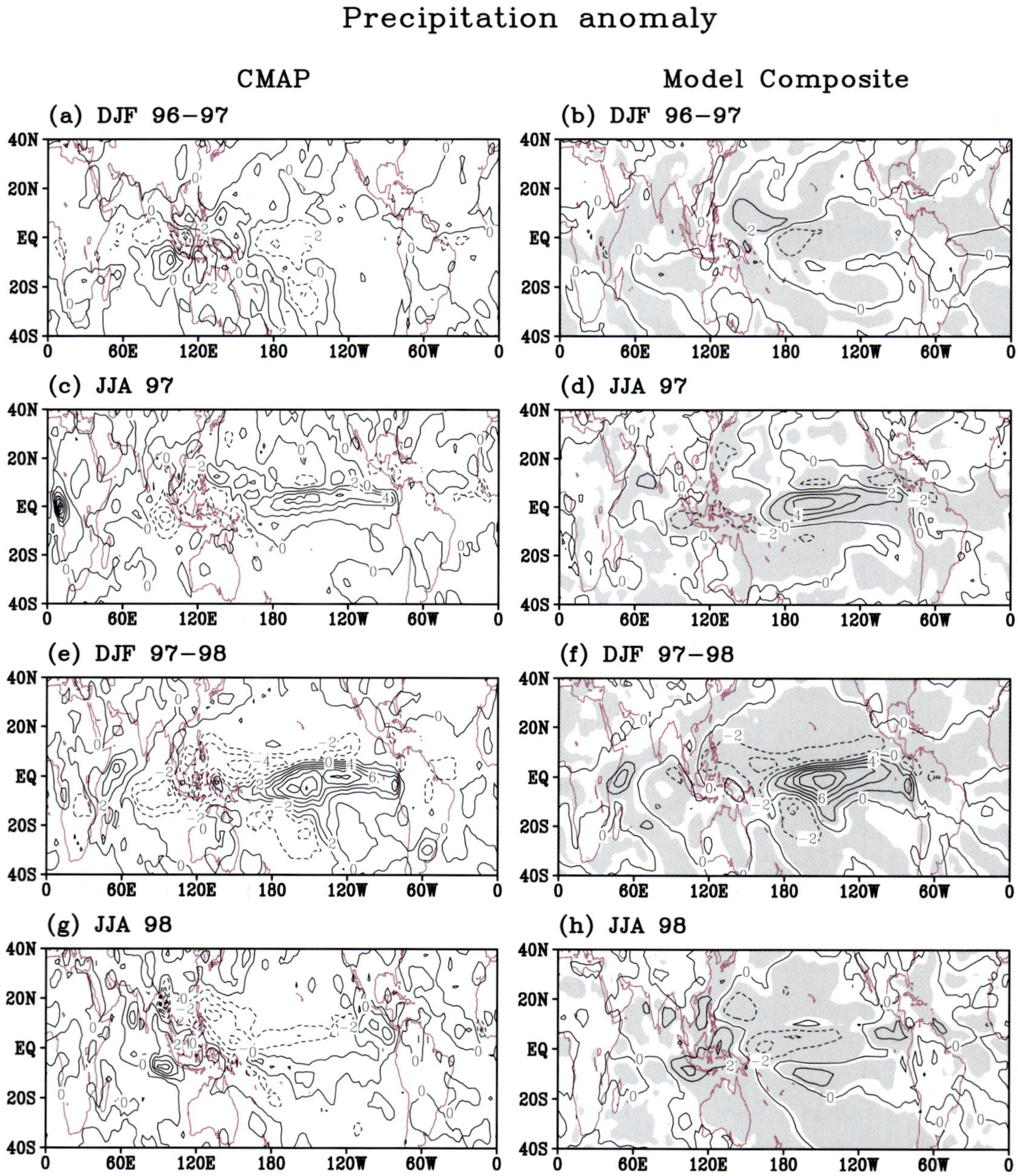

Figure 10. Precipitation anomalies for the two winter and two summer seasons in the period between September 1996 and August 1998. The left panels are the observed CMAP anomalies and the right panels are the model composites. Shading indicates the anomalies that are significant at the 99% level.

The overall performance of the GCMs in simulating the atmospheric anomalies associated with the tropical SST anomalies can be examined in terms of the Southern Oscillation index (SOI), which correlates negatively with the SST variations associated with El Niño events (Trenberth 1976; Horel and Wallace 1981). The SOI of each model is calculated by subtracting the sea level pressure (SLP) anomaly averaged over 125°E - 135°E and 5°S - 5°N from the SLP anomaly averaged over 145°W - 155°W and 5°S - 5°N based on the ensemble mean data of the 10 runs of each model. The model index is compared to the SOI obtained using the NCEP reanalysis data (Kalnay *et al.* 1996). Hereafter, the statistics from the reanalysis data will be referred to as the observed, although it is understood that they are not exactly the same as the observation. Fig. 9b shows the SOIs from each model and from the observation. A comparison of the characteristics of the observed SOI with those simulated by the models indicates that all the models sensitively respond to the tropical Pacific SST in a way similar to the observed, although each model produces a slightly different phase and amplitude of the index. This result indicates that the present GCMs as a whole are able to mimic the tropical atmospheric anomalies associated with the evolution of the recent El Niño.

It is expected that the rainfall distribution simulated by the models is closely related to the prescribed SST, as demonstrated by a number of previous studies (e.g., Horel and Wallace 1981; Ropelewski and Halpert 1987). To check this relationship, all model composites of seasonal-mean rainfall anomalies were calculated and then compared with the corresponding observations from the CMAP. The right panels in Fig. 10 show the precipitation anomalies of all model composite for the two winters and two summers in the period from September 1996 to August 1998. The corresponding CMAP precipitation anomalies are shown in the left hand side of Fig. 10. The winter mean precipitation observed for the winter of December 1996 – February 1997, when the tropical Pacific was in a normal state, shows positive anomalies over the Maritime continent and negative anomalies over the central tropical Pacific (Fig. 10a). The simulated anomalies for the corresponding period show a distribution similar to those of the observed, although the magnitudes of the simulated anomalies are much weaker, particularly over the Maritime continent. The spatial correlation between the two maps is 0.55 for the tropical region between 30°S and 30°N. Although similar, it may not be possible to say what is "forced" by the SST anomalies and what is simply due to the average of a small number of chance fluctuations. This issue is addressed by performing a simple *Student t*-test. The regions of statistical significance at the 99% level are highlighted by shading in the figures. The results of the *Student t*-test indicate that the large-scale precipitation anomalies simulated in the tropics are closely linked to the SST anomalies not only for the El Niño period (Figs. 10d and 10f) but also for the normal seasons (Figs. 10b and 10h). The results in Fig. 10 indicate that the precipitation anomalies, specifically in the tropical oceans are closely related to the SST anomalies, and this relationship is reproduced reasonably well by the model composite. However, the models underestimate the precipitation anomalies in the Maritime continent, particularly during the El Niño period.

The performance of each model in simulating the ENSO anomalies is illustrated in Fig. 11, which shows the ensemble mean precipitation anomaly of each model. Here the focus is on the simulation of the 1997-98 winter anomalies. For comparison, the observed CMAP precipitation anomaly is also shown in Fig. 11a. The *Student t*-test is performed for each model and the anomalies of significance level at 99% are shaded in the figures. Figures show that most models reproduce the precipitation anomalies in the tropical Pacific, which are significantly related to the prescribed SST anomalies. Whereas, the simulated precipitation anomalies in the Indian and East Asian regions are poorly related to the SST anomalies. It is particularly mentioned that the intensity of the anomalies in the equatorial Pacific and Maritime continent shows large intermodel differences. The ENSO related positive anomalies in the equatorial central Pacific (Horel and Wallace 1981) simulated by GEOS and IAP are very weak compared to the observed counterpart. On the other hand, COLA, NCAR, and SUNY models produce larger positive anomalies in the same region. All the models, except

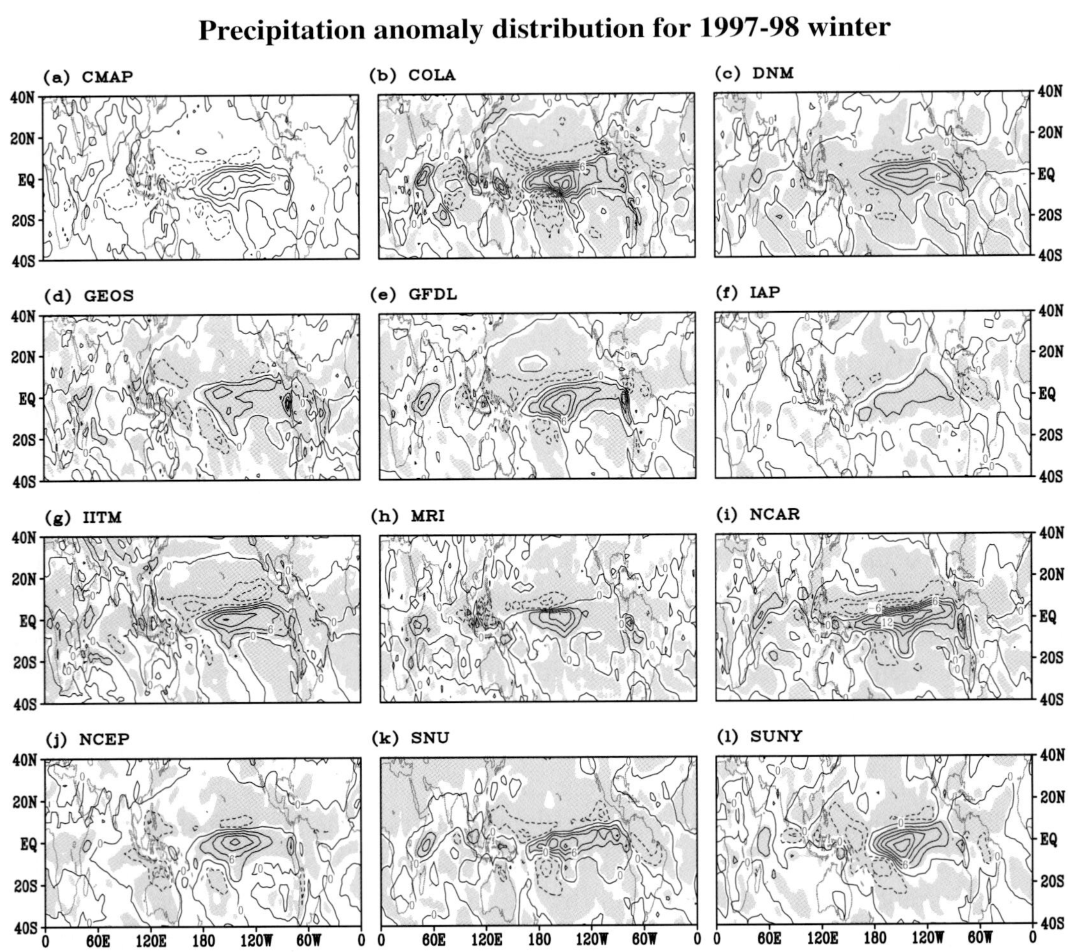

Figure 11. Distribution of precipitation anomaly during the 1997-98 winter for the CMAP (a), and for the ensemble mean of each model (b-l). Shaded area denotes anomalies that are significant at the 99% level.

GFDL, poorly simulate the observed negative anomalies over the western Pacific near Indonesia, which is also the region of greatest disagreement among the models.

It is also mentioned that not only the ensemble mean but also each individual run has a significantly large value of the correlation coefficient during the El Niño seasons (Kang *et al.* 2002a). The range of correlation values of the individual runs is considerably narrower for the El Niño seasons than for the normal seasons of 1996-97 winter and 1998 summer. Kang *et al.* (2002a) also show that, regardless of the season, the correlation value of the ensemble mean of each model is larger than that of any individual run with few exceptions, and the correlation value of the model composite is larger than that of the ensemble mean of individual model, with an exception of the GFDL 1997-98 winter simulation.

It is also interesting to examine how the anomalies of each model during the El Niño seasons are related to the accuracy of the model climatology. To address this issue, the

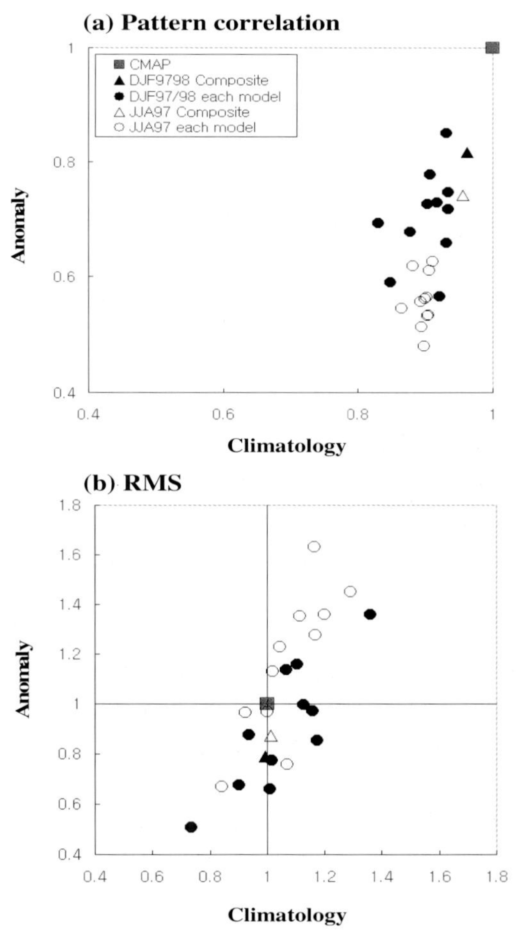

Figure 12. (a) Scatter plot of the pattern correlation of the ensemble anomalies of each model over the monsoon-ENSO region versus that of the corresponding climatology for the 1997 summer and 1997-98 winter seasons. (b) As in (a) except for the normalized rms. Open and dark circles are for the 1997 summer and 1997-98 winter, respectively

pattern correlation between the observed climatology and the climatology of each model for the monsoon-ENSO region (60°E - 90°W and 30°S - 30°N) were calculated and plotted against the pattern correlation of the anomalies. Fig. 12a shows a scatter diagram of the pattern correlation of climatology (x-axis) versus the pattern correlation of anomaly (y-axis). Clearly, the 1997-98 winter anomaly simulations are generally better than those of 1997 summer. Figure 12a illustrates that there is no clear relationship between the degree of accuracy of the climatology and that of the anomaly for both seasons, although the best model of the anomaly coincides with the best model of the climatology for each season. On the other hand, Fig. 12b shows that for both seasons, the intensity of simulated anomaly is generally proportional to the intensity of the corresponding climatology of each model. Here, the intensity of simulated anomalies over the region is measured in terms of the root-mean-square of all anomalies over the region, normalized with respect to the observed counterpart. Thus, the models with a larger (smaller) climatological mean precipitation than the observed generally simulate larger (smaller) El Niño anomalies. It implies that a simple bias correction of anomalies can be made based on the quality of the model climatology.

6. Monsoon Variability during 1997-98

In order to conveniently measure the Asian summer monsoon circulation anomalies, Webster and Yang (1992) proposed a circulation index, the Webster-Yang index (WYI), which is defined by the vertical zonal wind shear between 200 and 850 hPa (u200-u850) averaged over the region (40°E - 110°E, 0° - 20°N). The regions selected for the monsoon indices examined here are shown in Fig. 13. The WYI describes the broad-scale South Asian monsoon variability that is primarily driven by two convective heat sources, one lies in the Bay of Bengal and the other in the vicinity of Philippines. Wang and Fan (1999) found that the variations of the two convective heat sources reflect, respectively, the variations of the Indian summer monsoon (ISM) and the western North Pacific-East Asian summer monsoon (WNP-EASM); but they are not significantly correlated on the interannual time scales. Thus, Wang and Fan (1999) proposed two indices that quantify the variability of the ISM and WNP-EASM, respectively. The ISM index (IMI) is defined by the 850 hPa zonal wind averaged over (5°N - 15°N 40°E - 80°E) minus that averaged over (20°N - 30°N, 60°E - 90°E) (Fig. 14). This meridional shear of zonal winds depicts the intensity of the Indian monsoon trough and associated southwesterly monsoon. Similarly, the WNP-EASM index (WEI) can be defined by the 850hPa wind speed averaged over (5°N - 15°N, 100°E - 130°E) minus that over (20°N - 30°N, 110°E - 140°E). The IMI and WEI are highly representative of the dominant EOF modes of the low-level circulation anomalies over the ISM (30°E - 100°E, 0° - 30°N) and the WNP-EASM (100°E - 170°E, 0° - 40°N) regions with correlation coefficients of 0.72 and 0.88, respectively for the period of 1958-1997. McBride *et al.* (1995) proposed a circulation index measuring the Australian summer monsoon variability using 850hPa zonal wind anomalies averaged over (0° - 10°S, 120°E - 150°E), which is referred to as Australian monsoon index (ASI) in this paper. The aforementioned four monsoon indices provide a succinct description of the Asian-Australian summer monsoon variability.

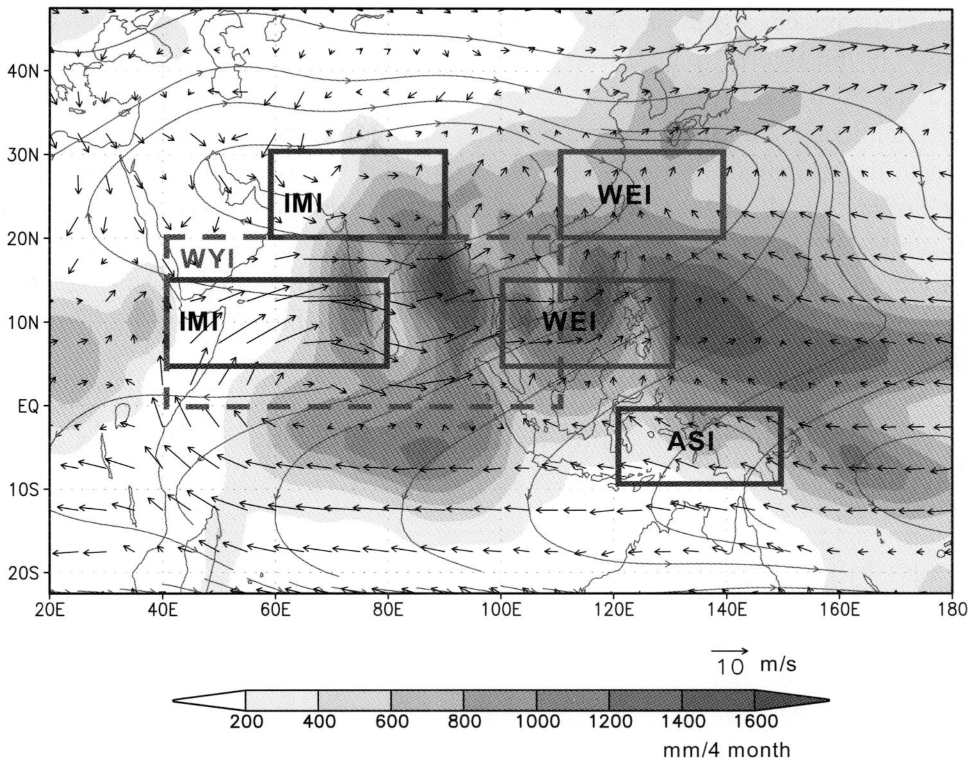

Figure13. Summer (JJAS) mean 850 hPa winds (*arrows*), 200 hPa streamlines (*thin lines*), and CMAP precipitation amount (*shading*). The solid-line boxes indicate locations in which three regional monsoon circulation indexes are defined: Indian monsoon index (IMI), western Pacific-East Asian monsoon index (WEI), and Australian monsoon index (ASI). The dashed-line box indicates the domain in which Webster-Yang index (WYI) is defined.

Fig. 14a compares the observed and simulated three indices (WYI, IMI, WEI) for JJA97 and the ASI index for the southern summer of DJF1997-98. The reanalysis shows that the broad-scale monsoon circulation for JJA97 is considerably weaker than normal as signified by a negative WYI. This large-scale circulation anomaly is simulated reasonably well by most of the models as well as the models' ensemble. Over the Indian monsoon region, both the reanalysis and model ensemble show near normal IMI. Over the western North Pacific and East Asia, the reanalysis and the model ensemble show a little discrepancy due to the considerable spreading among individual models. Figure 14b presents results for JJA98 during which the Pacific warming drastically decays from positive to negative SST anomalies in the NINO 3.4 region (Fig. 9a). The broad scale monsoon circulation (WYI) and IMI are normal but the WEI is extremely weak in the reanalysis. The model ensemble captures the weak WEI with the intensity being underestimated. The broad scale South Asian monsoon and the ISM are also reasonably well captured by the model ensemble, although the signs of anomalies are reversed. Again, the spread among individual models are considerably large. Also shown in the figures is that nearly all individual models and model ensemble mean reproduce the ASI realistically for the two consecutive austral summers (DJF96/97 and DJF97/98), although the model spread is large and the model ensemble mean tends to overestimate the magnitudes of anomalies. This indicates that the AGCMs perform well on simulation of the low-level zonal wind anomalies associated with the anomalous Walker cell during austral summer.

It is noticed that all models simulate realistic Australian and WNP-EA monsoon circulation anomalies throughout the entire two-year period (Fig. 14a and b). The monthly

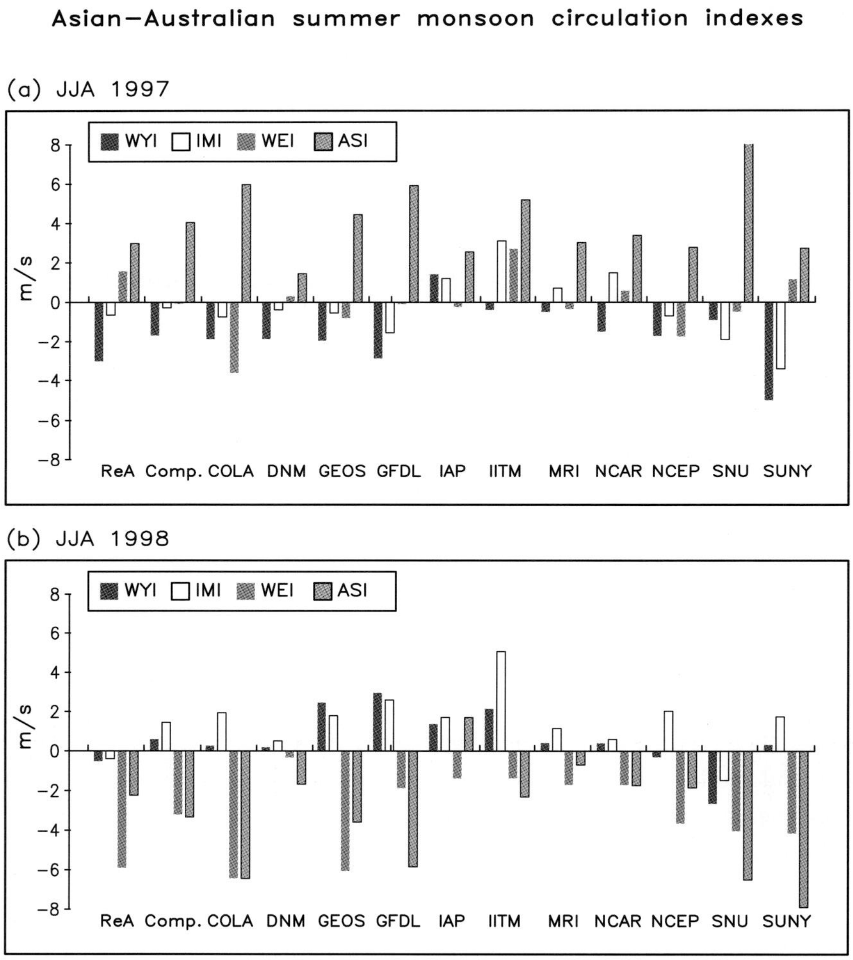

Figure 14. Comparison of the Asian summer monsoon circulation indexes derived from NCEP/NCAR reanalysis (ReA), the all-model ensemble mean (Comp), and from each model's ensemble for JJA 97 (a) and JJA 98 (b), respectively. The definitions for the broad-scale south Asian summer monsoon (WYI), Indian summer monsoon (IMI), the western Pacific-East Asian summer monsoon (WEI), and the Australian summer monsoon (ASI) are referred to section 2a. Acronyms for AGCMs are explained in the Introduction.

correlation coefficient between the all-model ensemble mean and the reanalysis reaches 0.73 for ASI and 0.65 for WEI (Table 2). The corresponding seasonal correlation coefficients are 0.91 and 0.75, respectively. In sharp contrast, the Indian monsoon circulation index simulated by the all-model ensemble is not significantly correlated with the observed counterpart, both for the monthly and seasonal means (Table 2). The broad scale South Asian tropical monsoon circulation is influenced by both the Indian and the Philippine convective heat source, thus the model's performance in terms of WYI is in between those of WEI and IMI. The IMI used here is highly correlated with the all-Indian rainfall index (Partharathy 1992) with a correlation coefficient of 0.72 for the 50-year period of 1948-1998. Thus, the result here supports the assessment made by Sperber and Palmer (1996) who showed that the JJAS averaged all-Indian rainfall indices from a suite of AMIP simulations for the period 1979-88 are not related to the corresponding observed counterparts. It is concluded that *the current AGCMs can reproduce monsoon circulation variations over the Australian and western Pacific-East Asian monsoon regions, but fails to simulate the Indian monsoon variability.*

Table 2. Correlation coefficients between the NCEP/NCAR reanalysis and the model ensemble means (for each individual model and the all-model ensemble (Comp)). The computation was done for four monsoon indexes and using (a) monthly and (b) seasonal mean data. ASI, IMI, WEI, and WYI denote, respectively, the Australian monsoon index, WNP-EA monsoon index, Indian monsoon index, and the Webster-Yang index. Acronyms for AGCMs are explained in the Introduction.

(a) Monthly mean

	Comp.	COLA	DNM	GEOS	GFDL	IAP	IITM	MRI	NCAR	NCEP	SNU	SUNY
ASI	0.65	0.63	0.57	0.55	0.41	0.29	0.44	0.51	0.60	0.60	0.64	0.43
IMI	0.08	-0.21	-0.19	0.06	-0.11	-0.46	-0.04	-0.11	0.04	0.23	-0.12	0.06
WEI	0.65	0.54	0.32	0.68	0.49	0.50	0.34	0.39	0.51	0.41	0.65	0.51
WYI	0.34	-0.35	0.11	-0.16	0.33	0.11	0.05	0.33	-0.02	0.53	-0.20	-0.05

(b) Seasonal mean

	Comp.	COLA	DNM	GEOS	GFDL	IAP	IITM	MRI	NCAR	NCEP	SNU	SUNY
ASI	0.95	0.98	0.84	0.88	0.83	0.44	0.92	0.68	0.85	0.89	0.94	0.89
IMI	-0.01	-0.65	-0.64	0.11	0.26	-0.54	-0.11	-0.10	-0.04	0.32	-0.13	0.25
WEI	0.75	0.66	0.37	0.86	0.57	0.74	0.47	0.74	0.57	0.53	0.85	0.75
WYI	0.43	-0.52	0.03	-0.37	0.55	0.00	0.25	0.42	-0.01	0.60	-0.20	-0.02

It has been known for recent years that El Niño weakens East Asian winter monsoon and brings warmer and wetter than normal weather to East Asian monsoon region (Zhang *et al.* 1996, Kang 1998, Lau and Nath 2000). These anomalous conditions can be well represented by the large-scale sea-level pressure anomaly patterns (Fig. 15). During DJF 97/98, an anomalous high pressure dominates in the WNP with two centers located over the Philippine Sea and Japan, respectively; while an anomalous low occupies Asian continent, indicating a weaker than normal Asian winter monsoon circulation (Fig. 15a). This large-scale pattern is well reproduced by the model ensemble mean. Furthermore, in both the reanalysis and the model ensemble, the high-pressure anomalies in the tropical WNP establishes in SON97 and the sea-level pressure anomaly pattern over Asian continent and the WNP persists into MAM 98. This result implies that the models are capable of simulating warmer and wetter than normal winter/spring in East and Southeast Asia and the suppressed convection over the Philippine Sea during JJA98. Thus, the impacts of ENSO on East Asian winter monsoon and spring climate, as well as suppressed WNP summer monsoon, are generally captured by the models.

It is conceivable that the skill of a model in simulation of monsoon anomaly may depend on the degree of accuracy of its simulated remote El Niño forcing. Hence, we examine whether the models that better reproduce El Niño rainfall anomalies also yield superior monsoon rainfall anomalies. Figure 16a shows the pattern correlation coefficient between the

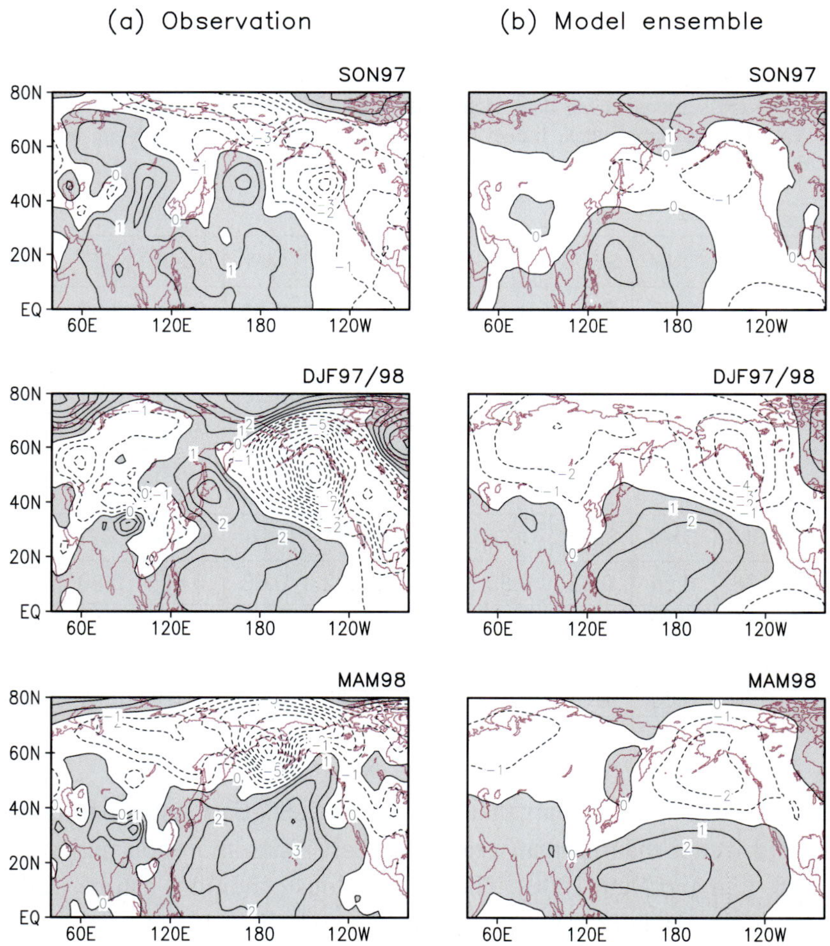

Figure 15. Sea-level pressure anomalies for the three consecutive seasons during 1997/98 El Niño obtained from observation (a) and all-model ensemble mean (b). The contour interval is 1 hPa.

area-mean precipitation over the El Niño region and that of the Monsoon region, produced by each model and the observation. In boreal winter (DJF97/98), the accuracy of simulating the El Niño rainfall pattern and the performance in simulating monsoon rainfall anomaly pattern are positively correlated, with the pattern correlation coefficient in the El Niño regions exceeding 0.6. On the other hand, during boreal summer (JJA 97), even though the pattern correlation over the El Niño region exceeds 0.6, there is virtually no correlation with the performance over the monsoon region. Thus, *a more accurate simulation of El Niño rainfall anomaly pattern does not warrant an improved simulation of the boreal summer monsoon rainfall anomaly pattern.* The season-dependent El Niño-monsoon relationship again

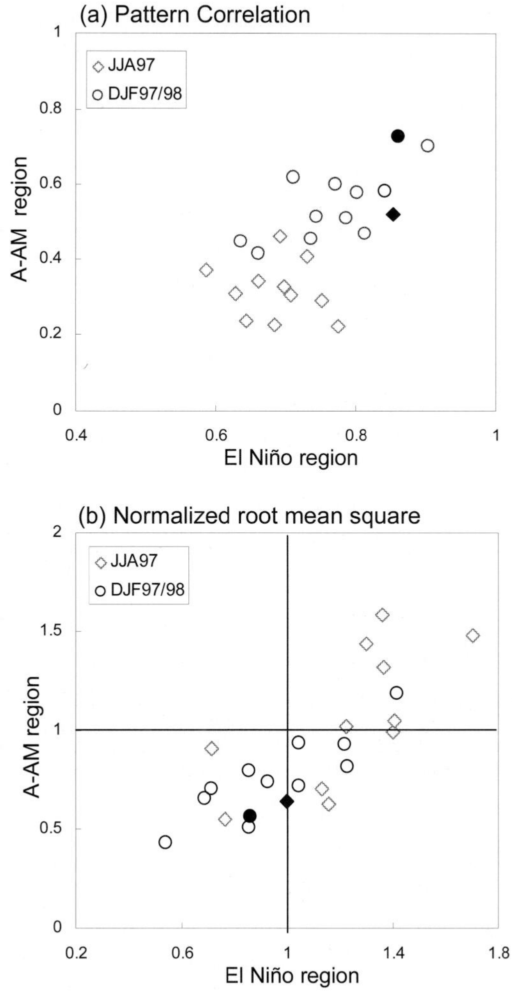

Figure 16. (a) Scatter diagram showing the relationship between the observed and simulated rainfall pattern correlation coefficients for the El Niño region (*abscissa*) versus those over the A-AM region (*ordinates*). (b) As in (a) except for the normalized root mean square that measures the averaged amplitude of an anomaly field. Different symbols represent different seasons. The empty and filled symbols represent, respectively, the correlation coefficient obtained from each individual model's ensemble experiments and from the all-model ensemble mean.

suggests a weaker impact of the El Niño forcing on the Asian summer monsoon than on the Australian summer monsoon. As far as the amplitude is concerned, the simulated variability in the A-AM region is highly correlated with that in the El Niño region, as indicated by Fig. 16b. Note also that the simulated monsoon rainfall anomalies during northern winter (the Australian summer monsoon) are weaker than the observed counterparts. This is a hint that the models and their ensemble mean tend to underestimate the effect of the remote forcing on the Australian summer monsoon.

7. Summary and Discussion

The present study examined the ability of 11 state-of-the-art GCMs to simulate the climatological variations of large-scale monsoon rainfall during boreal summer and the ENSO and Monsoon precipitation anomalies during the 1997-98 El Nino. All the model composites show the large-scale spatial patterns similar to the corresponding observations and are better than the best individual model results. The regional structures of the climatological mean precipitation simulated by individual models differ substantially from the observed. Most models produce excessive rainfall and overestimate the amplitude of the climatological variability in the Indian monsoon region. In the western Pacific, on the other hand, most of the models underestimate the climatological seasonal variation during the summer. All models simulate the spatial pattern of the observed El Niño precipitation anomalies in the tropical central Pacific reasonably well, although their amplitudes show large intermodel differences. On the other hand, most of the models have difficulty in simulating the negative anomalies over the Maritime continent during the El Niño.

A group of models including DNM, IAP, MRI, and NCAR fail to reproduce the observed large precipitation in the subtropical western Pacific. This group of models produces large precipitation confined to the equatorial region. The other group of models which include COLA, GEOS, IITM, SNU, simulate the maximum precipitation in the subtropical western Pacific but their large precipitation region is somewhat shifted to the north compared to the observed. These models tend to produce a stronger Indian monsoon compared to the previous group. None of the models reproduces the observed rain band in the region from the East China Sea to the mid Pacific, i.e., the Mei-yu, Changma, and Baiu rainbands. Most models simulate the rainband shifted toward the continental side along East Asia and a dry zone extending from the Sea of Okhotsk to the Korean peninsula. This situation is similar to results from a previous AMIP-I intercomparison (Lau *et al.* 1996). The poor simulation of the Mei-yu rainband indicates that that the model representation of physical processes leading to the summertime East Asian jetstream variability has not been improved since AMIP-I. However, it is also noted that simulation of the East-Asian rainbands may depend on the model resolution. Sperber *et al.* (1994) showed that the model with a higher resolution of about 1 degree (i.e. T106) was it possible to capture the East Asian rainbands.

Also examined are the climatological intraseasonal oscillations of precipitation over the Indian region and the East Asia-western Pacific monsoon region. The spatial distribution of the averaged amplitude of the *tropical* CISO component simulated by the models is similar to the observed. However, the present models simulate the *extratropical* CISO, which are much weaker than the observed. It is noted that, in the observations, the northward propagations of wet and dry events are evident in the extratropics between 15°N and 40°N during the summer. This extratropical character is largely missing in the model composite. Although several models such as COLA, GEOS, IITM, and SNU show the northward moving CISO components, their phases are shifted by 20-30 days such that the models simulate an early onset of the East Asian monsoon.

The performances of the AGCMs in simulating A-AM system show remarkable regional differences. All models simulate realistic Australian and western North Pacific-East Asian (WNP-EA) monsoon circulation anomalies throughout the entire two-year period of September 1996 - August 1998. The monthly (seasonal) correlation coefficient between the all-model ensemble and observation reaches 0.73 (0.91) for Australian monsoon index and 0.65 (0.75) for WNP-EA Monsoon index (Table 2) for a sample size of 264 (88). In sharp contrast, the Indian monsoon circulation index simulated by the all-model ensemble is not significantly correlated with the observed counterpart, both for the monthly and seasonal means (Table 2). Thus, in contrast to the WNP-EA monsoon and Australian monsoons, *the Indian monsoon variability is essentially not reproducible in the eleven AGCMs.* This is an indication of intrinsic chaotic nature of the Indian monsoon (Palmer 1994).

It is also noted that poor AGCM simulation of the monsoon rainfall anomalies over the oceans may be attributed in part to the deficiencies of the models' response to the local anomalous SST forcing, in particular in the western Pacific where air-sea interaction is active. The air-sea interaction influences not only atmospheric circulation and precipitation anomalies but also local SST. Atmospheric convection may cool down the local SST by reducing the solar radiation and stronger wind mixing. However, a stand-alone AGCM simulation would produce the convective anomalies over the warm oceans. In fact, the local correlations between AGCM simulated precipitation and prescribed SST anomalies are mostly positive over the western Pacific, whereas the local correlation between the observed precipitation and SST has negative values in some regions of the western Pacific. Therefore, even with a perfect AGCM, the forced response does not necessarily reproduce the observed atmospheric anomalies because of neglecting the feedback and rectification effects of the air-sea coupling. In this sense, consideration of the active monsoon-ocean interaction may be necessary for realistic simulation of Asian summer monsoon. Further studies of the nature of the monsoon-ocean interaction based on a coupled ocean-atmosphere GCM may hold a key for advancement of our understanding of the monsoon variability.

Acknowledgments

The author appreciates the participating GCM groups of the CLIVAR/Monsoon GCM intercomparison project, Ms. K. Jin, Ms. J.-Y. Lee, and Prof, Bin Wang for allowing the author to use their works. The author has been supported by the Monsoon project sponsored by Korea Meteorological Administration and Climate Environment System Research Center at Seoul National University.

References

Betts, A. K., 1986: A new convective adjustment scheme. Part I: Observational and theoretical basis. *Quart. J. Roy. Meteor. Soc.*, **112**, 677-691.

Bonan, G. B., 1998: The land surface climatology of the NCAR land surface model (LSM 1.0) coupled to the NCAR Community Climate Mode l(CCM3). *J. Climate*, **11**, 1307-1326.

Chou, M.-D., and M. J. Suarez, 1994: An efficient thermal infrared radiation parameterization for use in general circulation models. *NASA Tech. Memo. 104606*, **Vol. 3**, Goddard Space Flight Center, Greenbelt, MD 20771.

______, W. Ridgway, and M.-H. Yan, 1993: One-parameter scaling and exponential-sum fitting for water vapor and CO_2 infrared transmission functions. *J. Atmos. Sci.*, **50**, 2294-2303.

Deardorff, J. W., 1978: Efficient prediction of ground surface temperature and moisture, with inclusion of a layer of vegetation. *J. Geophys. Res.*, **83**, 1889-1903.

Fels, S. B., and M. D. Schwarzkopf, 1975: The simplified exchange approximation: A new method for radiative transfer calculations. *J. Atmos. Sci.*, **32**, 1475-1488.

Gadgil, S., S. Sajani, 1998: Monsoon precipitation in the AMIP runs. *Climate Dyn.*, **14**, 659-689.

Graham, N. E., T. P. Barnett, R. Wilde, M. Ponater, and S. Schubert, 1994: On the roles of tropical and midlatitude SSTs in forcing interannual to interdecadal variability in the winter Northern Hemisphere circulation. *J. Climate*, **7**, 1416-1441.

Gregory, D., and P. R. Rowntree, 1990: A mass flux convection scheme with representation of cloud ensemble characteristics and stability dependent closure. *Mon. Wea. Rev.*, **118**, 1483-1506.

Gruber, A., X. Su, M. Kanamitsu, and J. Schemm, 2000: The Comparison of Two Merged Rain Gauge–Satellite Precipitation Datasets. *Bull. Amer. Meteo. Soc.*, **81**, 2631-2644.

Harshvardhan, R. D., D. A. Randall, and T. G. Corsetti, 1987: A fast radiation parameterization for general circulation models. *J. Geophys. Res.*, **92**, 1009-1016.

Horel, J. D., and J. M. Wallace, 1981: Planetary-scale atmospheric phenomena associated with the Southern Oscillation. *Mon. Wea. Rev.*, **109**, 813–829.

Ingram, W. J., S. W. Ward, and J. Edward, 1996: Radiation. *Unified model documentation paper no. 23.*

Kalnay, E., M. Kanamitsu, R. Kistler, W. Collins, D. Deaven, L. Gandin, M. Iredell, S. Saha, G. White, J. Woollen, Y. Zhu, A. Leetmaa, B. Reynolds, M. Chelliah, W. Ebisuzaki, W. Higgins, J. Janowiak, K. C. Mo, C. Ropelewski, J. Wang, R. Jenne, and D. Joseph, 1996: The NCEP/NCAR 40-year reanalysis project. *Bull. Amer. Meteor. Soc.*, **77**, 437-471.

Kang, I.-S., S.-I. An, C.-H. Joung, S.-C. Yoon, and S.-M. Lee, 1989: 30-60 day oscillation appearing in climatological variation of outgoing longwave radiation around East Asia during summer. *J Korean Meteor. Soc.,* **25**, 149-160.

______, 1998: Relationship between El-Niño and Korean climate variability. *J. Korean Meteor. Soc.,* **34**, 390-396.

______, C.-H. Ho, Y.-K. Lim, and K.-M. Lau, 1999: Principal modes of climatological seasonal and intraseasonal variations of the Asian summer monsoon. *Mon. Wea. Rev.*, **127**, 322-339.

______, K. Jin, K.-M. Lau, J. Shukla, V. Krishnamurthy, S. D. Schubert, D. E. Wailser, W. F. Stern, V. Satyan, A. Kitoh, G. A. Meeh, M. Kanamitsu, V. Y. Galin, J.–K. Kim, A. Sumi, G. Wu, and Y. Liu, 2002a: Intercomparion of atomospheric GCM simulated anomalies associated with the 1997-98 El Niño. *J. Climate*, **15**, 2791-2805.

______, K. Jin, B. Wang, K.-M. Lau, J. Shukla, V. Krishnamurthy, S. D. Schubert, D. E. Wailser, W.

F. Stern, A. Kitoh, G. A. Meeh, M. Kanamitsu, V. Y. Galin, V. Satyan, C.–K. Park, and Y. Liu, 2002b: Intercomparion of the climatological variations of Asian summer monsoon precipitation simulted 10 GCMs. *Climate Dyn.*, **19**, 383-395.

Katayama, A., 1978: Parameterization of the planetary boundary layer in atmospheric general circulation models. *Kisyo Kenkyu Note*, **No. 134**, Meteorological Society of Japan, 153-200 (in Japanese).

Kiehl, J. T., J. J. Hack, G. Bonan, B. Boville, D. Williamson, and P. Rasch, 1998: The National Center for Atmospheric Research Community Climate Model (CCM3). *J. Climate*, **11**, 1131-1149.

Kitoh, A., K. Yamazaki, and T. Tokioka, 1988: Influence of soil moisture and surface albedo changes over the African tropical rain forest on summer climate investigated with the MRI GCM-I. *J. Meteor. Soc. Japan*, **66**, 65-86.

Krishnamurti, T. N., 1985: Summer monsoon experiment-A review. *Mon. Wea. Rev.*, **113**, 1590-1626.

Lacis, A. A., and J. E. Hansen, 1974: A parameterization for the absorption of solar radiation in the Earth's atmosphere. *J. Atmos. Sci.*, **31**, 118-133.

Lal, M, U. Cubasch, J. Perlwitz, and J. Waszkewitz, 1997: Simulation of the Indian monsoon climatology in the ECHAM3 climate model : Sensitivity to horizontal resolution. *Intl. J. Climatol.,* **17**, 847-858.

Lau, K.-M., and T. M. Li, 1984: The monsoon of East Asia and its global association – A survey. *Bull. Amer. Meteo. Soc.*, **65**, 114-125.

______, and S. Yang, 1996: The Asian monsoon and predictability of the tropical ocean-atmosphere system. *Quart. J. Roy. Meteor. Soc.*, **122**, 945-957.

______, G. J. Yang, and S. Shen, 1988: Seasonal and intraseasonal climatology of summer monsoon rainfall over East Asia. *Mon. Wea. Rev.*, **116**. 18-37.

______, J. H. Kim, and Y. Sud, 1996: Intercomparison of hydrologic processes in AMIP GCMs. *Bull. Amer. Meteo. Soc.*, **77**, 2209-2227.

Lau, N.-C., and M. J. Nath, 1994 : A modeling study of the relative roles of tropical and extratropical SST anomalies in the variability of the global atmosphere-ocean system. *J. Climate*, **7**, 1184-1207.

______, and M. J. Nath, 2000: Impact of ENSO on the variability of the Asian–Australian monsoons as simulated in GCM experiments. *J. Climate*, **13**, 4287–4309.

Manabe, S., J. Smagorinsky, and R. F. Strickler, 1965: Simulated climatology of a general circulation model with a hydrologic cycle. *Mon. Wea. Rev.*, **93**, 769-798.

Martin, M., 1999: The resolution of the Asian summer monsoon, and its sensitivity to horizontal resolution, in the UK meteorological office unified model. *Quart. J. Roy. Meteor. Soc.*, **125**, 1499-1525.

Matsuno, T., 1966: Quasi-geostrophic motions in the equatorial area. *J. Meteor. Soc. Japan,* **44**, 25-42.

McBride, J. L., N. E. Davidson, K. Puri, and G. C. Tyrell, 1995: The flow during TOGA COARE as diagnosed by the BMRC tropical analysis and prediction system. *Mon. Wea. Rev.*, **123**, 717–736.

Moorthi, S., and M. J. Suarez, 1992: Relaxed Arakawa-Schubert: A parameterization of moist convection for general circulation models. *Mon. Wea. Rev.*, **120**, 978-1002.

Nakajima, T., and M. Tanaka, 1986: Matrix formulation for the transfer of solar radiation in a plane-parallel scattering atmosphere. *J. Quant. Spectrosc. Radiat. Transfer*, **35**, 13-21.

Nakazawa, T., 1992: Seasonal phase lock of intraseasonal variation during the Asian summer monsoon. *J. Meteor. Soc. Japan*, **70**, 257-273.

Palmer, T. N., 1994: Chaos and predictability in forecasting the monsoons. *Proc. Indian Net. Sci. Acad.,* **60A**, 57-66.

Pan, H.-L., and L. Mahrt, 1987: Interaction between soil hydrology and boundary layer developments. *Boundary Layer Meteor.*, **38**, 185-202.

Parker, D. E., M. Jackson, and E. B. Horton, 1995: The GISST 2.2 sea surface temperature and sea-ice climatology. *CERT* **No. 63**, Hadley Centre for Climate Prediction and Research, Bracknell, UK.

Parthasarathy, B., R. R. Kumar, and D. R. Kothawale, 1992: Indian summer monsoon rainfall indices, 1871-1990. *Meteor. Mag.,* **121**, 174-186.

Reynolds, R.W., and T. M. Smith, 1994: Improved global sea surface temperature analyses using optimum interpolation. *J. Climate*, **7**, 929-948.

Ropelewski, C. F., and M. S. Halpert, 1987: Global and regional scale precipitation patterns associated with the El Niño/Southern Oscillation. *Mon. Wea. Rev.*, **115**, 1606-1626.

Schemm, J.–K., S. Schubert, J. Terry, S. Bloom, and Y. Sud, 1992: Estimates of monthly mean soil moisture for 1979-89, *NASA Tech. Memo*. **No. 104571**, Goddard Space Flight Center, Greenbelt, MD 20771.

Schwarzkopf, M. D., and S. B. Fels, 1991: The simplified exchange method revisited: An accurate, rapid method for computation of infrared cooling rates and fluxes. *J. Geophys. Res.*, **96**, 9075-9096.

Shi, G. Y.,1981: An accurate calculation and the infrared transmission function of the atmospheric constituents. Ph.D. Thesis, Dept. of Sci., Tohoku University of Japan. 191 pp.

Shibata, K., and T. Aoki, 1989: An infrared radiative scheme for the numerical models of weather and climate. *J. Geophys. Res.*, **94**, 14923-14943.

Shukla, J. M., and J. Fennessy, 1994: Simulation and predictability of monsoons. *Proc. of the Inter. Conference of Monsoon Variability and Prediction.* Tech. Rep. WCRP-84, WCRP, Geneva, Switzerland, 567-575.

Slingo, A., 1989: A GCM parameterization for the shortwave radiative properties of water clouds. *J. Atmos. Sci.*, **46**, 1419-1427.

______, and R. C. Wilderspin, 1986: Development of a revised longwave radiation scheme for an atmospheric general circulation model. *Quart. J. Roy. Meteor. Soc.*, **112**, 371-386.

Smith, R. N. B., 1990: Subsurface, surface and boundary layer processes, Unified Model Doc. Paper 24, Available at Nat. Met. Lib., Bracknell.

Soman, M. K., and J. M. Slingo, 1997: Sensitivity of the Asian summer monsoon to aspects of sea surface temperature anoamlies in the tropical Pacific Ocean. *Quart. J. Roy. Meteor. Soc.*, **123**, 309-336.

Sperber, K. R., S. Hameed, G. L. Potter, and J. S. Boyle, 1994: Simulation of the northern summer monsoon in the ECMWF model: Sensitivity to horizontal resolution. *Mon. Wea. Rev.*, **122**, 2461-2481.

______, and T. N. Palmer, 1996: Interannual tropical rainfall variability in general circulation model simulations associated with the Atmosphere Model Intercomparison Project. *J. Climate*, **9**, 2727-2750.

______, J. M. Slingo, and H. Annamalai, 2000: Predictability and the relationship between subseasonal and interannual variability during the Asian summer monsoon. *Quart. J. Roy. Meteor. Soc.*, **126**, 2545-2574.

Tanaka, M., 1992: Intraseasonal oscillation and the onset and retreat dates of the summer monsoon over East, Southwest and the western North Pacific region using GMS high cloud amount data. *J.*

Meteor. Soc. Japan, **70**, 613-629.

Tao, S. Y., and Y. H. Ding, 1981: Observational evidence of the influence of the Qinhai-Xizang (Tibet) plateau on the occurrence of heavy rain and severe storms in China. *Bull. Amer. Meteo. Soc.*, **62**, 23-30.

Tao, S,, and L. Chen, 1987: A review of recent research on the East Asian summer monsoon in China. *Monsoon Meteorology*. C. P. Chang and T. N. Krishnamurti, Eds, Oxford University Press, pp 60-92.

Trenberth, K. E., 1976: Spatial and temporal variations of the Southern Oscillation. *Quart. J. Roy. Meteor. Soc.*, **102**, 639-653.

______, 1997: The Definition of El Niño. *Bull. Amer. Meteor. Soc.*, **78**, 2771–2777.

Volodin, E. M., and V. N. Lykossov, 1998: Parameterization of heat and moisture transfer in the soil-vegetation system for use in atmospheric general circulation models: 1. Formulation and simulations based on local observational data. *Izvestiya Atmospheric and Oceanic Physics*, **34**, 405-416 (Translated from *Izvestiya AN. Fizika Atmosfery i Okeana*).

Wang, B., and X. Xu, 1997: Nothern hemisphere summer monsoon singularities and climatological intraseasonal oscillation. *J. Climate*, **10**, 1071-1085.

______, and Z. Fan, 1999: Choice of South Asian summer monsoon indices. *Bull. Amer. Meteo. Soc.*, **80**, 629-638.

______, and H. Lin, 2001: Rainy seasons of the Asian-Pacific monsoon defined using a single variable. *J. Climate*. (in press)

______, R. Wu, and K.-M. Lau, 2001: Interannual variability of Asian summer monsoom: Contrast between the Indian and western North Pacific summer monsoon. *J. Climate*, **14**, 4073-4090.

______, I.-S. Kang, and J.-Y. Lee, 2004: Asian-Australian monsoon anomalies during 1997/1998 El Nino: Results from ensemble simulation of eleven AGCMs. *J. Climate*. **17(4)**, 803-818.

Webster, P. J., and S. Yang, 1992: Monsoon and ENSO. *Quart. J. Roy. Meteor. Soc.*, **118**, 877-926.

______, V. O. Magana, T. N. Palmer, J. Shukla, R. A. Tomas, M. Yanai, and T. Yasunari, 1998: Monsoons: Processes, predictability, and the prospects for prediction. *J. Geophys. Res.*, **103**, 14451-14510.

Xie, P., and P. A. Arkin, 1997: Global precipitation: A 17-year monthly analysis based on gauge observation, satellite estimates and numerical model outputs. *Bull. Amer. Meteo. Soc.*, **78**, 2539-2558.

Xue, Y.-K., P. J. Sellers, J. L. Kinter II, and J. Shukla, 1991: A simplified biosphere model for global climate studies. *J. Climate*, **4**, 345-364.

Zachary, A. E., and D. A. Randall, 1999: Sensitivity of the simulated Asian summer monsoon to parameterized physical processes. *J. Geophys. Res.*, **104**, 12177-12191.

Zhang, G. J. and N. A. McFarlane, 1995: Sensitivity of climate simulations to the parameterization of cumulus convection in the Canadian Climate Centre general circulation model. *Atmos.-Ocean*, **33**, 407-446.

Zhang, R., A. Sumi, and M. Kimoto, 1996: Impact of El Niño on the East Asian monsoon: A diagnostic study of the '86/87 and '91/92 events. *J. Meteor. Soc. Japan*, **74**, 49-62.

Zhang, Y., K. R. Sperber, J. S. Boyle, M. Dix, L. Ferranti, A. Kitoh, K.-M. Lau, K. Miyakoda, D. Randall, L. Takacs, and R. Wetherald, 1997: East Asian winter monsoon: Results from eight AMIP models. *Climate Dyn.*, **13**, 792- 820.

9. SIMULATION STUDIES OF THE ASIAN MONSOON USING THE CCSR/NIES AGCM

AKIMASA SUMI AND MASAHIDE KIMOTO
Center for Climate System Research
University of Tokyo
4-6-1, Komaba, Meguro
Tokyo, 153-8904, Japan
E-mail: sumi@ccsr.u-tokyo.ac.jp

XUESHUN SHEN
Research Organization for Information
Science and Technology
2-2-54, Naka-Meguro, Meguro
Tokyo, 153-0061, Japan

Several numerical simulations of the Asian monsoon have been conducted at CCSR (Center for Climate System Research) since 1991. The results are summarized from a viewpoint of sensitivity to spatial resolution and the boundary conditions such as SST and land surface conditions.

T21, T42 and T106 results are compared to demonstrate the sensitivity to spatial resolution. Simulation performance is generally better, the higher the resolution, although it should be noted that increase in resolution does not automatically contribute to improvement in simulation. Care must be taken to match the spatial resolution to the actual physical processes involved.

Regarding the surface boundary conditions, it is concluded that the SST anomaly contributes to the inter-annual variability, as defined by the Webster-Yang Index, more than the land surface conditions. At the same time, contribution of sea surface temperature anomalies over the several regions is estimated. The contribution of both the tropical eastern Pacific and the Indian Ocean is noted. It is emphasized that an anti-cyclonic anomaly of 850 hPa over the tropical Western Pacific plays an important role in controlling the East Asian monsoon activity.

1. Introduction

The Center for Climate System Research (CCSR) of the University of Tokyo was established in 1991, where its objectives were (1) to develop a climate model, (2) to understand the mechanisms of climatic variability and (3) to promote collaboration studies with in Japan and with the other countries.

The Atmospheric General Circulation Model (AGCM) at CCSR has been under joint development with NIES (National Institute of Environmental Studies) since 1991 and the first version was completed in 1995 (Numaguti *et al.* 1997) and is denoted as CCSR/NIES AGCM hereafter. The Ocean GCM, which has been developed by the ocean group at CCSR, is denoted as the COCO (CCSR Ocean Component Model). The models have been coupled, creating the CCSR/NIES climate model.

The variability and predictability of the Earth climate system has been investigated using these models. Simulation of the Global Change due to the doubling of CO2 has been performed and the results were presented in the IPCC Assessment reports. These activities are summarized in Sumi (2000).

The Asian monsoon is one of the main themes which has been extensively investigated at CCSR and we summarize our results here. However, as the Asian monsoon has been investigated by many researchers from different viewpoints, it is necessary to explain the background and purpose of our research.

Prior to FGGE (the First GARP Global Experiment) held during 1978-79, the main focus was placed on seasonal forecasts and understanding of monsoon depressions. During the FGGE, two sub-projects were undertaken, namely WMONEX (Winter MONsoon EXperiment) (December 1978 - February 1979) and SMONEX (Summer MONsoon EXperiment) (June-August 1979), where the Asian monsoon was investigated in the context of the global circulation. Many interesting features were found in these experiments and are summarized in the book edited by Chang and Krishnamurti (1987). The particular importance of the Australian monsoon was noted in the context of the general circulation (Murakami and Sumi 1982a, b) and a significant 40-60 day fluctuation was found by Krishnamurti and Subrahmanyan (1982) and Murakami *et al.* (1984).

After FGGE, the climate variability associated with ENSO (El-Nino Southern Oscillation) attracted the world's attention. In this context, the annual variation of the Asian monsoon has been intensively investigated in association with the ENSO events during the TOGA (Tropical Ocean and Global Atmosphere)-period (1985-1994). In particular, the monsoon and ENSO relationship was pointed out by many researchers. For example, the global ENSO mode was pointed out by Yasunari (1987a, b) and the biennial oscillation of the Asian monsoon was presented by Meehl (1987). In their studies, the linkage between ENSO and the Asian monsoon is well noted, but the missing link is the interaction between the Atlantic disturbance's activity and the Eurasian snowfall, and the influence of the Eurasian snowfall on the Asian monsoon.

The linkage between the Eurasian snowfall and the intensity of the Asian monsoon is one of the main scientific issues in the GAME (GEWEX Asia Monsoon Experiment), which was conducted from 1996 until 2001 following the TOGA (1985-1994). The GAME was commenced in order to understand the role of the Asian monsoon in the context of the global Water and Energy cycle and improve our skill in predicting the Asian monsoon. The linkage was investigated in CCSR by Shen *et al.* (1998) and Shen and Kimoto (1999).

Besides these studies, the 1998 summer Asian monsoon showed very interesting features. There was a tremendously heavy rainfall in the Yangtze-River Valley and adjacent river valleys in China. There also was heavy rainfalls in Japan and Korea. These phenomena attracted our interest. At the same time, SST anomalies in the Indian Ocean were found by Webster *et al.* (1999) and Saji *et al.* (1999). The impact of the Indian Ocean on the East

Asian monsoon was studied using the numerical simulations conducted in the CCSR (Shen *et al.* 2001).

In this article, a description of the CCSR/NIES AGCM model will be given in section 2. In section 3, sensitivity of the simulation to horizontal resolution and the fluxes between atmosphere and land surface will be discussed in section 4. In section 5, the influence of the land surface condition and Sea Surface Temperature will be discussed. Summaries will be given in Section 6.

2. Model Description

The atmospheric general circulation model at CCSR has been developed jointly with NIES and is named CCSR/NIES AGCM (Numaguti *et al.* 1997). There follows a brief summary of the model characteristics.

It is a global spectral model with a horizontal resolution of T21, T42 and T106 (triangular truncation of the spherical harmonics at wavenumber 21, 42 and 106; the corresponding Gaussian grid distances are 5.6, 2.8 and 1.125 degrees, respectively). Twenty vertical levels are defined and a finite difference scheme with sigma coordinates is used. Prognostic variables of the model are streamfunction, velocity potential, air temperature, surface pressure and the mixing ratios of an arbitrary number of components, including water vapor and cloud liquid water. In addition, soil moisture, soil temperature, snow amounts and river water are predicted in the land surface model.

The physical processes include radiation, cumulus convection, large-scale condensation with prognostic cloud water, vertical diffusion and gravity wave drag. The radiative transfer code is based on the 2-stream, k-distribution method of Tanaka and Nakajima (1986). The diurnal cycle is included. The cumulus parameterization scheme is a simplified Arakawa-Schbert scheme (Moorti and Suarez 1992). The prognostic cloud water is computed using the scheme of Le Treut and Li (1991). Distinction between snow and rainfall is based on the ground surface temperature. Surface fluxes are based on the scheme of Louis (1985), and the closure model by Meller and Yamada (1974, 1982) is used, where the moisture effect is considered through the use of moist Richardson number. The effects of orographically excited gravity waves are parameterized following McFarlane (1987).

The parameterization of land surface processes is treated by adopting three layers for soil temperature and one for soil moisture. In the land surface sub-model, the surface skin temperature is implicitly determined by the surface energy balance in an infinitely thin layer. The ground temperature is prognostically calculated by a diffusion scheme and the ground wetness is predicted by the bucket model with a field capacity of 15 g cm-2 (Manabe *et al.* 1965). The snow accumulation is determined by the balance between snowfall, snowmelt and sublimation. When snow exists, the surface evaporation efficiency, surface albedo and roughness length are modified as an effect of the snow (Numaguti *et al.* 1987). The albedo and roughness length of the snow-free land surface are prescribed according to the 32 categories of surface type (Matthews 1983).

3. Sensitivity of the Asian monsoon Simulation to the Horizontal Resolution and the Fluxes between Atmosphere and Land Surface

Many numerical experiments have been conducted for the Asian monsoon research by using various AGCMs. Although many results are obtained, the results are usually model-dependent and it is difficult to obtain definitive results. Particularly as the Asian monsoon is strongly related to convective heating, many results become dependent on the parametarization scheme. Many studies have been conducted to investigate impacts of a parameterization scheme to the Asian monsoon prediction and simulation, but without definite conclusions. Other important factors are the horizontal and vertical resolutions, because orographic effects and convective heating are strongly influenced by resolution. Thus, the sensitivity of the Asian monsoon simulation results to horizontal resolution will be discussed with reference to many experiments conducted by CCSR in this section.

First, CCSR participated in the AMIP (Atmospheric Model Intercomparison Project) (Gates 1992), where a 10 year (1979-1989) simulation was conducted with reference to the observed sea surface temperatures (SST). In the AMIP run, a low resolution AGCM (T21) with 20 levels was used by Shen *et al.* (1988). In this paper, we concluded that the large-scale features of the Asian monsoon can be well simulated using a low resolution model, although there are many unsatisfactory aspects of the simulation results. Following the AMIP, the AMIP-2 projects were proposed, where a 15 year (1979-1994) simulation was conducted with observed SST. In the AMIP-2 simulation, a T42 model with 20 levels was used for simulation. Here, both results are compared to estimate the sensitivity to horizontal resolution of the Asian monsoon simulation with emphasis on the mean field and annual variation.

Fig. 1 (top) shows the 14-year average of June-July-August (JJA) precipitation of the AMIP-2 run using the CCSR/NIES AGCM (T42L20). For reference, a horizontal distribution of the 14 year average JJA precipitation of CMAP is shown in Fig. 1 (middle) with the 14 year average of NCEP re-analysis of wind fields at 850 hPa. The results based on T21 GCM are represented in Fig. 1 (bottom). Major maxima of precipitation over the oceanic regions, such as the Bay of Bengal, the Arabian Sea, the South China Sea and the Western Pacific are well simulated in both models, although small-scale features are better represented in T42 than T21. When they are compared with the CMAP observation, significant differences are noted in the Western Pacific ocean, the equatorial Indian Ocean and the mainland China. For example, the rainfall amount associated with ITCZ in the Northern Hemisphere is less than observed. This corresponds to the weaker easterlies there. The precipitation over these regions is considered to be coupled by the large-scale circulation and both fields will be improved at the same time. In the Indian Ocean, the rainfall maximum around 5S is noted by observation. This feature was not simulated in T21, but it can be simulated in the T42-run, although the magnitude is weaker. This rainfall amount may be related to disturbance activity.

Maximum rainfalls are noted south (30°N, 80°E) and east (35°N, 105°E) of the Tibetan Plateau in the simulation. This is considered to be related to the poor representation of the convective system around the high mountain region, where small-scale features are dominant.

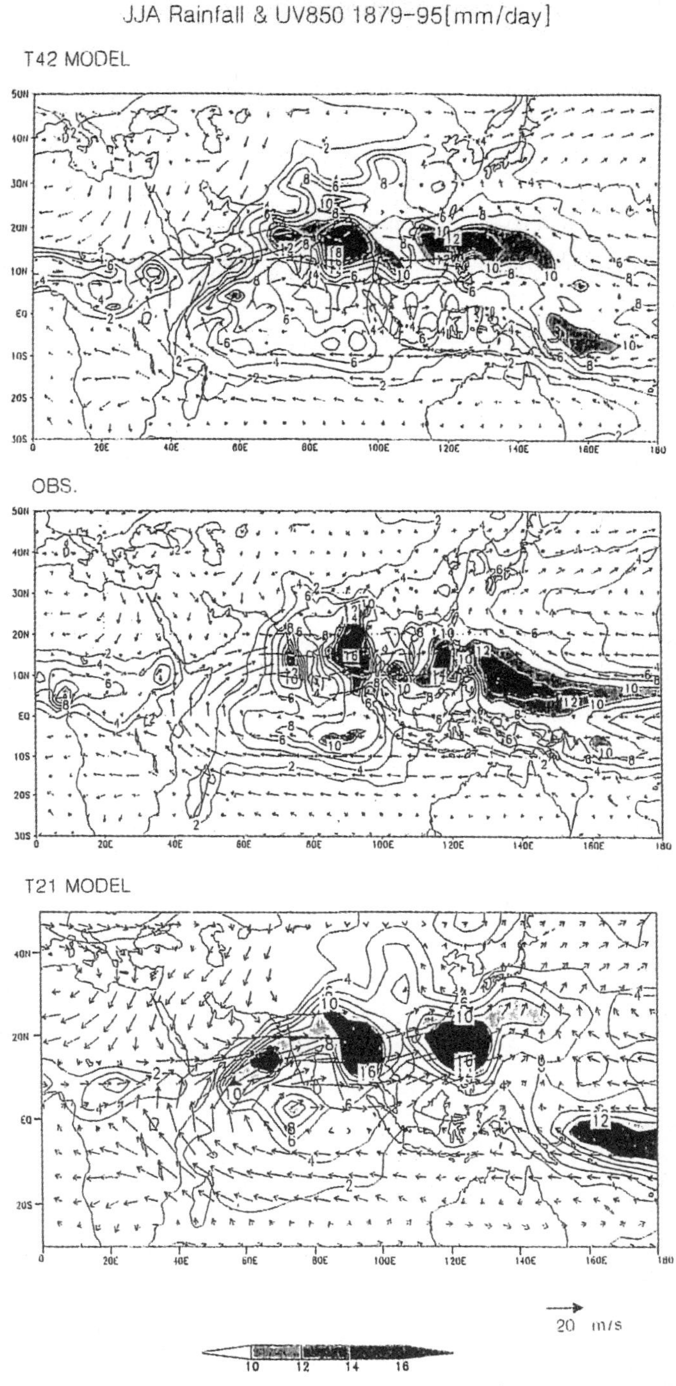

Figure 1. June, July, August (JJA) mean rainfall and 850 hPa winds. (Top) a simulation by T42L20, (middle) observed and (bottom) a simulation of T21L20. The observed rainfall climatology and 850 hPa flow are based on the Legates/MSU (Microwave Sounding Unit) and ECMWF analyses for 1980-88, respectively. Precipitation contours are plotted at an interval of 2mm /day. The unit of 850 hPa winds is m/s.

However, it should be noted that observed rainfall estimate over data-poor regions in the continent is not as good and it is difficult to estimate the rainfall amount over the land, using observations such as OLR and microwave satellite. Only the TRMM (Tropical Rainfall Measurement Mission) PR (Precipitation Radar) data can provide information about the rainfall amount over the land without in-situ observation. By contrast, the estimation of the precipitation over the ocean is well advanced, using satellite remote sensing such as AVHRR, TMI, and other sensors.

The Baiu front in the East China Sea and south of Japan is simulated better in the T42 run than the T21 run. This is considered to be due to increase of horizontal resolution. This is consistent to the results of Kawatani and Takahashi (2002), where the basic features of the Baiu front are well simulated when using the T106L20 CCSR/NIES AGCM.

The large-scale features of low-level atmospheric circulation, such as the easterly winds over in the southern Indian Ocean, the Somali jet off the Kenya peninsula and the southwesterly winds over the Arabian Sea to the western Pacific Ocean, are well simulated in the simulation by both models. However, there are several biases in the simulations. For example, we can note stronger south westerlies across the Indian sub-continent, strong northerlies in the Arabian Peninsula and excessive intrusion of westerlies in the South China Sea. These differences are based on the model insufficiency and dependence on horizontal resolution is small.

Why does the Asian monsoon vary year by year in spite of the large heat difference between land and ocean? The monsoon variability is one of the key issues in the monsoon studies. In the AMIP run, the interannual variation of the Asian monsoon is expected to be simulated because the observed SST is specified. To investigate the interannual variability, we used the monsoon circulation index by Webster and Yang (1992), which is a difference in the deviations of a zonal wind field from a mean field between 850 hPa and 200 hPa. However, there are arguments that Webster and Yang index cannot represent the Asian monsoon activity (Ailikun and Yasunari 1998; Goswami *et al.* 1999). It is very difficult to define the intensity of the Asian monsoon as one index (Miyakoda 2002). Thus we use the Webster Yang index as an indicator of the variation in the large-scale Asian monsoon circulation.

In Fig. 2 (a), the time series of the JJA-mean anomalies of the monsoon index in the T42 simulation is shown in association with the indices by NCEP analysis, with the same quantities based on T21 model until 1988 are also shown. Here, it is noted that an interannual variation of the index exists. T21 results correspond well with the NCEP results until 1988. T42 results and NCEP results show similar tendencies, although some discrepancies exist (for example, 1987 and 1992). It is also noted that discrepancies exist between T42 and T21. For example, a strong (weak) monsoon is simulated in T21 (T42) in 1984 and a weak (strong) monsoon in T21 (T42) in 1987. As both runs are bounded by the same SST distribution condition, the difference in the results is considered to be caused by the difference of horizontal resolutions. More precisely, the result is due to the difference in circulation change and the land surface condition in the two models. In other words, the inter-annual variation of the Asian monsoon is not completely controlled by the variation in the tropical ocean SST.

However, it should be noted that these simulations are based on one simulation run. If we run multiple simulations, the conclusion may be different.

Next, we address the sensitivity to the horizontal resolution by comparing the T42 and the T106 simulations of the 1998 Asian summer monsoon, when heavy rainfall occurred along the Yangtze River valley. In order to avoid the sensitivity to the initial conditions, we applied the ensemble method (3 samples) in this experiment. The AGCM was integrated from April 1, 3, and 15 until the end of September and the monthly means from May to September were compared.

The JJA ensemble-mean wind fields at 850 hPa, using T42 and T106 model, are shown in Fig. 3 (top) and Fig. 3 (bottom), respectively. For the large-scale features, it is very difficult to find difference between these two fields. But the intensity of the winds along the China coast and over the Arabian Bay is stronger in T106 than that it is in the simulation T42. It should be noted that the trough in the northern part of the Indian subcontinent is well represented in T106. This is because the effect of the Himalayas is better represented by the higher horizontal resolution. The JJA ensemble-mean precipitation fields, using the T42 and the T106 models, are shown in Fig. 4 (top) and Fig. 4 (bottom), respectively. Again, it is difficult to find differences in the largescale patterns, but many small scale differences can be

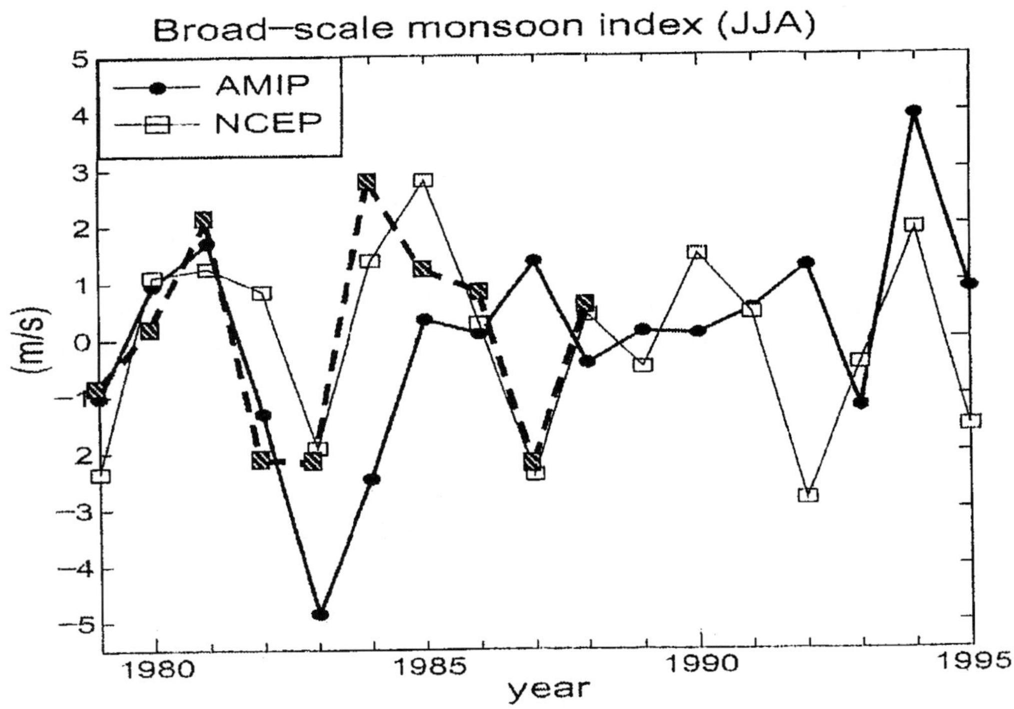

Figure 2. Time series of JJA-mean anomalies of Webster and Yang's monsoon index from 1979 to n1995. The index is defined as the zonal wind shear between 850 and 200 hPa levels averaged over (40°E - 110°E, 5°N - 20°N) region. The unit is m/s. The thick line represents T42L20, and thin line is an observed value based on NCEP analysis. For reference, the T21L20 is displayed (1979-1988) by a thick broken line.

seen. For example, the rainfall south of the Himalayas is well represented and the rain area extends north-westward in the north of the Indian subcontinent in the T106 simulation. The rain area in the T106 simulation is limited over the Bay of Bengal, but it is extended eastward into Thailand in the T42 simulation. The rainfall maximum corresponding to the ITCZ around (10°N, 160°E - 180°E) in the T106 run is smaller than that in the T42 run. For comparison, the JJA precipitation in 1998, estimated by the TRMM PR, is shown in Fig. 5. It should be noted that TRMM PR is the first space-born precipitation radar in the world, which can observe the precipitation over both land and ocean. In general, the rainfall amount in the simulation is larger than the observed value. Furthermore, the precipitation to the west of Sumatra and the maritime continent is not well simulated. These aspects are considered to be due to insufficient parameterization of the convective process.

In Fig. 6, the seasonal extent of the monthly rainfalls, averaged between 60°E and 80°E, is shown. The monthly rainfall estimate due to CMAP is also shown. This area corresponds to the Arabian Sea. Two maxima are noted, one is around 15°N and the other is around 5°S. The first one corresponds to the rainfall maximum, east of the mountain rage. The simulated rainfall is larger than the observed amount, although it is noted that the precipitation amount in T42 is larger than that in T106. In general, the T106 results are closer to observation than the T42 results. This is because the convection in the simulation tends to remain over the ocean due to the interaction between the monsoon westerly and the topography. On the

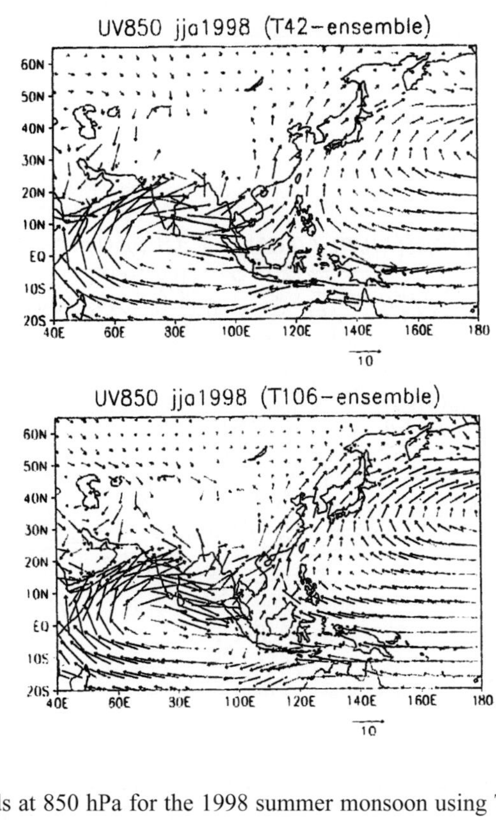

Figure 3. JJA-mean wind fields at 850 hPa for the 1998 summer monsoon using T42 ensemble simulations (top), and T106 ensemble simulations (bottom). Unit is m/s.

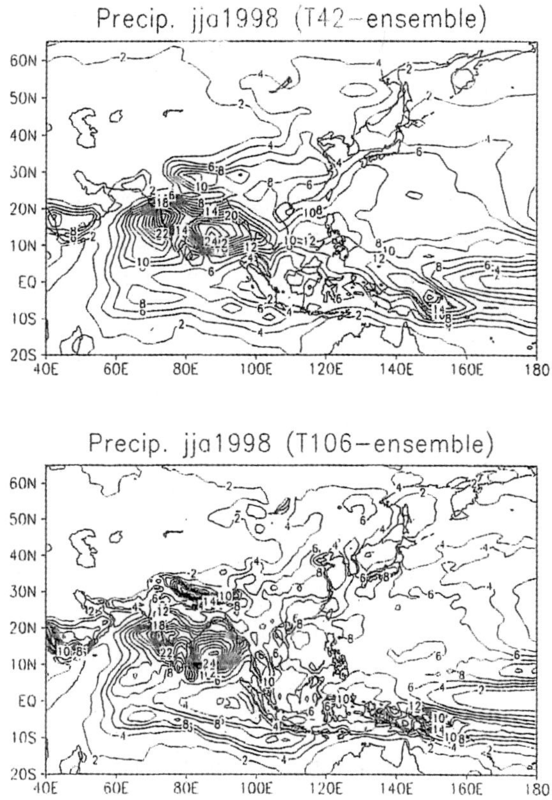

Figure 4. Same as Fig. 3, except for precipitation. Unit is mm/day.

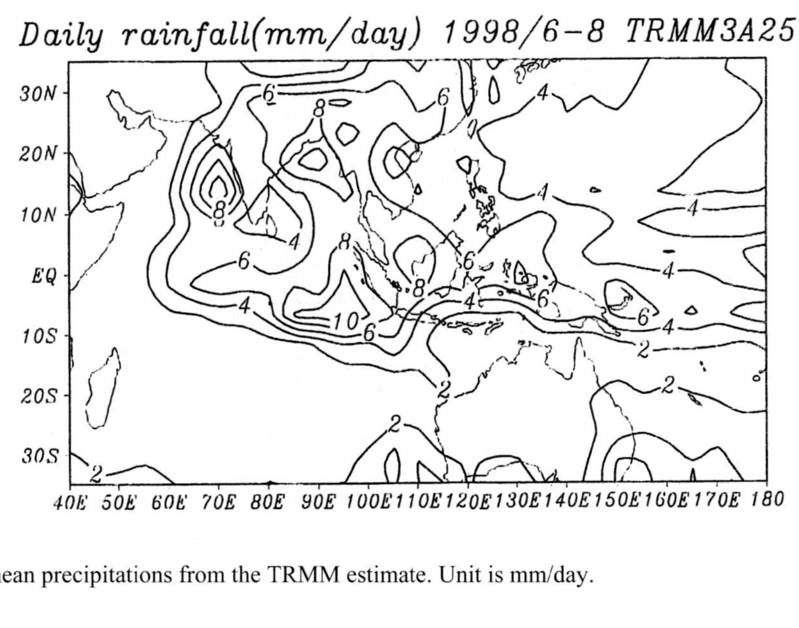

Figure 5. JJA-mean precipitations from the TRMM estimate. Unit is mm/day.

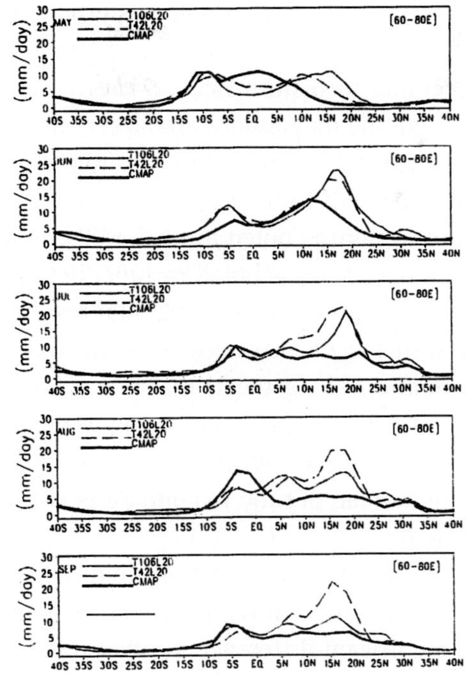

Figure 6. Meridional cross-section of monthly averaged rainfall between 60°E - 80°E from May until September, 1998. Unit is mm/day.

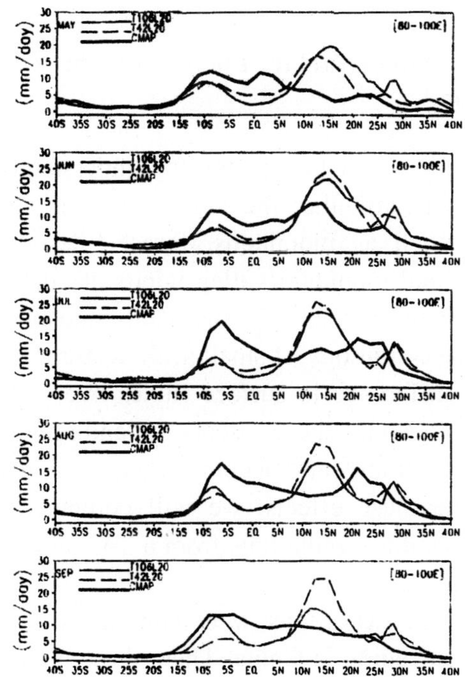

Figure 7. Same as Fig. 6, except for 80°E - 100°E.

contrary, the observed convection is associated with disturbances, which are mobile. Then, the precipitation field tends to be smoothed when we take a monthly average. The maximum around 5S is considered to be due to disturbances in the Indian Ocean. This maximum is represented in the simulations, but the magnitudes are larger, similarly to the above case.

In Fig. 7, the meridional distribution of monthly rainfalls averaged between 80°E and 100°E are shown. This area corresponds to the Bay of Bengal region. Again, the same feature in Fig. 7 is noted. The large maximum is noted around 15°N, which corresponds to the Bay of Bengal. By observation, this maximum is noted in June, but is seen in June- September in the simulation. This feature is insensitive to the horizontal resolution. This suggests that our model cannot simulate very well the disturbances over the Bay of Bengal. Small peaks at 30N correspond to the precipitation along the Himalayas. This is better represented in the T106 than the T42. Within this longitude the precipitation in the T42 is larger than that in the T106. Large maximum was noted around 10S from May to September by observation. This feature is captured by the simulations, but the magnitude is weaker. This is common to the characteristics discussed in Fig. 6.

In summary, it is concluded that the precipitation field is better represented in the T106 simulation. In particular, the precipitation along the Himalayas tends to become narrower in the T106 run, but there is no significant difference in other regions. In general, rainfall in the simulations tends to be steady in the Arabian Sea and the Bay of Bengal, but it is considered that the observed rainfall is associated with disturbances. The representation of disturbances or an organization of convection is poor in the simulations. This is considered to be the major issue in the model development. A new parameterization scheme may have to be introduced.

4. Sensitivity of the Estimation of Surface Fluxes over the Asian monsoon Regions to Horizontal Resolution

The land-atmosphere interaction is a key factor for understanding the mechanism of the Asian monsoon variability. In particular, the land surface is characterized by small-scale features of various kinds, such as mountains, forests, deserts and grasslands. Over a mountain-valley region, regional mountain valley winds are dominant and may have some impact on the land-atmosphere interaction. The land-atmosphere interaction is strongly dependent on the land surface conditions. If small scale features of land surfaces are taken in to account, there may be a difference in surface fluxes over a large region. Then, a main question is how to take into account these effects of land-atmosphere interaction over the heterogeneous land surface conditions into account when we apply large-scale models to the Asian monsoon. To examine these effects, we will compare the fluxes over the Asian monsoon region by using ensemble simulations from a T42 and T106 AGCM. Details of the simulations are given in section 2.

In GAME IOP, intensive observations were conducted over the Tibetan region, the Thailand, and the Huaihe-river region in China. Then, these three regions of Tibet (27.5°N - 35°N, 80°E - 100°E), Thailand (15°N - 20°N, 100°E - 105°E), and China (30°N - 35°N, 115°E - 120°E) are chosen for comparison of the surface fluxes.

(a) Tibetan Plateau

In Fig. 8, the monthly averaged rainfall for T42 and T106 (top, left), latent heat flux for T42 and T106 (middle, left), and sensible heat flux for T42 and T106 (bottom, left) are shown for the Tibetan area. In these experiments, three integrations were conducted at T42 and T106 resolutions, respectively. The mean and standard deviation are computed from three samples. The standard deviations are displayed in the figures. Although three samples are not really sufficient to evaluate the standard deviations, it is a useful measure to inspect the diversity of results. Generally, monthly rainfall increases as the season proceeds. In May and June, sensible flux is dominant, while latent heat becomes dominant in July and August. In general, the differences between T42 and T106 are small, but significant differences are noted in May and June.

In May and June, more rain is simulated in the T106 simulation than in the T42 simulation. This is considered to be due the increased contribution of the mountain range. For example, rainfall along the Himalaya in the T106 simulation is more than that in the T42 simulation (refer to Fig. 8). As the rainfall increases, the land surface gets wetter in the T106 simulation than in the T42 simulation. This is the reason why the evaporation is larger with T106 resolution than in the T42 simulation and the sensible heat flux is larger in T42 than in T106. After the onset of the monsoon, there is no difference between the two runs. Rainfall increases and evaporation becomes dominant. This phenomenon is considered to be a large-sale feature and the impact of the difference of horizontal resolution is considered to be small.

(b) Thailand

The rainfall over the Thai region for the T42 and T106 simulations (top, center), the evaporation for T42 and T106 simulations (middle, center) and the sensible heat flux for T42 and T106 simulation (bottom, center) are also shown in Fig. 8. In this region, the variance in the ensemble is large because rainfall intensity is large and is controlled by convective processes. In general, the rainfalls in the T42 run are larger than those in the T106 simulation. The difference in May, July, August and September is beyond the variance of the ensemble. It is considered that the rainfall over the Thai region is strongly influenced by the rainfall over the Bay of Bengal. In other words, the precipitation area intrudes into the Thai region from the Bay of Bengal to the Thai peninsula. As the mountain in Myanmar is well represented in the T106, the rainfall is simulated separately from the rain over the Bay of Bengal. On the contrary, the difference of the evaporation is small, although the evaporation in the T106 run is larger than in the T42 run in May, June and September. It is considered that because the topography in the Thai region is relatively flat, the effect of difference of horizontal resolution is small. In the case of sensible flux, differences are noted in both runs, but as the quantities are small, it is very difficult to draw a definite conclusion.

In the Thai region, surface fluxes were estimated using the eddy correlation method in 1998 during GAME-IOP (Toda *et al.* 2002). The observation is conducted in the complex land region (100°E, 17°N). The observed fluxes are shown in Table 1 In field observations, a large sensible heat flux is noted in June, but rapidly decreases in July, August and September. In contrast to this, the evaporation increases in July, August and September. These changes correspond to the change in land surface conditions associated with the onset of the rainy season. These changes are not well represented in the simulations. Besides, the latent heat flux in the model is larger than the observed flux. These differences may be caused by the

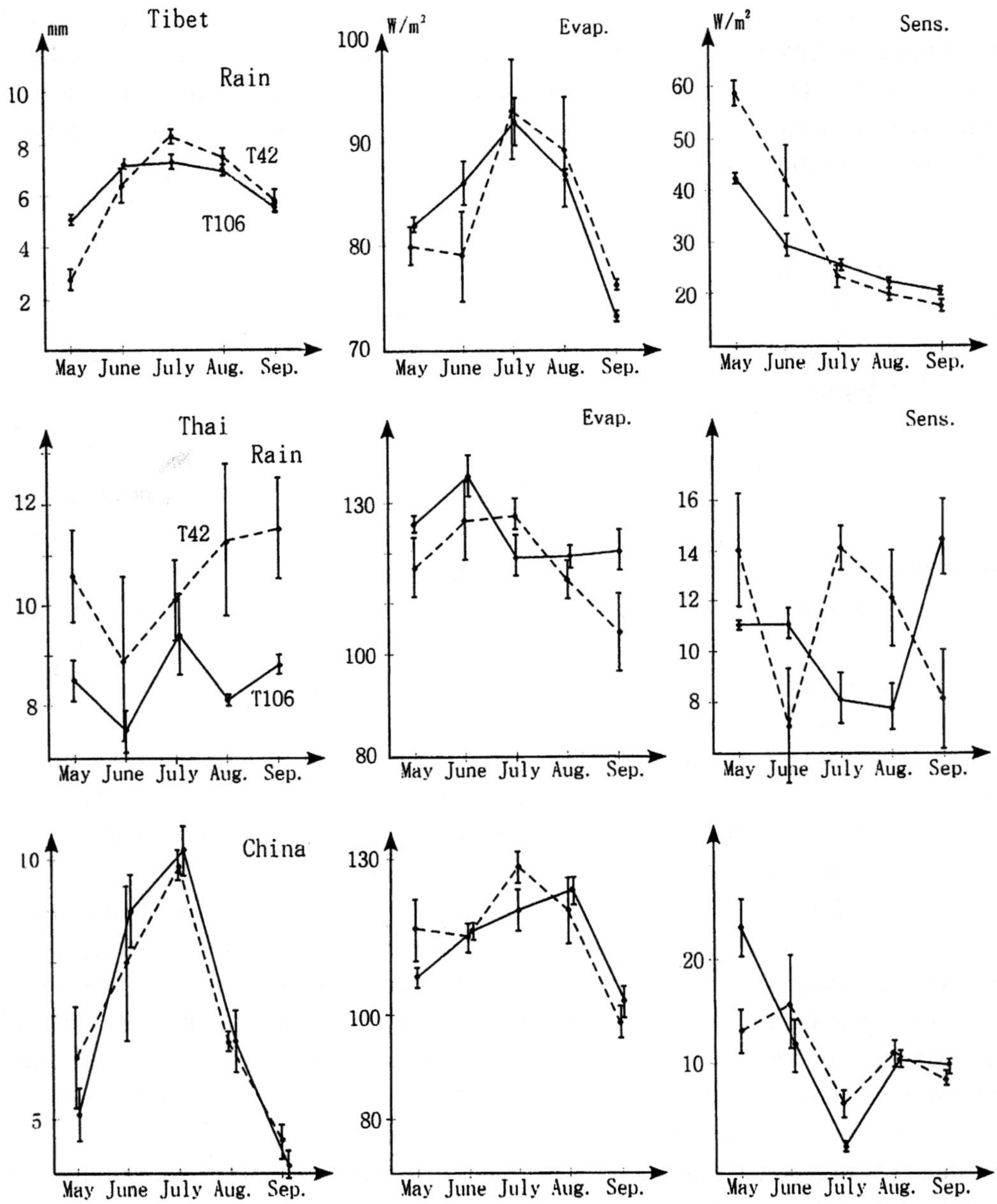

Figure 8. Monthly averaged rainfall(top), latent heat flux(middle) and sensible heat flux(bottom) over the Tibetan Plateau region(left),Thailand region(center) and China region(right). Unit is mm/day for rainfall and Wm-2 for fluxes. Full lines indicate T42 and broken lines, T106.

difference in the character of the fluxes. The flux is observed at a single point and the simulated values are based on the average over the large domain. Another possibility is that a land surface in the model tends to be wet in the pre-monsoon season. This aspect should be further investigated in the future.

(c) China

The same results for the China region are shown in the right of Fig. 8. Rainfall increases in June and July, which corresponds to the onset of the Mei-yu season. It is very difficult to find significant differences between the T42 and T106 simulations. This is because there is little variation in topography and land surface conditions between the two resolutions. If we wish to investigate the heterogeneous aspects of the land surface conditions, we need to use much a finer horizontal resolution.

Table 1. Observed sensible heat flux and latent heat flux at Sukhothai in the Thailand from June to September, 1998.

Month	Sensible Heat(W/m2)	Latent Heat(W/m2)
June,1998	72.3	48.1
July,1998	18.5	76.7
August,1998	19.3	92.7
September,1998	17.7	81.9

In summary, we could not find large differences in the rainfall, evaporation and sensible flux for the large-scale domain between the T42 and T106 simulation, except for the large-scale mountain area in the pre-monsoon period. It is considered that only the difference due to large-scale mountains can be represented with the T106 horizontal resolution.

5. Impacts of Land Surface Process and SST in the Simulation

It is considered that the Asian monsoon is controlled by internal dynamics and external forcing. The land surface condition and SST anomalies are considered to be the major external forcing. Their impact on the Asian monsoon has been investigated at CCSR using the CCSR/NIES AGCM. In this section, we will summarize our results.

An impact of the land surface conditions over the Eurasian continent was discussed by Shen *et al.* (1998). In their study, the effect of the land surface conditions in spring to the Asian summer monsoon was investigated by interchanging the land surface conditions between strong and weak monsoon years. For example, the model has been integrated from March, 1st. until the end of August in a strong monsoon year, but with the weak monsoon land surface initial conditions (snow and soil wetness) over the land area (0° - 60°N, 40°E - 150°E) in Eurasia. The other conditions, including SST distribution, are kept unchanged from the original 10 year run. In their study, nine permutations were available since three strong and three weak monsoon years were chosen.

Fig. 9 shows the height-latitude cross-section of the temperature fields in the ensemble mean between the experiment and the control run (i.e., original 10-year integration), averaged temporally over April and May and between 40°E and 110°E longitudes in these experiments. Fig. 9 (a) depicts the case where the land surface initial condition of strong monsoon years are replaced by that of weak monsoon years, while Fig. 9 (c) is for the opposite case. For comparison, the composite troposphere temperature difference in the strong and weak monsoon is displayed in Fig. 9 (b) and (d), respectively. When the initial snow and soil wetness in strong monsoon years are replaced by those in weak monsoon years as shown in Fig. 9 (a), it is found that the increased snow and soil wetness aid in cooling the atmosphere over the Asian landmass south of 50°N, but this effect is confined only in the middle and lower levels of troposphere. On the other hand, when the initial land surface conditions of weak monsoon years are replaced by those of strong monsoon years, the warming effect is confined mainly in the lower atmosphere. By comparing these results, it is found that the temperature fields are influenced by the change of the land surface conditions but the changes are limited to the lower layer (below 700hPa) over the continent.

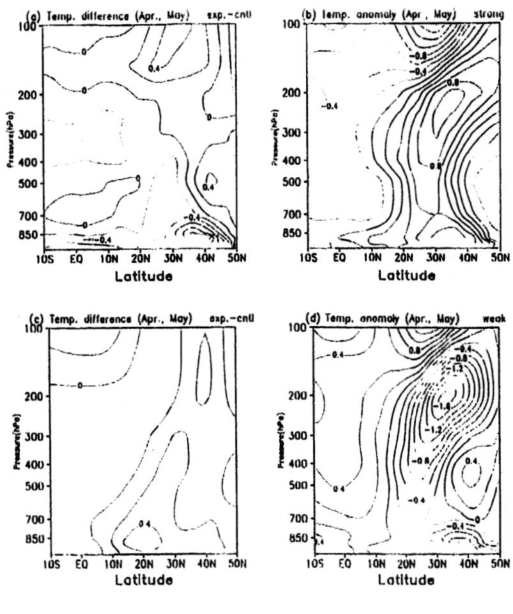

Figure 9. Latitude-height cross section of troposphere temperature changes averaged over April and May and over 40°E-110°E. (a) Temperature difference between ensemble mean of the experiment and original strong monsoon composite. (b) Temperature anomaly in the strong monsoon composite. (c) Temperature difference between ensemble mean of the experiment and original weal monsoon composite. (d) Temperature anomaly in the weak monsoon composite. The negative values are plotted with dash lines. The unit is ℃.

In Fig. 10, the results are summarized based on the Webster and Yang's monsoon index. Two solid curves indicate the weak and strong monsoon composites in the original 10 year simulation and two dashed lines define the ensemble mean of the experiments where the land surface conditions are interchanged. In both weak and strong monsoon cases, the impact of the land surface conditions are noted but are insufficient to change the weak monsoon to the strong monsoon and vice versa. This suggests that strong and weak monsoons are mainly controlled by other factors such as SST distribution. In Shen *et al.* (1998), it is concluded that the ENSO related anomalies play a significant role in determining a strong and weak monsoon. However, it should be noted that the monsoon index by Webster and Yang tends to be influenced by the ENSO related anomalies as described in section 3.

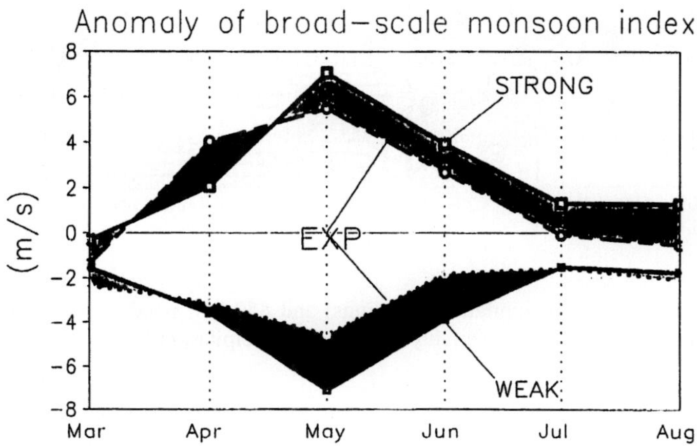

Figure 10. Anomalies of the broad-scale monsoon index for the composite weak and strong monsoons (solid lines) and the ensemble results (dashed lines) of the experiment by changing the land surface initial conditions. The unit of monsoon index is m/s.

Shen *et al.* (2001) investigated which area of the SST anomaly could exert an influence on the Asian monsoon of 1988. In summer 1998, cloudy and rainy weather was dominant over most of the East Asia. A record-breaking flood occurred over the Yangtze River valleys in China and lasted for almost three months. Heavy rainfall occurred in northern and eastern Japan in July and August, and hit the Korean peninsula in August. In Fig. 11, the anomalies of wind field at 850 hPa and precipitation are shown. Precipitation anomalies along Japan, in Korea, in the middle of China and south of Indonesia were noted. A strong easterly anomaly was also significant in the western Pacific region and the South China Sea. This anticyclonic anomaly in the 850 hPa wind field was pointed out by Zhang *et al.* (1995) to be related to precipitation anomalies in the southern part of China.

In summer 1998 the strong 1997/98 El Nino was decaying and a La Nina was developing. In Fig. 12, the SST anomalies of summer 1998 (June, July and August) are shown. Positive SST anomalies are found to the west of 160E in the western Pacific and a negative SST anomaly is noted in the central and eastern equatorial Pacific (160°E-120°W). These SST anomalies correspond to the growing La-Nina situation. Interestingly, the positive SST anomalies, especially south of the Indonesia, are noted in the Indian Ocean, as noted by Webster *et al.* (1999) and Saji *et al.* (1999). As these SST anomalies are considered to influence the atmospheric circulation over the monsoon region, they were investigated using the CCSR/NIES AGCM (Shen *et al.* 2001).

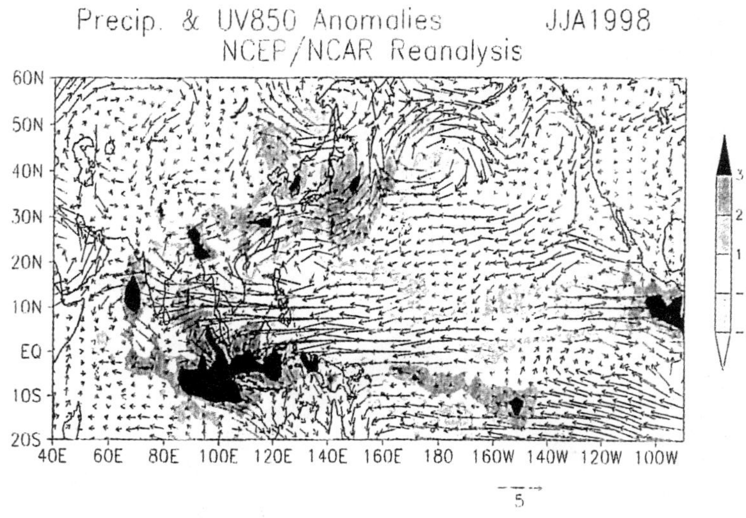

Figure 11. JJA-mean anomalies in precipitation (shadings) and 850 hPa winds (vectors) by the CMAP and the NCEP/NCAR re-analysis data sets, respectively. Units of precipitation and winds are mm/day and m/s, respectively.

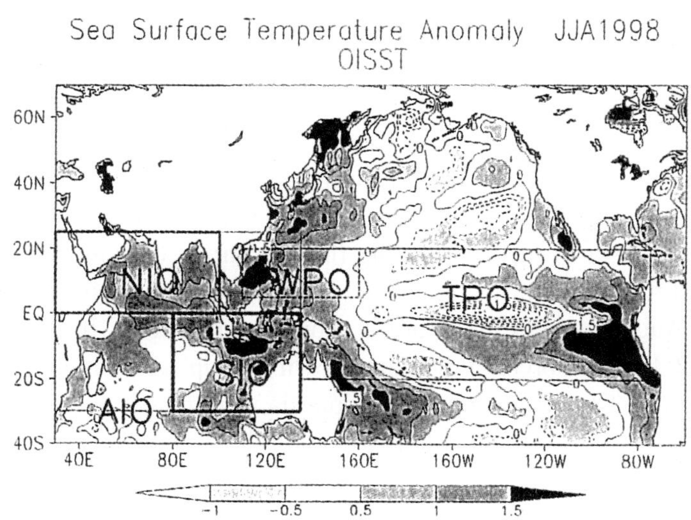

Figure 12. JJA-mean sea surface temperature anomaly in 1998 based on the NCEP 1-degree climatology. Contour interval is 0.5℃. The rectangles indicate the areas where regional sea surface temperature anomalies are utilized for the AGCM experiments.

To investigate the impacts of each SSTA, ensemble simulations for the summer 1998 monsoon were conducted, specifying different SST anomalies. T42 AGCM was used in these experiments. In these experiments, 10 samples are used, i.e., 10 integrations from April 21 to April 30 until the end of August 1998 were carried out using the observed SST.

Fig. 13 shows the simulated anomalies of precipitation and 850 hPa wind fields in the run with boundary conditions set by global SST anomalies (observed 1998 SST). Compared with Fig. 11, the major features of anti-cyclone and enhanced precipitation in the China and the Indian Ocean are depicted well in the simulations, although a heavy precipitation anomaly along the Yangtze River valley is not represented.

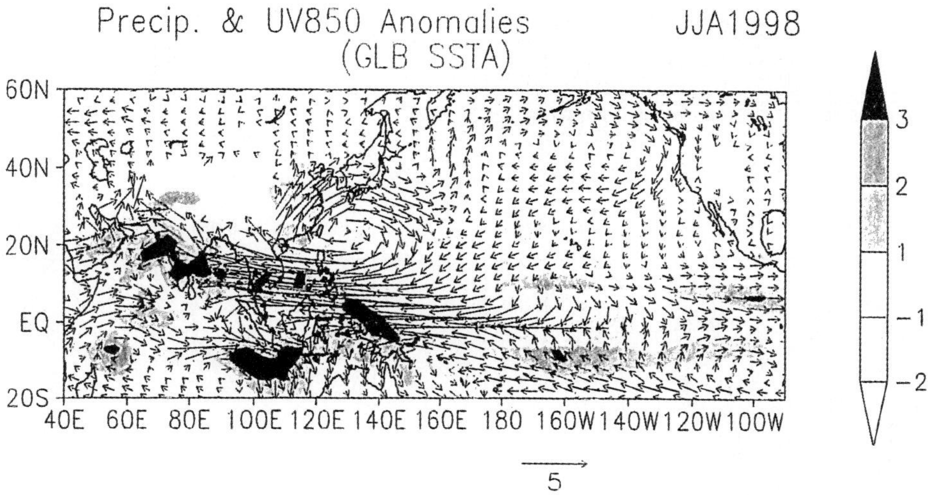

Figure 13. JJA-mean anomalies of precipitation (shadings) and 850 hPa winds (vectors) simulated by the AGCM ensemble integration using global SSTA (GLB). Unit of precipitation and winds are mm/day and m/s, respectively.

In order to investigate the impacts of regional SST anomalies, five regions were selected, i.e., (1) tropical and subtropical Indian Ocean (30°E - 135°E, 30°S - 25°N), which is referred to as AIO), (2) southeastern part of the Indian Ocean (80°E - 135°E, 30°S - EQ), which is referred to as SIO), (3) tropical Pacific Ocean (180°E - 75°W, 20°S - 20°N; TPO), (4) north Indian Ocean (25°N - EQ, 30°E - 100°E), and (5) western Pacific ocean (5°N - 20°N, 110°E - 160°E). These regions are presented in Fig. 12. In each model run, only the SST anomaly in the region was specified and SST distributions in the other regions were set to their climatological values.

The resultant of precipitation and 850 hPa wind anomalies are shown in Fig. 14. Compared with observation (Fig. 11), it is found that the low-level anti-cyclonic circulation anomaly and suppressed precipitation over the subtropical western Pacific are reproduced to an extent by these two regional SST anomaly cases (AIO and TPO), but WPO simulation cannot reproduce these anomalies. This suggests that the anomalies in the 1998 East Asian monsoon are not locally induced but remotely induced by the SST anomalies in the Indian Ocean and the tropical central and eastern Pacific Ocean. By comparing the NIO with the SIO, it can be determined whether the northern or the southern part of the Indian Ocean SST contributes to the anomalies over the East Asian monsoon region. Both NIO and SIO contribute to the anti-cyclonic circulation anomalies over the western Pacific region but only

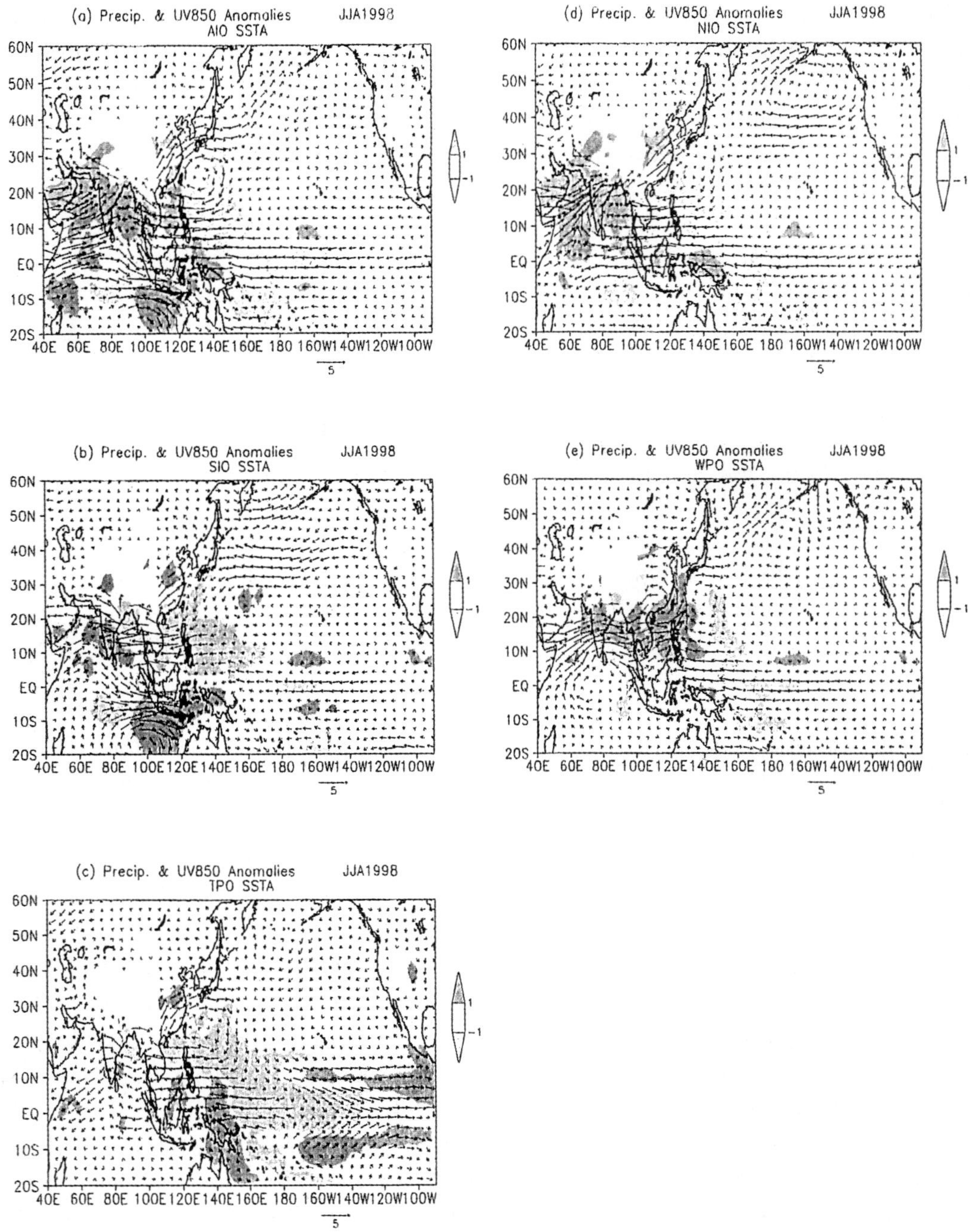

Figure 14. Same as Fig. 13, except for the regional SSTA. (a) is the result with the SSTA in the whole Indian Ocean (AIO), (b) is with the SSTA in the southern Indian Ocean (SIO), (c) is with the SSTA in the tropical Pacific Ocean, (d) is with the SSTA in the northern Indian Ocean (NIO) and (e) is with the SSTA in the western Pacific Ocean (WPO). Unit of precipitation and winds are mm/day and m/s, respectively.

SIO contributes an easterly anomaly and a precipitation anomaly south of the Indonesia. It should be noted that the strength of these anomalies is weaker than that observed in the run using global SST anomalies (sees Fig. 11). This is due to the non-linear effects of these SST anomalies.

In summary, the southeastern Indian Ocean and the equatorial eastern Pacific Ocean were most influential in forming the subsidence anomaly and associated anti-cyclonic anomaly at 850 hPa over the tropical western Pacific region. This is considered to be due to the weaker Hadley circulation by the Indian SST anomalies and weaker Walker circulation by the tropical Pacific SST. This persistent low-level anti-cyclonic anomaly is responsible for the 1998 East Asian monsoon.

6. Summary

The Asian monsoon has been intensively investigated by simulation using the CCSR/NIES AGCM. Main purpose of our study is to investigate the sensitivity of horizontal resolution and boundary conditions to the Asian monsoon simulation.

Initially, an AMIP-type simulation, i.e., the 10-year simulation of the Asian monsoon with the observed SSTs, was conducted and the year-to-year variation was investigated (Shen *et al.* 1998). In general, the large-scale features of the Asian monsoon are well simulated by the CCSR/NIES AGCM (T21L20), and the inter-annual variability of the Asian monsoon on a large-scale was well represented. However, it is considered that features exist which cannot be represented by the coarse resolution. Similarly, the AMIP-2 computation was conducted using T42L20 AGCM and 1998 Asian monsoon was simulated using T106 AGCM.

The sensitivity of the Asian monsoon simulation to horizontal resolution has been investigated. In general, performance at higher resolution (T106) is better than that at lower resolution (T42). Synoptic disturbances are better represented at the high-resolution (T106). In particular, rainfall along the Himalayas is well represented by T106, but small differences in the large-scale features of the monthly mean parameters is noted between the coarse resolution (T42) and the high-resolution (T106). A difference in the inter-annual variations of the Webster-Yang Monsoon Index is noted between T21 and T42 in the AMIP run. This suggests that the circulation in the tropics is not fully determined by the SST-distributions.

The sensitivity of fluxes between atmosphere and land was investigated. The fluxes are also compared with observations conducted in the GAME-IOP. With respect to the monthly mean fluxes in the 5-degree by 5-degree region there are little difference except in region of the Himalayas in the pre-monsoon season (May and June). However, this does not suggest that small-scale features are not important in the land-atmosphere interaction, but because the spatial resolution of the model in our experiments is insufficient. In future, we will use a higher resolution model for monsoon studies and study the role of the small-scale land-atmosphere interaction on the Asian monsoon activities.

The mechanism of the monsoon variability and the effects of the SSTA, together with the snowcover over the Eurasian continent have been investigated. In general, it is concluded that the snow over the Eurasian continent in the spring has an impact on the following Asian monsoon activities but not so strong as to change the intensity. It is concluded that the SSTAs surrounding the Eurasian continent have an impact on the Asian monsoon activity. Especially, the 1998 East Asian monsoon was strongly influenced by the SSTA in the Indian Ocean and the tropical Pacific Ocean. It is considered that they induced the anti-cyclonic anomaly at 850 hPa through the local Hadley and the Walker circulations. This anti-cyclonic anomaly correlates with the precipitation anomaly in the southern part of China.

References

Ailikun, B. and T. Yasunari, 1998: On the two indices of Asian summer monsoon variability and their implications. *Proceedings of the International Conference on Monsoon and Hydrologic Cycle.* Kyongju, Korea, 222-226.

Chang, C. P., and T. N. Krishnamurti, 1987: *Monsoon Meteorology*, Oxford University Press, 544pp.

Gates, W. L., 1992: AMIP, the Atmospheric Model Intercomparison Project. *Bull. Amer. Meteor. Soc.*, **73,** 1962-1970.

Goswami, B. N., V. Krishnamurthy, and H. Annamalai, 1999: A broad-scale circulation index for the interannual variability of the Indian summer monsoon. *Quart. J. Roy. Meteor. Soc.*, **125**, 611-633.

Hahn, D. J., and J. Shukla, 1976: An apparent relationship between Eurasian snowcover and Indian monsoon rainfall, *J. Atmos. Sci.*, **33**, 2461-2462.

Kawatani, Y., and M. Takahashi, 2003: Simulation of the Baiu Front in a high resolution AGCM. *J. Meteor. Soc. Japan*, **81**, 113-126.

Kimoto, M., X. Shen, A. Sumi, A. Numaguti, and J. Matsumoto, 1999: Simulation of the 1998 East Asian summer monsoon by the CCSR/NIES AGCM. *J. Meteor. Soc. Japan*, **79**, 741-757.

Krishnamurti, K. N., and D. Subrahmanyan, 1982: The 30 to 50 day mode at 850mb during MONEX. *J. Atmos. Sci.,* **39,** 2288-2095.

Le Treut, H., and Z.-X. Li, 1991: Sensitivity of an atmospheric general circulation model to prescribed SST changes: feedback effects associated with simulation of cloud optical properties. *Climate Dyn.*, **5**, 175-187.

Louis, J., 1979: A parametric model of vertical eddy fluxes in the atmosphere. *Bound. Layer Meteor.*, **17**, 187-202.

Manabe, S., J. Smagorinsky, and R. J. Strickler, 1965: Simulated climatology of a general circulation model with a hydrologic cycle. *Mon. Wea. Rev.*, **93**, 769-798.

Matthews, E., 1983: Global vegetation and land use: New high-resolution data bases for climate studies. *J. Climate. Appl. Meteor.*, **22**, 474-487.

McFarlane, N. A., 1987: The effect of orographically excited wave drag on the general circulation of the lower stratosphere and troposphere, *J. Atmos. Sci.*, **44**, 1775-1800.

Meehl, G., 1987: The annual cycle and interannual variability in the tropical Pacific and Indian Ocean region. *Mon. Wea. Rev.*, **115**, 27-50.

Meller, G. L., and T. Yamada, 1982: Development of a turbulence closure model for geophysical fluid problems, Rev. *Geophys. Atmos. Phys.*, **20**, 851-875.

Miyakoda, K., J. Kinter, and S. Yang, 2000: Analysis of the connection from the South Asian monsoon to ENSO by using precipitation and circulation indices. *COLA Technical Report 90*, 72pp.

Moorthi, S., and M. J. Suarez, 1992: Relaxed Arakawa-Schburt: A parameterization of moist convection for general circulation models. *Mon. Wea. Rev.*, **120**, 978-1002.

Murakami, T., T. Nakazawa, and J. He, 1984: On the 40-50 day oscillations during the 1979 northern hemisphere summer, Part I: Phase propagation. *J. Meteor. Soc. Japan*, **62**, 440-468.

______, and A. Sumi, 1982a: Southern Hemisphere summer monsoon circulation during the 1978-79 WMONEX, Part I. Monthly mean wind fields. *J. Meteor. Soc. Japan*, **60**, 638-648.

______, and ______, 1982b: Southern Hemisphere summer monsoon circulation during the 1978-79 WMONEX Part. II Onset, active and break monsoon. *J. Meteor. Soc. Japan*, **60**, 649-671.

Nakajima, T., and M. Tanaka, 1986: Matrix formulation for the transfer of solar radiation in a plane parallel scattering atmosphere. *J. Quart. Spectrosc. Radaiation Transfer*, **35**, 13-21.

Numaguti, A., S. Sugata, M. Takahashi, T. Nakajima, and A. Sumi, 1997: Study on the climate system and mass transport by a climate model. *CGER's Supercomputer Monograph Report*, **Vol. 3**, 91pp. (available at CGER/NIES)

______, M. Takahashi, T. Nakajima, and A. Sumi, 1997: Description of CCSR/NIES atmospheric general circulation model. *CGER's Supercomputer Monograph Report*, **Vol. 3**, 91pp. (available at CGER/NIES)

Saji, N. H., B. N. Goswami, P. N. Vinayachandran, and T. Yamagata, 1999: A dipole mode in the tropical Indian Ocean. *Nature*, **401**, 360-363.

Shen, X., M. Kimoto, and A. Sumi, 1998: Role of land surface processes associated with interannual variability of broad-scale Asian summer monsoon as simulated by the CCSR/NIES AGCM. *J. Meteor. Soc. Japan*, **76**, 217-236.

______, and ______, 1999: Influence of El Nino on the 1997 Indian summer monsoon. *J. Meteor. Soc. Japan*, **77**, 1023-1037.

______, ______, A. Sumi, A. Numaguti, and J. Matsumoto, 2001: Simulation of the 1998 east Asian summer monsoon by the CCSR/NIES AGCM. *J. Meteor. Soc. Japan*, **79**, 741-757.

Sumi, A., 2000: Climate Simulation Studies. *CCSR, General Circulation Model Development Past, Present and Future*, D. A. Randall, ed., Academic Press, 489-508.

Takahashi, M., 1996: Simulation of the stratospheric quasi-biennial oscillation using a general circulation model. *Geophys. Res. Lett.*, **23**, 661-664.

______, 1999: Simulation of the quasi-biennial oscillation in a general circulation model. *Geophys. Res. Lett.*, **26**, 1307-1310.

Toda, M., K. Nishida, N. Ohte, M. Tani, and K. Mushiake, 2002: Observation of energy fluxes and evapotranspiration over terrestrial complex land covers in the tropical monsoon environment. *J. Met. Soc. Japan*, **80**, 465-484.

Webster, P. J., A. M. Moore, J. P. Loschnigg, and R. R. Leben, 1999: Coupled ocean-atmosphere dynamics in the Indian Ocean during 1997-98, *Nature*, **401**, 356-360.

______, and S. Yang, 1992: Monsoon and ENSO: Selectively interactive systems. *Quart. J. Roy. Meteor. Soc.*, **118**, 877-926.

Yasunari, T., 1987a: Global structure of the El-Nino/Southern Oscillation. Part I. El Nino composites. *J. Meteor. Soc. Japan*, **65**, 67-80.

______, 1987b: Global structure of the El-Nino/Southern Oscillation. Part II. Time evolution. *J. Meteor. Soc. Japan*, **65**, 81-102.

Zhang, R., A. Sumi, and M. Kimoto, 1995: Impact of El-Nino on the East Asian monsoons: A diagnostic study of the 86/87 and 91/92 events. *J. Meteor. Soc. Japan*, **74**, 49-62.

Part IV

Synoptic and Mesoscale Processes

10. RESEARCH ON THE PHENOMENA OF MEIYU DURING THE PAST QUARTER CENTURY: AN OVERVIEW

GEORGE TAI-JEN CHEN

Department of Atmospheric Sciences
National Taiwan University
1, Sec. 4, Roosevelt Rd.,
Taipei City 106, Taiwan
E-mail: george@george2.as.ntu.edu.tw

Meiyu is a unique feature of East Asia and is one of the most important weather phenomena in China and vicinity. Numerous studies have been focused in the last quarter century on various aspects of the Meiyu system. The main purpose of this paper is to present an overview of the current understanding of the climatological characteristics of the Meiyu systems, as well as the structure and dynamics of the related features such as the Meiyu fronts, monsoon disturbances, low-level jets, mesoscale convective systems (MCSs), and orographically related disturbances during the Meiyu season.

1. Introduction

Over East Asia, seasonal rainfall over coastal China, Japan and Taiwan reaches a maximum during the transition from the northeast monsoon in winter to the southwest monsoon in summer. The rainfall during this period is called “Meiyu” (plum rain) in Chinese and “Baiu” in Japanese. The occurrence and variation of this distinct climatological system has important impacts to agriculture, water resources, and human activities.

The Meiyu (Baiu) rainy season starts concurrently with the East Asian summer monsoon onset in the South China Sea, which tends to occur before the South Asian monsoon development over the Bay of Bengal during late May and over the western coast of the Indian subcontinent in early June (Tao and Chen 1987, Fig. 1). The Meiyu season occurs in the period of mid-May to mid-June over South China and Taiwan and in the period of mid-June to mid-July over the Yangtze River Valley. Synoptically, the rainfall in Meiyu season is associated with the repeated occurrence of a front which develops in the deformation wind field between a migratory high to the north and the subtropical Pacific high to the south (G. Chen 1977b, 1983). Some investigators have considered the arrival of the Meiyu front in the South China Sea as a possible mechanism that contributes to the triggering of the monsoon onset (Chang and Chen 1995). Although the individual Meiyu front often moves slowly southeastward after its formation, the mean position of the front undergoes sequential northward shift between May and July depending upon the intensity and the position of the large-scale circulation in the subtropical latitudes (G. Chen 1988; Tao and Chen 1987). Figure 2 presents the annual mean frequency distribution of 850 hPa fronts in the Meiyu

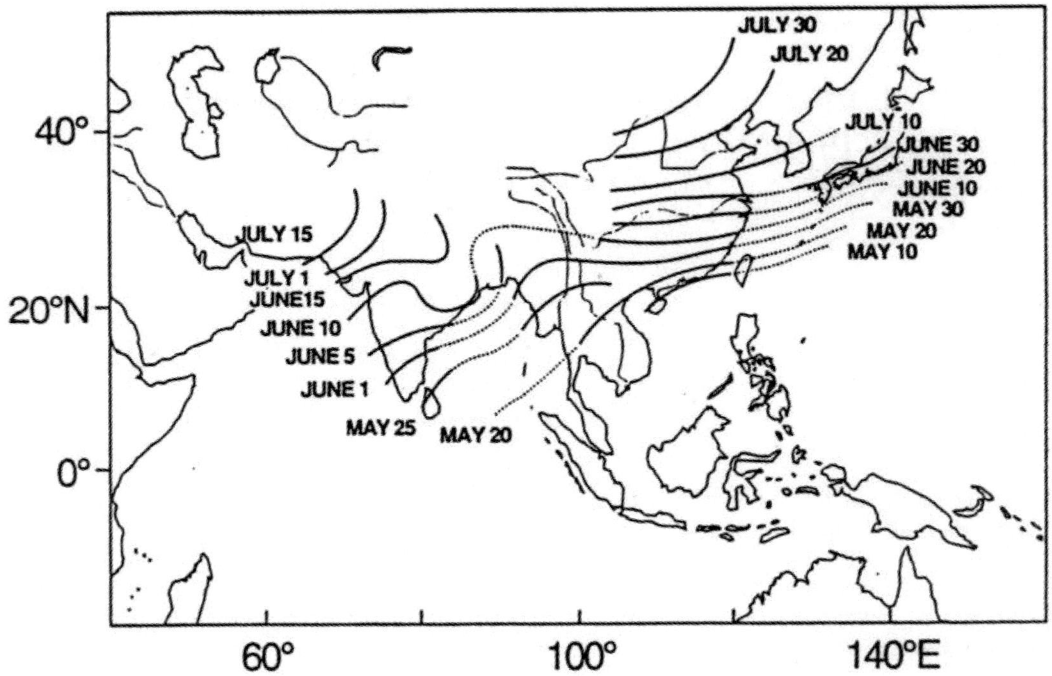

Figure 1. Mean onset date of the summer monsoon (from Tao and Chen 1987).

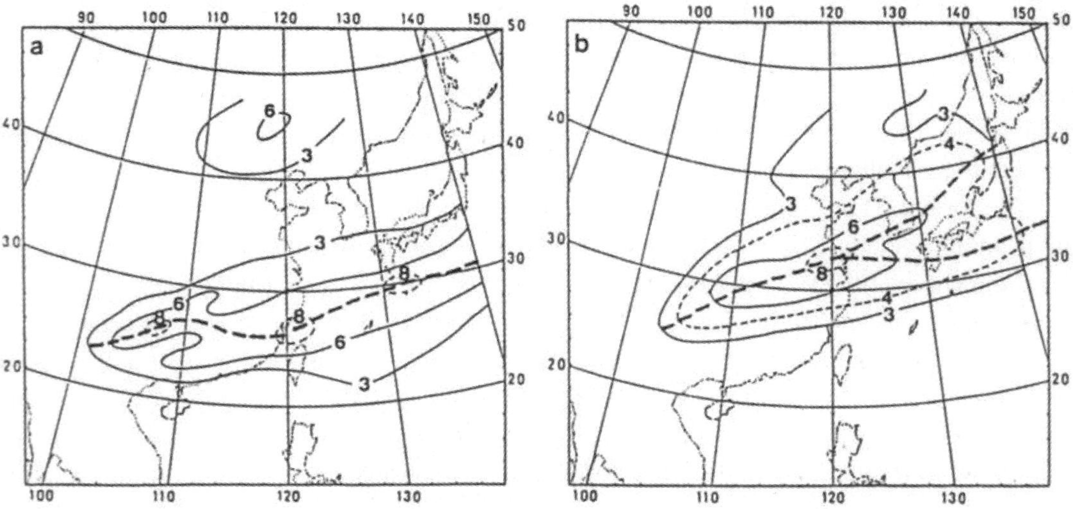

Figure 2. Annual mean (1975–1986) frequency distribution of 850 hPa fronts in (a) southern China and Taiwan Mei-Yu season (15 May–15 June) and (b) Yangtze River Valley Mei-Yu season (16 June–15 July). Front frequency is counted at 12 h intervals and analyzed at 1° lat ×1° long grid intervals. Heavy dashed line indicates maximum axis (from G. Chen 1988).

season of South China and Taiwan (mid-May to mid-June) and of the Yangtze River Valley (mid-June to mid-July) (G. Chen 1988). For the former case, the axis of maximum frequency, indicating the mean position of the Meiyu front, is oriented approximately in an east-west direction extending from southern Japan to southern China. The mean position shifts northward to Japan and central China in the Meiyu season of the Yangtze River Valley. The Meiyu front often moves southeastward slowly in the early stage of its lifetime and appears as a quasi-stationary front in the late stage with an average lifetime of 8 days.

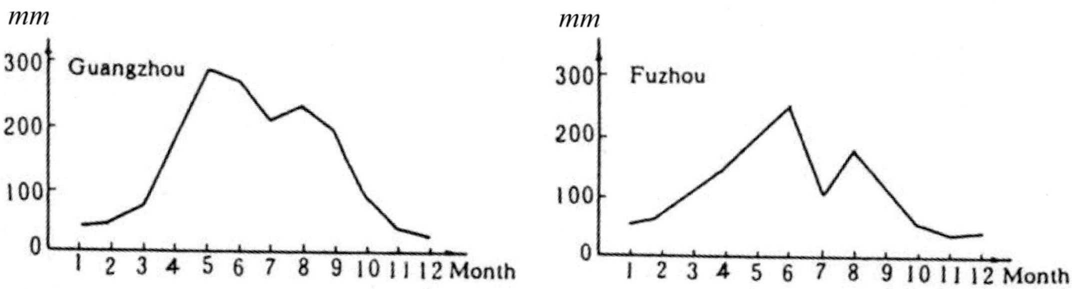

Figure 3. Long-term mean (1960–1979) annual range of rainfall amount at Guangzhou (23°N, 113°E) and Fuzhou (26°N, 119°E) in southern China (from Huang 1986).

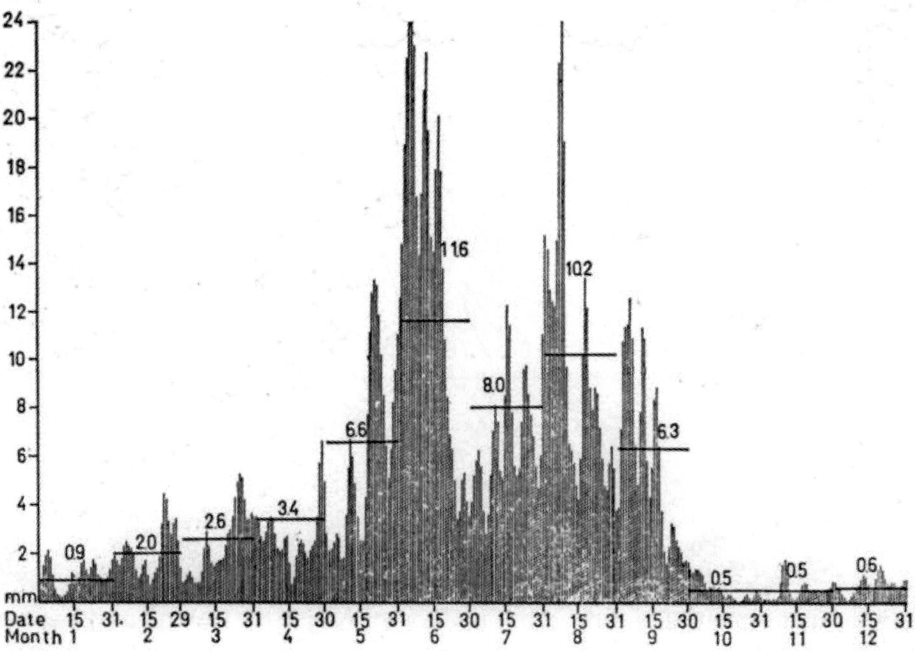

Figure 4. Long-term mean (1951–1970) daily rainfall amount (mm d^{-1}) from 1 Jan. to 31 Dec. at Taichung (24.1°N, 120.7°E) in central Taiwan. Mean daily rainfall amount in each month is indicated by the horizontal line (from G. Chen and Wu 1978).

The role of the Meiyu in seasonal rainfall distribution was studied in Taiwan (G. Chen and Wu 1978; S. Wang 1970) and in South China (Huang 1986). For example, the seasonal rainfall distribution at Guangzhou and Fuzhou in South China as presented in Fig. 3 and at Taichung in central Taiwan as presented in Fig. 4 clearly illustrates the primary rainfall peak during the Meiyu season of May-June and the secondary peak during the typhoon season of July-September. The time-latitude section of 10-day mean rainfall over eastern China analyzed by Lau *et al.* (1988) is presented in Fig. 5. The monsoon onset between 25°N and 30°N is indicated by the steep rise in precipitation starting from the first 10-day period of May. The most noteworthy feature is the northward propagation of Meiyu from southern to central China during the period of May-July. Outside the Meiyu season, strong subseasonal scale fluctuations with period of approximately 20 days appears to dominate the rainfall variation between 25°N and 40°N from the latter part of July to the end of September. This is an interesting feature, but will not be reviewed in this paper.

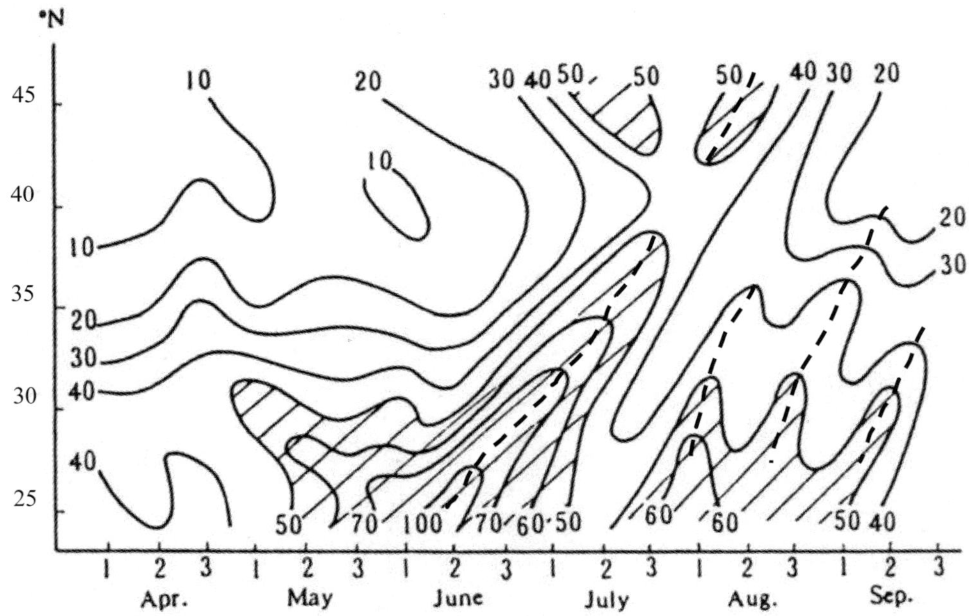

Figure 5. Time-latitude section of 10-day mean rainfall over eastern China (100-115°E). Isohyets analyzed at 10 mm (10 days)$^{-1}$ intervals. Regions of heavy rain (>50 mm) are shaded. The heavy dashed lines indicate northward propagation of the rainbands (from Lau *et al.* 1988).

To further understand the northward migration of Meiyu as well as the development of the southwest monsoon flow as the season proceeds, time-latitude sections of wind and vorticity at 850 hPa along eastern China coastal area are computed using the National Center for Environmental Predictions - National Center for Atmospheric Research (NCEP-NCAR) global atmospheric reanalysis dataset covering a 29-yr period from May to July 1968-1996 and are presented in Fig. 6. The Meiyu front remains quasi-stationary to the south of 30°N in southern China during the period of mid-May to mid-June and moves northward to the north of 30°N during the Meiyu season of the Yangtze River Valley in the period of mid-June to mid-July. To the south of Meiyu front, the southwest monsoonal flows increase gradually from mid-May and reach the peak values in late June to early July when Meiyu in the Yangtze River Valley reaches mature stage.

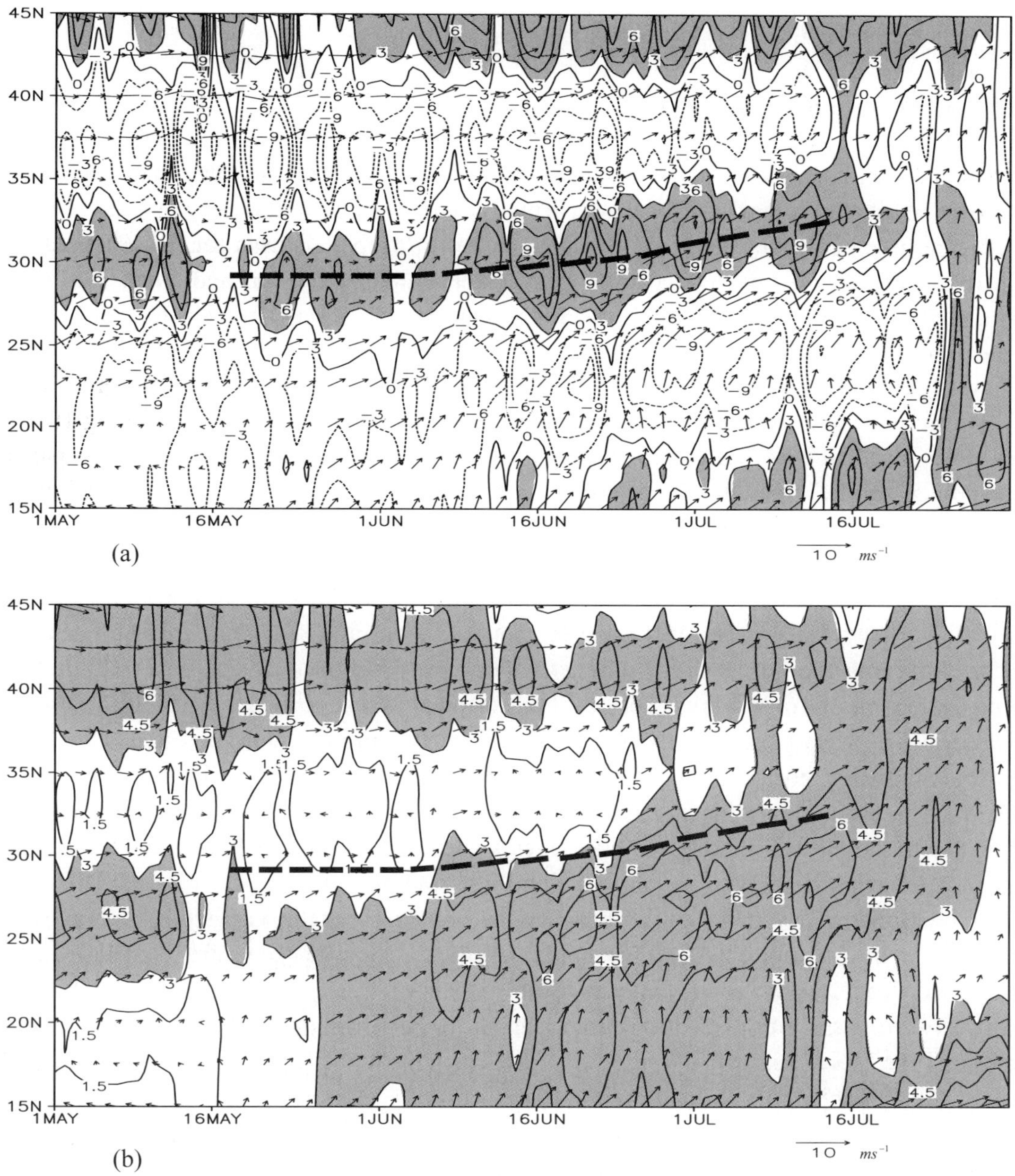

Figure 6. Time-latitude section of wind vectors and (a) vorticity (10^{-6} s^{-1}) and (b) isotach (m s^{-1}) at 850 hPa along 117.5-120°E from 1 May to 31 July in 1968–1996. Heavy dashed line indicates the frontal position as defined by the axis of vorticity maximum.

Because of its importance in East Asia, in the last quarter century numerous studies have been carried out to focus on various aspects of the Meiyu problem. The main purpose of this paper is to present an overview of the current understanding of the climatological characteristics of the Meiyu as well as the structure and dynamics of the related features such as the Meiyu fronts, monsoon disturbances, low-level jets, mesoscale convective systems (MCSs), and orographically related disturbances during the Meiyu season. For completeness, some of the earlier research as reviewed by G. Chen (1992) and Ding (1992) will also be included in this paper.

2. Climatological Characteristics

To avoid the confusion, it is worthwhile to point out that different terminology has been used for the same meteorological phenomenon, which occurs over different geographical regions in different time periods of the year. In China, the term "Meiyu" is used for the rainy season from mid-June to mid-July over the Yangtze River Valley (*e.g.*, Tao and Chen 1987). In Japan, the term "Baiu" is used both for the rainy season over Okinawa region from early May to mid-June and over the Japanese Main Islands from mid-June to mid-July (*e.g.*, Saito 1985). In Taiwan, on the other hand, the term "Meiyu" is used both for the rainy season over Taiwan and over South China from mid-May to mid-June (*e.g.*, G. Chen 1983, 1988; S. Wang 1970). Therefore, the "Meiyu" season over South China and Taiwan discussed in this paper corresponds to the "South China pre-summer rainy period" used by many Chinese meteorologists (*e.g.*, Ding 1992; Tao and Chen 1987), and the "pre-Meiyu" period used by Chang *et al.* (2000 a, b).

Figure 7 presents the distribution of 15-day mean stream function at 850 hPa in May-June (G. Chen 1994). The Meiyu over South China and Taiwan usually occurred concurrently with the onset of the summer southwest monsoon over the South China Sea during the period of 16-31 May (Phase I). The northwest advance of the subtropical ridge at the Phase III of Meiyu season (16-30 June) occurred concurrently with the establishment of the quasi-stationary position of Meiyu front over the Yangtze River Valley and Japan. To reveal the characteristics of large-scale circulations and the associated atmospheric conditions during the Meiyu season of the Yangtze River Valley and of South China and Taiwan, the same data set as used in Fig. 6 are also used in Figs. 8-11 and Fig. 13. The mean circulations during the Meiyu season over South China and Taiwan and over the Yangtze River Valley at 1000 and 850 hPa are presented in Figs. 8 and 9, respectively. The Meiyu front, as defined by the axis of vorticity maximum, extends from the ocean to the east of the Japan Islands southwestward passing through the Bashi Channel to the northern South China Sea at 1000 hPa during the Meiyu season of South China and Taiwan. At 850 hPa, it is located well to the north of the 1000 hPa position particularly to the west of 130°E, suggesting baroclinic nature of the system. During the Meiyu season of the Yangtze River Valley, the Meiyu front tends to be orientated approximately in an east-west direction with a smaller northward shift to 850 hPa from its 1000 hPa position particularly to the west of 120°E, indicating a semitropical disturbance with equivalent barotropic nature of the system as suggested by G. Chen and

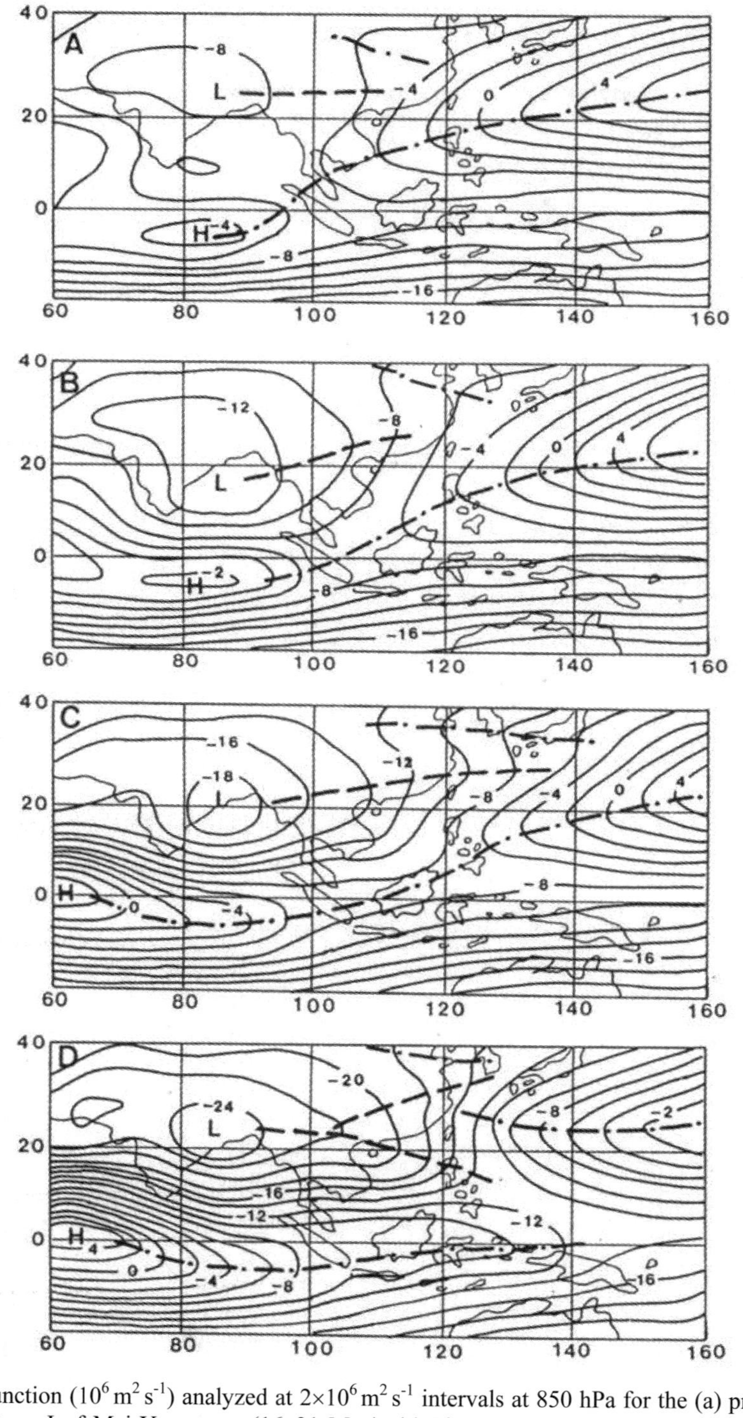

Figure 7. Stream function ($10^6\,m^2\,s^{-1}$) analyzed at $2{\times}10^6\,m^2\,s^{-1}$ intervals at 850 hPa for the (a) pre-Mei-Yu season (1–15 May), (b) Phase I of Mei-Yu season (16–31 May), (c) Phase II of Mei-Yu season (1–15 June), and (d) Phase III of Mei-Yu season (16–30 June) in 1981–1986. Dashed and dash-dotted lines indicate troughs and ridges, respectively. Cyclonic and anticyclonic centers are indicated by L's and H's, respectively (from G. Chen 1994).

Chang (1980). Note that the strong southwest monsoon flows prevail to the south of the Meiyu front at 850 hPa during the Meiyu season over the Yangtze River Valley, South China, and Taiwan. The seasonal rainfall maximum during the Meiyu period is caused by the organized distribution of upward motion and the moisture flux convergence along the Meiyu frontal zone as illustrated in Figs. 10 and 11. Satellite pictures usually revealed a long stratiform cloud band along the Meiyu front with vigorous convection embedded within the band (G. Chen 1978a,b). Figure 12 presents a case of Meiyu frontal cloud band with

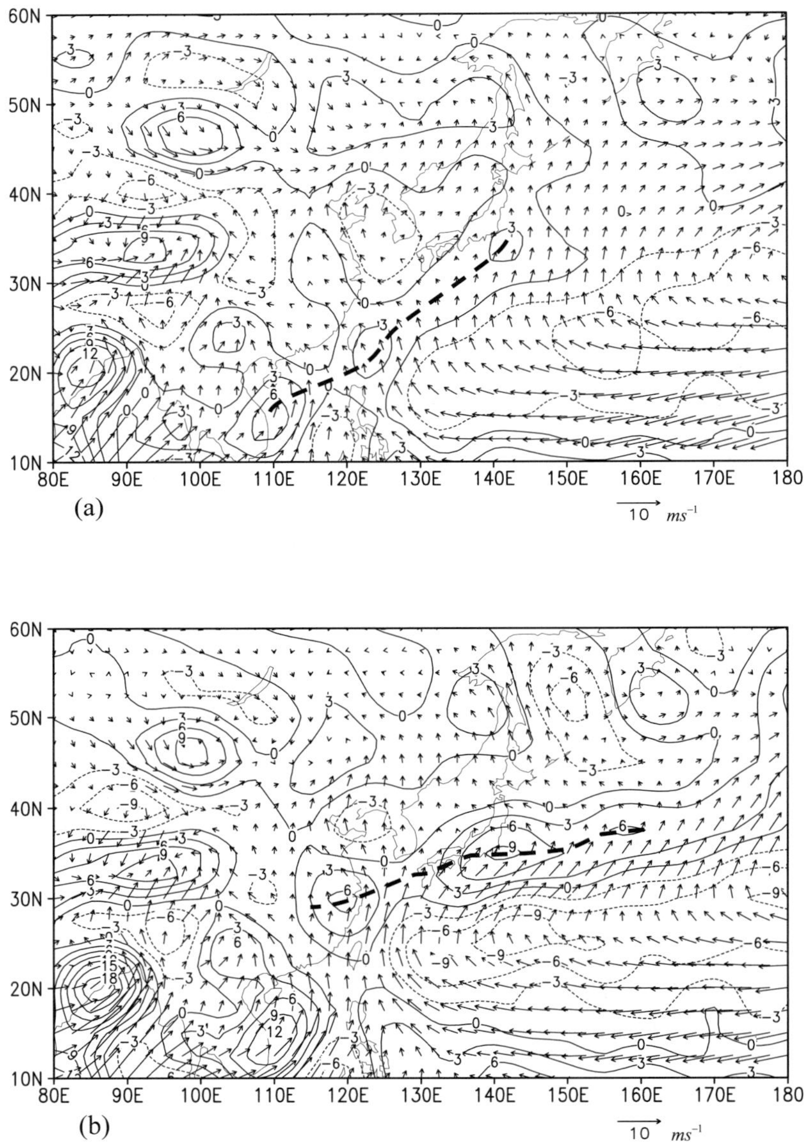

Figure 8. Vorticity (10^{-6} s^{-1}) and wind vectors (m s^{-1}) at 1000 hPa in (a) 16 May–15 June and (b) 16 June–15 July 1968-1996. Vorticity analyzed at 3×10^{-6} s^{-1} intervals with negative values dashed. Heavy dashed line indicates the frontal position as defined by the axis of vorticity maximum.

stratiform and convective clouds on 7 June 2003 as revealed by NOAA GOES9 satellite infrared imagery. The organized MCSs developed along and to the south of the Meiyu front over the southern China coast, the northern South China Sea, Taiwan and vicinity. The MCSs produced more than 300 mm daily rainfall over southern Taiwan on the same day in this case. The rainfall during the Meiyu season might be continuous or intermittent for several days to a few weeks and include frequent rainshowers and thunderstorms (G. Chen 1983, 1990c; Ding 1992; Hsu and Chi 1974). The convective activity develops under the potentially unstable atmosphere along the Meiyu front and to the south of the front as the equivalent potential temperature (θ_e) decreased with height in this environment (Fig. 13).

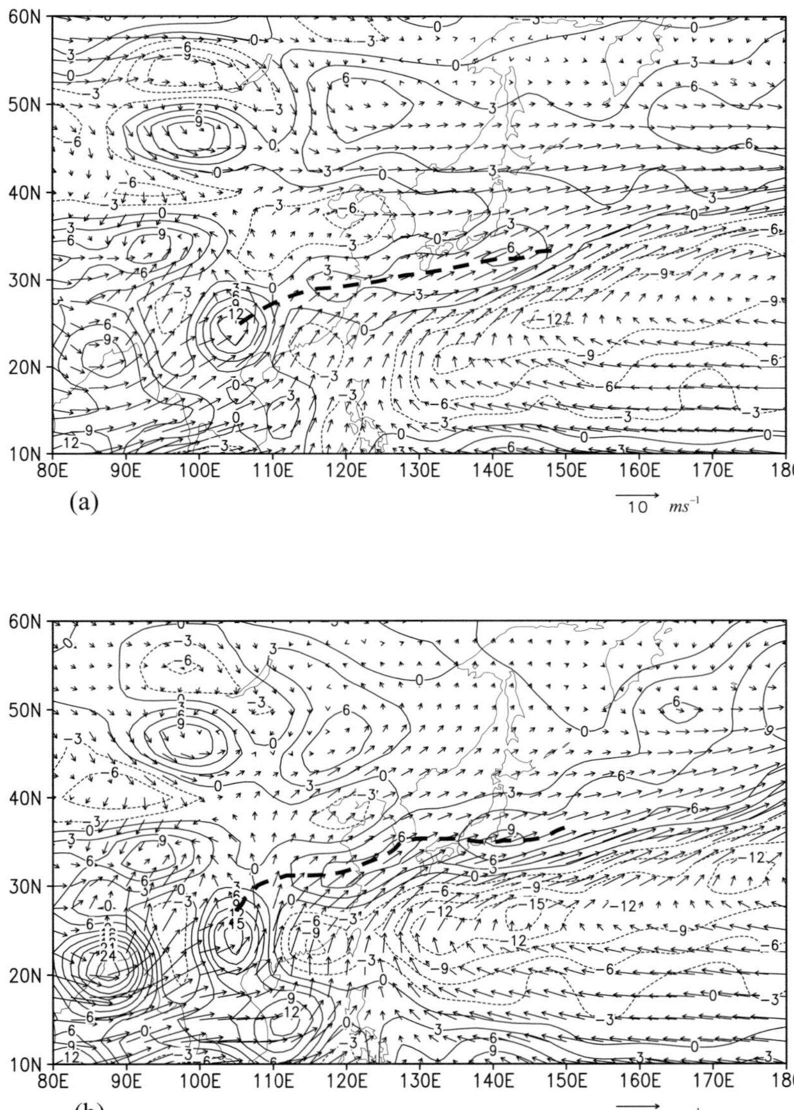

Figure 9. As in Fig. 8 except at 850 hPa. Heavy dashed line indicates the frontal position as defined by the axis of vorticity maximum.

Y. Wang (1992) studied the effects of blocking on the rainfall in the Meiyu season and found that the frequency of blocking anticyclones in the region of 50-70°N, 121-160°E (*i.e.* the Okhotsk Sea) was positively correlated with both the persistence of the Meiyu period and the rainfall amount during the Meiyu season in Central China. A statistical analysis in that study suggested that the existence of cold air in the lower layer over the area just to the north of the Meiyu front, associated with the blocking over the Okhotsk Sea, is an important factor in activating the Meiyu front in Central China and in enhancing the low-level southerly wind

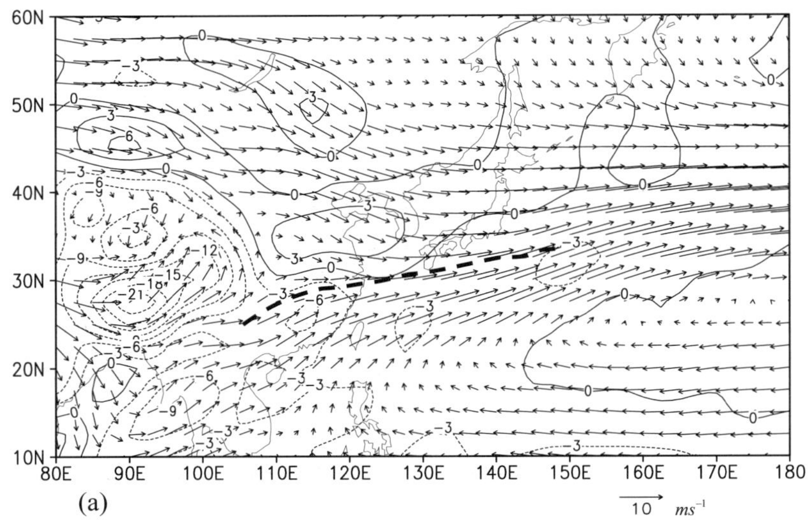

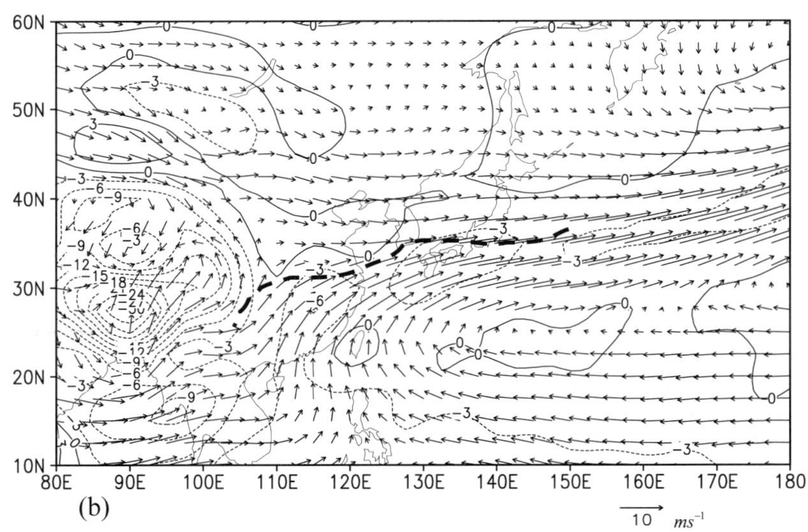

Figure 10. As in Fig. 8 except for vertical p-velocity (10^{-2} Pa s^{-1}) and wind vectors at 700 hPa. P-velocity analyzed at 3×10^{-2} Pa s^{-1} intervals with negative values dashed. Heavy dashed line indicates the frontal position at 850 hPa.

component in South China. It was also suggested that the cold air intrusion toward Central China in this situation is accompanied by the development of a baroclinic instability wave around 50°N/100-110°E. A high positive correlation between NINO3 SST (5°S-5°N, 90-150°W) and rainfall amount was found for the July Baiu over Japan (Tanaka 1997). The relationship between Meiyu and eastern equatorial Pacific SST is more complex because of the north-south space and time structure of Meiyu over China. Over the Yangtze River Valley the correlation between Meiyu rainfall and eastern equatorial Pacific SST is generally

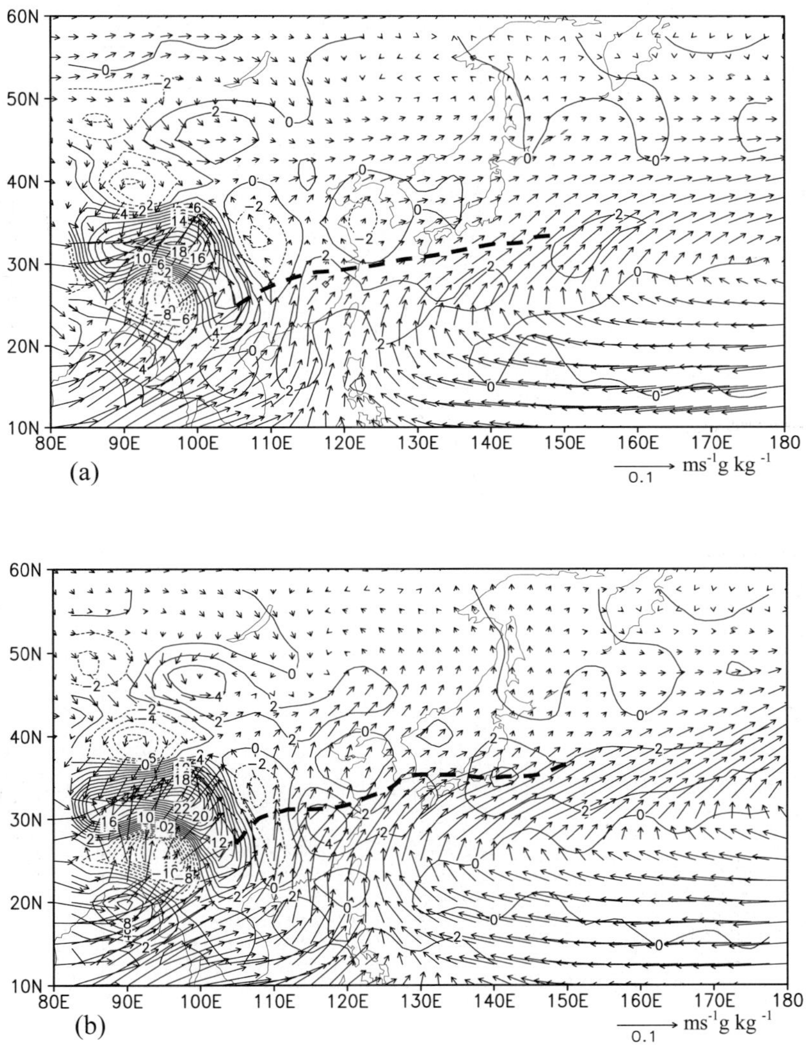

Figure 11. As in Fig. 8 except for moisture transport ($q\vec{V}$ arrows; m s^{-1}g kg^{-1}) and moisture flux convergence ($-\nabla\bullet q\vec{V}$; 10^{-5} s^{-1}g kg^{-1}) in 1000-850 hPa layer. Heavy dashed line indicates the frontal position at 850 hPa.

positive (*e.g.,* Chang *et al.* 2000b; Huang and Wu 1989; Shen and Lan 1995). For regions further south, the phase relationship becomes variable due to the variation of the location and strength of the western Pacific subtropical ridge that is affected by the tropical SST (Chang *et al.* 2000c). Further details can be found in other Chapter of this book.

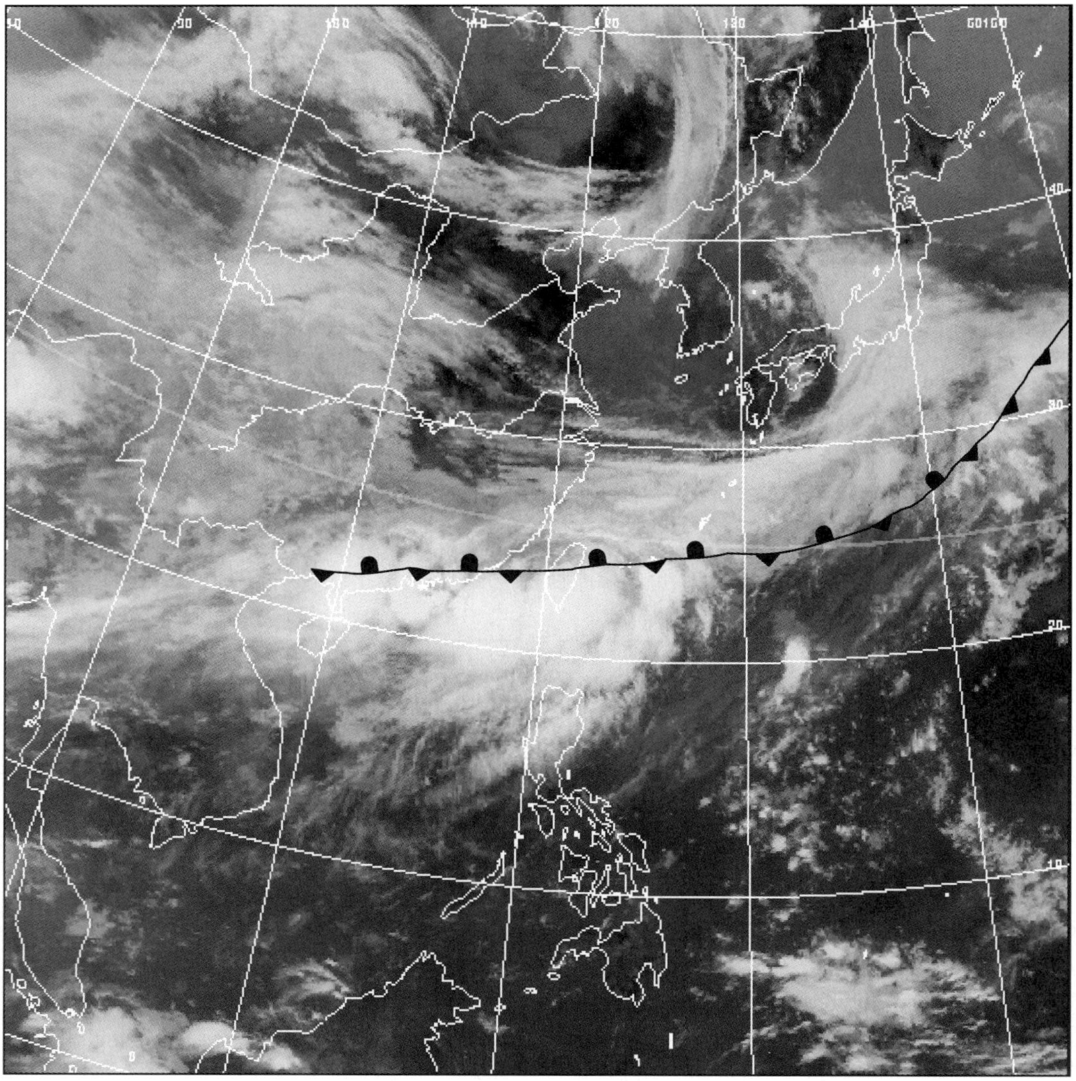

Figure 12. NOAA GOES9 infrared satellite image at 0023 UTC 7 June 2003. Surface Meiyu front is indicated.

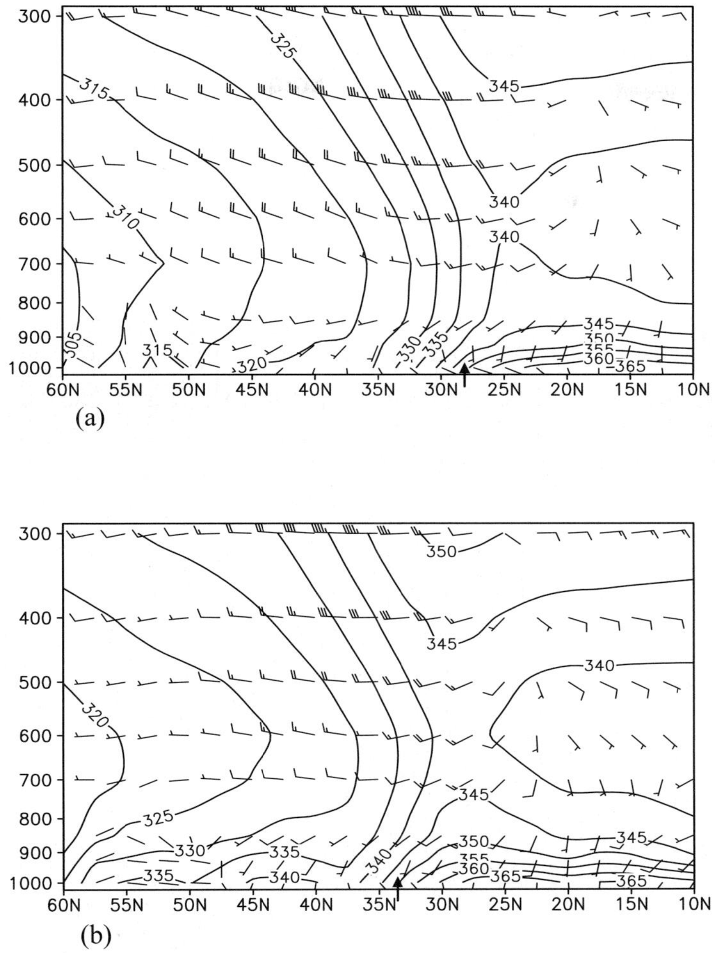

Figure 13. Height-latitude section for equivalent potential temperature (solid; K) and wind vectors along 120°E in (a) 16 May–15 June and (b) 16 June–15 July 1968–1996. Pennants represent 25 m s^{-1}, full wind barb 5 m s^{-1}, and half-wind barb 2.5 m s^{-1}. Heavy arrow indicates the 850 hPa frontal position as illustrated in Fig. 9.

By analyzing the vertical profiles of heating rate $Q1/C_p$ and drying rate $Q2/C_p$, it was found that the Meiyu rainfall was dominated by deep cumulus convection over South China (Johnson *et al.* 1993; Luo and Yanai 1984) and was associated with less deep convection and a higher percentage of stratiform precipitation over the Yangtze River Valley (Ding and Wang 1988; Johnson *et al.* 1993; Luo and Yanai 1984).

In summary, the surface Meiyu front as defined by the axis of vorticity maximum in the mean circulations in the Meiyu season of South China and Taiwan is located in the subtropical latitudes over East Asia extending from the ocean to the east of the Japan Islands southwestward passing through the Bashi Channel to the northern South China Sea. It shifts northward and is oriented in an east-west direction during the Yangtze River Valley Meiyu season. In the mean circulations, strong southwest monsoonal flows prevail in the lower troposphere to the south of the Meiyu front. The Meiyu front is characterized by the organized distribution of upward motion and the moisture flux convergence. The convective activity develops under the potentially unstable atmosphere along and to the south of the Meiyu front. It is also found that the Meiyu (Baiu) rainfall in Japan and the Yangtze River Valley is positively correlated to the eastern equatorial Pacific SST, whereas in South China and Taiwan this relationship becomes variable.

3. The Meiyu Front

G. Chen and Chi (1980a) analyzed the frontogenesis frequency during the Meiyu season of South China and Taiwan. They found two distinct areas of frontogenesis: one in middle latitudes to the north of 40°N and the other in the subtropical latitudes to the south of 35°N

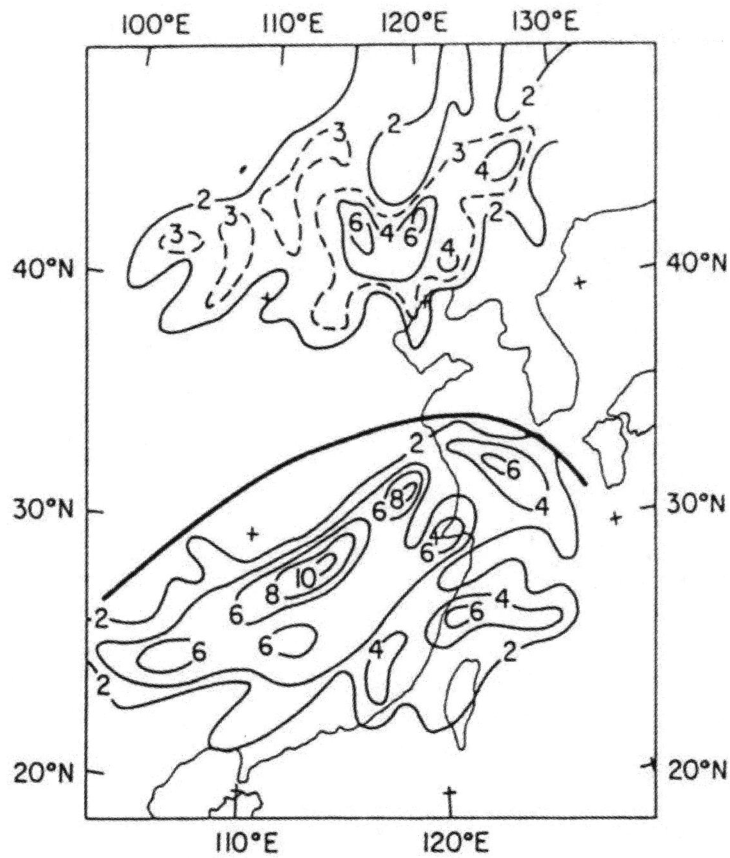

Figure 14. Frequency of frontogenesis at 1° lat × 1° long grid intervals during the Meiyu season of 15 May-15 June 1968-77 (1970 and 1975 excluded) over South China and Taiwan. The heavy solid line marks the boundary between the formation of the polar front and the Meiyu front (from G. Chen and Chi 1980a).

(Fig. 14). The heavy solid line in Fig. 14 marks the boundary between the formation of the polar front and Meiyu front. Among the Meiyu fronts analyzed in this period, 95% influenced the weather over Taiwan and South China during their lifetime, and all the Meiyu fronts that reached Taiwan and South China formed in the areas of 20-35°N and 100-130°E. The lifetime of these fronts ranged from 3 days to 22 days with an average of 8 days. There are about 4-5 frontal systems affecting Taiwan during the Meiyu season each year (G. Chen 1988; G. Chen and Chi 1978, 1980a). The stratiform and/or convective precipitation associated with the frontal system covers an area of 700 km wide across the front for both northern and southern Taiwan (G. Chen and Chi 1978).

G. Chen and Chang (1980) studied the synoptic-scale structure and dynamics over different sections of the Meiyu front. The results indicated that the structure of the eastern (near Japan) and central (the East China Sea) sections resembled a typical midlatitude baroclinic front with strong vertical tilt toward an upper level cold core and a strong horizontal temperature gradient. Whereas the western (southern China) section resembled a semitropical disturbance with an equivalent barotropic warm core structure, a weak horizontal temperature gradient, and a rather strong horizontal wind shear in the lower troposphere. The vorticity budget showed that generation of cyclonic vorticity by horizontal convergence was counteracted by cumulus damping in the eastern section and by boundary layer friction in the mountainous western section. Theoretical studies by Cho and Chen (1994, 1995) suggested that the frontogenetic process is initiated and maintained by the CISK mechanism through the interaction between the potential vorticity (PV) anomaly and the convection induced by Ekman layer pumping as the western section of a Meiyu front is characterized by a positive low-level PV anomaly. The scale contraction produced by the convergence flow with convection provides the basic frontogenetic forcing. The crucial role of the convective latent heating on the Meiyu frontogenesis was also demonstrated in a modeling case study by S. J. Chen *et al.* (1998) and recently in a PV diagnosis case study by G. Chen *et al.* (2003).

Using Taiwan Area Mesoscale Experiment (TAMEX) data, the mesoscale structure and new dynamical insights of the Meiyu front have been obtained. Case studies showed that the Meiyu front may possess appreciable baroclinity with shallow cold air within the lowest 1 km (Y. Chen and Hui 1990; Y. Chen *et al.* 1989; Trier *et al.* 1990). The most interesting feature of the front studied by Trier *et al.* (1990) during IOP 8 is its shallowness, with a depth ~1 km at the leading edge, tilting to a depth of 1.5 km over a distance of 5 km and then becoming quasihorizontal. The highly baroclinic shallow nature of this front strongly suggests that the density-current-type structure at the leading edge is an inherent property of this front. The major features of this density-current-type front include a wind shift occurring in the scale of 1 km, a convergence zone at the leading edge, and a steep slope with height of this wind shift followed by a more horizontally oriented frontal surface. It was suggested by Parsons *et al.* (1989) that the smaller and shallower frontal updraft in TAMEX system as compared to a midlatitude system was due in part to the smaller magnitude of the maximum low-level convergence from a shallower and slower moving Meiyu front. As the fronts moved southward over Taiwan during the Meiyu season, air-sea interaction was suggested as a

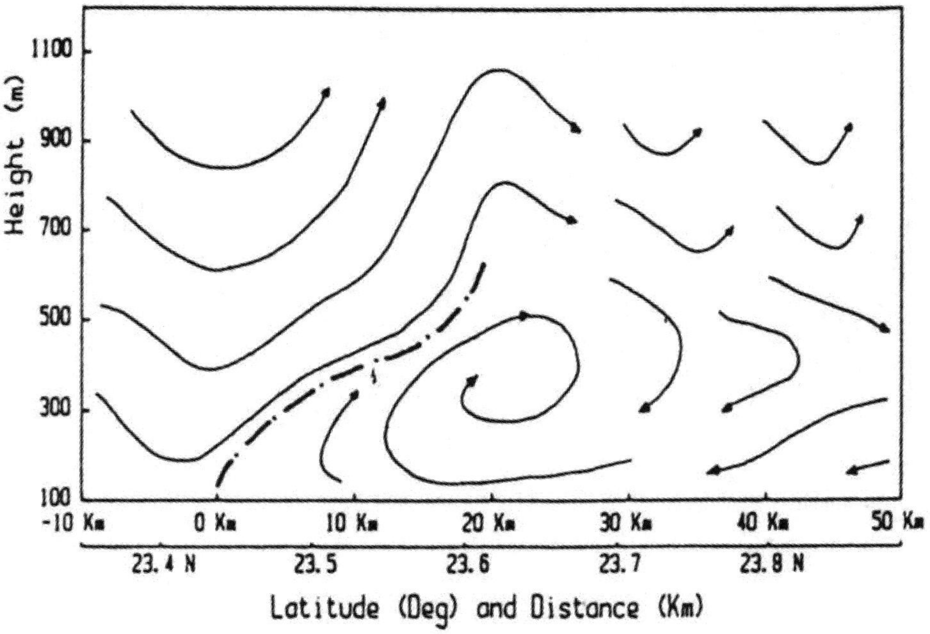

Figure 15. The relative airflow in the north-south cross section across the front east of Taiwan during 1139–1407 UTC 15 June 1987. The estimated frontal speed, 6 m s^{-1}, has been subtracted from the cross-front wind component. Dash-dotted line represents the leading edge of the cold air (from Y. L. Chen and Hui 1990).

possible mechanism for the slowness (G. Chen and Chi 1980a) and weakening (Y. Chen *et al.* 1989) of the fronts. An analysis of the Meiyu front presented by Trier *et al.* (1990) showed that the reduction of temperature contrast across front as the front moved southward was due to a warming and deepening of the postfrontal boundary layer and this warming could be explained by upward sensible heat flux.

The secondary circulation across the synoptic-scale Meiyu front was found to be thermally direct with warm air rising to the south and cold air sinking to the north of front (G. Chen and Chang 1980). Using composite technique, G. Chen and Chi (1978) was able to observe two secondary circulation cells associated with the Meiyu front in the northern Taiwan with a thermally direct circulation cell across the front and a reversed cell to the south of the front. Convective activity was enhanced over the area of ascending part of the cell and was suppressed over the area of descending part. For the TAMEX cases, the thermally direct circulation of the Meiyu front was deduced from a Doppler radar and P-3 aircraft observations (Y. Chen and Hui 1990; Jou *et al.* 1990; Trier *et al.* 1990). Figure 15 illustrates the thermally direct vortex circulation, derived from P-3 aircraft observations, at the leading edge of the front in the north-south cross section across a front to the east of Taiwan on 15 June 1987 (Y. Chen and Hui 1990). The slope at the leading edge of the cold air boundary was gentle; it rose to a depth of 800 m in a distance of 20 km. At the top of the cold air behind the vortex circulation a wavelike pattern was observed with saturated air in

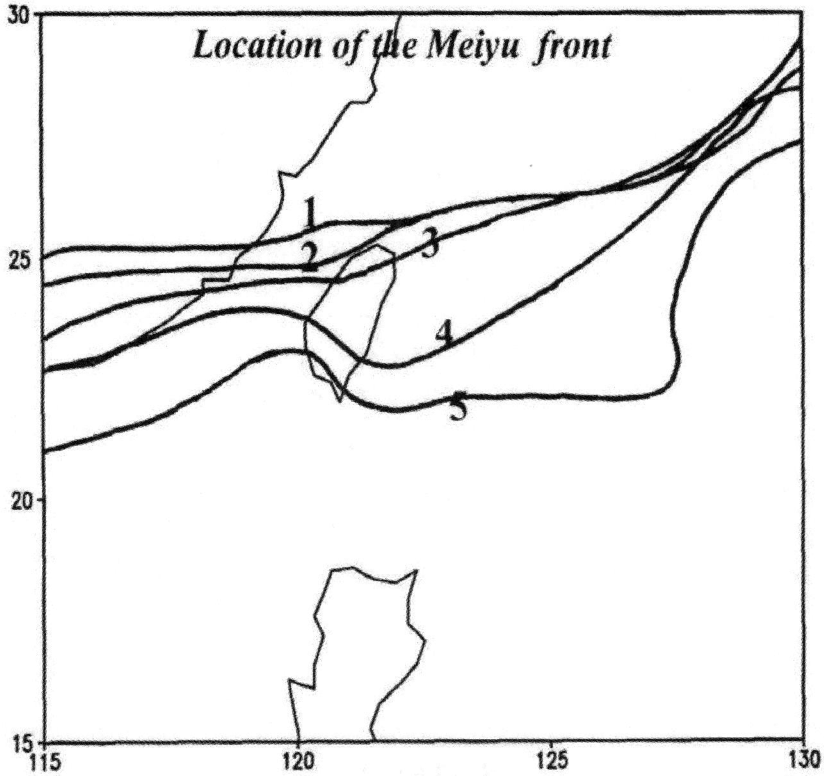

Figure 16. Location of the Meiyu front at (1) 0000 UTC 10 June, (2) 0000 UTC 11 June, (3) 1200 UTC 11 June, (4) 0000 UTC 12 June, and (5) 1200 UTC 12 June (5), 2000 (from Yeh *et al.* 2002).

the rising branches, which were partially driven by the buoyancy force due to latent heat release. Hor *et al.* (1998) studied the same front off the east coast of Taiwan and found the frontal edge possessed a well-defined density current structure. Their results also suggested that the propagation of the density current and the maintenance of the frontal system were due to the intense horizontal pressure gradient force from near the front in the cold core region and the moderate convective instability at the head of the system as well as the kinetic energy transport from the mean flow.

Another interesting feature of the Meiyu front was its modification due to the topography. Mesoscale analyses by G. Chen (1978b, 1979b) showed that the Meiyu front split into two branches, one to the east of Taiwan, and the other to the west, as the front approached the Taiwan Central Mountain Range (CMR) from the north. The eastern segment of the front often moved southward at a significantly faster speed than the western segment (G. Chen 1978b; S. Wang 1986). The frontal deformation with different propagation speeds to the east and west of the CMR was also observed in TAMEX (Y. Chen *et al.* 1989; Trier *et al.* 1990; S. Wang 1989) and in a recent case study using QuikSCAT oceanic winds by Yeh *et al.* (2002). Figure 16 presents the Meiyu front position under the influence of island topography over

Taiwan by incorporating the QuikSCAT oceanic surface winds (Yeh *et al.* 2002). It is clear that the front was distorted and the surface front to the east of Taiwan moved faster than that over southwestern Taiwan and the Taiwan Strait. This was related to the earlier arrival of the strong northerly winds along the east coast and over the ocean to the east of Taiwan without the obstruction by the island topography. Mannouji and Kurihara (1990) performed a numerical experiment for the frontal case on 24 June 1987 by using Japan Meteorological Agency (JMA) spectral model and illustrated the essential role of the CMR on the distortion of the front with a faster propagation speed on the east side only when mountain was included in the model. The strong ridging behind the leading edge of cold air along the east side of the CMR due to the blocking effect of the topography was quite similar to the case of cold air damming on the east slope of the Appalachian mountain chain in the eastern United States (Bell and Bosart 1988; Forbes *et al.* 1987). The acceleration of highly ageostrophic flow of cold air parallel to the CMR in response to the building pressure gradient was suggested to be responsible for the increase in propagation speed of the front along the east side of the mountain.

In summary, the Meiyu front affecting South China and Taiwan forms in the subtropical latitude, which is a distinct area from that for the formation of polar front in the Meiyu season. It resembles a semitropical disturbance with an equivalent barotropic warm core structure, a weak horizontal temperature gradient, a rather strong horizontal wind shear, and a positive low-level potential vorticity (PV) anomaly. Results from theoretical, modeling, and observational studies suggest that the Meiyu frontogenetic process is initiated and maintained by the CISK mechanism through the interaction between the PV anomaly and the convective latent heating. Some of the Meiyu fronts observed in TAMEX were characterized by the density-current-type structure at the leading edge with a thermally direct circulation cell across the front. The smaller and shallower frontal updraft in TAMEX system as compared to a midlatitude system was suggested due in part to the smaller magnitude of the maximum low-level convergence from a shallower and slower moving Meiyu front. Observational studies and numerical simulations with Taiwan topography revealed the frontal deformation with different propagation speeds to the east and west of the Taiwan Central Mountain Range. The acceleration of highly ageostrophic flow of cold air in response to the building pressure gradient was suggested to the responsible for the faster moving front to the east of Taiwan.

4. Low-Level Jet (LLJ)

One of the very interesting features accompanying a Meiyu front was the existence of a LLJ to the south or southeast of the 850 / 700 hPa trough (or shear line) (G. Chen 1977a, 1978a; Tao and Chen 1987). Figure 17 presents an example of a LLJ in a vertical cross section along 120°E on 11 June 1975 (G. Chen 1978a). In this case, LLJ was evident near 700 hPa in the vicinity of Taiwan (J=6) to the south of the Meiyu front. The 850 hPa Meiyu front/shear line was clearly indicated by the intersection point of zero lines of u and v (near J=7). It is clear that LLJ was a distinct feature of west-southwesterlies in contrast to the upper-level northwesterly jet.

A close relationship between extremely heavy rainfall and a LLJ in the 850-700 hPa layer during the Meiyu season was found in many observational case studies over various geographic locations (Akiyama 1973; G. Chen 1979a, 1983; G. Chen and Chi 1978; Matsumoto 1972; Ninomiya and Akiyama 1974; Tao and Chen 1987; Tsay and Chain 1987). The physical processes through which the existence of a LLJ leads to heavy rainfall was proposed by Matsumoto (1972) who suggested that the upward branch of the secondary circulation associated with the LLJ was responsible for the heavy rainfall. G. Chen's (1977a) study, on the other hand, suggested that the LLJ was the main mechanism for creating

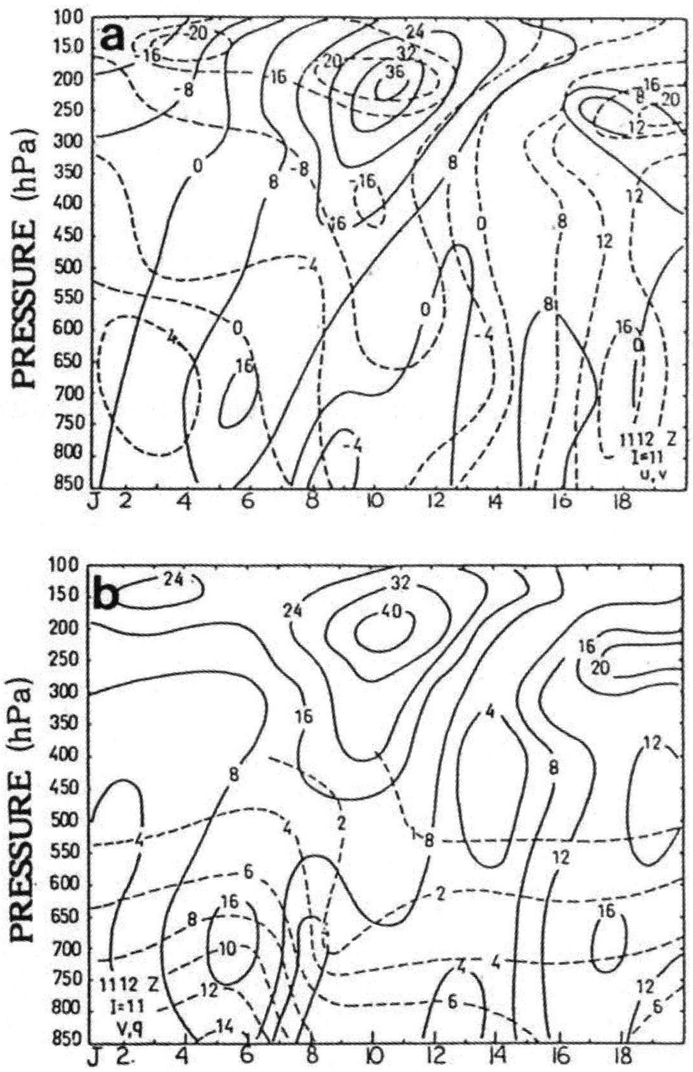

Figure 17. Vertical cross sections of (a) east-west wind (*u*; solid) and north-south wind (*v*; dashed) components in m s^{-1} and (b) total wind speed (solid) in m s^{-1} and mixing ratio (dashed) in g kg^{-1} along 120°E at 1200 UTC 11 June 1975. Grid indices (J) in increasing order are from south to north and the grid distance is 240 km (from G. Chen 1978a).

potential instability to the warm side of a Meiyu front and for generating the convective heavy rainfall by the associated convergence downstream. A numerical case study of mesoscale convective systems (MCSs) along the Meiyu front over the Yangtze River Valley by S. J. Chen *et al.* (2000) also showed that the LLJ transported warm moist air into the MCSs and the strong convergence existed on the left-front region of the LLJ. Numerical simulation of a TAMEX rainstorm case by Trier and Parsons (1995) suggested that the strong vertical shear associated with the LLJ provided a mechanism for producing greater local rainfall rates by allowing enhanced forcing for low-level updrafts in the mainly saturated ambient environment. G. Chen and Yu (1988) examined 35 heavy rainfall events over northern Taiwan in May and June during the years 1965-84. They found that there was an 84% chance that a LLJ with a maximum speed of at least 12.5 m s^{-1} would be present at 700 hPa 12-h prior to the start of the heavy precipitation. Conversely, when a LLJ was present over northern Taiwan, there was a 91% likelihood that heavy precipitation was in progress or would begin within 24-h. They also made a composite of vertical wind profiles for ten cases of heavy rainfall over northern Taiwan accompanied by a LLJ with an axis located within 1° latitude from Taipei at each of the three time periods: 12 hours before heavy rainfall, at time of heavy rainfall, and 12 hours after rainfall. Figure 18 presents the results. It seems clear

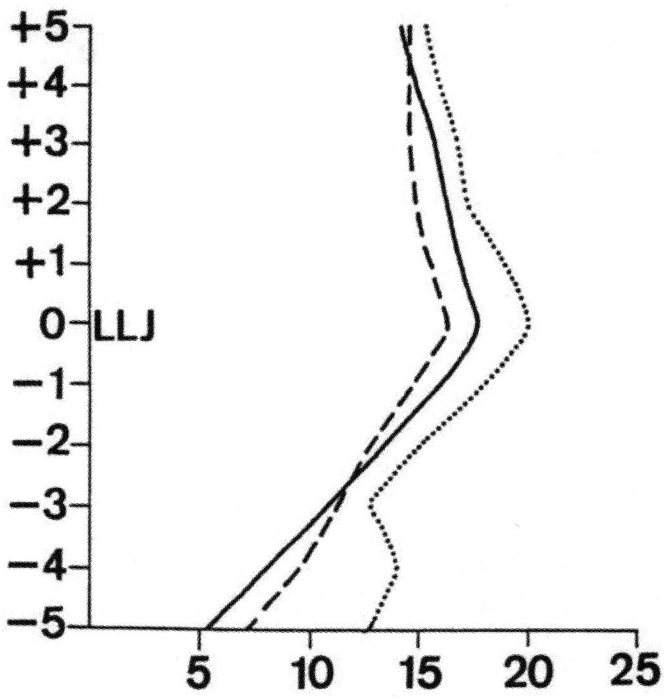

Figure 18. Composite vertical wind profiles for ten cases of heavy rainfall over northern Taiwan accompanied by a LLJ with axis located within 1° latitude from Taipei at each of the three time periods: 12 hours before heavy rainfall (dotted), at time of heavy rainfall (solid), and 12 hours after heavy rainfall (dashed). Wind speeds were composited with respect to the level of each LLJ. Abscissa is wind speed in m s^{-1} and ordinate is altitude at 300 m intervals above (+) and below (–) the LLJ level (0) (from G. Chen and Yu 1988).

from the composite profiles that vertical shear above and below LLJ level as well as the LLJ intensity decreased at and after the time of the start of extremely heavy rainfall.

Ageostrophic characteristics of the LLJ over Japan areas (Matsumoto *et al.* 1971) led to a hypothesis of downward momentum transport in the formation of a LLJ (Akiyama 1973; Matsumoto and Ninomiya 1969). Over Taiwan and China, on the other hand, a LLJ was usually observed prior to the major convection and decreased its intensity after the heavy rainfall (G. Chen and Chi 1978; G. Chen and Yu 1988; Ding 1992). Also the areas of MCS formation were usually determined by the location and intensity of a low-level jet (G. Chen 1977a; G. Chen and Pu 1985; S. J. Chen *et al.* 2000; Chiou and Liao 1984). Momentum budget study on a prefrontal convection case in TAMEX by Y. J. Lin *et al.* (1993) revealed that the vertical flux convergence and divergence of horizontal momentum by organized convection were largely responsible for weakening a low-level jet. Therefore, it seems reasonable that at least some of the LLJs over Taiwan and China were the cause rather than the effect of the convection.

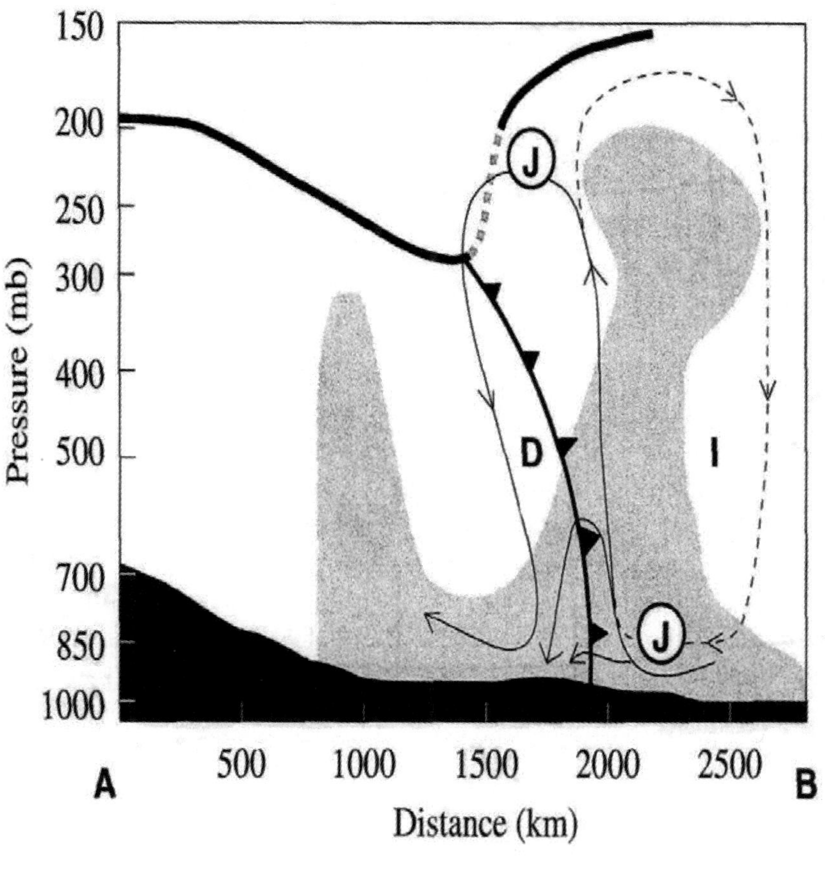

Figure 19. Schematic diagram showing the flow structure of an observed Meiyu front (from Y. L. Chen *et al.* 1994). The thin solid line depicts the direct (D) circulation while the thin dashed line depicts the indirect (I) circulation. The heavy solid line shows the frontal position. The character J denotes the jet position. The thick heavy line represents the tropopause boundary. Regions with relative humidity greater than 70% are shaded.

Observational study of G. Chen and Yu (1988) suggested that a LLJ might form to the south of the heavy rainfall area and proposed that the reversed secondary circulation to the south of Meiyu front as observed in the composite by G. Chen and Chi (1978) presumably driven by convective latent heating was a possible formation mechanism for a LLJ. This was also suggested both in theoretical study (Q. Chen 1982) and numerical experiments (S. J. Chen *et al.* 1998; S. J. Chen *et al.* 2000; Chou *et al.* 1990). Results of Q. Chen's study (1982) suggested that the existence of an unstable inertio-gravity wave in the ascending motion region caused a thermally direct circulation beneath the upper-level jet. Latent heating in the area of ascending motion accelerated the thermally direct circulation to the north and induced a reversed circulation to the south through the geostrophic adjustment process. It was concluded that the LLJ formed through the Coriolis acceleration of northward flow in the lower branch of this reversed circulation. Numerical study of Chou *et al.* (1990) simulated the formation of a LLJ to the south of the area of strong convection in a two-dimensional frontogenesis model and suggested the importance of cumulus convection and especially a slantwise structure in developing the reversed circulation and the LLJ. A weak thermally indirect circulation was observed to the south of the Meiyu front for TAMEX IOP 5 case as presented in Fig. 19 (Y. Chen *et al.* 1994). In that case, the LLJ was observed to the south of the Meiyu front within the lower return branch of the secondary circulation and the thermally direct circulation across the front was observed similar to that of G. Chen and Chi (1978) and G. Chen and Chang (1980). Numerical study of this LLJ case by Y. Chen *et al.* (1997) revealed that the LLJ developed through the Coriolis force acting on the cross contour ageostrophic winds in response to the increased pressure gradients related to the development of the cyclone and was enhanced by latent heating. Observational case studies (G. Chen *et al.* 1986; S. C. Lin 1988; S. C. Lin and Chiou 1985; S. C. Lin and Tsai 1989; Pu and Chen 1988) also showed that a LLJ tended to form or to intensify to the south of an MCS in South China. Thus, results of observational, theoretical, and numerical studies all suggested that a LLJ could form to the south of the convective heavy rainfall area presumably through the Coriolis acceleration of the lower branch of an induced secondary circulation. Over southern China, geostrophic forcing and lee cyclogenesis processes were also suggested as possible mechanism for the formation of LLJs in spring and Meiyu seasons (G. Chen and Pu 1985; X. Chen and Chen 1995; Y. Chen and Tseng 2000; Pu and Chen 1988). Besides, latent heating from the stratiform clouds rather than the cumulus heating was suggested to the responsible for a case of LLJ intensification over south China during TAMEX by a numerical study by Hsu and Sun (1994). Composite results of Tsay and Kau (1989) revealed that a LLJ tended to form to the south of the entrance region of an upper-level jet. They suggested that a strong vertical coupling existed between the upper-level jet and the LLJ.

Composite structure of the LLJ over South China during TAMEX obtained by G. Chen and Hsu (1997) revealed that the LLJ formed and intensified in association with a non-quasigeostrophic trough intensification process. It had a highly supergeostrophic nature. The distribution of divergence and vertical velocity (Fig. 20) was mainly characterized by an anti-symmetrical pattern upstream and downstream with upward motion caused by the along-jet ageostrophic wind speed convergence over the exit region. It also showed an anti-symmetrical pattern to a lesser degree on both the left and right hand sides. The characteristics of secondary circulation across the jet axis were quite different from those of the upper-level jet. A thermally indirect circulation across the jet axis was observed in the entrance and the jet core regions and a thermally direct circulation existed over the exit region (Fig. 21).

Analyses of TAMEX cases showed that the prevailing low-level southwesterly flows tended to be deflected upstream of the CMR due to the blocking effect of the mountain and the LLJ appeared to form to the northwest of Taiwan and to the southern tip of Taiwan (Li and Chen 1998; Shu 1991). These observed features of flow deflection and the formation of a LLJ locally to the northwest Taiwan were well captured in the model simulations by Huang and Raman (1990) and Mannouji and Kurihara (1990). The existence of the LLJ was found to be essential for maintaining the pre-frontal squall line by providing the strong vertical shear in the low troposphere (G. Chen and Chou 1993; Jou *et al.* 1990; T. Wang *et al.* 1990). The

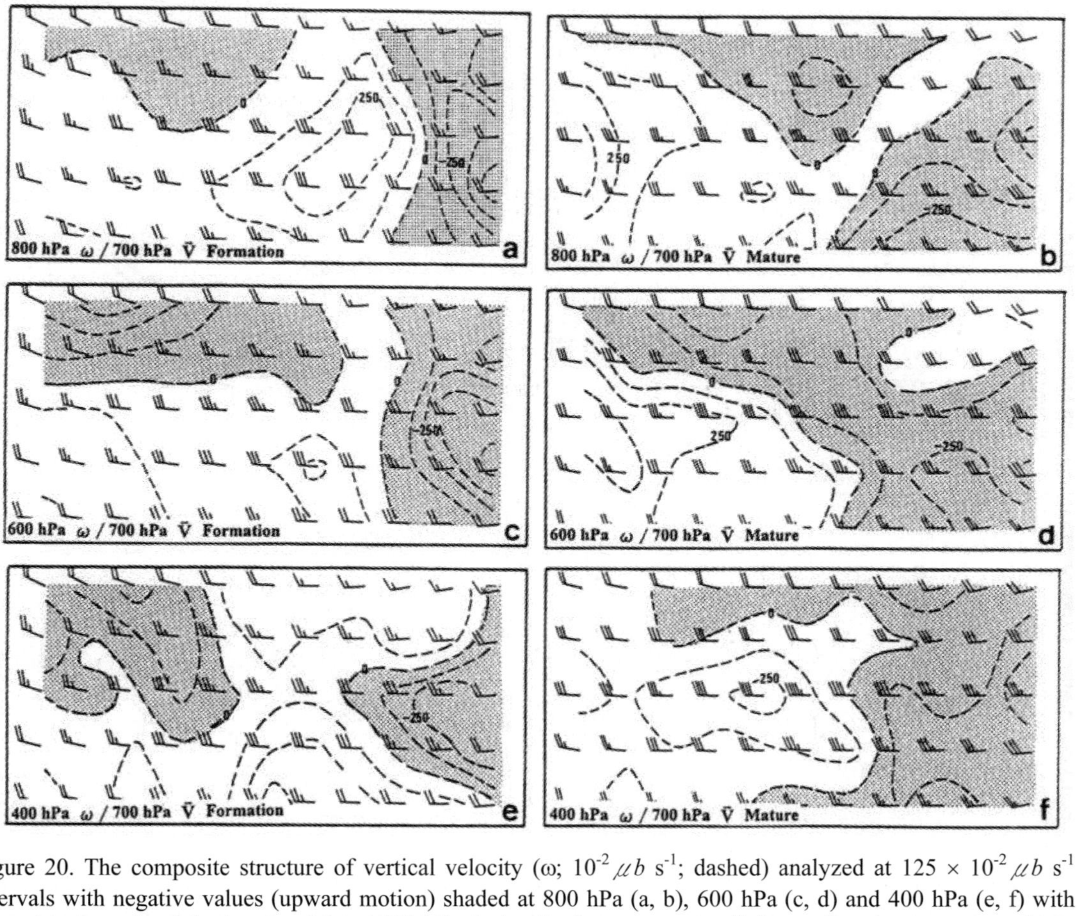

Figure 20. The composite structure of vertical velocity (ω; $10^{-2}\,\mu b\ s^{-1}$; dashed) analyzed at $125 \times 10^{-2}\,\mu b\ s^{-1}$ intervals with negative values (upward motion) shaded at 800 hPa (a, b), 600 hPa (c, d) and 400 hPa (e, f) with respect to the axis of the low-level jet at 700 hPa during its formation stage (left panel) and mature stage (right panel). The wind plots in each panel are for the formation (left) and mature (right) stages at 700 hPa (from G. Chen and Hsu 1997).

horizontal vorticity existed in the vertical shear tended to balance the horizontal vorticity produced by buoyancy gradients near the leading edge of the cold pool in those TAMEX squall line cases. This approximate balance was referred to as an optimal state for MCS development by Rotunno *et al.* (1988).

In summary, the existence of the LLJ to the south of the Meiyu front is crucial to the occurrence of convective heavy rainfall by providing strong low-level moisture flux convergence, vertical shear, and upward motion. The LLJ observed in TAMEX had a highly supergeostrophic nature with quite different characteristics of secondary circulation across the jet axis from those of the upper-level jet. A thermally indirect circulation was observed in the entrance and the jet core regions and a thermally direct circulation existed over the exit region. Results of observational, theoretical, and numerical simulation studies all suggested that a LLJ can form to the south of the convective heavy rainfall area through the Coriolis

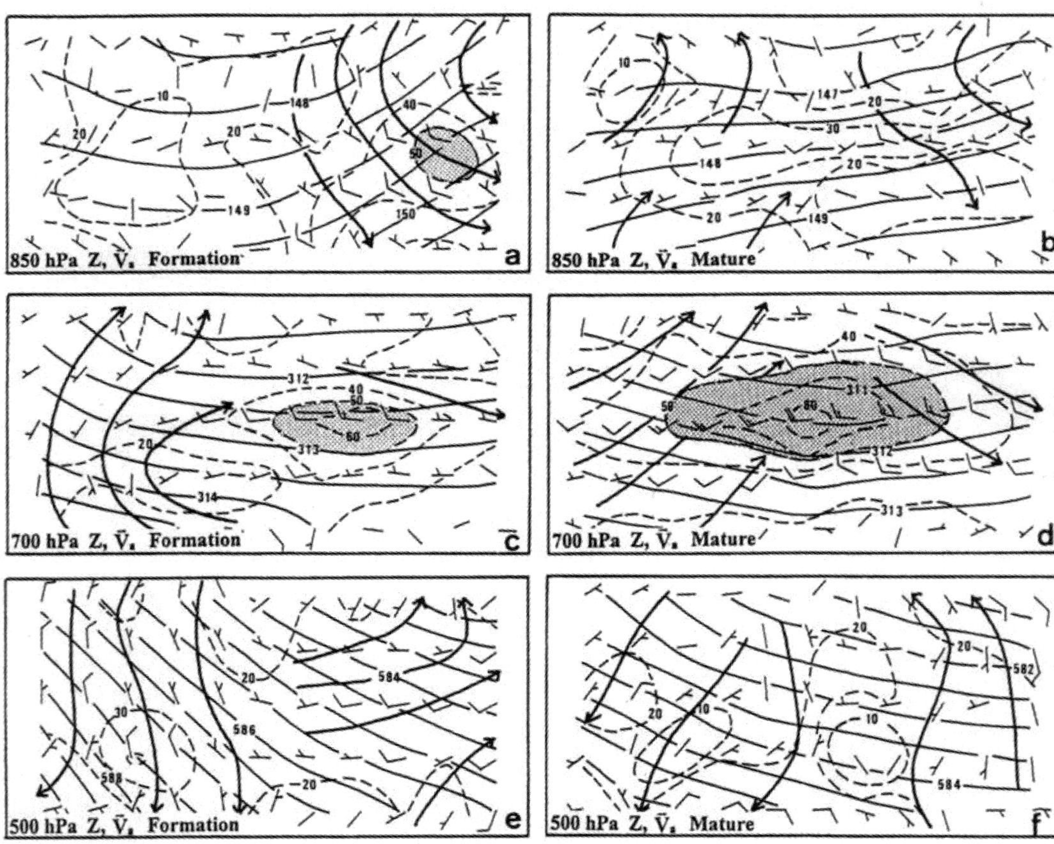

Figure 21. The composite structure of ageostrophic wind ($\vec{V}_a$) and geopotential height (solid; dam) at 850 hPa (a, b), 700 hPa (c, d) and 500 hPa (e, f) with respect to the axis of the low-level jet at 700 hPa (center line in each panel) during its formation stage (left panel) and mature stage (right panel). Wind plots with full barb and half barb represent 10 kt (5 m s^{-1}) and 5 kt (2.5 m s^{-1}), respectively. Dashed lines indicate the ratio of the ageostrophic wind to the geostrophic wind with the ratio greater than 50 % shaded. Heavy solid arrows illustrate the stream lines of the cross-contour ageostrophic flows (from G. Chen and Hsu 1997).

acceleration of the lower branch of an induced secondary circulation by convective latent heating.

5. Mesoscale Convective Systems (MCSs)

The heavy rainfall producers in the Meiyu season are the MCSs. There are different sizes of MCSs embedded in the Meiyu frontal cloud band. G. Chen *et al.* (1986) studied the climatology of the MCSs over western Pacific and South China in the Meiyu season of 1981-1983 using GMS satellite cloud pictures. Results showed that the duration of the Meiyu MCSs is similar to that observed over North America in warm season. The mean durations of the meso-α and meso-β MCSs were 14.6-h and 14.1-h, respectively. These are somewhat shorter than that obtained by Maddox (1980) of 16.5-h over the U.S.. They also found that the duration of MCS was positively correlated with its horizontal scale, which increased in warmer seasons. The MCSs in subtropical China tend to move southeastward over land and then recurve eastward or northeastward offshore. The afternoon maximum of the MCS formation over land was apparently due to solar heating. The early morning maximum of the MCS intensification was suggested to be due to the differential radiation effect between the cloudy and cloud-free area.

Since heavy rainfall in the Meiyu season is produced by the MCS embedded in the frontal cloud band, the environmental conditions are similar for the heavy rainfall and for the MCS (G. Chen 1977b; Chi and Chen 1988, 1989b, 1989c; Chiou and Chen 1989; Chu and Chen 1988; S. C. Lin 1988; S. M. Lin and Lin 1989). These synoptic case studies for the MCSs over Taiwan and South China suggested that there are at least six conditions favorable for the development of MCS. They are (1) warm advection in the lower troposphere, (2) low-level convergence over the low pressure and/or the front, (3) low-level jet, (4) short wave trough in the lower to middle troposphere, (5) the middle and upper troposphere diffluent flow and/or speed divergence, and (6) potential instability in the lower and middle troposphere.

Tsay and Chen (1980) pointed out that the large-scale upward motion to the south of the Meiyu front was generated mainly by vertical differential of the horizontal vorticity advection, Laplacian of the temperature advection, low-level frictional effect, and convective latent heating. However, the relative importance of these forcing terms was different over Japan, Taiwan and South China. The areas of this large-scale upward motion and the horizontal moisture flux convergence are the areas favorable for the development of MCSs (C. S. Chen 1990; G. Chen 1979a; G. Chen and Chang 1980; G. Chen and Tsay 1978; Chi and Chen 1989c; Kuo and Anthes 1982; Lee 1990; Wu and Chen 1988). Figure 22 presents the composite structure of environment conditions for the 12 cases of meso-α scale MCS (*i.e.* MCC) over South China selected in May-June 1981-1986 in their formation and mature stages (Wu and Chen 1988). The overall structure was quite similar to that for the midlatitude MCC in the North America as obtained by Maddox (1983). The MCCs form and intensify in the warm sector to the south of the Meiyu front/shear line. The strong warm advection and speed convergence (*i.e.* convergence due to the downstream speed decrease) in the lower-tropospheric southwesterlies, possible lifting mechanisms in the formation and intensification stages, prevail over the area of MCC. The MCC tended to form and to intensify on the cyclonic side of the LLJ exit region. Anticyclonic circulation and diffluent flow in the upper troposphere provided conditions favorable for the intensification of MCC. Case study of the MCSs over South China by Chiou and Chen (1989) revealed a warm core structure in middle and upper troposphere during the developing and mature stages.

Observational studies suggested that some mesoscale circulation systems are apparently responsible for the formation of the MCSs over Taiwan and South China. These included the frontal secondary circulation, low-level jet, upper-level jet streak, quasi-stationary mesolows, terrain induced flows, outflow boundaries, and local circulations (C. K. Chen 1979; C. S.

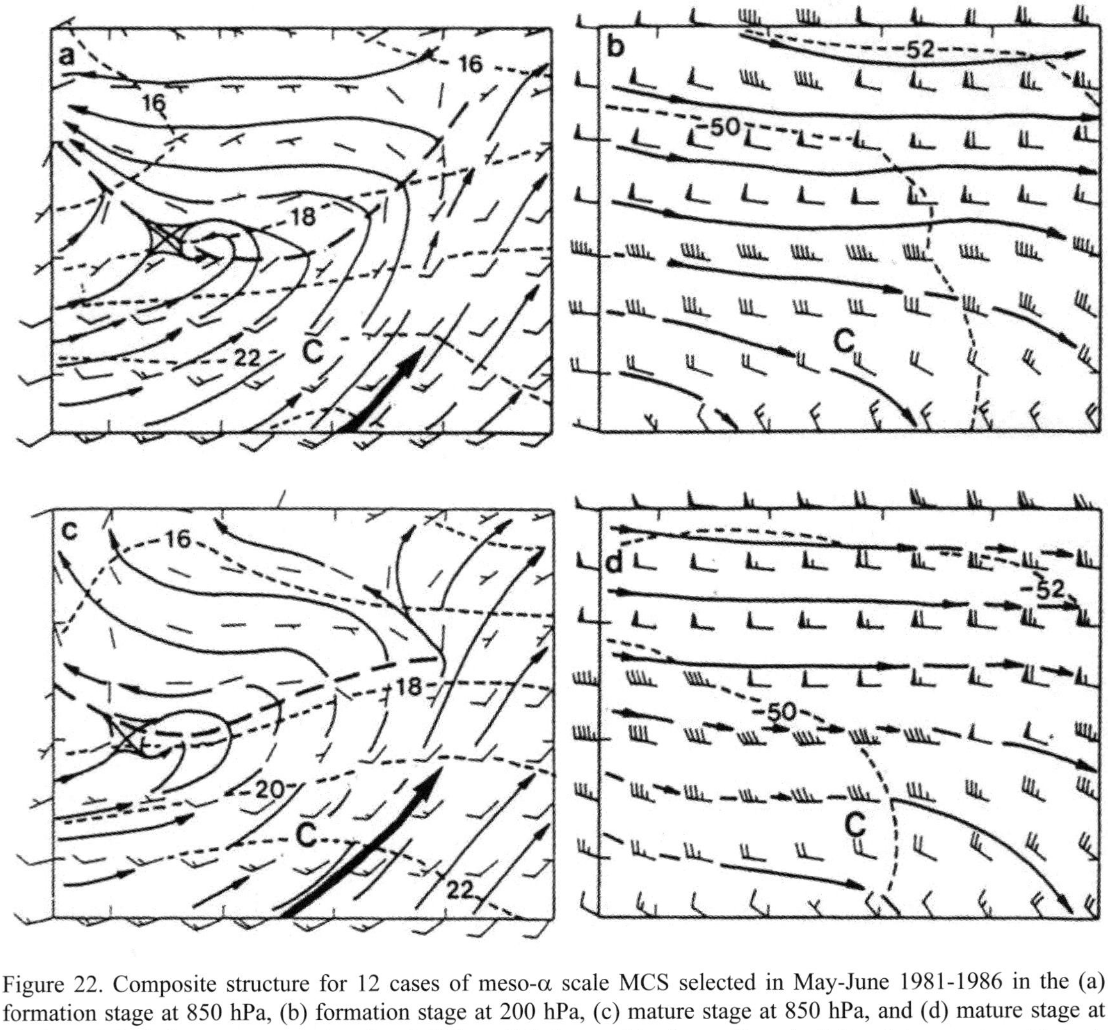

Figure 22. Composite structure for 12 cases of meso-α scale MCS selected in May-June 1981-1986 in the (a) formation stage at 850 hPa, (b) formation stage at 200 hPa, (c) mature stage at 850 hPa, and (d) mature stage at 200 hPa. Wind plots are conventional at 2.5° lat × 2.5° long grids. Streamlines (solid) and temperature (dashed, °C) are analyzed. Heavy arrow indicates LLJ position, heavy dashed line shows the Mei-Yu shear line, and C indicates the MCS position (from Wu and Chen 1988).

Chen 1989; G. Chen 1977a, 1978b, 1990b; G. Chen and Chi 1978, 1980b; G. Chen and Yu 1988; S. J. Chen *et al.* 1998, 2000; Chi and Chen 1989b; S. M. Lin and Lin 1988; Trier *et al.* 1990). The mesoscale circulation systems appeared to be important mechanisms for creating greater instability over a smaller area and providing stronger lifting necessary for the mesoscale convection. Under undisturbed conditions during TAMEX, on the other hand, wind direction below 3 km height was suggested to be very important for occurrence of afternoon precipitation in mountain area in Taiwan (C. S. Chen and Lin 1997). Using 2-D cloud model to study a precipitation system in northeastern Taiwan during TAMEX IOP 10, C. S. Chen and Lin (1996) was able to demonstrate that the magnitude of the wind speed was important for the formation of convective cells. A numerical study of the MCSs along the Meiyu front over the Yangtze River Valley by S. J. Chen *et al.* (2000) suggested that barotropic processes might be responsible for triggering these MCSs along the Meiyu front.

Climatological aspects of convective rainfall in Taiwan were strongly modulated by the local circulations in Meiyu season (G. Chen *et al.* 2002; Bresch and Johnson 1990). The convective rainfall occurred mainly in the daytime hours (0900-2000 LST), indicating the importance of solar heating effect. Besides, the occurrence time of maximum frequency appeared to correlate well with the inland distance of the station. The coastal stations had a peak frequency in the late morning and the inland stations in the afternoon. This seems to indicate the importance of see breeze in modulating the convective rainfall. Results of G. Chen *et al.* (2002) also suggested that the land-sea breeze circulations prevailed and the peak frequency of convective rainfall occurred when the sea breeze started. Note that majority of the heavy rainfall events over northern Taiwan occurred at night (G. Chen 1985), whereas the convective rainfall occurred mainly in the daytime hours on the western side of the CMR (G. Chen *et al.* 2002). This would suggest that the characteristics of the MCSs in heavy rainfall were different from the MCSs produced by the sea breeze and solar heating effect. A recent case study of IOP 3 rainfall event over northwestern Taiwan by Yeh and Chen (2002) suggested that the convergence responsible for the local rainfall maximum off the northwestern coast was a combination of the synoptic-scale forcing, orographic effects, and the feedback of convection.

In TAMEX, Doppler radar and P-3 aircraft observations provided a great opportunity for studying the general characteristics and internal structure of the MCSs. Jorgensen and LeMone (1989) and Jorgensen *et al.* (1991) studied the characteristics of vertical velocity and kinematic structure of oceanic convection in the vicinity of Taiwan. They observed that the updrafts and cores in TAMEX were comparable in size and strength with those measured in GATE and hurricanes but much weaker than those measured in continental thunderstorms. They also found that the oceanic convection in TAMEX was dominated by warm rain coalescence processes, similar to that in GATE and hurricanes, and that a large fractional rainout occurred below the freezing level. Momentum budget study for the oceanic convection during TAMEX by LeMone and Jorgensen (1991) revealed that the positive momentum flux at upper levels was consistent with the updrafts' eastward tilt. It was countergradient and not consistent with previous observations, but was suggested in numerical simulation of systems in an environment similar to that for this system. The internal structure and dynamics of MCSs in different TAMEX cases were studied extensively by various investigators (*e.g.,* Akaeda *et al.* 1991; Chang and Yoshizaki 1991; Jou *et al.* 1990; P. L. Lin *et al.* 1989; Y. J. Lin *et al.* 1990, 1992, 1993; Tao *et al.* 1991; Teng *et al.* 2000; T. Wang 1988, 1989; T. Wang *et al.* 1990). Case study of a north-south oriented pre-frontal squall line on 17 May 1987 (IOP 2) by T. Wang *et al.* (1990) showed that many structural features were similar to those for a fast-moving tropical squall line. Figure 23 presents an east-west vertical cross section over northern part of this squall line. The convective updraft (≥ 9 m s^{-1}) was tilted toward the west in the lower layer (0-4 km) heading toward the trailing stratiform region and rather strong and deep rear-to-front flow came from the back of the leading edge. A density-current-type structure was evident at the leading edge as shown by a remarkable change in the u component. The environmental conditions observed during TAMEX squall lines were found with characteristics between tropical and midlatitude squall lines (G. Chen and Chou 1993). The inflow ahead of the squall lines was deeper and stronger

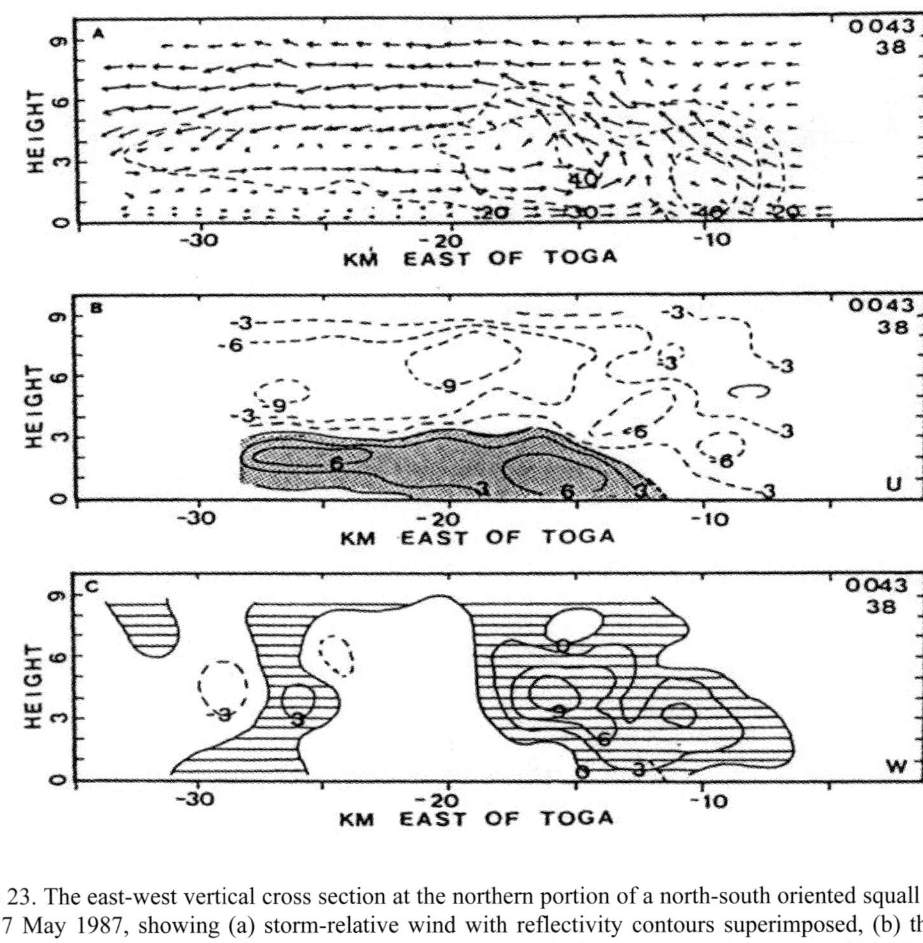

Figure 23. The east-west vertical cross section at the northern portion of a north-south oriented squall line at 0043 LST 17 May 1987, showing (a) storm-relative wind with reflectivity contours superimposed, (b) the cross-line component (u), and (c) vertical velocity (w). Contour intervals for Z, u, v, and w are 10 dBZ, 3 and 3 m s^{-1}, respectively. Negative values are dashed with positive values either shaded or hatched. The heavy dashed line signifies the guest front (from T. Wang *et al.* 1990).

below 400 hPa, and the CAPE was higher during the mature stage as compared to the intensifying stage. The squall-line collapse was correlated with decreasing CAPE and low-level inflow ahead of the lines.

In summary, favorable conditions for the development of MCSs and the occurrence of heavy rainfall in the Meiyu season were identified such as warm advection in the lower troposphere, low-level convergence over the low pressure and/or the front, low-level jet, short-wave trough in the lower to middle troposphere, the middle and upper troposphere diffluent flow and/or speed divergence, and potential instability in the lower and middle troposphere. The mesoscale forcings for triggering MCSs included the frontal secondary circulation, low-level jet, upper-level jet streak, quasi-stationary mesolows, terrain-induced flows, outflow boundaries, and local circulations. The updrafts and cores of MCSs in TAMEX were comparable in size and strength with those measured in GATE and hurricanes but much weaker than those measured in continental thunderstorms. The oceanic convection in TAMEX was dominated by warm rain coalescence processes, similar to that in GATE and hurricanes, and that a large fractional rainout occurred below the freezing level. The structural feature of some squall lines in TAMEX was similar to that for a fast-moving tropical squall line. The environmental conditions during the TAMEX squall lines were found with characteristics between tropical and midlatitude squall lines. The front-to-rear inflow was deeper and stronger and the CAPE was higher during the mature stage as compared to the intensifying stage. The squall-line collapse was correlated with decreasing CAPE and low-level front-to-rear inflow.

6. Disturbances in Meiyu Season

Over China, the heavy rainfall during the Meiyu period are mainly generated by the meso-α- and meso-β-scale disturbances which are embedded within and propagated along the Meiyu cloud and rain band or frontal zone with horizontal length scale of several thousand kilometers (Ding 1992). The meso-α-scale systems during the Meiyu period may be classified into two types: the Yangtze River Valley shear line and the low-level vortex. The Yangtze River Valley (112-120°E, 30-35°N) shear line is the major synoptic system, which generated heavy rainfalls in this region.

There were at least two kinds of low-level vortices that generated heavy rains during the Meiyu season. One was the SW (southwest) vortex. It was generated on the lee-side of the Tibetan Plateau and tended to be stationary if there was no upper-level trough to steer it out of the Sichuan Basin. It could produce heavy rainfall locally in Sichuan Basin. Once it is steered out and moves eastward, it moves along the Meiyu shear line in most cases and moves northeastward or southeastward in some others. Another kind of low vortex is the intermediate-scale cyclone which formed along the Meiyu front with a horizontal scale of 1000-3000 km. Results of a case study of the heavy rain event in 23-25 June 1983 over the Yangtze River Valley by Ma and Bosart (1987) revealed that a quasi-stationary frontal boundary, separating very warm and moist tropical Pacific air from slightly cooler but still moist air, served to focus the rains in a relatively narrow latitudinal band. The situation was consistent with those obtained in the previous studies for the occurrence of heavy rains in the Meiyu season over the Yangtze River Valley (Ding 1992; Tao and Chen 1987; Tao and Ding 1981).

The Sichuan flood catastrophe event of 11-15 July 1981 was found to be directly related to the extreme development of a long-lived SW vortex over the basin (S. J. Chen and Dell'Osso 1984; Kuo *et al.* 1986, 1988). It was found that the formation of the SW vortex

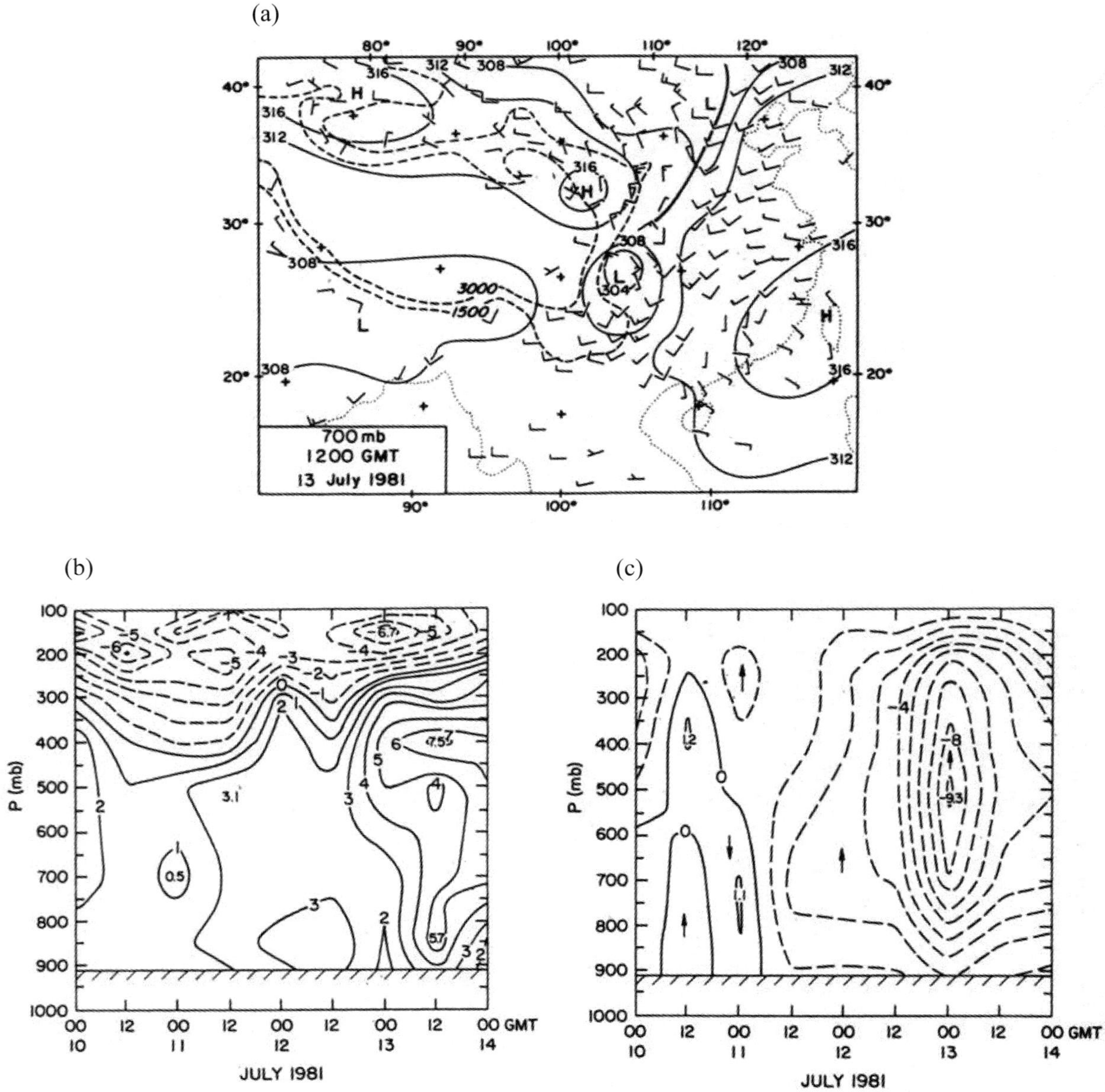

Figure 24. (a) The 700 mb analyses at 1200 UTC 13 July 1981. One full barb = 10 m s^{-1}. Solid contours are geopotential height (dam). The time-height section of (b) relative vorticity (ζ), and (c) vertical motion (ω) averaged over an area of 400 km × 400 km following the center of the SW vortex. The unit for relative vorticity is 10^{-5} s^{-1} and that for ω is μb s^{-1} (from Kuo *et al.* 1986).

was a consequence of the blocking of the southwesterly monsoon flow by the mesoscale mountain range (located to the southeastern corner of the Tibetan Plateau) and was not influenced by diabatic processes. However, the latent heat was essential for the development of the SW vortex (Kuo *et al.* 1986, 1988). Figure 24 presents the SW vortex of this case on the 700 hPa level at 1200 UTC 13 July when it reached a maximum intensity and the time-height sections for relative vorticity and vertical motion in 10-14 July. It is clear that the SW vortex was a shallow, low-level vortex during the early stages of its development, with major vorticity confined below 700 hPa. By 1200 UTC 13 July, the vortex showed strong cyclonic vorticity from the surface up to 250 hPa. Even at the mature stage, a vorticity maximum was clearly shown at 850 hPa. Upward motion gradually amplified as the SW vortex increased its intensity. By 0000 UTC 13 July, a strong upward motion was established at 500 hPa with a maximum intensity of -9.3 $\mu b\ s^{-1}$. For the low vortex along the Meiyu front, it was also found that the latent heat played a major role for its development (Chang *et al.* 2000a; Zhao *et al.* 1982). In one of the very strong intensification cases, Chang *et al.* (1998) found that the vertical coupling processes compounded the diabatic heating effect that was sustained by tropical moist air, leading to the strong intensification.

Due to the observational spatial data limitations in China, very little work has been done on meso-β-scale systems. The Meiyu experiment over the middle and lower reaches of the Yangtze River (1980-1983) for the first time provided an opportunity for studying this system on the horizontal scale of 25-250 km, by using the denser network of the upper-air and surface observations. The major findings have been summarized in the monograph by Zhang (1990). It was found that the meso-β-scale systems occurred in advance of the forward tilting minor wave trough which was located near the Meiyu cloud and rain bands, on the right side of the upper-level jet, and the left side of the low-level jet. In general, this system was associated with the mesoscale shear line.

In summary, the Yangtze River Valley shear line and low-level vortex are the meso-α-scale systems, which generate heavy rainfalls in the Yangtze River Valley Meiyu season. One kind of the low-level vortex is the SW (southwest) vortex, which occurs in Sichuan Basin on the lee-side of the Tibetan Plateau. The formation of the SW vortex is due to the local topographical blocking of the southwest monsoon flow and the latent heat is essential for the development of the SW vortex. Another kind of low-level vortex is the intermediate-scale cyclone (1000-3000 km), which forms along the Meiyu front and the latent heat release also plays a major role for its development. It was also found that the vertical coupling processes compounded the diabatic heating effect could lead to the strong intensification of the Meiyu frontal cyclone.

7. Orographically Related Disturbances

The lower tropospheric mesolows often form over southeastern, southwestern, and northwestern Taiwan when the Meiyu front passes over Taiwan or its vicinity (G. Chen 1978b, 1990a; Y. L. Lin 1993). Figure 25 presents the frequency distribution of the mesolow appearance in the Taiwan Meiyu season of 1972-1977. The mean lift span of mesolows was about 12-15-h. Climatological analysis by G. Chen (1978b) showed that the rainfall and mesolow over western Taiwan were positively correlated and the converse was true for the mesolow over eastern Taiwan. G. Chen and Chi (1980b) observed a close relationship between the heavy rainfall and mesolow over northwestern Taiwan. It was suggested that the mesolow over northwestern Taiwan probably served as a mechanism for producing heavy rainfall through enhanced southwesterlies. A case study by G. Chen (1979b) showed that mesolow formation and the accompanied wind changes over southwestern Taiwan were closely related to the enhancement of convective rainfall. This close relationship again existed in another study by G. Chen (1990b) using 18 cases of heavy rainfall events accompanying mesolows on the west side of the CMR in 1983-1987 Meiyu season. In that study the rainfall rate and radar echo intensified substantially when and after the mesolow formed. This intensification was perhaps due to the increase of local convergence, which was caused by the increase of pressure gradient and low-level winds to the south of mesolow.

A case study by G. Chen (1979b) showed that a mesolow on the west coast of Taiwan had a horizontal scale of 200 km and was characterized by strong cyclonic vorticity, horizontal convergence and boundary layer upward motion. G. Chen and Yu (1990) found the mesolow over southwestern Taiwan was quite shallow (limited in the lowest 1.6 km) with a horizontal scale of 150 km and was responsible for determining the intensity of heavy rainfall in southern Taiwan. In TAMEX field phase, a mesolow east of Taiwan was observed by P-3 aircraft on 16 May 1987. Figure 26 presents the structure as analyzed using *in situ* observations from the aircraft coupled with the conventional data collected in the IOP (Kuo and Chen 1990). The analysis shows two surface lows on the east side of the island. The southern low, which took an elongated shape, was located very close to the mountain over southeastern Taiwan where was the most favorable area for mesolow formation and appearance as illustrated in Fig. 25. The other center (the primary mesolow) was located east of Hualien, to the northeast of the first low-pressure center. The central pressure of the lows was 1007 hPa, ~2 -3 hPa lower than the synoptic-scale pressure field. In addition to these two low-pressure centers, a mesohigh was observed on the windward side of the CMR (over southwestern Taiwan). The high/low pressure couplet across the CMR is consistent with the theoretical results of flow blocking by mesoscale mountains (Smith 1982). Indeed, the wind field analysis (Fig. 26b) at the flight level (~900 hPa) clearly revealed the low-level blocking and flow splitting upstream of the CMR. A strong horizontal wind shear existed to the east of Taiwan between the weak northerly flow and the strong southwesterly flow. On the eastern part of this shear line, a closed cyclonic circulation associated with the mesolow was observed. An analysis of the thermodynamic structure showed that the southwesterly flow was generally warmer and drier than the northerly flow (Fig. 26c and 26d). This cyclonic mesolow appeared to have a warm and dry core structure indicating the importance of subsidence in its formation.

In addition to the lower tropospheric mesolow, a mid-level cyclonic vortex was observed in TAMEX to the east of the CMR (G. Chen and Liang 1992). It had a horizontal scale of about 100-200 km and a life span of more than 24-h. Analyses showed that this vortex had a cold, moist core in the middle and lower troposphere (900-550 hPa; 1-5 km) and tilted slightly northwestward towards the mid-tropospheric cold air in the early stage. It developed upward to the 300 hPa (~10 km) warm region and downward to 950 hPa (~0.5 km) in the later stage. It was proposed that the blocking effect coupled with the vertical vortex stretching process might be responsible for the formation of this vortex.

One of the very important issues related to the mesolow is the formation mechanism in different geographic areas to the southeast, southwest and northwest of Taiwan. For the mesolow over southeastern Taiwan, subsidence warming coupled with vertical vortex stretching process is one possible mechanism as suggested by observations which indicated a warm and dry lower troposphere and less rainfall in the mesolow area (G. Chen 1978b, 1990a; Kuo and Chen 1990; S. Wang 1989; S. Wang and Chen 1990) and by the model simulations (Kuo *et al.* 1989; Y. L. Lin 1989, 1990; Mannouji and Kurihara 1990; Sun *et al.* 1991). Low-level blocking coupled with flow separation is another mechanism suggested by observations (S. Wang 1989; S. Wang and Chen 1990) and by model studies (Huang and Raman 1990; Sun and Wu 1989). Numerical study by Soong *et al.* (1989), on the other hand, suggested

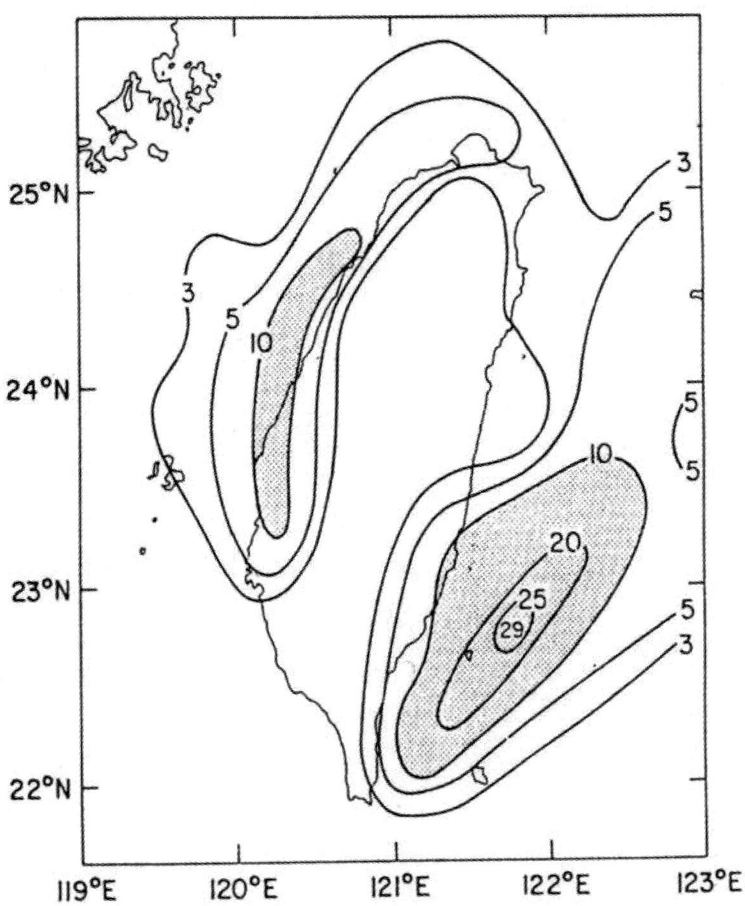

Figure 25. Frequency distribution of mesolow appearance near Taiwan (inside 0.5° lat × 0.5° long boxes) during the Mei-Yu season (15 May to 18 June) of axis years from 1972 to 1977 (from G. Chen 1978).

that the diabatic heating and the associated sea breeze circulation could also be a mechanism for the formation of mesolow over southeastern Taiwan. Recently, diagnostic case studies of mesolow formation over southeastern Taiwan by C. Wang and Chen (2002, 2003) revealed that mesolow could form under different Froude number (Fr) flow regimes as presented in Fig. 27. The top panel illustrates the processes found in type L when both Fr and Rossby

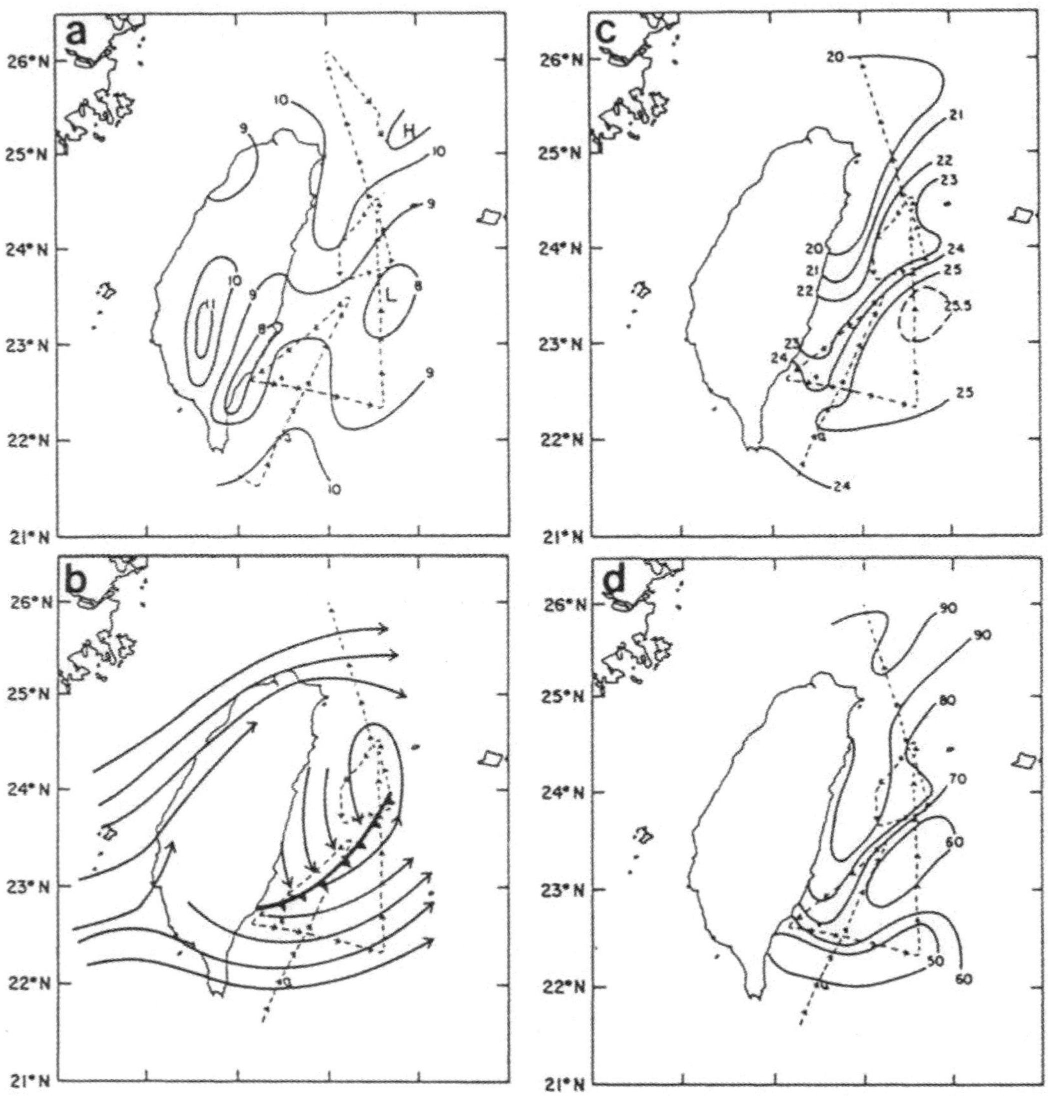

Figure 26. Analysis at 1800 UTC 16 May 1987: (a) sea-level pressure (hPa), (b) 900-hPa streamline, (c) 900-hPa temperature (°C), and (d) 900-hPa relative humidity (%). Dashed arrows indicate P-3 flight tracks (from Kuo and Chen 1990).

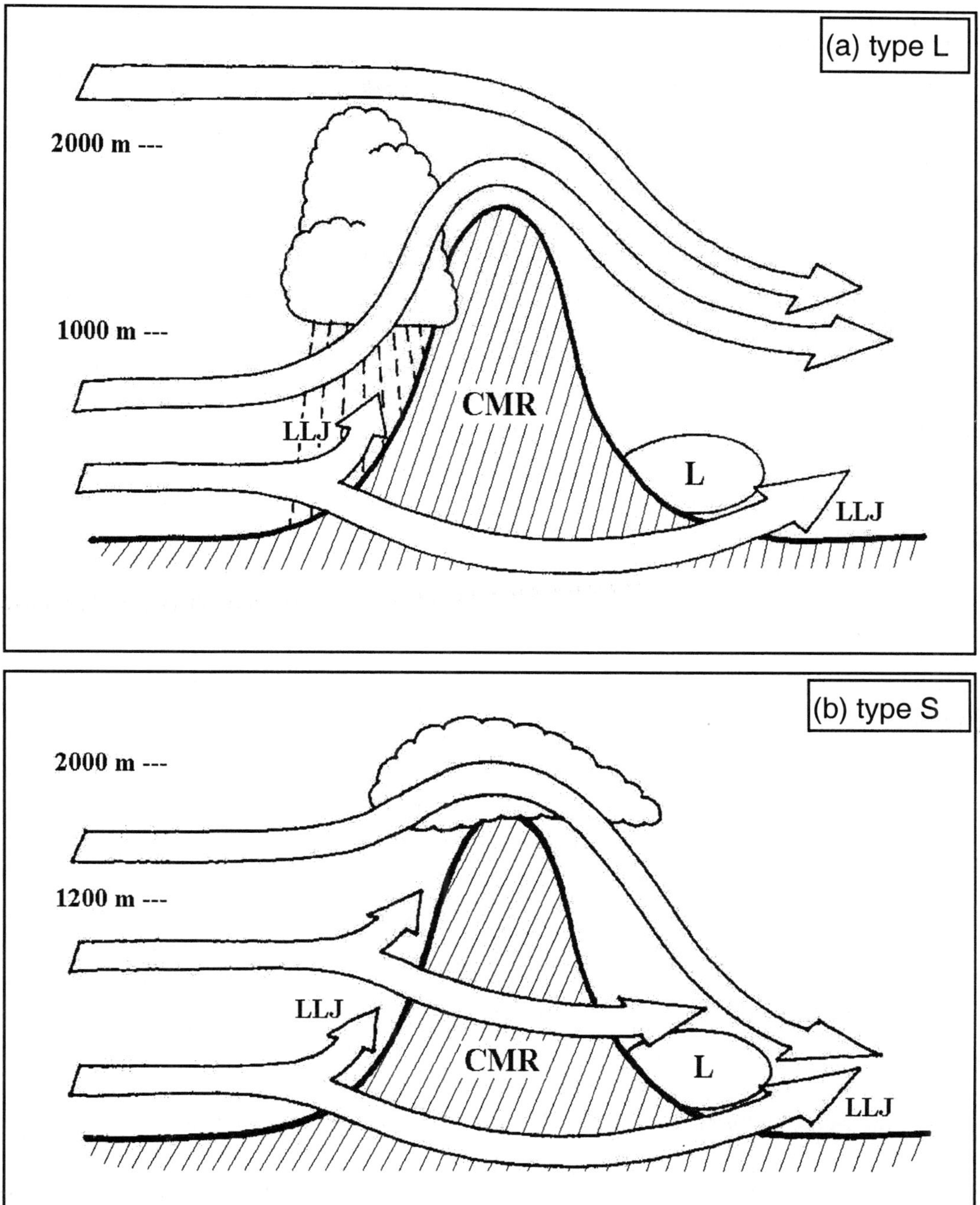

Figure 27. Schematic illustration of processes found (a) in the type L case (adopted from C. Wang and Chen 2003), and (b) in the type S case. Thick solid line and hatching represent the southern Central Mountain Range (CMR) of Taiwan, and open arrows represent trajectories of airflow at different levels. Locations of cloud, precipitation, low-level jet (LLJ), and leeside mesolow are also indicated.

number (Ro) were relatively large ($Fr \geq 0.5$, $Ro \geq 2.0$). Only air at the lowest level was blocked by the CMR and split upstream, producing LLJs both to the northwest and southeast of the terrain. Flow at higher elevations (above about 800 m), was forced to climb along the windward slope and produced heavy precipitation, then subsided at the leeside. Air parcels at yet higher levels above the terrain also subsided directly at lee. The sinking led to mesolow formation through adiabatic warming, and the bottom of the sinking was near 1000 m. In type S when Fr and Ro were relatively small ($Fr \leq 0.2$, $Ro \leq 1.0$), a thick layer of low-level air (up to ~1400m) was blocked by the CMR and also split upstream due to stronger Coriolis deflection (Fig. 27b). Air parcels above 1400 m rose at the windward side, but the shorter distance of uplifting and lower initial moisture content produced nearly no rain, and latent heating contributed little in raising θ values during the ascent. However, due to strong upstream blocking, air parcels subsided at the leeside directly from an elevation of about 2000 m to nearly 500 m, considerably closer to the surface than that in type L.

For the mesolow over northwestern Taiwan, a modeling study by Chern and Sun (1989) suggested that the formation is due to subsidence warming and lack of cold advection under the prevailing northeasterly flow to the north of Meiyu front. On the other hand, results of a non-linear semi-geostrophic model by Y. L. Lin (1989) indicated that a mesolow forms over northwestern Taiwan under the prevailing southwesterly barotropic flow with a diabatic cooling effect. Increasing convergence produced by diabatic cooling appears to be responsible for the formation of upstream low in this case. For the mesolow over southwestern Taiwan, observational studies suggested that both the convective latent heating (G. Chen 1990b; G. Chen and Yu 1990) and lack of cold air due to the flow blocking by the CMR (Y. Chen and Hui 1990; Y. Chen *et al.* 1989) were possible formation mechanism. Numerical study by Y. L. Lin (1989) using a non-linear semi-geostrophic model, on the other hand, suggested that a possible mechanism appears to be the diabatic cooling effect, which increases convergence over the mesolow area under the prevailing northwesterly baroclinic flow.

Besides the mesolows, the orographically induced mesoscale circulations were often observed due to the blocking and deflecting effects of the CMR on the prevailing flow under relatively undisturbed conditions. From synoptic experience, the deflecting effect of the prevailing westerly or southwesterly flow can produced local convergence in northwestern Taiwan and divergence in southwestern Taiwan. In the TAMEX IOP-13 case (Chi and Chen 1989a), the MCS intensified in northwestern Taiwan where the frontal forcing and the local convergence produced by the deflecting flow seemed to dominate. As the MCS moved over southwestern Taiwan, it weakened in the area of local divergence produced by deflecting flow. A case study of IOP-8 by Trier *et al.* (1990) showed that the splitting flow produced by the influence of the island barrier was felt at least 150 km from the island. The deceleration of the low-level prefrontal westerlies results in a splitting of flow around the island produced higher surface pressure and divergent wind field in the vicinity of the southwest coast. The flow deflecting and splitting features under the prevailing southwesterlies have been well captured in the model simulations. The importance of the CMR in generating these features was clearly demonstrated in a modeling study by Mannouji and Kurihara (1990). Huang and

Raman (1990) found that a strong southwesterly Taiwan due to the flow deflection. This is supported by an observational study (Shu 1991) in that the southwesterly LLJ tended to intensify in northwestern Taiwan during TAMEX. Another interesting feature in the model simulation is that a line of convergence appeared to form to the northeast of Taiwan due to the blocking and splitting effects of the CMR (Soong *et al.* 1989). However, the observational study of this feature is hampered by the lack of data over the ocean.

A splitting of the postfrontal flow at the northern tip of the island was observed in TAMEX IOP-8 case by Trier *et al.* (1990) and IOP-9 case by Y. Chen and Hui (1990) who estimated the Froude number to be 0.3. This value is within the theoretical limit for flow blocking by obstacles. The cold air stacking over the windward slope led to the pressure increase, which was more significant on the northeast coast than on the northwest coast. The splitting ensured that the shallow, cold air mass never reached the higher elevations of central Taiwan. In fact, a couplet of mesolow in southwestern Taiwan and a mesoscale ridge on the east side of the CMR (G. Chen and Liang 1992; Y. Chen *et al.* 1989) was primarily due to these blocking and splitting effects. The blocking effect over the southern Taiwan Strait off the south-western coast using P-3 aircraft *in situ* measurements (Y. Chen and Hui 1990) indicated that the depth of cold air was about 450 m to the south of 23°N in the immediate vicinity of southwestern coast. However, this depth increased to about 1 km in the region where the northeasterlies were not obstructed by the island barrier. The shallowness of the cold air in the lee area was suggested by Chen and Hui to be due to the lateral spreading of the cold air as it moved around the island.

In summary, the lower tropospheric shallow mesolows form over southeastern, southwestern, and northwestern Taiwan when the Meiyu front passes over Taiwan or its vicinity. The southeastern mesolow can be associated either with the heavy rain or no rain situation over the windward side and is generally accompanied by the clear sky due to the subsidence over the lee-side mesolow area. The enhanced southwesterlies and local convergence during the formation of northwestern and southwestern mesolows are suggested to be responsible for the occurrence of heavy rainfall over and to the south of the mesolows. Formation mechanisms of mesolows over different areas are studied and proposed. These include the subsidence warming coupled with vertical vortex stretching, low-level blocking coupled with flow separation, diabatic heating coupled with sea breeze circulation, latent heating coupled with subsidence warming, and blocking coupled with subsidence warming for the southeastern mesolow. The formation mechanisms proposed for the northwestern mesolows include subsidence warming coupled with the lack of cold advection and southwesterly barotropic flow coupled with diabatic cooling. Whereas for the southwestern mesolows, convective latent heating and northwesterly baroclinic flow coupled with diabatic heating are being proposed. In addition to the formation of the mesolows, the upstream flow blocking, deflecting, and splitting also occur due to the topographical effect of the Taiwan CMR. These mesoscale flow features are responsible for the formation of barrier jet and the enhancement of rainfall over northwestern Taiwan and for the suppression of rainfall over southwestern Taiwan.

8. Concluding Remarks

Meiyu (or Baiu) is a unique regional weather and climate phenomenon occurring during late spring and early summer over East Asia and the western North Pacific. Because of the seasonal rainfall maximum associated with Meiyu, it is very important to agriculture, water resources, and human activities in China and Taiwan. The heavy rainfall and the associated flash flooding occurred in the Meiyu season often caused tremendous property damage and loss of life. Therefore, one of the key issues in meteorological community is to improve, through better understanding, the forecasting of heavy rainfall events that lead to flash floods. The major objective of the Taiwan Area Mesoscale Experiment (TAMEX) in 1983-1993 was to study various scientific problems relevant to the heavy rainfall events. What we have learned about Meiyu in the past 25 years can be summarized in the following.

Climatologically, the mean circulations possess strong southwest monsoonal flows in the lower troposphere to the south of the Meiyu front which is characterized by the organized distribution of upward motion and the moisture flux convergence. The convective activity develops under the potentially unstable atmosphere along and to the south of the Meiyu front. The Meiyu (Baiu) rainfall in Japan and the Yangtze River Valley is positively correlated to the eastern equatorial Pacific SST, whereas in South China and Taiwan this relationship becomes variable.

The Meiyu front affecting South China and Taiwan forms in the subtropical latitude, which is a distinct area from that for the formation of polar front in the Meiyu season. It resembles a semitropical disturbance with an equivalent barotropic warm core structure, a weak horizontal temperature gradient, a rather strong horizontal wind shear, and a positive low-level potential vorticity (PV) anomaly. The CISK mechanism has been suggested for the Meiyu frontogenesis through the interaction between the PV anomaly and the convective latent heating. Some of the Meiyu fronts observed in TAMEX were characterized by the density-current-type structure at the leading edge with a thermally direct circulation cell across the front. The frontal deformation occurs with different propagation speeds to the east and west of the Taiwan Central Mountain Range. The acceleration of highly ageostrophic flow of cold air in response to the building pressure gradient has been suggested to the responsible for the faster moving front to the east of Taiwan.

The existence of the LLJ to the south of the Meiyu front is crucial to the occurrence of convective heavy rainfall by providing strong low-level moisture flux convergence, vertical shear, and upward motion. The LLJ observed in TAMEX had a highly supergeostrophic nature with quite different characteristics of secondary circulation across the jet axis from those of the upper-level jet. A thermally indirect circulation was observed in the entrance and the jet core regions and a thermally direct circulation existed over the exit region. An LLJ can form to the south of the convective heavy rainfall area through the Coriolis acceleration of the lower branch of an induced secondary circulation by convective latent heating.

Concerning the initiation and development of MCSs and heavy rainfalls, favorable environmental conditions and mesoscale forcings in the Meiyu season are identified and summarized. The updrafts and cores of MCSs in TAMEX were comparable in size and strength with those measured in GATE and hurricanes but much weaker than those measured in continental thunderstorms. The oceanic convection in TAMEX was dominated by warm rain coalescence processes, similar to that in GATE and hurricanes. The structural feature of some squall lines in TAMEX was similar to that for a fast-moving tropical squall line. The environmental conditions during the TAMEX squall lines were found with characteristics between tropical and midlatitude squall lines.

The Yangtze River Valley shear line and low-level vortex are the meso-α-scale systems, which generate heavy rainfalls in the Yangtze River Valley Meiyu season. One kind of the low-level vortex is the SW (southwest) vortex, which occurs in Sichuan Basin on the lee-side of the Tibetan Plateau. The formation of the SW vortex is due to the local topographical blocking of the southwest monsoon flow and the latent heat is essential for the development of the SW vortex. Another kind of low-level vortex is the intermediate-scale cyclone (1000-3000 km), which forms along the Meiyu front and the latent heat release also plays a major role for its development. The vertical coupling processes compounded the diabatic heating effect could lead to the strong intensification of this kind of frontal cyclone.

The mesoscale features which are related to the topographical effects of the Taiwan CMR and are important factors for the formation and enhancement of MCSs and rainfalls include mesolows over northwestern, southwestern, and southeastern Taiwan, and the flow blocking, deflecting, and splitting upstream of the CMR. Some of the heavy rainfall events over northwestern Taiwan are due to the formation of barrier jet and the local convergence as produced by the blocking and deflecting effects under the prevailing low-level southwesterlies.

We have attempted to give an overview of the recent research results on the climatological characteristics of Meiyu and the associated structure and dynamics of the synoptic and mesoscale features. It is clear that our knowledge is far from complete. More research efforts will be needed to conduct observational, theoretical, and modeling studies in order to improve our understanding of the Meiyu phenomenon.

Acknowledgments

Comments and suggestions to improve the manuscript from Professor C. P. Chang and the reviewer are highly appreciated. Thanks are to Professor Z. H. Jiang, Miss Mei-Lan Chen, and Mr. An-Hsiang Wang for preparing this manuscript. This paper is partially supported by NSC91-2119-M-002-029 and NSC91-2111-M-002-030.

References

Akaeda, K., T. Yokoyama, A. Tabata, M. Ishihara, and H. Sakakibara, 1991: Evolution of the kinematic structure within a meso-β-scale convective system in the growing and mature stages. *Mon. Wea. Rev.*, **119**, 2664-2676.

Akiyama, T., 1973: Ageostrophic low-level jet stream in the Baiu season associated with heavy rainfalls over the sea area. *J. Meteor. Soc. Japan*, **51**, 205-208.

Bell, G. D., and L. F. Bosart, 1988: Appalachian cold-air damming. *Mon. Wea. Rev.*, **116**, 137-161.

Bresch, J. F., and R. H. Johnson, 1990: Rainfall and vertical motion associated with TAMEX precipitation systems. *Proc. Workshop on TAMEX Scientific Results, NCAR*, 24-26 September, 84-89. [Available from Mesoscale and Microscale Meteor. Div., NCAR, P.O. Box 3000, Boulder, CO 80307]

Chang, C. P., and G. T. J. Chen, 1995: Tropical circulations associated with southwest monsoon onset and westerly surges over the South China Sea. *Mon. Wea. Rev.*, **123**, 3254-3267.

______, S. C. Hou, H. C. Kuo, and G. T. J. Chen, 1998: The development of an intense East Asian summer monsoon disturbance with strong vertical coupling. *Mon. Wea. Rev.*, **126**, 2692-2712.

______, L. Yi, and G. T. J. Chen, 2000a: A numerical simulation of vortex development during the 1992 East Asian summer monsoon onset using the Navy's regional model. *Mon. Wea. Rev.*, **128**, 1604-1631.

______, Y. Zhang, and T. Li, 2000b: Interannual and interdecadal variations of the East Asian summer monsoon and the tropical sea-surface temperatures. Part 1: Relationships with Yangtze River Valley rainfall. *J. Climate*, **13**, 4310-4325.

______, ______, and ______, 2000c: Interannual and interdecadal variations of the East Asian summer monsoon and the tropical sea-surface temperatures. Part 2: Meridional structure of the monsoon. *J. Climate*, **13**, 4326-4340.

Chang, C. Y., and M. Yoshizaki, 1991: A numerical study of the mesoscale convective system observed over Okinawa Island in June 1987. *Mon. Wea. Rev.*, **119**, 2724-2733.

Chen, C. K., 1979: An analysis of the relationship between the low level jet stream and the heavy rainfall during the Mei-Yu season in Taiwan. *Atmos. Sci.*, **6**, **1**, 29-37 (in Chinese with English abstract).

Chen, C. S., 1989: The relationship between low-level lifting and the formation of new convection in Mei-Yu season. *Papers Meteor. Res.*, **12**, 1-16.

______, 1990: A numerical study of the terrain effects on a squall line. *Terrestrial Atmos. Oceanic Sci.*, **1**, 73-90.

______, and C. Y. Lin, 1996: A study of a precipitation system in northeastern Taiwan during TAMEX IOP#10. *Meteor. Atmos. Phys.*, **59**, 185-200.

______, and ______, 1997: A preliminary study of the formation of precipitation systems under undisturbed conditions during TAMEX. *Meteor. Atmos. Phys.*, **64**, 83-105.

Chen, G. T. J., 1977a: An analysis of moisture structure and rainfall for a Meiyu regime in Taiwan. *Proc. Natl. Sci. Counc.*, **1**, **11**, 1-21.

______, 1977b: A synoptic case study on mean structure of Mei-Yu in Taiwan. *Atmos. Sci.*, **4**, 38-47.

______, 1978a: The structure of a subtropical Mei-Yu system in Southeast Asia. *Sci. Rep., Dept. of Atmos. Sci. Natl. Taiwan Univ.*, **2**, 9-23. [Available from the Dept. of Atmos. Sci. Natl. Taiwan Univ., Taipei, Taiwan, R.O.C.]

______, 1978b: On the meso-scale systems for the Mei-Yu regime in Taiwan. *Proc. Conf. Severe Weather in Taiwan Area*, NSC and Academia Sinica, 150-157 (in Chinese with English abstract). [Available from the Dept. of Atmos. Sci. Natl. Taiwan Univ., Taipei, Taiwan, R.O.C.]

______, 1979a: On the moisture budget of a Mei-Yu system in southeastern Asia. *Proc. Natl. Sci. Counc.*, **3**, **1**, 24-32.

______, 1979b: Mesoscale analysis for a Mei-Yu case over Taiwan. *Papers Meteor. Res.*, **2**, 63-74.

______, 1983: Observational aspects of the Mei-Yu phenomena in subtropical China. *J. Meteor. Soc. Japan*, **61**, 306-312.

______, 1985: Feasibility study of "A Severe Regional Precipitation Observation and Analysis Experiment". *Natl. Sci. Counc., Sci. and Tech. of Disaster Prevention Program, Tech. Rep.* 73-42, 32 pp (in Chinese with English abstract). [Available from the Dept. of Atmos. Sci., Natl. Taiwan Univ., Taipei, Taiwan, R.O.C.]

______, 1988: On the synoptic-climatological characteristics of the East Asian Mei-Yu front. *Atmos. Sci.*, **16**, 435-446 (in Chinese with English abstract).

______, 1990a: On the climatological characteristics of the mesolow in Taiwan Mei-Yu season. *Atmos. Sci.*, **18**, 73-84 (in Chinese with English abstract).

______, 1990b: Study of rainfalls and radar echoes in the heavy rainfall events accompanied by mesolow in Mei-Yu season. *Atmos. Sci.*, **18**, 213-228 (in Chinese with English abstract).

______, 1990c: Overview of Mei-Yu research in Taiwan. *East Asia and Western Pacific Meteorology and Climate*, P. Sham and C. P. Chang, Eds., World Scientific Publishing Co., 14-37.

______, 1992: Mesoscale features observed in the Taiwan Mei-Yu season. *J. Meteor. Soc. Japan*, **70**, 497-516.

______, 1994: Large-scale circulations associated with the East Asian summer monsoon and the Mei-Yu over South China and Taiwan. *J. Meteor. Soc. Japan*, **72**, 959-983.

______, and C. P. Chang, 1980: The structure and vorticity budget of an early summer monsoon trough (Mei-Yu) over southeastern China and Japan. *Mon. Wea. Rev.*, **108**, 942-953.

______, and S. S. Chi, 1978: On the meso-scale structure of Mei-Yu front in Taiwan. *Atmos. Sci.*, **5**, **1**, 35-47 (in Chinese with English abstract).

______, and ______, 1980a: On the frequency and speed of Mei-Yu front over southern China and the adjacent areas. *Papers Meteor. Res.*, **3**, **1&2**, 31-42.

______, and ______, 1980b: On the mesoscale rainfall and mesolow in the Mei-Yu season in Taiwan. *Atmos. Sci.*, **7**, 39-48 (in Chinese with English abstract).

______, and H. C. Chou, 1993: General characteristics of squall lines observed during TAMEX. *Mon. Wea. Rev.*, **121**, 726-733.

______, and Y. S. Hsu, 1997: Composite structure of a low level jet over southern China observed during the TAMEX period. *J. Meteor. Soc. Japan*, **75**, 1003-1018.

______, and C. Y. Liang, 1992: A midlevel vortex observed in the Taiwan Area Mesoscale Experiment (TAMEX). *J. Meteor. Soc. Japan*, **70**, 25-41.

______, and C. P. Pu, 1985: A case study of the formation of a low level jet over subtropical China and heavy precipitation in northern Taiwan. *Atmos. Sci.*, **12**, 23-32 (in Chinese with English abstract).

______, and C. Y. Tsay, 1978: A synoptic case study of Mei-Yu near Taiwan. *Papers Meteor. Res.*, **1**, 25-36.

______, C. C. Wang, and S. C. S. Liu, 2003: Potential vorticity diagnostics of a Mei-Y u front case. *Mon. Wea. Rev.*, **131**, 2680-2696.

______, ______, and J. S. Yang, 2002: Characteristics of spatial and temporal distribution of convective precipitation in Taiwan Mei-Yu season. *Atmos. Sci.*, **30**, 83-98 (in Chinese with English abstract).

______, and C. C. Wu, 1978: On the climatological characteristics of five cities in Taiwan. *Atmos. Sci.*, **5**, **2**, 1-16 (in Chinese with English abstract).

______, C. W. Wu, and S. S. Chi, 1986: Climatological aspects of the mesoscale convective systems over subtropical China and the western North Pacific during Mei-Yu season of 1981-1983. *Atmos. Sci.*, **13**, 33-45 (in Chinese with English abstract).

______, and C. C. Yu, 1988: Study of low-level jet and extremely heavy rainfall over northern Taiwan in the Mei-Yu season. *Mon. Wea. Rev.*, **116**, 884-891.

______, and______, 1990: Role of Mei-Yu front and mesolow on the heavy rainfall events: Two cases in TAMEX phase I (1986). *Atmos. Sci.*, **18**, 129-147 (in Chinese with English abstract).

Chen, Q., 1982: The instability of the gravity-inertia wave and its relation to low-level jet and heavy rainfall. *J. Meteor. Soc. Japan*, **60**, 1041-1057.

Chen, S. J., and L. Dell'Osso, 1984: Numerical prediction of the heavy rainfall vortex over Eastern Asia monsoon region. *J. Meteor. Soc. Japan*, **62**, 730-747.

______, Y. H. Kuo, W. Wang, Z. Y. Tao, and B. Cui, 1998: A modeling case study of heavy rainstorms along the Mei-Yu front. *Mon. Wea. Rev.*, **126**, 2330-2351.

______, W. Wang, K. H. Lau, Q. H. Zhang, and Y. S. Chung, 2000: Mesoscale convective systems along the Meiyu front in a numerical model. *Meteor. Atmos. Phys.*, **75**, 149-160.

Chen, X. A., and Y. Chen, 1995: Development of low-level jet during TAMEX. *Mon. Wea. Rev.*, **123**, 1695-1719.

Chen, Y. L., X. A. Chen, S. Chen, and Y. H. Kuo, 1997: A numerical study of the low-level jet during TAMEX IOP 5. *Mon. Wea. Rev.*, **125**, 2583-2604.

______, ______, and Y. X. Zhang, 1994: A diagnostic study of a low-level jet during TAMEX IOP 5. *Mon. Wea. Rev.*, **122**, 2257-2284.

______, and N. B. F. Hui, 1990: Analysis of a shallow front during Taiwan Area Mesoscale Experiment. *Mon. Wea. Rev.*, **118**, 2649-2667.

______, and S. F. Tseng, 2000: Comments on "The intensification of the low-level jet during the development of mesoscale convective systems on a Mei-Yu front". *Mon. Wea. Rev.*, **128**, 495-506.

______, Y. X. Zhang, and N. B. F. Hui, 1989: Analysis of a surface front during the early summer rainy season in Taiwan. *Mon. Wea. Rev.*, **117**, 909-931.

Chern, J. D., and W. Y. Sun, 1989: Lee cyclogenesis and interaction between front and mountain. *Proc. Workshop on TAMEX Preliminary Scientific Results*, *Taipei*, 22-30 June, 343-346. [Available from the Dept. of Atmos. Sci., Natl. Taiwan Univ., Taipei, Taiwan, R.O.C.]

Chi, S. S., and G. T. J. Chen, 1988: A diagnostic case study of the environmental conditions associated with mesoscale convective complexes: 27-28 May 1981 Case. *Atmos. Sci.*, **16**, 14-30 (in Chinese with English abstract).

______, and ______, 1989a: A study on the MCS case of TAMEX IOP-13. *Proc. Workshop on TAMEX Preliminary Scientific Results*, *Taipei*, 22-30 June, 14-21. [Available from the Dept. of Atmos. Sci., Natl. Taiwan Univ., Taipei, Taiwan, R.O.C.]

______, and ______, 1989b: Case study of the MCSs and rainfall during TAMEX/Phase I. *Atmos. Sci.*, **17**, 59-75 (in Chinese with English abstract).

______, and______, 1989c: A moisture budget analysis of two MCC cases during Taiwan Mei-Yu season. *Papers Meteor. Res.*, **12**, 143-157.

Chiou, T. K., and S. Y. Liao, 1984: A study of mesoscale convective system in the southern China and its vicinity. *Atmos. Sci.*, **11**, 85-100 (in Chinese with English abstract).

______, and C. S. Chen, 1989: The structure of mesoscale convective systems in southern China during the Mei-Yu period. *Papers Meteor. Res.*, **11**, 195-217.

Cho, H. R., and G. T. J. Chen, 1994: Convection and Mei-Yu front. *Terrestrial Atmos. Oceanic Sci.*, **5**, 121-136.

______, and ______, 1995: Mei-Yu frontogenesis. *J. Atmos. Sci.*, **52**, 2109-2120.

Chou, L. C., C. P. Chang, and R. T. Williams, 1990: A numerical simulation of the Mei-Yu front and the associated low-level jet. *Mon. Wea. Rev.*, **118**, 1408-1428.

Chu, C. M., and C. S. Chen, 1988: The analysis of environmental conditions associated with mesoscale convective systems in SE China. *Atmos. Sci.*, **16**, 349-371 (in Chinese with English abstract).

Ding, Y. H., and X. F. Wang, 1988: An analysis of distribution of apparent heat sources and sinks over the middle reaches of Yangtze River during the Meiyu season in 1983. *Tropical Meteorology*, **4**, 134-145 (in Chinese).

______, 1992: Summer monsoon rainfalls in China. *J. Meteor. Soc. Japan*, **70**, 373-396.

Forbes, G. S., R. A. Anthes, and D. W. Thomson, 1987: Synoptic and mesoscale aspects of an Appalachian ice storm associated with cold-air damming. *Mon. Wea. Rev.*, **115**, 564-591.

Hor, T. H., M. H. Chang, and B. J. D. Jou, 1998: Mesoscale structure of air flow in a Mei-Yu front leading edge observed by aircraft off the east coast of Taiwan during TAMEX IOP 9. *J. Meteor. Soc. Japan*, **76**, 473-496.

Hsu, M. T., and S. S. Chi, 1974: On the analysis of Mei-Yu in Taiwan. *Meteor. Bull.*, **20**, **4**, 25-44 (in Chinese with English abstract).

Hsu, W. R., and W. Y. Sun, 1994: A numerical study of a low-level jet and its accompanying secondary circulation in a Mei-Yu system. *Mon. Wea. Rev.*, **122**, 324-340.

Huang, C. Y., and S. Ramam, 1990: Numerical simulation of Taiwan island circulations: diurnal modification. *Proc. Workshop on TAMEX Scientific Results*, *NCAR*, 24-26 September, 199-203. [Available from Mesoscale and Microscale Meteor. Div., NCAR, P.O. Box 3000, Boulder, CO 80307]

Huang, R., and Y. Wu, 1989: The influence of ENSO on the summer climate change in China and its mechanism. *Adv. in Atmos. Sci.*, **6**, 21-32.

Huang, S. S., 1986: *The Presummer Heavy Rainfall over South China*. Guangdong Science and Technology Press, 243 pp.

Johnson, R. H., Z. Wang, and J. F. Bresch, 1993: Heat and moisture budgets over China during the early summer monsoon. *J. Meteor. Soc. Japan*, **71**, 137-152.

Jorgensen, D. P., and M. A. LeMone, 1989: Vertical velocity characteristics of oceanic convection. *J. Atmos. Sci.*, **46**, 621-640.

______, ______, and B. J. D. Jou, 1991: Precipitation and kinematic structure of an oceanic mesoscale convective system. Part I: Convective line structure. *Mon. Wea. Rev.*, **119**, 2608-2637.

Jou, B. J. D., J. S. Hong, and S. M. Deng, 1990: Dual-Doppler radar analysis of a convective (Mei-Yu) frontal rainband. *Atmos. Sci.*, **18**, 239-264 (in Chinese with English abstract).

Kuo, Y. H., and R. A. Anthes, 1982: Numerical simulation of a Mei-Yu system over southeastern Asia. *Papers Meteor. Res.*, **5**, 15-36.

______, and J. W. Bao, and S. Chen, 1989: Structure and dynamics of the mesolow observed during TAMEX IOP-2. Part II: mesoscale numerical simulations. *Proc. Workshop on TAMEX Preliminary Scientific Results*, *Taipei*, 22-30 June, 339-342. [Available from the Dept. of Atmos. Sci., Natl. Taiwan Univ., Taipei, Taiwan, R.O.C.]

______, and G. T. J. Chen, 1990: Taiwan Area Mesoscale Experiment: An overview. *Bull. Amer. Meteor. Soc.*, **71**, 488-503.

______, L. Cheng, and R. A. Anthes, 1986: Mesoscale analyses of Sichuan flood catastrophe, 11-15 July 1981. *Mon. Wea. Rev.*, **114**, 1984-2003.

______, ______, and J. W. Bao, 1988: Numerical simulation of the 1981 Sichuan flood. Part I:

Evolution of a mesoscale southwest vortex. *Mon. Wea. Rev.*, **116**, 2481-2504.

Lau, K. M., G. J. Yang, and S. H. Shen, 1988: Seasonal and intraseasonal climatology of summer monsoon rainfall over East Asia. *Mon. Wea. Rev.*, **116**, 18-37.

Lee, C. S., 1990: The moisture and energy budgets over Taiwan Strait during TAMEX. *Atmos. Sci.*, **18**, 111-127 (in Chinese with English abstract).

LeMone, M. A., and D. P. Jorgensen, 1991: Precipitation and kinematic structure of an oceanic mesoscale convective system. Part II: Momentum transport and generation. *Mon. Wea. Rev.*, **119**, 2638-2653.

Li, J., and Y. Chen, 1998: Barrier jets during TAMEX. *Mon. Wea. Rev.*, **126**, 959-971.

Lin, P. L., T. Wang, and C. C. Yeh, 1989: Doppler observational study of a long-lived rainband in TAMEX IOP 13. *Papers Meteor. Res.*, **12**, 91-119.

Lin, S. C., and T. K. Chiou, 1985: Objective scale separation technique and its application on the mesoscale convective system diagnosis. *Papers Meteor. Res.*, **8**, **2**, 69-94.

______, 1988: The life cycle and structure of a mesoscale convective system occurring in the southern China area during Mei-Yu season. *Papers Meteor. Res.*, **11**, 1-26.

______, and C. M. Tsai, 1989: Kinetic energy budgets of a mesoscale convective system occurring in the Mei-Yu season. *Atmos. Sci.*, **17**, 187-209 (in Chinese with English abstract).

Lin, S. M., and M. S. Lin, 1988: A case study of mesoscale convective systems during the TAMEX phase I. *Atmos. Sci.*, **16**, 163-174 (in Chinese with English abstract).

Lin, Y. J., T. Wang, R. W. Pasken, H. Shen, and Z. S. Deng, 1990: Characteristics of a subtropical squall line determined from TAMEX dual-Doppler data. Part 2: Dynamic and thermodynamic structure and momentum budgets. *J. Atmos. Sci.*, **47**, 2382-2399.

______, R. W. Pasken, and H. W. Chang, 1992: The structure of a subtropical prefrontal convective rainband. Part I: Mesoscale kinematic structure determined from dual-Doppler measurements. *Mon. Wea. Rev.*, **120**, 1816-1836.

______, H. W. Chang, and R. W. Pasken, 1993: The structure of a subtropical prefrontal convective rainband. Part II: Dynamic and thermodynamic structures and momentum budgets. *Mon. Wea. Rev.*, **121**, 1671-1687.

Lin, Y. L., 1989: A study of the flow over a mesoscale mountain with diabatic heating. *Proc. Workshop on TAMEX Preliminary Scientific Results, Taipei*, 22-30 June, 327-333. [Available from the Dept. of Atmos. Sci., Natl. Taiwan Univ., Taipei, Taiwan, R.O.C.]

______, 1990: Formation mechanisms of Taiwan mesolows during the Mei-Yu season. *Proc. Workshop on TAMEX Scientific Results, NCAR*, 24-26 September, 174-181. [Available from Mesoscale and Microscale Meteor. Div., NCAR, P.O. Box 3000, Boulder, CO 80307]

______, 1993: Orographic effects on airflow and mesoscale weather systems over Taiwan. *Terrestrial Atmos. Oceanic Sci.*, **4**, 381-420.

Luo, H. B., and M. Yanai, 1984: The large-scale circulation and heat sources over the Tibetan Plateau and surrounding areas during the early summer of 1979. Part II: Heat and moisture budgets. *Mon. Wea. Rev.*, **112**, 966-986.

Ma, K. Y., and L. F. Bosart, 1987; A synoptic overview of a heavy rain event in southern China. *Wea. Forecasting*, **2**, 89-112.

Maddox, R. A., 1980: Mesoscale convective complexes. *Bull. Amer. Meteor. Soc.*, **61**, 1374-1387.

______, 1983: Large-scale meteorological conditions associated with midlatitude, mesoscale convective complexes. *Mon. Wea. Rev.*, **111**, 1475-1493.

Mannouji, N., and K. Kurihara, 1990: A numerical experiment of TAMEX IOP #13 by the spectral limited area model of the JMA. *Proc. Workshop on TAMEX Scientific Results, NCAR*, 24-26 September, 121-126. [Available from Mesoscale and Microscale Meteor. Div., NCAR, P.O. Box 3000, Boulder, CO 80307]

Matsumoto, S., 1972: Unbalanced low-level jet and solenoidal circulation associated with heavy rainfalls. *J. Meteor. Soc. Japan*, **50**, 194-203.

______, and K. Ninomiya, 1969: On the role of convective momentum exchange in the mesoscale gravity wave. *J. Meteor. Soc. Japan*, **47**, 75-85.

______, K. Ninomiya, and S. Yoshizumi, 1971: Characteristic features of "Baiu" front associated with heavy rainfall. *J. Meteor. Soc. Japan*, **49**, 267-281.

Ninomiya, K., and T. Akiyama, 1974: Band structure of mesoscale echo cluster associated with low-level jet stream. *J. Meteor. Soc. Japan*, **52**, 300-313.

Parsons, D. B., S. B. Trier, and J. B. Klemp, 1989: Numerical simulations of frontal updrafts observed in subtropical and middle latitudes. *Proc. Workshop on TAMEX Preliminary Scientific Results, Taipei*, 22-30 June, 6-13. [Available from the Dept. of Atmos. Sci., Natl. Taiwan Univ., Taipei, Taiwan, R.O.C.]

Pu, C. P., and G. T. J. Chen, 1988: A preliminary analysis of the low-level jet and mesoscale convective systems over subtropical China during Mei-Yu season: The case of June 1-3, 1983. *Meteor. Bull.*, **34**, 285-297(in Chinese with English abstract).

Rotunno, R., J. B. Klemp, and M. J. Weisman, 1988: A theory for strong, long-lived squall lines. *J. Atmos. Sci.*, **45**, 463-485.

Saito, N., 1985: Quasi-stationary waves in mid-latitudes and the Baiu in Japan. *J. Meteor. Soc. Japan*, **63**, 983-995.

Shen, S., and K. M. Lau, 1995: Biennial oscillation associated with the East Asian monsoon and tropical sea surface temperatures. *J. Meteor. Soc. Japan*, **73**, 105-124.

Shu, I. S., 1991: A study of low-level jet in Taiwan area during TAMEX. M.S. Thesis, Dept. Atmos. Sci., Natl. Taiwan Univ., 89 pp (in Chinese). [Available from the Dept. of Atmos. Sci., Natl. Taiwan Univ., Taipei, Taiwan, R.O.C.]

Smith, R. B., 1982: Synoptic observation and theory of orographically disturbed wind and pressure. *J. Atmos. Sci.*, **39**, 60-70.

Soong, S. T., M. Mathur, and W. K. Tao, 1989: The effect of topography and diurnal heating cycle on airflow and precipitation in Taiwan. *Proc. Workshop on TAMEX Preliminary Scientific Results, Taipei*, 22-30 June, 304-309. [Available from the Dept. of Atmos. Sci., Natl. Taiwan Univ., Taipei, Taiwan, R.O.C.]

Sun, W. Y., and C. C. Wu, 1989: Numerical simulation of mesoscale circulation in Taiwan and surrounding area. *Proc. Workshop on TAMEX Preliminary Scientific Results, Taipei*, 22-30 June, 173-176. [Available from the Dept. of Atmos. Sci., Natl. Taiwan Univ., Taipei, Taiwan, R.O.C.]

______, J. D. Chern, C. C. Wu, and W. R. Hsu, 1991: Numerical simulation of mesoscale circulation in Taiwan and surrounding area. *Mon. Wea. Rev.*, **119**, 2558-2573.

Tanaka, M., 1997: Interannual and interdecadal variations of the western north Pacific monsoon and Baiu rainfall and their relationship to the ENSO cycles. *J. Meteor. Soc. Japan*, **75**, 1109-1123.

Tao, S. Y., and Y. H. Ding, 1981: Observational evidence of the influence of the Qinghai Xizang (Tibet) Plateau on the occurrence of heavy rain and severe convective storms in China. *Bull. Amer. Meteor. Soc.*, **62**, 23-30.

______, and L. X. Chen, 1987: A review of recent research on the East Asian summer monsoon in China. *Monsoon Meteorology*, C. P. Chang and T. N. Krishnamurti, Eds., Oxford University

Press, 60-92.

Tao, W. K., J. Simpson, and S. T. Soong, 1991: Numerical simulation of a subtropical squall line over the Taiwan Strait. *Mon. Wea. Rev.*, **119**, 2699-2723.

Teng, J. H., C. S. Chen, and T. Wang, 2000: Orographic effects on a squall line system over Taiwan. *Mon. Wea. Rev.*, **128**, 1123-1138.

Trier, S. B., and D. B. Parsons, 1995: Updraft dynamics within a numerically simulated subtropical rainband. *Mon. Wea. Rev.*, **123**, 39-58.

______, ______, and T. J. Matejka, 1990: Observations of a subtropical cold front in a region of complex terrain. *Mon. Wea. Rev.*, **118**, 2449-2470.

Tsay, C. Y., and G. T. J. Chen, 1980: Dynamic processes for vertical motion in a Mei-Yu season. *Papers Meteor. Res.*, **3**, 67-77.

______, and B. F. Chain, 1987: A composite study of low level jet and its relationship with heavy rainfall in Taiwan area during Mei-Yu season. *Atmos. Sci.*, **15**, **1**, 1-16 (in Chinese with English abstract).

______, and W. S. Kau, 1989: The coupling of upper-level and low-level jet streaks during Taiwan heavy rainfall period in Mei-Yu season. *Papers Meteor. Res.*, **12**, 17-32.

Wang, C. C., and G. T. J. Chen, 2002: Case study of the leeside mesolow and mesocyclone in TAMEX. *Mon. Wea. Rev.*, **130**, 2572-2592.

______, and ______, 2003: On the formation of leeside mesolows under different Froude number flow regime in TAMEX. *J. Meteor. Soc. Japan*, **81**, 339-365.

Wang, S. T., 1970: On the Plum Rain in Taiwan. *Quart. J. Meteor.*, **44**, 12-20 (in Chinese with English abstract).

______, 1986: Observational analysis of the interaction between fronts and orography in Taiwan during the late winter monsoon season. *Preprints*, *International Conference on Monsoon and Mesoscale Meteorology*, November 4-7, Taipei, 123-135. [Available from the Meteor. Soc. of the R.O.C., 64 Kung Yuan Rd., Taipei, Taiwan, R.O.C.]

______, 1989: Observational study of the orographically induced disturbances during TAMEX. *Proc. Workshop on TAMEX Preliminary Scientific Results, Taipei*, 22-30 June, 279-286. [Available from the Dept. of Atmos. Sci., Natl. Taiwan Univ., Taipei, Taiwan, R.O.C.]

______, and G. T. J. Chen, 1990: On the formation and structure of mesolows in TAMEX. *Proc. Workshop on TAMEX Scientific Results*, *NCAR*, 24-26 September, 166-173. [Available from Mesoscale and Microscale Meteor. Div., NCAR, P.O. Box 3000, Boulder, CO 80307]

Wang, T. C. C., 1988: The radar analysis of two precipitation systems during 1986 Mei-Yu season. *Papers Meteor. Res.*, **11**, 63-94.

______, 1989: The formation of new convection at the intersection of two line echo systems during TAMEX IOP#2. *Papers Meteor. Res.*, **12**, 75-90.

______, Y. J. Lin, R. W. Pasken, and H. Shen, 1990: Characteristics of a subtropical squall line determined from TAMEX dual-Doppler data. Part Ⅰ: Kinematic structure. *J. Atmos. Sci.*, **47**, 2357-2381.

Wang, Y., 1992: Effects of blocking anticyclones in Eurasia in the rainy season (Meiyu/Baiu season). *J. Meteor. Soc. Japan*, **70**, 929-950.

Wu, C. W., and G. T. J. Chen, 1988: Composite structure of environment conditions associated with MCCs in Taiwan Mei-Yu season. *Proc. Conference on Weather Analysis and Forecasting, Central Weather Bureau*, Taipei, 95-106 (in Chinese with English abstract). [Available from the Central Weather Bureau, 64 Kung Yuan Rd., Taipei, Taiwan, R.O.C.]

Yeh, H. C., G. T. J. Chen, and W. T. Liu, 2002: Kinematic characteristics of a Mei-yu front detected by the QuikSCAT oceanic winds. *Mon. Wea. Rev.*, **130**, 700-711.

Yeh, H. C., and Y. Chen, 2002: The role of offshore convergence on coastal rainfall during TAMEX IOP 3. *Mon. Wea. Rev.*, **130**, 2709-2730.

Zhang, B. C., 1990: *A study of the Meiyu rainstorms over the Yangtze Valley*. China Meteorological Press, 269 pp (in Chinese).

Zhao, S. X., X. P. Zhou, K. S. Zhang, and S. H. Liu, 1982: Numerical simulation experiment of the formation and maintenance of meso-scale low. *Scientia Atmospherica Sinica*, **6**, 109-119 (in Chinese with English abstract).

11. LARGE- AND MESOSCALE FEATURES OF MEIYU-BAIU FRONT ASSOCIATED WITH INTENSE RAINFALLS

KOZO NINOMIYA

Frontier Research System for Global Change
236-0001 Yokohama, Japan
E-mail: nmiya@jamstec.go.jp

This review paper discusses the multi-scale features, from the large scale to the mesoscale, of the Meiyu-Baiu front observed in 1-10 July 1991 based on the author's recent studies. Main data utilized are ECMWF reanalysis data and Geostationary Meteorological Satellite infrared data. During this period, the Meiyu-Baiu front with intense rainfalls extended from the foot of the eastern Tibetan Plateau to the Japan Islands in association with the westward extension of the North Pacific subtropical anticyclone, development of a blocking ridge and two cold lows in the northern latitudes (45°-60°N). A strong low-level jet stream, nearly moist neutral stratification, and a strong gradient of the specific humidity and that of equivalent potential temperature characterize the Meiyu-Baiu frontal zone. The weak thermal gradient in the Meiyu region of frontal zone is due to the cooling in the frontal precipitation zone. The relatively large thermal gradient in the Baiu region of frontal zone is sustained between the tropical maritime airmass and the polar maritime airmass. The strong meridional convergence of the moisture flux sustains the large precipitation in the frontal zone. The differential advection of the equivalent potential temperature generates convective instability against the stabilizing effect of the cumulus convection, and thus, a moist neutral stratification is sustained during the period of intense rainfalls. The strong low-level convergence in the frontal zone is accompanied by the northward strong acceleration along the northwestern rim of the westward protruding subtropical anticyclone.

A large diurnal variation of the convective clouds is observed over 90°-100°E, whereas the eastward passages of subsynoptic-scale and meso-α-scale cloud systems are evident in the frontal zone to the east of ~110°E. These cloud systems formed in the heavy rainfall area over the Continent, and develop during the passage along the Baiu frontal zone, where the significant baroclinity is present within a nearly moist neutral layer in the lower troposphere. A few meso-α-scale cloud systems develop along the trailing portion of the preceding sub-synoptic-scale cloud system. They form a family of cloud systems in the Meiyu-Baiu front.

1. Introduction

The summer monsoon rainfall over the East Asia is identified as Meiyu in China and Baiu in Japan, respectively. The intense rainfalls concentrate in a west-to-east elongated rain zone, which is called by the Meiyu-Baiu frontal rain zone.

Many studies have been made on the Meiyu-Baiu frontal rain zone from various viewpoints (see review papers, e.g., Ninomiya and Murakami 1987; Tao and Chen 1987; Ding 1991; Ninomiya and Akiyama 1992). Roughly speaking, previous studies can be classified into four categories according to the scales discussed. Studies in the first category have mainly discussed the East Asia monsoon as a subsystem of the planetary scale monsoon circulation. These studies paid attention to the variations of the whole monsoon system, the influence of Tibetan Plateau, the relation between the Indian monsoon and the East Asia monsoon, 30-50 day oscillations, and relation to the El Nino Southern Oscillation (ENSO).

The studies in the second category paid attention to the structure and formation process of the Meiyu-Baiu front and influences of large-scale circulation systems on the front. Several articles (e.g., Matsumoto *et al.* 1970 and 1971; Akiyama 1973 and 1975; Ninomiya 1984, 1999 and 2000; Ninomiya and Muraki 1986; Ninomiya and Murakami 1987) have shown that the Baiu front is characterized by a low-level jet stream, strong moisture gradient and the moist neutral stratification. The important influences of the moisture transport along the western rim of the Pacific subtropical anticyclone, the confluence/convergence of the Indian westerly monsoon with the Pacific trade easterly, and the mid-high latitude circulation systems on the Baiu front were also pointed out in these papers. Many authors (e.g., Tao and Chen 1987; Ding 1991) also pointed out that the confluence/convergence of the monsoon westerly and the easterly winds over the South China Sea is an important condition for the occurrence of the intense Meiyu rainfalls. They also pointed out that "the Tibetan Plateau vortices" (Ding 1991) are the important rain-producing circulation systems over China.

The studies in the third category were made with interests on the mesoscale precipitation systems in the frontal zone. Many authors (e.g., Matsumoto and Tsuneoka 1970; Matsumoto and Ninomiya 1971; Akiyama 1984, 1989 and 1990; Ninomiya 1978; Ninomiya *et al.* 1988) show various mesoscale precipitation systems develop under the influence of larger circulation systems. Ninomiya and Akiyama (1992) and Shibagaki *et al.* (1997) further stress the multi-scale features of the Meiyu-Baiu rainfalls.

Papers of the fourth category (e.g., Akiyama 1978; Ninomiya and Yamazaki 1979; Takeda and Iwasaki 1987; Kato 1998; Shibagaki *et al.* 2000) studied the fine inner structure of mesoscale cloud systems and rainband. However, they did not make detailed studies of the associated larger circulation systems.

For all the aforementioned studies, multi-scale features, from the large scale to the mesoscale, of the Meiyu-Baiu cloud and precipitation systems have not been fully understood yet, since the subjects of these studies diverge for a wide range of interests.

The present review paper will discuss the multi-scale features of the Meiyu-Baiu front based on the author's recent studies (Ninomiya 2000 and 2001; Ninomiya and Shibagaki 2003) on the west-to-east elongated Meiyu-Baiu front observed on 1-10 July 1991. We will focus our attentions on the multi-scale features of the Meiyu-Baiu front, "cloud system family" (train of cloud systems) along the front, and influence of the large-scale circulation systems on the front and the cloud systems.

The main data used for these studies are the European Center for Medium Range Forecasts (ECMWF) re-analysis data (vertical velocity is included), which are recorded in a 2.5° latitude-longitude grid system, and the cloud-top equivalent blackbody temperature (TBB) obtained from the infrared radiation (IR) observations from the Geostationary Meteorological Satellite (GMS) of Japan Meteorological Agency (JMA), which is averaged over 1° latitude-longitude areas. The data from the Global Analysis (GANL) of JMA are also utilized in Section 3.

2. Overview of the East Asia Monsoon in July 1991

On the monthly averaged TBB maps for June and July of each year, the Meiyu-Baiu cloud zone appears as an elongated cloud zone from the eastern foot of the Tibetan Plateau to the Japan Islands, and then into the Northwestern Pacific. However, the cloud images indicate significant temporal variations, from the wave-pattern with large south-north amplitude (Fig. 1A) to the west-east elongated cloud band (Fig. 1B), in response to the changes in the large-scale circulations. The cloud distributions in the Meiyu-Baiu season in 1968 (Akiyama 1973), 1972 (Akiyama 1975), 1979 (Ninomiya 1989), and 1982 (Akiyama 1989) have been documented in detail. In these four years, an elongated quasi-stationary cloud zone was formed in 1-10 July 1968, 7-15 July 1972, 21-30 June 1979, and 16-25 July 1982. These dates indicate that an elongated stationary cloud zone tends to appear in the peak period of the Baiu season. During these periods, large precipitation occurred in the cloud zone, and cloud systems of sub-synoptic-scale and meso-α-scale were formed successively along the cloud zone.

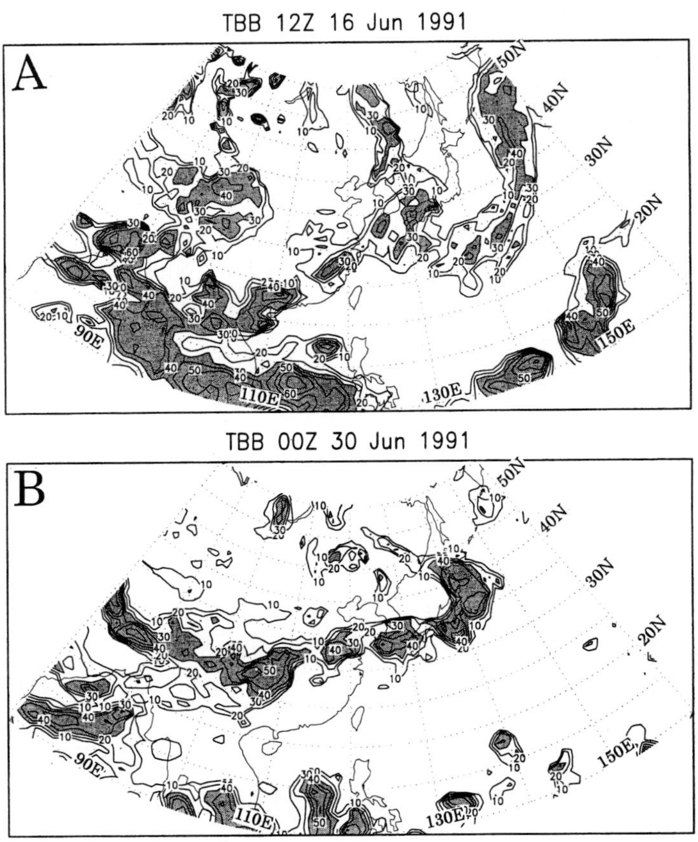

Figure 1. A: TBB map at 12 UTC 16 June 1991. The isopleths are given at 10°C intervals and the minus sign of TBB is omitted. The stippling indicates an area where TBB is less than −30°C. This map shows a typical example of meandering Meiyu-Baiu cloud zone. B: TBB map at 00 UTC 30 June 1991. This map shows the typical features of the west-to-east elongated Meiyu-Baiu cloud zone.

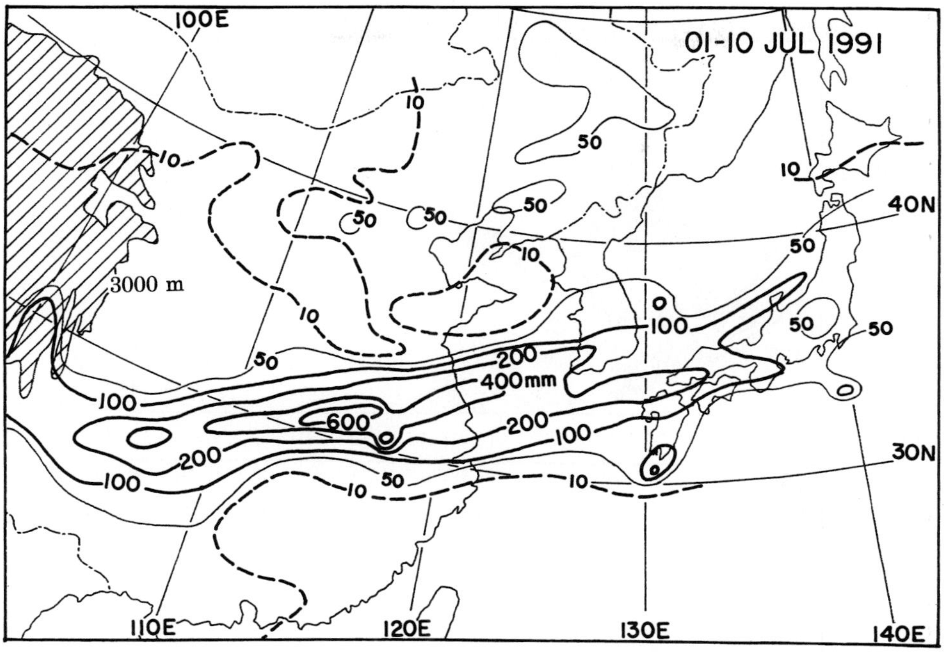

Figure 2. The 10-day total precipitation (mm) over East Asia during 1-10 July 1991. The hatched area indicates an area higher than 3000 m above the sea level.

The quasi-stationary Meiyu-Baiu precipitation zone stayed over the Yangtze River basin during the 13-day period between 30 June and 12 July 1991 (Ding *et al.* 1993; Ninomiya 2000). The map of the total precipitation during 1-10 July 1991 (Fig. 2) shows a rain band of ~300 km width and ~4000 km length that extends from the eastern foot of the Tibetan Plateau to the Japan Islands. The precipitation is concentrated within a narrow rain band with a maximum total precipitation of ~600 mm during the 10 days. It is notable that an area of very small precipitation is present close to the southern edge of the rain band.

Maps of 10-day averaged "cloud top equivalent black body temperature (TBB)" and the topography of the East Asia are shown in Figs. 3A and 3B, respectively. Five major cloud areas are identified in Fig. 3A. The first one (C-1, the Meiyu-Baiu frontal cloud zone) is the quasi-stationary, long elongated cloud zone from the eastern foot of the Tibetan Plateau to the western North Pacific. The second (C-2) extends from the northern parts of Thailand toward the southern China, where it merges with the southwestern tip of the Meiyu cloud zone. The third (C-3) extends from the southern rim of the Tibetan Plateau toward its southeastern foot, where it merges with the Meiyu cloud zone. Large diurnal variations of convective cloud occur in the C-2 and C-3 regions. Area C-4 is the cloud area around Bangladesh, while C-5 is the cloud zone over the Bay of Bengal.

To overview the temporal variations of the Baiu cloud zone during the period, the latitude-time section of TBB along the 129.5°E meridian is presented in Fig. 4. The Baiu cloud zone (30°-35°N) is clearly separated from the ITCZ (0°-10°N) and the northern latitude cloud zone (45°-60°N) by areas of small cloud amount.

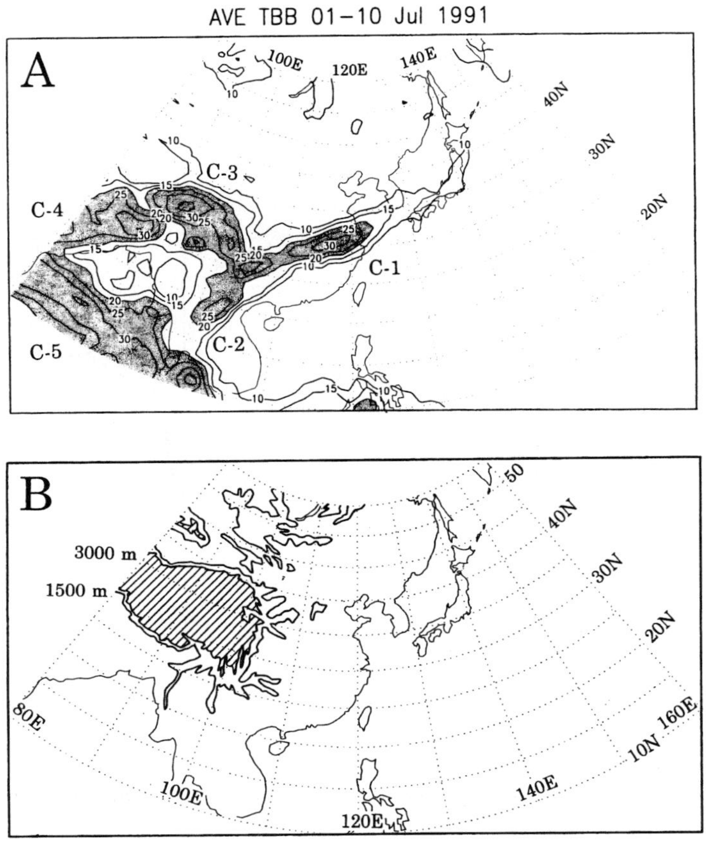

Figure 3. A: The 10-day averaged TBB for 1-10 July 1991. The isopleths are given at 5°C intervals and the minus sign of TBB is omitted. The stippling indicates an area where TBB is less than −20°C. In this map, C-1, C-2, C-3, C-4 and C-5 indicate five major cloud areas. B: Orography of East Asia. The shading shows the area higher than 3000 m above sea level and the solid line indicates the 1500 m contour.

Successive 12 hourly TBB maps (not shown) indicate that several Meiyu-Baiu cloud systems with low TBB values propagate eastward successively with a speed of ~1000 km d^{-1} along the cloud zone during the period between 29 June and 12 July 1991. In Fig. 4, cloud systems on 29 June, 1, 4 and 7 July exhibit relatively wide north-south extensions, while other cloud systems have smaller north-south extensions. Detailed discussion of these cloud systems will be given in Sections 9 and 10.

The 10-day averaged 500-hPa geopotential-height is shown in Fig. 5. The location of the low-level jet stream at 850 hPa, the axes of the maximum wind at 500 hPa, and the subtropical jet stream at 200 hPa are also indicated in this map. The westward extension of the Pacific subtropical anticyclone, two cold lows, and a blocking ridge over northern latitudes (45°-60°N) characterize the large-scale situation during this period. The Meiyu-Baiu cloud zone is located to the north of the low-level jet stream and to the south of the subtropical jet stream. These features are generally found during the active phase of the Meiyu-Baiu front.

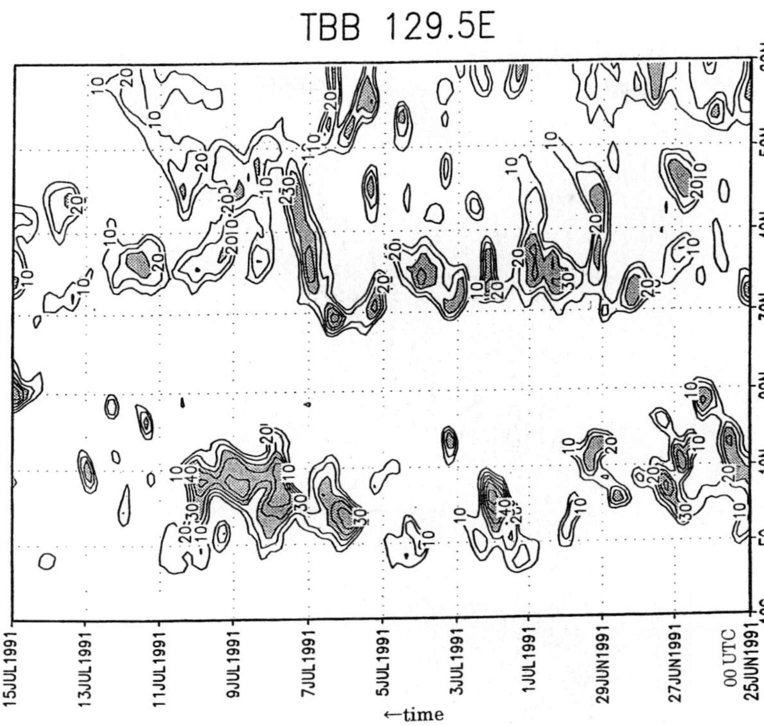

Figure 4. The latitude-time section of TBB along 129.5°E meridian. The isopleths are given at 10°C intervals and the minus sign of TBB is omitted. The stippling indicates an area where TBB is less than −30°C.

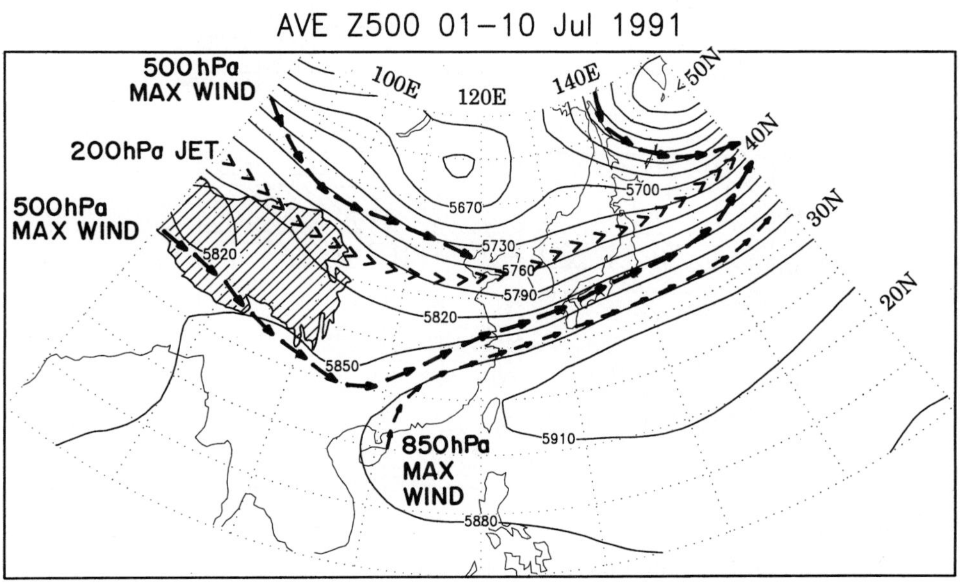

Figure 5. The 10-day averaged geopotential height at 500 hPa for 1-10 July 1991. The height contours are given at 30 m interval. The maximum wind axes at 850, 500 and 200 hPa are indicated.

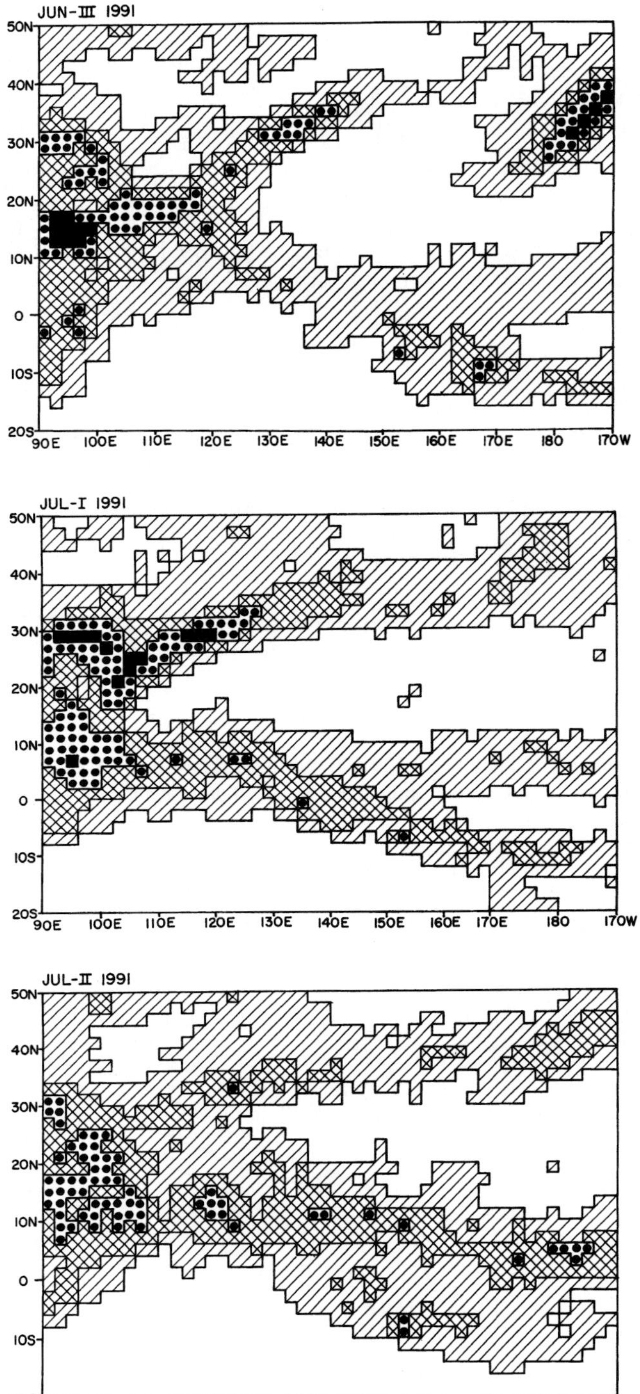

Figure 6. Distribution of 10-day averaged Ch (amount of cloud whose top is higher than the 400 hPa level) for 21-30 June (top), 1-10 July (middle) and 11-20 July 1991 (bottom). The black area, stippled area, dense-hatched area and thin-hatched areas indicate the area of Ch more than 0.7, 0.5, 0.3 and 0.1, respectively. The blank area indicates the area with Ch less than 0.1, which is defined as a clear area in the present report.

3. Seasonal Changes of the North Pacific Subtropical Anticyclone

The seasonal variation of the Pacific subtropical anticyclone in June and July 1991 will be discussed in relation to the variation of the Indian westerly monsoon. The variation of the subtropical anticyclone is examined by using "Ch" cloud cover that is the amount of cloud higher than 400 hPa, which is estimated by TBB observations. Figure 6 shows the maps of 10-day averaged Ch for 21-30 June, 1-10 July and 11-20 July 1991. In these maps, "the clear area" is defined as Ch values smaller than 0.1, which corresponds to the subsiding area of the subtropical anticyclone. The clear area protrudes westward during 1-10 July, and simultaneously the active convective region, which forms along the northwestern rim of the clear area, moves westward and is significantly intensified.

Figure 7 presents longitude-time section of the 10-day averaged **Q**, which is the vertically integrated moisture transport from the 1000 to 300 hPa level, along 20°N. It is evident that the westward extension of the Pacific subtropical anticyclone at 20°N occurs simultaneously with the westward retreat of the Indian westerly monsoon. Thus, the Meiyu-Baiu intense rainfalls have close relation with the 30-50-day west-east oscillations of both the Indian westerly monsoon and the Pacific subtropical anticyclone, although detailed discussion on this subject is not given in the present review article.

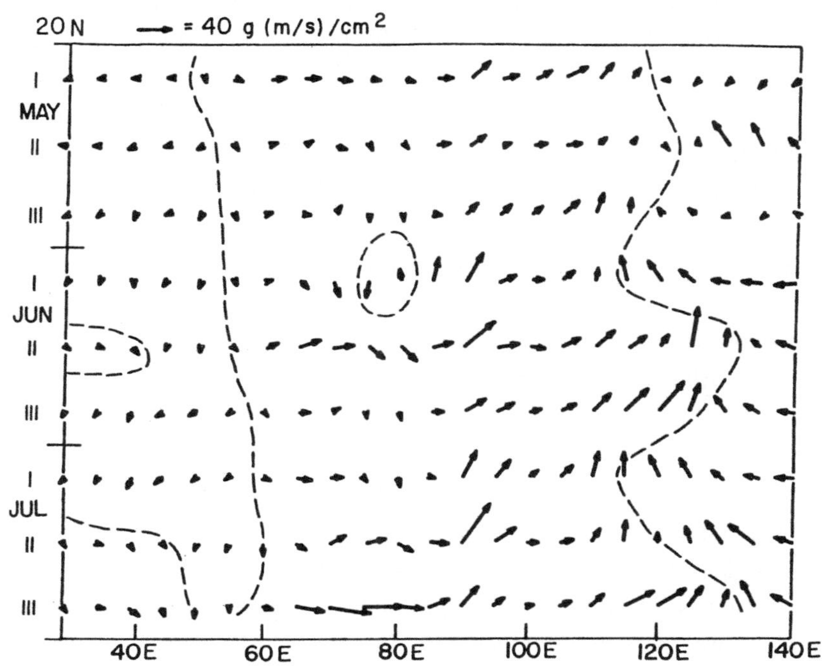

Figure 7. The longitude-time section of the 10-day averaged vertically integrated (1000-300 hPa) moisture transport along 20°N for 1-10 July 1991. A dashed line indicates the boundary between the westerly and easterly moisture transport.

4. Large-Scale Circulation Systems around the Meiyu-Baiu Front

The 10-day averaged wind velocity at 850 and 700 hPa is presented in Figs. 8A and 8B, respectively. In the lower troposphere, the Indian monsoon westerlies and the Pacific subtropical anticyclone are dominant circulations in the lower latitudes. At 850 hPa, the monsoon westerly from the Bay of Bengal is split into two branches over the Indochina Peninsula. The northern branch becomes southwesterly toward the cloud area C-3. The southern branch extends toward the South China Sea, where it meets with the easterly trade wind over the South China Sea and yields strong southerly winds. Cloud areas C-1 and C-2 form in association with these strong southerly winds. This southerly wind turns into the southwesterly along the northwestern rim of the Pacific subtropical anticyclone. This strong southwesterly, which extends along the south side of the Meiyu-Baiu cloud zone (C-1), is identified as the Meiyu-Baiu low-level jet stream.

The averaged wind field at 500 hPa (not presented) is almost the same as that at 700 hPa. The Pacific subtropical anticyclone predominates not only in the lower troposphere, but also in the middle troposphere. The middle latitude strong wind belt at 700 hPa, which is the lower portion of the polar jet stream, is in the ~45°N zone over the Asian Continent. Around 110°E, the north-south distance between the polar jet stream and the low-level jet stream is ~2000 km. These two jet streams merge gradually, and form a strong wind zone along ~35°N in the 130°-160°E zone.

In the northern latitudes, two predominant quasi-stationary cut-off lows are over the Bering Sea (~50°N, 160°E) and over Siberia (~50°N, 115°E), respectively. A blocking ridge is between these cold lows. The northerly flow invades southward on the southwestern sides of these cold lows. Another zone of northerly wind is along the eastern rim of the Tibetan Plateau. Owing to the influence of these northerly winds, confluence and convergence of the tropical air with the polar air occurs along the north side of the Meiyu-Baiu cloud zone in the 100°-115°E.

The wind field at 200 hPa (Fig. 8C) is predominated by the Tibetan anticyclone, which is centered near ~27°N, 85°E. The eastern part of the Tibetan anticyclone merges with the west-east oriented narrow anticyclonic zone over ~27°N, which forms over the Meiyu-Baiu rainfall zone. At 200 hPa, the subtropical jet stream with a maximum wind velocity of ~40 m s^{-1} is along ~40°N, while the easterly jet with wind velocity of ~25 m s^{-1} is elongated at ~17°N.

The 10-day averaged temperature T at 850 hPa, specific humidity q at 700 hPa, and equivalent potential temperature θ_e at 700 hPa are presented in Fig. 9. At 850 hPa (Fig. 9A), a warm area higher than 18°C is present over the Continent to ~50°N, which is due to the strong summer insolation. A strong thermal gradient over the Continent is thus sustained near 45°-50°N, which approximately coincides with the latitude of the polar jet stream in 700-500 hPa (Fig. 5). An especially warm area higher than 24°C spreads to the north of the Tibetan Plateau. The small specific humidity (Fig. 9B) and subsidence (Fig. 10C) in this area suggest that this high temperature is partially due to the subsidence warming. It is noted that weak thermal gradient in the Meiyu frontal zone is not due to the continent-ocean thermal

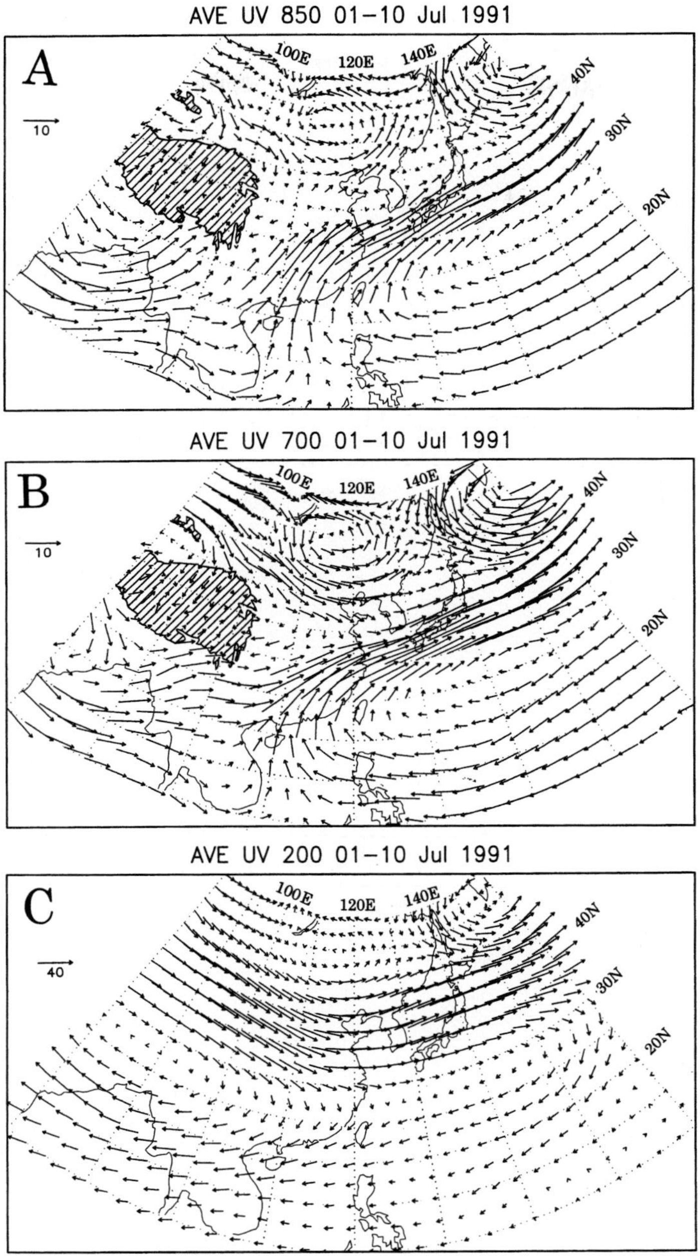

Figure 8. The 10-day averaged wind velocity (m s^{-1}) at (A) 850 hPa, (B) 700 hPa, and (C) 200 hPa for 1-10 July 1991.

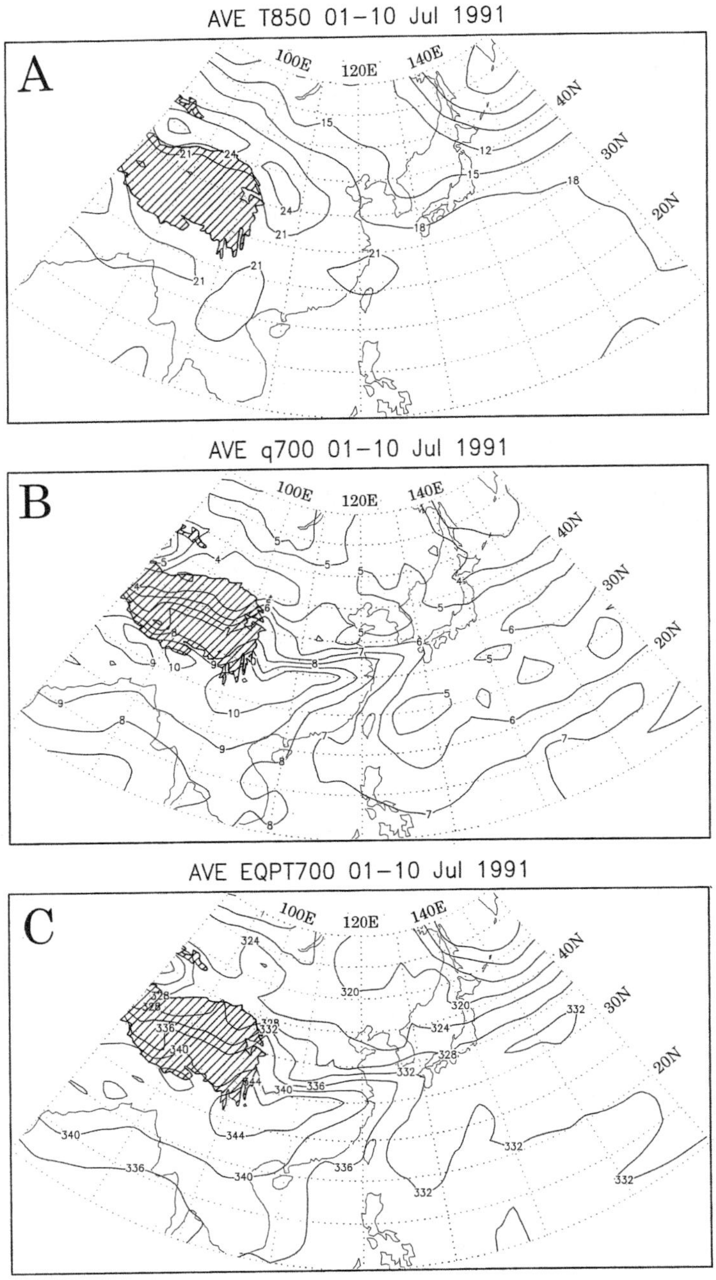

Figure 9. The 10-day averages for 1-10 July 1991 of (A) temperature at 850 hPa (°C), (B) specific humidity at 700 hPa (g kg^{-1}), and (C) equivalent potential temperature θ_e at 700 hPa (K).

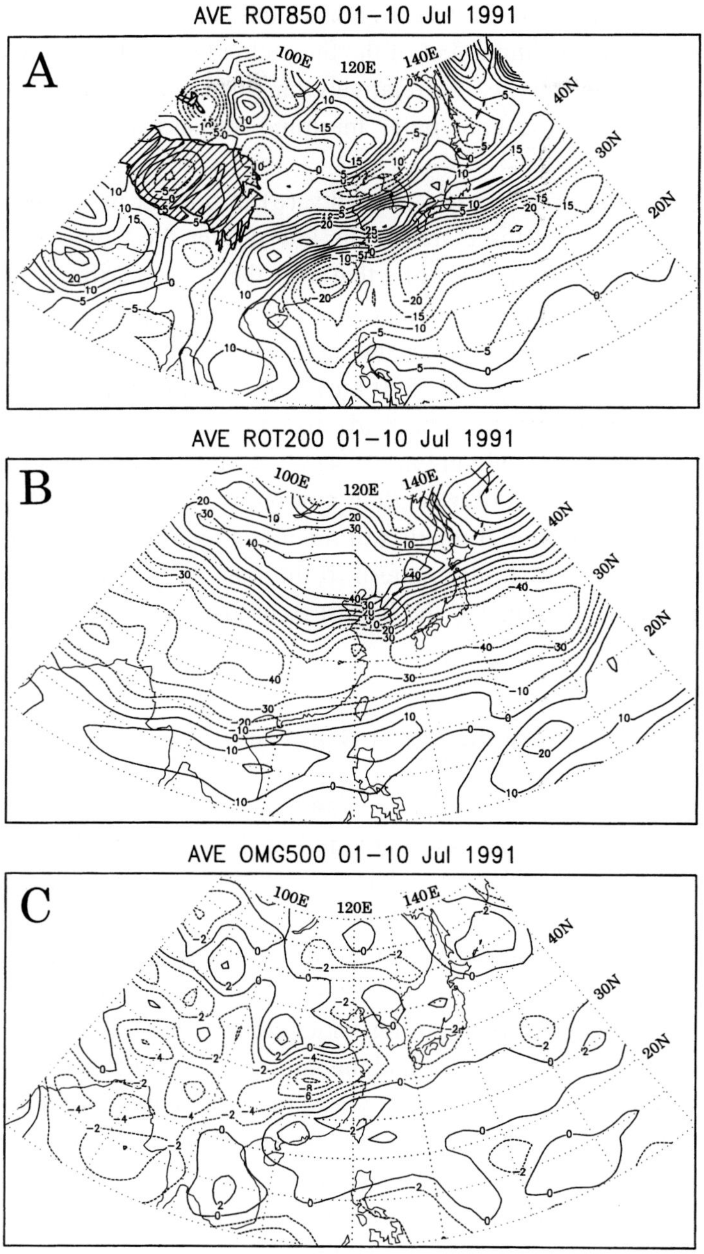

Figure 10. The 10-day averages for 1-10 July 1991 of (A) vorticity at 850 hPa (10^{-6} s^{-1}), (B) vorticity at 200 hPa (10^{-6} s^{-1}), and (C) vertical-p-velocity at 500 hPa (hPa h^{-1}).

contrast, but is mainly due to the relatively cool frontal zone, which results from the cooling effect of precipitation and shading effect of the thick cloud layer. Over the Yellow Sea, Japan Sea and the western North Pacific, the relatively large thermal gradient of the front in 35°-40°N is sustained between the cool polar maritime airmass and warm tropical maritime airmass.

At 700 hPa (Fig. 9B), low specific humidity (q) is present to the north of the Tibetan Plateau and over the polar maritime areas. The low q over the Continent is due to the low relative humidity in the subsiding air, while that over the polar maritime airmass is due to the low temperature. An area with very large q (~10 g kg^{-1}) protrudes eastward from the southeastern foot of the Tibetan Plateau as the moist tongue along the intense rainfall zone.

The largest q gradient of ~5 g kg^{-1} $(500\ km)^{-1}$ appears along the Meiyu front near 100°-120°E. Therefore, the Meiyu front has been identified as the "moisture front" or "subtropical front" in several articles (e.g., Ninomiya 1984). It is noted that the area of high specific humidity coincides with the area of ascending motion, which indicates the important role of the vertical advection of q for the formation of the moist tongue at 700 hPa. The map of θ_e at 700 hPa (Fig. 9C) indicates a similar pattern to the map of q. The largest θ_e of ~344 K is associated with the moist tongue, and the largest θ_e gradient of ~15 K $(500\ km)^{-1}$ appears in the Meiyu frontal zone.

5. Circulation Associated with the Meiyu-Baiu Front

The 10-day averaged relative vorticity ζ at 850 and 200 hPa are shown in Figs. 10A and 10B. The vertical velocity ω at 500 hPa is presented in Fig. 10C. At 850 hPa, an arch-shaped zone of large positive vorticity extends from Vietnam to the Yangtze River Basin and farther eastward to the western North Pacific. The cloud zones C-1 and C-2 coincide with this maximum vorticity zone. Strong anticyclonic vorticity appears in the northern part of the Pacific subtropical anticyclone. The especially large anticyclonic vorticity over the southeastern China is owing to both strong anticyclonic curvature of the stream along the westward protruding Pacific anticyclone and the large lateral wind shear.

At 200 hPa, a zone with large cyclonic vorticity is elongated along the northern side of the subtropical jet stream axis. The west-east elongating ridge axis of the upper anticyclone is recognized as a zone of strong anticyclonic vorticity along ~30°N. Figure 10C shows strong upward motion in the Meiyu-Baiu cloud zone. An especially large upward velocity of ~ -10 hPa h^{-1} appears over the intense rainfall area of the Meiyu front.

The meridional-vertical cross sections along 115°E of the 10-day averaged zonal wind velocity (u), meridional wind velocity (v), vertical velocity (ω), and equivalent potential temperature (θ_e) are presented in Fig. 11. These cross sections cut cross the zone of intense rainfall near 29°-33°N. In Fig. 11A, the maximum of u around 700 hPa at ~27°N is the Meiyu-Baiu low-level jet stream. The core of the subtropical jet stream is at ~200 hPa near 37°N, while the easterly jet stream is at ~100 hPa and 17°N.

In Fig. 11B, the largest southerly winds appear in 925-600 hPa over ~27°N, which is to the south of the intense rainfall zone, while a relative maximum northerly wind appears in

700-500 hPa over 37°N, which is to the north of the intense rainfall zone. The dominant meridional current in the upper troposphere is the northerly flow at 150 hPa and ~25°N, which indicates the outflow from the upper-level anticyclone formed over the intense rainfall area. Thus, the strong low-level meridional convergence capped by the strong upper-level meridional divergence appears in the intense rainfall zone. Strong ascending motion is concentrated within the intense rainfall zone, and significant subsidence is on both the northern and southern sides of the ascending motion (Fig. 11C).

In Fig. 11D, a thick layer with a nearly moist neutral lapse rate extends from 925 to 300 hPa within the intense rainfall zone near ~30°N. Convectively unstable stratification is present in both the subtropical and higher latitudes. In these regions, convection is inhibited by the subsidence. The convective instability is released selectively only in the Meiyu frontal zone, where the ascending motion is strong.

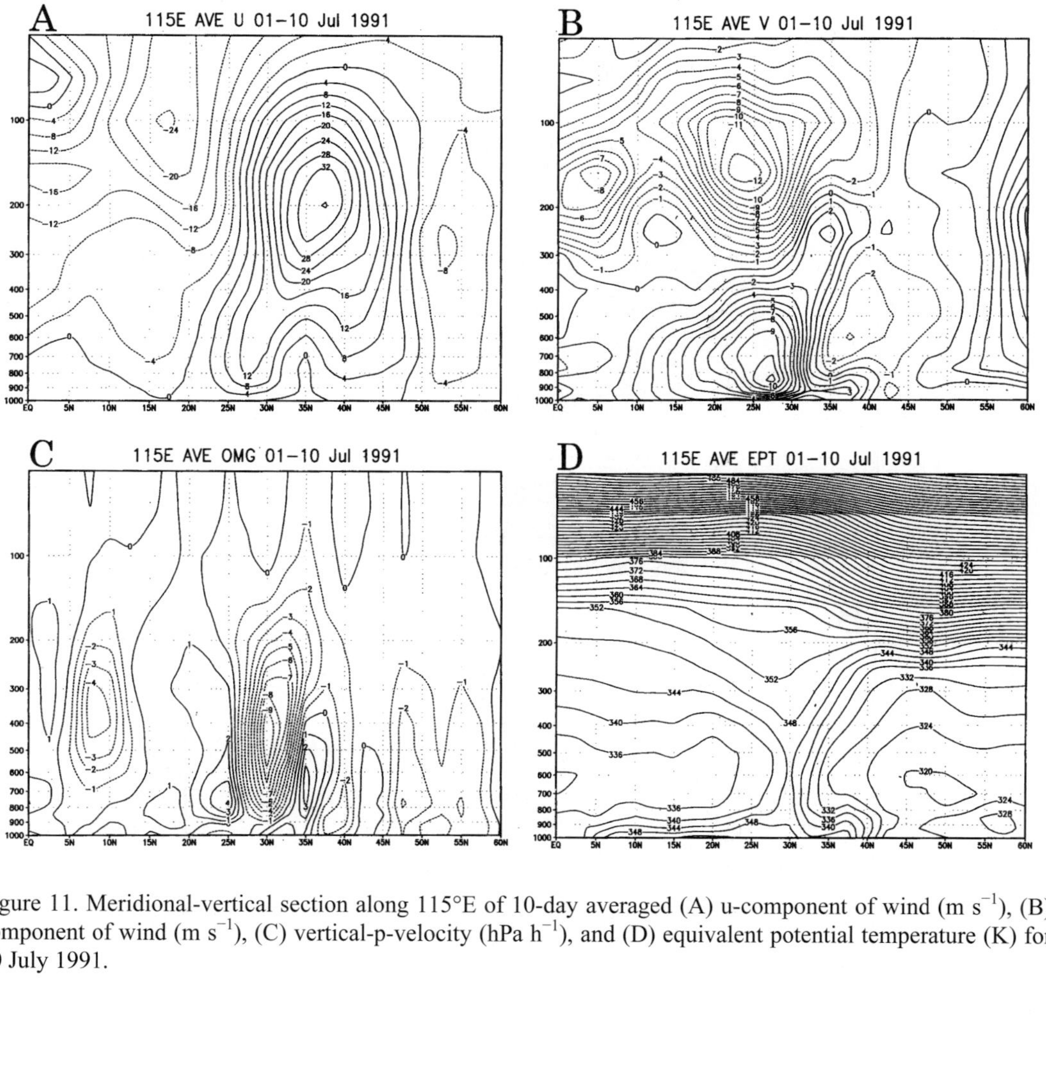

Figure 11. Meridional-vertical section along 115°E of 10-day averaged (A) u-component of wind ($m\ s^{-1}$), (B) v-component of wind ($m\ s^{-1}$), (C) vertical-p-velocity ($hPa\ h^{-1}$), and (D) equivalent potential temperature (K) for 1-10 July 1991.

6. Influence of the Subtropical Anticyclone on the Meiyu-Baiu Front

As already shown in Fig. 10C and Fig. 11C, strong ascending motion appears within the Meiyu-Baiu frontal cloud zone, while strong subsidence appears on the south side of the cloud zone. The 10-day averaged divergence at 850 hPa (Fig. 12A) has a dipole of strong convergence (in the cloud zone) and divergence (to the south of the cloud zone) over southern China in association with the westward-protruding Pacific subtropical anticyclone. Although alternate interpretations may be given for the dipole of low-level convergence and divergence, here we will relate the ageostrophic wind velocity (u_a,v_a) to the acceleration along the rim of the subtropical anticyclone by using the following equations of motion,

$$\frac{\partial u}{\partial t}+u\frac{\partial u}{\partial x}+v\frac{\partial u}{\partial y}+\omega\frac{\partial u}{\partial p}=f v_a \qquad (1)$$

and,

$$\frac{\partial v}{\partial t}+u\frac{\partial v}{\partial x}+v\frac{\partial v}{\partial y}+\omega\frac{\partial v}{\partial p}=-f u_a \qquad (2).$$

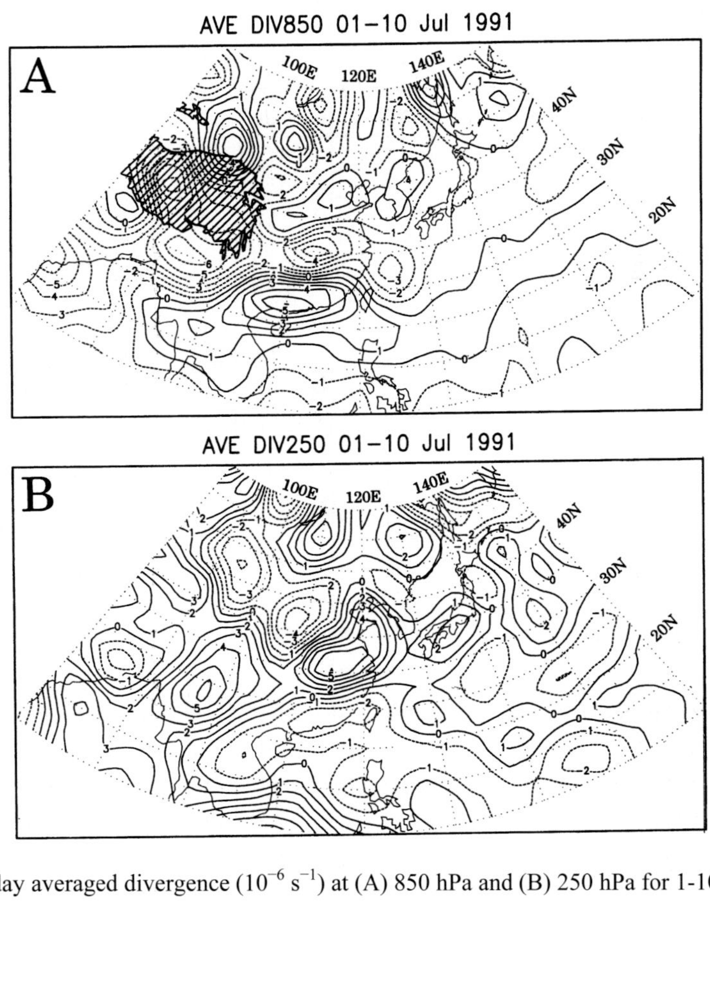

Figure 12. The 10-day averaged divergence (10^{-6} s^{-1}) at (A) 850 hPa and (B) 250 hPa for 1-10 July 1991.

The 10-day averaged ageostrophic wind velocity at 850 hPa (Fig. 13A) indicates that the large ageostrophic wind from the subtropical anticyclone toward the frontal zone is associated with the large acceleration along the periphery of the anticyclone, where both the anticyclonic curvature and horizontal wind shear are large. Consequently, a dipole of large convergence and divergence is associated with the westward-protruding subtropical anticyclone. Aforementioned ageostrophic wind and the dipole of convergence and divergence are present not only at 850 hPa but in the deep layer from 925 hPa to 500 hPa (not shown).

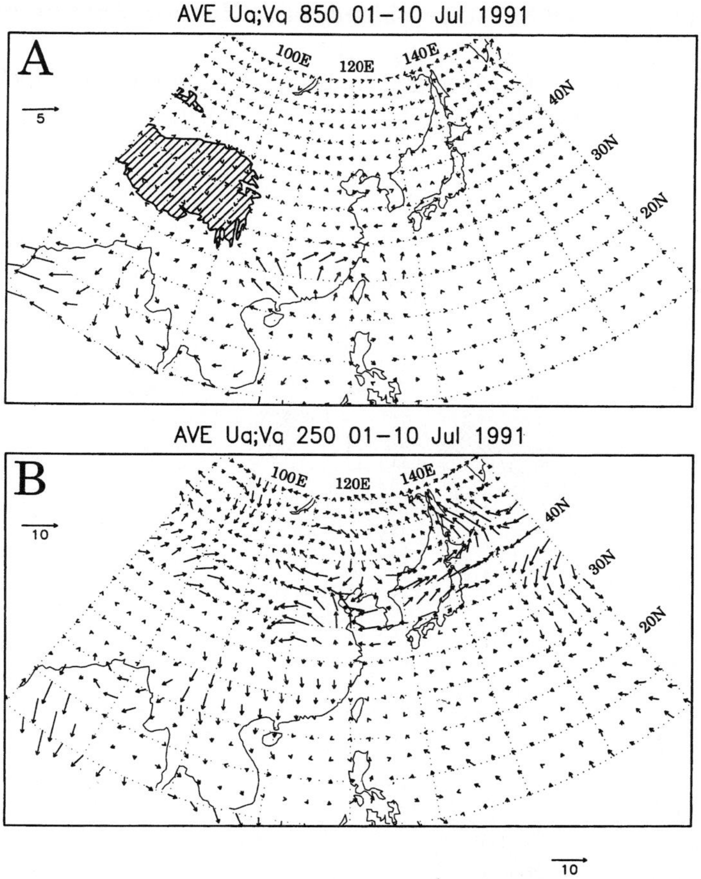

Figure 13. The 10-day averaged ageostrophic wind velocity (m s^{-1}) at (A) 850 hPa and (B) 250 hPa for 1-10 July 1991.

The 10-day averaged divergence at 250 hPa (Fig. 12B) has strong high-level divergence and convergence over the low-level convergence and divergence areas, respectively. The 10-day averaged ageostrophic wind velocity at 250 hPa evaluated by equations (1) and (2) is presented in Fig. 13B. The strong divergence of the ageostrophic wind near ~30°N, 115°E is associated with the acceleration in the anticyclonic flow with strong curvature along the rim of the eastward-protruding Tibetan anticyclone. The ageostrophic wind from the anticyclone became weaker over the southern coast of China, and therefore, the strong upper-level convergence appears over the low-level divergence area to the south of the Meiyu rainfall zone.

The influence of the intense precipitation on the upper-level divergence should be also noted. The strong condensation heating in the intense rainfall zone (see Fig. 14 in Section 7) will produce strong upper-level divergence over the strong ascending motion in the Meiyu rainfall zone, and enhance the upper-level anticyclone.

7. Moisture Transport and Moisture Balance in the Intense Precipitation Zone

Figures 14A presents the 10-day averaged moisture transport $\boldsymbol{V}$q at 850 hPa, and Fig.14B shows the horizontal moisture flux divergence $\nabla \cdot \boldsymbol{V}q$ at 850 hPa. The moisture transport is

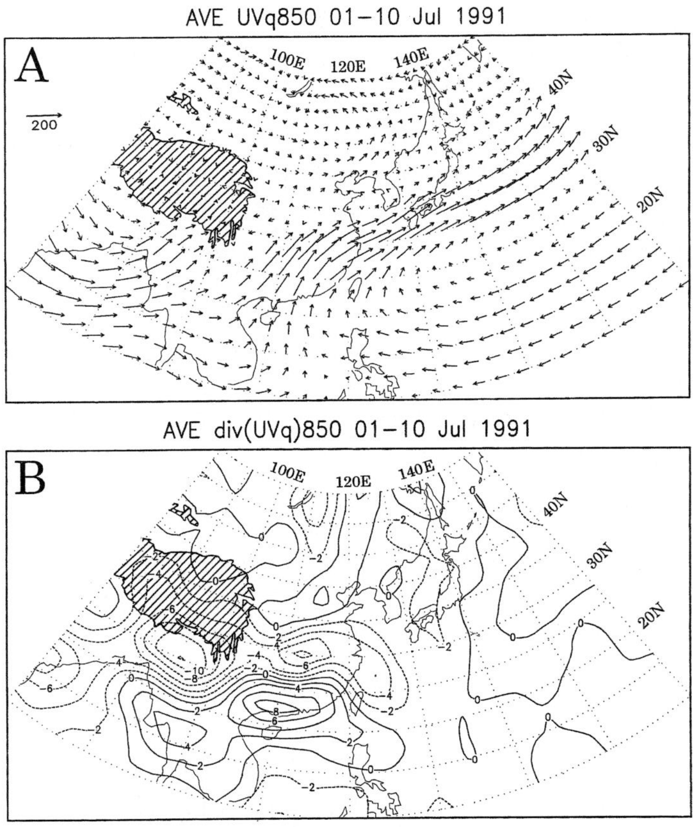

Figure 14. The 10-day averaged (A) moisture flux Vq ($g\ kg^{-1}\ m\ s^{-1}$) at 850 hPa and (B) moisture flux divergence div Vq ($10^{-5}\ g\ kg^{-1}\ m\ s^{-1}$) at 850 hPa for 1-10 July 1991.

characterized by the westerly moisture transport in the monsoon zone (10-20°N and 80-105°E), and anticyclonic moisture flow around the North Pacific subtropical anticyclone. The confluence of the moist monsoon westerlies with the moist trade easterlies yields very strong southerly moisture flow toward the intense precipitation zone.

The strong moisture flux convergence near ~25°N, 95°E is due to the blocking effect of the Himalayan Mountains, while the large convergence in the Meiyu frontal zone is associated with the large influx along the northwestern rim of the subtropical anticyclone. Strong moisture flux divergence exists near ~22°N, 110°E, where strong subsidence takes place.

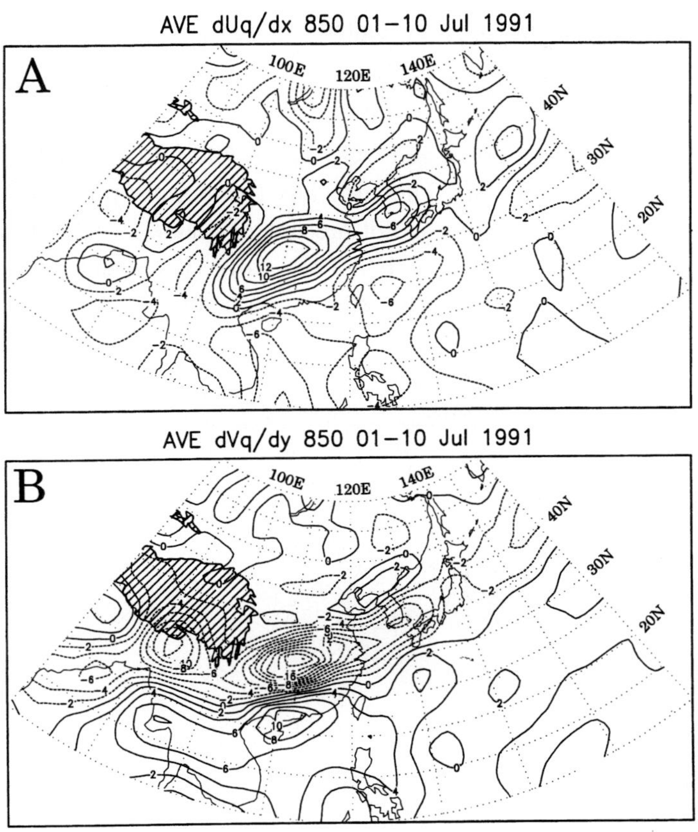

Figure 15. The10-day averaged (A) ∂ uq / ∂ x (10^{-5} g $kg^{-1}s^{-1}$) at 850 hPa and (B) ∂ vq / ∂ y (10^{-5} g $kg^{-1}s^{-1}$) at 850 hPa for 1-10 July 1991.

The 10-day averaged zonal moisture flux divergence ($\partial uq/\partial x$) and the meridional moisture flux divergence ($\partial vq/\partial y$) at 850 hPa are presented in Figs. 15A and 15B respectively. These terms are actually calculated in the λ-ϕ coordinate system. The zonal moisture flux convergence is very strong over the South China Sea (monsoon confluence region), where the Indian monsoon westerlies meet the trade easterly winds. This strong zonal moisture flux convergence is accompanied by the strong meridional moisture flux divergence over the monsoon confluence region, and the large southerly moisture flux toward the frontal zone. Thus, the monsoon confluence zone plays the role of the channel transporting the moisture from the tropical oceanic region toward the Meiyu-Baiu frontal zone (Ninomiya 1999).

On the contrary, the meridional moisture flux convergence is very strong within the intense rainfall zone over China. Large parts of the meridional moisture flux convergence are consumed as precipitation, but the remaining parts are transported eastward, and large zonal moisture flux divergence is found over the Meiyu rainfall zone. Thus, the Meiyu frontal zone plays the role of transporting the moisture from the monsoon confluence region to the Baiu frontal zone.

The apparent latent heat sink Q_M is evaluated by

$$Q_M = -L\left(\frac{\partial q}{\partial t} + u\frac{\partial q}{\partial x} + v\frac{\partial q}{\partial y} + \omega\frac{\partial q}{\partial p}\right) \qquad (3).$$

Figure 16 shows the 10-day averaged ($-Q_M/c_p$) at 600 hPa. The zone of large ($-Q_M/c_p$) extends over the intense rainfall zone. The vertically (from the surface to 300 hPa) integrated apparent moisture sink ($-Q_M/L$, figure is not presented) reaches ~35 mm d^{-1} at the center of the intense rainfall zone (~30°N, 115°E). This value is consistent with the observed precipitation averaged over the width of the precipitation zone.

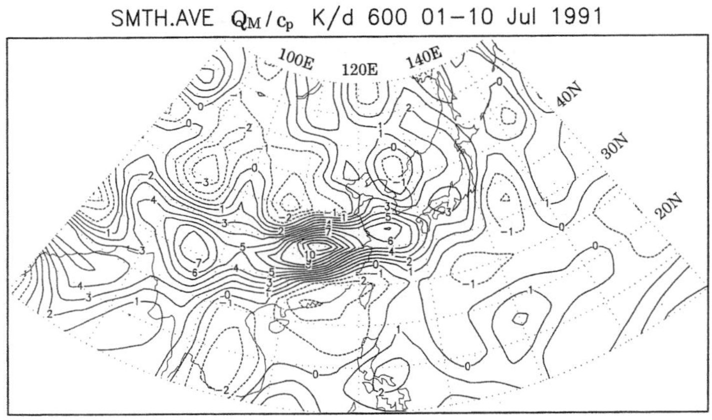

Figure 16. The 10-day averaged $-Q_M/c_p$ (K d^{-1}) at 600 hPa for 1-10 July1991. Horizontal smoothing has been applied in this map to eliminate small irregular patterns.

8. Vertical Stratification in and around the Meiyu-Baiu Frontal Zone

The vertical cross-section of equivalent potential temperature along 115°E (Fig. 11D) indicates a nearly moist neutral layer within the Meiyu frontal zone. The horizontal distribution of the vertical stratification is next examined. Figure 17A shows the 10-day averaged $-(\partial \theta_e/\partial p)$ evaluated for the layer between 850 and 500 hPa. Whereas the oceanic areas north of ~35°N have convectively stable stratification, the oceanic areas south of ~30°N, and the continental area south of ~45°N, are characterized by a convectively unstable stratification. In these areas, the subsiding motion inhibits the cumulus convection, and therefore the unstable stratification is maintained.

It is important that the Meiyu-Baiu frontal zone, which is associated with intense convective precipitation, is not characterized by the strong convective instability, but by nearly moist neutral stratification. This indicates the release of the convective instability associated with the cumulus convection.

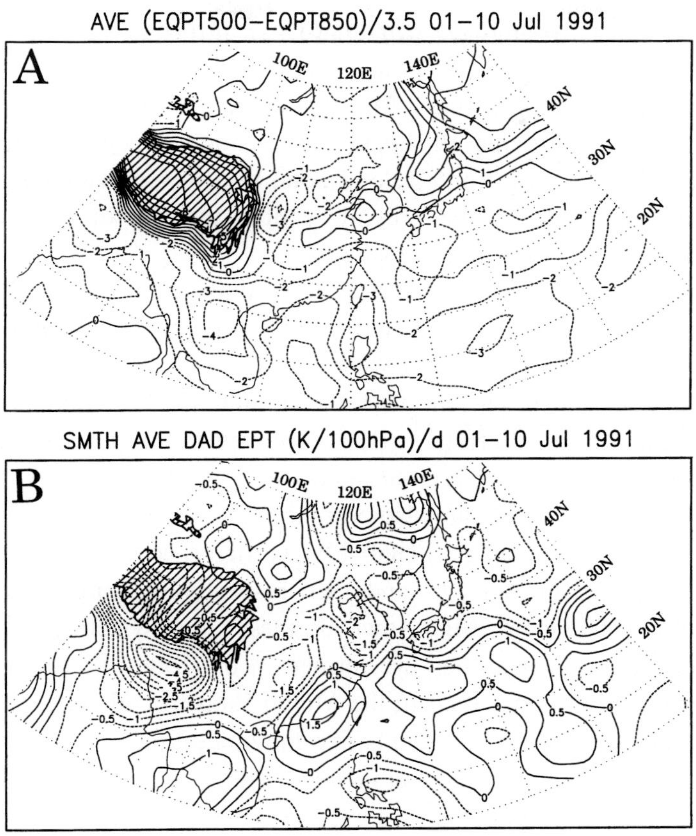

Figure 17. The 10-day averaged (A) vertical stability $-\partial \theta_e/\partial p$ (K (100 hPa)$^{-1}$) in 850-500 hPa and (B) differential advection of θ e, $\partial\,[u\partial \theta_e/\partial x + v\partial \theta_e/\partial y + \omega\partial \theta_e/\partial p]/\partial p$ (K (100 hPa)$^{-1}$ d^{-1}) in 850-500 hPa for 1-10 July 1991.

For the sustenance of the strong convective precipitation for more than 10 days, some large-scale process must generate convective instability against the stabilizing effect of the convective clouds. The local time change of convective stability due to the differential advection of θ_e is expressed by

$$[-\partial(\partial\theta_e/\partial p)/\partial t]_{ADV} = \partial[(u\partial\theta_e/\partial x + v\partial\theta_e/\partial y + \omega\partial\theta_e/\partial p)]/\partial p \quad (4).$$

The 10-day averaged value of $[-\partial(\partial\theta_e/\partial p)/\partial t]_{ADV}$ evaluated for the layer between 850 and 500 hPa is shown in Fig. 17B. Areas of negative differential advection (generation of convective instability) are present over the Meiyu-Baiu frontal zone, which indicates that the differential advection generates successively convective instability against the release of the instability by the convective clouds. As the result of these two processes, the large precipitation and nearly moist neutral stratification are maintained within the frontal precipitation zone.

9. Meiyu-Baiu Frontal Disturbances and Cloud-Precipitation Systems

As pointed out by many authors (e.g., Matsumoto and Tsuneoka 1970; Ninomiya and Akiyama 1971 and 1972; Akiyama 1984 and 1989; Ninomiya *et al.* 1988; Ninomiya and Akiyama 1992), the Meiyu-Baiu frontal cloud zone is a row of cloud systems of various scales aligned along the front. In the present section, features of the cloud-precipitation systems in the Meiyu-Baiu frontal zone are studied.

As a typical example, the TBB and the relative vorticity at 500 and 850 hPa at 00 UTC 1 July 1991 are presented in Figs. 18A, B and C, respectively. The Meiyu-Baiu cloud zone appears as the chain of cloud systems on the subsynoptic-scale and mesoscale. Two cloud systems of subsynoptic-scale are present near ~130°E and ~180°. Cyclonic vorticity cores at 850 hPa (Fig. 18A) and 500 hPa (Fig. 18B) accompany these cloud systems. However, a smaller cloud system is not necessarily associated with a vorticity core. A few cores of large vorticity associated with cloud areas are also present in the northern latitudes.

The longitude-time section of ζ_{850} (850 hPa vorticity) along 35°N, and that of TBB along 33.5°N are presented in Figs.19A and 19B, respectively. Large diurnal variations of clouds are seen in 90°-105°E (Fig. 19B). However, these cloud systems, which develop in the afternoon, are not sustained, since they are formed in the zone of anticyclonic circulation. Significant diurnal variation is not seen in the Meiyu-Baiu frontal zone, while the eastward propagation of the cloud systems is evident. The longitude-time section in Fig. 19A shows eastward propagation of the core of ζ_{850} associated with cloud systems. However, Fig. 19B shows finer structure of cloud systems as compared with ζ_{850}, because the TBB data have higher resolution. The wavelength of the major disturbances in Fig. 19B is estimated to be ~2000 km, which falls on the border between macro-β- and meso-α-scale, according to the scale classification by Orlanski (1975). Therefore, these disturbances are identified as subsynoptic-scale Meiyu-Baiu frontal disturbances in the present report. Some authors (e.g., Matsumoto *et al.* 1972; Ninomiya and Akiyama 1971 and 1972; Ninomiya and Yamazaki 1979; Yoshizumi 1977) classified them as medium-scale disturbances.

The longitude-time section of ζ_{500} (500 hPa vorticity) along 50°N (figure is not presented) shows that the propagation of disturbances in the northern latitudes is quite different from that in the Meiyu-Baiu frontal zone. The disturbances in the northern latitudes tend to remain at the almost same longitude, which indicates very slow propagation. This is due to the development of the blocking ridge over 130°-140°E.

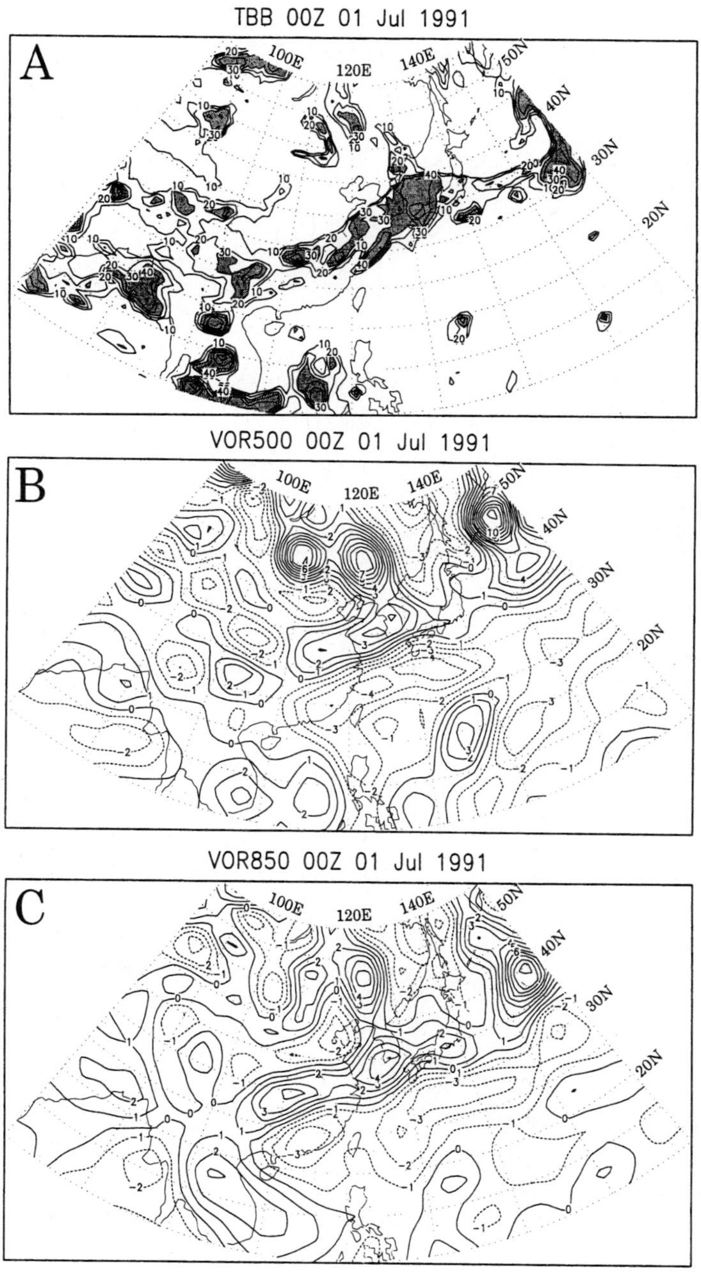

Figure 18. (A) Maps of TBB at 00 UTC 1 July 1991. The isopleths are at 10°C intervals, and the minus sign of TBB is omitted. The stippling indicates an area where TBB is less than −30°C. Maps of (B) ζ_{500} (500 hPa relative vorticity, 10^{-6} s^{-1}) and (C) ζ_{850} (850 hPa relative vorticity, 10^{-6} s^{-1}) at 00 UTC 1 July 1991.

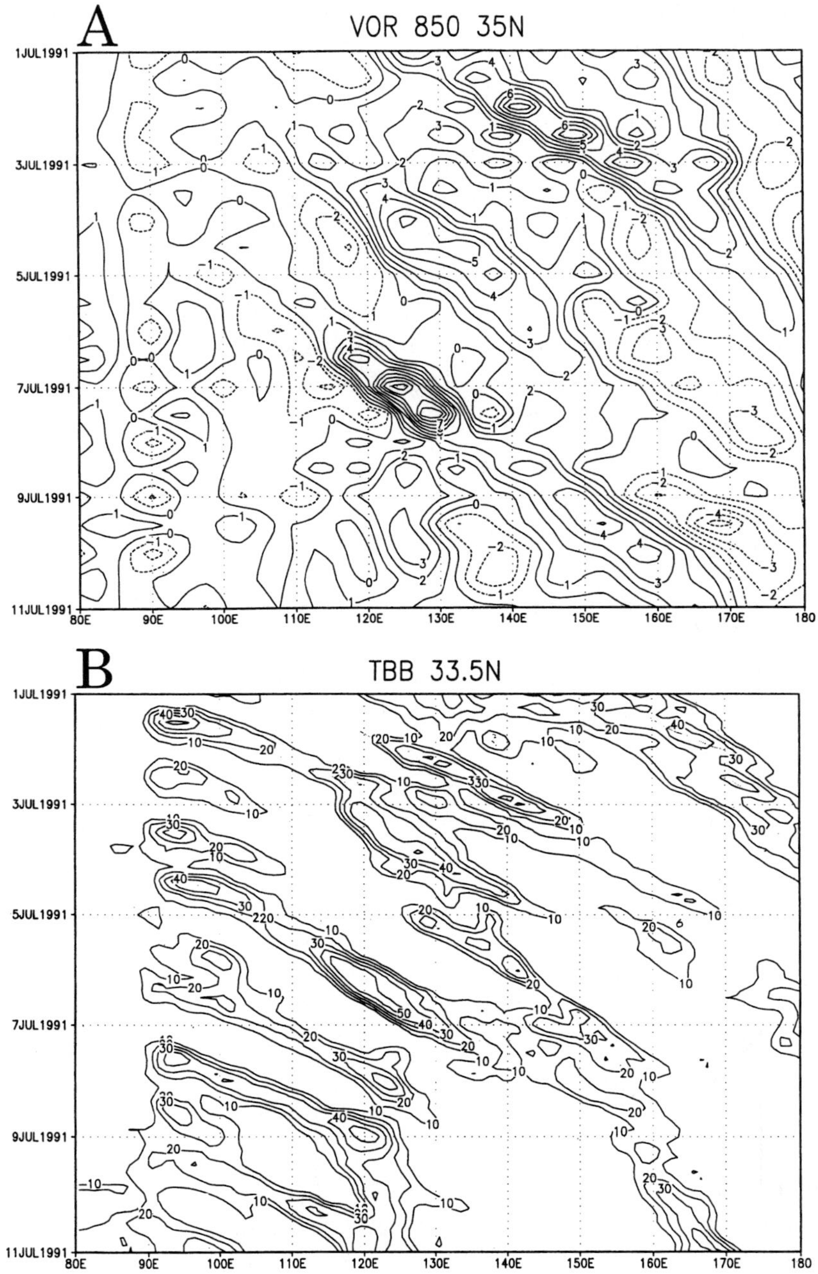

Figure 19. (A) The longitude-time section of ζ_{850} (850 hPa relative vorticity, unit in 10^{-6} s^{-1}) at 35°N. (B) The longitude-time section of TBB at 32.5°N. The isopleths are at 10°C intervals, and the minus sign of TBB is omitted.

Figures 20A and 20B show the locations of vorticity cores at 500 and 850 hPa derived from successive 12-hourly maps. Two paths of the disturbances are found. The southern path extends along the Meiyu-Baiu frontal zone, while the northern path is along the southern rim of the cold low over Siberia. When the disturbances in the northern latitudes move southward, the coupling of the northern disturbance with the Meiyu-Baiu disturbance results in the formation of a north to south oriented short-wave trough and a northward extension of the cloud/precipitation zone (Ninomiya 2001).

We next examine temporal variations of $\boldsymbol{V}_{850}$ (850 hPa wind velocity), ω_{500} (500 hPa vertical velocity) and T_{850} (850 hPa temperature) associated with the passage and development of the disturbances. To show the variations within the 10-day period, the standard deviations of these quantities are calculated (Fig. 21). For example, the standard deviation of temperature T is defined at each grid point as

$$\text{STD of T} = [\Sigma\ (T - T_m)^2 / N\,]^{1/2} \qquad (5).$$

Here, T_m is the average value within the 10-day period, and N is the number of samples (in this case, N = 20; 12-hourly data over 10 days).

Large STDs of $\boldsymbol{V}_{850}$ (Fig. 21A) and ω_{500} (Fig. 21B) are seen in the Meiyu-Baiu frontal zone. In 115°-160°E, the area of maximum STDs of $\boldsymbol{V}_{850}$ and ω_{500} coincide with the zone of maximum vorticity at 850 hPa. It is notable that STD of ω_{500} already attains its maximum in the intense precipitation area over China, while STD of $\boldsymbol{V}_{850}$ begins to increase gradually on the downstream side of the intense precipitation area. The STD of $\boldsymbol{V}_{850}$ increases eastward and has a maximum of ~8 m s^{-1} over the Japan Islands. This indicates that the Meiyu-Baiu frontal disturbances, which are generated in the vicinity of the intense precipitation zone as convective systems, develop into subsynoptic-scale circulation systems as they propagate eastward. As the propagation speed of these disturbances is ~1000 km d^{-1}, the disturbances seem to reach a developed stage within ~1 day after the generation in the intense precipitation region.

The distribution of STD of T_{850} (Fig. 21C) has different features from STD of $\boldsymbol{V}_{850}$ and STD of ω_{500}. The STD of T_{850} is very small in the intense precipitation region over China, where the horizontal thermal gradient is also small. The area of relatively large STD of T_{850} in the Baiu frontal zone in 135-160°E indicates the development of the disturbances related with the baroclinic instability.

As shown in Fig. 19B, a few mesoscale cloud systems occur between the successive subsynoptic-scale cloud systems. These meso-α-scale cloud systems are accompanied by vorticity maximum at 850 and 500 hPa. The horizontal spacing between them is 500-1000 km. Even smaller mesoscale cloud systems are not necessarily associated with the significant vorticity maximum, and are not sustained beyond ~20 hours.

According to Eddy's baroclinic instability model (Gill 1982), the maximum growth rate of the baroclinic wave, σ_{max}, is given by

$$\sigma_{max} = 0.310\,(f / N)(\,\partial U_g / \partial z)\,, \qquad (6)$$

at the wave number defined by

$$N\,k\,H = 0.803\,f \qquad (7).$$

Here N, k, H, f, and ($\partial U_g / \partial z$) are the Brunt-Vaisalla frequency ($N^2 = g\partial \ln \theta / \partial z$), wave number in the direction of the thermal wind vector, half depth of the layer in which the disturbances exists, Coriolis parameter, and the vertical shear of the geostrophic wind velocity, respectively. Ninomiya (2000) estimated the wavelength of the most unstable baroclinic wave and its growth rate from the 10-day averaged quantities in the lower troposphere within 125°-140°E. He obtained a wavelength of ~3100 km, and the maximum growth rate of ~ $5.4 \times 10^{-6}\ s^{-1}$, which corresponds to the e-folding time of ~48 hours. This estimate of the most unstable baroclinic wave coincides approximately with the wavelength of the subsynoptic-scale frontal disturbances.

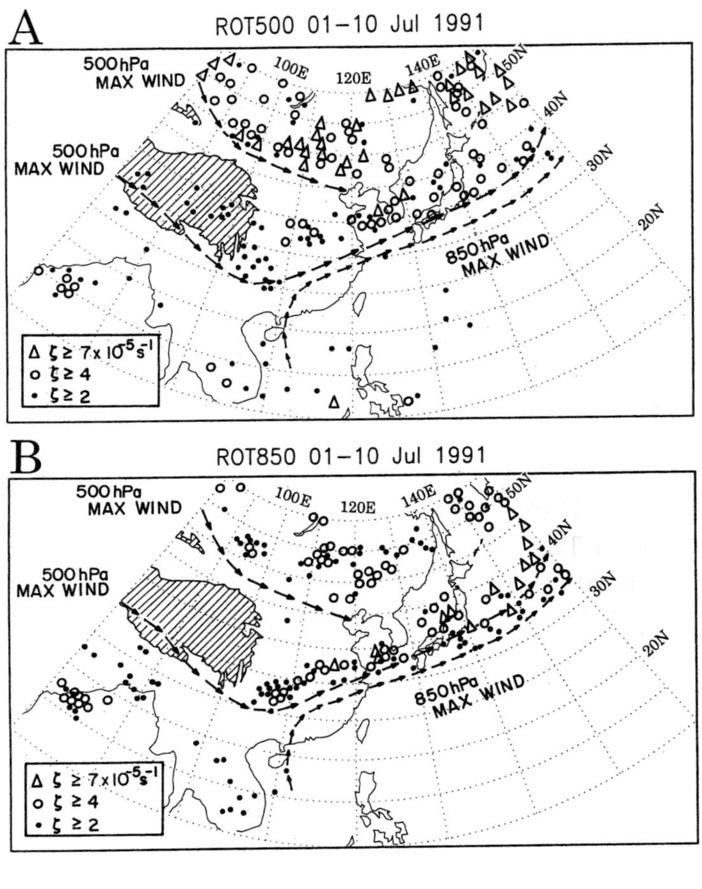

Figure 20. Locations of vorticity cores at (A) 500 hPa and (B) 850 hPa during the 10-day period.

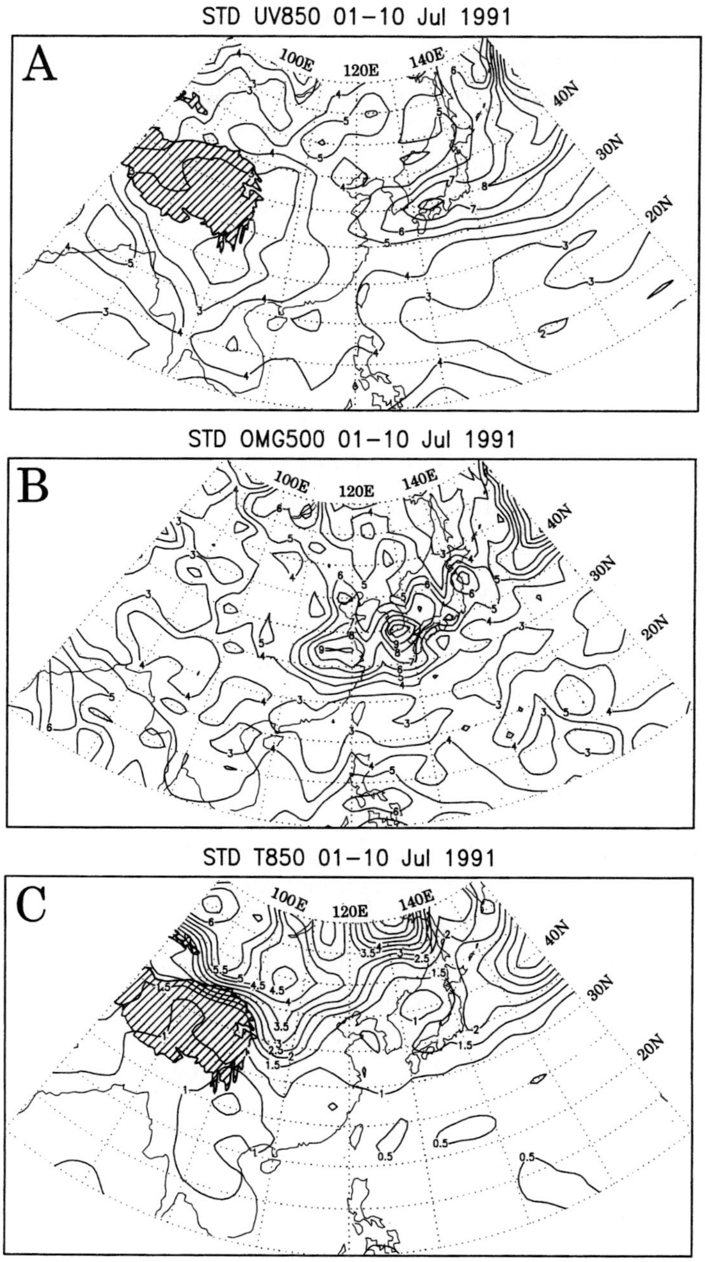

Figure 21. Standard deviations of (A) wind velocity at 850 hPa (m s^{-1}) , (B): vertical-p-velocity at 500 hPa (hPa d^{-1}), and (C) temperature at 850 hPa (K) for 1-10 July 1991.

For the mesoscale frontal cloud systems, the equivalent potential temperature θ_e should be used for the evaluation, since the cloud zone is very moist. When N is replaced by N* ($N^{*2} = g\, \partial \ln\theta_e / \partial z$), the wavelength of the most unstable wave is ~600 km and its growth rate is 2.8 x 10^{-5} s^{-1}, which corresponds to the e-folding time of ~10 hour. These values are consistent, at least qualitatively, with the observed spacing between the meso-α-scale cloud systems in the cloud zone, although the observed features of the mesoscale cloud system family are too complicated to be interpreted from the linear theory. It should be mentioned that the Meiyu-Baiu front and cloud systems interact. While the front provides environmental conditions favorable for the cloud systems, the successive development of the cloud systems works to sustain the front.

It is also emphasized that the mesoscale cloud systems begin to develop over the area of intense precipitation around ~115°E, where the horizontal thermal gradient is very weak. This fact suggests the important role of condensation heating for the generation of the mesoscale convective systems, as theoretically discussed by Tokioka (1973).

10. Family of Mesoscale Cloud Systems

As indicated in Figs. 18 and 19, the Meiyu-Baiu cloud zone contains a train of cloud systems of subsynoptic-scale and mesoscale.

Each subsynoptic-scale cloud system is associated with a respective subsynoptic-scale frontal depression. There are a few mesoscale sub-systems within a subsynoptic-scale cloud system. In general, active convective clouds are favored in its southwestern portion. Sometimes, significant banded structures of clouds are found.

The subsynoptic-scale subsidence and cold advection occurring on the backside of the depression sustain a narrow frontal zone after the passage of the depression. The weak convectively stable layer and relatively strong thermal gradient are also sustained in the long trailing portions of the depressions. Thus, one series of the cloud zone with a length of 2000-3000 km will consist of a family of subsynoptic-scale cloud system, and meso-α-scale cloud systems that are aligned along the trailing portion of the preceding subsynoptic-scale cloud system.

The low TBB value and sharp boundary of the low TBB area of the meso-α-scale cloud systems indicate that the main part of them consists of active convective clouds. They tend to appear on satellite cloud images as either oval-shaped cloud clusters or leaf-shaped cloud systems with a cloud band in its southwestern tail. A fine inner structure is usually seen.

These meso-α-scale cloud systems are classified into three types according to their lifetime and the magnitude of the associated relative vorticity.

Type-A meso-α-scale cloud systems are characterized by the cyclonic circulation in the lower and middle troposphere. The magnitude of the relative vorticity is less than 4 x 10^{-5} s^{-1}, which is significantly smaller than that of the subsynoptic-scale depression. They do not necessarily develop into a frontal depression. They form within the moist and convectively unstable airmass over the southern China and develop as they propagate along the Baiu front. They weaken as they propagate into the stable airmass to the east of 150°E. The lowest TBB is about −60°C. The horizontal spacing between the subsynoptic-scale cloud systems and/or meso-α-scale cloud systems is 500-1000 km.

Type-B meso-α-scale cloud systems are characterized by a cyclonic circulation that is confined to the lower troposphere. They form within the moist and convectively unstable airmass over southern China and weaken when they propagate into the stable airmass. Although the lowest TBB is about −50°C, the extent of the cloud system is only about 200 km.

Type-C meso-α-scale cloud systems are characterized by a short life of less than 12 hour. They form over the southern China during the afternoon due to heating by strong insolation. Many of them form within the region of low-level (850 hPa) anticyclonic circulation, which extends to the south of the Meiyu front, and are not sustained beyond 12 hour.

11. Summary and Conclusion

Many previous authors have studied the Meiyu-Baiu front from various viewpoints. However, the multi-scale features, from large-scale to mesoscale, of whole frontal system have not been fully understood yet. The present review discusses the Meiyu-Baiu front and associated cloud systems observed during 1-10 July 1991 based on the author's recent studies, which utilized mainly ECMWF re-analysis data and the Geostationary Meteorological Satellite IR data. Important features found in the present review are:

(1) The west-to-east elongated Meiyu-Baiu cloud zone associated with intense rainfalls appears only in a limited period of the Meiyu-Baiu season.

(2) The west-to-east elongated Meiyu-Baiu front with intense precipitation is sustained under the characteristic large-scale environmental conditions, which are associated with a westward protruding Pacific subtropical anticyclone and strong confluence of the Indian monsoon westerlies with the Pacific trade easterly winds around the South China Sea. The large-scale features in the northern latitudes are characterized by development of the blocking ridge over 130°-140°E, and two cold lows to the east and west of the blocking ridge.

(3) A dipole of the low-level convergence in the Meiyu-Baiu front and divergence to the south of the front is the most significant feature in the west-to-east elongated Meiyu-Baiu rainfall zone. This dipole is related with the strong ageostrophic wind in the western periphery of the westward protruding Pacific anticyclone.

(4) The west-to-east elongated Meiyu-Baiu frontal cloud zone consists of subsynoptic-scale and meso-α-scale cloud systems aligned along the cloud zone. These cloud systems are categorized into the subsynoptic-scale and meso-α-scale cloud systems based on the horizontal extent of the cloud area, duration, and the magnitude of the relative vorticity associated with the cloud system.

(5) During the 10-day period, four subsynoptic-scale cloud systems, which where associated with Meiyu-Baiu frontal depressions, formed with a time interval of 2-3 days, under the influence of the westerly troughs in the middle latitudes that had wavelengths of 2000-3000 km. The subsynoptic-scale cloud system consists of mesoscale convective systems and widespread clouds in its southern and northeastern portions, respectively.

(6) The subsidence and cold advection occurring on the back of the subsynoptic-scale depressions sustain a narrow frontal zone after the passage of the depressions. The nearly moist neutral stratification and relatively strong thermal gradient are also sustained in the long trailing portion of the depressions.

(7) A few meso-α-scale cloud systems develop in the long trailing portion of the preceding subsynoptic-scale depression. They form a family of Meiyu-Baiu frontal cloud systems. The low temperature at cloud top and the sharp boundary of the cold cloud area indicate that the main part of the meso-α-scale cloud system consists of active convective clouds. While some of the cloud systems have features of oval-shaped cloud clusters, some others have features of leaf-shaped cloud systems with a cloud band in its southwestern portion.

(8) Meso-α-scale cloud systems are categorized into three types based on the horizontal extent of the cloud area, duration, and the magnitude of the relative vorticity associated with the cloud system. Type-A and Type-B meso-α-scale cloud systems are identified for about 20-30 hours, although their inner structures exhibit shorter time-α-scale variations. The Type-A meso-α-scale cloud systems are accompanied by cyclonic vorticity in the lower and middle troposphere, while the Type-B meso-α-scale cloud systems are only accompanied by low-level vorticity. The shorter-lived Type-C meso-α-scale cloud systems are not accompanied by a cyclonic vorticity maximum.

(9) The west-to-east elongated Meiyu-Baiu cloud zone consists of a few cloud system families. A "cloud system family" consists of a preceding sub-synoptic-scale cloud system, and a few meso-α-scale cloud systems that are aligned along the trailing portion of the preceding subsynoptic-scale cloud system.

On the basis of the aforementioned features, a conceptual model of the meso-scale cloud system family in the Meiyu-Baiu front is proposed in Fig. 22. In this model the cloud zone consists of a few cloud system families, each of which consists of two parts: a sub-synoptic-scale cloud system associated with a sub-synoptic-scale Meiyu-Baiu frontal depression (indicated by S), and a few meso-α-scale cloud systems (indicated by α). The latter are aligned along the trailing portion of the preceding sub-synoptic-scale cloud system. Cold lows and a midlatitude blocking ridge and the Pacific subtropical anticyclone all have strong influences on the Meiyu-Baiu cloud systems. Rows of large and small arrows in Fig. 22 indicate the 500-hPa and 850-hPa maximum wind axes, respectively. The short-wave trough that propagates along the northern maximum wind zone becomes coupled with the short-wave trough in the Meiyu-Baiu frontal zone under the influence of the cold low over Siberia, leading to the development of a sub-synoptic-scale frontal depression. Subsequently, a few meso-α-scale cloud clusters form along the trailing portion of the preceding sub-synoptic-scale cloud system.

Although the multi-scale structure of the Meiyu-Baiu cloud zone has been evident from the satellite cloud observation, there are yet no solid physical reason available to explain why a few kinds of cloud systems coexist in the frontal zone. Studies about the inner fine structure in each cloud system utilizing radar observations are also needed to further understanding of the Meiyu-Baiu cloud systems.

The results of the various previous papers should be compared with the present paper. Some of the previous papers did not clearly describe the environmental situation around the precipitation/cloud systems. Such cases cannot be compared with the present case. The cases in 16-25 July 1982 (Akiyama 1989) have similar features as in the present study. It is not clear whether the cases in 5-8 July, 1968 (Matsumoto and Tsuneoka 1970; Akiyama 1978) are subsynoptic-scale or meso-α-scale system, since the wavelength is about 1000 km. It should be also noted that the precipitation/cloud systems associated with a cold vortex aloft (Ninomiya and Akiyama 1974; Yoshizumi 1975), and a large-scale extratropical cyclone (Matsumoto and Ninomiya 1971), are another type of cloud system that differs from these in the quasi-stationary Meiyu-Baiu front.

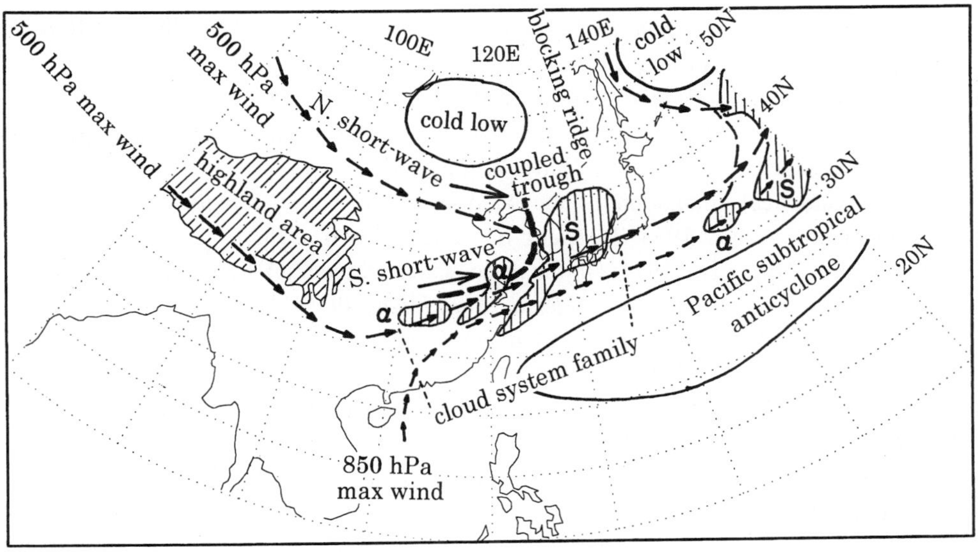

Figure 22. Conceptual model of the Meiyu-Baiu frontal cloud zone.

Acknowledgments

The present study has been partially down in 1997-2000 at the Center for Climate System Research (CCSR), the University of Tokyo. The GMS data are prepared by a "Cooperative research on the tropical convective activity and variability of large-scale circulation" project between the Meteorological Research Institute (MRI) and CCSR. About this, the present author expresses his sincere thanks to Dr. N. Yamazaki of MRI.

References

Akiyama, T., 1973: The large-scale aspects of the characteristic features of the Baiu front. *Pap. Met. Geophy.*, **24**, 157-188.

______, 1975: Southerly transversal moisture flux into the extremely intense rainfall zone in the Baiu season. *J. Meteor. Soc. Japan*, **53**, 304-316.

______, 1978: Meso-scale pulsation of convective rain in medium-scale disturbances developed in Baiu front. *J. Meteor. Soc. Japan*, **56**, 267-283.

______, 1984: A medium-scale cloud cluster in a Baiu front. Part 1: Evolution process and fine structure. *J. Meteor. Soc. Japan*, **62**, 485-504.

______, 1989: Large, synoptic and meso-scale variations of the Baiu front during July 1982. Part 1: Cloud features. *J. Meteor. Soc. Japan*, **67**, 57-81.

______, 1990: Large, synoptic and meso-scale variations of the Baiu front during July 1982. Part 2: Frontal structure and disturbances. *J. Meteor. Soc. Japan*, **68**, 559-574.

Ding, Y. -H., 1991: *Monsoon over China*, Kluwer Academic Pub., 419 pp.

______, 1993: *Study on the persistent heavy rainfall over the Yangtze River Valley and Huaihe River Basin in 1991.* China Meteorological Press. 255 pp.

Gill, A. E., 1982: *Atmosphere-ocean dynamics*. Academic Press. 602 pp.

Iwasaki, H., and T. Takeda, 1993: Structure and behavior of meso-scale cloud clusters traveling over the Baiu-frontal zone. *J. Meteor. Soc. Japan*, **71**, 733-747.

Kato, T., 1998: Numerical simulation of the band-shaped torrential rain observed over southern Kyushu, Japan on 1 August 1993. *J. Meteor. Soc. Japan*, **76**, 97-128.

Matsumoto, S., and K. Ninomiya, 1971: On the meso-scale and medium-scale structure of a cold front and the relevant vertical circulation. *J. Meteor. Soc. Japan*, **49**, 648-662.

______, ______, and S. Yoshizumi, 1971: Characteristics of the Baiu front with heavy rainfall. *J. Meteor. Soc. Japan*, **49**, 267-281.

______, and Y. Tsuneoka, 1970: Time-lapse composite echo pattern of wave disturbances embedded in the Baiu front. *J. Meteor. Soc. Japan*, **48**, 189-203.

______, S. Yoshizumi, and M. Takeuchi, 1970: On the structure of the Baiu front and associated intermediate-scale disturbances in the lower atmosphere. *J. Meteor. Soc. Japan*, **48**, 479-491.

Ninomiya, K., 1978: Heavy rainfalls associated with frontal depression in Asian subtropical humid region. *J. Meteor. Soc. Japan*, **56**, 253-266.

______, 1984: Characteristics of Baiu front as a predominate subtropical front in the summer Northern Hemisphere. *J. Meteor. Soc. Japan*, **62**, 880-894.

______, 1989: Cloud distribution over East Asia during Baiu period in 1979. *J. Meteor. Soc. Japan*, **62**, 880-894.

______, 1999: Moisture balance over China and the South China Sea during the summer monsoon in 1991 in relation to the intense rainfalls over China. *J. Meteor. Soc. Japan*, **77**, 737-751.

______, 2000: Large- and meso-α-scale characteristics of Meiyu-Baiu front associated with intense rainfalls in 1-10 July 1991. *J. Meteor. Soc. Japan*, **78**, 141-157.

______, 2001: Large λ-shaped cloud zone formed around July 6, 1991 with pole-ward moisture transport from intense rainfall area in Meiyu-Baiu front. *J. Meteor. Soc. Japan*, **79**, 805-813.

______, and T. Akiyama, 1971: The development of medium-scale disturbance in the Baiu front. *J. Meteor. Soc. Japan,* **49**, 663-677.

______, and ______, 1972: Medium-scale echo clusters in the Baiu front as revealed by multi-radar composite echo maps. *J. Meteor. Soc. Japan*, **50**, 558-569.

______, and ______, 1974: Band structure of meso-scale echo clusters associated with low-level jet stream. *J. Meteor. Soc. Japan*, **52**, 300-313.

______, and ______, 1992: Multi-scale features of Baiu, the summer monsoon over Japan and the East Asia. *J. Meteor. Soc. Japan*, **70**, 467-495.

______, ______, and M. Ikawa, 1988: Evolution and fine structure of a long-lived meso-α-scale convective system in Baiu frontal zone. Part 1: Evolution and meso-β-scale characteristics. *J. Meteor. Soc. Japan*, **66**, 331-350.

______, and T. Murakami, 1987: The early summer rainy season (Baiu) over Japan. *Monsoon Meteorology*, C.-P. Chang and T. N. Krishnamurti, Eds., Oxford Univ. Press, 93-121.

______, and H. Muraki, 1986: Large-scale circulations over East Asia during Baiu period of 1979. *J. Meteor. Soc. Japan*, **64**, 409-429.

______, and Y. Shibagaki, 2003: Cloud system families in the Meiyu-Baiu front observed in 1-10 July 1991. *J. Meteor. Soc. Japan*, **81**, 193-209.

______, and K. Yamazaki, 1979: Heavy rainfalls associated with frontal depression in Asian subtropical humid region (II). *J. Meteor. Soc. Japan*, **57**, 399-413.

Orlanski, J., 1975: A rational subdivision of scales for atmospheric process. *Bull. Amer. Met. Soc.*, **56**, 527-530.

Shibagaki, Y., M. D. Yamanaka, H. Hashiguchi, A. Watanabe, H. Uyeda, Y. Maekawa, and S. Fukao, 1997: Hierarchical structures of vertical velocity variations and precipitation clouds near the Baiu frontal cyclone center observed by the MU and meteorological radars. *J. Meteor. Soc. Japan*, **75**, 569-596.

______, ______, S. Shimizu, H. Uyeda, A. Watanabe, Y. Maekawa, and S. Fukao, 2000: Meso-β to γ-scale wind circulations associated with precipitating cloud near Baiu front observed by the MU and meteorological radars. *J. Meteor. Soc. Japan*, **78**, 69-91.

Takeda, T., and H. Iwasaki, 1987: Some characteristics of meso-scale cloud clusters observed in East Asia between March and October 1980. *J. Meteor. Soc. Japan*, **65**, 507-513.

Tao, S., and L. Chen, 1987: A review of recent research on the East Asia summer monsoon over China. *Monsoon Meteorology*, C. -P. Chang and T. N. Krishnamurti, Eds.,Oxford Univ. Press, 50-92.

Tokioka, T., 1973: A stability study of medium-scale disturbances with inclusion of convective effects. *J. Meteor. Soc. Japan*, **51**, 1-10.

Yoshizumi, S., 1975: Development of an instability line under a cut-off low. *Pap. Met. Geophy.* , **26**, 167-180.

______, 1977: On the structure of intermediate–scale disturbances on the Baiu front. *J. Meteor. Soc. Japan*, **55**, 107-120.

12. OCEANIC EAST ASIAN MONSOON CONVECTION: RESULTS FROM THE 1998 SCSMEX

RICHARD H. JOHNSON AND PAUL E. CIESIELSKI

Atmospheric Science Department
Colorado State University
Fort Collins, CO 80523-1371
E-mail: johnson@atmos.colostate.edu, paulc@atmos.colostate.edu

THOMAS D. KEENAN

Bureau of Meteorology Research Centre
Melbourne, Australia
E-mail: T.Keenan@bom.gov.au

Observations from the May-June 1998 South China Sea Monsoon Experiment (SCSMEX) have been used to determine the characteristics of the onset of the summer monsoon over the northern South China Sea (SCS). The SST gradually increased over the northern SCS during SCSMEX, interrupted by slight cooling following monsoon onset until early June when the warming resumed. Surface sensible and latent heat fluxes increased after onset, but then decreased in June as a result of warm, moist air advecting over cooler water near the south China coast. The heating and moistening rates and vertical eddy flux of total heat during the early-June active period were greater than those observed during the May monsoon onset active period, indicating more vigorous deep convection as the monsoon ensued, a finding supported by Tropical Rainfall Measuring Mission (TRMM) precipitation radar data. Analysis of ground-based radar data reveals that lower and middle level vertical shears exerted a dominant control over the structure and orientation of mesoscale convective systems over the northern SCS. The findings are consistent with those of LeMone *et al.* for the equatorial western Pacific, except two new organizational modes have been identified: shear-parallel bands for strong low-level shear and weak midlevel shear when the air is dry aloft, and shear-parallel bands for strong shears in both layers when the shear vectors are in the same direction. Midlatitude influences likely contributed to these two additional modes.

1. Introduction

The East Asian Monsoon (EAM) is characterized by distinctly different winter and summer circulation patterns, with transitions between the two occurring on a relatively rapid time scale. During the winter monsoon, cold air at low levels flows off the Asian continent southwestward over the South China Sea (SCS), eventually converging into deep convective systems over the Indonesian-Malaysian maritime continent. In contrast, the summer monsoon features southwesterly flow of moist air across the SCS feeding deep convective systems over the regions of China, Japan, and Korea. The focus of this paper is on the summer component of the East Asian Monsoon. It is during this period that the 1998 South China Sea Monsoon

Experiment (SCSMEX; Lau *et al.* 2000b) was conducted, providing new insights into convective systems over the northern SCS around the time of the onset of the EAM.

The onset of the broadscale Asian summer monsoon is concurrent with the reversal of the temperature gradient south of the Tibetan Plateau (Flohn 1957; Yanai *et al.* 1992; Li and Yanai 1996). The initial onset occurs in early to mid-May over the SCS area (Tao and Chen 1987). In early spring, convection in this region begins to shift northward from the equator and monsoon rains become anchored over the Borneo-Malaysia-Indo-China land bridge from mid-April until mid-May. Then convection appears to rapidly jump across the SCS during mid-to-late May and establish itself over southern China and the northern SCS (Hirasawa *et al.* 1995; Lau and Yang 1997; Lau *et al.* 1998). The subsequent arrival in June of the rains over central China and Japan marks the beginning of the Meiyu and Baiu rainy seasons over those areas, respectively.

In May-June 1998 SCSMEX was conducted to investigate processes associated with the onset of the East Asian Monsoon. SCSMEX provides a first opportunity to study the detailed characteristics of convection and circulation changes over the northern SCS during the EAM onset. Multiple observation platforms (e.g., soundings, Doppler radar, ships, wind profilers, etc.) were deployed during SCSMEX to explore these changes. In this paper we will report on radar and atmospheric sounding studies of the organization of convection, as well as the vertical profiles of heating and moistening, over the northern SCS during the 1998 onset of the EAM.

1.1. *Pre-SCSMEX Studies of Latent Heating Distributions*

During the summer monsoon, the convective areas of southern China and the northern SCS areas provide a significant contribution to the total precipitation and diabatic heat source in the Asian monsoon system (Luo and Yanai 1984; Yanai and Tomita 1998). Several studies of latent heating distributions have been carried out over mainland China. Atmospheric budget studies for the Yangtze Valley show the existence of both convective and stratiform precipitation during the Meiyu, with considerable interannual variability (Luo and Yanai 1984; Kato 1985; Ding and Wang 1988; Ding and Hu 1992). In a study of three summer monsoon seasons, Johnson *et al.* (1993) found bands of maximum heating and moistening characteristic of deep convection just inland and parallel to the coastline of southern China during the pre-Meiyu and Meiyu rainy periods.

In contrast, very little is known about the part of the monsoon heat source that extends over the northern SCS. Satellite studies (e.g., Lau and Yang 1997) indicate a sudden increase in high cloudiness during onset around mid-May in the northern SCS, although there is considerable year-to-year variability. However, details of these cloud fields and their changes with the onset are not known.

In this study, we plan to extend these earlier works by examining the evolution of large-scale and regional-scale flows, precipitation patterns, and tropospheric heating and moistening for the 1998 summer monsoon onset using SCSMEX sounding data. Precipitation estimates based on atmospheric budgets will be compared to those obtained from satellite measurements, with particular emphasis on Tropical Rainfall Measuring Mission (TRMM) observations (Kummerow *et al.* 2000).

1.2. *Pre-SCSMEX Studies of the Organization of Convection*

It has long been known that the organization of tropical convection is influenced predominantly by the vertical shear and convective available potential energy or CAPE (e.g., Moncrieff and Green 1972). Various observational studies in the eastern Atlantic and northern Australia have confirmed the strong influence of environmental winds and CAPE on the structure and orientation of convective bands (e.g., Barnes and Seickman 1984; Alexander and Young 1992; Keenan and Carbone 1992). Recently, LeMone *et al.* (1998) investigated the organization of convection over the western Pacific warm pool using aircraft data from the Tropical Ocean Global Atmosphere Coupled Ocean-Atmosphere Response Experiment (TOGA COARE). They found that vertical shear in the low-to-mid- troposphere was key in determining the orientation of convective bands, while CAPE influenced their depth and longevity. Their results have been recently supported by numerical simulations of convection in shear by Robe and Emanuel (2001).

In this study, the relationship between convection and vertical wind shear is further explored using radar reflectivity data from the Bureau of Meteorology Research Centre (BMRC) C-POL radar on Dongsha Island in the northern SCS and proximity sounding data. The findings will be related to those determined by LeMone *et al.* (1998) and others for various regions in the Tropics.

2. Data and Analysis Procedures

2.1. *Sounding Data*

During the May-June 1998 SCSMEX, a sounding network was established in the SCS and surrounding region to investigate the onset and development of the East Asian monsoon. Preliminary results from the SCSMEX sounding network have been reported in Ding and Liu (2001). The SCSMEX network was contained within the larger domain of the GEWEX Asian Monsoon Experiment (GAME). Sounding sites within the SCSMEX/GAME domain and the May-June 1998 mean sea surface temperature (SST) are shown in Fig. 1. The mean SST field was derived from daily values obtained from the TRMM Microwave Imager (TMI) SST product (Wentz *et al.* 2000). Figure 1 contains information concerning the vertical resolution and frequency of data receipt for all of the sites during the 47-day Intensive Observing Period (IOP) (5 May-20 June). The highest sounding frequency and resolution were in the vicinity of the SCS (up to 4 launches per day and 5-hPa vertical resolution) decreasing to 1-2 times per day and operational vertical resolution (~30-100 hPa) away from the SCS. Two special sounding networks are indicated by the polygons in Fig. 1, one in the northern SCS – the Northern Enhanced Sounding Array or NESA – surrounding the dual-Doppler array (the TOGA radar aboard R/V *Shiyan #3* and the BMRC C-POL radar on Dongsha Island) and the other surrounding the R/V *Kexue #1* in the southern SCS - the Southern Enhanced Sounding

Array. These networks were designed to determine and contrast the properties of convection in two distinct oceanic regions of the Asian monsoon. However, this study will focus only on results for the NESA.

Figure 1 shows that a strong north-south SST gradient existed across the NESA during SCSMEX. Cooler waters near the south China coast, left over from the winter monsoon, existed throughout the IOP, although the gradient weakened slightly throughout the period. A tongue of cooler water extended through the Taiwan Straits towards *Shiyan #3*, which reduced the surface fluxes at this location following monsoon onset (to be discussed later).

Sounding quality control procedures and other aspects of the sounding data analysis are described in Johnson and Ciesielski (2002).

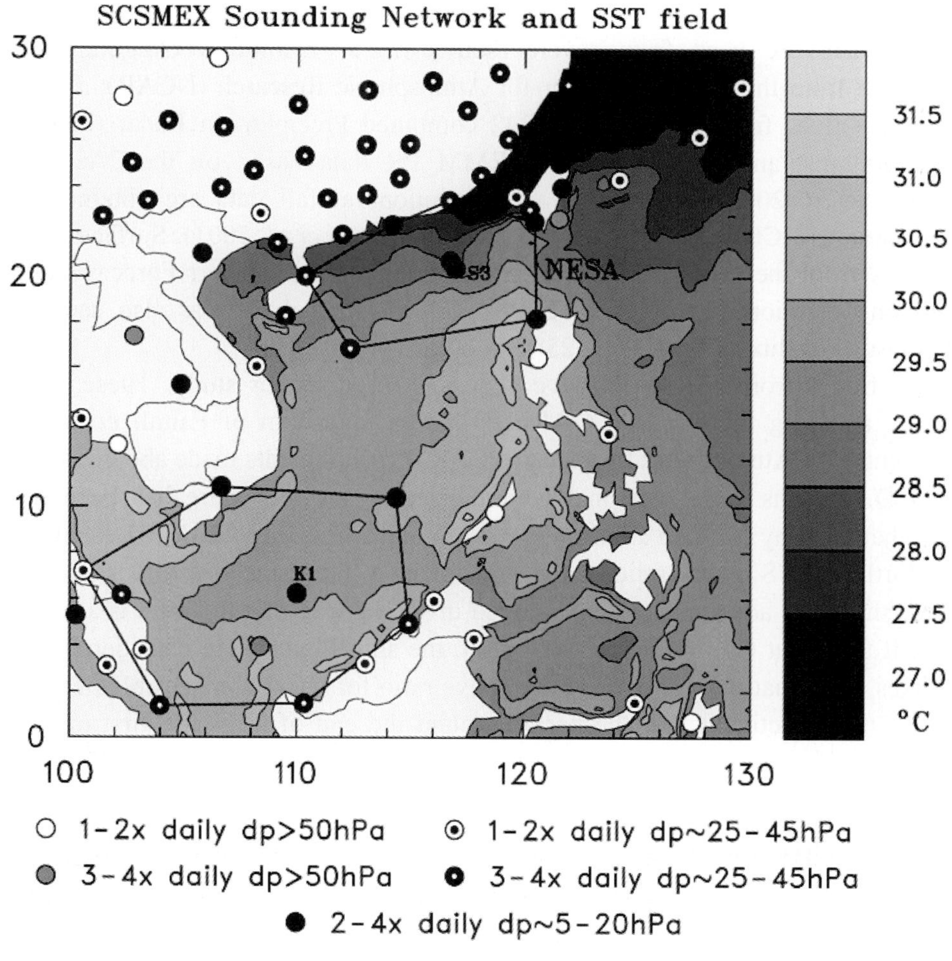

Figure 1. SCSMEX/GAME sounding network and SST distribution (scale to right) for May-June 1998. Frequency of observations and vertical resolution are indicated by dots with various shadings (*dp* refers to vertical resolution; *S3* to R/V *Shiyan #3*; *K1* to R/V *Kexue #1*). Sounding polygons in the northern and southern South China Sea are also indicated; NESA refers to Northern Enhanced Sounding Array.

2.2. *Other Data Sources*

Ground-based radar data are obtained from the BMRC, 5-cm dual-polarimetric Doppler radar (C-POL) that was located at Dongsha Island (20.7°N, 116.7°E) in the northern SCS. Animations of base-scan reflectivity fields were used to determine the organizational modes and propagation characteristics of the convection.

In the preparation of a gridded dataset, GAME Reanalysis data are used to supplement data from the sounding network to fill in voids and lessen the impacts of topography on northern SCS budgets. The GAME reanlaysis employs a three-dimensional optimum interpolation scheme and the Japan Meteorological Agency (JMA) forecast model; has T213 (0.5625 degree) horizontal resolution and 30 vertical layers; and uses satellite cloud-drift winds, aircraft reports and wind profiler observations in addition to the enhanced SCSMEX/GAME sounding network (Yamazaki *et al.* 2000).

Satellite data are used from several sources: 2.5° resolution Outgoing Longwave Radiation data from the National Center for Atmospheric Research (NCAR), 1°–resolution daily rainfall values from the TRMM 3B42 combined Precipitation Radar (PR) and TMI rainfall algorithms, and high-resolution TRMM PR data based on the 2A25 algorithm (Kummerow *et al.* 2000). In addition, 1°–resolution rainfall data are obtained from the Global Precipitation Climatology Project (GPCP; Huffman *et al.* 2001). Surface rainfalls are also obtained from the European Centre for Medium Range Weather Forecasts (ECMWF) 1°–resolution operational analyses. The JMA/GAME Reanalysis was also used to obtain rainfall and surface flux estimates at 1.25° resolution for the NESA.

Surface fluxes from *Shiyan #3* have also been used in our study. These fluxes were obtained by applying the TOGA COARE bulk flux algorithm of Fairall *et al.* (1996) to National Center for Atmospheric Research (NCAR) measurements made aboard *Shiyan #3* (J. Pinto and D. Parsons 2001, personal communication). A procedure has been developed similar to that used by Lin and Johnson (1996) to adjust the JMA/GAME Reanalysis fluxes over the northern SCS by the ratio of the ship fluxes to the reanalysis fluxes at a grid point nearest the ship. The adjustment is constrained in such a way that the ratio is always greater than zero. It turns out that after monsoon onset, the ship fluxes were considerably less than model fluxes, such that the averages of the above ratio for latent and sensible heat fluxes are 0.50 and 0.76, respectively. For the period around the end of May and first of June when *Shiyan #3* was off station, these mean values were used to adjust the fluxes.

3. Large-Scale Evolution of the Flow and Convection during Onset

The evolution of the flow accompanying the onset of the 1998 summer monsoon has been discussed in detail by Ding and Liu (2001). They noted that the onset occurred first over the northern SCS around 15 May and was influenced significantly by the southward propagation of cold air from midlatitudes, as often occurs (Chang and Chen 1995; Chan *et al.* 2000). Over the central and southern SCS, the onset occurred slightly later, around 20 May, and was characterized by an eastward retreat of the western Pacific subtropical ridge and the establishment of southwesterlies across the SCS (Ding and Liu 2001). The rapid large-scale circulation changes at 850 and 200 hPa at this time are shown in Fig. 2. From 19 to 21 May the 850-hPa subtropical ridge quickly withdrew from the SCS and from 19 to 22 May the 200-hPa anticyclone center shifted northward to near the head of the Bay of Bengal.

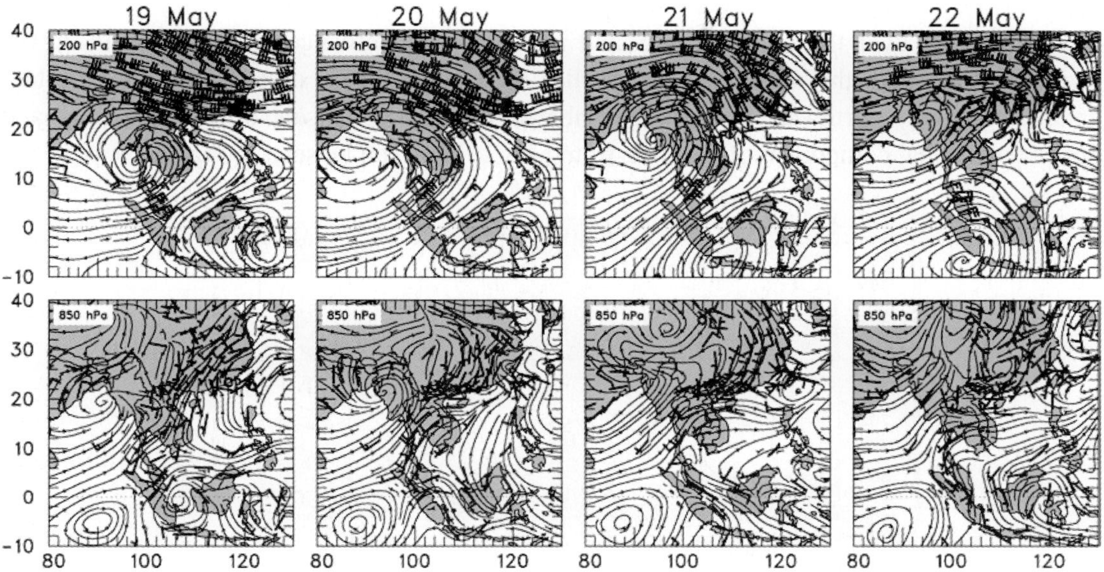

Figure 2. Streamline analyses at 200 hPa on 19-22 May 1998 (top panels) and at 850 hPa (bottom panels). Wind barbs at the sounding sites are indicated.

4. Atmospheric Variability over the Northern South China Sea during the SCSMEX IOP

Precipitation over the NESA has been computed from the moisture budget using the gridded dataset following Yanai *et al.* (1973):

$$\langle Q_2 \rangle = L(P - E), \tag{1}$$

where $(Q_2 \equiv -L(\partial\overline{q}/\partial t + \overline{\mathrm{v}} \cdot \nabla\overline{q} + \overline{\omega}\partial\overline{q}/\partial p)$ is the apparent moisture sink, q the specific humidity, L the temperature-dependent latent heat of vaporization, P precipitation rate, E evaporation rate, overbar denotes a horizontal average, $\langle\ \rangle \equiv 1/g\int_{p_T}^{p_s}(\ )dp$, p_T is the tropopause pressure and p_s the surface pressure. Surface evaporation E over the NESA is based on the GAME reanalysis values adjusted by measurements from *Shiyan #3* as described in Section 2. We also compute the apparent heat source Q_1, defined as $c_p\left[\left(\partial\overline{T}/\partial t + \overline{v}\cdot\nabla\overline{T} + (p/p_o)^{\kappa}\,\overline{\omega}\partial\overline{\theta}/\partial p\right)\right]$, where $\kappa = R/c_p$, R is the gas constant, and

c_p the specific heat at constant pressure for moist air.

In order to view changes over the northern SCS, gridded fields have been averaged over the NESA. Time series of *u*, *v*, and relative humidity for May and June over the NESA are shown in Fig. 3. The onset of low-level westerlies in mid-May can be seen to be related to an upper-tropospheric westerly wind maximum, further indicating a connection between the monsoon onset over the northern SCS and midlatitude disturbances (Ding and Liu 2001). Eventually in June, the upper-level westerlies were replaced by easterlies as the 200-hPa anticyclone built northward toward the Tibetan Plateau. Reversals of the low-level and upper-level zonal component of the flow from May to June can be seen in Fig. 4.

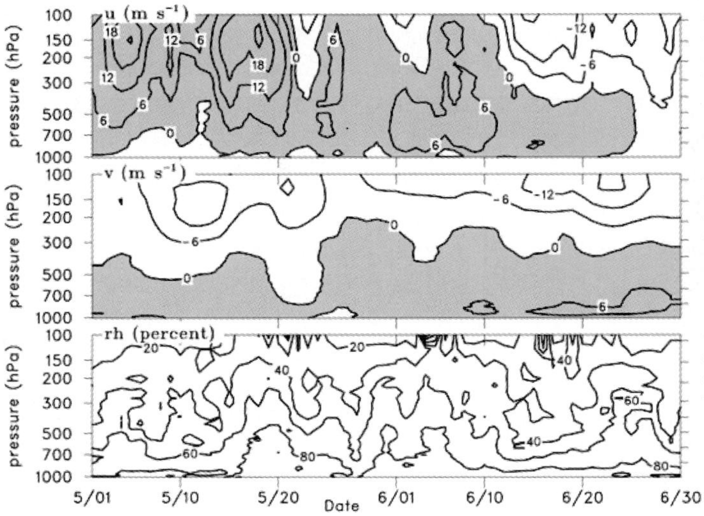

Figure 3. Time series of *u, v* and relative humidity over the NESA during May and June 1998. Positive values of wind are shaded.

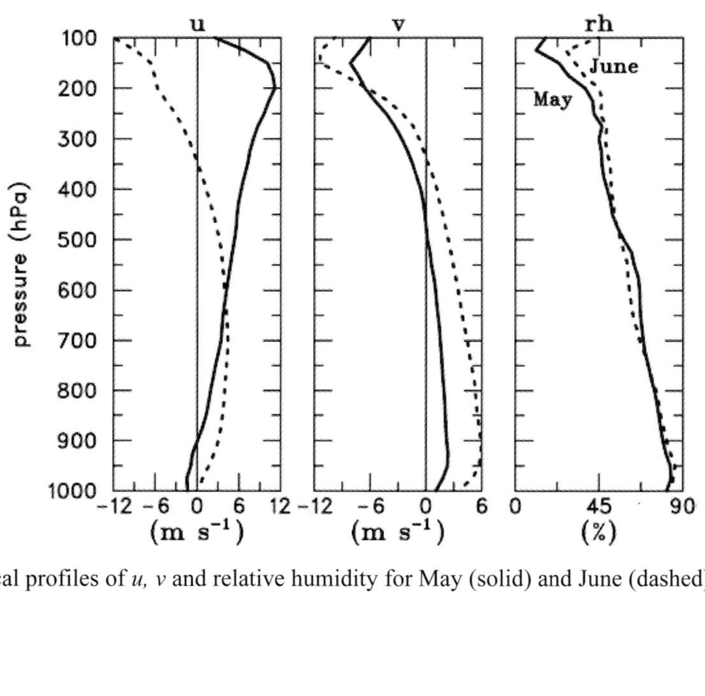

Figure 4. Vertical profiles of *u, v* and relative humidity for May (solid) and June (dashed) 1998.

The meridional winds in Figs. 3 and 4 show that low-level southerlies strengthened during June as the summer monsoon advanced and heavy rains shifted northward toward the Yangtze Valley. Although the relative humidity varied significantly on ~week to ten-day time periods (Fig. 3), the mean profiles for May and June were almost the same (Fig. 4).

Time series of divergence, vertical motion, Q_1, and Q_2 are shown in Fig. 5. Two extended periods of deep convection are inferred, one in mid-May and the other in early June. Strong upward motion, peaking in the mid-to-upper troposphere can be seen at these times. During these periods, peaks in upper-level divergence occurred between 125 and 150 hPa, with a deep layer of convergence in the low-to-mid- troposphere. Q_1 and Q_2 displayed maxima characteristic of deep convection: peaks in Q_1 in the mid- to upper troposphere and peaks in Q_2 in the mid- to lower troposphere (Luo and Yanai 1984). Strong downward motion was diagnosed prior to the monsoon onset on 15 May, between the two convectively active periods, and again in mid-to-late June. The periods of upward and downward motion, corresponding to *apparent* convective drying (positive Q_2) and moistening (negative Q_2), also correspond to periods of *actual* moistening and drying, respectively, in the mid- to upper troposphere (see relative humidity panel in Fig. 3). The dry period over the northern SCS in mid-June marked the first stage of the Meiyu season, when heavy rainfall shifted northward to the region south of the Yangtze Valley (Ding and Liu 2001).

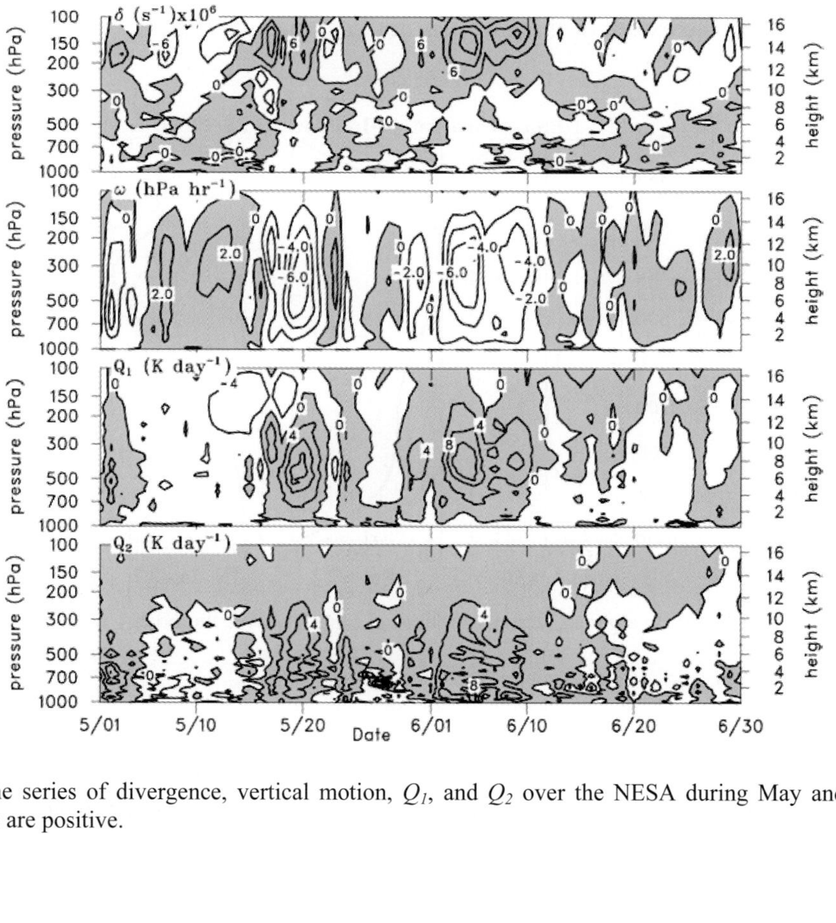

Figure 5. Time series of divergence, vertical motion, Q_1, and Q_2 over the NESA during May and June 1998. Shaded values are positive.

Averages of divergence, vertical motion, Q_1, and Q_2 for the NESA are shown in Fig. 6 for the 37-day period from the monsoon onset on 15 May through 20 June.[a] There is strong low-level convergence, upper-level divergence peaking at 150 hPa, and mid-level convergence. The latter feature may be evidence of "melting-layer convergence" in deep convective systems near the 0°C level arising from extensive regions of stratiform precipitation (Mapes and Houze 1995; Johnson *et al.* 1996). The mid-level convergence gives rise to an inflection in the ω profile between 600 and 700 hPa, suggesting the upper peak in ω in Fig. 6 may reflect an important contribution from the stratiform components of the convective systems. The mean Q_1 and Q_2 profiles are indicative of deep convection, with the shape of Q_1 resembling that determined for the western Pacific (Yanai *et al.* 1973) and the eastern Atlantic (Thompson *et al.* 1979). The negative values of Q_1 near the surface may indicate the effects of rainfall evaporation. The shape of Q_2 also resembles that determined in the above studies, except below 850 hPa. The Q_2 peak between 900 and 950 hPa does not have a counterpart in the other studies. Although it may be related to condensation near cloud base, it is more likely a consequence of errors in the low-level vertical motion over the NESA. Unlike the deep Tropics, horizontal gradients of low-level specific humidity exist in this region (in association with the horizontal gradients of SST, Fig. 1). Horizontal advection of q not properly compensated by vertical advection of q may yield errors in low-level values of Q_2. The low-level distribution of Q_2 could also be partly affected by moisture biases from the various sounding systems of the region which could introduce spurious low-level moisture advection; however, the extent of this effect is not known.

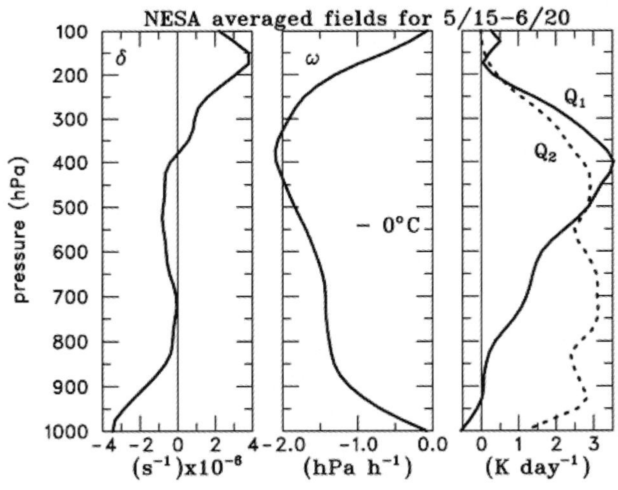

Figure 6. Vertical profiles of divergence, vertical motion, Q_1, and Q_2 over the NESA from 15 May through 20 June 1998.

---

[a] Errors encountered in computing divergence in the pre-onset period are partially but not totally eliminated by replacement of winds at Laoag, Philippines, by GAME Reanalysis winds below 700 hPa (Johnson and Ciesielski 2002). Note in Fig. 5 unrealistic values of $Q_1 < -4$ K day^{-1} in the upper troposphere prior to onset in mid-May, so this period of the IOP is excluded from the averages in Fig. 6.

There is an interesting contrast between the heat and moisture budgets for the SCSMEX NESA and those for the western Pacific warm pool. In particular, note that throughout most of the column $Q_2 > Q_1$ (Fig. 6). However, the opposite was found to be true for the warm pool (Johnson and Ciesielski 2000, their Fig. 3). Since

$$\langle Q_R \rangle = \langle Q_1 \rangle - \langle Q_2 \rangle - S - LE \,, \tag{2}$$

where $< Q_R >$ is the net radiative heating rate, S the surface sensible heat flux, and LE the latent heat flux, this difference implies, assuming that the surface fluxes are comparable, that net tropospheric radiative cooling rate over the SCSMEX region is greater than that over the western Pacific warm pool. Indeed, this is found to be the case, as will be discussed later.

The relationship between SST, surface air temperature T_o, CAPE, vertical motion at 400 hPa, and OLR during May and June over the NESA is shown in Fig. 7.[b] During the IOP, there is a high positive correlation between ω and OLR ($r = 0.81$ for daily-averaged values), indicating the enhanced sounding network is doing a good job of depicting the vertical motion field over the northern SCS. The correlation falls dramatically after the IOP in late June when the enhanced sounding network was terminated. There was a general increase in the SST and surface air temperature throughout the two-month period, interrupted by brief periods of cooling during the late-May and early-June convective episodes presumably due to rainfall, cloud shading, enhanced evaporation, and vertical mixing and horizontal transport in the upper ocean. Lau and Yang (1997) show for a ten-year period that the SST over the entire SCS exhibits a broad maximum during May and June, with a gradual cooling thereafter until September. The surface temperature fell during the convective periods, with a corresponding reduction in CAPE, as observed in the eastern Atlantic (Thompson *et al.* 1979) and elsewhere during deep convection. Further reduction in CAPE occurred after convection abated (around May 25) due to the passage of a weak cold front from the China mainland. The surface specific humidity, on the other hand, increased during the convective periods. Presumably, dry air from aloft (Fig. 3) was entrained into the mixed layer during the inactive and break periods to reduce the surface specific humidity at those times, although advection of drier air from the mainland could have contributed to the reduction in late May.

The difference between SST and T_0 (Fig. 7) and the time series of wind speed (Fig. 8) indicate that the surface sensible heat flux should exhibit a peak in late May. This result is consistent with the JMA/GAME Reanalysis sensible heat flux time series (Fig. 8) showing a peak in late May. The latent heat fluxes peaked at the same time (Fig. 8), which is consistent with the minimum in q_0 at that time (Fig. 7). The adjusted-JMA fluxes (by the NCAR measurements from *Shiyan #3*) also show a peak in late May, but their amplitudes after the onset of the monsoon are considerably reduced. Data from the NCAR measurements aboard *Shiyan #3* indicate markedly reduced surface fluxes after onset. Tentatively, we explain these as a consequence of increased surface southerly flow over the NESA (Figs. 3 and 4)

[b] CAPE is computed assuming pseudo-adiabatic ascent using mean thermodynamic conditions in the lowest 60 hPa, the approximate depth of the mixed layer.

advecting warmer, moister air over cooler water (Fig. 1). In fact, the adjusted sensible heat fluxes actually became negative in mid-June as the southerly component of the flow strengthened over the NESA (Fig. 3).

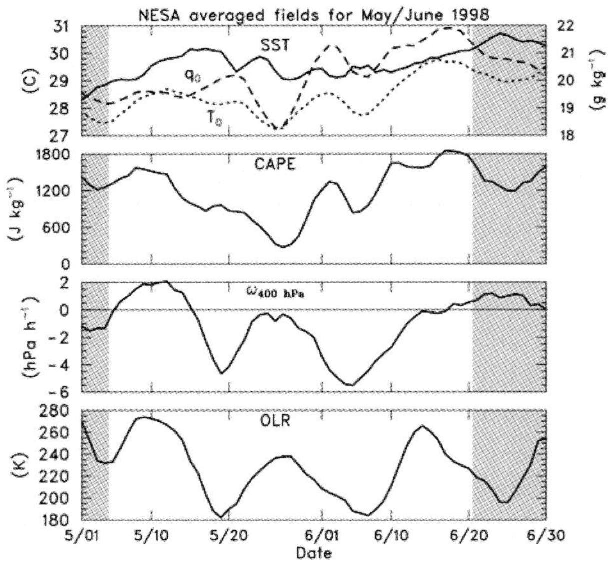

Figure 7. Time series of SST (solid), surface air temperature T_0 (dotted), surface specific humidity q_0 (dashed), CAPE, vertical motion at 400 hPa, and OLR over the NESA for May and June 1998. Shaded times are outside the SCSMEX IOP.

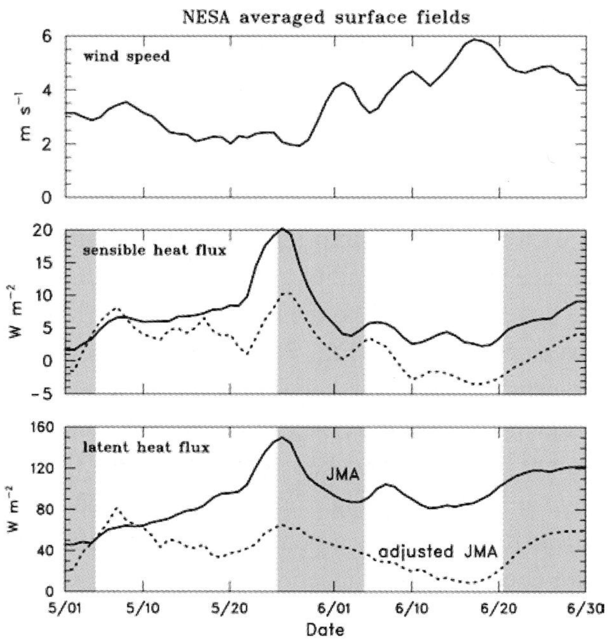

Figure 8. Time series of wind speed and adjusted (dashed line) and unadjusted (solid line) JMA/GAME Reanalysis sensible and latent heat fluxes. Adjusted fluxes are based on measurements of bulk fluxes from *Shiyan #3* (see text for explanation). Times when *Shiyan #3* was off station are shaded.

5. Precipitation, Heating and Moistening Rates over the Northern South China Sea during Active Periods of the Monsoon

A time series of precipitation rate P from the ECMWF, GAME Reanalysis, GPCP, and TRMM 3B42 datasets is shown in the upper panel of Fig. 9. In the lower panel, the moisture budget results are shown both with and without the replacement of Laoag, Philippines (station in southeast corner of NESA in Fig. 1) sounding winds below 700 hPa with GAME Reanalysis winds, and also with both the Laoag wind replacement and using the adjusted JMA/GAME Reanalysis surface fluxes. The replacement of low-level winds at Laoag with model winds is motivated by occurrence of significant flow-blocking in the vicinity of Laoag, which is aliased onto larger scales by the objective analysis scheme. The large, unrealistic negative precipitation rates in the pre-onset period, when low-level easterlies prevailed, are reduced considerably by the Laoag replacement, while after onset (when low-level westerlies ensued and terrain-blocking effects over the NESA lessened) the differences are generally reduced. When both changes are made, the rainfall rates are reduced throughout most of the period.

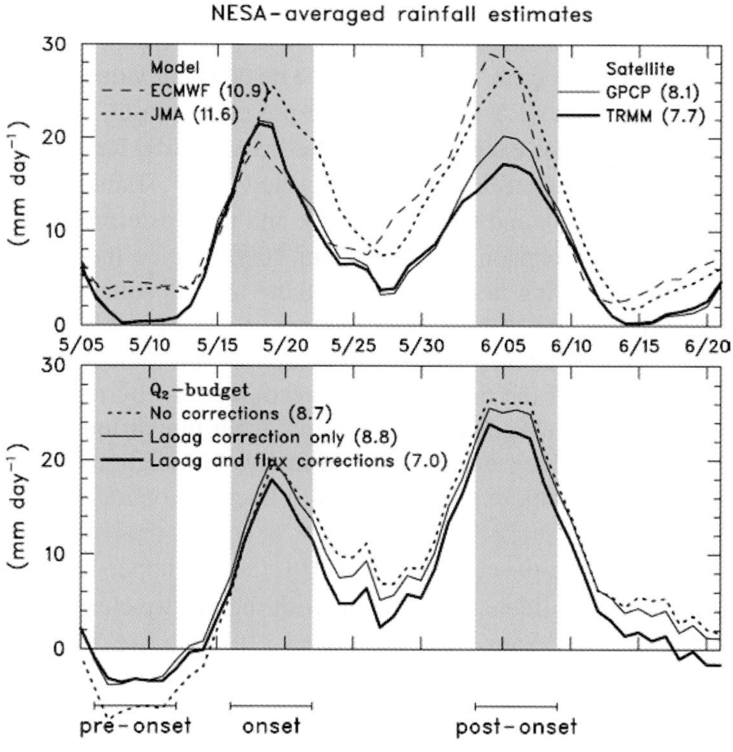

Figure 9. (top panel) Time series of rainfall rate for model (JMA/GAME Reanalysis, ECMWF) and satellite products (GPCP, TRMM 3B42). (bottom panel) Time series of rainfall rate from moisture (Q_2) budget for three cases: (a) no adjustments to winds at Laoag and surface fluxes, (b) replacement of Laoag winds below 700 hPa by JMA/GAME Reanalysis winds, and, (c) both (b) and adjustment of JMA/GAME surface fluxes by direct measurements at *Shiyan #3*. IOP-mean values are in parentheses. Pre-onset, onset, and post-onset periods are denoted by shading.

Overall, there is good agreement among model, budget, and satellite precipitation time series in Fig. 9. The 47-day mean rainfall rates from the various moisture budget estimates, 6.8-8.6 mm day^{-1}, bracket the GPCP (8.1 mm day^{-1}) and TRMM (7.7 mm day^{-1}) estimates, but average about 25% smaller than the model estimates, 10.9 and 11.6 for the ECMWF and GAME Reanalysis, respectively. The time series indicate two ~10-day rainy periods separated by a break at the end of May. The moisture budget estimates are slightly smaller than the TRMM and GPCP estimates during the first rainy period, and conversely during the second rainy period. We use these time series to define three 7-day periods – pre-onset, onset, and post-onset (Fig. 9) – for further analysis.

Mean profiles of divergence, ω, Q_1, Q_2 , and vertical eddy flux of total heat F for these three periods are shown in Fig. 10. Here F is defined by

$$F \equiv -\frac{\overline{h'\omega'}}{g} = S + LE + \frac{1}{g}\int_{p_s}^{p}\left(Q_1 - Q_2 - Q_R\right)dp, \tag{3}$$

where h = $c_p\,T + Lq + gz$ is the moist static energy. In computing F, a constant Q_R is assumed (since its vertical profile is not known) so that $F = 0$ at 100 hPa. At the surface, $F = S + LE$, where the surface fluxes from the JMA/GAME Reanalysis have been adjusted by the NCAR-measured fluxes from the *Shiyan #3*. The constant values of Q_R that give $F = 0$ at 100 hPa are –1.6, –1.2, and –1.8 K day^{-1} for the pre-onset, onset, and post-onset periods, respectively. These cooling rates exceed those of the JMA/GAME Reanalysis (–0.96, –0,20, and –0.16 K day^{-1} for the three successive periods); the GARP Atlantic Tropical Experiment (GATE), –1.1 K day^{-1} from Cox and Griffith (1979); and the western Pacific TOGA COARE estimate of ~ –0.5 K day^{-1} (Johnson and Ciesielski 2000). Since there is no clear physical explanation for such large cooling rates, we expect there are still some remaining errors in the budgets for the SCSMEX region. Considering the sensitivity of the budgets over the NESA to low-level winds around Luzon, and the other significant topography surrounding the region, it is likely that some of the discrepancy is accounted for by topographic effects. Part may also be due to biases in humidity measurements from the various sounding systems of the region, which may introduce errors in Q_2. Finally, although the sounding data were quality controlled, possible errors (e.g., in the winds) from the operational sites in the region may have eluded detection.

During the undisturbed pre-onset period (Fig. 10), there is upper-level convergence, low-level divergence, and deep subsidence, consistent with the mostly clear skies and high values of OLR (Fig. 7). Radar data from Dongsha Island (20.7°N, 116.7°E) during this period indicated no deep convection, with only widely scattered precipitating cumulus congestus clouds. Q_1 is negative at all levels with values in excess of –2.5 K day^{-1} in the upper troposphere. Since Q_1 approximately equals the net radiative heating rate Q_R during this undisturbed period, we conclude from the large cooling rates that excess subsidence is still being diagnosed at this time despite the adjustment of the low-level winds at Laoag. Q_2 exhibits moistening in the lower troposphere, indicative of shallow nonprecipitating cumulus, but it is also excessive since it extends too high to exclusively represent nonprecipitating clouds.

The profiles of divergence and vertical motion during the May onset and June post-onset periods (Fig. 10) are dramatically different from the pre-onset period. Low-level convergence, upper-level divergence and strong upward motion occur during both periods. Deep convergence extends to 300-500 hPa, similar to that found in the western Pacific (Reed and Recker 1971; Yanai *et al.* 1973). Some of this convergence is presumably associated with downdrafts and some with melting-induced convergence in the stratiform precipitation regions (Mapes and Houze 1995). The Q_1 and Q_2 profiles are characteristic of deep convection during the onset and post-onset periods, with a greater separation of the two curves in the post-onset period, suggesting more vigorous deep convection at that time. This conclusion is supported by profiles of F (Fig. 10), which indicate much stronger eddy heat fluxes during the second active period. The occurrence of more intense convection during the second active period is consistent with higher surface specific humidity q_o at that time, which also led to slightly higher CAPEs (Fig. 7).

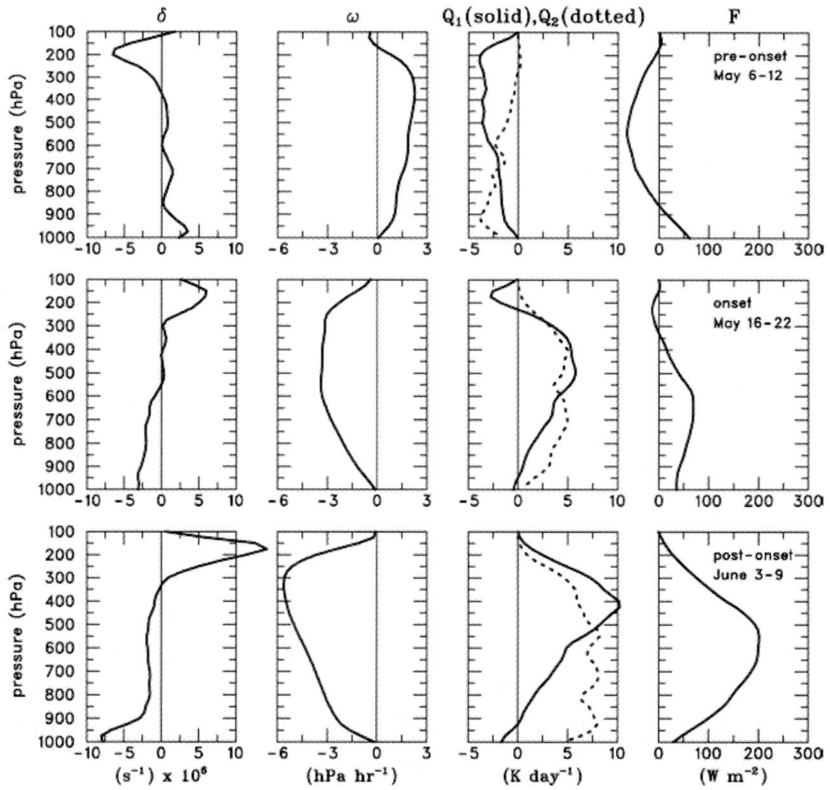

Figure 10. Vertical profiles of divergence, vertical motion, Q_1, Q_2, and total energy flux F for the undisturbed pre-onset period (top panel), the onset convective period (middle panel), and the post-onset convective period (bottom panel).

More vigorous convection during the post-onset period is also consistent with TRMM precipitation radar (PR) reflectivity profiles for the two periods derived from the 2A25 algorithm (Kummerow *et al.* 2000). Mean reflectivity profiles for stratiform and convective precipitation for thirteen overpasses in each of the periods are shown in Fig. 11. Although the number of overpasses is limited and uncertainties in TRMM-based, weekly mean reflectivity profiles are large (Steiner and Houze 1998), the NESA area is very large (~5×10^5 km^2) and there are over 70,000 individual profiles comprising each of the means. For the times of the overpasses, the budget and PR rainfall rates for the onset period are 13.2 and 11.9 mm day^{-1} for the NESA, and for the post-onset period are 22.8 and 24.7 mm day^{-1}, respectively. Higher reflectivities are observed in both the stratiform and convective profiles for the post-onset period, supporting the budget findings of stronger convection at that time.

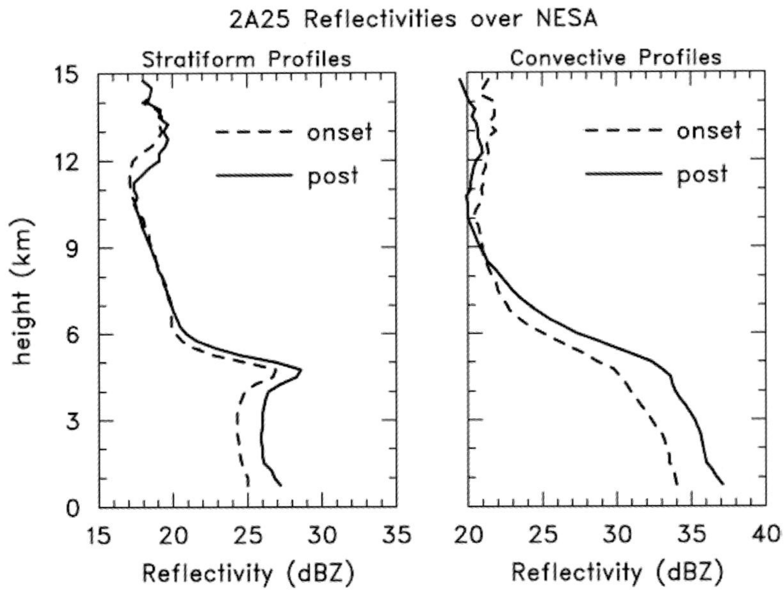

Figure 11. Vertical profiles of stratiform and convective radar reflectivity from TRMM precipitation radar (PR) using algorithm 2A25 (Kummerow *et al.* 2000) for onset (16-22 May 1998) and post-onset (3-9 June 1998) periods.

The horizontal distributions of rainfall from the moisture budget, GPCP, TRMM 3B42 combined algorithm, and the JMA/GAME Reanalysis for the onset and post-onset periods are shown in Figs. 12 and 13. For the onset period (Fig. 12) the rainfall patterns from the budget, GPCP and TRMM are in reasonably good agreement, with greatest amounts in the western half of the NESA. The average budget rainfall rate for the NESA (13.2 mm day^{-1}) is about 25% less than the GPCP and TRMM 3B42 estimates (18.2 and 17.7 mm day^{-1}, respectively). The JMA/GAME Reanalysis does not agree as well with the other two, showing a broad rainfall maximum in the south-central part of the NESA and a considerably greater onset mean (22.9 mm day^{-1}). The positioning of the primary rainfall maximum away from the south

China coastline in the onset period (Fig. 12) may be related to the SST gradient across the region (Fig. 1), i.e., warmer water to the south.

For the post-onset period (Fig. 13) the agreement between the budget, GPCP, and TRMM 3B42 rainfall patterns is also reasonably good with multiple maxima present within and near the NESA. Both the budget and GPCP show two primary maxima, one south of Taiwan and another along the south coast of China (although the positions of the latter differ). The positions of a weaker maximum near Hainan also differ somewhat. The TRMM 3B42 and JMA/GAME show a single elongated maximum, although their positions generally coincide with the envelope of the strongest maxima for the budget and GPCP. The mean values for the period range from 16.4 mm day^{-1} for the TRMM 3B42 up to 25.9 mm day^{-1} for the JMA/GAME, with the budget and GPCP values falling in between.

The generally good agreement between the independent budget, GPCP, and TRMM rainfall patterns for both convectively active periods lends confidence to the ability of the sounding network to describe the heating and moistening distributions reasonably well during SCSMEX. However, integral constraints applied to the budgets that yield estimates for $\langle Q_R \rangle$ indicate that some systematic errors still exist, likely due to topographic effects and biases in the sounding data.

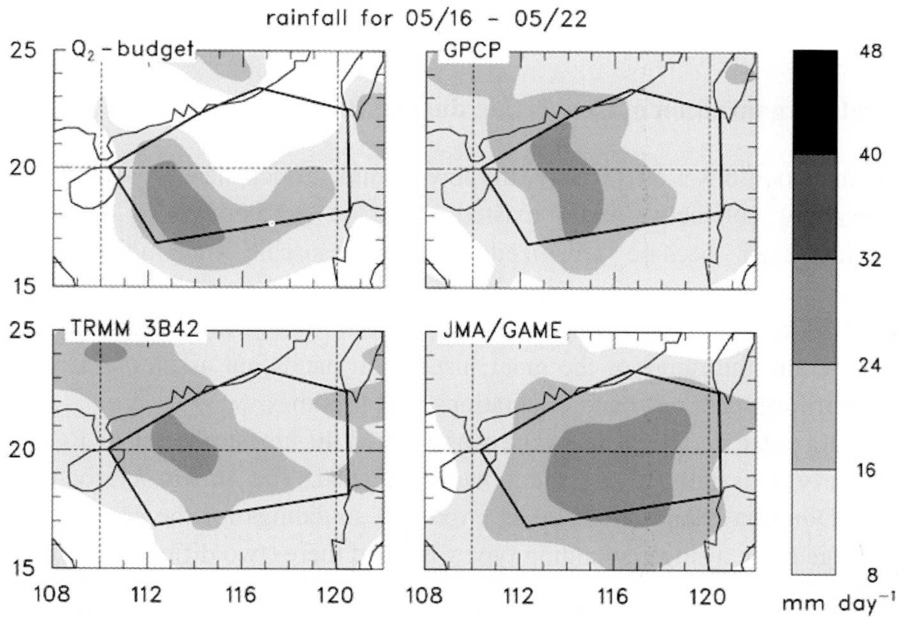

Figure 12. Horizontal distributions of rainfall rates (scale on right) for the 16-22 May 1998 onset period from the moisture budget, the GPCP, the TRMM 3B42 algorithm, and the JMA/GAME Reanalysis. Numbers in parentheses are NESA-averaged rainfall rates for the period.

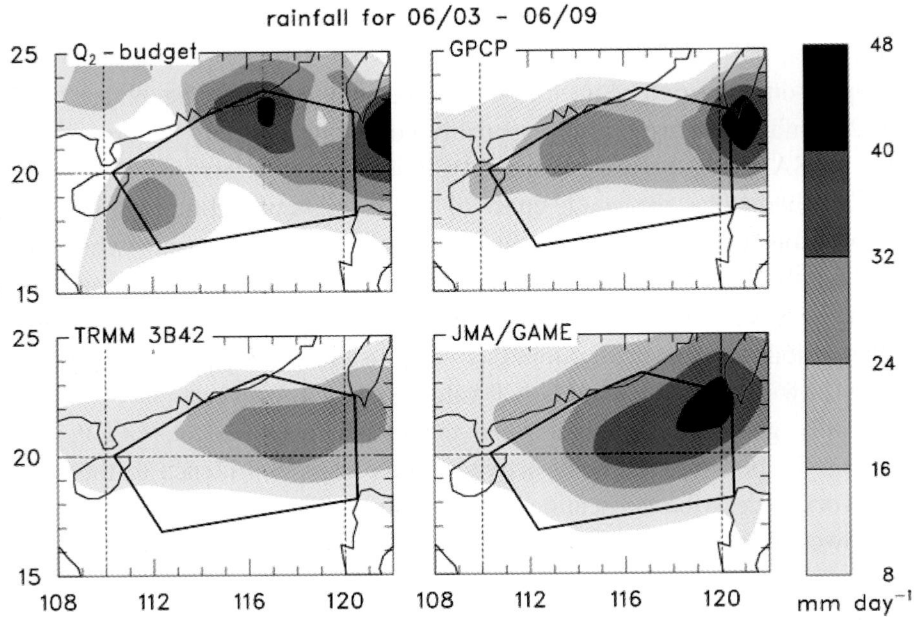

Figure 13. Horizontal distributions of rainfall rates (scale on right) for the 3-9 June 1998 post-onset period from the moisture budget, the GPCP, the TRMM 3B42 algorithm, and the JMA/GAME Reanalysis. Numbers in parentheses are NESA-averaged rainfall rates for the period.

6. Mesoscale Organization of Convection during Monsoon Onset

Prior to the monsoon onset on 15 May, the low-to-mid- troposphere was very dry and only scattered, shallow convective cells existed over the northern SCS. Following onset, convection increased, became organized on the mesoscale and moistened the entire troposphere as the large-scale circulation changed to produce convergence over the region. For a period of ten days following monsoon onset (15-25 May), precipitation occurred continuously within the range of the radar, usually in bands but assuming a wide range of convective/stratiform patterns, band orientations, and lifetimes.

The C-POL radar on Dongsha Island was located 40 km northwest of R/V *Shiyan #3*, where six-hourly Vaisala-GPS soundings were launched. The *Shiyan #3* soundings, rather than those at Dongsha Island, are used as proximity soundings for the convection in the C-POL radar range since a single sounding type was used there (two different sounding systems were used at Dongsha). During most of the 15-25 May period, sufficient CAPE (~1000 J kg^{-1}, Fig. 7) and minimal convective inhibition (CIN) existed to support deep convection in a background of large-scale convergence. However, as will be seen later, drying aloft occurred toward the end of this period, which suppressed deep convection. *Shiyan 3* wind data were used to compute vertical shears in the low- and mid- troposphere.

Time series of the vertical shear in the lower (1000-800 hPa) and middle (800-400 hPa) levels are shown in Fig. 14. Shading denotes periods when the shear exceeded thresholds defined by LeMone *et al.* (1998) for shear-perpendicular convection based on the low-level shear (> 4 m s^{-1}) and shear-parallel convection based on midlevel shear (> 5 m s^{-1}) when the low-level shear is weak. Layer shear directions, used to classify line orientations, were computed but are not shown in Fig. 14. Finally, the bottom panel in Fig. 14 shows a time series of the 600-300 hPa mean relative humidity.

Animations of the C-POL radar base-scan reflectivity within ~200 km radius of Dongsha Island have been used to determine the dominant modes of convective organization at six-hour intervals (e.g., 00-06, 06-12 UTC, etc.). Lines were classified as shear-parallel or shear-perpendicular if the line orientations were within 30° of specific shear vectors for the low- and mid- troposphere illustrated in Fig. 15. This figure, from LeMone *et al.* (1998), is a summary of the findings from TOGA COARE, but supplemented by new results from SCSMEX (two new modes **2r** and **4c**, discussed below).

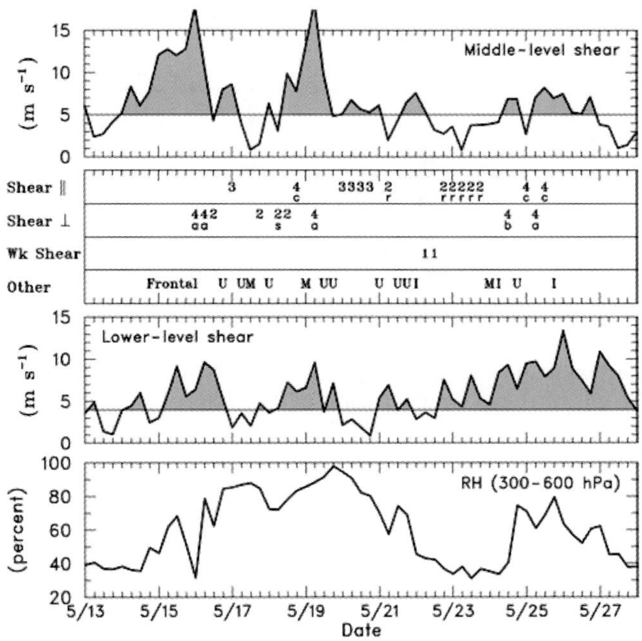

Figure 14. Time series of shear vector magnitude in the low levels (1000-800 hPa) and middle levels (800-400 hPa), and 600-300 hPa relative humidity, based on six-hourly sounding data from *Shiyan #3*. Periods with shear exceeding 5 m s^{-1} in the middle levels and 4 m s^{-1} in the lower levels are shaded. Convective organization in relation to the shear in the two layers indicated in second panel from top, with numbers and letters referring to organizational modes depicted in Fig. 15. Those cases not fitting any of those modes are denoted a **Other: Frontal** = convection influenced by frontal system; **U** = Unclassifiable; **I** = Isolate convection.

LeMone *et al.* (1998) found that the orientation of primary convective bands in mesoscale convective systems is perpendicular to the shear in the lowest 200 hPa when the shear magnitude exceeds 4 m s^{-1} (upper- and lower-right frames in Fig. 15). Secondary lines parallel to the low-level shear were found in some cases ahead of the primary band (**2s** in Fig. 15). In the absence of strong low-level shear, lines form parallel to the 800-400 hPa shear when its magnitude exceeds 5 m s^{-1} (lower-left frame). When the vertical shear exceeds the thresholds in both layers and the shear vectors are not in the same direction (lower-right frame), the primary band is normal to the low-level shear (**4a** or **4b**). For midlevel shear normal to the low-level shear, the primary band remains two dimensional (**4a**). Trailing secondary bands parallel to the midlevel shear occur if the midlevel shear is opposite the low-level shear (**4b** in Fig. 15). If the shear is weak in both layers, convection develops in arcs along outflow boundaries (**1**).

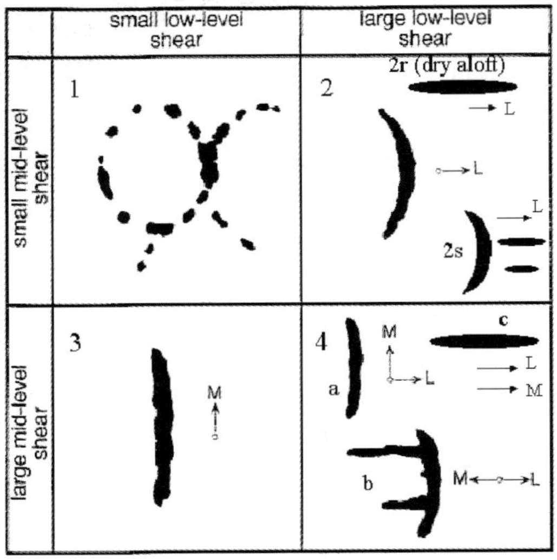

Figure 15. Schematic depiction from LeMone *et al.* (1998) of convective structures for given vertical shears in the lower troposphere (1000-800 hPa) and at middle levels (800-400 hPa) based on COARE observations, but modified to include results from SCSMEX (modes **2r**, **2s**, and **4c** added). Length of schematic convective bands is ~100-300 km; line segments in upper-left frame are up to 50 km length. Cutoff between "strong" and "weak" shear for lower layer (1000-800 hPa) is 4 m s^{-1} and for middle layer (800-400 hPa) is 5 m s^{-1}. Arrows marked **L** and **M** are shear vectors for lower and middle layers, respectively. See text for description of convective modes.

For the majority of the 15-25 May period, the orientation of SCSMEX convective bands fit the classification scheme of LeMone *et al.* (1998). Modes identified at six-hour intervals are denoted by numbers and letters in the second panel of Fig. 14 taken from Fig. 15 (e.g., **4a**, **2s**, **3**, etc.) and placed into shear-parallel, shear-perpendicular, and weak-shear categories. Two additional modes of convection have been identified from analysis of SCSMEX C-POL radar data (Fig. 15): shear-parallel bands (**2r**) for strong low-level shear and weak midlevel shear when the air is dry aloft, and shear-parallel bands (**4c**) for strong shears in both layers when the shear vectors are in the same direction. Both of these new modes reflect the influence of midlatitude effects, as will be discussed.

From Fig. 14 it is evident that a nearly equal number of shear-parallel and shear perpendicular convective bands occurred over the northern SCS during the monsoon onset period of SCSMEX. The first convection on 15 May was influenced by a frontal system that moved south from the southern China coast and led to a temperature drop of 4°C and shift to northerly wind at Dongsha Island. Subsequent convection was mostly nonfrontal, although a shift to northerly wind and some cooling occurred again on 19 May. The results show that, in general, shear-perpendicular convection was common when the low-level shear exceeded 4 m s^{-1}. This finding is consistent with the idea that convection organizes to establish a balance between the cold pool and low-level shear when the low-level shear is sufficiently large (Rotunno *et al.* 1988). An exception occurred on 23 May when the middle troposphere became extremely dry (bottom panel, Fig. 14), thus preventing deep convection and their associated downdrafts and leading to shallow, shear-parallel lines [**2r**, where **r** indicates convection likely associated with longitudinal roll vortices in the boundary layer (Kuettner 1971; LeMone 1973)]. Isolated convective cells (**I**) also occurred during this period, likely due to a suppression of deep convection by the dry air aloft. A number of unclassifiable cases (**U**) occurred throughout the period, many of which exhibited several modes of organization within the radar domain.

Most of the shear-perpendicular bands were of the **4** variety since midlevel shear was also generally strong, although two **2**'s also occurred. This control on convective organization was influenced by the general trend toward decreasing midlevel shear and increasing low-level shear throughout the 15-25 May period (Fig. 14) as the transition to summer occurred - the midlevel jet pulled northward and low-level monsoon flow strengthened.

The strong midlevel westerly flow during mid-May occurred in association with a high amplitude upper-level trough that moved into southern China around the time of monsoon onset (Fig. 2). This trough contributed to the cold-front passage at Dongsha Island on 15 May and illustrates the importance of midlatitude influence on the onset of the East Asian summer monsoon (Chang and Chen 1995; Lau *et al.* 2000a). With strong westerly shears at low- and midlevels during this period, shear-parallel bands developed on four occasions (**4c** in Fig. 14). The origin and nature of these bands are uncertain, but their alignment approximately along the low-level shear vector makes boundary-layer rolls a candidate mechanism. This point was made for category **3** shear-parallel bands by LeMone *et al.* (1998). It is possible that weak downdrafts during this period prevented shear-perpendicular bands from developing. LeMone *et al.* did not observe similarly oriented low- and midlevel shear vectors (category **4c** convection) in the TOGA COARE region. However, LeMone (1983) identified a convection line on 14 September 1974 during the GARP Atlantic Tropical Experiment (GATE) having characteristics similar to **4c**.

7. Summary and Conclusions

The characteristics of convection over the northern South China Sea (SCS) have been investigated using data from the May-June 1998 South China Sea Monsoon Experiment (SCSMEX). The main results from this study are as follows:

- The onset of the monsoon over the northern SCS occurred on 15 May in association with a rapid retreat of the western Pacific subtropical high. The onset was followed by a week-to-ten-day period of deep convection, a ~ten-day break period, and then another week-to-ten-day convectively active period in early June.

- SST gradually increased over the northern SCS during the IOP, interrupted by slight cooling following monsoon onset and during the early-June convective period. Surface sensible and latent heat fluxes increased just after monsoon onset, while CAPE decreased as a result of vertical mixing by deep convection and brief incursions of cool air from the China mainland. Surface fluxes then decreased in June as a result of warm, moist air advecting over cooler water near the south China coast, although CAPE increased due to an increase in the boundary-layer specific humidity and temperature.
- The mean apparent heat source Q_1 during convective periods resembled that determined for other tropical oceanic regions with a peak near 400 hPa. The apparent moisture sink Q_2, also resembled profiles for other tropical regions, except for larger values below 900 hPa. However, the latter feature may be associated with errors in the low-level vertical motion field and/or errors in the humidity field over this data-sparse and topographically influenced region. The budgets indicate an IOP-mean net tropospheric radiative heating rate of ~ -1.5 K day^{-1}. This cooling rate is greater than other tropical estimates by at least 0.5 K day^{-1}, suggesting that some errors in the budgets still remain (e.g., due to topographic effects, moisture biases at the sounding sites, other errors in the sounding data, sparse sounding network).
- The temporal and spatial variabilities of rainfall rates throughout the SCSMEX IOP from independent estimates – moisture budget and satellites (TRMM and GPCP) – agree well. Agreement with models (GAME Reanalysis and ECMWF operational analysis) is not quite as good in terms of spatial patterns and intensity (the models' mean rainfall rate for the 47-day Intensive Observing Period (IOP) is about 25% greater than the other estimates).
- The heating and moistening rates and vertical eddy flux of total heat during the early-June active period were greater than those observed during the May monsoon onset active period, indicating more vigorous deep convection during the June period. This finding is supported by vertical reflectivity profiles from the TRMM precipitation radar.
- Analysis of radar data from the BMRC C-POL radar on Dongsha Island from 15 to 25 May – a ten-day period following the onset of the East Asian summer monsoon over the northern SCS – reveals that lower and middle level vertical shears exert a dominant control over the structure and orientation of mesoscale convective systems in this region. The findings are consistent with those of LeMone *et al.* (1998) for TOGA COARE, except two new organizational modes have been identified: shear-parallel bands (**2r** in Fig. 15) for strong low-level shear and weak midlevel shear when the air is dry aloft, and shear-parallel bands (**4c**) for strong shears in both layers when the shear vectors are in the same direction. Midlatitude influences likely contributed to these two additional modes by producing strong westerlies (in the

case of **4c**) during the passage of a strong upper-level trough and mid-tropospheric drying (in the case of **2r**) following passage of a trough.

Several important aspects of the convection in this region were not addressed in this paper. One aspect has to do with the diurnal cycle. Both satellite and radar observations during the experiment clearly indicated a strong diurnal cycle of convection over the northern SCS apparently linked to land and sea breezes along the southern coast of China. Further work is also underway to investigate this aspect of monsoon convection during SCSMEX. A second aspect concerns the organizational modes of convection during the convectively active June post-onset period. The organization of convection during this period is currently being explored.

Acknowledgments

This research has been supported by the National Aeronautics and Space Administration under Grant No. NAG5-9665 and by the National Oceanic and Atmospheric Administration under Grant No. NA67RJ0152. We thank Drs. Dave Parsons and James Pinto for providing the surface flux data from *Shiyan #3*, Chelle Gentemann for providing the TMI SST data, and Peggy LeMone and an anonymous reviewer for their helpful comments.

References

Alexander, G. D., and G. S. Young, 1992: The relationship between EMEX mesoscale precipitation feature properties and their environmental characteristics. *Mon. Wea. Rev.*, **120**, 554-564.

Barnes, G. M., and K. Seickman, 1984: The environment of fast- and slow-moving tropical mesoscale convective cloud lines. *Mon. Wea. Rev.*, **112**, 1782-1794.

Chan, Johnny C. L., Y. Wang, and J. Xu, 2000: Dynamic and thermodynamic characteristics associated with the onset of the 1998 South China Sea summer monsoon. *J. Meteor. Soc. Japan*, **78**, 367-380.

Chang, C.-P., and G. T. J. Chen, 1995: Tropical circulations associated with Southwest Monsoon onset and westerly surges over the South China Sea. *Mon. Wea. Rev.*, **123**, 3254-3267.

Cox, S. K., and K. T. Griffith, 1979: Estimates of radiative divergence during Phase III of the GARP Atlantic Tropical Experiment: Part II. Analysis of Phase III results. *J. Atmos. Sci.*, **36**, 586-601.

Ding, Y.-H., and J. Hu, 1988: The variation of the heat sources in East Asia in the early summer of 1984 and their effects on the large-scale circulation in East Asia. *Adv. in Atmos. Sci.*, **6**, 171-180.

______, and X.-F. Wang, 1988: An analysis of the distribution of apparent heat sources and sinks over the middle reaches of Yangtze River during the Meiyu season in 1983. *Tropical Meteorology*, **4**, 134-145. (in Chinese).

______, and Y. Liu, 2001: Onset and evolution of the summer monsoon over the South China Sea during SCSMEX field experiment in 1998. *J. Meteor. Soc. Japan*, **79**, 255-276.

Fairall, C. W., E. F. Bradley, D. P. Rogers, J. B. Edson and G. S. Young, 1996: Bulk paramaterization of air-sea fluxes for TOGA COARE. *J. Geophys. Res.*, **101**, C2, 3747-3764.

Flohn, H., 1957: Large-scale aspects of the "summer monsoon" in South and East Asia. *J. Meteor. Soc. Japan*, 75th Ann. Vol., 180-186.

Hirasawa, H., K. Kato, and T. Takeda, 1995: Abrupt change in the characteristics of the cloud zone in subtropical East Asia around the middle of May. *J. Meteor. Soc. Japan*, **73**, 221-239.

Huffman, G. J., R. F. Adler, M. M. Morrissey, D. T. Bolvin, S. Curtis, R. Joyce, B. McGavock, and J. Susskind, 2001: Global precipitation at one-degree daily resolution from multisatellite observations. *J. Hydrometeor.*, **2**, 36-50.

Johnson, R. H., and P. E. Ciesielski, 2000: Rainfall and radiative heating rate estimates from TOGA-COARE atmospheric budgets. *J. Atmos. Sci.,* **57**, 1497-1514.

______, and ______, 2002: Characteristics of the 1998 summer monsoon onset over the northern South China Sea. *J. Meteor. Soc. Japan*, **80**, 561-578.

______, Z. Wang, and J. F. Bresch, 1993: Heat and moisture budgets over China during the early summer monsoon. *J. Meteor. Soc. Japan*, **71**, 137-152.

______, P. E. Ciesielski, and K. A. Hart, 1996: Tropical inversions near the 0°C level. *J. Atmos. Sci.,* **53**, 1838-1855.

Kato, K., 1985: On the abrupt change in the structure of the Baiu front over the China continent in late May of 1979. *J. Meteor. Soc. Japan*, **63**, 20-36.

Keenan, T. D., and R. E. Carbone, 1992: A preliminary morphology of precipitation systems in tropical northern Australia. *Quart. J. Roy. Meteor. Soc.,* **118**, 283-326.

Kuettner, J. P., 1971: Cloud bands in the earth's atmosphere: Observations and theory. *Tellus*, **23**, 404-426.

Kummerow, C. and Co-Authors, 2000: The status of the Tropical Rainfall Measuring Mission (TRMM) after two years in orbit. *J. Appl. Meteor.*, **39**, 1965-1982.

Lau, K.-M., Y.-H. Ding, J.-T. Wang, R. H. Johnson, T. D. Keenan, R. Cifelli, J. Gerlach, O. Thiele, T. Rickenbach, S.-C. Tsay, and P.-H. Lin, 2000b: A report of the field operations and early results of the South China Sea Monsoon Experiment (SCSMEX). *Bull. Amer. Meteor. Soc.,* **81**, 1261-1270.

______, K.-M. Kim, and S. Yang, 2000a: Dynamical and boundary forcing characteristics of regional components of the Asian summer monsoon. *J. Climate,* **13**, 2461-2482.

______, H.-T. Wu, and S. Yang, 1998: Hydrologic processes associated with the first transition of the Asian Summer Monsoon: A pilot satellite study. *Bull. Amer. Meteor. Soc.,* **79**, 1871-1882.

______, and S. Yang, 1997: Climatology and interannual variability of the Southeast Asian Summer Monsoon. *Adv. in Atmos. Sci.,* **14**, 141-162.

LeMone, M. A., 1973: The structure and dynamics of horizontal roll vortices in the planetary boundary layer. *J. Atmos. Sci.,* **30**, 1077-1091.

______, 1983: Momentum transport by a line of cumulonimbus. *J. Atmos. Sci.*, **40**, 1815-1834.

______, E. J. Zipser, and S. B. Trier, 1998: The role of environmental shear and thermodynamic conditions in determining the structure and evolution of mesoscale convective systems during TOGA COARE. *J. Atmos. Sci.,* **55**, 3493-3518.

Li, C., and M. Yanai, 1996: The onset and interannual variability of the Asian summer monsoon in relation to land-sea thermal contrast. *J. Climate*, **9**, 358-375.

Lin, X., and R. H. Johnson, 1996: Heating, moistening and rainfall over the western Pacific warm pool during TOGA COARE. *J. Atmos. Sci.,* **53**, 3367-3383.

Luo H.-B., and M. Yanai, 1984: The large-scale circulation and heat sources over the Tibetan Plateau and surrounding areas during the early summer of 1979. Part II: Heat and moisture budgets. *Mon. Wea. Rev.*, **112**, 966-989.

Mapes, B. E., and R. A. Houze, Jr., 1995: Diabatic divergence profiles in western Pacific mesoscale convective systems. *J. Atmos. Sci.,* **52**, 1807-1828.

Moncrieff, M. W., and J. S. A. Green, 1972: The propagation and transfer properties of steady convective overturning in shear. *Quart. J. Roy. Meteor. Soc.*, **98**, 336-352.

Reed, R. J., and E. E. Recker, 1971: Structure and properties of synoptic-scale wave disturbances in the equatorial western Pacific. *J. Atmos. Sci.,* **28**, 1117-1133.

Robe, F., and K. A. Emanuel, 2001: The effect of vertical wind shear on radiative-convective equilibrium states. *J. Atmos. Sci.,* **58**, 1427-1445.

Rotunno, R., J. B. Klemp, and M. L. Weisman, 1988: A theory for strong, long-lived squall lines. *J. Atmos. Sci.,* **45**, 463-485.

Steiner, M., and R. A. Houze, Jr., 1998: Sensitivity of monthly three-dimensional radar-echo characteristics to sampling frequency. *J. Meteor. Soc. Japan*, **76**, 73-95.

Tao, S. Y. and L. X. Chen, 1987: A review of recent research on the east Asian summer monsoon in China. *Monsoon Meteorology III*, eds. by C.-P. Chang and T. N. Krishnamurti, , Oxford University Press, 60-92.

Thompson, R. M., Jr., S. W. Payne, E. E. Recker and R. J. Reed, 1979: Structure and properties of synoptic-scale wave disturbances in the intertropical convergence zone of the eastern Atlantic. *J. Atmos. Sci.,* **36**, 53-72.

Wentz F. J., C. Gentemann, D. Smith, and D. Chelton, 2000: Satellite measurements of sea surface temperature through clouds. *Science*, **288**, 847-850.

Yamazaki, N., H. Kamahor, A. Yatagai, K. Takahashi, H. Ueda, K. Aonashi, K. Kuma, Y. Takeuchi, H. Tada, Y. Fukutomi, H. Igarashi, H. Fujinami, Y. Kajikawa, 2000: On the release of GAME Reanalysis products. *Tenki*, **47**, 659-663 (in Japanese).

Yanai, M., S. Esbensen, and J. H. Chu, 1973: Determination of bulk properties of tropical cloud clusters from large-scale heat and moisture budgets. *J. Atmos. Sci.,* **30**, 611-627.

______, C. Li, and Z. Song, 1992: Seasonal heating of the Tibetan Plateau and its effects on the evolution of the Asian summer monsoon. *J. Meteor. Soc. Japan*, **70**, 319-351.

______, and T. Tomita, 1998: Seasonal and interannual variability of atmospheric heat sources and moisture sinks determined from NCEP-NCAR reanalysis. *J. Climate,* **11**, 463-482.

Part V

Interactions with Other Circulations

13. MONSOON-RELATED TROPICAL CYCLONES IN EAST ASIA

RUSSELL L. ELSBERRY
Department of Meteorology
Graduate School of Engineering and Applied Sciences
Naval Postgraduate School
Monterey, CA93943, USA
E-mail: Elsberry@nps.edu

After monsoons, the tropical cyclone is the major economic and social weather event that impacts the people of East Asia. In many aspects, the large-scale circulations established by the monsoon that extend to the adjacent ocean areas control when and where tropical cyclones will form and where the cyclones will move. This chapter will explore recent understandings of these monsoon-related effects on tropical cyclone formation and motion in the East Asia region. Seasonal displacements of the East Asia monsoon circulation are accompanied by latitudinal displacements of monsoon trough-related tropical cyclone formations over the western North Pacific. Intraseasonal variations between active and inactive monsoon troughs lead to favorable and unfavorable environmental conditions for tropical cyclone formation. Some theoretical and observational studies are beginning to provide more insight on specific aspects of tropical cyclone formation in the monsoon environment, but more studies are required. A recent modeling study by H.-C. Kuo and colleagues provides some new insights on the important problem of why tropical cyclone formation is favored near the eastern end of the monsoon trough. The roles of the upstream waves in the easterlies, the convergence effect in the zone between the equatorial westerlies and the easterlies, and the nonlinear effects leading to wave energy accumulation and axisymmetrization of the waves into vortices are clarified by that study. The environmental influence of the monsoon circulation on tropical cyclone motion has been described in terms of synoptic patterns and regions. The typical monsoon trough situation is included in the Standard synoptic patterns, and the reverse-oriented monsoon trough is included in the Poleward synoptic pattern. A monsoon Gyre synoptic pattern is unique to the western North Pacific. In each pattern, synoptic regions are defined that have characteristic tracks, so that a change in pattern/region combination is accompanied by a track change. This pattern/region change may occur via various tropical cyclone-environment interactions, including a sudden poleward track change when a tropical cyclone interacts with a monsoon gyre.

1. Introduction

The monsoon is the dominant weather phenomenon throughout East Asia. The social and economic impacts of the monsoon are of unquestioned importance. However, the large-scale monsoon circulation patterns also extend offshore over the adjacent ocean areas and are a contributing factor to where and when tropical cyclones will form. The environmental flow patterns associated with the monsoon circulations also provide the first-order effect on where the tropical cyclone will move. Thus, it is appropriate to discuss tropical cyclones in this volume.

The tropical cyclone is a cyclonic circulation that forms over warm tropical oceans and has a central convective region of heavy precipitation and high winds. Although tropical cyclone designations vary among forecast centers, the convention used in this chapter will be that of the Joint Typhoon Warning Center (JTWC) Honolulu, Hawaii: Tropical Depressions (TD) with a definite closed surface circulation and maximum winds between 25 kt (12.5 m/s) and 33 kt (16.5 m s^{-1}); Tropical Storms (TS) with maximum winds of at least gale-force (34 kt, or 17 m s^{-1}); and Typhoons (TY) with maximum winds equal or greater than 64 kt, or 32 m s^{-1}. Whereas the JTWC designations are based on 1-minute averaged winds, the other national meteorological centers in East Asia use 10-minute average winds. Thus, some confusion may occur as to the tropical cyclone intensities issued by the different forecast centers. However, an agreement has recently been reached that the tropical cyclone will not be named until the Regional Specialized Meteorological Center, Typhoon Center, Tokyo decides that gale-force winds exist near the center. A common name list has also been adopted with names provided by the various countries.

The special threat of a tropical cyclone comes from the combination of high winds and heavy precipitation near the center. In addition to the potential wind damage to structures and vegetation, the winds can build high ocean surface waves and create storm surges of more than 5 m elevation. A storm surge during November 1970 with an estimated maximum level of 9 m at Chittagong in Bangladesh cost 350,000 lives. Even as recently as 1991, about 140,000 people died in a storm surge generated by a cyclone that struck Bangladesh.

Increases in coastal population and industry raise the risk factor from tropical cyclone strikes. In recent years, it has become evident that heavy precipitation-induced flooding associated with tropical cyclones is also an important and difficult forecasting problem. This flooding is not limited to coastal areas and may occur with slowly moving tropical storms. Thus, the cyclone continues to be a danger well inland from the coast, especially when the storm circulation interacts with orography.

A primary focus of this chapter is how the East Asia monsoon circulation contributes to tropical cyclone formation. *Formation* may be defined from a physical basis as having occurred when a warm-core, cyclonic circulation in the tropics with convection and winds concentrated near the center has been established throughout the troposphere, since a warm-core cyclone throughout the troposphere is a characteristic of the mature tropical cyclone. In the case of formations from the warm-core monsoon depressions and gyres, a concentrated core of central convection and winds must be created, rather than just having gale-force winds at outer radii. In the case of cold-core systems with maximum vorticity in the mid-troposphere, the cyclonic vortex must be extended downward. From an energetics viewpoint, a tropical cyclone can be said to have formed when the maximum cyclonic vorticity (winds) are present at the top of the atmospheric boundary layer and are strong enough that the low-level cyclonic wind field will be able to enhance the oceanic heat and moisture sources and establish a self-intensifying feedback that will lead to a mature tropical cyclone – in the absence of inhibiting effects such as excessive vertical wind shear. This latter self-intensifying stage following formation would be referred to as *intensification.*

From an operational viewpoint, formation is typically defined as the time when either a tropical depression or a (named) tropical storm (see definitions above) exists. No consensus exists among the operational centers, and even so the varying use of one-minute verses ten-minute surface wind averaging times might still lead to discrepancies as to the time of tropical cyclone formation. Perhaps an operational definition might be developed if all centers would use the same satellite pattern recognition technique; however, this has not occurred, perhaps because of real differences in how tropical cyclones form in different environments.

The second focus of this chapter will be on how the monsoon circulations affect tropical cyclone motion. Different environmental flows (defined as synoptic patterns and regions) are associated with different characteristic tracks, and a transition from one synoptic pattern/region combination to another brings about a track change. Some recent advances in the ability to forecast tropical cyclone tracks can be attributed in part to improved numerical weather prediction guidance, which includes the ability to correctly forecast monsoon-related circulations.

2. Favorable Environmental Conditions for Tropical Cyclone Formation

The six environmental factors originally defined by Gray (1968, 1975) will be condensed to four favorable (but not sufficient) conditions for tropical cyclone formation (i) Sea-surface temperatures (SSTs) exceeding 26°C; (ii) convective instability in the lower troposphere, with large values of relative humidity in the middle troposphere; (iii) large cyclonic absolute vorticity in the lower troposphere and anticyclonic relative vorticity in the upper troposphere; and (iv) weak vertical shear of the horizontal winds. As discussed by Gray (1975), the first two "thermodynamic conditions" are met over large areas of the tropical oceans and vary only slowly in time.

The requirement for SST > 26°C is related to the ocean being the ultimate energy source for the tropical cyclone formation. Since the surface air is in near-equilibrium (within 1°C) with the SST over most of the tropical ocean during the tropical cyclone season, so the surface air temperature is ≥ 26°C. Given the typical upper-tropospheric thermal structure in the tropics, an air parcel near the surface must have a temperature of at least 26°C if it is to be lifted to saturation, ascend approximately along a moist adiabat, and still be buoyant to near the tropopause. That is, a buoyant parcel ascent of saturated air in deep convection will only occur for sufficiently high surface air temperatures. Creation of warm air aloft is favorable for lowering the central pressure relative to the environmental value, and further warming will follow if the ascending air is forced to subside in dry air immediately adjacent to the cloud.

Convective instability refers to an environmental thermal structure in which lifting of a saturated air parcel will cause it to ascend with a higher temperature than that of the environment. If the air is not saturated, the ascent will be along the dry adiabat, which will be negatively buoyant relative to the environment, and the parcel will return to the original level. The low-level tropical atmosphere is not saturated, and a lifting mechanism must exist to bring the cloud parcel to saturation. However, the boundary layer air over the tropical ocean is sufficiently warm and moist that lifting of only about 60 mb (600 m) is required. This lifting may be accomplished by frictional or dynamical convergence on various horizontal scales.

A consensus exists that the mid-tropospheric relative humidity in the center of a developing tropical cyclone is higher than the average environmental values. However, the explanations for the moist mid-tropospheric differ, as do the proposed mechanisms for achieving the moistening. Gray (1998) states that the precipitation rates (and thus latent heat release) are higher with a moist mid-troposphere. Gray hypothesizes that an extreme horizontal convergence of moisture into a 50-km "extreme convection" region would be required to raise the water vapor contents such that an additional 20-25% humidity increase [to near saturation] would occur in a short time.

Bister and Emanuel (1997) propose two mechanisms for eliminating the normal dry mid-tropospheric air and establishing high θ_E values in the boundary layer. In their conceptual model, the evaporation of rain from the stratiform rain area of a mesoscale convective system (MCS) will eventually moisten the lower tropospheric air. In their numerical model, continued rain over 18-36 h is required for this step. A crucial part of their hypothesis is that this continued rain will extend the mid-tropospheric vortex down to the top of the boundary layer, and that introduction of these stronger winds will initiate larger air-sea fluxes of heat and moisture, which will re-establish high θ_E values at the surface. In their model, deep convection erupts in a concentrated region of the cold (but humid) air, and a warm-core vortex develops where there was previously a cold-core vortex. However, many remotely-sensed and *in situ* observations of MCSs in the western North Pacific during the Tropical Cyclone Motion (TCM) field experiments in 1992 and 1993 did not support the Bister and Emanuel scenario of continued stratiform rain leading to the downward extension of the vortex to the surface.

Whereas the SST threshold and convective instability condition are commonly satisfied over the tropical oceans, the two dynamical environmental conditions for tropical cyclone formation vary significantly on synoptic time scales. Consequently, they may be expected to be more related to observed periods of tropical formation or of inactivity.

2.1. *Favorable Environment Vorticity Condition*

A long list of Colorado State University studies (see Gray 1998 for references) searched for environmental features revealed by composites of rawinsondes that would distinguish between developing and non-developing systems. After exhaustive tests, one primary difference is that developing storms have low-tropospheric environmental winds with larger cyclonic relative vorticity than is observed in non-developing disturbances. In a cylindrical coordinate system centered on the storm, the tangential wind components (v_θ) are larger positive (cyclonic) in a deep tropospheric layer (Fig.1) for developing storms than for non-developing storms. The explanation for why an environment with positive absolute vorticity is favorable for cyclogenesis is found in the vorticity tendency equation,

$$\frac{d\zeta}{dt} = -\zeta_a \nabla \bullet V, \qquad (1)$$

where ζ and ζ_a are the relative and absolute vorticities and $\nabla \cdot V$ is the divergence. A spinup of cyclonic vorticity (cyclogenesis) occurs in regions of horizontal convergence with large values of absolute vorticity. For the same cloudy-area ascent with compensating low-level convergence, the storm needs to be at some distance from the equator so the Coriolis parameter contributes to the absolute vorticity. For the same latitude (Coriolis parameter), increased tangential winds at 300-600 km from the center in Fig. 1 represent enhanced cyclonic relative vorticity that increases the absolute vorticity term in the vorticity generation term.

Climatological studies (e.g., McBride 1995) indicate that perhaps 75% or more of the named tropical storms over the globe form in a monsoon trough environment. A somewhat idealized monsoon trough, as might appear in a time-averaged wind field, is given in Fig. 2 to illustrate a favorable environmental wind structure as in Fig. 1. More specific monsoon circulation structures related to tropical cyclone formation will be presented later. By definition, the monsoon trough (lowest surface pressure) is displaced more than about 8° latitude from the equator, so that lower-tropospheric westerly winds are found equatorward of the monsoon trough. An alternate source for westerlies along the equator is to have near-equatorial troughs in both hemispheres (see Fig. 11 below), and these troughs may be only 5°-8° lat. from the equator. Easterly (trade) winds are found poleward of the monsoon trough between the trough and the subtropical high pressure region farther poleward. Thus, the monsoon trough as in Fig. 2 provides a favorable cyclonic relative vorticity environment for tropical cyclone formation.

An additional favorable environmental wind feature from the composite rawinsonde studies is a stronger anticyclonic circulation in the upper troposphere above developing storms (Fig. 1). This condition is also satisfied in the monsoon trough, as indicated in the

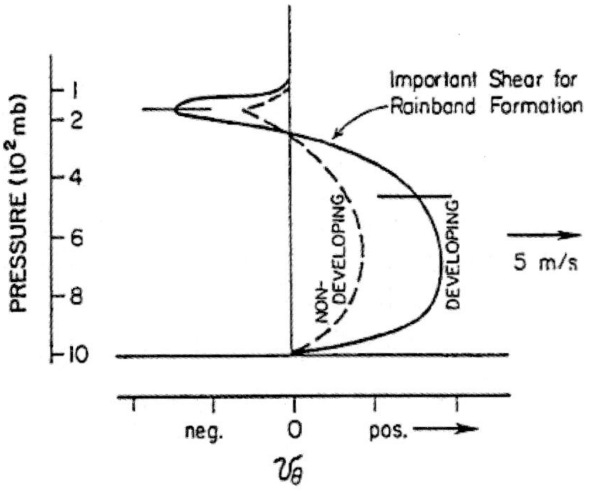

Figure 1. Vertical distribution of the tangential wind at a radius of about 300 to 600 km in storm-centered rawinsonde composites by Colorado University researchers (Gray 1998) comparing developing (solid line) versus non-developing (dashed) tropical storms. Positive (negative) tangential winds are cyclonic (anti-cyclonic) in this Northern Hemisphere example.

Figure 2. Schematic of a Northern Hemisphere monsoon trough displaced more than 8° latitude from the Equator so that Equatorial Westerlies are found on the equatorward side and tradewind easterlies on the poleward side. The convergence of warm, moist air from both sides makes the monsoon trough a favorable region for cloud clusters (Gray 1998).

(Northern Hemispheric) schematic in Fig. 3. Consider first the vertical profiles of the zonal wind on the right side of Fig. 3. As indicated by Fig. 2, lower tropospheric (here at 1 km) westerlies (easterlies) are found on the south (north) side of the monsoon trough. Low surface pressures are found in the monsoon trough because it is the warmest column. From the hypsometric equation, warmer columns have a greater thickness between two pressure surfaces. Thus, a high pressure and anticyclonic wind circulation are found aloft owing to the deep tropospheric warm column above the monsoon trough. The change from cyclonic shear of the horizontal winds across the monsoon trough in the lower troposphere (1 km) to anticyclonic wind shear in the upper troposphere (12 km) is shown to the left in Fig. 3. That is, low-level equatorial westerlies (tradewind easterlies) to the south (north) of the trough in Fig. 3 have upper-tropospheric easterlies (westerlies) aloft.

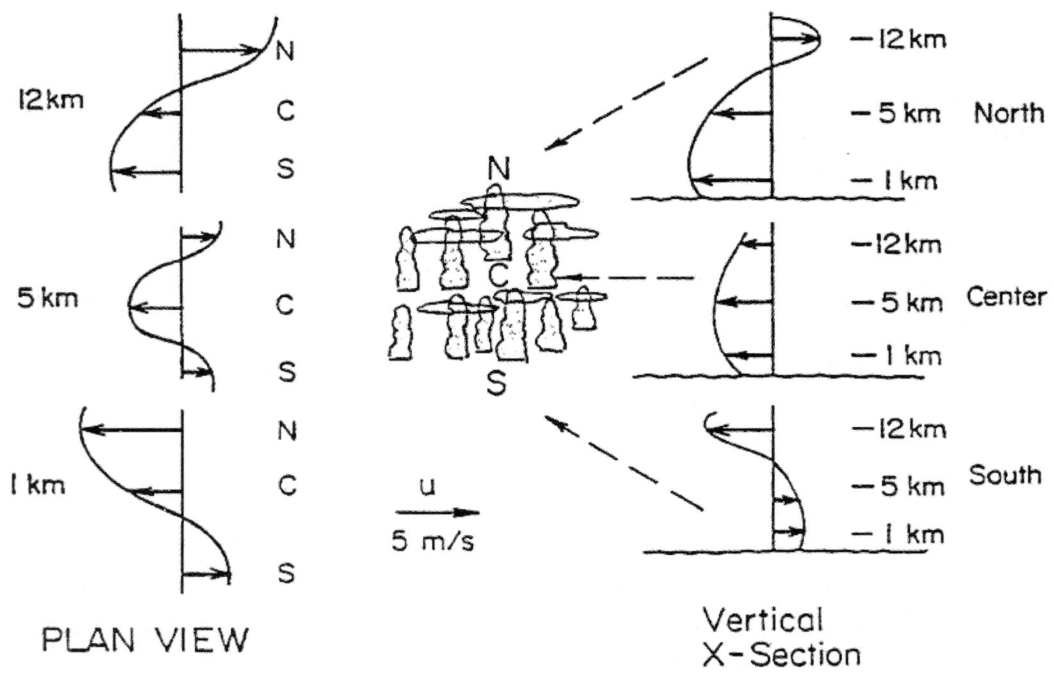

Figure 3. Plan and cross-section views of the typical arrangement of the zonal winds to the North (N) and South (S) of a pre-tropical cyclone cloud cluster in a Northern Hemisphere monsoon trough as in Fig. 2 with the center located at C (Gray 1998).

The schematics in Figs. 2 and 3 for the monsoon trough are representative of the favorable tropical cyclone formation environmental condition of lower-tropospheric cyclonic vorticity and upper-tropospheric anticyclonic vorticity. Specific monsoon-related and other synoptic circulations that satisfy this favorable environmental vorticity condition will be presented below.

2.2. *Minimum Vertical Wind Shear Condition*

It has been long observed by forecasters and researchers that tropical cyclone formation is inhibited in the presence of vertical wind shear (e.g., Riehl 1954; Palmen 1956; Simpson *et al.* 1969; Atkinson 1971). In operational situations, the vertical wind shear is often calculated between the 850 mb and 200 mb levels, because these are the levels with the best data coverage of cumulus and cirrus cloud-drift winds, respectively, from animation of geostationary satellite imagery. The Cooperative Institute for Meteorological Satellite Studies at the University of Wisconsin website has a real-time display of vertical wind shear throughout the tropics that is primarily based on satellite-derived wind vectors.

A recent study by Zehr (1992) of western North Pacific disturbances suggested a critical 850-200 mb wind difference of 12 m s^{-1} separates the developing from the non-developing disturbances. Elsberry and Jeffries (1996) suggest some caution in using operational wind analyses for such a vertical shear calculation. They found the cirrus-level outflow from a developing disturbance may be "fighting-off" the encroaching upper-tropospheric flow, but that the operational analysis did not have the required mesoscale wind vectors, or perhaps the horizontal resolution, to represent this effect. Consequently, the 850-200 mb wind shear from the operational analysis exceeded the Zehr (1992) threshold value, even though the disturbance continued to develop.

One explanation given for this minimum environmental vertical wind shear condition is that too large of a difference between lower-tropospheric winds and upper-tropospheric winds will not allow heat (and moisture) to accumulate in the column to form a warm core and lower the surface pressure. Consider the vertical wind shear in the monsoon trough environment illustrated on the right side of Fig. 3, which may serve as examples of this "ventilation of heat" concept. From thermal wind considerations with the warmest air over the monsoon trough and colder air to the north, the vertical wind shear is westerly to the north. On the south side of the monsoon trough, the vertical wind shear is reversed (easterly), because the warmest column is over the monsoon trough and cooler columns are found to the south. Specifically, lower-tropospheric equatorial westerlies change to upper-tropospheric easterlies on the south side of the monsoon trough (see lower-right wind profile in Fig. 3). In either of these regions in which the winds reverse from the lower to the upper troposphere, heating generated by a cloud cluster would be advected in opposite directions in the lower and upper portions of the tropospheric column. This "ventilation" of the column if the vertical wind shear is too large it is thought to prevent accumulation of the warm air in the column. *Without the ventilation effect, the warm air aloft is kept above the disturbance, and the resulting surface pressure falls will contribute to growth (cyclonic spinup) of the lower-tropospheric disturbance.*

Whereas large westerly (easterly) vertical wind shear is found to the north (south) of the monsoon trough (see wind profiles on right side of Fig. 3), the vertical wind shear must pass through zero in the region of the monsoon trough (see middle-right profile in Fig. 3). Thus, disturbances in the monsoon trough meet the minimum vertical wind shear condition that is favorable for tropical cyclone formation (McBride and Zehr 1981), and this is another reason why more than 75% of the global tropical cyclones form in the monsoon trough.

3. Tropical Cyclone Formation in a Monsoonal Environment

The four favorable environmental conditions described in section 2 clearly apply in the monsoonal environment. This theme of relating tropical cyclone formation to favorable environmental conditions in monsoon regions will be further developed in this section as a number of variations on that theme will be presented. As in section 2, the emphasis will be on the dynamical aspects of lower- (upper-) tropospheric cyclonic (anticyclonic) relative vorticity in the environmental flow, with the presumption that stronger dynamical forcing leads to more tropical cyclone formations given that the other three favorable environmental conditions are also met. Where large-scale environmental conditions establish extended areas of favorable conditions in monsoonal regions, simultaneous (or a sequence of) tropical cyclone formation may occur in the region. By contrast, inactive monsoon regimes also tend to be inactive tropical periods (Harr and Elsberry 1995). Because of these active and inactive periods of the monsoon environment, active and inactive tropical cyclone periods also tend to be clustered (Fig. 4). As the capability to predict the large-scale environment with numerical models is extended to longer periods, it may become possible to anticipate active and inactive tropical cyclone periods, which could be useful in scheduling maritime activities that would be impacted by a tropical cyclone.

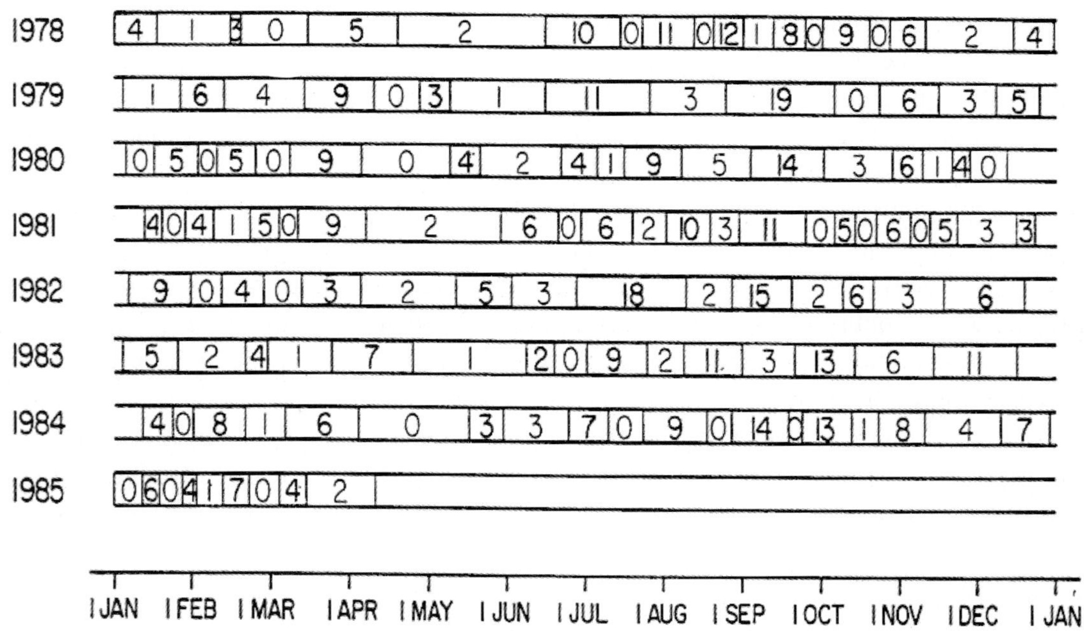

Figure 4. Tendency of the global tropical cyclone formations to be clustered into active and inactive periods. Numbers indicate named tropical cyclone formations within arbitrarily selected periods (provided by W. Gray).

3.1. *Enhancement in the Typical Monsoon Trough*

Although these examples of the typical monsoon trough and other variations of the monsoon environment will be taken from the western North Pacific, similar situations will be found in other monsoonal regions, and will have similar effects on tropical cyclone formations.

3.1.1. Seasonal Variations

The seasonal migration of the typical monsoon trough in the East Asian area is indicated in Chapter 1 of this volume. As the East Asia summer monsoon is established in June, the monsoon trough extends typically southeastward toward its more normal latitude over the tropical ocean (near the latitude of maximum SST closer to the equator). With the low pressure in the monsoon trough displaced more than 8-10° lat. from the equator, equatorial westerlies with warm, moist conditions are established on the equatorward side of the monsoon trough (Fig. 2). As the summer monsoon advances poleward over East Asia through August, the western anchor of the monsoon trough over the western North Pacific is also displaced poleward. Southward retreat of the summer monsoon over east Asia during September and October is also accompanied by an equatorward displacement of the western North Pacific monsoon trough. These seasonal displacements indicate the monsoon trough-related tropical cyclone formation areas will be similarly displaced.

3.1.2. Active and Inactive Monsoon Troughs

Harr and Elsberry (1995) have defined six recurrent large-scale circulation patterns via a cluster analysis of the time coefficients from a vector Empirical Orthogonal Function (EOF) representation of the 700-mb anomalous wind circulations. Cluster 1 (Fig. 5a) is typical of an active monsoon trough. Notice the large-scale nature of this circulation pattern with an active monsoon trough that extends from India into the western North Pacific. Just as a long band of enhanced equatorial westerlies are found on the equatorward side, a long band of enhanced tradewind easterlies are found on the poleward side of the active trough. Furthermore, stronger than normal westerlies are found on the poleward side of the subtropical anticyclone. Although such a description does not establish cause-and-effect relationships, it appears that an active monsoon trough over the western North Pacific may be related to large-scale (i.e., East and South Asia) monsoon circulations.

Harr and Elsberry (1995) Cluster 4 (not shown) is almost the opposite pattern of Cluster 1 with an anomalously weak monsoon trough and subtropical ridge, which is characteristic of an inactive tropical cyclone regime. Cluster 3 (Fig. 5b) also has some of the characteristics of an inactive monsoon regime with an anomalous anticyclonic circulation in the South China Sea and equatorial easterlies across the southern Philippine Sea and into the Indian Ocean. Monsoon trough-related tropical cyclone formations would not be favored in such inactive monsoon conditions, although other types of tropical cyclone formations (e.g., the baroclinic type) could still occur.

One phenomenon that may contribute to such an active monsoon trough is the intraseasonal Madden-Julian Oscillation (MJO). The MJO has a baroclinic mixed Kelvin and

Rossby wave structure in the Indian and western Pacific Oceans that propagates eastward at about 5 m s^{-1}. Farther eastward, a Kelvin-type wave propagates eastward at phase speeds of 10-12 m s^{-1}, and anomalous winds at 850 mb can extend to the Americas. The active MJO periods (characteristic repeat times of 30-60 days) coincide with anomalous equatorial westerlies, and thus low-level cyclonic wind shear across the active monsoon trough. Thus, it is not surprising that Liebmann *et al.* (1994) found more frequent tropical cyclones in the

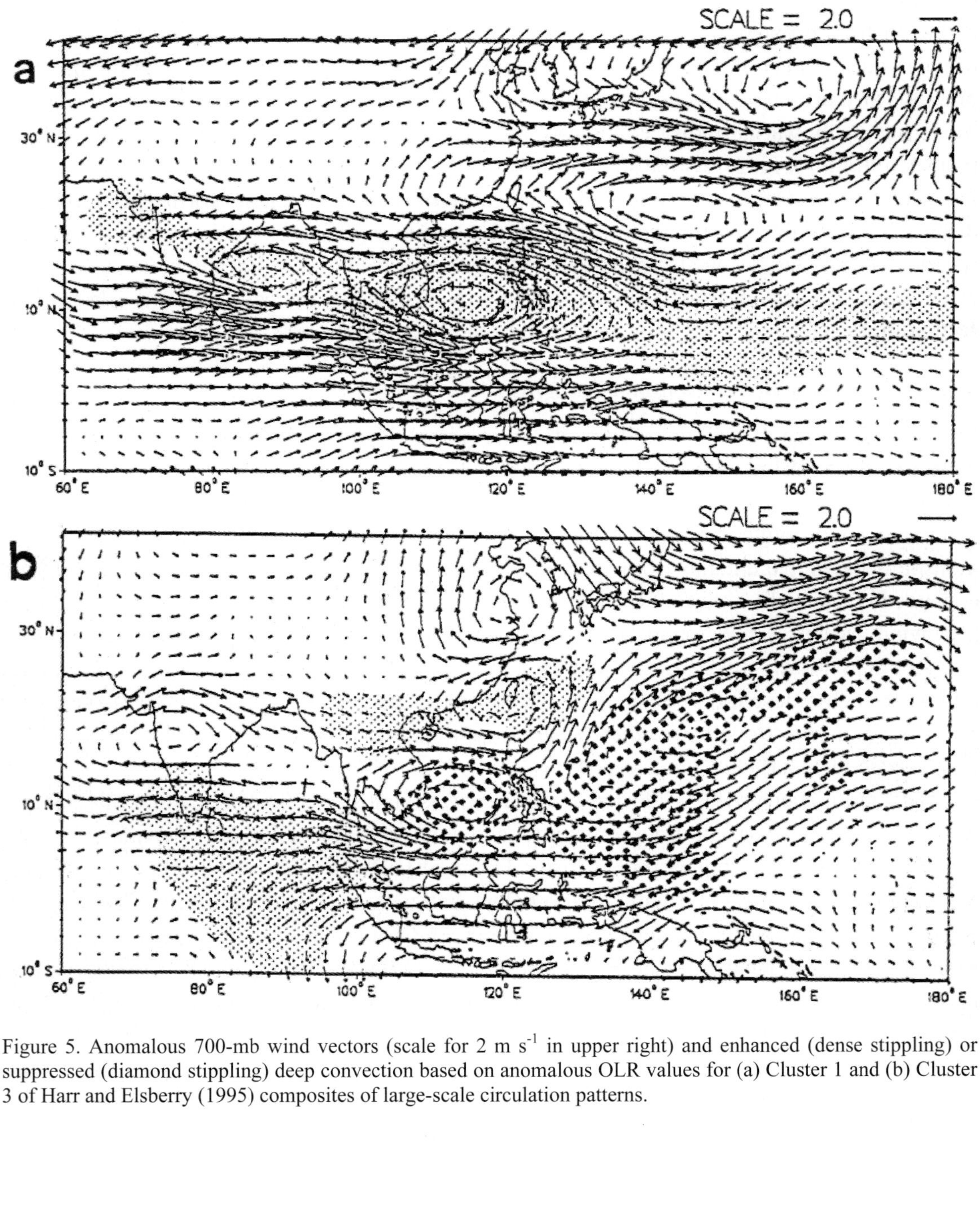

Figure 5. Anomalous 700-mb wind vectors (scale for 2 m s^{-1} in upper right) and enhanced (dense stippling) or suppressed (diamond stippling) deep convection based on anomalous OLR values for (a) Cluster 1 and (b) Cluster 3 of Harr and Elsberry (1995) composites of large-scale circulation patterns.

western North Pacific during the active MJO phase. However, they noted that increased tropical cyclone activity was also associated with active monsoon troughs not associated with the MJO.

M. Lander (e-mail 1 Sep 2000 on Tropical Storms Bulletin Board) defines an active monsoon trough as one that extends well eastward into the western North Pacific basin (at least to 150°E). He specifies a period of about 20 days between active and inactive monsoon troughs. This period is closer to the frequency of active tropical cyclone formations in Fig. 4, and indeed Lander includes a requirement for tropical cyclone formations as an integral part of his active monsoon trough definition. Lander's conceptual model is that the monsoon trough will be drawn poleward as the tropical cyclones move poleward. In the example in Fig. 6, Typhoons Kirogi and Kai-Tak have formed in the monsoon trough and a new disturbance is starting to spinup between the two typhoons. In this case, TY Kirogi is moving poleward more rapidly than Kai-Tak under the influence of the subtropical anticyclone to the east of Kirogi. This relative displacement can lead to the monsoon trough assuming a reverse-oriented (i.e., southwest-to-northeast) monsoon trough orientation. As the tropical cyclones and monsoon trough move poleward, the normal monsoon trough flow is replaced by easterlies with subsidence and drier air aloft that are unfavorable for tropical cyclone formation. According to Lander, the approximate 20-day monsoon trough cycle reflects the time for the monsoon trough to become re-established along 10°N with its associated equatorial westerlies and begin the process again, and this 20-day cycle may be independent of the MJO that tends to have maximum amplitude during Northern Hemisphere winter rather than during the typhoon season.

Figure 6. Infrared satellite image at 2300 UTC 6 July 2000 indicating TY Kai-Tak in the South China Sea and TY Kirogi near 28°N, 136°E with a new cyclonic circulation starting near 20°N, 133°E.

3.1.3. Idealized ITCZ Breakdown into Disturbances

In a barotropic model by Guinn and Schubert (1993), a zonally elongated monsoon trough with low-level easterlies on the poleward side and low-level westerlies on the equatorward side is represented as a zonal strip of high Potential Vorticity (PV) (Fig. 7). The associated (negative) meridional gradient of Potential Vorticity (PV) on the poleward side will support a PV wave that will propagate eastward relative to the easterly flow. Similarly, the positive meridional PV gradient on the equatorward side will support a westward-propagating PV wave relative to the equatorial westerlies. Guinn and Schubert propose these counter-propagating waves may have the same phase speed relative to the earth such that a phase-locking will be favorable for mutual growth of the waves. The subsequent ITCZ breakdown leads to a pooling of PV into disturbances (Fig. 7) that are hypothesized to break away and form tropical cyclones. Although Guinn and Schubert show a satellite image in which multiple disturbances with a wavelength similar to Fig. 7 form on an elongated ITCZ, moist processes are clearly required for intensification to a tropical cyclone.

Guinn and Schubert do not address how the initial ITCZ formed, which also must have involved moist processes. Rather, their purpose was to indicate that a sharp PV gradient associated with easterlies (westerlies) on the poleward (equatorward) side of an ITCZ could result in spontaneous growth of perturbations on time scales of about 5 days.

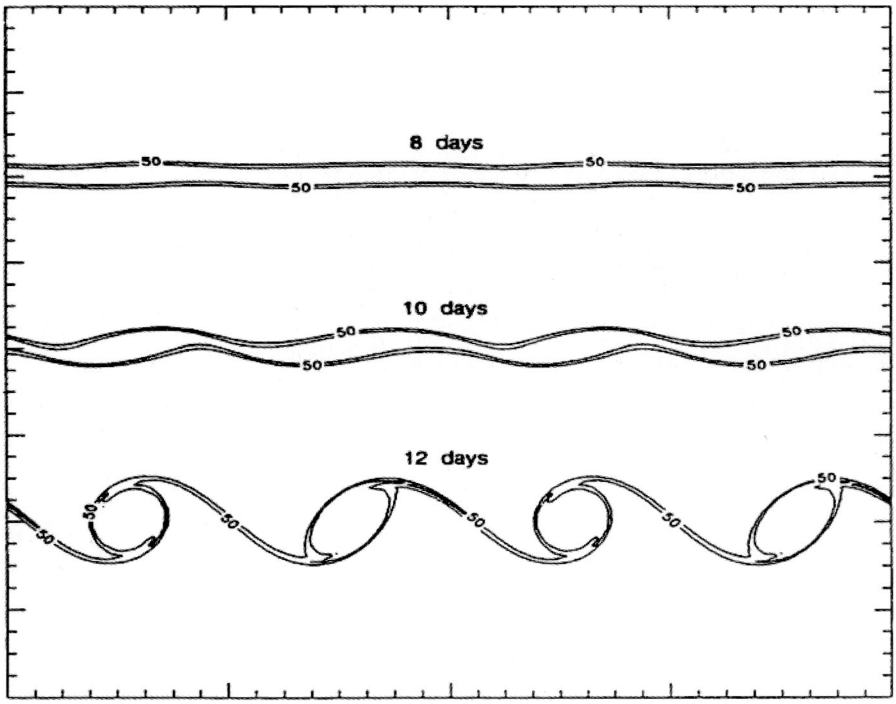

Figure 7. Schematic of breakdown of the ITCZ into discrete vortices as simulated by Guinn and Schubert (1993) model. Contours are of vorticity 8, 10, and 12 days into the simulation. Tick marks along the boundaries are at 400 km.

Ferreira and Schubert (1997) propose a mechanism for developing an environmental flow instability that would produce disturbances that might serve as seedlings for tropical cyclone formation. Continued release of latent heat in an extended band such as along the ITCZ will generate cyclonic PV below the maximum in heating, and anticyclonic PV above the heating maximum. As in Guinn and Schubert (1993), Ferreira and Schubert propose that the maximum in low-level PV from such a heating can create a Charney-Stern instability that can produce a breakdown into disturbances that might subsequently become tropical cyclones. Notice this *in situ* development is only for the early stages and can not explain tropical cyclone formation without the inclusion of moist processes.

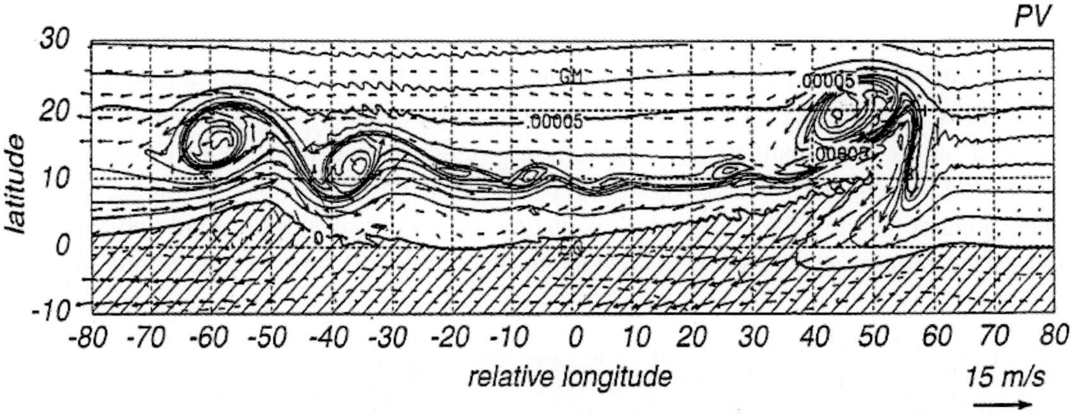

Figure 8. Perturbations of PV after 10 days in the Ferreira and Schubert (1997) simulation with a zonally oriented, irregularly shaped mass sink imposed for 5 days. Notice the dual vortices on the left (representing western North Pacific), minimal perturbations in the central region, and a larger vortex on the right (representing eastern North Pacific). All growing vortices are tilted to grow via extraction of zonal flow kinetic energy.

Ferreira and Schubert (1997) simulate the formation and breakdown of an ITCZ by imposing a mass sink in a (one-layer) shallow-water model. This mass sink represents convection, and as air parcels converge into the mass sink region, a cyclonic vorticity region is produced. After about 5 days, a horizontal shear layer with easterlies (westerlies) of about 5.5 m s^{-1} (12 m s^{-1}) on the poleward (equatorward) side is produced when Ferreira and Schubert placed an elongated mass sink along 10°N. Thus, a region with reversed meridional gradient of absolute vorticity is created on the northern edge as in the Guinn and Schubert (1993) idealized vortex strip. In the Ferreira and Schubert simulation, perturbations on the PV strip amplified with a wavelength of about 2070 km. These perturbations (Fig. 8) have a horizontal tilt such that they grow via barotropic conversion of energy from the mean flow as in the Guinn and Schubert simulation. A zonally elongated mass sink produces a more intense PV anomaly on the poleward side, so the vortices that emerge are several hundred kilometers poleward of the center of the mass sink (simulated ITCZ), which is consistent with the favored location of tropical cyclone formations that occur well poleward of the primary cloud band equatorward of the monsoon trough. However, tropical cyclones in the western North and South Pacific and South Indian Oceans tend to form along the axis of the monsoon trough. Since the maximum convection in the monsoon trough is within the equatorial westerlies, the tropical cyclone then does form poleward of what would be the mass sink in this model. These simulations of ITCZ breakdown do not imply barotropic instability is the sole process, because an actual breakdown would be expected to involve interplay of barotropic, baroclinic, and moist physical processes.

3.1.4. Example of Monsoon Trough-Related Tropical Cyclone Formation

The tropical cyclone formation in the monsoon trough often occurs in a synoptic-scale circulation called a *monsoon depression*. The Joint Typhoon Warning Center (JTWC 1994) defines the monsoon depression as a large cyclonic vortex (outer closed isobar with diameter of order 600 n mi or 1000 km) with a loosely organized cluster of deep convection (Fig. 9) that lacks a distinct cloud system center, and has a low-level, light-wind core of perhaps 100 n mi (185 km) diameter that is partially surrounded by a band of gales. Transition to a TC structure thus requires development of a central region of deep convection and a core of accelerated winds. This distinction between monsoon depressions and tropical cyclones will be retained in this chapter.

The practice at the JTWC has been to distinguish the tropical gale winds on the outer region of the monsoon depression as a separate phenomenon from the gale winds near the

Figure 9. Visible satellite image at 0530 UTC 26 October 2000 of a monsoon depression that later developed into TS Xangsane. Observations at Palau Island in the lower-central region included gusts as high as 44 kt that likely indicates sustained gale winds at a large radius from the center (provided by Mark Lander).

center that would justify a designation as a tropical storm. Thus, a swath of gale winds may have existed on the equatorward and eastern sides long before a tropical storm warning is issued. This practice is prudent since not all monsoon depressions acquire the tropical cyclone characteristics. For those depressions that do achieve central convection and gale winds that merge with the outer gales, the first tropical storm warning may have a large radius of 35 kt winds. Indeed, one of the characteristics of tropical cyclones that form from monsoon depressions is that they have stronger outer winds, and thus are larger in "size." If the tropical

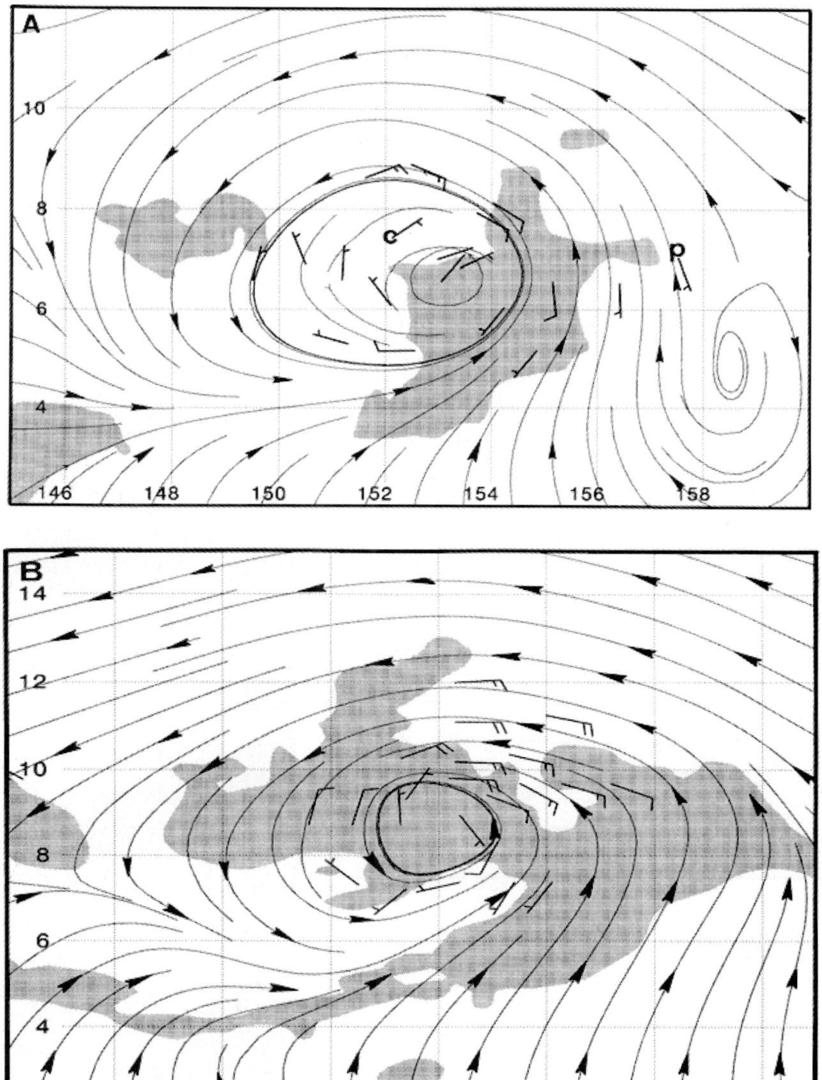

Figure 10. Wind structure at 850 mb of the TS Robyn (a) before and (b) after a central core of high winds developed and merged with the outer gale winds in the monsoon depression to form a large tropical cyclone (Harr *et al.* 1996a).

gale warnings have been placed accurately, the subsequent tropical storm warnings that merge the outer gale winds with the newly developed central gale winds should serve the purpose of adequately preparing the customers for the threat of high winds. However, an educational process may be required to ensure that the customers understand that the tropical gale warning in these situations contains an imminent threat of much higher winds as would be implied if it was a tropical cyclone advisory.

Harr *et al.* (1996a) have used aircraft and satellite data to document the evolution of a monsoon depression into Typhoon Robyn during 1993. Bands of enhanced equatorial westerlies on the south side and enhanced easterlies on the north side (not shown) simultaneously appeared to approach the formation location from the west and from the east, respectively, and eventually wrapped around the east and west, respectively, sides of the monsoon depression (Fig. 10a). That is, gale-force winds existed at several hundred kilometer radius before the central-core convection developed. The inner-core winds later accelerated and merged with this ring of gale-force winds to form a large tropical cyclone (Fig. 10b).

3.1.5. Eastward Extension of the Monsoon Trough

Another variation of the normal monsoon trough is an eastward extension in conjunction with a "westerly wind burst" along the equator (Fig. 11a). As the wind burst translates eastward, strong cyclonic environmental wind shear may be introduced in the lower troposphere on either side of the equator, which is favorable for twin-tropical cyclone formations (Fig. 11b).

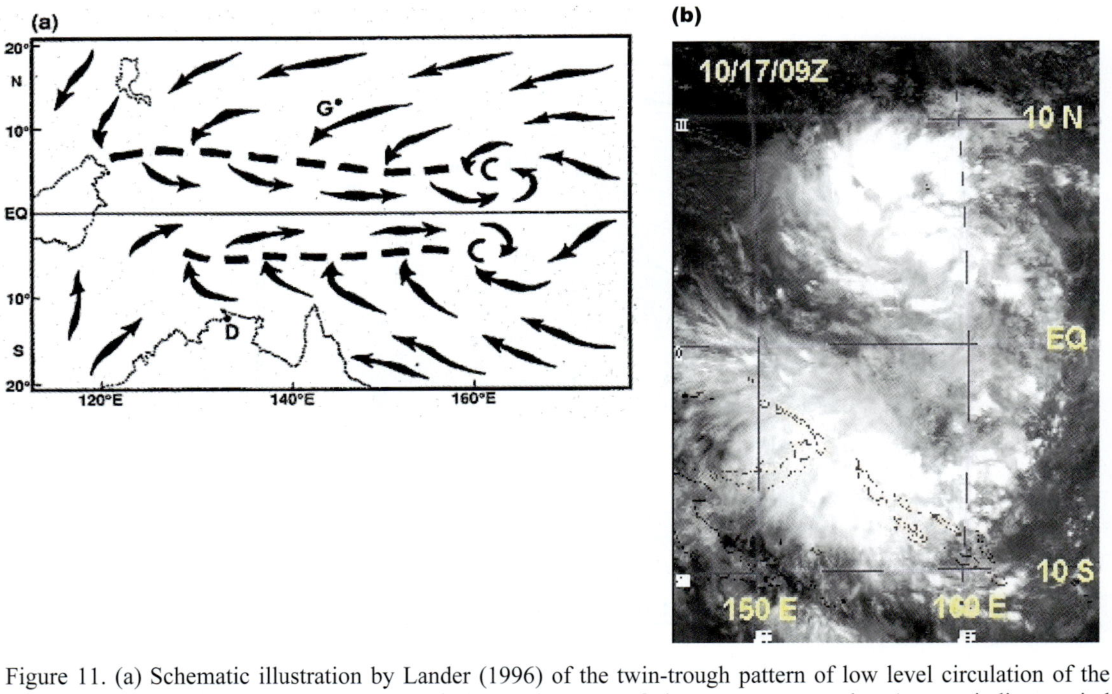

Figure 11. (a) Schematic illustration by Lander (1996) of the twin-trough pattern of low level circulation of the western North Pacific. Bold dashed lines indicate the axes of the monsoon troughs. Arrows indicate wind direction. The locations of Guam (G) and Darwin (D) are indicated. Bold Cs show cyclone circulations. (b) Satellite infrared image at 0900 UTC 17 October 2001 indicating a case of twin tropical cyclones attempting to form on adjacent sides of the Equator in response to an equatorial westerly wind burst.

In the case in Fig. 11b, Quikscat observations on the previous day indicated 45 kt westerly surface winds along the Equator. Such a pattern is likely to occur during the transition season(s) between summer and winter when a near-equatorial trough is present in both hemispheres. As each tropical cyclone develops, it will break out of the trough into the easterly flow on the poleward side, and thus move westward.

Such an equatorial westerly wind burst may be produced in association with a MJO. The most pronounced MJO cases tend to occur during the Northern Hemisphere winter over the eastern Indian Ocean and western North Pacific. A common representation of wave patterns associated with the MJO deep convection that is translating eastward is the Gill pattern in response to an equatorial heat source. That is, an equatorially trapped Kelvin wave (predominantly zonal wind perturbation) propagates eastward from the convective area, and two moist Rossby waves emanate from the convection and propagate westward in both hemispheres. The Climate Diagnostic Center website (www.cdc.noaa.gov/map/clim/olr_modes/mapanim2.html) has an animated gif file of the tropical strip outgoing longwave radiation mapped on several modes (MJO, equatorial Kelvin, Rossby waves). In some cases (Fig. 12), it appears the positive vorticity centers of these moist Rossby waves are the loci about which each of the twin vortices forms and grows. Some people have assumed that the westerly wind burst along the Equator is created in response to the twin low-pressure regions. According to Lander (1990), the sequence is that the westerly wind burst occurs first, a convective blow-up is triggered, and the moist Rossby waves emanate from the convection and initiate the twin vortices. However, Hartten (1996) found only 8% of synoptic-scale westerly wind bursts had associated concurrent cyclones in both hemispheres.

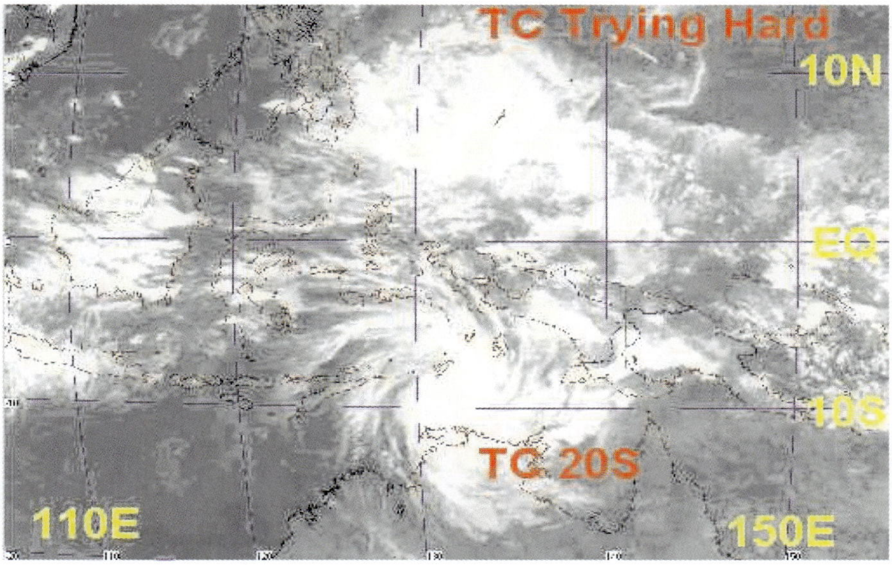

Figure 12. Satellite infrared image at 1200 UTC 16 April 2001 of the twin convective regions that gave rise to TC Alistair in the Southern Hemisphere and later became TC02W in the Western Hemisphere. These twins followed the development of a large equatorial cloud cluster on the Equator that was associated with a westerly wind burst (provided by Mark Lander).

3.2. *Other Monsoonal Environments*

3.2.1. Reverse-Oriented Monsoon Trough

Although the northwest-southeast orientation of the monsoon trough in the schematic in Fig. 2 is the normal situation during most months of the typhoon season, a reverse-oriented monsoon trough exists a significant fraction of the time in the western North Pacific (Fig. 13a). That is, the orientation is southwest to northeast with one or more cyclonic vortices embedded in the trough. In the satellite image in Fig. 13b, multiple convective regions or tropical cyclones are present within the monsoon trough, much like the schematic in Fig. 13a. This trough is in a break in the subtropical anticyclone, with the axis of the anticyclone farther poleward to the west (over Asia) than to the east. Instead of low-latitude equatorial

Figure 13. (top) Conceptual model of a reverse-oriented (with respect to climatology) monsoon trough with streamlines indicating boundary-layer flow in the tropical western North Pacific at 0000 UTC 17 August 1981 (adapted by Lander 1994). Island locations shown are Okinawa (O); Iwo Jima (I.J.); Guam (G); Wake (W); and Tarawa (T). (bottom) Annotated satellite image of a reverse-oriented monsoon trough and ridge to the southeast.

westerlies as in Fig. 2, a band of easterlies that extends to the Equator approaches the anticyclone axis before turning poleward. The subsidence and lower-tropospheric anticyclonic vorticity to the east are unfavorable to tropical cyclone formation. However, the southwesterlies (northeasterlies) to the east (west) of the reverse-oriented monsoon trough represents cyclonic relative vorticity (and low-level convergence) that is favorable for tropical cyclone formation. These reverse trough-related TCs may have an initial eastward motion, and generally have a poleward track orientation, which constitutes about 30% of all tropical cyclone forecasts in the western North Pacific (Carr *et al.* 1995). In many cases, the reverse-oriented monsoon trough may be a result of two (or three) tropical cyclones forming in an east-west trough and the eastern-most tropical cyclone moves poleward more rapidly as in Fig. 6. If a stable southwest-northeast oriented ridge (indicated in Fig. 13b) forms from the combination of the beta-induced peripheral anticyclones to the southeast of the cyclones, the tropical cyclones will have a poleward and eastward steering flow.

The large-scale environmental conditions that may lead to persistent reverse-oriented monsoon trough situations are not known. In some typhoon seasons, these conditions are present over a significant fraction of the typhoon season. Consequently, poleward-oriented tracks are favored that are generally more difficult to forecast (Webb 1996). In the very anomalous 1998 typhoon season in which an all-time low of 18 named storms occurred, the reverse-oriented monsoon trough pattern as in Fig. 13a persisted. No tropical cyclones formed east of 145°E where easterlies were continually present because the monsoon trough never became established east of this longitude. How this anomalous situation might be related to the rapid diminishing of the intense El Nino of 1997-98 and establishment of a La Nina is unknown.

3.2.2. Mei-Yu Monsoon Trough

Lander (1996) notes that a similar reverse orientation of a trough exists in late spring when the Mei-yu (Chinese for plum rains) front is active (Fig. 14). This persistent zone of disturbed weather is typically positioned across southern China and Taiwan around mid- to late-May. During June and July, this boundary between the anticyclone over Asia and the subtropical ridge moves poleward to lie across northern China through northern Japan. Occasionally, this Mei-yu frontal zone is transformed into a major convective cloud band that includes an extension of monsoon southwesterly flow into subtropical latitudes (Fig. 14). Tropical cyclone formation may then occur at unusually high latitudes along the axis of the Mei-yu monsoon trough. A cyclonically curved perturbation on the Mei-yu front in Fig. 14b had several characteristics of a tropical cyclone as might apply in the conceptual model in Fig. 14a. However, the satellite image indicates the convection is experiencing vertical wind shear.

3.2.3. Monsoon Gyre

Approximately once every two years, the low-level circulation in the tropical western North Pacific takes the form of a large low-pressure area (about 2500 km diameter of the outer closed isobar). On the satellite imagery, a persistent band of deep convective cloud extends from the southwest around the eastern periphery in a "fish-hook." An example of such a *monsoon gyre* is shown in Fig. 15. Notice that the sea-level center is north of the normal monsoon trough axis for July through September (prime months of gyre occurrence), and that a high-pressure ridge separates the monsoon gyre from the low-pressure area over Asia.

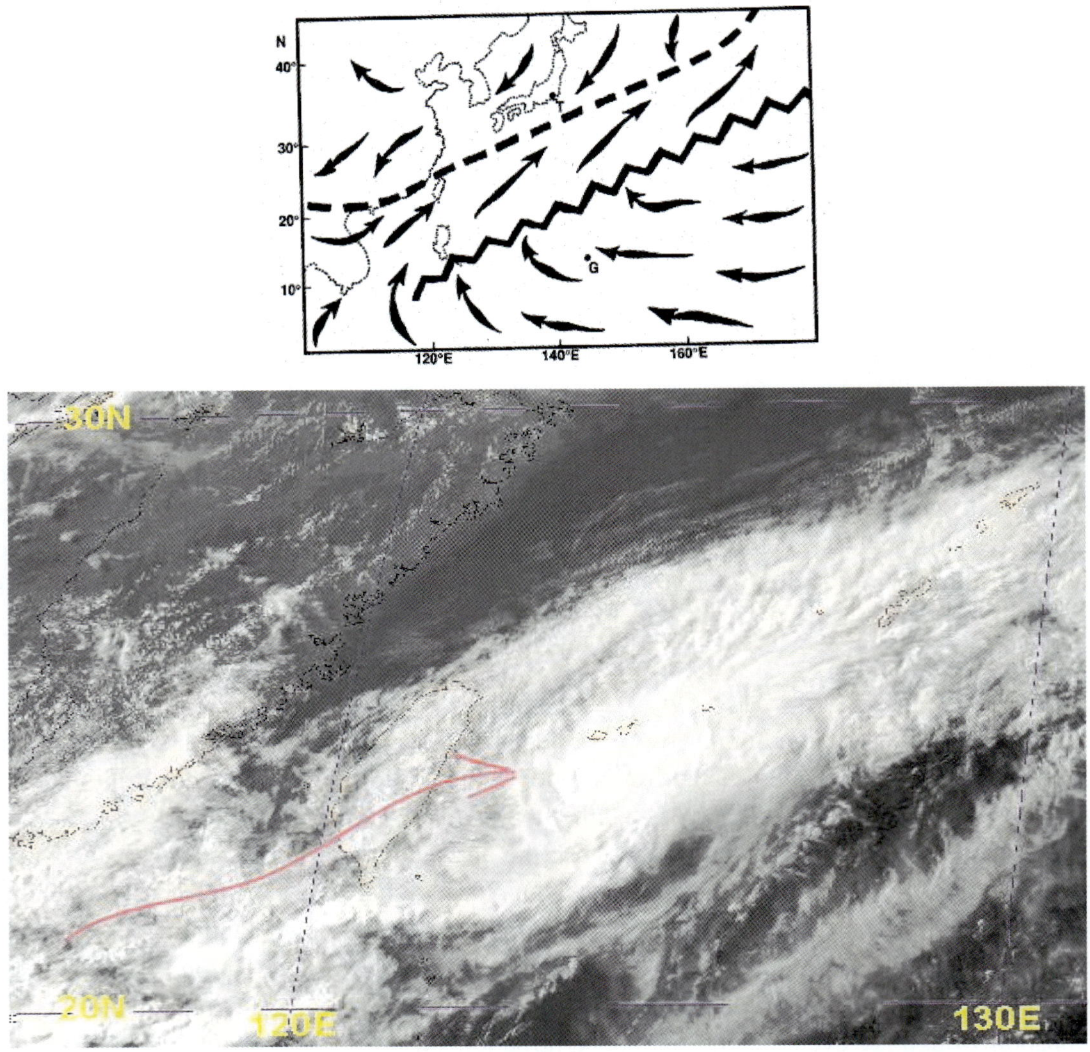

Figure 14. (top) Schematic by Lander (1996) of the active Mei-yu pattern of the low-level circulation of the western North Pacific. Bold dashed line indicates the axis of the Mei-yu/monsoon trough. Arrows indicate wind direction. The locations of Guam (G) and Tokyo (T) are indicated. A ridge of high pressure (zig-zag line) is associated with easterly wind flow throughout the low latitudes. Southwesterly monsoon flow occurs in the South China Sea and extends northeastward to the south of the Mei-yu monsoon trough. When this flow pattern occurs during the summer, tropical cyclones rarely form at low latitude but may form at unusually high latitudes along the trough axis. (bottom) Satellite infrared image at 0730 UTC 31 May 2001 with a cyclonic perturbation on the Mei-yu front that has moved from the South China Sea across Taiwan (arrow).

Sequential genesis of small tropical cyclones tends to occur in conjunction with such monsoon gyres. These tropical cyclones tend to emerge from the downstream end of the peripheral cloud band to the southeast, which is also a region of lower-tropospheric cyclonic vorticity and convergence. In this example (Fig. 15), the midget tropical cyclone Ellie and a Tropical Depression (TD 13) have formed in the southeast quadrant and rotated cyclonically around the gyre. Harr *et al.* (1996b) describe the formation of the midget TS Ofelia in a monsoon gyre during 1993 based on aircraft and satellite data. The August 1991 monsoon gyre in Fig. 15 subsequently was the site of central convection that expanded outward to encompass the gyre wind field. Such a giant tropical cyclone will have a tendency to translate poleward and westward, and thus vacate the location.

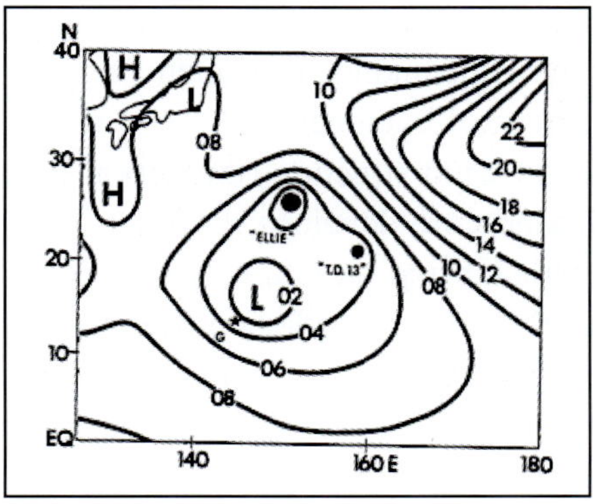

Figure 15. (Left) A monsoon gyre sea-level pressure contours at 2-hPa intervals for 0000 UTC 14 August 1991. Tropical cyclone Ellie and Tropical Depression 13 (T. D. 13) are embedded in the outer circulation of a monsoon gyre whose low pressure center is located near Guam (the star labeled G) (Lander 1994). (Right) Satellite image on 5 August 1999 illustrating the "fish-hook" character of the monsoon gyre cloud pattern with embedded circulations.

In summary, a variety of monsoon circulations may provide favorable environmental conditions for tropical cyclone formation. In all cases, the maximum cyclonic vorticity is near the top of the boundary layer in these warm-core systems, which typically have anticyclonic circulations aloft. Other favorable thermodynamic criteria of sufficiently high SSTs and conditional instability with warm, moist air are generally met because these monsoonal formations are over the low-latitude tropical oceans. Two exceptions are the developments in the central core of a monsoon depression or a monsoon gyre, which are generally regions of drier, subsiding air that is less cloudy than in the peripheral rings or bands of cloud. Although vertical wind shear may be an inhibiting environmental factor, the fact that the other favorable environmental conditions in section 3 are generally met helps to understand why such a large fraction of the global tropical cyclones form in a monsoonal environment.

Three special initial tropical cyclone wind structures are noted depending on the monsoonal environment in which formation occurred. First, the tropical cyclone that forms in a monsoon depression will merge with the outer wind gales and will tend to have a larger wind structure than a tropical cyclone that forms in a shearline monsoon trough (e.g., a reverse-oriented monsoon trough as in Fig. 13a). Second, tropical cyclones that form in the peripheral cloud band of a monsoon gyre tend to have a small wind structure (midget) at the initial time. Third, in the rare circumstance that a central wind core develops in a monsoon gyre and subsumes its large outer circulation, a giant typhoon wind structure is expected. Although these different outer wind structure may evolve as the tropical cyclone moves into a different environment, or experiences physical processes that cause outer wind structure changes, these tendencies for certain initial tropical cyclone sizes are helpful in distinguishing among the variety of tropical cyclone characteristics that are possible in a monsoonal environment.

3.3. *Tropical Cyclone Formation at Eastern End of Monsoon Circulation*

Several studies have suggested that tropical cyclone formation within the monsoon trough circulation is favored at the eastern end. In the idealized schematic of the typical monsoon trough (Fig. 2), a favorable low-level convergence and cyclonic vorticity would be found at the confluence between the equatorial westerlies and the tradewind easterlies. This confluence region is also a favored region of deep convection (low values of outgoing long-wave radiation or cold cloud tops) in the active monsoon cluster composite (Fig. 5a) of Harr and Elsberry (1995). Zehr (1992) suggests a "wind surge" in the equatorial westerlies at the eastern end would be a favored region for tropical cyclone formation as it would enhance convergence against the easterlies. The strengthening equatorial westerlies prior to the development of TY Robyn (Fig. 10) might be interpreted by Zehr as a wind surge. The formation of the midget TS Ofelia was at the eastern end of the monsoon trough where easterly winds were meeting and undercutting the equatorial westerlies (Fig. 16). Thus, it is not surprising that Briegel and Frank (1997) found a significant number of the monsoon trough-related formations in the western North Pacific during 1988-89 were near the eastern end of the trough (Fig. 17). Eleven (17) of the 17 (29) tropical cyclones that formed in the monsoon trough during 1988 (1989) were near the eastern end.

Ritchie (1995) and Holland (1995) have also emphasized the favored eastern-end location in conjunction with a wave-energy accumulation mechanism. They propose that a westward-propagating wave in the easterly trade flow will lead to an accumulation of wave energy at the eastern end of the monsoon trough. Whereas numerous studies (e.g., Wallace 1971) have documented upper-tropospheric waves, differing opinions exist as to the occurrence of waves in the low-level easterlies of the western North Pacific. Recent studies are summarized by Sobel and Bretherton (1999). Although Ritchie (1995) suggested 10-15% of the western North Pacific tropical cyclogenesis cases during 1990-92 could be attributed to easterly waves, Briegel and Frank (1997) found only one case during 1988-89 of an easterly wave that could be related to a monsoon trough-related formation as in Fig. 17a.

Another interesting aspect of these eastern-end formations is that another tropical cyclone tends to form in the wake of a previous cyclone as it moves northwestward along the northwest-southeast oriented monsoon trough. This next tropical cyclone tends to form near the re-established confluence region between the equatorial westerlies and tradewind easterlies as in Fig. 17a. An example of sequential tropical cyclone formation during the Tropical Cyclone Motion (TCM-90) field experiment is given in Fig. 18. Notice the cyclonic circulation in the right-central region of the 9/3 analysis grows and moves northwestward through the 9/10 analysis and then leaves the analysis domain. A new cyclonic circulation becomes evident in the lower-right region of the 9/7 analysis where equatorial westerlies have become re-established to the west and easterlies are present on the eastern edge of the analysis (similar to the monsoon depression in Fig. 10). This new cyclonic circulation grows in areal extent and translates to the northwest throughout the remainder of the period. Formation of a third cyclonic circulation is suggested at the bottom right area of the 9/16 analysis.

Chang *et al.* (1996) documented northwestward-propagating wave patterns with a period of 8-9 days when the western North Pacific monsoon trough had its typical Northwest-Southeast orientation. These waves had a zonal wavelength of 28° longitude, had a warm core, and had a poleward moisture flux with a trailing anticyclone that was dry. Thus, Chang *et al.* suggested these waves might be the initial disturbances from which tropical cyclones may form. Takayabu and Nitta (1993) had earlier documented "tropical depression-type" disturbances in an off-equatorial wave train that extended northwestward from the Equator. They proposed that mixed Rossby-gravity waves propagating westward along the Equator might trigger the TD-type disturbances. Because the waves in the Chang *et al.* (1996) study had a period of 8-9 days, and those in the Takayabu and Nitta study had a period of 3-5 days, these may not be the same waves.

Kuo *et al.* (2001) have proposed that nonlinear barotropic processes on a beta plane may explain both why a tropical cyclone tends to form at the eastern end of the monsoon trough and tends to trigger another cyclonic development about 8 days later in the convergence area between equatorial westerlies and tradewind easterlies. Kuo *et al.* use the idealized background flow with a convergence zone between westerlies and easterlies (Fig. 17b) to explain some of the physics, and then utilize a more realistic monsoonal flow (see later Fig. 20a) to illustrate the proposed evolution.

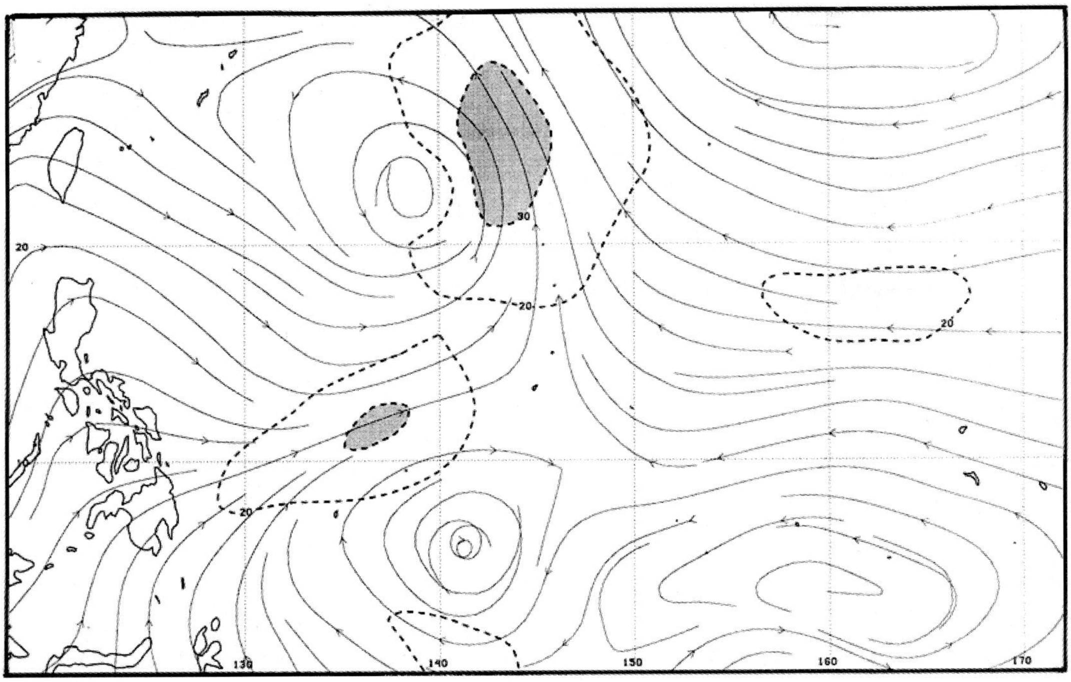

Figure 16. Formation region of the midget TS Ofelia at the eastern end of a monsoon trough where the 850 mb tradewind easterlies were meeting the southwesterly flow in the gyre (Harr *et al.* 1996b). TY Nestor to the north near 22°N, 138°E had previously formed in a similar convergence region.

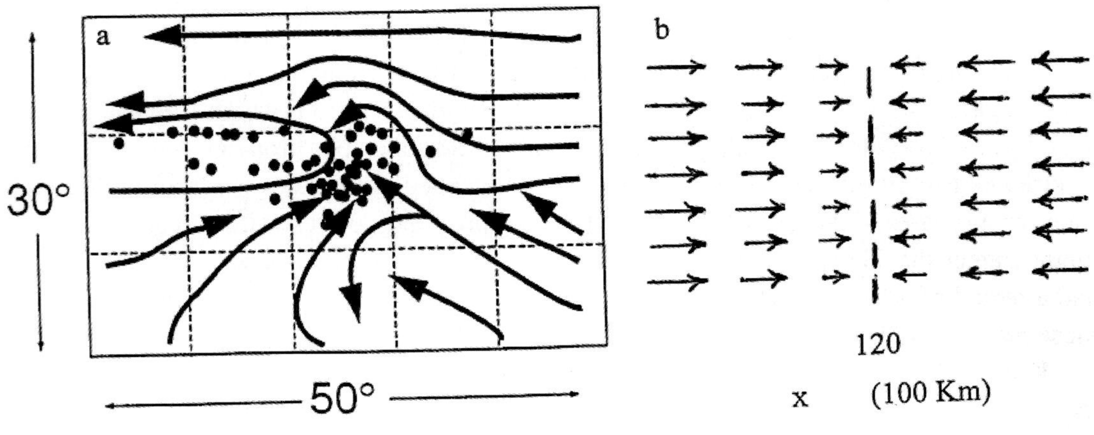

Figure 17. (a) Locations relative to an idealized monsoon trough of the monsoon-related tropical cyclogenesis during 1998-89 as defined by Briegel and Frank (1997). (b) Schematic of idealized convergence of zonal westerlies and easterlies at x = 12,000 km in the model of Kuo *et al.* (2001).

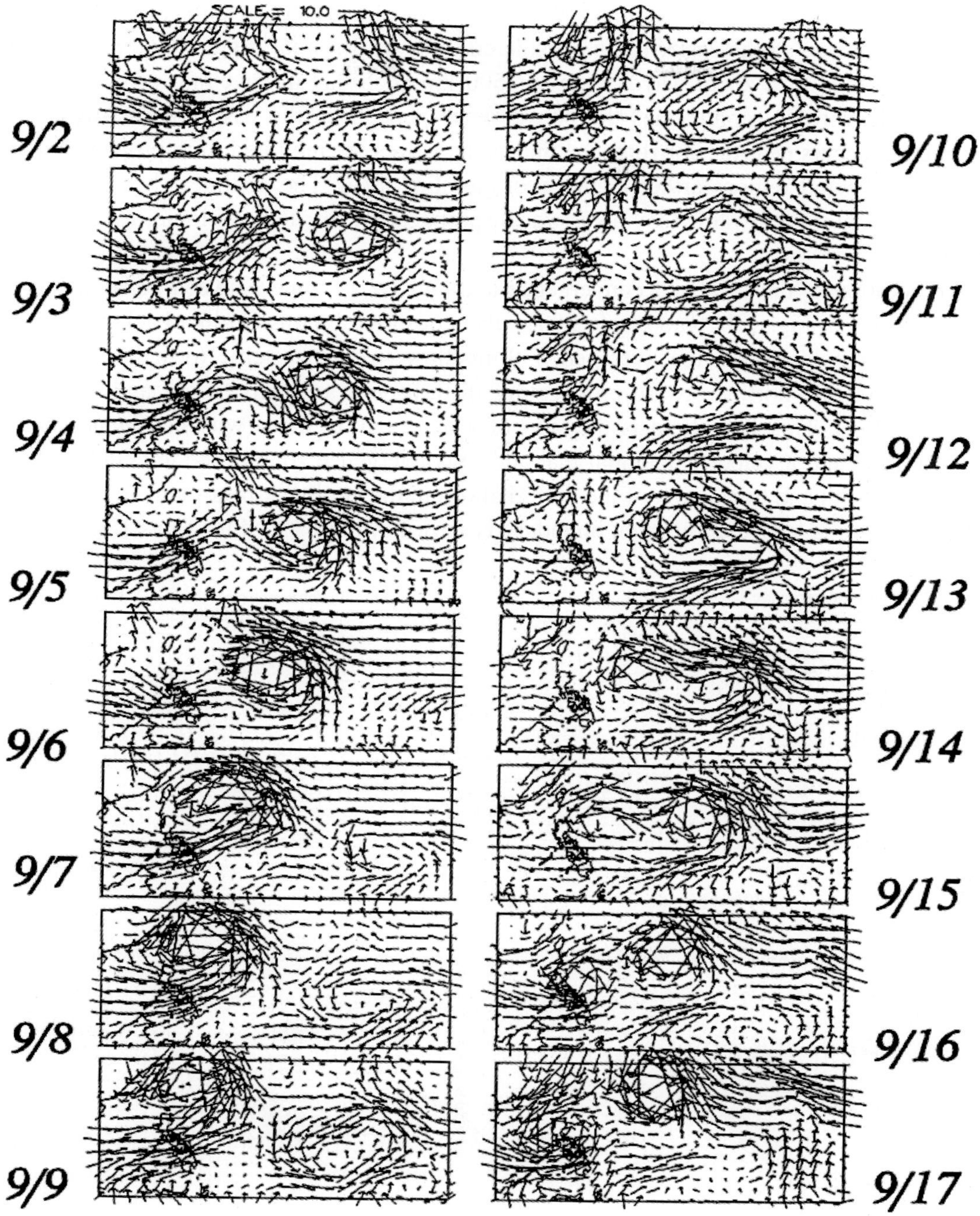

Figure 18. Daily analyses of 850-mb winds during the Tropical Cyclone Motion (TCM-90) field experiment during 2 September (labeled 9/2) – 17 September 1990 (Chang *et al.* 1996).

As in the Ritchie (1995), Holland (1995), and Sobel and Bretherton (2000) studies, the presence of waves in the easterlies is considered to be an essential trigger. Kuo *et al.* (2001) initiate waves with a zonal wavelength of 2500 km such that a wavetrain of cyclonic and anticyclonic vorticity approaches the convergence zone from the east. An east-west scale contraction occurs near the convergence line as the waves are inhibited from propagating into the westerly wind region farther to the west (see Fig. 17b). The nonlinear beta effect then leads to a northwestward (southwestward) propagation of the cyclonic (anticyclonic) circulations.

In addition to the idealized convergence and confluence zone between zonal westerlies and easterlies in Fig. 17b, Kuo *et al.* (2001) also construct a monsoon-like background flow (Fig. 19a) that resembles the averaged 850-mb flow in the western North Pacific during periods when the northwestward-propagating disturbances are active (Lau and Lau 1990, their Fig. 17b). This flow field is the sum of a deformation field (with confluence but no convergence) and a velocity potential that adds a convergence zone in the left center of Fig. 19a. The growth and translation of the perturbations in the nonlinear beta-plane model when a 2500 km wave is forced at the right boundary is shown in Figs. 19b-d for Days 18-26. As the

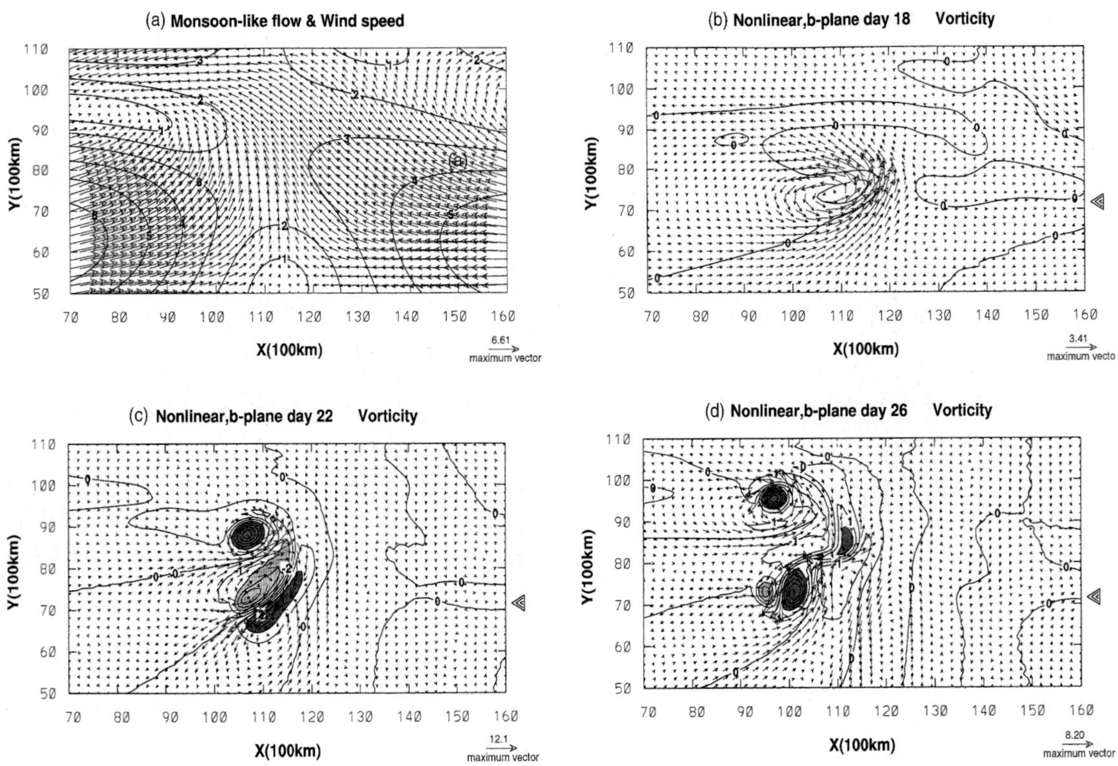

Figure 19. (a) Wind vectors of the monsoon-like background flow with isopleths of wind speed (m s^{-1}), and forecast winds and vorticities at (b) Day 18, (c) Day 22, and (d) Day 26 for the Kuo *et al.* (2001) nonlinear beta-plane calculation with a forcing zonal wavelength of 2500 km. The vorticity isopleth interval is 10^{-5} s^{-1}, with heavy shading for values > 2 x 10^{-5} s^{-1} and light shading for values < -2 x 10^{-5} s^{-1}. The central latitude of the east boundary forcing (right edge arrow) is at y = 7200 km.

wave approaches the convergence zone (Fig. 19b), the cyclonic background flow (Fig. 19a) adds to the wave vorticity to produce a cyclonic vortex. As in the nonlinear, beta-plane calculation, the cyclonic vortex at Day 22 is propagating westward and poleward with a Rossby dispersion of a trailing anticyclone and cyclone couplet to the southeast.

Finally, the cyclonic vortex in the trailing Rossby wave wake is intensifying at Day 26 (Fig. 19d), which is eight days after the initial cyclonic vortex is formed (Fig. 19b). Kuo *et al.* conclude that successive formations of new disturbances may thus be explained by barotropic processes. Kuo *et al.* demonstrate by removing the imposed convergence component and leaving just the confluence of easterlies and westerlies that actual convergence is necessary for the cyclogenesis at the eastern end of the monsoon trough. Although no diabatic effects are included in this model, this study suggests the early stages of tropical cyclogenesis at the eastern end of the monsoon trough may have an important contribution from nonlinear barotropic processes.

Whereas the approach of a 2500-km wave in the easterlies is essential in this model, it should be noted that many cases of multiple tropical cyclone outbreaks along the monsoon trough occur in which a wave in the easterlies is not evident. Indeed, real-time monitoring of the outgoing long-wave radiation (OLR) with filtering of the signal to highlight specific regions of the wavenumber-frequency domain indicate that mixed Rossby-gravity waves may be the result of the formation rather than the cause.(see http://www.bom.gov.au/bmrc/clfor/cfstaff/matw/maproom/OLR_modes/). Whereas waves in the easterlies might be present upstream of the western North Pacific, evidence has not been found for such a contribution with the tropical cyclone formations in the monsoon troughs in the Southern Hemisphere.

3.4. *Concluding Remarks*

The key conclusion from this discussion is that the East Asia monsoon circulation has an important contribution to where and when tropical cyclone formation will occur over the western North Pacific. The seasonal displacements of the East Asia monsoon circulations are accompanied by displacements of where monsoon trough-related tropical cyclone formations occur. Intraseasonal variations of active and inactive monsoon trough conditions are also favorable and unfavorable formation conditions for tropical cyclone formation. Some theoretical studies of how the monsoon trough instabilities contribute to the cyclone formations are progressing from barotropic models to more comprehensive physical representations. Few studies with *in situ* observations of monsoon depressions leading to tropical cyclone formation have been accomplished. The Harr *et al.* (1996a, b) studies suggest a key role for mesoscale convective systems in determining when and how a monsoon depression will be converted to a tropical cyclone. Clearly, more such observational studies, and accompanying numerical model predictions, are necessary for progress in understanding the specific conditions for formation. The recent study by Kuo *et al.* (2001) has provided some valuable insights on tropical cyclone formation at the eastern end of the monsoon trough. Their barotropic model simulations need to be followed with more complete physical representations in baroclinic models, and with real-data prediction studies to improve our ability to forecast tropical cyclone formation.

As new understandings have been achieved in the role of East Asia monsoon circulations on tropical cyclone formation, attempts have been made to apply these ideas to other tropical basins. The western North Pacific conceptual models would seen to have the most applicability in the Southern Hemisphere, and especially in the eastern South Indian Ocean and western South Pacific. Although the Australian monsoon effects are not so extensive as the East Asian monsoon, a number of analogous circulation occur that favor tropical cyclone formation. Clearly the equatorial westerly wind bursts that contribute to twin tropical cyclone formations (section 3.a.5) apply to both hemispheres. One special semi-permanent circulation feature is the South Pacific Convergence Zone (SPCZ). Some of the features of the reverse-oriented monsoon trough of the western North Pacific would seem to apply to tropical cyclone formation in the SPCZ, but the upper-tropospheric troughs in the Southern Hemisphere have an important modulating effect (Bannister *et al.* 1997; Reader *et al.* 1999, 2000).

Some of the East Asia monsoon trough effects also would seem to apply in the North Indian Ocean. It is well known that tropical cyclone formations in that region are confined to the pre- and post-monsoon periods since the vertical wind shear is so large during the monsoon as to prevent tropical cyclone formations. As the monsoon trough moves poleward from near- equatorial latitudes in the pre-monsoon period, favorable environmental conditions for tropical cyclone formation are present. Likewise, favorable conditions are again established when the monsoon trough recedes to near-equatorial latitudes in the fall. An evaluation of how these monsoonal circulations affect tropical cyclone motion in the North Indian Ocean is in progress at the Naval Postgraduate School.

Application of the East Asia monsoon circulation conceptual models to the eastern North Pacific and the Atlantic tropical cyclone formations has aroused more controversy on the Tropical Storms Bulletin Board. The eastern North Pacific certainly has a monsoon trough with cross-equatorial flow to the south. The controversy is the extent to which this monsoonal type ITCZ contributes to the tropical cyclone formation compared with the wave disturbances in the easterlies, or the topographically induced flows downstream from Mexico and Central America. Even more controversy is generated by those attempting to apply East Asian monsoon concepts over Africa to explain African wave generation and maintenance and the downstream tropical cyclone formations over the Atlantic. Whereas some aspects of the environmental flows are similar, more attention needs to be given to the differences if we are to achieve a better understanding of how monsoon circulations around the globe affect tropical cyclone formation.

4. Tropical Cyclone Motion in a Monsoonal Environment

4.1. *Some Basic Tropical Cyclone Motion Concepts*

Understanding of the tropical cyclone motion has been advanced considerably in the last decade (Elsberry 1995). One of the important advancements is a better understanding of the relative roles of the environment (first order) and of the asymmetric circulations (second order) in tropical cyclone motion. The focus in this section is how the East Asia monsoon circulations affect the environmental flow around western North Pacific tropical cyclones. A second important advancement is the recognition that the tropical cyclone can in certain circumstances affect its environment in such a way as to change its motion. A few examples of such tropical cyclone-environment interactions in the monsoon environment will be given.

Many definitions of the environmental flow affecting tropical cyclone motion have been proposed (Elsberry 1995). Some areal averaging or horizontal filtering is clearly necessary to remove the tropical cyclone circulation. The appropriate layer over which to average in the vertical to define a steering flow is dependent on the tropical cyclone intensity (Elsberry 1995). Whereas averaging just over a lower tropospheric layer is appropriate for tropical depressions, the layer average should include middle-tropospheric winds for a tropical storm, and the upper- tropospheric winds to 300 mb for typhoons. For many qualitative (non-quantitative) purposes of defining the important synoptic circulations contributing to the environmental steering, it is adequate to use the 500 mb streamlines and isotachs as a representative level. Knowledge of the recent track positions and that the isotach maximum will be to the right (Northern Hemisphere) of the steering flow are also important factors in establishing qualitatively the steering flow from operational analyses and dynamical model predictions.

4.2. *Monsoon Circulations and the Tropical Cyclone Environment*

One example of the environmental flow affecting tropical cyclone motion is the anomalous 700-mb winds associated with active and inactive monsoon circulations in Fig. 5. Just as these anomalies affect tropical cyclone formation locations, the subsequent tracks will also differ. For example, a cyclone forming in the active monsoon trough (Fig. 5a) will have a strong westward steering flow that is more likely to advect the cyclone toward China. Recall that the only favorable formation location when the monsoon is inactive (Fig. 5b) is off the central China coast. However, the steering flow will then be more poleward with a preferred track toward Japan.

Carr and Elsberry (1994) and Carr *et al.* (1995) have defined the environment structure around western North Pacific cyclones in terms of synoptic patterns and regions (Fig. 20, updated by Carr and Elsberry 1999). Three of these synoptic patterns (Standard, Poleward, and Midlatitude) have been found in the Southern Hemisphere, eastern and central North Pacific, and Atlantic basins. The unique synoptic pattern in the western North Pacific is the monsoon Gyre (G) pattern (Fig. 20, right-center). It is significant that only three common synoptic patterns plus one special pattern in each of these tropical cyclone basins is adequate to define the environment structure in more than 4500 cases examined so far.

The synoptic regions defined within the synoptic patterns in Fig. 20 are associated with characteristic track motions (Carr *et al.* 1995). Consequently, a change from one synoptic pattern/region combination to another pattern/region combination necessarily involves a track change. Establishing the pattern/region from the present synoptic analysis is then the first step in the track forecast process because that specifies the synoptic circulations that are determining the steering flow over the tropical cyclone. The track forecast problem then can be cast in terms of whether the same synoptic circulations will continue to determine the steering flow, or whether the environment structure will change and a track change will be occurring.

The monsoon circulations primarily affect the three synoptic patterns in the lower half of Fig. 20, and each of these patterns will now be discussed. The Standard (S) synoptic pattern applies for approximately 50% of the cases examined (Carr *et al.* 1995). Although for convenience the monsoon trough is illustrated as being east-west oriented in Fig. 20, the pattern may be re-oriented relative to the typical west-northwest to east-southeast monsoon trough orientation as appropriate to the daily situation. The key regions relative to the monsoon trough are the Equatorial Westerlies (EW) on the equatorward side, in which case the steering flow is eastward, and the Tropical Easterlies (TE) region, which has a westward steering flow. Tropical cyclones closer to the col region between the two subtropical anticyclone cells will be in either the Poleward Flow (PF) region with a northwestward steering flow, or the Equatorward Flow (EF) region with a south of west steering flow.

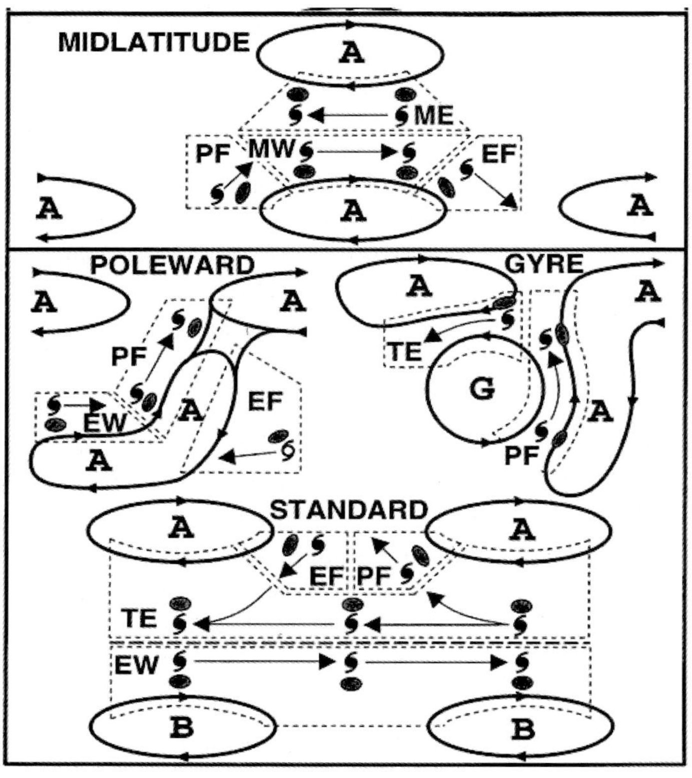

Figure 20. Synoptic patterns (Standard, Poleward, Gyre, and Midlatitude) and synoptic regions (Equatorial Westerlies-EW; Tropical Easterlies-TE; Poleward Flow-PF; Equatorward Flow-EF; Midlatitude Easterlies – ME; and Midlatitude Westerlies-MW) as defined by Carr and Elsberry (1999) for western North Pacific tropical cyclones.

The Poleward (P) synoptic pattern in Fig. 20 is a manifestation of the reverse-oriented monsoon trough discussed in section 3.b.1. The tropical cyclones in the EW region of the P pattern will have a predominantly eastward steering current. A more common occurrence is for the cyclone to form and move poleward in the PF region. The occurrence of poleward-moving tropical cyclones in the P/PF pattern/region is much more frequent in the western North Pacific than in the other basins mentioned above. Although the reasons for this high frequency are unknown, it must be related to the East Asia monsoon in some way. That is, the reverse-oriented monsoon trough must be a stable mode of the East Asia monsoon circulation, because in some years (e.g., 1998) this synoptic pattern seems to be quite persistent. In those years, tropical cyclone formation is favored in the extreme western North Pacific and not in the normal low-latitude monsoon trough locations, and the tracks are more poleward and shorter.

As mentioned above, the Gyre (G) synoptic pattern in Fig. 20 is unique to the western North Pacific. As described in section 3.b.3, small tropical cyclones tend to form in the southeast quadrant of the gyre. Since the large-scale monsoon gyre circulation is the steering flow for these cyclones, they tend to more poleward on a cyclonically curved path in the G/PF pattern/region. A bifurcation of tracks occurs in the northeast quadrant of the gyre. Smaller tropical cyclones continuing to have the gyre circulation as a steering flow will move westward in the TE region. Other cyclones more under the influence of the anticyclonic circulation to the east will move through the subtropical ridge into the Midlatitude (M) synoptic pattern in Fig. 20, and thus recurve into the midlatitudes.

In summary, the various modes of the monsoon circulation over the western North Pacific as represented by the Standard (typical monsoon trough), Poleward (reverse-oriented monsoon trough), and Gyre synoptic patterns can lead to diverse tropical cyclone tracks.

4.3. *Monsoon-Related Tropical Cyclone-Environment Transformations*

Carr and Elsberry (1994, 1999) describe a number of tropical cyclone-environment interactions by which the environment structure (synoptic pattern and/or region) is changed, and thus the track is changed. For example, a reverse-oriented monsoon trough (see section 3.b.1) may be formed if two (or three) tropical cyclones form in an east-west oriented trough and the eastern cyclone moves more rapidly poleward (e.g., TY Kirogi in Fig. 6) to establish a southwest-northeast orientation of the low-pressure trough. If the trailing peripheral anticyclones to the southeast of each cyclone also merge, then a Poleward (P) synoptic pattern as in Fig. 20 will be formed. The track change will then be toward the pole and eastward along the reverse-oriented monsoon trough.

Carr and Elsberry (1995) have described another transformation called monsoon gyre-tropical cyclone interaction that is accompanied by a sudden poleward track change. The precursor to this event is a monsoon surge of gale-force winds and deep convection along the eastern periphery of the monsoon gyre. Carr and Elsberry (1995) used a non-divergent, barotropic model to test the hypothesis that the sudden track change is initiated by a binary interaction of the tropical cyclone and monsoon gyre. Depending on the positions of the tropical cyclone in the northern (Fig. 21a) or the southern (Fig. 21b) quadrant of the monsoon gyre, the track changes may have a cyclonic loop or just a period of slowed westward motion before rapidly turning poleward. Carr and Elsberry (1995) show tracks and Naval Operational Global Atmospheric Prediction System analyses that resemble the barotropic model simulations. They conclude that the observed track changes are manifestations of the interaction of a monsoon gyre and a nearby tropical cyclone, and that barotropic processes provide a first-order description of the interaction.

4.4. *Requirements for Improved Tropical Cyclone Tracks*

The descriptions above demonstrate that the East Asia monsoon circulation has an important role in the track of the tropical cyclone. Variations in the monsoon circulation thus will be revealed in tropical cyclone tracks on interannual, seasonal, intraseasonal, and synoptic time scales. What are the requirements then to improve tropical cyclone track warnings to say 5 days?

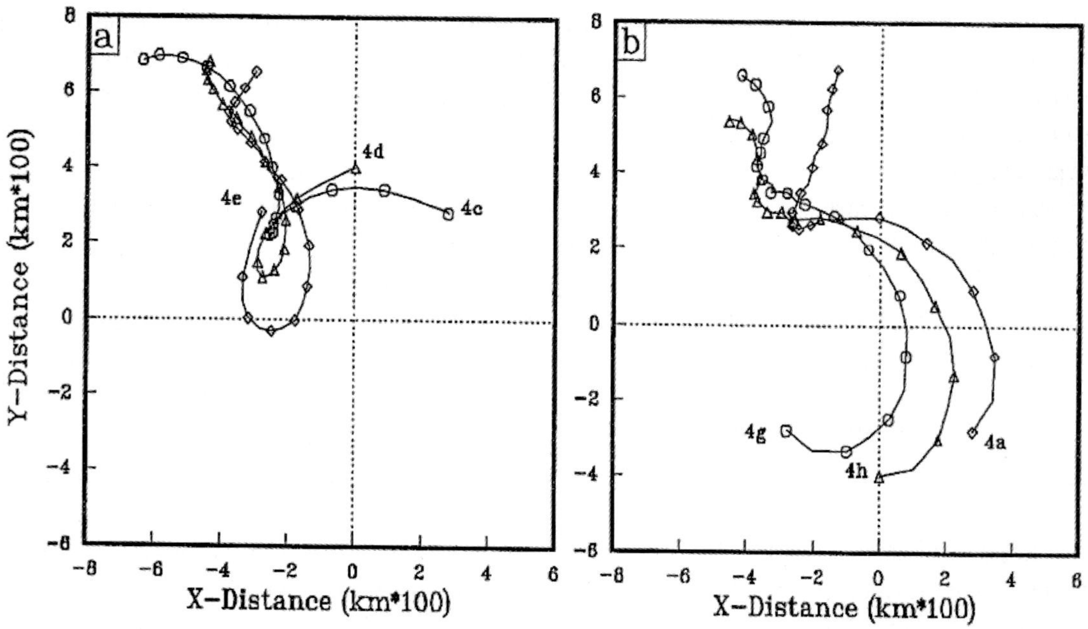

Figure 21. Simulated tropical cyclone tracks in the barotropic model of Carr and Elsberry (1995) with a monsoon gyre at the origin and three initial tropical cyclone positions 400 km from the gyre center in the (a) northern and (b) southern quadrants of the gyre. Track positions are each 6 h to 96 h.

Since the environmental flow is the first-order effect on tropical cyclone motion, observing, analyzing, and predicting the environmental flow is of highest importance. In many developing nations, the rawinsonde launches have been curtailed or abandoned because of the high cost. Some of this degradation in observational coverage has been offset by new, or better use of existing, satellite systems. For example, the water-vapor winds from animation of geostationary satellite imagery have supplemented the data coverage in regions with no cumulus or cirrus targets for the satellite cloud-drift winds. The Quikscat has provided much improved coverage of the surface wind vectors compared to earlier scatterometer systems. The numerical weather prediction centers have also developed better techniques for extracting thermal and moisture information in the tropics from the polar-orbiting satellite radiances. A number of planned satellite instrument deployments is expected to increase the data flow by a factor of ten or more.

The importance of the ocean and land fluxes for the monsoon circulation prediction is well known. Thus, the prediction models must accurately represent these physical processes over both East Asia and the adjacent oceans. Given the rugged orography and the complex vegetation types over East Asia, this will not be an easy task.

The inclusion of the location, structure, and recent motion of the tropical cyclone in the initial conditions of the numerical models was an important achievement in producing viable track forecasts (Elsberry 1995). A similar model initial specification for monsoon depressions, which are so important in East Asia, has not been achieved. Use of the tropical cyclone structure specification is not satisfactory since gale-force winds on the periphery of the monsoon depressions are not represented by synthetic tropical cyclone observations. Without adequate observations of these gale winds from other sources, the structure of the monsoon depression vortex will not be adequately specified, which may lead to inaccurate track forecasts when the system is transformed into a tropical cyclone.

Another opportunity for improving tropical cyclone warnings is to provide the forecasters with better tools for utilizing the model track guidance. This approach has been quite successful at the JTWC, who have successfully implemented the Systematic Approach Forecasting Aid (SAFA) developed at the Naval Postgraduate School (Carr *et al.* 2001). The JTWC 72-h track errors have been reduced from about 300 n mi (555 km) during the 1990s to 163 n mi (302 km) during the 2002 typhoon season. The SAFA is an information management system that systematically guides the forecaster through an evaluation of the dynamical model track guidance. The goal is to detect likely erroneous track forecasts based on conceptual models developed from careful study of a historical data base of cases with large 72-h track errors. By knowing when to accept the consensus of the dynamical model guidance, and when value can be added by a selective consensus after eliminating erroneous guidance, the JTWC forecasters have achieved a quantum improvement in forecasting.

5. Concluding Remarks

The objective of this chapter has been to describe recent understandings of how the East Asia monsoon circulation affects both tropical cyclone formation and motion over the western North Pacific. These influences occur on interannual, seasonal, intraseasonal, and synoptic time scales. As improvements are made in East Asia monsoon observations, analysis, and prediction, these improvements will have a positive downstream impact on the western North Pacific tropical cyclone prediction and warnings. Improved tropical cyclone warnings will provide opportunities to reduce the destructive impacts of these cyclones on the economy and lives of people in East Asia.

Acknowledgments

This review has benefited from many years of collaboration with Les Carr, Patrick Harr, and Mark Boothe at the Department of Meteorology, Naval Postgraduate School. This research has been supported by the U. S. Office of Naval Research and Space and Naval Warfare Command. Patrick Harr, Mark Boothe, Kevin Cheung, Bill Frank, and an anonymous reviewer provided helpful comments on the manuscript, which was skillfully prepared by Mrs. Penny Jones.

References

Atkinson, G. D., 1971: *Forecaster's guide to tropical meteorology.* Tech. Rep. 240, United States Air Force, Air Weather Service, 350 pp.

Bannister, A. J., M. A. Boothe, L. E. Carr, III, and R. L. Elsberry, 1997: *Southern Hemisphere application of the Systematic Approach to tropical cyclone track forecasting. Part I. Environment structure characteristics.* Tech. Rep. NPS-MR-98-001, Naval Postgraduate School, Monterey, CA 93943-5114, 96 pp.

______, ______, ______, and ______, 1998: *Southern Hemisphere application of the Systematic Approach to tropical cyclone track forecasting. Part II. Climatology and refinement of meteorological knowledge base.* Tech. Rep. NPS-MR-98-004, Naval Postgraduate School, Monterey, CA 93943-5114, 69 pp.

Bister, M., and K. Emanuel, 1997: The genesis of Hurricane Guillermo: TEXMEX analyses and a modeling study. *Mon. Wea. Rev.*, **125**, 2662-2682.

Briegel, L. M., and W. M. Frank, 1997: Large-scale influences on tropical cyclogenesis in the western North Pacific. *Mon. Wea. Rev.*, **125**, 1397-1413.

Carr, L. E., III, M. A. Boothe, S. R. White, C. S. Kent, and R. L. Elsberry, 1995: *Systematic and integrated approach to tropical cyclone track forecasting. Part II. Climatology, reproducibility, and refinement of meteorological data base.* Tech. Rep. NPS-MR-95-001, Naval Postgraduate School, Monterey, CA 93943-5114, 96 pp.

______, and R. L. Elsberry, 1994: *Systematic and integrated approach to tropical cyclone track forecasting. Part I. Approach overview and description of meteorological basis.* Tech. Rep. NPS-MR-94-002, Naval Postgraduate School, Monterey, CA 93943-5114, 273 pp.

______, and ______, 1995: Monsoonal interactions leading to sudden tropical cyclone track changes. *Mon. Wea. Rev.*, **123**, 265-289.

______, and ______, 1999: *Systematic and integrated approach to tropical cyclone track forecasting. Part III. Traits knowledge base for JTWC track forecast models in the western North Pacific.* Tech. Rep. NPS-MR-99-002, Naval Postgraduate School, Monterey, CA 93943-5114, 227 pp.

______, ______, and J. E. Peak, 2001: Beta test of the Systematic Approach expert system prototype as a tropical cyclone track forecasting aid. *Wea. Forecasting*, **16**, 355-368.

______, and R. T. Williams, 1989: Barotropic vortex stability to perturbations from axisymmetry. *J. Atmos. Sci.*, **46**, 3177-3191.

Chang, C.-P., J.-M. Chen, P. A. Harr, and L. E. Carr, III, 1996: Northwestward-propagating wave patterns over the tropical western North Pacific during summer. *Mon. Wea. Rev.*, **124**, 2245-2266.

Elsberry, R. L., 1995: Tropical cyclone motion. *Chap. 4, Global Perspectives on Tropical Cyclones.* Tech. Doc. WMO/TD-No. 693, World Meteorological Organization, Geneva, Switzerland, 106-197.

______, and R. Jeffries, 1996: Vertical wind shear influences on tropical cyclone formation and intensification during TCM-92 and TCM-93. *Mon. Wea. Rev.*, **124**, 1374-1387.

Ferreira, R. N., and W. H. Schubert, 1997: Barotropic aspects of ITCZ breakdown. *J. Atmos. Sci.*, **54**, 261-285.

Gray, W. M., 1968: Global view of the origin of tropical disturbances and storms. *Mon. Wea. Rev.*, **96**, 669-700.

______, 1975: *Tropical cyclone genesis*. Dept. of Atmos. Sci., Paper No. 232, Colorado State University, Ft. Collins, CO, 121 pp.

______, 1998: The formation of tropical cyclones. *Meteor. Atmos. Phys.*, **67**, 37-69.

Guinn, T. A., and W. H. Schubert, 1993: Hurricane spiral bands. *J. Atmos. Sci.*, **50**, 3380-3408.

Harr, P. A., and R. L. Elserry, 1995: Large-scale circulation variability over the tropical western North Pacific. Part I: Spatial patterns and tropical cyclone characteristics. *Mon. Wea. Rev.*, **123**, 1225-1246.

______, ______, and J. C.-L. Chan, 1996: Transformation of a large monsoon depression to a tropical storm during TCM-93. *Mon. Wea. Rev.*, **124**, 2625-2643.

Harr, P. A., M. S. Kalafsky, and R. L. Elsberry, 1996b: Environmental conditions prior to formation of a midget tropical cyclone during TCM-93. *Mon. Wea. Rev.*, **124**, 1693-1710.

Hartten, L. M., 1996: Synoptic setting of westerly wind bursts. *J. Geophys. Res.*, **101**, 16997-17019.

Holland, G. J., 1995: Scale interaction in the western Pacific monsoon. *Meteor. Atmos. Phys.*, **56**, 57-79.

JTWC, 1994: *Annual Tropical Cyclone Report*, Joint Typhoon Warning Center, 425 Luapele Road, Pearl Harbor, HI 96860-3130, 337 pp.

Kuo, H.-C., J.-H. Chen, R. T. Williams, and C.-P. Chang: 2001: Rossby waves in zonally opposing mean flow: Behavior in Northwest Pacific summer monsoon. *J. Atmos. Sci.*, **58**, 1035-1050.

Lander, M. A., 1990: Evolution of the cloud pattern during the formation of tropical cyclone twins symmetrical with respect to the equator. *Mon. Wea. Rev.*, **118**, 1194-1202.

______, 1994: Description of a monsoon gyre and its effects on the tropical cyclones in the western North Pacific during August 1991. *Wea. Forecasting*, **9**, 650-654.

______, 1996: Specific tropical cyclone track types and unusual tropical cyclone motion associated with a reverse-oriented monsoon trough in the western North Pacific. *Wea. Forecasting*, **11**, 170-186.

Lau, K.-H., and N.-C. Lau, 1990: Observed structure and propagation characteristics of tropical summertime synoptic scale disturbance. *Mon. Wea. Rev.*, **118**, 1888-1913.

Liebmann, B., and H. H. Hendon, 1990: Synoptic-scale disturbances near the equator. *J. Atmos. Sci.*, **47**, 1463-1479.

______, ______, and J. D. Glick, 1994: The relationship between tropical cyclones of the western Pacific and Indian Oceans and the Madden-Julian oscillation. *J. Meteor. Soc. Japan*, **72**, 401-412.

McBride, J. L., 1995: Tropical cyclone formation. *Chap. 3, Global Perspectives on Tropical Cyclones*, Tech. Doc. WMO/TD No. 693, World Meteorological Organization, Geneva, Switzerland, 63-105.

______, and R. M. Zehr, 1981: Observational analysis of tropical cyclone formation. Part II: Comparison of non-developing and developing systems. *J. Atmos. Sci.*, **38**, 1132-1151.

Palmen, E., 1956: A review of knowledge on the formation and development of tropical cyclones. *Proc. Trop. Cyclone Symp.*, Bureau of Meteorology, Brisbane, Australia, 213-232.

Reader, G., M. A. Boothe, R. L. Elsberry, and L. E. Carr, III, 1999: Southern Hemisphere application of the Systematic Approach to tropical cyclone track forecasting. Part III: Updated environmental structure characteristics. Tech. Rep. NPS-MR-00-004, Naval Postgraduate School, Monterey, CA 93943-5114, 73 pp.

______, ______, ______, and ______, 2000: Southern Hemisphere application of the Systematic Approach to tropical cyclone track forecasting: Part IV: Sources of large track errors by dynamical models. Tech. Rep. NPS-MR-00-004, Naval Postgraduate School, Monterey, CA 93943-5114, 51 pp.

Riehl, H., 1954: *Tropical meteorology*. McGraw-Hill, NY, 392 pp.

Ritchie, E. A., 1995: Mesoscale aspects of tropical cyclone formation. Ph.D. dissertation, Monash University, 167 pp.

Simpson, R. H., N. Frank, D. Schideler, and H. M. Johnson, 1969: Atlantic tropical disturbances of 1968. *Mon. Wea. Rev.*, **97**, 240-255.

Smith, G. B., and M. T. Montgomery, 1995: Vortex axisymmetrization: Dependence on azimuthal wavenumber or asymmetric radial structure changes. *Quart. J. Roy. Meteor. Soc.*, **121**, 1615-1650.

Sobel, A. H., and C. S. Bretherton, 1999: Development of synoptic-scale disturbances over the summertime Northwest Pacific. *J. Atmos. Sci.*, **56**, 3106-3127.

Takayabu, Y. N., and T. Nitta, 1993: 3-5 day period disturbances coupled with convection over the tropical Pacific Ocean. *J. Meteor. Soc. Japan*, **71**, 221-246.

Wallace, J. M., 1971: Spectral studies of tropospheric wave disturbances in the tropical western Pacific. *Rev. Geophys. Space Phys.*, **9**, 557-611.

Webb, B. H., 1996: Evaluations of the Northwest Pacific tropical cyclone track forecast difficulty and skill as a function of environmental structure. M. S. Thesis, Naval Postgraduate School, Monterey, CA 93943-5114, 100 pp.

Zehr, R. M., 1992: Tropical cyclogenesis in the western North Pacific. NOAA Tech. Rep. NESDIS 61, Department of Commerce, Washington, DC, 181 pp.

14. FORMATION OF THE SUMMERTIME SUBTROPICAL ANTICYCLONES

GUOXIONG WU, YIMIN LIU AND PING LIU
State Key Lab of Atmospheric Sciences and Geophysical Fluid Dynamics (LASG)
Institute of Atmospheric Physics
Chinese Academy of Sciences
Beijing, 100029 China
E-mail: gxwu@lasg.iap.ac.cn, lym@lasg.iap.ac.cn, liup@lasg.iap.ac.cn

The monthly mean data set of the NCEP/NCAR reanalysis is employed to investigate the relation between the subtropical anticyclone and diabatic heating. Criteria for defining the location and intensity of the zonal mean subtropical anticyclone are given to study its characteristics. Comparison between the Hadley circulation and the subtropical anticyclone is made. Results show that the maximum convergence of meridional mass flux occurs in the subtropics, resulting in the formation of the subtropical anticyclone. In the free atmosphere, the ridgeline of the subtropical anticyclone deviates completely from the sinking arm of the Hadley cell, with the former being located equatorward of the latter. Due to friction, the subtropical anticyclone in the planetary boundary layer coincides with the sinking arm of the Hadley cell that extends vertically from the planetary boundary layer to tropopause. It is stressed that either in the free atmosphere or in the planetary boundary layer, descent cannot be considered as a mechanism for the formation of the subtropical anticyclone. The theories of thermal adaptation of the atmosphere to external thermal forcing are employed to understand the formation of subtropical anticyclone in the three dimensional domain. Numerical experiments are designed to verify these theories. Results show that strong land-surface sensible heating in the summer subtropics generates not only the strong surface cyclone and upper layer anticyclone over the continent, but also the strong lower layer equatorward flow along the western coast of the continent, breaking the zonal symmetric anticyclone belt along the subtropics and forming the surface anticyclones over the oceans to the west. On the other hand, strong deep convective latent heating generates upper tropospheric equatorward flow that closes off the upper tropospheric subtropical anticyclone to the west of the heating, and lower tropospheric poleward flow that closes off the lower tropospheric subtropical anticyclone to its east. Radiative cooling along the subtropics in the two hemispheres coincides well with the tropospheric sinking and the location of subtropical anticyclone in the planetary boundary layer. Over the eastern oceans in the subtropics, such radiative cooling is strong with its maximum occurring at the top of the planetary boundary layer. It generates strong surface equatorward flow and shifts the centers of the oceanic subtropical anticyclones, which is forced by the land-surface sensible heating, towards the eastern parts of the oceans. In the boreal summer, the formation of the strong South Asian High (SAH) and the North American anticyclone in the upper troposphere and the subtropical anticyclone over the western Pacific (SAWP) in the middle and lower troposphere is therefore considered partly as a result of the convective latent heating associated with the Asian monsoon, but affected by orography and the surface sensible heating over the continent. On the other hand, the formation of the subtropical anticyclones at the surface over the eastern North Pacific and North America is mainly due to the strong surface sensible heating over continents, but affected by radiative cooling over the eastern North Pacific. It is shown that the different kinds of diabatic heating over each continent and the adjacent oceans along the summer subtropics are organized in a quadruplet pattern LOSECOD that forces a specific zonal asymmetric circulation pattern. The global summer subtropical heating and circulation can then be viewed as "mosaics" of such a quadruplet heating and circulation patterns, respectively. The fundamental importance of the land-sea distribution in forming the summertime subtropical circulations is then proved.

1. Introduction

Along the subtropics of the Northern and Southern Hemispheres there exist belts of high pressure or geopotential height. These are usually referred to as the belts of subtropical anticyclone. The existence of mountains, air-sea interaction, land surface processes, land-sea contrast, sea-ice and snow cover etc. changes the energy budget of the atmosphere, and breaks the belts into discrete subtropical anticyclones. Such subtropical anticyclones link the westerly systems to the north and tropical weather systems to the south, affect the atmospheric circulation and water-vapor transfer in the surroundings, and influence the weather and climate not only in the adjacent regions and along the subtropics, but also over the world. In the summer half year, the subtropical anticyclones cover huge area in the lower and upper troposphere. For instance, the sea surface subtropical anticyclone over the Northern Pacific in the boreal summer alone covers about twenty to twenty five percent of the Northern Hemisphere. Therefore, anomalies in either the size or intensity of the subtropical anticyclone are usually associated with abnormal regional or global circulations and weather and climate disasters, causing great damages in social economy and human life.

The severe weather and climate such as torrential rainfall, typhoon, the persistent Meiyu front, and floods and draughts in China during the summer months are accompanied with prominent anomalies in the South Asian High (SAH) in the upper troposphere and the subtropical anticyclone over the western Pacific (SAWP) in the lower troposphere. Therefore, subtropical anticyclone and its dynamics have long been the subjects of meteorological studies. Many important results have been obtained. These studies can be cataloged into two groups (Liu and Wu 2000), i.e., synoptic analysis and diagnosis of physical mechanisms. In synoptic analysis, the early studies on the SAWP of Tao *et al.* (1962a, 1962b), Tao (1963), Huang *et al.* (1962) and Huang (1963) reveal the seasonal variation in its intensity and location, its structure in association with temperature, specific humidity, wind field, vertical velocity, meridional circulation, vorticity and divergence, and its association with the distribution of summer rainfall in China. Ye *et al.* (1958a, 1958b) pointed out that from winter to summer, the abrupt northward movement of the subtropical anticyclone in the Asian monsoon area is accompanied with the abrupt changes in circulation patterns. Other studies showed that, the movement of subtropical anticyclone possesses low frequency variabilities, including quasi- two-week (Tang *et al.* 1982; Zhou and Yang 1986; Bi 1989; Yu and Qian 1992) and 30 to 60 day (Xie and Ye 1994) oscillations, and interannual and interdecadal variations (Yan *et al.* 1990). The activities of the SAWP are also closely linked to the weather and climate anomalies in Korea and Japan (e.g., Kurihara and Tsuyuki 1987; Kurihara 1989; Nikaidou 1988). In diagnostic studies, many factors have been proposed to explain the variation and formation of the subtropical anticyclone. These include the interactions in circulation between upper and lower troposphere and between middle and high latitudes and subtropics (Tao and Zhu 1964; Zhang and Wei 1990; Liao and Zhao 1992), the impact of the Tibetan Plateau (Krishnamurti 1973; Gao 1977; Ye and Gao 1979; Ji *et al.* 1984; Luo *et al.* 1986, Wu and Zhang 1998; Ye and Wu 1998; Wu *et al.* 2002a), the influence of the east Asian monsoon (Li and Luo 1988; Yu and Wang 1989; Nikaidou 1989; Qian and Yu 1991; Hoskins 1996; Wu *et al.* 1999, Liu *et al.* 2001; Rodwell and Hoskins 2001), the effects of sea surface temperature, the upper ocean current and thermal contain over the western Pacific (Nitta 1987; Fu and Teng 1988; Huang and Li 1988; Pu and Yu 1993), and the effects of sea ice (Fan 1986; Peng 1987; Wu 1997). Besides, the long-term variations of the subtropical anticyclone are found to be coordinated with the variations of solar radiation (Xu 1986; Jiang and Wang 1989). The seasonal variations of the subtropical anticyclone are found to be associated with the rotation of the earth (Peng *et al.* 1982; Qian 1991).

However, due to the limitations in available data and development of our sciences, our knowledge on subtropical anticyclone is still incomplete. Its formation mechanism is unclear, and its forecast is still unsatisfied even in short- and medium-range operations. By the middle of 1990's, the NCEP/NCAR reanalysis data set (Kalnay *et al.* 1996) became available, and a climate system model of LASG/IAP GOALS was developed coupling the five climate sub-systems together (Wu and Zhang *et al.* 1997; Zhang *et al.* 2000). Conditions for pursuing climate study had been improved tremendously in China. To advance our understanding on the subtropical anticyclone, in 1995 the National Natural Science Foundation of China (NSFC) decided to set up a Key Project entitled "Formation and variation of subtropical anticyclone". From 1996 to 1999, more than ten scientists were organized to carry out intensive studies. Traditional theories on the formation of the subtropical anticyclone were examined. Data diagnosis and numerical experiments were designed to get new insights. By the end of 1999, the project was completed and many new research results had been obtained. Some of them were presented in the Session of "Subtropical anticyclone dynamics" during the IUGG 8th Assembly that was held in Birmingham, England in 1999. Some others were published in Chinese journals (Wu *et al.* 2002b). The material presented in this Chapter condenses these research results, including theoretical study, data diagnosis, and numerical modeling. In Section 2, the general concepts and dynamics of the zonal mean subtropical anticyclone are discussed. The contrasts in observation as well as dynamics between the zonal mean subtropical anticyclone and Hadley cell are made in Section 3. To understand the three-dimension features of the subtropical anticyclone, some relevant dynamics are presented in Section 4. The dynamics includes the adaptation of the atmospheric circulation to external thermal forcing and the importance of the vertical heating profile in determining the circulation patterns in the subtropics.

Webster (1972) used a two-layer linear model to show that latent heating is an important local source in maintaining the stationary circulations in low latitudes. Based on another two-layer linear model, Egger (1978) attempted to separate the contributions of sensible heating and latent heating to the maintenance of the summertime subtropical anticyclones, and obtained a similar conclusion. However due to the coarse vertical resolution, their models greatly underestimate the impacts of the sensible heating in the planetary boundary layer, and this conclusion needs to be re-examined. Furthermore, the surface subtropical anticyclones over the North Pacific and North Atlantic in these models have unrealistic shapes, and their centers are biased to the east, rather asymmetric to their central meridional axis. Chen *et al.* (2001) interpreted such a "tilting" as a result of the meridional shear of the zonal wind. However the simulated anticyclone following this line was presented as a parallelogram with its center biased westward, much unlike the observations. All these aspects imply that the synthetic diabatic effects on the maintenance of the summertime subtropical anticyclone are still unclear. Sections 5 to 7 are therefore devoted to study the contributions to the summertime subtropical anticyclone configuration by surface sensible heating, radiative cooling and deep convective condensation heating, respectively. Because in the real atmosphere such individual thermal forcing is closely linked and well organized, the relation between the synthetic heating and the summertime subtropical anticyclones are investigated in Section 8. Discussions and conclusions are presented in Section 9.

2. Dynamics of the Zonal Mean Subtropical Anticyclone

2.1. *General Concepts*

In the free atmosphere in a meridionally steady state $(\partial v / \partial t = 0)$, the zonal mean flow can be described by using the geostrophic relation, and the meridional momentum equation can be simplified as:

$$f\rho u + \partial p / \partial y = 0. \tag{1}$$

From this relation, the equation of the divergence of meridional mass flux ($\partial(\rho v) / \partial y = 0$) in a steady state can be obtained as:

$$\partial(f\rho u)/\partial y + \partial^2 p / \partial y^2 = 0, \tag{2}$$

the symbols used here are conventional in meteorology. Equation (2) implies that under the constraint of geostrophic relation, the convergence of mass flux due to the exertion of the Coriolis force upon the zonal flow ($-\partial(f\rho u) / \partial y = 0$) is balanced by the divergence of mass flux produced by the pressure gradient force ($-\partial^2 p / \partial y^2 = 0$), as depicted schematically in Figure 1a.

Within the planetary boundary layer, friction impacts need to be considered and equations (1) and (2) are respectively modified as:

$$f\rho u + \partial p / \partial y = -\rho k v \tag{3}$$

$$\partial(f\rho u)/\partial y + \partial^2 p / \partial y^2 = -k\partial(\rho v) / \partial y, \tag{4}$$

where k is a friction coefficient and a linear drag is assumed. In this layer, both the mixed layer and Ekman spiral theories indicate that the horizontal wind has a component directed from higher pressure toward lower pressure (Holton 1992). Thus the surface subtropical anticyclone is accompanied with strong horizontal divergence of mass flux ($\partial(\rho_0 v)/\partial y > 0$) within the boundary layer. Since the mass transport into the boundary layer at the top of the Ekman layer D_e is equal to the divergence of the cross-isobaric mass transport in the layer, i.e.,

$$\rho_0 w(D_e) = -\frac{\partial}{\partial y}\int_0^{D_e} \rho_0 v dz, \tag{5}$$

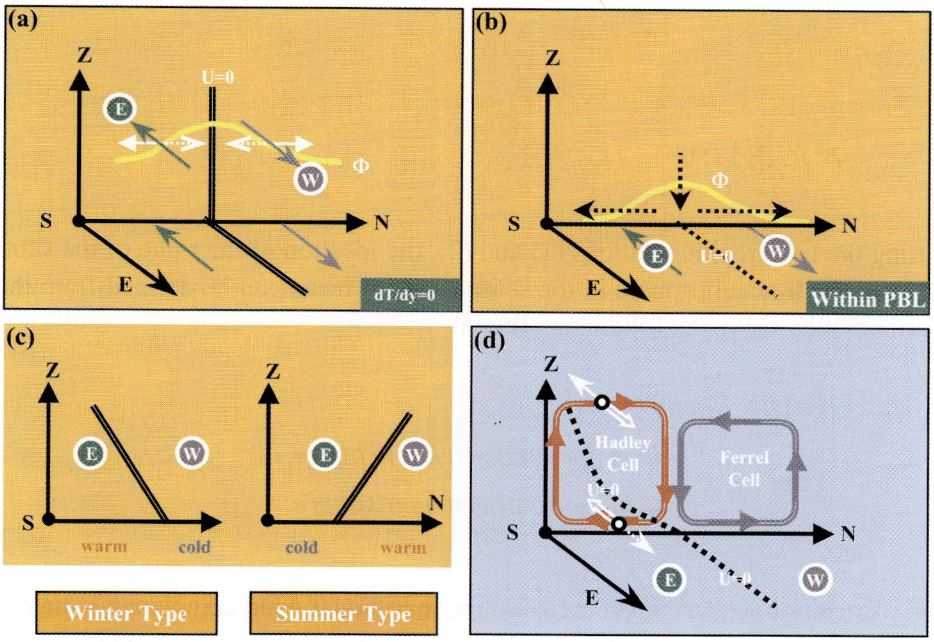

Figure 1. Schematic diagram showing the mechanism for the maintenance of the zonal mean subtropical anticyclone and the Hadley Cell. (a) In the north-south direction at the ridgeline of subtropical anticyclone, the convergence of meridional mass flux owing to the inertial effects of the Earth's rotation (*-fρu*, dotted white arrow) is balanced by its divergence due to the pressure gradient force (*-∇Φ*, solid white arrow). (b) In the planetary boundary layer, the cross-isobaric flow that diverges from the subtropical anticyclone outwards is balanced by the descent at the top of the planetary boundary layer into the layer. (c) According to the thermal wind relation, the ridge of subtropical anticyclone tilts with increasing height towards beneath warmer region, forming westerly shear in winter and easterly shear in summer when crossing the ridgeline upwards in the Asian monsoon area. (d) The inertial torques associated with the horizontal branches of the Hadley Cell (blank white arrow) are balanced by the generation of angular momentum due to friction at the surface (dotted arrow), and by the divergence of angular momentum from tropics to mid-latitudes in the upper troposphere (solid arrow).

the location of the subtropical anticyclone in the planetary boundary layer is then characterized by strong descent at the top of the Ekman layer. It is worthwhile to point out that, although at a steady state the two terms in (5) balance each other, the descent at the top of the planetary boundary layer cannot be used as a mechanism to explain the formation of subtropical anticyclone. This is because both the descent and the cross-isobaric flow in association with the anticyclone are secondary non-divergent circulations, and do not contribute to the mass built-up in the layer, as depicted schematically in Figure 1b.

2.2. *Location*

The belt of subtropical anticyclone is referred to a zone along the subtropics where the air pressure accumulates a maximum in the meridional direction. This means that along the ridge of the subtropical anticyclone,

$$\partial p / \partial y = 0, \tag{6}$$

and

$$\partial^2 p / \partial y^2 < 0. \tag{7}$$

By using the geostrophic relations (1) and (2), the location of the ridge of the subtropical anticyclone in the free atmosphere in the sense of zonal mean can be defined from the zonal wind distribution by using the following criteria:

$$\begin{cases} (a)\, u = 0;\ and \\ (b)\, \partial u/\partial y \begin{cases} > 0, in\ the\ northern\ hemisphere, \\ < 0, in\ the\ southern\ hemisphere. \end{cases} \end{cases} \tag{8}$$

In the planetary boundary layer, because the meridional wind component vanishes at the latitude where the zonal mean center of the subtropical anticyclone is located, equation (3) becomes equivalent to equation (1). Because, according to the Ekman theory, the meridional component of pressure gradient force is greater than that of the frictional force, the sign of the first term on the left hand side of (4) is determined by the sign of the second term. Therefore the criteria (8) for identifying the location of the center of subtropical anticyclone in the free atmosphere are also valid in the planetary boundary layer.

The vertical variation in the location of the subtropical anticyclone can be understood by using the thermal wind relation:

$$\partial u / \partial p < (\alpha / f\theta) \partial\theta / \partial y. \tag{9}$$

For simplicity, p-coordinate system is adopted in (9) in which, θ is potential temperature and α, the specific volume of the air. Such relation between vertical wind shear and horizontal temperature gradient along the ridgeline of the subtropical anticyclone can be demonstrated schematically in Figure 1c. According to this relation, if the tropical region is warmer than the mid-latitude region, across the ridgeline ($u=0$) of the subtropical anticyclone upwards there should be a westerly shear. On the contrary, if the tropical region is colder than the mid-latitude region, across the ridgeline upwards there should be an easterly shear. The latter is applicable in the Asian monsoon region in the boreal summer when the land surface of the southern part of Asia becomes warmer than the sea surface of the North Indian Ocean. In other words, the ridgeline of the subtropical anticyclone tends to tilt towards warmer latitudes with increasing height.

2.3. *Intensity*

The intensity of the subtropical anticyclone can be measured by the convergence or accumulation of the meridional mass flux

$$Int. = \partial(f\rho u)/\partial y = f\rho\partial u/\partial y + \rho u\beta. \tag{10}$$

Here density ρ is assumed to be independent of y in the vicinity of the ridgeline of the subtropical anticyclone. Formula (10) together with (2) indicates that the intensity of the subtropical anticyclone can be measured by $-\partial^2 p/\partial y^2$. In the p-coordinate system it is equivalent to $-\rho\partial^2\Phi/\partial y^2$. Since the density changes little horizontally across the ridgeline of the subtropical anticyclone, for simplicity $-\partial^2\Phi/\partial y^2$ can also be used as an intensity index. Further more, if the Legender polynomial or sinusoidal function is assumed for Φ, $-\partial^2\Phi/\partial y^2$ is proportional to Φ, and geopotential height can also be used as an intensity index of the subtropical anticyclone for comparison purpose.

According to (10), the following reasoning can be obtained:

(a) Zonal mean relative vorticity, i.e., $\partial u/\partial y$, can be used as a measure of the intensity of zonal mean subtropical anticyclone only at the same level $\rho \approx const.$ and at the same latitude $f \approx const.$

(b) For the same meridional gradient of zonal wind, the intensity of the subtropical anticyclone increases with decreasing height and/or increasing latitude.

According to (b), if $\partial u/\partial y$ along the ridgeline of a subtropical anticyclone keeps unchanged at a level, it will be intensified during its poleward moving, but weakened during its equatorward moving.

3. Distribution of the Zonal Mean Subtropical High and the Hadley Circulation

The eighteen-year monthly means ranging from 1980 to 1997 of the NCEP/NCAR reanalysis are employed for the following analysis. In Figure 2, the heavy dashed curves denote the $u=0$ isopleth, and the shading indicates the deviation of geopotential height from its value at the equator and at the same level, i.e.,

$$\phi(y,p)=\Phi(y,p)-\Phi(0,p), \tag{11}$$

where $\Phi(0,p)$ represents the geopotential height at the equator and at pressure p. Since $\Phi(0,p)$ is independent of y, $\partial\Phi(0,p)/\partial y=\partial^2\Phi(0,p)/\partial y^2=0$. According to the discussions in Sections 2.2 and 2.3, $\phi(y,p)$ is equivalent to $\Phi(y,p)$ when the distribution and intensity of the zonal mean subtropical anticyclone are studied. The advantage of using $\phi(y,p)$ is that, by removing the large amount $\Phi(0,p)$ that is irrelevant to our present study, the much smaller deviation field $\phi(y,p)$ can demonstrate the three-dimension structure of the subtropical anticyclone much clearer (Liu 1999). In Figure 2, the isopleth $u=0$ coincides almost everywhere with the maximum $\phi(y,p)$ at the same level. Since easterly exists in the tropical region between the two $u=0$ isopleths, whereas westerly exists outside the region, it becomes evident from Figure 2 that the criteria (8) is adequate in defining the location of subtropical anticyclone at least in so long as the zonal mean situation is concerned. The two ridgelines in the two hemispheres are approximately symmetric to the equator: they are at about 30° latitude near the surface, and tilt equatorward with increasing height. The equatorward tilting of the ridgeline of the subtropical anticyclone is due to the fact that the temperature in tropics is warmer than in mid-latitudes in annual mean. In Figure 2, significant

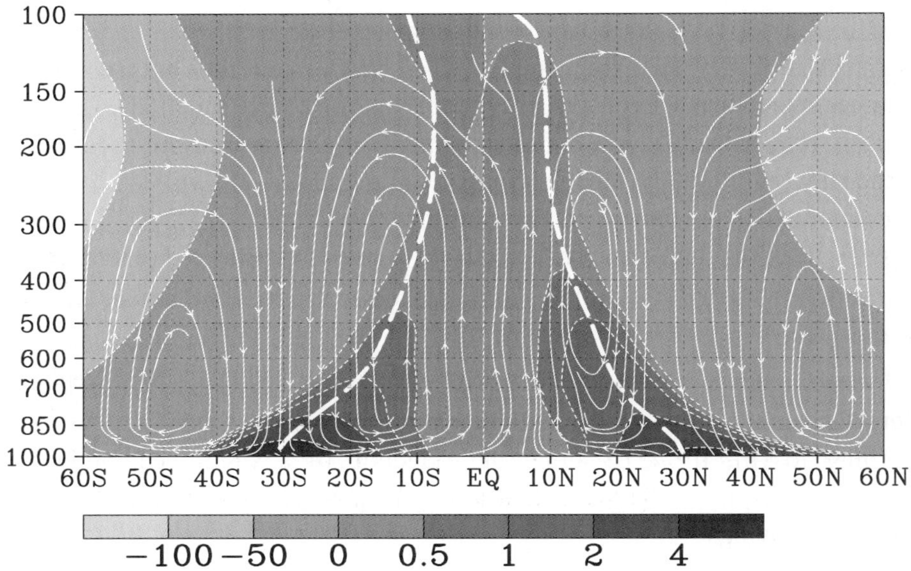

Figure 2. Annual mean distributions calculated from the NCEP/NCAR Reanalysis of the meridional deviation of the zonal mean geopotential height from its equatorial value at the same level (shading, bar labeled in 10 gpm), the mean meridional circulations (light streamline with vectors), and the ridgeline of subtropical anticyclone identified by the curve $u=0$ (heavy dashed curve).

difference between the two hemispheres in the domain of positive geopotential height deviation can also be observed below 850 hPa. Particularly at 1000 hPa, although its poleward rim in the Southern Hemisphere is bounded by 44°S, in the Northern Hemisphere it extends northward to around 60°N. This high extension can be attributed to the existence of the huge and strong Siberia High in the boreal winter.

The deviation fields $\phi(y,p)$ along the ridgelines in the two hemispheres are almost symmetric to the equator as well: they are about 50 gpm near the surface, and decrease with increasing height. At 500 hPa, it is 10 gpm in the Northern Hemisphere, but only 6 gpm in the Southern Hemisphere. At about 200 hPa, $\phi(y,p)$ becomes very weak. The weakening in intensity of the subtropical anticyclone with increasing height is in accordance with the equatorward tilting of its ridgeline. This is because as the ridgeline is approaching the equator with increasing height, both the Coriolis parameter and air density become smaller. According to (8), for the same zonal wind shear, there is less convergence of the mass flux along the ridgeline, leading to the weakening in intensity.

Traditionally, the formation of the subtropical anticyclone is attributed to the dynamical effects of the subsidence in the sinking arm of the Hadley circulation (Peixoto and Oort 1991). However, the strongest Hadley cell occurs in the winter hemisphere and is accompanied with the weakest subtropical anticyclone, whereas the weakest Hadley cell occurs in the summer hemisphere and is accompanied with the strongest subtropical anticyclone, as pointed out by Hoskins (1996). The subsidence does not well match the existence of the subtropical anticyclone. To explore this mismatch further, in Figure 2 is also shown the 18-year mean of the meridional circulation. It is apparent that the sinking arms of the two Hadley cells do not coincide with the ridgelines of subtropical anticyclone except at the surface. They are vertically located over a vast latitude span, usually to the poleward side of the subtropical anticyclone in the free atmosphere. As well known, the mean meridional circulation is forced only when the geostrophic and hydrostatic balances are destroyed by diabatic heating and/or eddy transfer of heat and momentum (Eady 1949; Kuo 1956; Holton 1992). In return, the generation of the mean meridional circulation is to restore such geostrophic and hydrostatic balances. Thermodynamically, the adiabatic cooling/heating of the vertical motion of the two vertical arms of a Hadley cell compensates the atmospheric diabatic heating/cooling. An equilibrium thermal state of the tropical atmosphere is then maintained. As long as the conservation of angular momentum is concerned, the positive inertial torque ($f\rho v$) of the upper branch of the Hadley cell balances the eddy transfer of angular momentum from the tropics to mid-latitudes (Figure 1d); whereas the negative inertial torque of its lower branch balances the positive surface frictional torque that is negatively proportional to the surface easterly zonal wind (Wu 1988). Since the surface domain of the lower branches of the two Hadley cells is associated with that of the surface easterlies, the two latitudes where the zonal wind vanishes bound the domain. This then explains why at the earth's surface the sinking arm of the Hadley cell coincides well with the ridge of subtropical anticyclone as shown in Figure 2. Away from the surface, however, the ridgeline of subtropical anticyclone deviates from the almost vertically located sinking arm of the Hadley cell.

From the above discussions, we see the contrast in the formation dynamics between the subtropical anticyclone and the Hadley circulation. The formation of subtropical anticyclone is associated with the convergence or accumulation of meridional mass flux. This can be understood by employing the v-momentum equation. Although the vertical tilting of the ridgeline of the subtropical anticyclone is determined by the thermal structure of the atmosphere, the ultimate forcing of the formation of the subtropical anticyclone is the rotation of the earth. On the other hand, the Hadley circulation is thermally driven. Its maintenance mechanism includes the requirement of conservation of angular momentum that can be understood by using the u-momentum equation (Held and Hou 1980; Lindzen and Hou 1988). Only because the generation and destruction of the atmospheric angular momentum due to the surface zonal frictional torque are separated by the nil zonal wind, does the ridgeline of subtropical anticyclone at the surface coincide with the sinking arm of the Hadley cell (Figure 1d). Due to the surface frictional dissipation, the descending air at the location of this surface subtropical anticyclone compensates the surface divergent meridional mass fluxes, and does not contribute to the mass built-up along the surface high. In other words, the descent at the surface subtropical high is a result, rather than the cause of the formation of the surface high, as was discussed in Section 2.1 and presented in Figure 1b.

The above discussions do not mean that the formations of the subtropical anticyclone and the Hadley circulations are separated. Although the subtropical anticyclone exists because of the meridional mass flux convergence due to Coriolis turning, the existence of the zonal mean tropical easterlies and mid-latitude westerlies in the first place is closely associated with the Hadley cell. A simple Hadley cell model of Held and Hou (1980) clearly shows that the Hadley circulation must exist at least within the subtropics in order to maintain the zonal winds and an associated temperature field consistent with angular momentum conservation. This model produces a westerly jet at the subtropical tropopause and surface easterly in the tropics. The ridgeline of the subtropical anticyclone defined by $u=0$ then tilts equatorward (his Figure 3) following the thermal wind relation. It is therefore separated from the descending arm of the Hadley cell and located to the equatorward side of the descending arm, as is observed from Figure 2 and represented in Figure 1d. This model predicts the importance of several parameters in determining the distributions of the Hadley cell and the zonal winds, including the gravitational acceleration, the rotation rate and radius of the Earth, the height of the tropopause, and the fractional change in potential temperature from equator to pole.

4. Thermal Adaptation and *PV*- Forcing

In the Northern Hemisphere in winter, westerlies dominate the mid-latitude and subtropics in the upper troposphere. Mountain forcing plays an important role in the formation of the circulation patterns (Charney and Elliason 1949; Bolin 1950; Yeh 1950; Rodwell and Hoskins 2001). On the other hand in the summer subtropics, the upper tropospheric westerly is weak, and thermal forcing more importantly influences the circulation (Hoskins 1991; Chen 2001). Before we discuss the formation of the subtropical anticyclones in the three-dimension domain, in this section we show how the atmosphere responds to a given thermal forcing.

4.1. *Intensity*

The Ertel potential vorticity (Ertel 1942) P, which is abbreviated as potential vorticity in the following context, is defined as:

$$P = \frac{1}{\rho}\vec{\zeta}_a \cdot \nabla\theta, \tag{12}$$

where $\vec{\zeta}_a$ is the three-dimension absolute vorticity. In the θ coordinate, since $\nabla_\theta \theta \equiv 0$, (12) becomes

$$P = \frac{1}{\rho}\vec{\zeta}_a \cdot \vec{n}\frac{\partial\theta}{\partial z} \tag{13}$$

where $\vec{n}$ is a unit vector along the gradient of θ, Δz is the thickness between two θ surfaces. At the θ surface, if a unit area ds is defined as $ds=dxdy$, then the mass contained in a unit box is $dm=\rho dxdydz$. Following the discussion of Hoskins (1991) and integrating (13) over an area S and between two adjacent θ surfaces θ_1 and θ_2 lead to

$$\frac{1}{S}\int P dm = \int_{\theta_2}^{\theta_1} \zeta_{an} d\theta, \tag{14}$$

where ζ_{an} is the absolute vorticity component along the $\vec{n}$ direction and averaged over S. (14) shows that the mass weighted integral of P per unit area equals to the θ weighted integral of absolute vorticity ζ_{an}. Suppose in a statically stable atmosphere, there are three horizontally located surfaces, and an external diabatic heating source Q within area S that increases monotonically with height in the lower layer ($\theta_1 - \theta_2$), but decreases monotonically with height in the upper layer ($\theta_2 - \theta_3$) (Figure 3a). T, A and B are the three parcels initially located, respectively, at the θ_3, θ_2 and θ_1 surfaces (Figure 3a). Thermodynamically the heating at θ_2 causes the surface sink downwards, increasing stability and P below and decreasing stability and P above the surface (Figure 3b). Dynamically the heating-induced ascent of parcel A results in convergence in the lower layer and divergence in the upper layer (Figure 3c). Since no penetration of P across θ_2 is permitted (Haynes and McIntyre 1987), both P and dm increase in the lower layer while decrease in the upper layer. According to (14), cyclonic vorticity in the lower layer and anticyclonic vorticity in the upper layer (Figure 3d) are then generated (Hoskins 1991; Wu and Liu 2000).

To get the vorticity budget associated with such heating induced circulation change, a geometric coordinate system is adopted for the following discussion, and the flux form of the following vorticity equation is employed (Hynes and McIntyre 1987; Wu and Liu 2000):

$$\frac{DW}{Dt} = \vec{F}_\zeta \cdot \nabla\theta + \vec{\zeta}_a \cdot \nabla Q, \tag{15}$$

where $\frac{DW}{Dt} = (\frac{\partial}{\partial t} + \nabla \cdot \vec{V})W$, $\vec{F}_\zeta$ is the three-dimension dissipation of vorticity, and

$$W = \rho P = \vec{\zeta}_a \cdot \nabla \theta, \tag{16}$$

is amount of potential vorticity per unit volume. Divide W into its vertical and horizontal components, i.e.:

$$\begin{cases} W = W_v + W_h, \\ W_v = (f + \zeta_z)\theta_z, \\ W_h = \nabla \times \vec{V} \cdot \nabla_h \theta, \end{cases} \tag{17}$$

and assume the following characteristic values for large scale atmospheric motions:

$$\begin{cases} \Delta Z \sim 10^3 \sim 10^4\,\mathrm{m}, \Delta\mathrm{x}, \Delta\mathrm{y} \sim 10^6\ \mathrm{m}, \Delta\theta \sim 10\mathrm{K}, \Delta\mathrm{V} \sim 10\mathrm{m\,s^{-1}}, \\ w \sim 10^{-3}\ \mathrm{m\,s^{-1}}, \Delta\mathrm{F} \sim 10^2\ \mathrm{W\,m^{-2}} \sim 10\,\mathrm{J\,s^{-1}\ m^{-2}}, \end{cases} \tag{18}$$

where F is heat flux of the atmosphere, and usually has the magnitude of 10^2 W m^{-2}. Thus, $\mathrm{Q} = -\frac{\theta}{\mathrm{C_p T \rho}}\frac{\partial \mathrm{F}}{\partial \mathrm{z}} = -\frac{1}{\mathrm{C_p \rho}}(\frac{\mathrm{P_0}}{\mathrm{P}})^{\mathrm{R/C_p}}\frac{\partial \mathrm{F}}{\partial \mathrm{z}} \sim 10^{-4} \sim 10^{-5}\ \mathrm{K\,s^{-1}}$, where $C_p \sim 10^3$ J K^{-1} Kg^{-1}, $\rho \sim 10^0$ Kg m^{-3}, $(p_0/p)^{R/Cp} \sim 10^0$. The magnitude of W is then estimated as

$$W = W_v + W_h = (f + \zeta)\frac{\partial \theta}{\partial z} + \nabla \times \vec{V} \cdot \nabla_h \theta \tag{19}$$
$$\sim 10^{-6}\ \mathrm{K\,s^{-1}\ m^{-1}}\ (10^0 \sim 10^{-1} + 10^{-1} \sim 10^{-2}).$$

If a total potential vorticity unit is defined as 1 TU=10^{-6} K s^{-1} m^{-1}, then

$$W = W_v + W_h \sim (10^0 \sim 10^{-1})\,\mathrm{TU} + (10^{-1} \sim 10^{-2})\,\mathrm{TU}. \tag{20}$$

This means that, when the slope of θ surface is small $(\leq 10^{-3})$, the vertical component W_v is one order of magnitude larger than the horizontal component W_h.

If the left hand side of (15) is expanded and ΔZ is taken as 10^3 m, the magnitude of each term of (15) can be estimated as:

$$\underset{\text{(a)}}{\frac{\partial W}{\partial t}} + \underset{\substack{\text{(b)}\\10^{-11}}}{\nabla_\mathrm{h} \cdot \vec{V}_h W} + \underset{\substack{\text{(c)}\\10^{-12}}}{\frac{\partial}{\partial z} wW} = \underset{\substack{\text{(d)}\\10^{-11}}}{\vec{F}_\zeta \cdot \nabla \theta} + \underset{\substack{\text{(e)}\\10^{-11}}}{\vec{\zeta}_a \cdot \nabla Q} \tag{21}$$

This is to say, the horizontal divergence of horizontal flux of W is one order of magnitude large than the vertical divergence of its vertical flux. In other words, the vertical divergence of the vertical flux of W is not large enough in compensating the generation of potential vorticity induced by diabatic heating.

Now let us consider the vorticity budget in the lower part $(z_1 - z_2)$ of the heated column (Figure 4a). In this layer $\dfrac{\partial Q}{\partial z} > 0$, and $\vec{\zeta}_a \cdot \nabla Q > 0$. According to Figure 3, cyclonic circulation is generated in the lower layer. At a steady state, $\partial W / \partial t$ vanishes.

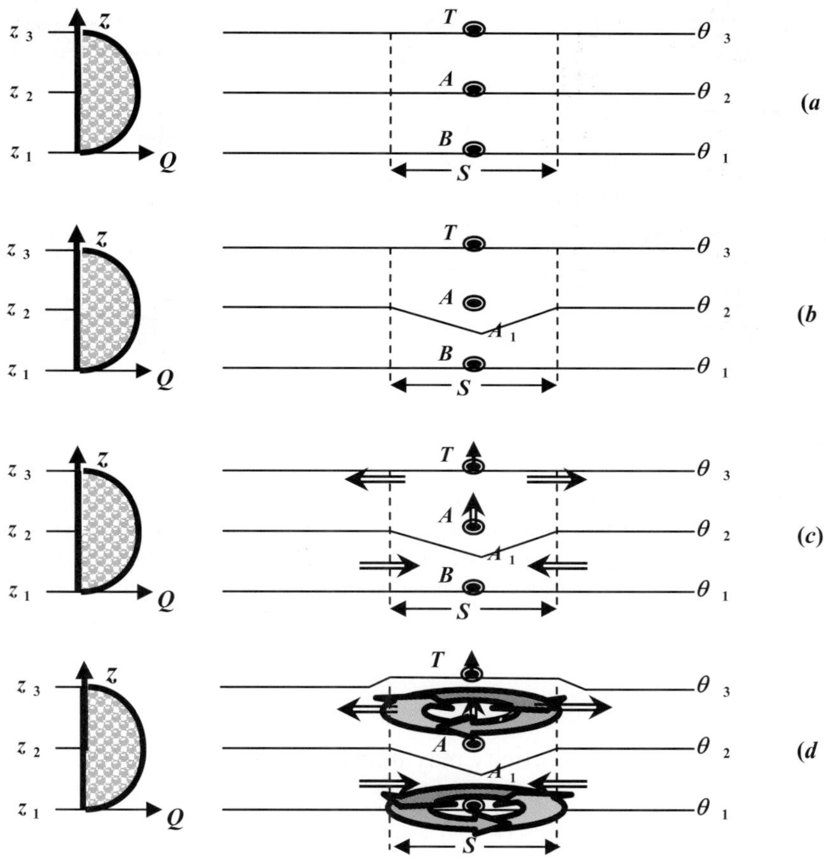

Figure 3. Schematic Diagram showing the thermal adaptation of the atmospheric circulation (arrow) to an external heating Q. Left panel indicates the vertical distribution of heating with height within the area S. (a) Initial state. T, A and B denote the parcels initially at the θ_3, θ_2 and θ_1 surfaces, respectively; (b) Thermal impact: heating increases θ within S, θ_2 surface descends, and potential vorticity increases in the lower layer but decreases in the upper layer; (c) Dynamic impact: heating causes ascent of parcel A, air between two θ surfaces converges in the lower layer but diverges in the upper layer; (d) Due to the combined effects of (b) and (c) and according to (4.30), cyclonic circulation is excited in the lower layer but anticyclonic circulation in the upper layer. (Please also refer to Figure 2 of Hoskins 1991).

Then since there is convergence in the interior of *S*, term (b) in (21) becomes (−) and cannot balance term (e). Although term (c) is (+), but is too small and cannot balance term (e) as well. Therefore near the surface the heating induced generation of vorticity must be compensated to some degree by the effect of friction, i.e.,

$$\vec{F}_{\zeta} \cdot \nabla \theta \cong -\vec{\zeta}_a \cdot \nabla Q. \tag{22}$$

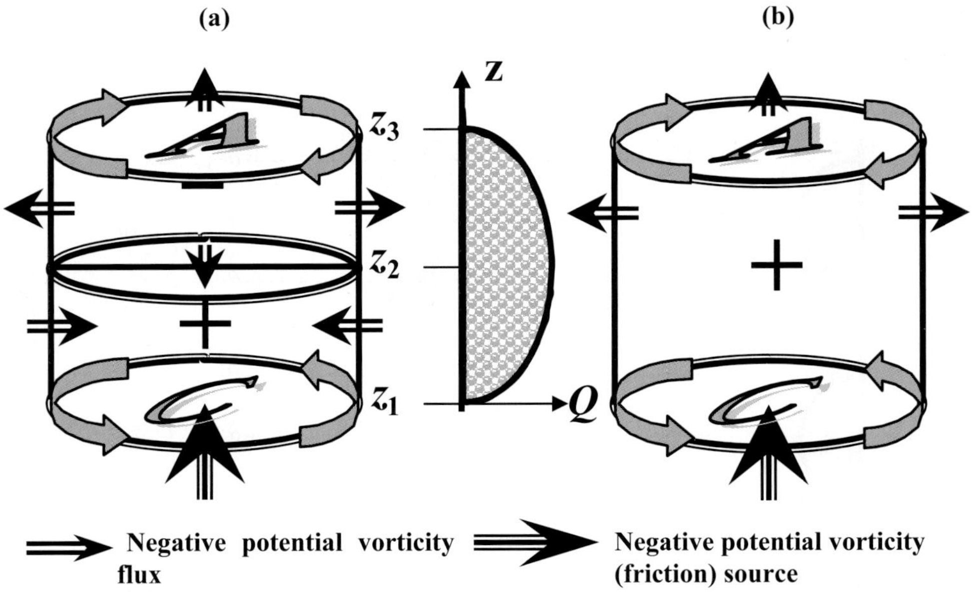

Figure 4. Budget of total potential vorticity per unit volume *W* during atmospheric thermal adaptation. (+) and (-) indicate, respectively, the source and sink of *W*; arrow indicates the direction of flux of negative *W*. (a) In the lower layer of a two-layered model, generation of positive *W* due to heating is balanced mainly by frictional dissipation and by cross-boundary flux; negative *W* generation in the upper layer resulting from the decreasing heating profile is balanced by cross-boundary negative *W* flux. (b) In a column model, the negative *W* generation due to friction at the bottom is compensated for partly by the heating-generated positive *W* $(\zeta_z \partial Q/\partial z)$, but mainly by the cross-boundary negative *W* flux.

In such case, the dissipation rate of vorticity may then be estimated as

$$F_z \cong -(f + \zeta_a)\frac{\partial Q}{\partial z} / \frac{\partial \theta}{\partial z} < 0, \tag{23}$$

which is as large as 10^{-9} s^{-1}. Therefore a heating rate of 100 W m^{-2} in the boundary layer would force a surface cyclonic vorticity of 10^{-5} s^{-1} within 2 to 3 hours if there is no vorticity exchange. However, the growth rate of the surface cyclone is limited. This is because the heating causes the intersection of the θ surface with the earth's surface, resulting in a $\nabla_h \theta$ pointing inward along the heating boundary Γ. At the same time, the cyclonic circulation in the lower layer and the anticyclonic circulation in the upper layer create a vertical wind shear, resulting in a horizontal vorticity pointing outward along Γ. Therefore, along the boundary Γ of the heating area S,

$$W_h = \nabla \times \vec{V} \cdot \nabla_h \theta < 0. \qquad (24)$$

Assuming the horizontal scale of the boundary $(\Delta x, \Delta y)$ be $10^4 \sim 10^5$ m, and its vertical scale ΔZ, 10^3 m, and referring to (18), we can estimate the magnitude of W along the boundary as:

$$W_\text{v} \sim 10^{-1} - 10^0 \text{TU}, \ \ W_\text{h} \sim 10^0 - 10^1 \text{TU}, \text{ at the boundary } \Gamma. \qquad (25)$$

It means that along the boundary,

$$|W_h| > |W_v|, \qquad (26)$$

and

$$W = W_v + W_h < 0. \qquad (27)$$

This implies that symmetric instability appears at the boundary of the heating region S. Integrating (21) from z_1 to z_2 over the area S and employing the Gaussian's law,

$$\int_s \nabla_h \cdot (\vec{V}A) ds = \oint_\Gamma (A\vec{V}) \cdot \vec{n} dl,$$

lead to

$$\int_{z_1}^{z_2} \text{d}z \oint_\Gamma W\vec{V} \cdot \vec{n} \text{d}l + \int_S W_2 w_2 \text{d}s = \int_{z_1}^{z_2} \int_S \left[F_z \theta_z + (f + \zeta_z) \frac{\partial Q}{\partial z} \right] \text{d}s\text{d}z \qquad (28)$$

$$\text{(b)} > 0 \qquad \text{(c)} > 0 \qquad \text{(d)} < 0 \qquad \text{(e)} > 0$$

where Γ is positive along an counterclockwise direction, and $\vec{n}$ is the unit vector perpendicular to Γ and pointing outward. Since there is lower layer convergence across the boundary ($\vec{V}_\Gamma \cdot \vec{n} < 0$), and since along the boundary W is negative (27), the sign of term (b) in (28) becomes positive. This means that, due to the intersection of the θ surface with the earth's surface and the occurrence of symmetric instability along the boundary, negative vorticity is pumped across the boundary into the heating area to dilute the generation of positive vorticity due to diabatic heating. As a result, the signs of the terms (b) to (e) in (28) become +, +, - and +, respectively. Therefore, the heating induced positive vorticity in the lower layer is balanced by the surface friction dissipation, transgression at the upper boundary and dilution of negative vorticity flux across the transverse boundary. A steady state is then maintained (Figure 4a).

As to the vorticity balance in the upper layer ($z_2 - z_3$), since $\partial Q / \partial z < 0$, negative vorticity generation and anticyclonic circulation are anticipated (Figure 4a). In this layer, friction dissipation can be ignored, and the generation of negative vorticity should be balanced mainly by the horizontal divergence of vorticity flux. This is indeed the case. Because in this layer $\nabla_h \theta$ and the horizontal vorticity vector possess opposite signs as they do in the lower layer, the discussions presented from (24) to (27) for the lower layer are still applicable for the upper layer. Since the air is divergent in the layer, $\vec{V}_\Gamma \cdot \vec{n} < 0$, integration of (21) from z_2 to z_3 over S then leads to

$$\underset{(b)<0}{\int_{z_2}^{z_3} \mathrm{d}z \oint_\Gamma W\vec{V} \cdot \vec{n} \mathrm{d}l} - \underset{(c)<0}{\int_S W_2 w_2 \mathrm{d}s} = \underset{(e)<0}{\int_{z_2}^{z_3} \int_S (f + \zeta_z) \frac{\partial Q}{\partial z} \mathrm{d}s\mathrm{d}z} \tag{29}$$

(29) indicates that, in the upper layer the generation of negative vorticity due to the decreasing of the diabatic heating with height is partly diluted by the upward transfer of positive vorticity across the lower boundary from below, but mainly by the outward transport of negative vorticity across the lateral boundary Γ of the heating area (Figure 4a). Here we also observe the importance of the symmetric instability along the boundary in the maintenance of vorticity balance within the heating area.

If the integration of (10) is taken from z_1 to z_3 over S, the vorticity balance over the whole column can then be estimated, and the result is the summation of (28) and (29):

$$\underset{(b)<0}{\int_{z_1}^{z_3} \mathrm{d}z \oint_\Gamma W\vec{V} \cdot \vec{n} \mathrm{d}l} = \int_{z_1}^{z_3} \int_S \left[\underset{(d)<0}{F_z \theta_z} + \underset{(e)>0}{\zeta_z \frac{\partial Q}{\partial z}} \right] \mathrm{d}s\mathrm{d}z \tag{30}$$

The term of vertical transfer (c) vanishes. The large portion ($f\partial Q/\partial z$) of the generation term is cancelled; only the small portion ($\zeta_z \partial Q/\partial z$) in (e) is remained. The term of friction dissipation (d) keeps unchanged. As to term (b), its sign should be the same as that of its upper layer. This is partly because the upper layer divergence should overwhelm the lower layer convergence so as to maintain the surface low, and partly because, judging from (28) and (29), the generation rate that is approximated as ($f\partial Q/\partial z$) in the upper layer is not dissipated by friction as it is in the lower layer. As a whole, the (b), (d) and (e) terms in (30) possess the signs of -, - and +, respectively. This is to say that for the total column integration, the negative vorticity pumped into the column due to surface friction is partly canceled by the positive vorticity generation due to the heating. The rest is pumped out of the column in the upper layer, affecting the circulations outside the heating region (Figure 4b).

4.2. *Numerical Experiments*

Numerical experiments are designed to examine the thermal adaptation theory established in the preceding section. The model used is the climate model of Global-Ocean-Atmosphere-Land System (GOALS hereafter) which is developed at LASG, IAP (Wu and Zhang *et al.* 1997; Zhang *et al.* 2000) and has participated the Atmospheric Model Inter-comparison Program (AMIP), Coupled Model Inter-comparison Program (CMIP) and Task I of the Inter-Governmental Program for Climate Changes (IPCC 2001). The atmospheric component has nine levels in the vertical and rhomboidally truncated at wave number 15 in the horizontal (Wu *et al.* 1996). The ocean component (Zhang *et al.* 1996) has 20 layers in the vertical with

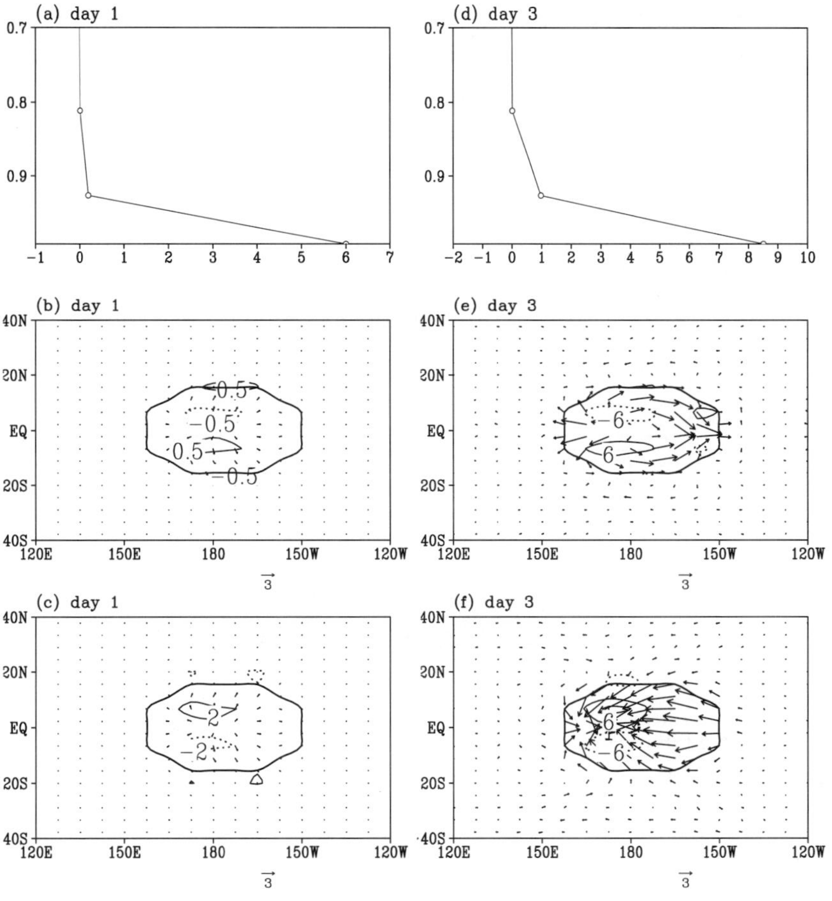

Figure 5. Numerical experiment on the thermal adaptation of the atmospheric flow (arrow) and vorticity (contour, 10^{-6} s^{-1}) to a prescribed surface sensible heating at Day 1 (a, b and c) and Day 3 (d, e and f) at the upper level σ=0.664 (b and e) and lower level σ=0.991 (c and f). (a) and (d) present the heating rate (K d^{-1}) at Day 1 and Day 3, respectively. Solid and dashed curves denote positive and negative values respectively, and the heavy solid curve bounds the heating region, which is more than 1 W m^2.

horizontal resolution of 4-degree latitude by 5-degree longitude. The land component uses the SSiB model (Xue 1991; Liu and Wu 1997). The *K*-distribution scheme developed by Shi (1981) is employed for representing the radiation processes.

For the present purpose, the ocean and land components are switched off, and an aqua planet is assumed. The climatic July and zonal-mean sea surface temperature is imposed as the lower boundary condition. The solar angle is fixed by its value on July 15. Other variables, including CO_2, aerosol, cloud amount and atmospheric variables, all are prescribed to their corresponding July zonal means. The initial wind and horizontal gradient of temperature $\nabla_h T$ are set to zero. For simplicity, an axial symmetric surface heating source is imposed at the equator centered at 176°W and bounded in the region of 11°S-11°N and 160°E-150°W. The intensity of the heating source is 100 W m^{-2} at the center, and decreases gradually in cosine form towards zero at the boundary. The heating region bounded by the 1 W m^{-2} heating is shown in Figure 5 by heavy curves. The setting of 100 W m^{-2} for the experiment is in reference to the observations: over the equatorial Africa and Latin America, the surface sensible heat flux in July is commonly above this value. To concentrate on the atmospheric response to such imposed surface diabatic heating, condensation heating and the sensible heating over other parts of the world are switched off from the thermodynamic equation.

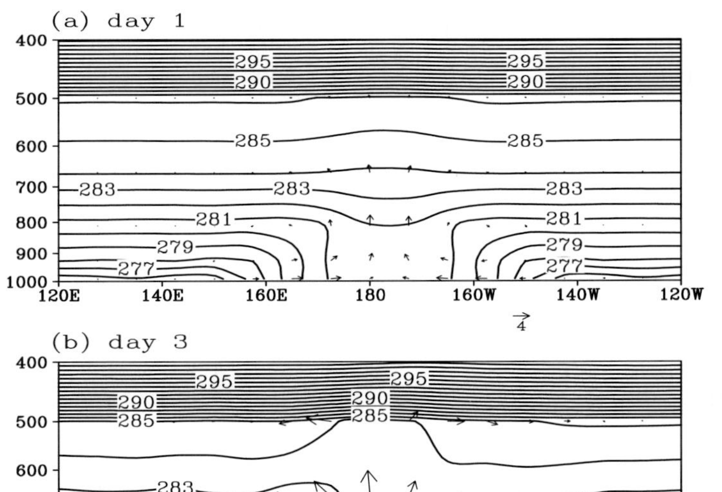

Figure 6. Vertical cross-sections at the equator of potential temperature (K) and velocity (unit: *u* in m s^{-1}, ω in –40 Pa s^{-1}) based on the numerical experiment of thermal adaptation of the atmosphere to a prescribed surface sensible heating. (a) Day 1; (b) Day 3.

Figures 5a and 5d show the vertical distributions of diffusive sensible heating over the heating region on Day 1 and Day 3 of the model integration. The maximum heating is near the surface with an intensity of 8 K d^{-1} on Day 3. The heating decreases with height and approaches zero near 800 hPa. Figures 5c and 5f present the wind and vorticity fields at the σ=0.991 level on Day 1 and Day 3. The diabatic heating results in cyclonic vorticity and horizontal convergence in the lower layer, but anticyclonic circulation and divergence at the upper level σ=0.664 (Figures 5b and 5e). The intensity of the vorticity is in the order of 10^{-6} s^{-1} outside the equator. Figure 6 demonstrates the vertical sections of potential temperature and wind vector composed by the zonal and vertical velocity on Day 1 and Day 3. After one day of heating, the isentropic surfaces become concave. Air converges in the lower layer, penetrates the isentropic surfaces over the heating region, and diverges at the upper level σ= 0.664. These processes get intensified at Day 3. All these results are in agreement with those shown in Figure 3.

The spectacular results occur at the western as well as eastern boundaries of the heating region, i.e., around 165°E and 157°W, where the isentropic surfaces intersect the earth's surface. According to Figures 5 and 6, the magnitudes of those terms relevant to the calculation of W and along the boundary can be estimated as follows:

$$(f+\zeta)\sim 10^{-6}-10^{-5}\ \mathrm{s}^{-1}, \qquad \frac{\Delta\theta}{\Delta z}\sim 10^{-3}-10^{-4}\ \mathrm{K\,m}^{-1},$$
$$\left(\frac{\Delta u}{\Delta Z},\frac{\Delta v}{\Delta Z}\right)\sim 10^{-2}\ \mathrm{s}^{-1}, \qquad \left(\frac{\Delta\theta}{\Delta X},\frac{\Delta\theta}{\Delta Y}\right)\sim 5\times 10^{-6}\ \mathrm{K\,m}^{-1}. \tag{31}$$

It is evident that even for a region with the horizontal scale of 4×10^6 m and an averaged heating rate of 64 W m^{-2} ($200/\pi$), the heating-induced horizontal vorticity W_h (the vertical wind shear) along the boundary is about three orders of magnitude larger than the vertical vorticity W_v. According to (31), the magnitudes of W_v and W_h can be estimated as:

$$W_v=(f+\zeta)\theta_z\sim 10^{-2}-10^{-4}\ \mathrm{TU}$$
$$W_h=\nabla\times\vec{V}\cdot\nabla_h\theta\sim 10^{-1}-10^{-2}\ \mathrm{TU} \tag{32}$$

It is apparent that along the boundary of the heating region, $|W_h|>|W_v|$. Since $W_h<0$, the validity of (26) and (27) and the occurrence of symmetric instability (W<0) along the boundary are proved. It is worthwhile to note that the sign of vorticity along the heating boundary is opposite to that in the interior. This is in favor of the inward transfer of negative vorticity flux along the boundary to dilute the positive vorticity generation due to heating in the lower layer. All these results support our analysis on the vorticity balances as presented in Figure 4. The only difference is that the magnitude of W shown in (32) is smaller than the previous scale analysis. This is because in this numerical experiment, the heating source is placed at the equator where f vanishes. Since the equatorial location of the symmetric heat source can generate a Gill-type atmospheric response (Gill 1980), such an experiment design can illustrate the thermal adaptation of the atmospheric circulation to an external forcing more clearly. In another experiment in which the center of the sensible heating source is located at 35°N, similar results are obtained and an asymmetric Gill type circulation are also induced (Figures 3.11- 3.14, Liu 1988; Wu *et al.* 2002b).

The theory developed in Section 4.1 describes how the steady circulation can be maintained in response to a heating profile; whereas the numerical experiment shows that when the maximum heating is close to the surface, strong surface cyclone and deep upper layer anticyclone are generated. If this happens along the subtropics, the symmetric anticyclone belt will then be broken, and enclosed anticyclone systems will be formed.

4.3. *PV-Forcing and Subtropical Anticyclone*

Based on the potential vorticity equation (15) and scale analysis, the relation between the diabatic heating Q and the forced atmospheric circulation along the subtropics and in steady state can be expressed as the following Sverdrup balance (Liu *et al.* 2001):

$$\vec{V}\cdot\nabla\zeta+\beta v\approx\theta_z^{-1}(f+\zeta)Q_z\quad(\theta_z\neq 0)\tag{33}$$

A similar relation to (33) at a steady state can also be obtained by combining the approximated thermodynamic equation

$$w\approx Q\theta_z^{-1},\quad \theta_z\neq 0,$$

and the vorticity equation

$$\vec{V}\cdot\nabla\zeta+\beta v=f\frac{\partial w}{\partial z}.$$

In the lower troposphere in the summer subtropics, or in the upper troposphere in the deep convection occasions, the vorticity advection is weak, and (33) can be further simplified to

$$\beta v\approx\theta_z^{-1}(f+\zeta)Q_z\quad(\theta_z\neq 0)\tag{34}$$

(34) implies that the thermally forced circulation along the subtropics depends strongly on the vertical profile of the heating. Since f is positive in the Northern Hemisphere but negative in the Southern Hemisphere, for a statically stable atmosphere, a heating increasing with altitude generates poleward flow, whereas a heating decreasing with altitude produces equatorward flow. This provides another basis for understanding the formation of the subtropical anticyclones in summer.

4.4. *Subtropical Anticyclone Forced by Surface Sensible Heating*

During boreal summer along subtropics, land-surface sensible heat flux over the western continents usually exceeds 100 W m^{-2} which amounts to a heating rate (Q) of 10^{-5} K s^{-1}. For a large-scale atmospheric system such as subtropical anticyclone, the order of magnitude for θ_z is estimated as $\theta_z \sim 10^{-2}$ K m^{-1}. Then the orders of magnitude of the forcing terms on the right hand sides of (33) and (34) are estimated to 10^{-10} s^{-2}. Following the discussions presented in Section 3.1 and the studies of Haynes and McIntyre (1987) and Hoskins (1991), such strong surface heating will generate a cyclone at the surface and an anticyclone aloft (lower part of Figure7a). At the same time, since the maximum sensible heating appears near the land surface, Q_z is negative. According to (34), the forced northerly in the lower layers is estimated to 10^0 m s^{-1}. This means that, in response to a surface sensible heat flux of 100 W m^{-2}, the equatorward winds of several m s^{-1} will be forced over the heating region in the lower layers with a thickness of about one kilometer. The symmetric subtropical flow is then broken and the lower layer anticyclone is generated to the west of the heating region, as shown schematically in the lower part of Figure 7a.

In the upper layers in subtropics, due to the vertical westerly wind shear, the zonal advection of vorticity overwhelms the β effect as demonstrated in Figure 9, and (33) can be approximated as

$$\partial\zeta / \partial x \approx (f+\zeta)u^{-1}Q_z\theta_z^{-1} < 0, \quad (\theta_z \neq 0) \tag{35}$$

For a horizontal heating scale of 10^6-10^7 m, and a westerly flow of 10^1 m, a negative relative vorticity of the order -10^{-5} s^{-1} should appear downstream of the westerlies, contributing to the formation of the upper layer subtropical anticyclone, as shown schematically by the upper part of Figure7a.

4.5. *Subtropical Anticyclone Forced by Deep Convective Condensation Heating*

Along the subtropics, maximum latent heating associated with deep convection usually occurs at a height between 300 and 400 hPa, where the heating rate due to condensation can be estimated as Q~10^{-5}~10^{-4} K s^{-1}. Using (34), the order of magnitude of the diabatic term $\theta_z^{-1}(f+\zeta)Q_z$ is also 10^{-10} s^{-2} below the maximum heating layer ($z = Z_M$), but 10^{-9} s^{-2} above this layer. However, unlike in the case of sensible heating, the large latent heating rate at $z = Z_M$ results in strong westerlies to its north and easterlies to its south. Thus, even in the upper troposphere in the subtropics, vorticity advection above a deep convection area is usually small. The response of the atmosphere to deep convective latent heating can then be approximated to

$$v \approx (f+\zeta)\beta^{-1}Q_z\theta_z^{-1}\begin{cases} <0 & (z > Z_M) \\ >0 & (z < Z_M) \end{cases} \quad (f > 0, \theta_z \neq 0) \tag{36}$$

(36) shows that poleward flow is thermodynamically generated below Z_M, while equatorward flow in the upper troposphere. Therefore, in the upper troposphere, subtropical anticyclone is formed to the west of the deep convection region as shown in the upper part of Figure 7b, while lower tropospheric subtropical anticyclone is formed to its east as shown in the lower part of Figure 7b. At the surface, cyclonic circulation should be generated due to the condensational heating as in the case of sensible heating (refer to Figure 13e).

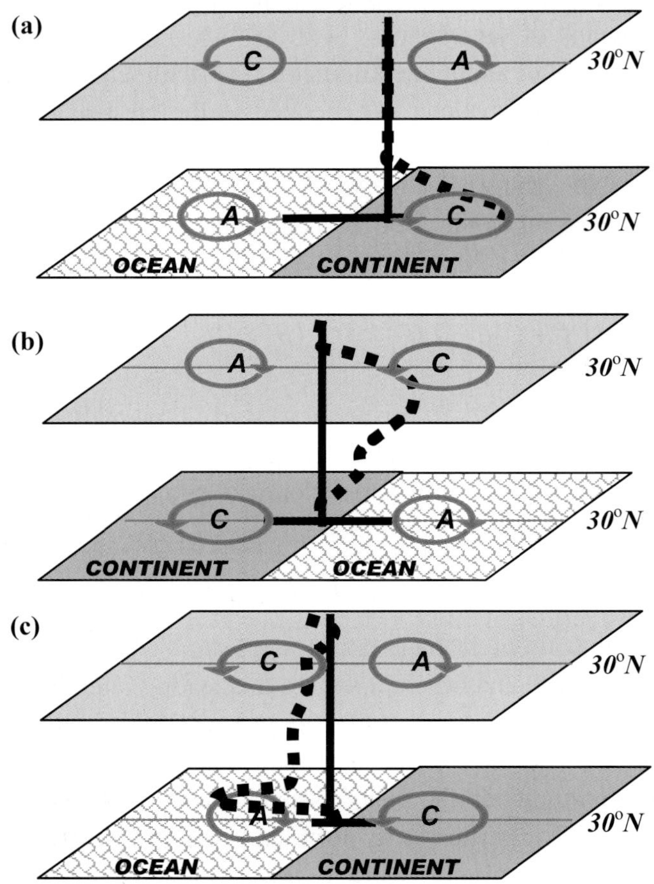

Figure 7. Schematic diagram indicating the summertime subtropical atmospheric response to the vertically differential diabatic forcing of (a) sensible heating, (b) deep condensation heating, and (c) longwave radiative cooling. "*A*" denotes anticyclone, "*C*" denotes cyclone, and the profiles at the centers of each panel indicate the dominating heating (adopted from Liu *et al.* 2003).

4.6. *Subtropical Anticyclone Forced by Long-Wave Radiative Cooling*

In summer in the subtropics radiative cooling overwhelms heating over the eastern oceans and on the poleward side of the western oceans (Wu and Liu 2003). In particular, it is stronger than -100 W m^{-2} over the eastern coastal region of each ocean basin. The maximum long-wave radiative cooling (LO) is usually below 850 hPa with an intensity of about – 6 K d^{-1}. Following a similar argument, such radiative cooling can produce a vorticity forcing of $+10^{-10}$ s^{-2} in the upper troposphere and -10^{-9} s^{-2} in the lower troposphere. Poleward flow of several m s^{-1} in the upper layer and stronger equatorward flow in the lower troposphere are then forced. Such radiation-induced meridional flows reinforce those circulation patterns generated by sensible heating (Figure 7a), and favor the formation of the subtropical anticyclone to the west of the radiative cooling in the lower troposphere but to its east in the upper troposphere, as shown in Figure 7c. We may therefore propose that, the strong negative vorticity forcing near the surface due to LO over the eastern oceans contributes to the eastward shift of the surface oceanic subtropical anticyclone, resulting in the asymmetric configuration of the surface subtropical anticyclones over oceans.

5. Sensible Heating and Subtropical Anticyclone

To investigate the influence of the summertime land surface sensible heating on the formation of the subtropical anticyclone, a simple idealized perpetual July experiment SH-0 is designed by using the GOALS climate model. In this experiment, the Earth is considered as an aqua-planet. The cloud distribution in all the experiments in this section is prescribed by using remote sensing climate data so that the cloud-radiation feedback is switched off. This allows us to isolate the impacts of the imposed thermal forcing on the atmospheric circulation. The simple experiment adopts a perpetual July integration in which the solar angle is fixed by its value on July 15. It uses only the July mean zonal symmetric climate forcing, including sea surface temperature, sea ice, cloud amount and atmospheric constituencies. The model is integrated for 24 months. The outputs of the last 12 months are then averaged for analysis. To mimic the surface sensible heating in the boreal subtropics in summer, two subtropical heating sources are imposed on the aqua planet. The first source is imposed in the region between 120°W and 90°W, and between 24.4°N and 46.6°N, with a heating maximum of 150 W m^{-2} located along 120°W and decreasing in sinusoidal form to zero along 90°W mimicking the distribution of surface sensible heat flux over the subtropical North America. The second source is imposed in the region between 0° and 105°E and between 24.4°N and 46.6°N. The sensible heating has a maximum of 150 W m^{-2} along the Greenwich Meridian, and decreases in sinusoidal form to zero at 105°E mimicking the distribution of surface sensible heat flux over the Africa-Eurasian area, as shown in Figure 8c. The model-induced condensation heating is not allowed to warm the atmosphere.

Results show that such an imposed surface sensible heating produces surface highs and 500 hPa lows over North Pacific and North Atlantic, but surface lows and 500 hPa highs over the two heating regions (Figures 8a and b). It can be seen from the vertical cross-section along 30°N that the main atmospheric response to such a surface sensible heating generates cyclonic circulation below 700 hPa and anticyclonic circulation above over the heating region, as is anticipated from the thermal adaptation theory stated in Section 4 and presented schematically in Figure 3d. In lower layers, strong northerlies are generated over the western heating regions (b). Surface highs are then produced to the west of the heating regions. These results can be well explained in terms of the Sverdrup balance as depicted in Figure 7a. It is important to note that with the imposed surface sensible heating alone, the prominent vertical out-of-phase feature in geopotential height observed along the summer subtropics (Figure 8d) is well captured in this simple experiment even in the absence of orography, and the generated pattern of geopotential height deviation at 1000 hPa (Figure 8b) is already similar to that in the observation (Figure 11c). This implies the fundamental importance of land surface sensible heating in the formation of the summertime surface subtropical anticyclones.

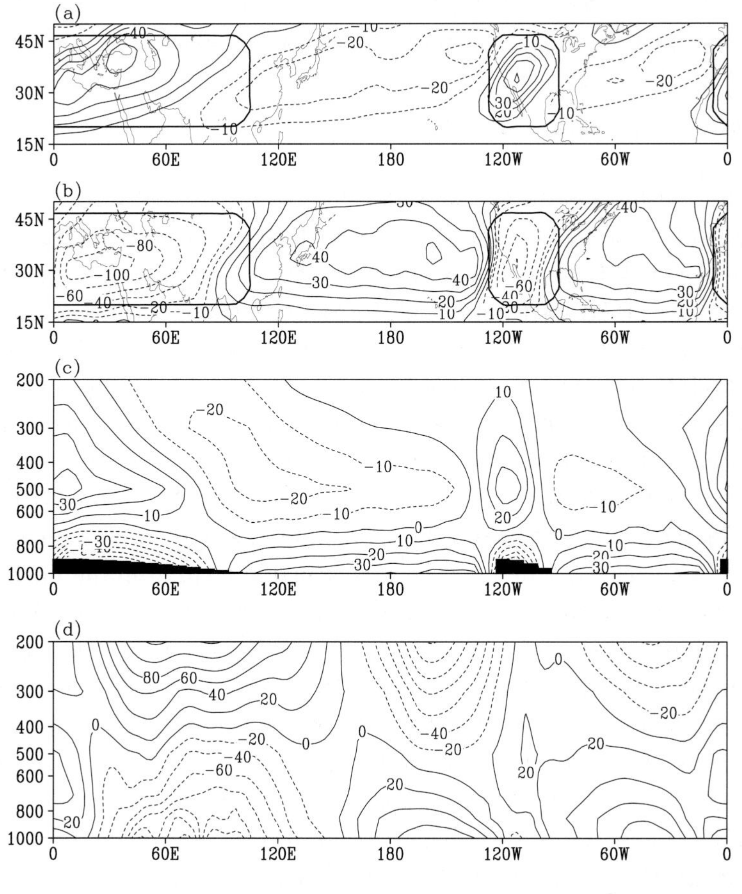

Figure 8. The July- mean zonal deviation of geopotential height (units: gpm) at (a) 500 hPa and (b) 1000 hPa, and (c) its vertical cross-section at 30°N in the perpetual-July idealized sensible-heating experiment SH-0 (adopted from Figure 8, Liu *et al.* 2003). (d) The same as (c) but in the observation. The heavy curves in (a) and (b) bound the region where heating is more than 1 W m^{-2}. Shading in (c) indicates the surface sensible heating region.

To see the relative importance of vorticity advection and β effect in the formation of the subtropical anticyclones over Northwestern Pacific and North America in the lower troposphere, the NCEP/NCAR reanalysis data from 1980 to 1995 are retrieved to evaluate the terms shown in the vorticity equation (33). Because the elevation of the ridge of the Rockies Mountains is about 3 km, 500 hPa is chosen to represent the lower troposphere. Panels (a) and (b) in Figure 9 show, respectively, the July mean distributions of the zonal asymmetric geopotential height at 1000 hPa and 500 hPa. At the surface over the eastern North Pacific (a), a positive center of more than 90 gpm is found to the west of North American. At 500 hPa the positive center is above North America (b), just down-stream of the maximum vorticity forcing as shown in (d). The distributions in the region of the surface sensible heat flux and the resultant vorticity forcing near the surface are shown in (c) and (d). It is prominent that the July surface sensible heat flux over the oceans is weak, and its maximum appears along the western coast of North America with intensity of about 150 W m^{-2} near 30°N (c). The corresponding vorticity forcing is negative, with maximum intensity of -3×10^{-10} s^{-2} located at about 35°N (d). At 500 hPa, the vorticity advection ($-\vec{V}\cdot\nabla\zeta$) is of positive sign with an intensity of $1.5\text{-}2.0\times10^{-10}$ s^{-2} above the surface forcing (e), and several times greater than the term $-\beta v$ (*f*). It is evident that the horizontal vorticity advection is a dominant factor in balancing the external vorticity forcing along the subtropics at higher levels in the lower troposphere. As a whole, the general features observed from Panels (a) to (f) are similar to those shown in the schematic diagram Figure7a, indicating the possible linkage between the surface sensible heating and the formation of the subtropical anticyclone via the β effect and vorticity advection. The general distributions shown in Panels (e) and (f) are well reproduced by the GOALS model (figures not shown).

To verify further the impacts of the surface sensible heating on the subtropical anticyclone, a pair of numerical experiments is designed. To focus on land surface processes, the oceanic component of the GOALS is switched off and replaced by a prescribed climate distribution of SST, which is based on observation and possesses only seasonal cycle. Then the model is integrated for 12 model years. Results from the last 10 years are extracted for analysis and defined as the Climate (CLI) run. The second numerical experiment is similar to the CLI run except the removal of surface sensible heat flux throughout the experiment, which is defined as the no-sensible-heat flux (nSH) run. Since the summer rainfall in the subtropical region over Northwestern America is limited in both the CLI run and nSH run (Liu 1998), the difference in rainfall over the area between CLI and nSH is not significant. Therefore, the differences between the two runs can be considered as the result of the surface sensible heating alone, and defined as SH-1. In Panels (g) and (h) of Figure 9 are shown the distributions of geopotential height in SH-1 at 1000 hPa and 500 hPa, respectively. These are close to the observation shown in (a) and (b) and the corresponding model simulations, and in good agreement with the schematic diagram shown in Figure 7a. We can therefore reach a conclusion that the formation of the subtropical anticyclone over Northeast Pacific and North America in summer is, to a great extent, forced by the land-surface sensible heating over the North American continent.

The main discrepancy between observation (a) and experiment (g) exists near the longitude of the anticyclone center: in the experiment the center is shifted towards the dateline, whereas in the observation it is closer to the western coast of North America. This is because in such an experiment design, the impacts of radiation cooling are excluded, which will be further explored in the following Section.

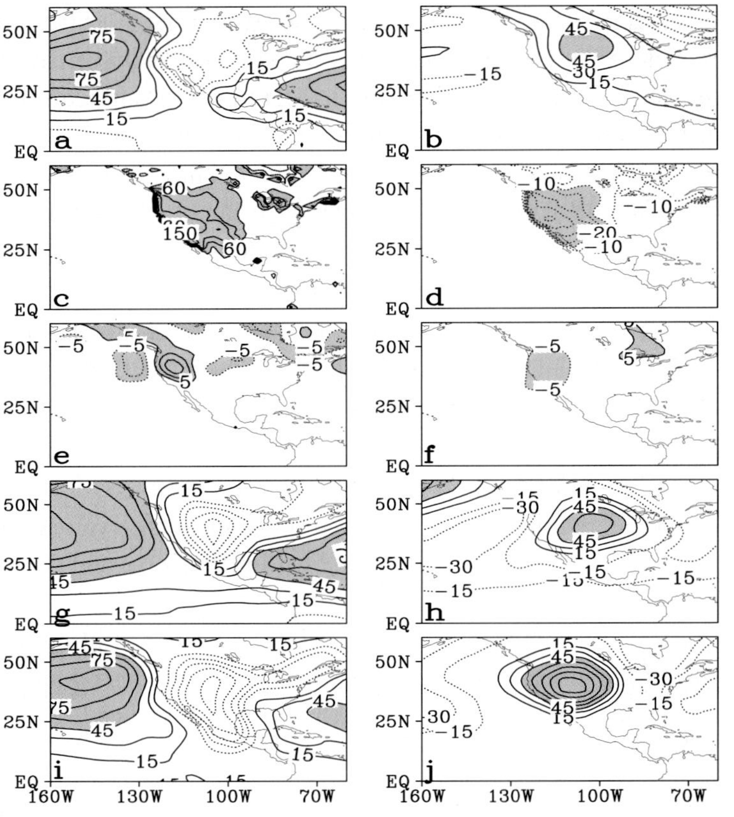

Figure 9. Distributions of July means calculated from the NCEP/NCAR reanalysis of the zonal deviation of geopotential height (gpm) at (a) 1000 hPa and (b) 500 hPa, (c) surface sensible heat flux (W m^{-2}), (d) vorticity forcing (10^{-11} s^{-2}), and vorticity advection at 500 hPa (10^{-11} s^{-2}) of (e) $-\boldsymbol{V}\cdot\nabla\zeta$ and (f) $-\beta v$. The zonal deviations of geopotential height (gpm) obtained from the GCM experiment SH-1 are shown in (g) for 1000 hPa and (h) for 500 hPa. Those obtained from the GCM experiment SH-2 are shown in (i) for 1000 hPa and (j) for 500 hPa.

6. Radiative Cooling and Subtropical Anticyclone

Upon employing the NCEP/NCAR reanalysis data, the geographical distribution of the column averaged radiative heating rate for July is calculated. Short-wave radiation is affected less by atmospheric constituencies and more zonal distributed except over high orography where the shortened atmospheric column gains less radiative heating. The heating rate is always positive except in the polar night area, but usually less than 1°C per day with the maximum heating of more than 0.9 K d^{-1} appearing in the subtropics of the summer hemisphere. By contrast, long-wave radiation always cools the atmospheric column, and is strongly asymmetrically distributed particularly along the subtropics in the summer hemisphere. Except in the mid-latitude area in winter, such long-wave cooling rate is more than 1 K d^{-1}. The areas of maximum cooling stronger than –2 K d^{-1} are found over the eastern parts of Pacific, Atlantic, and South Indian Oceans, and in the winter hemisphere over the ocean sectors along the subtropics. Less long-wave cooling is observed over high latitudes where the air temperature is low and over the tropics where deep convection occurs. The July distributions of total radiative heating, i.e., the sum of the short-wave and long-wave radiations, are shown in Figure 10a. They are negative everywhere, and the patterns resemble those of long-wave radiative cooling. This is because the amount of long-wave radiative cooling is stronger than that of short-wave radiative heating, and because the distribution of shortwave radiative heating is mainly zonal symmetric. The cooling belt more than -1.4°C d^{-1} meanders around the tropics and subtropics in the Southern Hemisphere. In the Northern Hemisphere such a strong cooling is confined only to the eastern Pacific and Atlantic Oceans and North Africa.

To see how the radiative cooling affects the subtropical circulation, in Figure 10b are shown the vertical profiles of different kinds of heating. The site selected for plotting is located at (124°W, 30°N), just within the strong radiative cooling region over the eastern North Pacific which is marked as L in Figure 10a. A sensible heating rate of 3.5 K d^{-1} is near the surface, but no deep condensation heating is observed. Strong long-wave radiative cooling of –6.5 K d^{-1} appears in the layer between 0.5 to 1.5 km (σ~0.85 to 0.95) above sea level, in correspondence with the existence of the *in situ* low stratus clouds (Rodwell and Hoskins 2001). The profile of the total heating resembles the profiles of the long-wave radiative cooling, with the maximum cooling near the 1.5 km altitude. Since the total heating decreases with altitude below this level, but increases with altitude above, based on (34), strong anticyclonic northerly near the surface and weaker cyclonic southerly in the upper troposphere should be forced.

In Figure 9g, although the surface sensible heating over continents can generate the surface anticyclones over the ocean sectors, the center of the anticyclone over the North Pacific is shifted to the central Pacific compared with the observation (Figure 9a). This may be due to lack of radiative cooling over the eastern North Pacific in this experiment according to the above discussions. To verify this hypothesis, another experiment is designed. It is similar to the climate run CLI but with the removal of latent heating in the thermodynamic equation, and labeled as SH-2. It can be considered as the atmospheric response to surface sensible heating and radiative cooling. As shown in Figures 9i and 9j, the zonal asymmetry of the surface oceanic anticyclone appears, and the simulated surface subtropical circulation (i) is closer to that in CLI. The forced anticyclone over North America is too strong because the model generates too strong sensible heating over the eastern part of North America. The anticyclone center over North Pacific is located at 145°W both in SH-2 and in CLI, in agreement with the observation (Figure 9a). Results from the above experiments then prove that it is mainly the radiative cooling over the eastern oceans that breaks the symmetry of

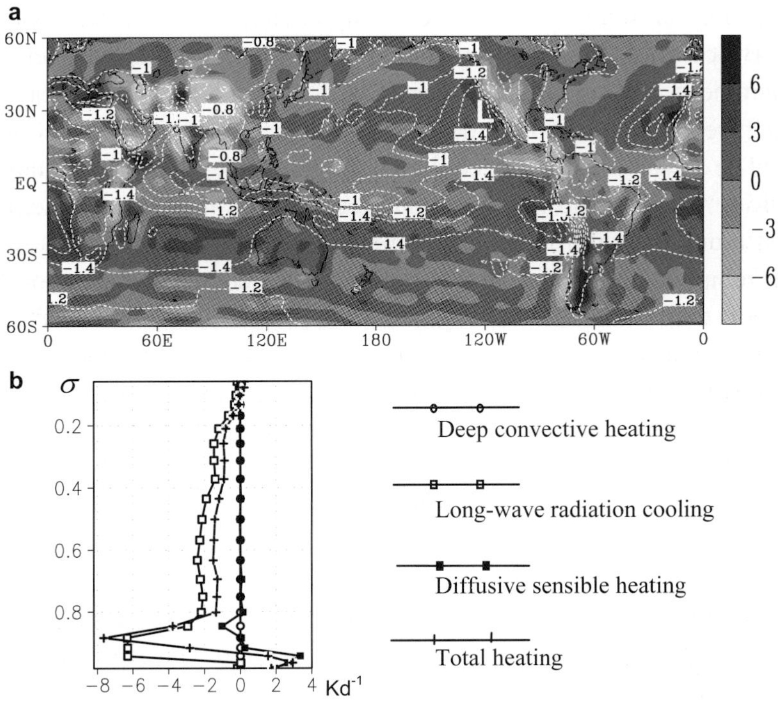

Figure 10. (a) Distributions of the July-mean column averaged total radiative heating (contours, unit: K d^{-1}) and the "vertical velocity" at 925 hPa (shading, unit: 10^{-2} hPa s^{-1}). (b) The vertical profiles of different kinds of heating at the site L (124°W, 30°N) as shown in (a).

the surface anticyclones over the oceans, and shifts their centers eastward towards the western coasts of the continents.

For comparison, the field of "vertical motion" at 925 hPa is also plotted in Figure 10a. The distributions of the vertical motion at levels below 850 hPa are similar and not shown here. It is clear that strong descent at the top of and within the planetary boundary layer is usually accompanied with the strong column integrated radiative cooling, and in coordination with the ridgeline of the surface subtropical anticyclone (Figure 11c). This is basically the same as in the zonal mean case (Figure 2), and can also be interpreted by the frictional impacts within the planetary boundary layer. Such descent is compensated for by the cross isobaric divergence at the ridgeline of the subtropical anticyclone, and does not contribute to the *in situ* mass build up.

The distribution of the vertical velocity at 925 hPa is in general similar to that at 500 hPa (Figure 11b). The main discrepancy exists over the western oceans in the Northern Hemisphere. Although the ridgeline of the surface subtropical anticyclone over the western Pacific is characterized by near surface descent, the region over the western subtropical

Pacific at 500 hPa is dominated by ascent. So it is over the western Atlantic. This can be understood by considering the thermal structure in the lower troposphere where the θ surface tilts upwards with increasing latitude. In adiabatic case, the southerly in the western part of a subtropical anticyclone should slide upward on the declined θ surfaces, while the northerly in its eastern part should slide downward. During the summer season, the monsoon latent heating to the west of the anticyclone and the radiative cooling to its east intensify the ascent over the western ocean and the descent over the eastern ocean. This is what we observed from Figure 11b. We are now facing another question: in eastern Asia, the approaching of an anticyclone is usually accompanied with clear sky, is this in conflict with the ascent feature of the middle tropospheric anticyclone? Notice that within the planetary boundary layer below these ascent regions, descent occurs along the ridgeline (Figure 10a). Since most of the water vapor in the atmosphere concentrates in the surface layer, descending motion and

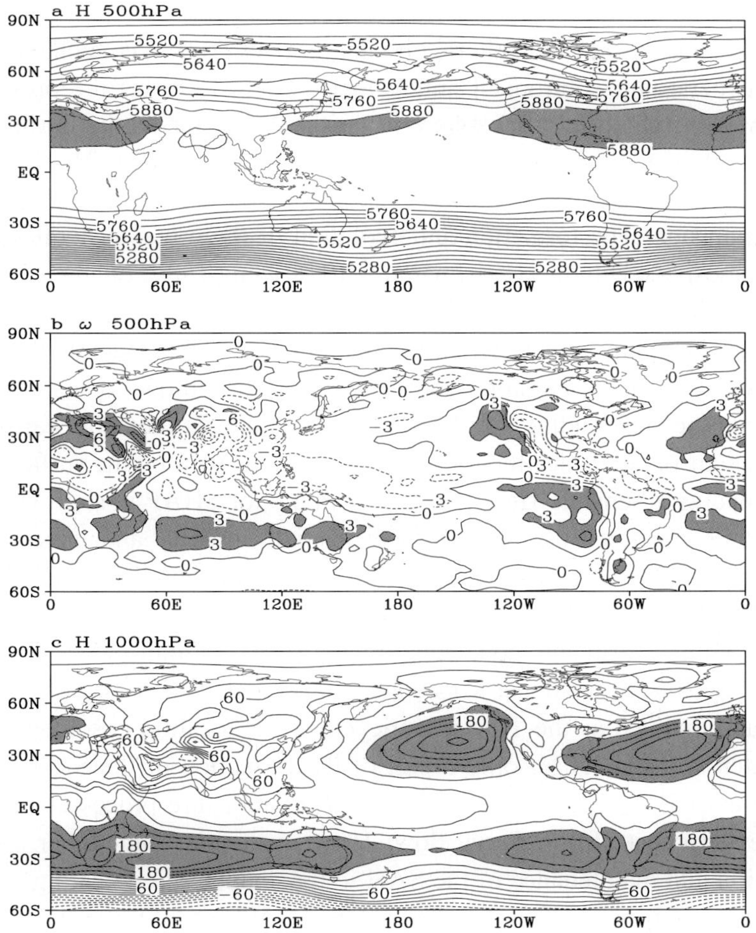

Figure 11. The distributions of the July-mean geopotential height at (a) 500 hPa and (c) 1000 hPa, and (b) ω at 500 hPa calculated from the NCEP/NCAR reanalysis. The contour intervals are 40 gpm in (a), 3×10^{-2} Pa s^{-1} in (b), and 20 gpm in (c).

the associated surface divergent flow prohibit the water vapor convergence and cloud formation in the layer along the ridgeline. Furthermore, since the adiabatic warming associated with descent decreases the relative humidity of the atmosphere, the feature of descent at the ridgeline of the surface subtropical anticyclone then explains why the weather in a region is usually clear when a surface anticyclone is moving towards the region.

7. Condensation Heating and Subtropical Anticyclone

The discussions presented in Sections 2 and 3 have shown that the classical descending theory does not work for the formation of the zonal mean anticyclone belt. This is also the case even in the three-dimension space. Figure 11 clearly shows that the lower layer subtropical anticyclones over the western oceans are associated with ascending motion in the middle troposphere. This cannot be interpreted in terms of sensible heating or radiative cooling. It is in fact due to the condensation heating associated with the Asian monsoon, as we will see below.

In the boreal summer, the atmospheric circulation over East Asia is characterized by the existence of two persistent subtropical anticyclone systems. One is the pronounced South Asian High (SAH) in the upper troposphere just over the region to the north of the Bay of Bengal; and the other, the subtropical anticyclone over the western Pacific (SAWP). The seasonal variations of these two systems are closely linked to the onset and withdrawal of the Asian summer monsoon. The formation and evolution of these two systems have long been the subjects of atmospheric studies (Krishnamurti 1973; Peng 1987; Zhang *et al.* 1995). After early spring, the strong elevated surface sensible heating of the Tibetan Plateau persistently warms up the air column aloft at a rate of 2 to $4^{o}C$ per day (Ye and Gao 1979; Wu and Zhang 1998). Strong ascent of the air column over the Tibetan Plateau due to such surface sensible heating then sucks the surrounding airs from below and expels those outwards in the upper layer, acting as an air pump. Since the air pump is mainly driven by the surface sensible heat flux, it was defined as a Sensible Heat Driven Air Pump, or SHAP in short (Wu *et al.* 1997; 2002a). The SHAP generates the near surface positive vorticity and upper-layer negative vorticity, and regulates the atmospheric circulations in the surrounding area. These processes and the mechanical forcing of the Tibetan Plateau are essential in maintaining the huge upper layer anticyclone with a warm and moist core (Ye *et al.* 1957; Flohn 1957; Ye and Wu 1998). However, the anticyclone in the upper troposphere generated in such a way is mainly over the Plateau, and usually named as the Tibetan anticyclone. Its location is somehow to the north of the SAH center. It accounts for only a part of the formation of the SAH. The numerical experiment results of Li and Luo (1988) show that the moisture processes can enhance the SAH and the development of meridional flow. There must exist some other mechanism that links the formation and maintenance of the SAH to the condensation heating in the monsoon area. On the other hand, most of the studies on the SAWP at 500 hPa were devoted to its impacts on the surrounding weather and climate (e.g. Huang and Yue 1962; Tao and Chen 1987; Samel *et al.* 1999).

In 1987, Nitta conducted a correlation analysis and found that there exists a wave train of geopotential height, which emanates from the heat source region near Philippines and propagates through Japan to North America. This wave train appears to be generated when convective activities over the Philippine Sea become intense. Huang and Li (1989) further showed that, during the northward propagation of these quasi-stationary planetary waves the SAWP shifts northward and gets intensified. Despite these great efforts the mechanism of the formation of the SAWP and its anomaly is still unclear in general. It is, however, interesting to notice that the movement of the SAH and that of the SAWP are not separated. The study of Tao and Zhu (1964) indicates that, the eastward migration of the 100 hPa SAH is usually followed by the westward movement of the 500 hPa SAWP. It then suggests the existence of some common mechanism that links the movement of one system to another.

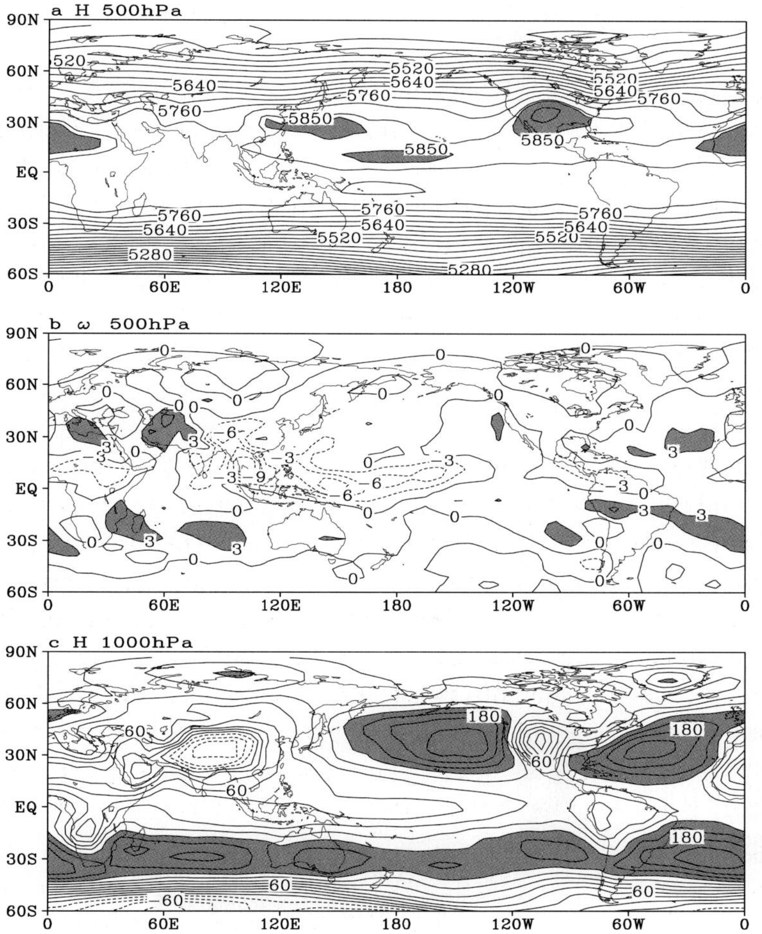

Figure 12. The same as Figure 9, but for the results derived from the IAP/LASG GAOLS climate model and averaged over a 16-year period.

To understand these, the GOALS model is employed again to conduct a series of numerical experiment. Figure 12 shows the circulation derived from the climate simulation CLI as described in Section 5. Compared with Figure 11, it can be seen that the space distributions of the subtropical anticyclones and vertical velocity are simulated reasonably well. The general features of the subtropical anticyclones over the western Pacific, Northern America, and Northern Africa at 500 hPa (Figure 12a), the two low-level oceanic anticyclones at 1000 hPa (Figure 12c), and the ascendant motion (Figure 12b) over the western parts of the 1000 hPa subtropical anticyclones are similar to the reanalysis (Figure 11). The apparent discrepancy in depicting the observed climate is over the Equatorial

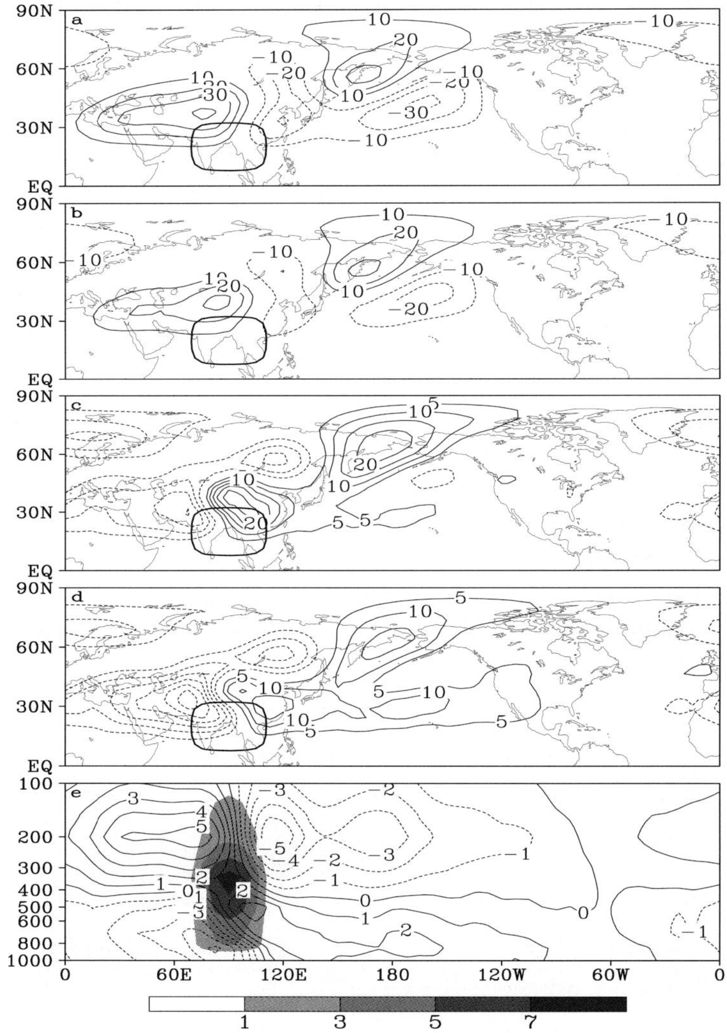

Figure 13. Distributions in the perpetual-July idealized latent-heating experiment LH-0 of the zonal deviation of geopotential height (gpm) at (a) 200 hPa, (b) 300 hPa, (c) 500 hPa and (d) 700 hPa, and of (e) the vertical cross-section of stream function (10^6 m^2 s^{-1}) at 30°N. The boxes in (a) to (d) indicate the region where heating is more than 1 K d^{-1} at the level σ=0.336. Shading in (e) indicates the vertical profile of the heating (units: K d^{-1}).

Middle Pacific, where the simulated geopotential height is too low at 1000 hPa and too high at 500 hPa, and the ascending over there is too strong.

In the real world, the atmospheric response to condensation heating may also be affected by other factors, such as orography, radiation and sensible heating, etc. A better understanding of the problem is sought through idealized experiments. To concentrate on the summer circumstances, the perpetual July experiment is again employed. An idealized "monsoon" heating, which mimics the three-dimension condensation heating distribution over the Bay of Bengal as obtained from the NCEP/NCAR reanalysis data, is introduced into the aqua-planet model integration (Liu *et al.* 2001). The boxes shown in Figure 13 indicate the elongated region where such heating is more than 1 K d^{-1} at the maximum heating level (σ=0.336). It is centered at 90°E and 20°N. The vertical profile of the heating is presented by the shading in Figure 13e. It possesses a maximum of 8 K d^{-1} at about 336 hPa and is modeled on what is observed from the reanalysis data. This experiment is labeled as LH-0, and its results are shown in Figure 13. Here the values of geopotential height are subtracted by their zonal means. Above the maximum heating level, high pressure appears on the western side of the heating source, and low pressure appears on its eastern side (Figures 13a and b). Below this level, the pressure pattern is reversed. High pressure appears on the eastern side of the heating source, and low pressure appears on its western side (Figure 13c and d). The vertical distribution of the stream function at 30°N (Figure 13e) is in good agreement with the above theoretical analysis: it is the vertical gradient of the condensation heating that produces a reverse circulation between the upper and lower troposphere, with positive deviation appearing to the west of the heating region in the upper troposphere, but to the east in the lower troposphere. This result is also in accordance with the numerical results of Hoskins and Karoly (1981) for a deep elliptical heat source at 15°N perturbing the Northern Hemisphere winter zonal flow.

Usually, it merits careful attention to conduct such a kind of idealized experiment because of the multi-scale nature of the atmospheric system and the scale-interaction issues. For the present study, these are not the major concerns since the dynamics involved is mainly linear. As stated by formula (34), the response of the meridional wind v to the forcing F_Q $(=\theta_z^{-1}(f+\zeta)\partial Q/\partial z)$ is roughly linear, i.e., $v \propto \lambda F_Q$. Each component of the source spectrum will force an in-phase v component. Therefore, for a stationary atmosphere, the forced strongest Southerly/Northerly should always be located just below/above the heating center, as shown in Figures 13a-d.

7.1. *Responses of the Zonal Symmetric Circulation to the 'Real' Latent Heating*

The above results show that the subtropical anticyclones can result from an idealized condensation heating at the subtropics. To make the study even close to reality, in this section a prescribed forcing of the July condensation latent heating is derived from the climate simulation (CLI) and saved at each grid point of the model. This forcing is considered as a 'real latent heating', and the atmospheric response to the forcing is investigated. Another perpetual July experiment LH-1 is designed for the present purpose. As used in the idealized experiment LH-0, in LH-1 no orography is used and the surface sensible heating is turned off.

The aforementioned "real" July latent heating is imposed as the sole external thermal forcing. Initiated with the same zonal-mean July circulation as in LH-0, this perpetual July experiment (LH-1) is also integrated for 24 months. Similarly, the zonal deviation results averaged over the last 12 months are employed for the following analysis.

The plot of precipitation shown in Figure 14a reflects the distribution of vertically integrated heating. The result shows that in the subtropics over the Asian monsoon, more than 2 mm d^{-1} covers from the eastern Tibetan Plateau to the west of the dateline. The steady responses of geopotential heights in LH-1 (Figure 14b-c) are similar to those in the idealized experiment LH-0. At 200 hPa along 30°N, there exist positive height deviations to the west of the heating, ranging from the Tibetan Plateau to the lower reaches of the Yangtze River with a maximum as high as 80 gpm. The negative height is over the central and eastern Pacific with a center of –60 gpm. Similar to Figures 13, Rossby wave excited by the heating propagates northeastward. In mid-high latitudes, it is still equivalent barotropic in nature, but with strength twice stronger than that in Figure 13. Moreover, the response centers shift eastward compared with those in LH-0. Such results are related to the fact that in LH-1, the imposed latent heating is stronger, and the region of the heating shifts eastward as well. At 500 hPa, the pattern in the subtropics is opposite to that at 200 hPa. Over the western Pacific, the high pressure appears to the east of the condensation heating with the strength of 20 gpm.

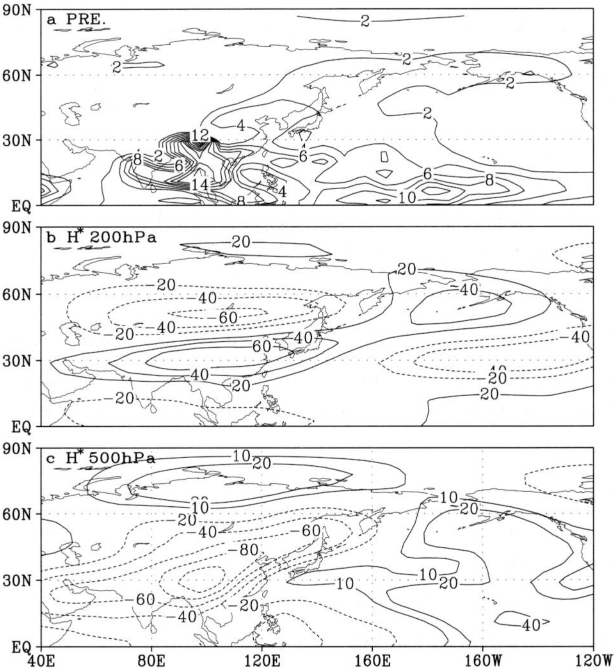

Figure 14. Distributions in the perpetual-July latent-heating experiment LH-1 of (a) precipitation (mm d^{-1}), and the zonal deviation of geopotential height (gpm) at (b) 200 hPa, and (c) 500 hPa.

On the other hand, in a contrast experiment in which the condensation heating is removed, there exists merely an almost zonal symmetric high belt over the subtropics at each level (figures not shown). It is clear that the prescribed July condensation heating can break the zonal symmetric anticyclone belts into isolated anticyclones and cyclones. It contributes to the formation of the SAH and tropical upper-tropospheric trough (TUTT) in the upper troposphere (200 hPa) and the SAWP in the middle troposphere (500 hPa). Using a linear two-layer model, Webster (1972) introduced a latent heating distribution for JJA as a prescribed diabatic heating to the integration, which was based on the estimates of cloudiness and also possesses a maximum over China and western Pacific along the boreal subtropics (his Figure 2c). Strong northeasterly at 250 hPa and southwesterly at 750 hPa was generated over this heating area, and a strong upper layer subtropical anticyclone from Japan to the eastern North Africa and a weaker low layer subtropical anticyclone over the western Pacific were produced (his Figure 15). Similar features are also presented in the linear model outputs of Lin (1983, Figure 9). They agree well with those shown in Figures 13 and 14.

7.2. *Comparison between Condensation Heating and Sensible Heating*

The next experiment (LH-2P) is almost the same as LH-1 but with orography included during the integration. The height deviation and condensation heating shown in Figure 15b indicate that with orography and land-sea distribution included the subtropical anticyclone in the middle and lower troposphere and the SAH in the upper troposphere can be produced to some extent by condensation latent heating alone. These results are consistent with the theoretical analysis presented in Section 4.5, and with the results from the idealized latent heating experiment LH-0 and the model-prescribed latent heating experiment LH-1. Comparing Figure 15b with Figure 14b, we see that the inclusion of the Tibetan Plateau in LH-2P causes the subtropical anticyclone at 200 hPa to shift westward to approach the observed location, just in phase with the plateau.

To reveal the relative importance between condensation heating and sensible heating in the formation of the subtropical anticyclone, another pair of perpetual July experiment is designed (Liu *et al.* 2003). The first one is similar to LH-2P but with the inclusion of sensible heating. It is indeed a perpetual July control experiment and labeled as CON-P. The second one is the same as the CON-P but with the removal of latent heating in the thermodynamic equation. It can be considered as an experiment with sensible heating as an external forcing, and labeled as SH-2P. The cross-sections along 30°N of the zonal deviation of geopotential height for these experiments are shown in Figure 15. The vertical out-of-phase feature in SH-2P (a) is similar to that in the idealized experiment SH-0 (Figure 8c), and enhanced due to the inclusion of orography. In LH-2P (b), the Sverdrup balance pattern that was obtained from the idealized run LH-0 (Figure 13d) is prominent over either the Asian monsoon region or the North American monsoon region: the local maximum positive geopotential height is observed to the west of the maximum heating in the upper layer and to its east in the lower layer, not at the surface. Besides, surface cyclone exists just below the heating.

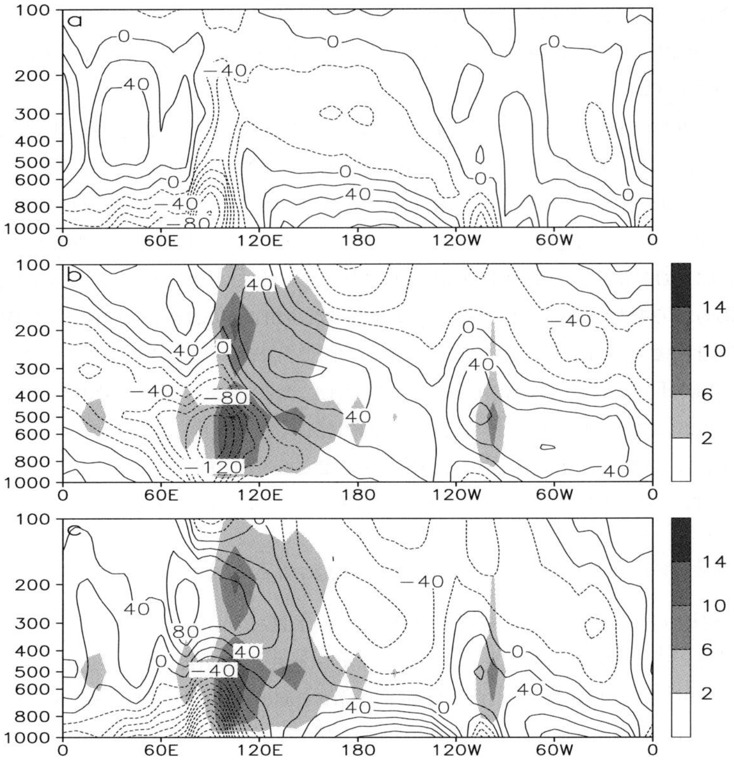

Figure 15 Cross-sections of the zonal deviation of geopotential height (unit: gpm) along 30°N in the perpetual July experiments (a) SH-2P, (b) LH-2P and (c) CON-P. Shading indicates the July distribution of precipitation in the GOALS climate simulation (CLI) that is taken as a latent heating source in these experiments (adopted from Liu *et al.* 2003).

It is interesting to compare the results of SH-2P and LH-2P with those of CON-P. The vertical out-of-phase feature in CON-P is well simulated in SH-2P particularly over the oceans. Surface continental cyclones and oceanic anticyclones in CON-P are also presented well in SH-2P. The magnitude of the surface anticyclone over Pacific in SH-2P is over 80 gpm and accounts for about 80% of its counterpart in CON-P (>100 gpm), and that over Atlantic is more than 60 gpm and accounts for more than 70% of its counterpart in CON-P (>80 gpm). The contributions to these surface anticyclones from condensation heating become secondary. In these aspects, sensible heating is more important than latent heating in maintaining the circulations. Over the two monsoon regions, however, the geopotential height fields in CON-P bear the strong impacts of the Sverdrup balance that is driven by condensation heating. As a result, a geopotential height center with 60 gpm at 200 hPa over Asia and another center with similar strength at 500 hPa over North America produced by condensation heating in LH-2P are rather close to their counterparts in CON-P, and the subtropical anticyclone at 500 hPa over western Pacific in CON-P owes its existence mainly to the condensation heating. It is important to note that the sensible heating and condensation heating act together to strengthen the anticyclone over Asia and the troughs over oceans in the upper troposphere, as well as the oceanic subtropical anticyclones and the huge Asian cyclone near the surface.

In LH-2 (Figure 15b), the SAH center is only 60 gpm. The SAWP strength is as high as 40 gpm at 500 hPa. The strength of the Subtropical Anticyclone over the Northern Pacific (SANP) is less than 20 gpm. Thus in LH-2, the forced SAH and SANP are too weak, and the SAWP is too strong. Since the geopotential height deviation forced by sensible heating is out of phase with that forced by condensation heating (Figure 7), when these two kinds of forcing act together, some compensation in the forced geopotential height may be anticipated. Therefore, the too strong SAWP at 500 hPa and too weak SAH and SANP in LH-2 must owe to the neglect of the effects of sensible heating. This is indeed the case. Figure 15c shows that the sensible heating strengthens the SAH in the upper troposphere, reduces the geopotential height over oceans and weakens the SAWP, and intensifies greatly the SANP and the subtropical anticyclone over the North Atlantic. Consequently, the cross-section of the deviation circulation pattern shown in Figure 15c is closer to that from the reanalysis data (Figure 8d).

It becomes clear that the sensible heating and condensation heating in summer play different roles in maintaining the subtropical circulations. Together with radiative cooling, they are all important in forming the summertime subtropical anticyclones. Although the magnitude of latent heating associated with the summer monsoon is stronger than sensible heating, the sensible heating over land surface is more fundamental in maintaining the summertime subtropical anticyclones near the surface.

The above conclusion is in contradiction to those drawn by Webster (1972) and Egger (1978) who attempted to separate the contributions of sensible and latent heating to the maintenance of the summertime subtropical anticyclones and stressed the dominant importance of latent heating. However as shown in Figure 5, the substantial part of sensible heating is concentrated below 800 hPa. Such a very low layer heating exerts profound impacts on the circulations not only in the lower layers but also in the upper layers as demonstrated in Figures 8, 12 and 15. On contrary, the sensible heating above 800 hPa is negligible compared with other heating. Therefore it may not be appropriate to evaluate the impacts of sensible heating on the atmospheric circulation by only using the kinds of two-layer circulation model in which the lower layer is at either 800 or 750 hPa. Chen *et al.* (2001) introduced two latent heating sources, respectively, over Asia and North America to a linear quasi-geostrophic (QG) model (Chen 2001) to study the atmospheric responses to the summertime diabatic heating, and interpreted the surface oceanic subtropical anticyclones in the Northern Hemisphere as a remote response of Rossby waves forced by the large-scale heating source over Asia, and the circulations over the two continents as a local response to monsoon latent heat release in the mid-troposphere. As discussed by Liu *et al.* (2003), the fact that the QG model lacks the mechanism of the heating-adjusted static stability and vorticity forcing can lead to an exaggerated surface response. In addition, the observed monsoon rainfall is located over Eastern China to western Pacific and over eastern North America to western Atlantic, whereas the imposed latent heating sources for their experiment were located between 50° and 130°E and between 116° and 94°W, too far to the west of the observations. These make it difficult to compare the current study with their results.

8. Synthetic Heating and the Summertime Subtropical Anticyclones

In previous sections, the impacts of individual thermal forcing on the summertime subtropical circulations have been studied. However, in real atmosphere such individual thermal forcing is not isolated. Instead, they are well organized. Wu and Liu (2003) showed that in the summer subtropics over each continent and its adjacent oceans, the local dominant heating is organized in a pattern with long-wave radiative cooling (LO) located to the west of the

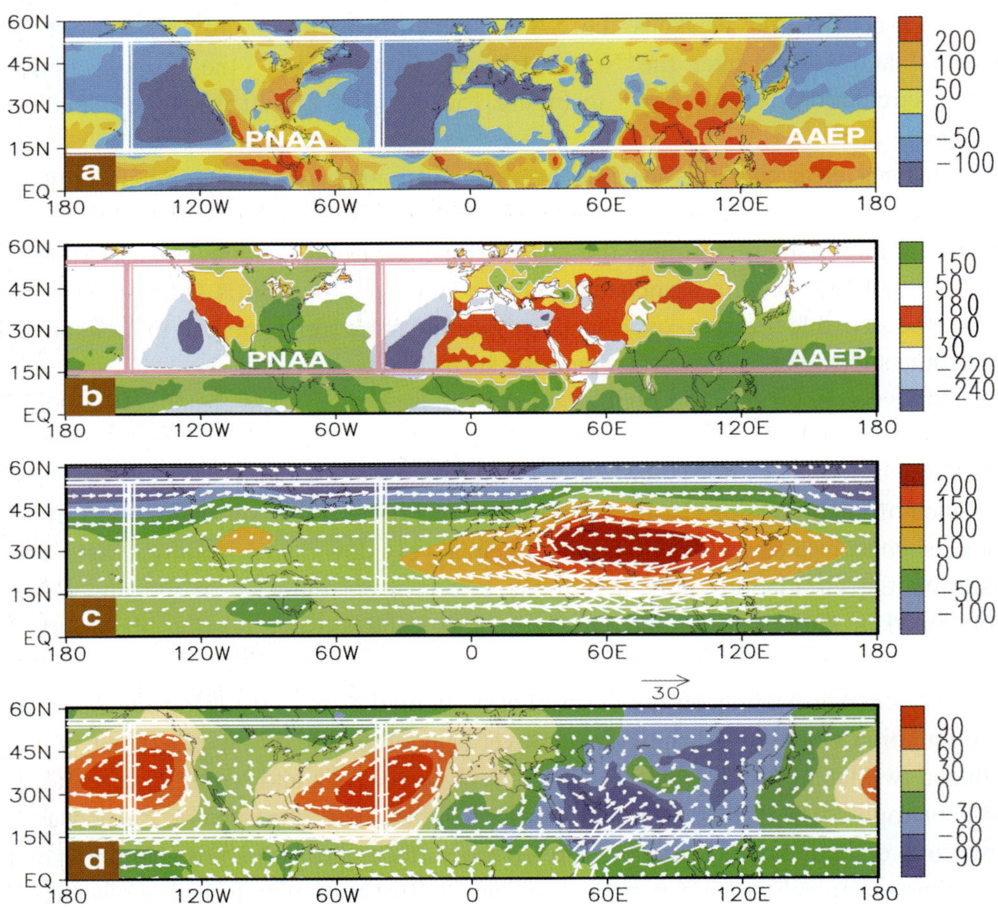

Figure 16. July-mean distributions in the Northern Hemisphere of the column-integrated (a) total heating and (b) main local heating in units of W m^{-2}, and the wind vector and deviation of geopotential height from the equatorial zonal-mean at (c) 100 hPa and (d) 1000 hPa with units of gpm. The heating distributions (a and b) demonstrate a mosaic of the quadruplet LOSECOD heating patterns, and the circulations (c and d) also demonstrate a mosaic of the circulation pattern with anticyclonic circulations over the continent in the upper troposphere and over the adjacent oceans in the lower troposphere (adopted from Wu and Liu 2003).

continent, sensible heating (SE) and condensation heating (CO) over the western and eastern continent, and the double dominant heating (LO and CO) to the east of the continent. These then compose a LOSECOD heating quadruplet pattern. This heating quadruplet forces a circulation pattern with the subtropical anticyclonic circulations over the continent in the upper troposphere and over the adjacent oceans in the lower troposphere. By employing the reanalysis data of NCEP/NCAR from 1980 to 1997, they demonstrated further that the global summer subtropical heating and circulation may be viewed as "mosaics" of such a quadruplet heating and circulation patterns, respectively. These are illustrated in Figures 16 and 17 (Figures 3 and 4, Wu and Liu 2003).

In the northern subtropics there are two big continents. By setting the lateral boundaries at the longitudes over the oceans where SE is negligible and the surface meridional wind components vanish, the subtropical belt can be divided into two regions, i.e., the Pacific-North America-Atlantic (PNAA) region and the Atlantic-Africa-Eurasia-Pacific (AAEP) region. The distribution of the July-mean column-integrated total heating and main local heating over the entire northern subtropics are respectively presented in Figures 16a and 16b. The LOSECOD heating quadruplet over AAEP or PNAA can be identified. Strong LO lobes are located over the eastern North Atlantic and North Pacific. SE lobes occupy the western and central AAEP and western PNAA areas. CO lobes are found over the East Asian monsoon region and eastern North America. Although deep condensation heating of more than 50 W m^{-2} extends from the East China Sea to Japan and from Florida Peninsular northeastward (Figure 16b), radiative cooling overwhelms the heating over western Pacific and Atlantic (Figure 16a).

The July-mean circulations of the Northern Hemisphere at 100 and 1000 hPa are shown in Figures 16c and 16d, respectively. Along the subtropics over AAEP and PNAA, the aforementioned circulation pattern can also be detected, and is well coordinated with the LOSECOD quadruplet. The 100 hPa anticyclonic circulations appear over the two continents, in correlation with the SE and CO heating lobes, whereas the surface anticyclonic circulations appear over the two oceans, in correlation with the *in situ* column integrated cooling. These can be well explained by the atmospheric thermal adaptation to the heating/cooling, as stated in Section 4. Furthermore, when the circulation patterns over AAEP and PNAA are placed side by side, the two troughs at 100 hPa and the two strong subtropical anticyclones at 1000 hPa appear just at the joined edges.

However, such anticyclonic circulations near the surface (Figure 16d) are strongly asymmetric about their central meridional axis. That is, over the LO lobe, the equatorward flow is strongly developed, whereas over the D lobe, only is the southwesterly observed just in coordination with the band of deep condensation heating over the western oceans (Figure 16b). Furthermore, the meridional winds over the LO and SE lobes to the west are equatorward near the surface but poleward in the upper troposphere, whereas they are poleward near the surface but equatorward in the upper troposphere over the CO and D lobes to the east. The correspondence between the profile of heating Q and meridional wind v can be interpreted by employing the Sverdrup balance (34). This relation implies that in a steady state and in the absence of zonal advection, the negative (positive) vorticity generation due to the decrease (increase) with height of a heating Q must be balanced by positive (negative) planetary vorticity advection that is brought in by meridional winds from high (low) latitudes. Since the total heating in the LO and SE lobes decreases with height rapidly in the lower layer, but increases with height in the deep upper layer (Figure 1c, Wu and Liu 2003), the *in situ* strong near surface equatorward flow and weaker upper layer poleward flow should be generated. A similar argument applies in the CO and D lobes. The axially asymmetric surface anticyclones are then produced.

It becomes apparent that eastern portion of each strong oceanic surface subtropical anticyclone is substantially affected by radiative cooling and continental sensible heating, whereas its western part is to a great extent affected by radiative cooling as well as condensation heating associated with the summer monsoon.

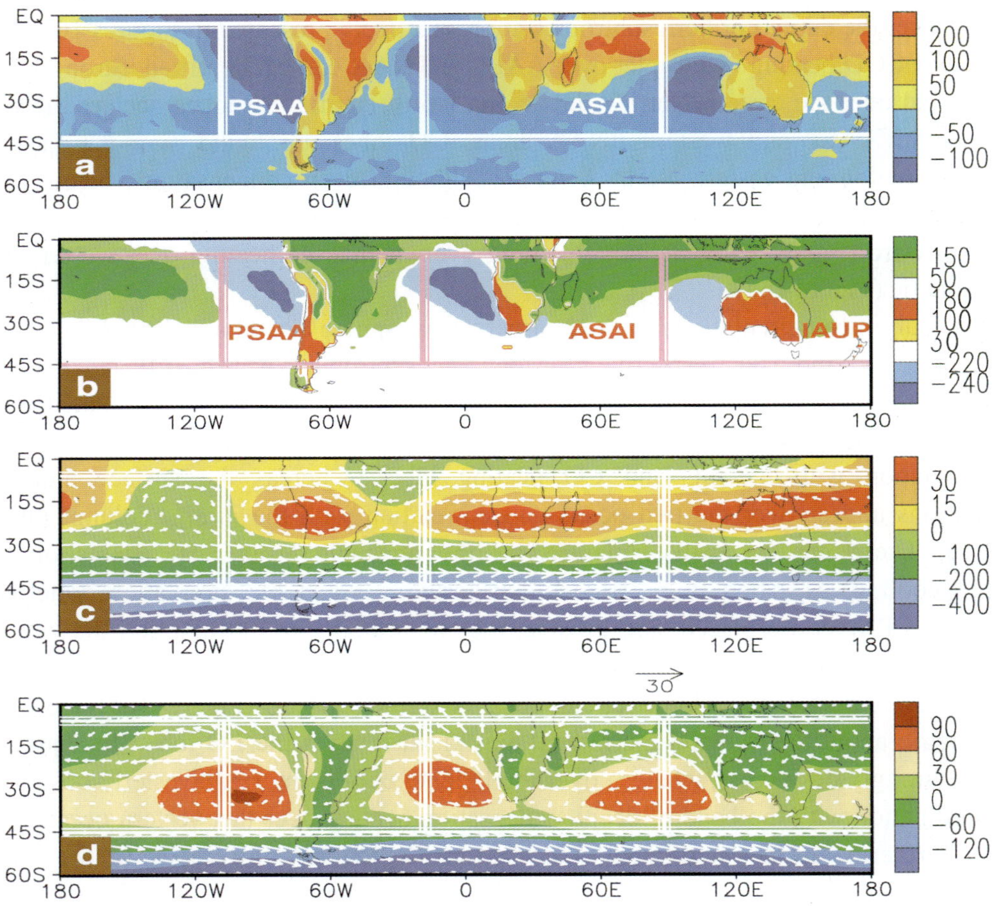

Figure 17. Same as in figure 16 for the January- mean distributions in the Southern Hemisphere (adopted from Wu and Liu 2003).

In the southern subtropics there are three continents. By selecting the three longitudes of 110°W, 20°W, and 90°E as boundaries, the southern subtropics can be divided into three regions: the Pacific-South America-Atlantic (PSAA), Atlantic-South Africa-Indian Ocean (ASAI), and Indian Ocean-Australia-Pacific (IAUP) regions, as shown in Figure 17. The distribution of the January-mean total heating is presented in Figure 17a, and the main local heating, in Figure 17b. The LOSECOD quadruplet can also be identified between 15° and 35°S in each region, with the central heating lobes SE and CO located over the continent, and radiative cooling over the oceans.

The January-mean circulations of the Southern Hemisphere at 100 and 1000 hPa are shown in Figures 17c and 17d, respectively. Again, the aforementioned circulation pattern is prominent over each continent. The intensities of the three upper tropospheric anticyclones (greater than 30 gpm) are similar (Figure 17c). When the three circulation patterns are tiled side by side, troughs in the upper troposphere and anticyclones near the surface are found at the three joined edges, whereas the upper level anticyclones appear over the continents, as is the case in the Northern Hemisphere.

9. Discussions and Conclusions

Traditionally, subtropical anticyclone is considered as a result of atmospheric descent, and regarded as a giant master that controls the movement of the surrounding weather systems such as typhoons, fronts, torrential rains and westerly troughs. Through theoretical investigation and numerical experiment, our studies show that the atmospheric descent cannot be used as a mechanism to interpret the formation of the subtropical anticyclone, and there exists interaction between the subtropical anticyclone and the latent heating associated with monsoon. In the summer time, diabatic heating along the subtropics plays important roles in the formation of the subtropical anticyclone. Strong surface sensible heating over continents breaks the zonal symmetric subtropical anticyclone belt, forming surface lows over continents and highs over oceans. Because the strong radiative cooling over the eastern oceans produces anticyclonic vorticity and strong equatorward flow near the surface, the center of the oceanic subtropical anticyclone is shifted eastward towards the western coast of the continent. Because this book concentrates on the eastern Asian monsoon, much material presented here is contributed to the study of the subtropical anticyclone over the region. It is shown that the latent heat release of the Asian summer monsoon contributes significantly to the formation of the South Asian High in the upper troposphere and the anticyclone over the western Pacific in the middle and lower troposphere. The orographic forcing of the Tibetan Plateau and the surface sensible heating over the land surface also have strong impacts on their locations and intensities.

The findings that the Asian monsoon rainfall can significantly affect the formation of the subtropical anticyclone over the western Pacific are of great importance. It raises an urgent need for change in the methodology of short-term climate prediction. For many decades in meteorology, people predicted the behaviors of the subtropical anticyclone over the northwestern Pacific in summer months by using different statistical means and data collected several months before. Then the prediction of a distribution of rainfall anomalies is induced. Results from this study indicate that this method is inadequate, because it uses a false cause-effect relation between rainfall and the variation of subtropical anticyclone. For the prediction of the anomalous monsoon rainfall, we have to seek other external forcing mechanisms, such as sea surface temperature anomalies in some oceanic regions, the Eurasian snow cover in winter months, and the thermal status of the Tibetan Plateau, etc. Continuous efforts oriented in this direction will lead us to a better climate prediction in the future.

The discussions presented in this Chapter concentrate on the climate time-scale. The short-term variation of the subtropical anticyclone is more complicated, and the interactions between the subtropical anticyclone and the surrounding weather systems as well as the global circulation adjustment should be considered. These are under study and new results are anticipated.

Acknowledgments

This study is supported jointly by Natural Science Foundation of China under the project number 49635170, 40135020 and 40221503, and by the Chinese Academy of Sciences under the grant number ZKCX2-SW-210. The authors thank Dr. Kevin Hamilton for helpful comments on the manuscript.

References

Bi, M. Y., 1989: Oscillation of the subtropical anticyclone over the western Pacific in summer. *Acta Meteorologica Sinica,* **47**, 468-474.

Bolin, B., 1950: On the influence of the earth's orography on the general character of the westerlies. *Tellus*, **2**, 184-195.

Charney, J. G., and A. Ellassen, 1949: A numerical method for predicting the perturbations of the middle-latitude westerlies. *Tellus*, **1**, 38-54.

Chen, P., 2001: Thermally forced stationary waves in a quasigeostrophic system. *J. Atmos. Sci.,* **58**, 1585-1594.

Chen, P., M. P. Hoerling, and R. M. Dole, 2001: The origin of the subtropical anticyclones. *J. Atmos. Sci.,* **58**, 1827-1835.

Eady, E. T., 1949: Long waves and cyclone waves. *Tellus,* **1**, 33-52.

Egger, J., 1978: On the theory of planetary standing waves: July. *Beitr. Phys. Atmos.*, **51**, 1-14.

Ertel, H., 1942: Ein neuer hydrodynamische wirbdsatz. *Meteorology. Z. Braunschweig*, **59**, 277-281.

Fan, Z. F., 1986: Interaction between the subtropical anticyclone in the Northern Hemisphere and the sea- ice over the Northern Pole. *Sci. Bull.,* **31**, 286-289.

Flohn, H., 1957: Large-scale aspects of the "summer monsoon" in South and East Asia. *J. Meteor. Soc., Japan* (75th Ann. Vol.), 180-186.

Fu, Z. B., and X. L. Teng, 1988: Climate anomalies in summer over China and El Nino/Southern Oscillation. *Atmos. Sci.,* Special Issue, 133-141.

Gao, Y. F., 1977: Impacts on the climate over China of the land- sea contrast and the Tibetan Plateau. *Collection of Tibetan Plateau Meteorology*, Gao Youxi, Ed., Science Press, Beijing, 34-46.

Gill, A. E., 1980: Some simple solutions for heat-induced tropical circulation. *Quart. J. Roy. Meteor. Soc.,* **106**, 447-662.

Haynes, P. H., and M. E. McIntyre, 1987: On the evolution of vorticity and potential vorticity in the presence of diabatic heating and frictional or other forces. *J. Atmos. Sci.,* **44**, 828-841.

Held, I. M., and A. Y. Hou, 1980: Nonlinear axially symmetric circulations in a nearly inviscid atmosphere. *J. Atmos. Sci.*, **37**, 515-533.

Holton, J. R., 1992: *An introduction to dynamic meteorology.* 3rd edition, Academic Press. Inc., New York, 117-140.

Hoskins, B. J., 1991: Towards a PV-θ view of the general circulation, *Tellus*, **43**AB, 27-35.

______, 1996: On the existence and strength of the summer subtropical anticyclones. *Bull. Amer. Meteor. Soc.,* **77**, 1287-1292.

______, and D. J. Karoly, 1981: The steady linear response of a spherical atmosphere to thermal and orographic forcing. *J. Atmos. Sci.*, **38**, 1179-1196.

Huang, S. S., 1963: Longitudinal movement of the subtropical anticyclone and its prediction. *Acta Meteorologica Sinica,* **33**, 320-332.

______, and Z. H. Yue, 1962: On the structure of the subtropical highs and same associated aspects of the general circulation of atmosphere. *Acta Meteorologica Sinica*, **31**, 339-359.

Huang, R. H. and L. Li, 1989: Numerical simulation of the relationship between the anomaly of the subtropical high over East Asia and the convective activities in the western tropical Pacific. *Adv. in Atmos. Sci.*, **6**, 202-214.

______, and W. J. Li, 1988: Impacts of the heat source anomaly over the western equatorial Pacific on the subtropical anticyclone over eastern Asia in summer and the associated mechanism. *Atmos. Sci.,* Special Issue, 107-116.

IPCC, 2001: Chap. 8, Climate Change 2001: The Scientific Basis. *Contribution of Working Group I to the Third Assessment Report of the Intergovernmental Panel on Climate Change*, Houghton, J. T. etc., Eds., Cambridge University Press, Cambridge and New York, 881pp.

Ji, L., R. J. Shen, and Y. X. Chen, 1982: A numerical experiment on the dynamic and thermal effects of the Qinghai-Xizang (Tibet) Plateau in Summer. *Proc. of the first Sino-Amer. Workshop on Mountain Meteorology,* Science Press, 355-370.

Jiang Q. R, and B. N. Wang, 1989: Solar radiation and the seasonal movement of the subtropical anticyclone. *Tropical Meteor.,* **5**, 289-294.

Kalnay, E., and co- authors, 1996: The NCEP/NCAR 40- year reanalysis project. *Bull. Amer. Meteor. Soc.,* **77**, 437-471.

Krishnamurti, T. N. *et al.*, 1973: Tibetan High and upper tropospheric tropical circulation during northern summer. *Bull. Amer. Meteor. Soc.*, **54**, 1234-1249.

Kuo, H. L., 1956: Forced and free meridional circulations in the atmosphere. *J. Meteor.,* **13,** 561-568.

Kurihara, K. 1989: A climatological study on the relationship between the Japanese summer weather and the subtropical high in the western northern Pacific. *Geophy. Mag.*, **43**, 45-104.

______, and T. Tsuyuki, 1987: Development of the barotropic high around Japan and its association with Rossby wave-like propagations over the North Pacific: Analysis of August 1984. *J. Meteor. Soc. Japan*, **65**, 237-246.

Li, M. C., and Z. X. Luo, 1988: Effects of moist process on subtropical flow patterns and multiple equilibrium states. *Sci. Sinica(B)*, **31**, 1352-1361.

Liao, Q. S., and Z. G. Zhao, 1992: Anomaly in the latitude location of the subtropical anticyclone over the northwestern Pacific in July and August and its impacts on the weather over China. *Long-term Weather Forecasts and the Sun- Earth Relation,* Zhang and Huang, Eds., Ocean Press, Beijing, 131-139.

Lin, B. D., 1983: The behavior of stationary waves and the summer monsoon. *J. Atmos. Sci.*, **40**, 1163-1177.

Lindzen, R. S., and A. Y. Hou, 1988: Hadley circulations for zonally averaged heating centered off the equator. *J. Atmos. Sci.*, **45**, 2416-2427.

Liu, H., and G. X. Wu, 1997: Impacts of land surface on climate of July and onset of summer monsoon: A study with an AGCM plus SsiB. *Adv. in Atmos. Sci.,* **14**, 289-308.

Liu, P., 1999: Interannual variations of the zonal mean subtropical anticyclone and the subtropical anticyclone over western Pacific and their association with the anomaly in sea surface temperature, Ph.D. Thesis, Ins. Atmos. Phys., 166pp.

Liu, Y. M.., 1998: Impacts of spatially non-uniform diabatic heating on the formation of subtropical anticyclone in boreal summer, Ph.D. Thesis, Ins. Atmos. Phys., 135pp.

______, and G. X. Wu, 2000: Review on the study of the subtropical anticyclone and several relevant fundamental problems. *Acta Meteorologica Sinica*, **58**, 500-512.

______, ______, H. Liu, and P. Liu, 2001: Dynamical effects of condensation heating on the subtropical anticyclones in the Eastern Hemisphere. *Climate Dyn.*, **17**, 327-338.

______, ______, and R. C. Ren, 2003: Relation between the subtropical anticyclone and diabatic heating. *J. Climate*, in press.

Luo, M. X., B. Z. Zhu, and R. J. Sheng, 1986: Numerical experiments on the impact of large- scale orography on the formation of subtropical anticyclone in the Northern Hemisphere in summer. *Atmos. Sci.,* **10**, 258-264.

Nikaidou, Y. 1988: Effects of high SST anomaly over the tropical western Pacific on climates predicted in 4-month integrations of the global spectral model T42. *Research Activities in Atmosphere and Ocean Modeling*, G. J. Boer, Ed., WMO/ TD, **263**, **7**, 19—20.

______, 1989: The PJ-like north-south oscillations found in 4-month integrations of the global spectral model T42. *J. Meteor. Soc. Japan,* **67**, 687—604.

Nitta, T., 1987: Convective activities in the tropic western Pacific and their impact on the Northern Hemisphere summer circulation. *J. Meteor. Soc. Japan,* **65**, 373—390.

Peixoto, J. P., and A. H. Oort, 1991: *Physics of climate.* American Institute of Physics Press, New York, 520pp.

Peng, G. B., 1987: Connections of the west Pacific subtropical high and snow hydroclimatical regions in China with Antarctic ice-snow indices. *Meteor. Atmos. Phy.*, **37**, 61—71.

______, W. Lu, and Y. Z. Yin, 1982: Response of the climate and circulation to the variation of the rotation rate of the Earth. *Acta Meteorologica Sinica,* **40**, 209-218.

Pu, S. Z., and H. L. Yu, 1993: Impacts on subtropical anticyclone of the thermal structure and circulation in the upper layer of the tropical western Pacific. *Acta Oceano. Sinica,* **15**, 31-43.

Qian W. H., 1991: Impacts of the Earth's rotation rate on the meridional movement of the subtropical anticyclone. *Acta Meteorologica Sinica,* **49**, 239-243.

Qian, Z. C., and S. H. Yu, 1991: Mid- term variation in condensation heating over East Asia and the quasi- two- week oscillation of the subtropical anticyclone over the western Pacific. *Tropical Meteor.*, **7**, 259-267.

Rodwell, M. R., and B. J. Hoskins, 2001: Subtropical anticyclones and monsoons. *J. Climate,* **14**, 3192-3211.

Samel, A. N., W. C. Wang, and X. Z. Liang, 1999: The monsoon rainfall over China and relationships with the Eurasian circulation. *J. Climate*, **12**, 115-131.

Shi, G. Y., 1981: An accurate calculation and the infrared transmission function of the atmospheric constituents. Ph.D. Thesis, Dept. of Sci., Tohoku University of Japan, pp 191.

Tang, M. M., S. G. Lu, and S. S. Huang, 1982: Variation in location and intensity of the subtropical anticyclone over the western Pacific. *Proceedings of Tropical Circulation and System.* Ocean Press, Beijing, 81-104.

Tao, S. Y., and S. Y. Xue, 1962a: Circulation characteristics in association with persistent summer drought and flood in the Zhangjiang- Huihe reaches. *Acta Meteorologica Sinica,* **32**, 1-18.

______, ______, and Q. Y. Guo, 1962b: Meridional and longitudinal circulation pattern in summer in the tropical and subtropical regions over East Asia. *Acta Meteorologica Sinica,* **32**, 91-102.

______, 1963: *On the summer synoptic systems in the subtropics over China.* Science Press, Beijing.

______, and F. K. Zhu, 1964: Variation of summer circulation pattern at 100 hPa over South Asia and its relation with the movement of the subtropical anticyclone over western Pacific. *Acta Meteorologica Sinica,* **34**, 385-394.

______, and L. Chen, 1987: A review of recent research on the East Asia summer monsoon in China. *Monsoon Meteorology*, Krishnamuri, Ed., Oxford University Press, 60-92.

Webster, P. J., 1972: Response of the tropical atmosphere to local, steady forcing. *Mon. Wea. Rev.,* **100**, 518-541.

Wu, B. Y., 1997: Variation of the sea- ice in North Pole and its impacts on global climate. Ph. D. thesis, Institute of Atmospheric Physics, Chinese Academy of Sciences.

Wu, G. X., 1988: Roles of the mean meridional circulation on atmospheric budgets of angular momentum and sensible heat. *Atmos. Sci.,* **12**, 11-24.

______, H. Liu, Y. C. Zhao, and W. P. Li, 1996: A nine-layer atmospheric general circulation model and its performance, *Adv. in Atmos. Sci.,* **13**, 1-18.

______, W. P. Li, H. Guo, and H. Liu, 1997: Sensible heating-driving air pump of the Tibetan Plateau and the Asian summer monsoon. *Memorial Volume of Prof. Zhao JZ*, Ye DZ, Ed., Science Press, Beijing, 116-126.

______, and X. H. Zhang *et al.*, 1997: The LASG global ocean- atmosphere- land system model GOALS/LASG and its simulation study. *Appl. Meteor.,* **8**, 15-28.

______, and Y. S. Zhang, 1998: Tibetan Plateau forcing and the timing of the monsoon onset over South Asia and the South China Sea. *Mon. Wea. Rev.*, **126**, 913-927.

______, Y. M. Liu, and P. Liu, 1999: Spatially inhomogeneous diabatic heating and its impacts on the formation and variation of subtropical anticyclone, I. Scale analysis. *Acta Meteorologica Sinica,* **57**, 257-263.

______, and Y. M. Liu, 2000: Thermal adaptation, overshooting, dispersion, and subtropical anticyclone. I: thermal adaptation and overshooting. *Chinese J. Atmos. Sci.*, **24**, 433-446.

______, L. Sun, Y. M. Liu, H. Liu, S. F. Sun, and W. P. Li, 2002a: Impacts of land surface processes on summer climate. *East Asia and Western Pacific Meteorology and Climate*, C. P. Chang *et al.* Eds., World Scientific, 64-76.

______, J. F. Chou, Y. M. Liu, and J. H. He, 2002b: *Dynamics of the Formation and Variation of the Subtropical Anticyclones.* Beijing, Chinese Science Press, 314pp.

Wu, G. X., and Y. M. Liu, 2003: Summertime quadruplet heating pattern in the subtropics and the associated atmospheric circulation. *Geophys. Res. Lett.*, **30**, 1201.

Xie, A., and Q. Ye, 1994: Low frequency oscillations identified from OLR and the occurrence of Typhoon over western Pacific. *Appl. Meteor.*, **5**, 143-150.

Xu, Q., 1986: Coupling oscillation between solar activity and the intensity of subtropical anticyclone in the Northern Hemisphere. *Atmos. Sci.*, **10**, 204-211.

Xue, Y. K., P. J. Sellers, J. L. Kinter, and J. Shukla, 1991: A simplified biosphere model for global climate studies. *J. Climate*, **4**, 345-364.

Yan, Z. W. *et al.*, 1990: On the abrupt change of the summer climate in the Northern Hemisphere in 1960's. *Sci. Sinica*, **23**, 97-103.

Ye, D. Z. (also Yeh TC), S. W. Lo and B. C. Chu, 1957: On the heat balance and circulation structure in the troposphere over the Tibetan Plateau and its vicinity. *Acta Meteorologica Sinica*, **28**, 108-121.

______, and B. Z. Zhu, 1958a: *Some fundamental problems of the general circulation of the atmosphere.* Science Press, Beijing, 159pp.

______, S. Y. Tao, and M. C. Li, 1958b: Abrupt seasonal change of the general circulation in June and October. *Acta Meteorologica Sinica,* **29**, 249-263.

______, and Y. X. Gao, 1979: *Meteorology over the Tibetan Plateau.* Science Press, Beijing, 278pp.

______, and G. X. Wu, 1998: The role of the heat source of the Tibetan Plateau in the general circulation. *Meteor. Atmos. Phys.*, **67**, 181-198.

______, 1950: The circulation of high troposphere over China in winter of 1945-46. *Tellus*, **2**, 173-183.

Yu, S. H., and Z. C. Qian, 1992: Mid- term variation of the summer circulation over East Asia and possible mechanism. *Appl. Meteor.*, **3**, 114-119.

______, and S. L. Wang, 1989: Circulation mechanism leading to the mid- term variation of the subtropical anticyclone over western Pacific. *Acta Oceano. Sinica*, **11**, 370-377.

Zhang, R, H. S. Shi, and S. H. Yu, 1995: A study of non-linear stability of the western Pacific subtropical high. *Chinese J. Atmos. Sci.,* **19**, 687-700.

Zhang, X. G., and F. Y. Wei, 1990: Relation between northern polar vortex and subtropical anticyclone. *Collection of Long- term Weather Forecasts.* Meteor. Press, Beijing, 226-231.

Zhang, X. H., K. M. Chen, X. Z .Jin, W. Y. Lin, and Y. Q. Yu, 1996: Simulation of thermohaline circulation with a twenty-layer oceanic general circulation model, *Theor. and Appl. Climatol.,* **55**, 65-88.

______., G. Y. Shi, H. Liu, and Y.Q. Yu. 2000: *IAP Global Ocean-Atmosphere-Land System Model.* Science Press, Beijing, New York, 252pp.

Zhou, J. Y., and D. S. Yang, 1986: Oscillation of the subtropical circulation system in summer and the spectrum analysis of rainfall over China. *Tropical Meteor.,* **2**, 195-203.

15. MAINTENANCE AND SEASONAL TRANSITIONS OF THE STATIONARY WAVES DURING EAST ASIAN MONSOON

MINGFANG TING[a]
Lamont-Doherty Earth Observatory of Columbia University
61 Rt. 9W, Palisades, NY 10964
E-mail: ting@ldeo.columbia.edu

RENU JOSEPH
Department of Atmospheric Sciences
University of Illinois at Urbana-Champaign
Urbana, Illinois 61801, USA

The dominant stationary wave features during the East Asian monsoon season, from April to October, are examined in this study using the NCEP/NCAR reanalysis and the linear and nonlinear stationary wave models. The study focuses on the stationary wave maintenance and seasonal transition and their relations to the monsoon diabatic heating. Monsoon rainfall indices over China, the stationary wave and diabatic heating indices over the monsoon region are constructed for easy comparisons. Four stationary wave features, i.e., the Tibetan anticyclone, the East Asian trough, the subtropical high, and the monsoon trough, are examined in detail. All four stationary wave centers go through distinct seasonal cycle during this seven-month period, with the Tibetan anticyclone, the subtropical high and the monsoon trough intensifying from April to July, and weakening from August to October, while the East Asian trough weakens from April to August and re-establishes itself from August to October. The linear and nonlinear stationary wave models indicate that the effect of the heating is the most dominant in contributing to the stationary wave seasonal cycle. Further decomposition of the heating shows that the monsoon heating is the one that causes the seasonal variation of the stationary wave centers except the East Asian trough.

1. Introduction

Stationary waves, or the zonally asymmetric circulation features, are an important part of the Asian monsoon system. For example, the upper tropospheric easterly jet and the westerly monsoon flow in summer are the results of the strong Tibetan anticyclone in the upper troposphere and the monsoon low at lower troposphere over South Asia, respectively. Many monsoon indices are based on the zonal or meridional winds at the upper and lower troposphere that are intimately related to the stationary waves (Webster and Yang 1992; Goswami *et al.* 1999). Another important stationary wave feature during the Asian summer monsoon, the Pacific subtropical high, has long been recognized by the East Asian monsoon countries as an important index for summer monsoon prediction (Huang and Yue 1962; Tao and Chen 1987; Samel *et al.* 1999). The forcing mechanisms for the stationary wave features and their anomalies, particularly the Tibetan anticyclone and the western Pacific subtropical

[a] This work started while the lead author was at University of Illinois, Urbana-Champaign, IL., Urbana.

high, during Asian summer monsoon has been studied extensively in the past (Krishnamurti 1973; Nitta 1987; Huang and Li 1989; Ting 1994; Hoskins 1996; Liu *et al.* 2001; Rodwell and Hoskins 2001). Yet many unanswered questions remain. In particular, it is not clear whether the monsoon heating associated with the rainfall is directly responsible for the stationary wave maintenance and whether the presence of Tibetan Plateau essential for the stationary wave maintenance.

In this chapter, the maintenance of the stationary wave features and their seasonal development during Asian monsoon will be examined in detail using a linear and a nonlinear stationary wave model and the NCEP/NCAR reanalysis. In the next section, the linear and the nonlinear models will be introduced. In section 3, the main features of the stationary waves associated with the East Asian monsoon onset and withdrawal are discussed using the NCEP/NCAR reanalysis. Section 4 presents the main results of the linear and nonlinear model diagnosis of the stationary waves and their seasonal cycle. A summary of the main findings is presented in section 5.

2. Model Description

The models that will be used include a linear and a nonlinear stationary wave model. The linear model is the same as that in Ting and Held (1990) and later used in Wang and Ting (1999) and Ting *et al.* (2001) for studying the stationary wave maintenance and seasonal cycle in the NCEP/NCAR reanalysis and the GFDL GCM. The linear model has a rhomboidal 30 horizontal resolution and 14 unevenly spaced sigma levels in the vertical. The basic equations consist of the steady state version of those for vorticity, divergence, temperature, and surface pressure plus the hydrostatic balance and the mass continuity equation. The model equations are linearized about the zonal mean basic state based on the NCEP/NCAR reanalysis monthly mean data. The forcings for the linear model include the diabatic heating, orography, transients, and the stationary nonlinearity. The diabatic heating is derived from the NCEP/NCAR reanalysis data as a residual in the thermodynamic equation. The transients, including the transient momentum and heat flux convergences, are computed from the four-times daily data of the reanalysis. The stationary nonlinearity, which represents the stationary nonlinear wave-wave interactions, is not a physical forcing of the system. It consists of the terms ignored during the linearization of the equations. The nonlinear terms are computed from the monthly mean reanalysis data and included in the linear model to represent the effect of nonlinearity in a diagnostic sense. The dissipations in the atmosphere are parameterized in the linear model as Rayleigh friction and Newtonian cooling, same as those used in Ting *et al.* (2001).

The nonlinear model is the same as that used in Ting and Yu (1998) and Ting *et al.* (2001). It is a time-dependent, fully nonlinear model based on the primitive equations in σ-coordinates. The horizontal and vertical resolution of the nonlinear model is exactly the same as that in the linear model. The basic model variables are deviations from a prescribed zonal mean basic state taken from the NCEP/NCAR reanalysis. The steady state solution of the nonlinear model is obtained by time integration and suppressing the model-generated transients. A quasi-steady state can be reached after about 30 days of integration. The stationary wave forcings in the nonlinear model include diabatic heating, orography, and transients. These forcings are exactly the same as that in the linear model, except that the stationary nonlinearity is no longer needed as a forcing in the nonlinear model, since the nonlinearity is already included in the model formulation. We also employ the same damping in the nonlinear model as those in the linear model, same as in Ting *et al.* (2001).

The basic state and forcings used in the linear and nonlinear models are derived from the NCEP/NCAR reanalysis (Kalnay *et al.* 1996) from 1948 to 1999 (52 years). The transient heat and momentum fluxes are computed based on the four times daily data and the diabatic heating is derived as a residual in the thermodynamic energy equation. More details on the diabatic heating calculations based on the reanalysis can be found in Wang and Ting (1999).

3. Stationary Wave Features during East Asian Monsoon

The climatological mean onset of the East Asian monsoon is identified as in early May over South China, and becomes progressively later northwestward to late July over Northern China and Korea (Tao and Chen 1987; Tanaka 1992; Lau and Yang 1997). The climatological mean retreat of the East Asian monsoon is in September, marking the end of the Northern China and Korean rainy season. There are significant fluctuations from year to year of the onset and retreat dates of the East Asian monsoon; the onset can be as early as April and retreat in October. To illustrate the relevant stationary wave features during the East Asian monsoon development, the stationary waves for the seven-month period, from April to October, are examined.

Figures 1 and 2 illustrate the upper (σ=0.171, approximately 171 mb) and lower (σ=0.866, approximately 866 mb) level stationary wave streamfunction for the seven-month period in the Asian monsoon region. The dominant changes in the upper level stationary waves during this period are the decrease and eastward retreat of the East Asian trough, a prominent stationary wave feature in Northern wintertime, from April to May, and a significant enhancement and westward and northward expansion of the Tibetan anticyclone from April to July. Accompanying the Tibetan anticyclone enhancement, the oceanic troughs over both the Pacific and Indian Ocean at upper level intensifies significantly as well. The intensity of the Tibetan anticyclone relates directly to the intensity of the easterly jet in the southern flank of the high, as is clearly seen on the streamfunction field in Fig. 1. It has been shown that the summer rainfall over China is related to the strength and location of the Tibetan anticyclone (Murakami and Matsumoto 1994). The intensity of this anticyclone is not only important for the East Asian monsoon, but also crucial for the Indian monsoon establishment. From August to October in the upper level (Fig. 1), the East Asian trough re-establishes itself while the Tibetan anticyclone weakens substantially.

At the lower level (Fig. 2), the most dominant stationary wave development from April to July is the intensification and widening of the trough over South Asia, and a simultaneous enhancement of the Indian Ocean anticyclone and the western Pacific subtropical high. From August to October, the monsoon trough weakens and shrinks in its horizontal extent. The link between the Pacific subtropical high and the East Asian monsoon has been explored extensively in the past (e.g. Tao and Chen 1987; Wang and Zhang 1992; Murakami and Matsumoto 1994; Nitta and Hu 1996; Samel *et al.* 1995, 1999; Rodwell and Hoskins 2001). The westward ridging of the subtropical high from April to August (Fig. 2), along with the intensified monsoon low, causes the enhanced southerly flow from tropical western Pacific to East Asia, bringing warm, moist air and heavy precipitation into China, Japan and Korea.

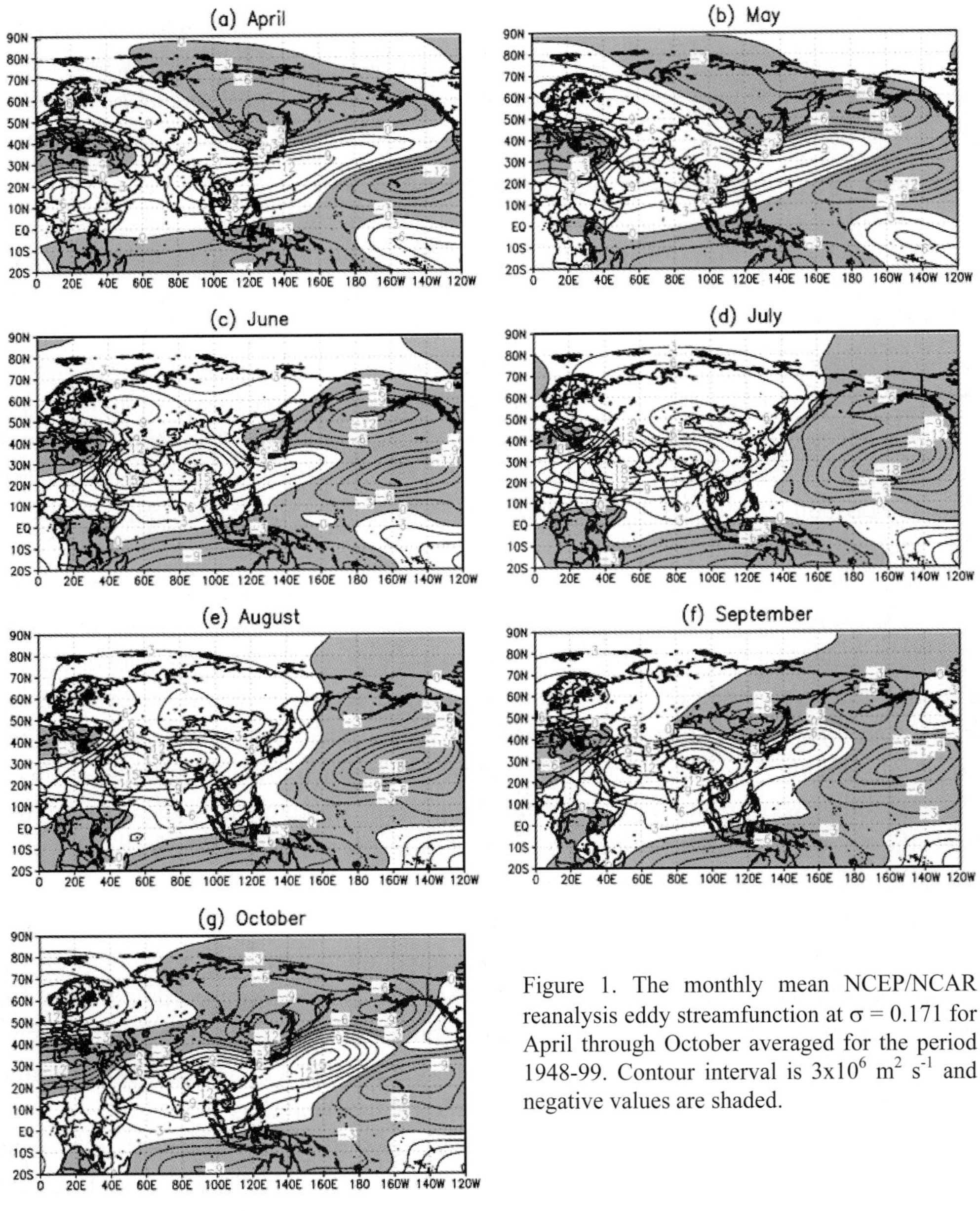

Figure 1. The monthly mean NCEP/NCAR reanalysis eddy streamfunction at $\sigma = 0.171$ for April through October averaged for the period 1948-99. Contour interval is 3×10^{6} m^2 s^{-1} and negative values are shaded.

Figure 3 shows the vertical cross section of the stationary wave features at 30°N from April to October. One can see clearly the maximum amplitude at upper and lower levels, showing a distinct baroclinic structure with the upper and lower level streamfunctions reversed in sign, particularly for the three-month summer, June to August. The baroclinic structure is an indication that the stationary waves are forced by tropical diabatic heating.

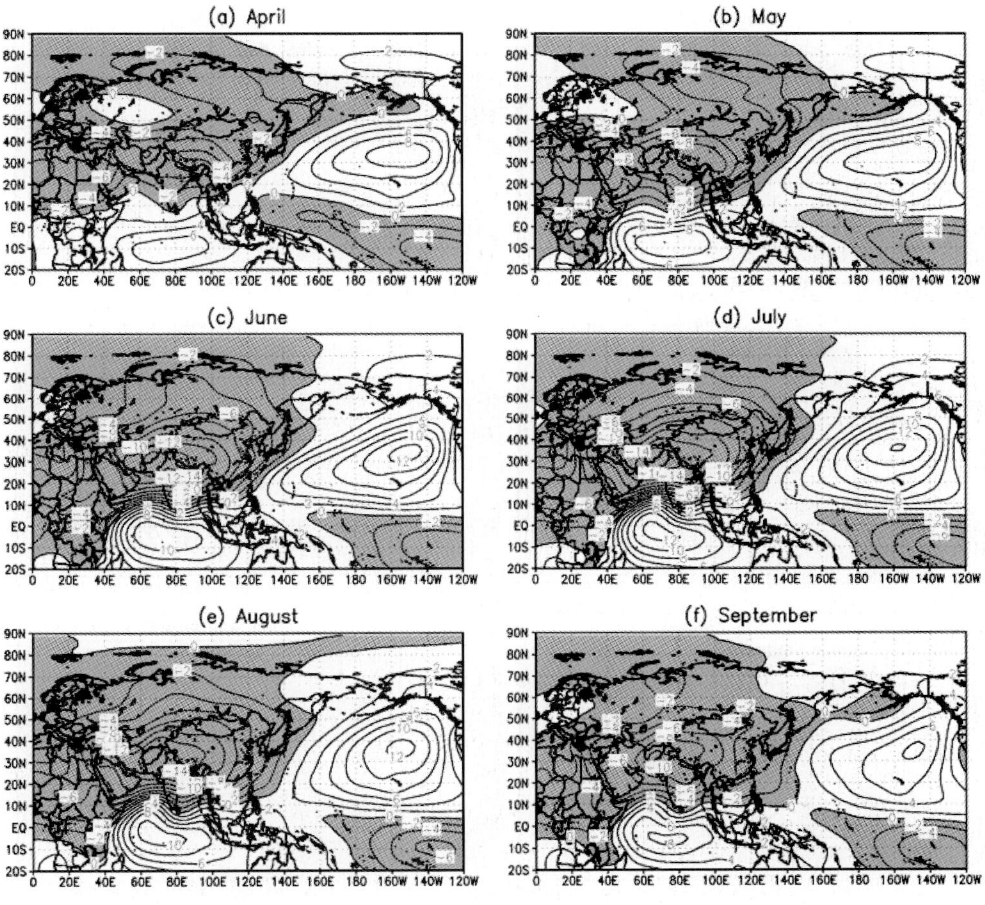

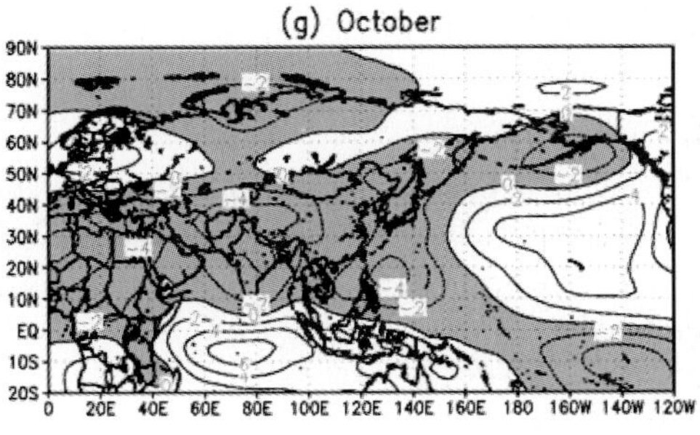

Figure 2. Same as Fig. 1, but for σ=0.866 and contour interval is $2x10^6$ m^2 s^{-1}.

In order to illustrate the seasonal cycle of the stationary wave features more concisely, we constructed four stationary wave indices, which are area-averaged stationary wave streamfunction for the Tibetan anticyclone at upper level (20°E - 120°E, 10°N - 50°N), the east Asian trough at upper level (100°E - 160°E, 40°N - 70°N), the monsoon low at lower level (40°E - 120°E, 15°N - 65°N), and the subtropical high at lower level (160°E - 120°W, 10°N - 60°N). Figure 4b illustrates the regional stationary wave indices along with the

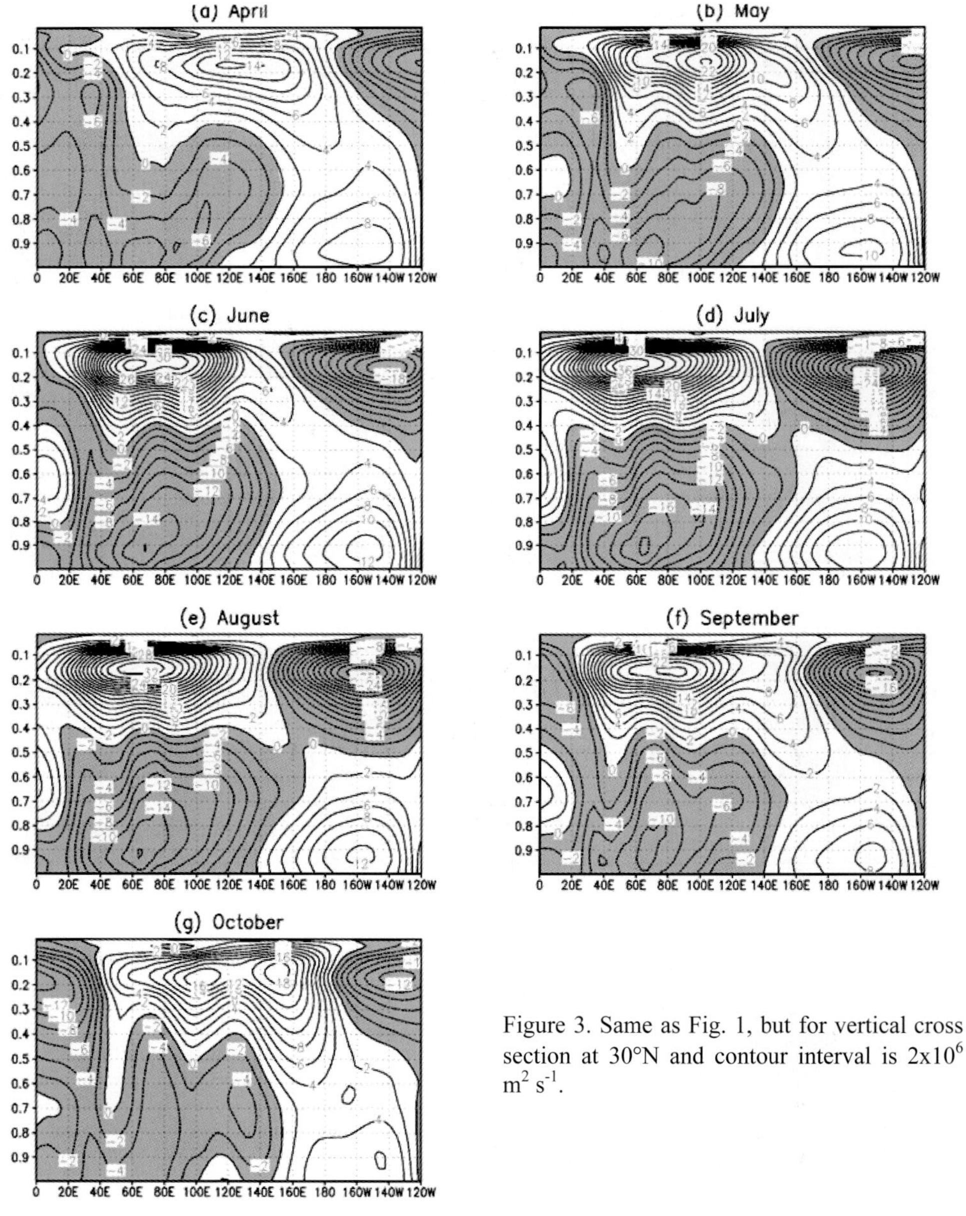

Figure 3. Same as Fig. 1, but for vertical cross section at 30°N and contour interval is 2×10^6 $m^2 s^{-1}$.

precipitation indices (Figure 4a) averaged over all available rain gauge stations for Southern China (South of 27°N and east of 110°E), the Yangtze River valley (107°E - 113°E and 27°N - 31°N, 113°E - 123°E and 29°N - 35°N), and Northern China (103°E - 113°E, 35°N - 42°N) using the rainfall observations. The monthly mean rainfall over South China shows a peak in

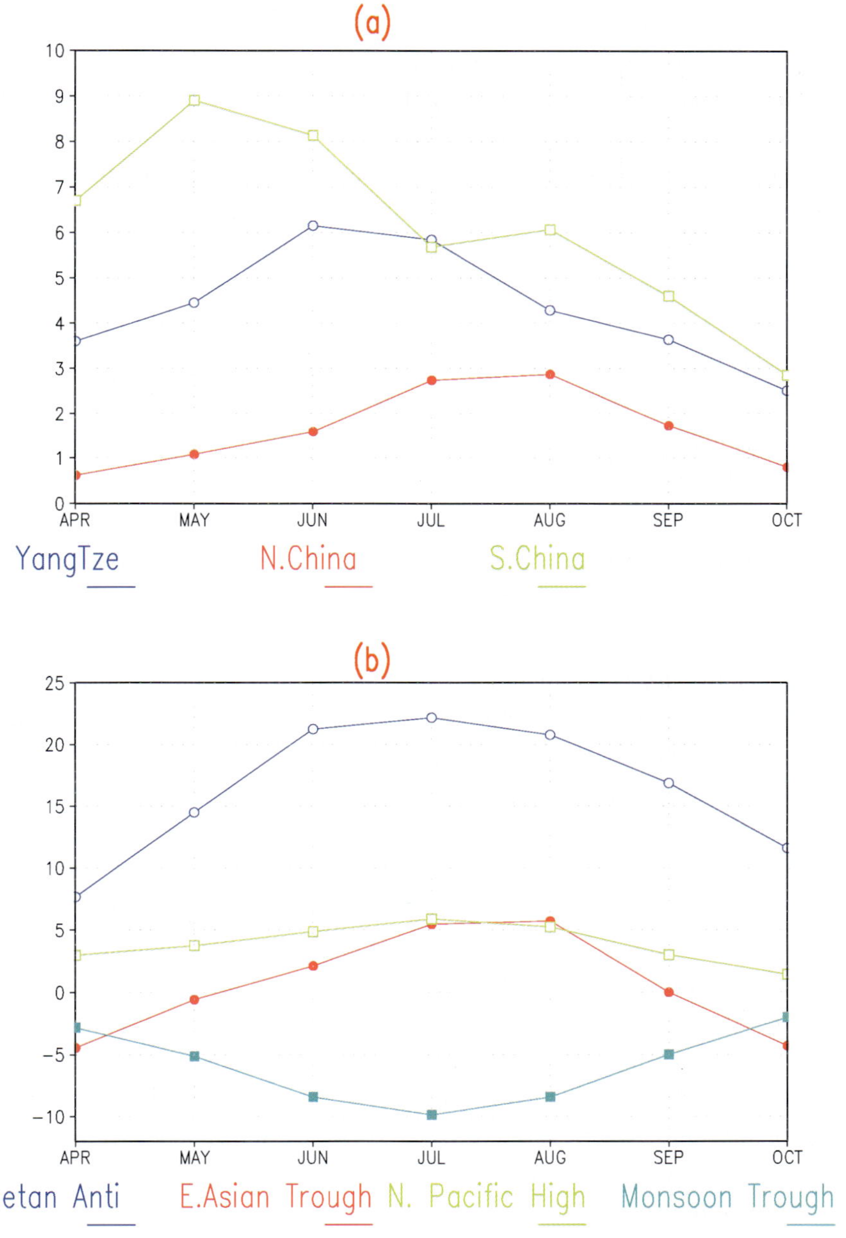

Figure 4. (a) Monthly precipitation indices averaged over all available rain gauge stations for South China (south of 27°N and east of 110°E), Yangtze River Valley (107°E - 113°E and 27°N - 31°N, 113°E - 123°E and 29°N - 35°N), and North China (103°E - 113°E and 35°N - 42°N). (b) Regional stationary wave indices for Tibetan anticyclone, East Asian Trough, North Pacific subtropical high, and Monsoon trough (see text for details).

May, showing the monsoon onset there in early May, and the Yangtze River Valley has a maximum in June and July, and the North China in August, indicating the progressively northward moving monsoon rain band. Comparing to the stationary wave indices, it is clear that the stationary wave development lags the rainfall peaks, with the exception of the Northern China rainfall. This suggests that the stationary waves may be forced responses of the monsoon heating, rather than the cause of the monsoon onset. There is, however, a similar seasonal cycle between the Tibetan anticyclone and the Yangtze River valley precipitation indices, and between the East Asian trough and the Northern China precipitation indices. This correspondence may be an indication that interaction between monsoon precipitation and the stationary wave features is important. This chapter does not try to establish the cause and effect between the stationary wave seasonal cycle and that of the East Asian monsoon rainfall (diabatic heating). The focus in this chapter is to address the following questions:

- What are the forcing mechanisms for the seasonal development of the above-noted stationary wave features?
- If diabatic heating is the dominant forcing mechanism, is the East Asian monsoon heating directly responsible for these developments?

In the next section, the maintenance and seasonal development of the stationary wave features accompanying the East Asian summer monsoon will be further examined using a linear and a nonlinear model.

4. Maintenance and Seasonal Developments of the Stationary Waves

4.1. *Effect of the Total Stationary Wave Forcing*

The observed stationary wave features for the seven months were reproduced using the linear and nonlinear stationary wave models by subjecting the models to the NCEP/NCAR reanalysis zonal mean basic state and the zonally asymmetric forcing (i.e. diabatic heating and transients derived from the NCEP/NCAR reanalysis plus orography). In the case of the linear model, the nonlinear terms ignored in the linearization of the basic equations are computed based on the reanalysis stationary waves and added into the linear model as the stationary nonlinearity forcing (see Wang and Ting 1999; Ting *et al.* 2001). The vertically integrated diabatic heating derived from the NCEP/NCAR reanalysis for April through October is shown in Fig. 5. The monsoon features are well captured in the total diabatic heating field. There is a large increase of diabatic heating in Southern China associated with monsoon onset there from April to May, consistent with the precipitation index in Fig. 4a. From May to June, there is a clear indication of the onset of Indian monsoon, the dramatic increase in precipitation over the Yangtze River valley and Japan. From June to July, both the Indian monsoon and the East Asian monsoon intensified and shifted further north in the case of the East Asian monsoon.

The four stationary wave indices based on the linear and nonlinear model responses to the total stationary wave forcings from April to October are shown in Fig. 6, along with the indices for the corresponding NCEP/NCAR reanalysis. Both the linear and the nonlinear models capture the essential features of the observed stationary wave seasonal cycle during this period. The Tibetan anticyclone (Fig. 6a) increases its strength from April to July and decreases from July to October; the East Asian trough (Fig. 6b) weakens its strength from April to August and then re-establishes itself in September and October; the Pacific subtropical high (Fig. 6c) intensifies from April to July and then weakens from July to October; the monsoon low (Fig. 6d) deepens from April to July and then weakens from July to October. The most obvious discrepancy between the linear and nonlinear model responses and the reanalysis is the underestimation of the amplitude of all the centers in both the linear

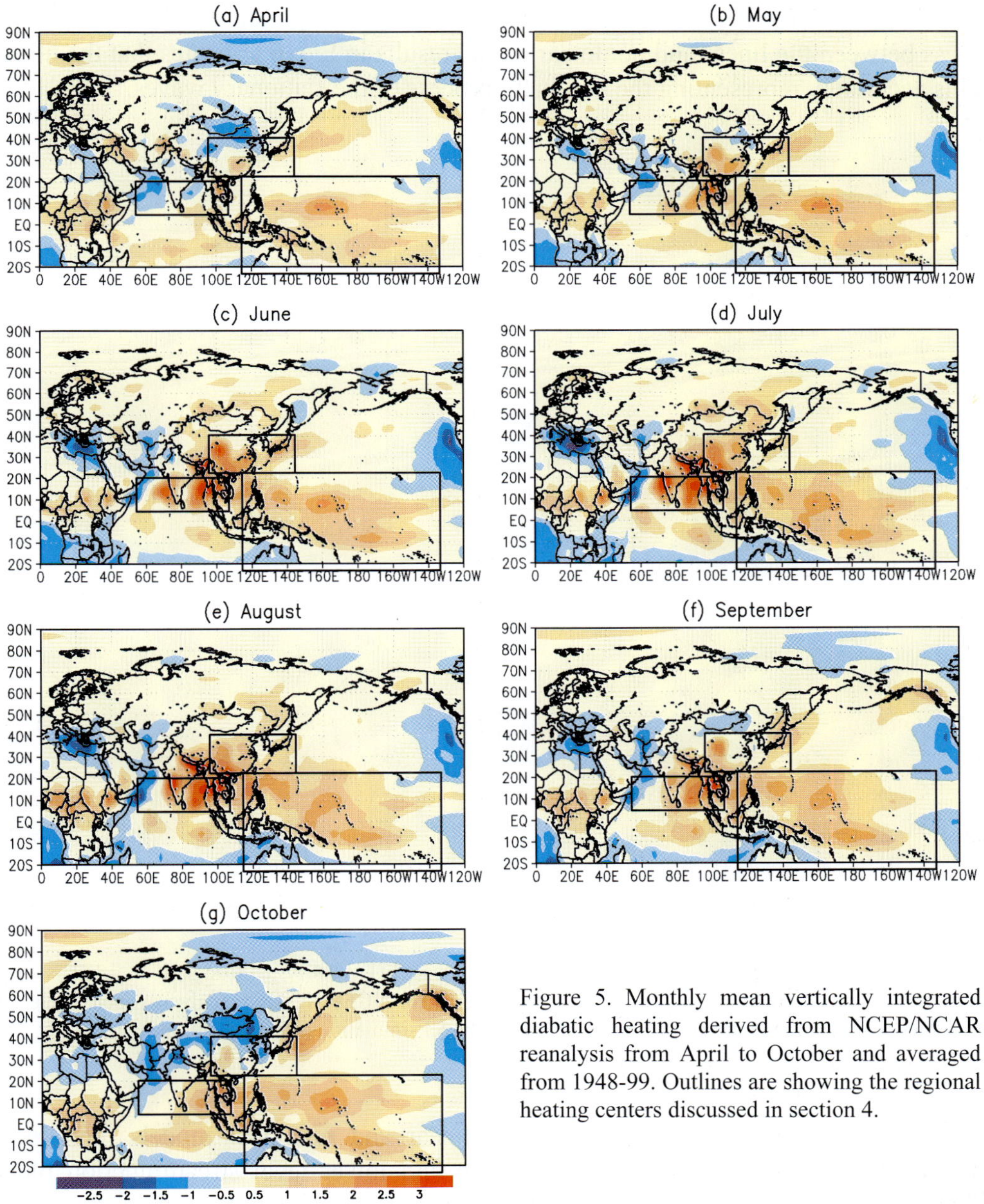

Figure 5. Monthly mean vertically integrated diabatic heating derived from NCEP/NCAR reanalysis from April to October and averaged from 1948-99. Outlines are showing the regional heating centers discussed in section 4.

and nonlinear models. This is particularly true for the North Pacific subtropical high in the nonlinear model. Note that the linear model, when including the stationary nonlinearity forcing, should give a perfect simulation of the observed stationary waves if the stationary wave forcings and dissipation parameterizations in the linear model are accurate (see Ting *et al.* 2001). The amplitude discrepancies in Fig. 6 are thus an indication of the inaccuracies in the forcings and dissipations used in the linear model. In this case, most of the inaccuracies in the forcings likely come from the diabatic heating, which is derived as a residual from the reanalysis thermodynamic equation. Annamalai *et al.* (1999) has shown that the NCEP/NCAR reanalysis underestimate the tropical precipitation over the Oceans. The closeness between the linear and nonlinear model results in Fig. 6 indicates that the nonlinear model is accurately representing the nonlinear process of the stationary waves.

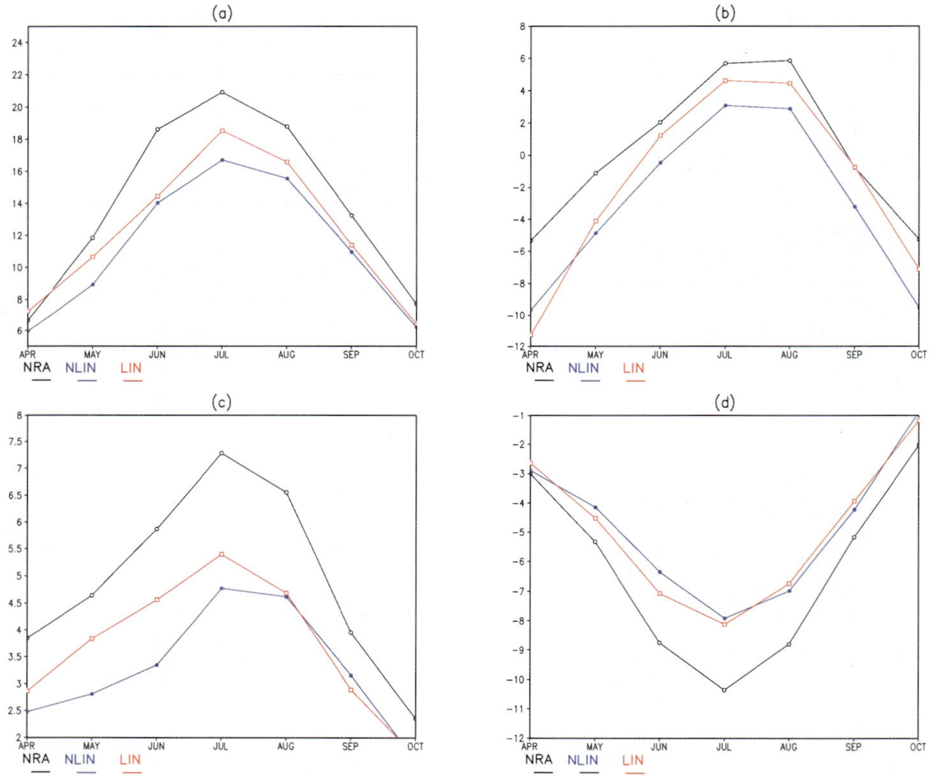

Figure 6. Stationary wave indices for Tibetan anticyclone at σ=0.171, East Asian trough at σ=0.171, subtropical high at σ=0.866 level, and Monsoon trough at σ=0.866 level for NCEP/NCAR reanalysis (open circle, black), nonlinear model simulation (solid circle, blue), and linear model simulation (open square, red).

To get a better picture of how well the linear and nonlinear models reproduce the stationary waves, we show in Fig. 7 the upper and lower level streamfunction plots for NCEP/NCAR reanalysis (top panels), linear model simulations minus the reanalysis (middle panels), and the nonlinear model simulations minus the reanalysis for July. The top panels have been shown in Fig. 1d and Fig. 2d and are plotted here again for easy comparison. The differences between the linear and the nonlinear model simulations and the NCEP/NCAR reanalysis are relatively small and show an underestimation of the stationary wave amplitude in the stationary wave models. The underestimation of the stationary wave amplitude in the linear and nonlinear models in Fig. 7 is consistent with that in the index plot (Fig. 6). Given the imperfect reanalysis stationary wave forcings, particularly errors in the derived heating fields, both the linear and nonlinear models simulate the observed stationary wave features fairly well. Thus the linear and nonlinear models are useful tools for elucidating the individual forcing contributions to the stationary wave features. Given the fact that the stationary nonlinearity forcing is not a physical forcing term in the linear model, the nonlinear model gives a more physical picture of the role of each forcing. In the rest of the text, results will be shown mainly for the nonlinear model with the notion that conclusions derived based on the nonlinear model are confirmed by the linear model as well.

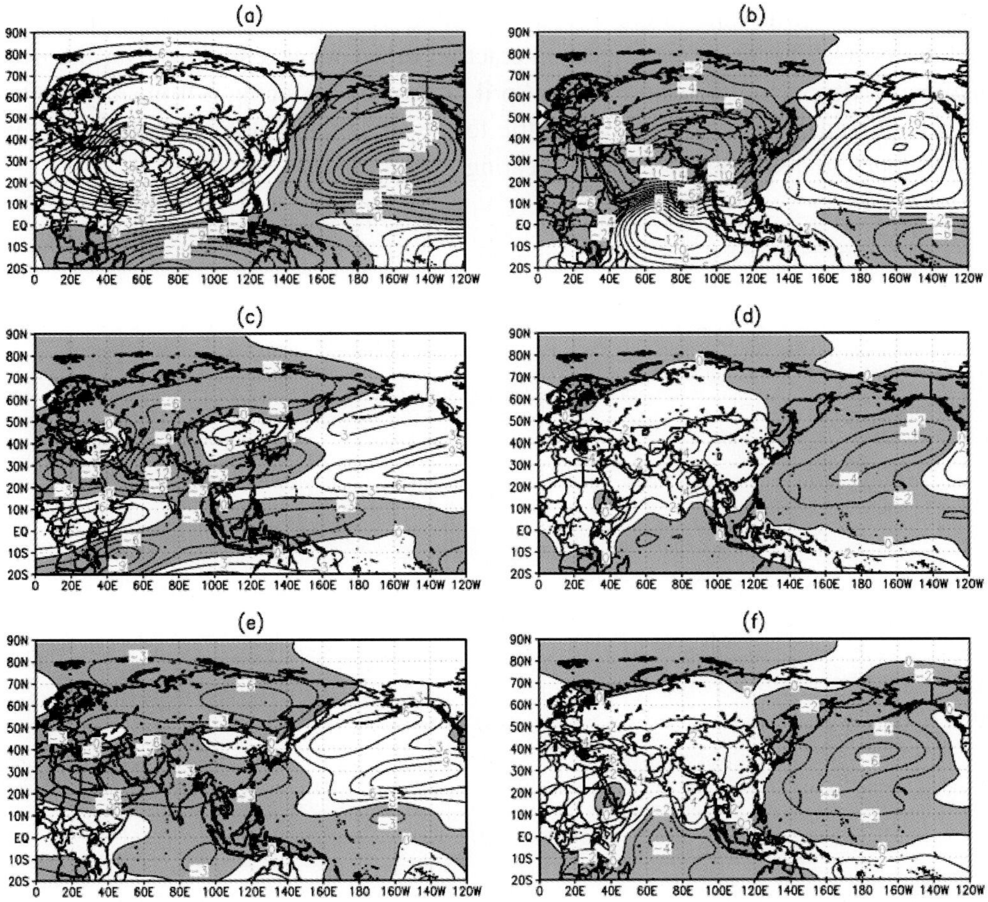

Figure 7. Eddy streamfunction at σ=0.171 (left) and σ=0.866 (right) levels in July for NCEP/NCAR reanalysis (a, b), linear model simulation with stationary nonlinearity forcing minus reanalysis (c, d), and nonlinear model simulation minus reanalysis (e, f). Contour interval is 3×10^6 m^2 s^{-1} for 0.171 level and 2×10^6 m^2 s^{-1} for 0.866 level. Negative values are shaded.

4.2. *Effect of the Individual Stationary Wave Forcing*

The nonlinear model streamfunction responses to global diabatic heating, orography, and transients are shown in Fig. 8 for the four stationary wave indices. Also included in Fig. 8 is the nonlinear effect that is not included in the nonlinear responses to heating, orography, and transients. This nonlinear effect is the result of nonlinear interactions among flows forced by heating, orography, and transients. The indices based on the nonlinear model responses to the total forcings are reproduced in Fig. 8 for comparison purposes. Figure 8 shows clearly that almost all the dominant stationary wave features are forced by diabatic heating, in particular, the seasonal variation of the Tibetan anticyclone at the upper level and the monsoon trough and North Pacific subtropical high at the lower level. The change in the strength of the East Asian trough at the upper level, however, is not due to the effect of heating alone, although heating is the most dominant of all effects. The effect of orography contributes to the seasonal weakening of the East Asian trough from April to July and the intensification from August to October, but the amplitude of the orographic contribution is rather small. Another forcing that has significant amplitude contribution to the East Asian trough in April, September and October is the nonlinear interaction term, but it tends to offset the observed seasonal cycle, deepening the trough from April to July and reducing it from July to October. The effects of transients contribute very little to the seasonal cycle of the stationary waves, although it tends to dissipate the effect of heating in most cases.

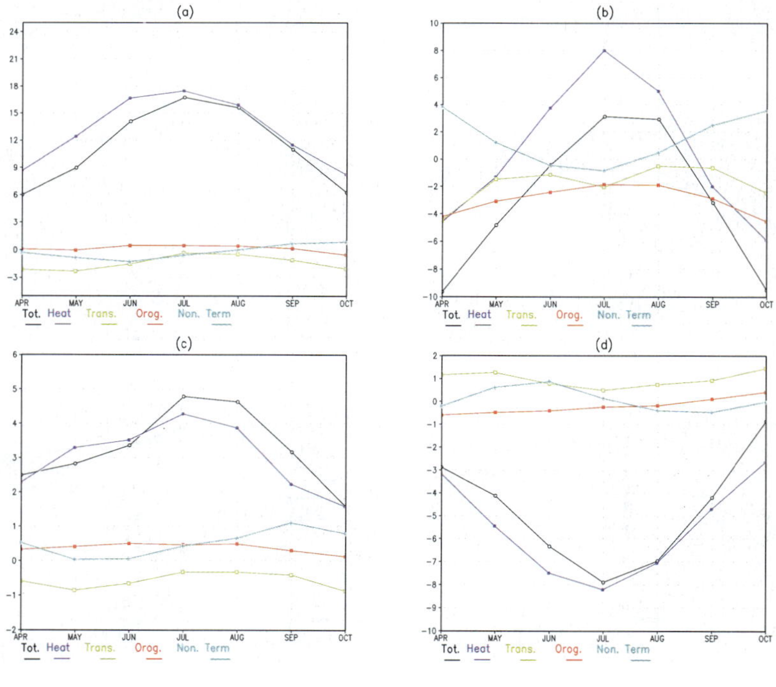

Figure 8. Stationary wave indices based on nonlinear model responses to total forcing (open circle, black), diabatic heating (solid circle, blue), orography (solid square, red), transients (open square, light green), and nonlinear interaction terms (plus sign, dark green) for Tibetan anticyclone (a), East Asian trough (b), subtropical high (c), and Monsoon trough (d).

The dominance of the diabatic heating effect is further illustrated in Fig. 9, where the nonlinear model responses to diabatic heating, orography, transients, and nonlinear interactions are shown for the upper and lower levels in July. The amplitude of the responses to orography, transients, and the nonlinear interactions are negligible compared to that of the

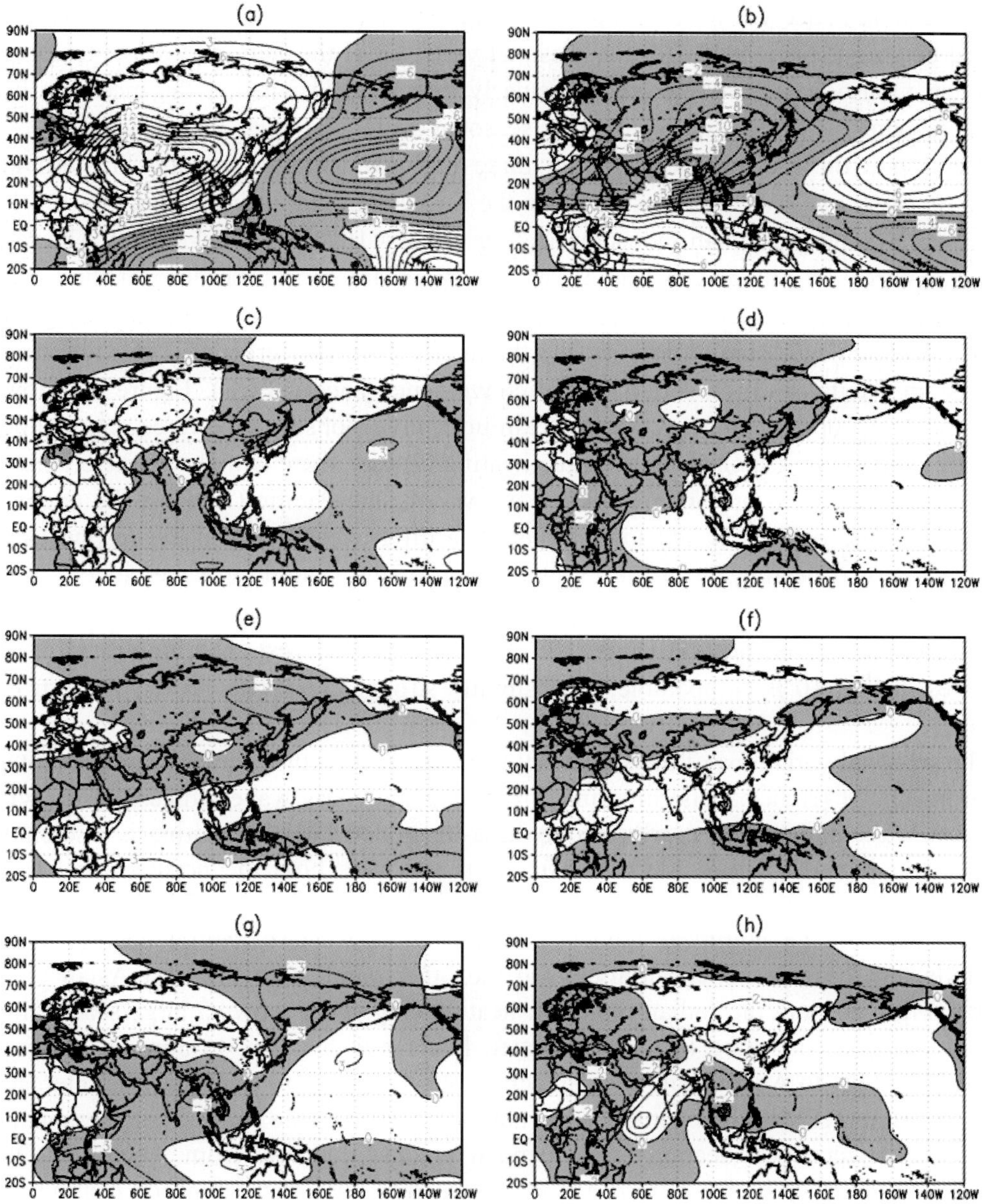

Figure 9. Eddy streamfunction at σ=0.171 (left) and σ=0.866 (right) in July for nonlinear model responses to diabatic heating (a, b), orography (c, d), transients (e, f), and nonlinear interaction terms (g, h). Contour intervals are $3x10^6$ m^2 s^{-1} at σ=0.171 level and $2x10^6$ m^2 s^{-1} at σ=0.866 level, and negative values are shaded.

response to diabatic heating in July. In other months, the effects of transients and orography may be stronger than that in July, but they are generally much smaller compared to the effect of heating, except for the case of East Asian trough in early and late part of the seven-month period (Fig. 8).

4.3. *Effect of the Regional Diabatic Heating*

Given the dominance of the diabatic heating effect, it is natural to ask which part of the diabatic heating is responsible for the seasonal variation of the stationary wave features over the Asian monsoon region. We are particularly interested in whether the monsoon heating itself is responsible for the stationary wave seasonal cycle.

We divide the diabatic heating into three regional features, i.e. the Western Pacific heating (114°E - 133°W, 22.5°S - 22.5°N), the East Asian monsoon heating (95°E - 144°E, 18°N - 40°N), and the Indian monsoon heating (54°E - 107°E, 4.5°N - 20°N). The regional heating features are outlined in Fig. 5. These regions are chosen because of their different characteristics, including the regional monsoon features of the Indian (including Bay of Bengal) and the East Asian monsoons, and the intertropical convergence (ITCZ) heating over the western Pacific and the south Pacific convergence zone (SPCZ). Figure 10 shows the nonlinear model upper and lower level streamfunction responses to the three regional heating centers in July. The responses to all three heating centers share similar characteristics, with an anticyclone over Asian landmasses and a low over the Pacific Ocean at the upper level, and the opposite polarity at the lower level. The differences between the responses to the different regional heating lie in the exact location of the stationary wave centers, with the response to Western Pacific heating (Figs. 10a, b) located further to the east and the response to the Indian monsoon heating (Figs. 10e, f) further to the west. The amplitude of all the responses in Fig. 10 is comparable, with slightly stronger responses to the western Pacific heating.

The seasonal variations of the stationary waves forced by the regional heating centers are shown in Fig. 11 (right panels) in terms of regional stationary wave indices. Along with the stationary wave indices, the corresponding area-integrated diabatic heating is also shown in Fig. 11 (left panels). The western Pacific heating covers both the northern and southern hemisphere, thus not showing a typical seasonal cycle in Fig. 11a. The stationary wave indices based on the nonlinear model response to the western Pacific heating has large amplitude, but little seasonal variation. On the other hand, both the East Asian monsoon heating (Fig. 11c) and the Indian monsoon heating (Fig. 11e) shows a large seasonal cycle, indicating a sharp increase in area-integrated heating from April to June, a rather small change from June to August, and then a decrease from August to October. It is interesting to notice the asymmetry in the intensification and decaying stages of the heating seasonal cycle for the Indian monsoon heating, with a much gradual decrease from August to October compared to the steep increase from April to June. Both the Indian monsoon heating and the East Asian monsoon heating contribute to the seasonal cycle of all the stationary wave features of interest. For the Tibetan anticyclone (open circle in Fig. 11), the Indian monsoon heating contributes to the intensification in early part of the summer (April to June), whereas the response to Asian monsoon heating tends to increase from April to August. In the decay phase of the Tibetan anticyclone, the response to East Asian monsoon shows a much sharper decrease than that of the Indian monsoon heating. For the monsoon trough at lower level (solid square in Fig. 11), both the responses to Indian monsoon and Asian monsoon heating show similar seasonal cycle, but the amplitude is much larger for the response to Asian monsoon heating. The contribution to the subtropical high seasonal cycle (open square) is comparable for the responses to Indian monsoon heating and that to the East Asian monsoon heating. Interestingly, the seasonal cycle of the East Asian trough (solid circle) is opposite due to the Indian monsoon heating and that to the East Asian monsoon heating. Thus the seasonal cycle of the East Asian trough cannot be explained by the three regional heating centers considered here. Further calculation shows that the amplitude seasonal cycle of the East Asian trough is due to heating further north of the monsoon.

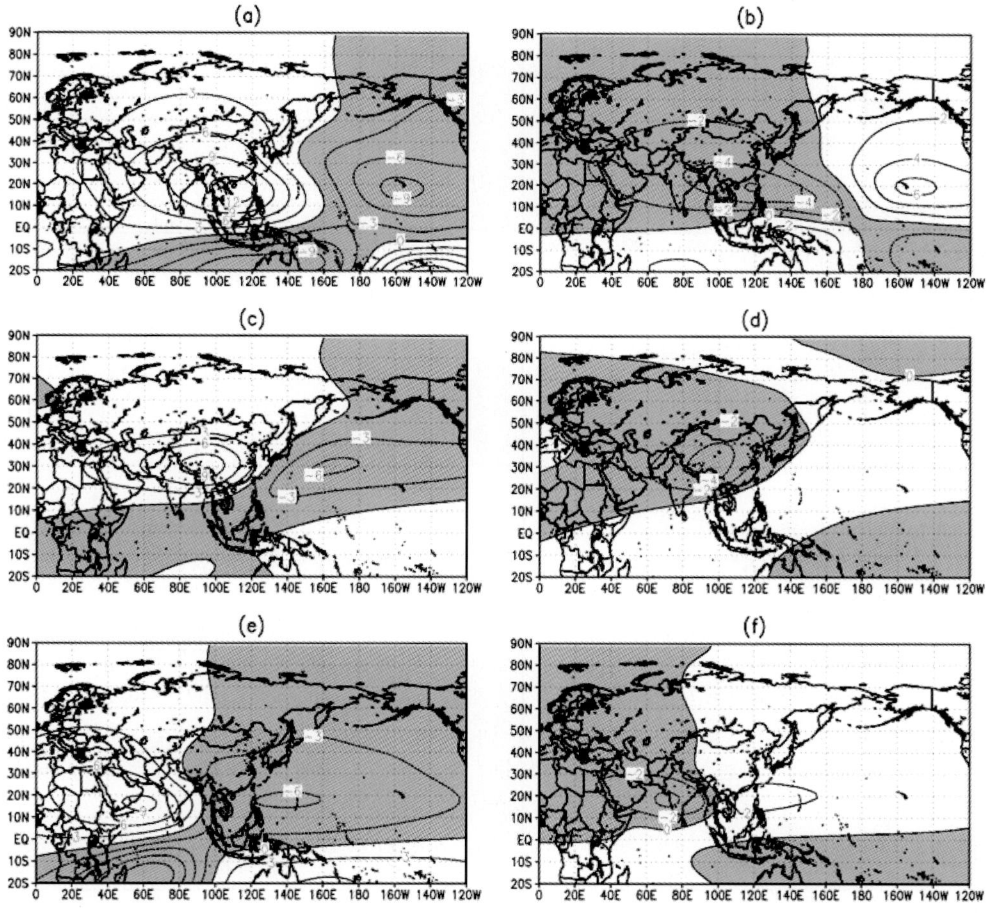

Figure 10. Eddy streamfunction at σ=0.171 (left) and σ=0.866 (right) in July for nonlinear model responses to regional diabatic heating over western tropical Pacific (a, b), East Asian monsoon region (c, d), and Indian monsoon region (e, f). Contour intervals are 3×10^6 m^2 s^{-1} at σ=0.171 level and 2×10^6 m^2 s^{-1}.

Although the seasonal variation of the stationary wave features is not directly related to the western Pacific heating, the amplitude of the stationary waves due to the western Pacific heating is the largest among all due to other regional heating centers, particularly during the beginning of the monsoon season (April and May). Thus the interannual variability of the western Pacific heating could strongly influence the amplitude of the stationary wave features discussed here. Given the importance of these stationary wave features in the monsoon establishment, interannual variability of the tropical western Pacific heating may further influence the Indian monsoon and the East Asian monsoon variability, as found by many researchers of the relation between El Nino and Asian monsoons (Webster and Yang 1992; Meehl 1993, 1994).

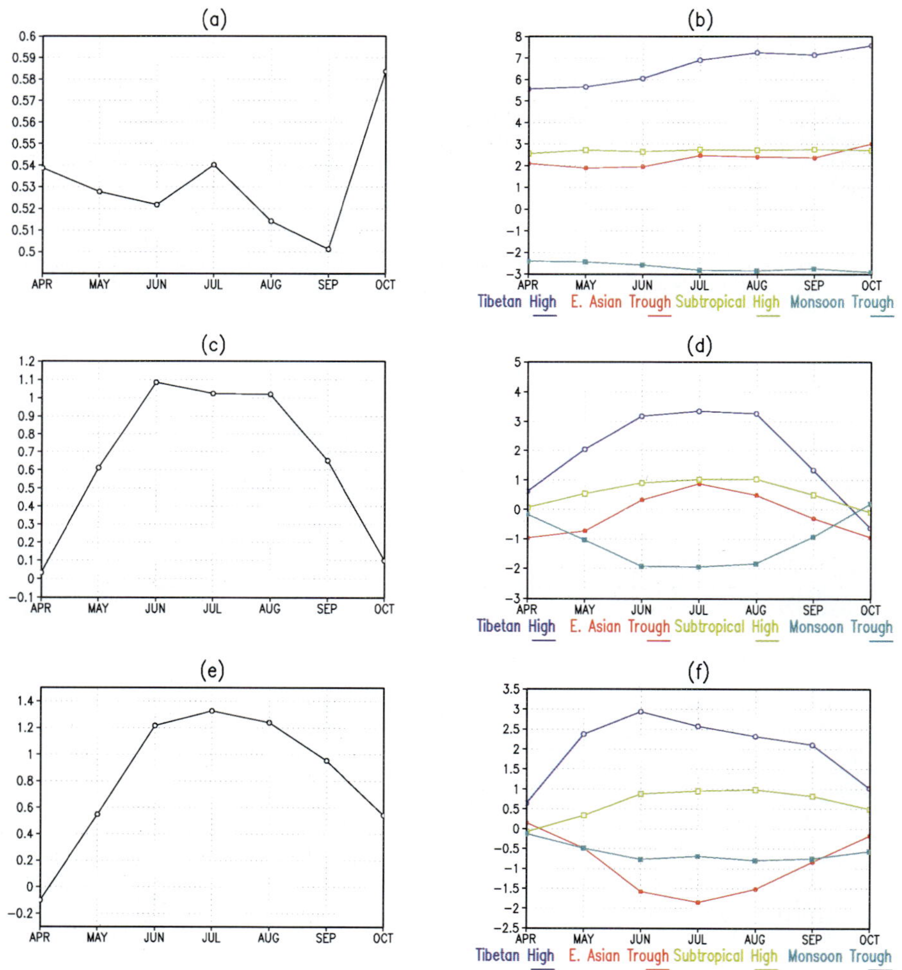

Figure 11. Regional heating indices averaged over western tropical Pacific (a), East Asian monsoon region (c), and Indian monsoon region (e), and regional stationary wave indices for nonlinear model streamfunction responses to these heating centers (b, d, f) for Tibetan anticyclone (open circle, blue), East Asian trough (solid circle, red), North Pacific subtropical (open square, light green), and Monsoon trough (solid square, dark green).

Figure 11 suggests that the seasonal variation of the stationary waves is largely accounted for by the seasonal variation of the monsoon diabatic heating. This suggests that the feedback between the monsoon heating and the stationary waves is the main cause for the seasonal variation of the stationary waves. Results in Fig.10 (c-f) indicate that while the responses to regional monsoon heating centers is global, the location of the maximum amplitude is local; the effect of Indian monsoon heating (Fig.10e, f) is limited to the western part of the domain, while the effect of East Asian monsoon (Fig.10c, d) is more to the east. This may provide support for the earlier findings that the two monsoon systems can vary independently and show distinctively different regimes (Yang and Gutowski 1992; Lau *et al.* 2000; Qian and Yang 2000; Wang *et al.* 2001).

4.4. *Effect of the Zonal Mean Flow Seasonal Cycle*

During the Asian monsoon season, the zonal mean circulation goes through a seasonal cycle as well, which is characterized by the northward shift of the zonal mean jet stream as the season progresses. Whether the stationary wave seasonal cycle is linked in any way to the zonal mean flow seasonal change or entirely due to the change in the stationary wave forcing fields is not entirely clear.

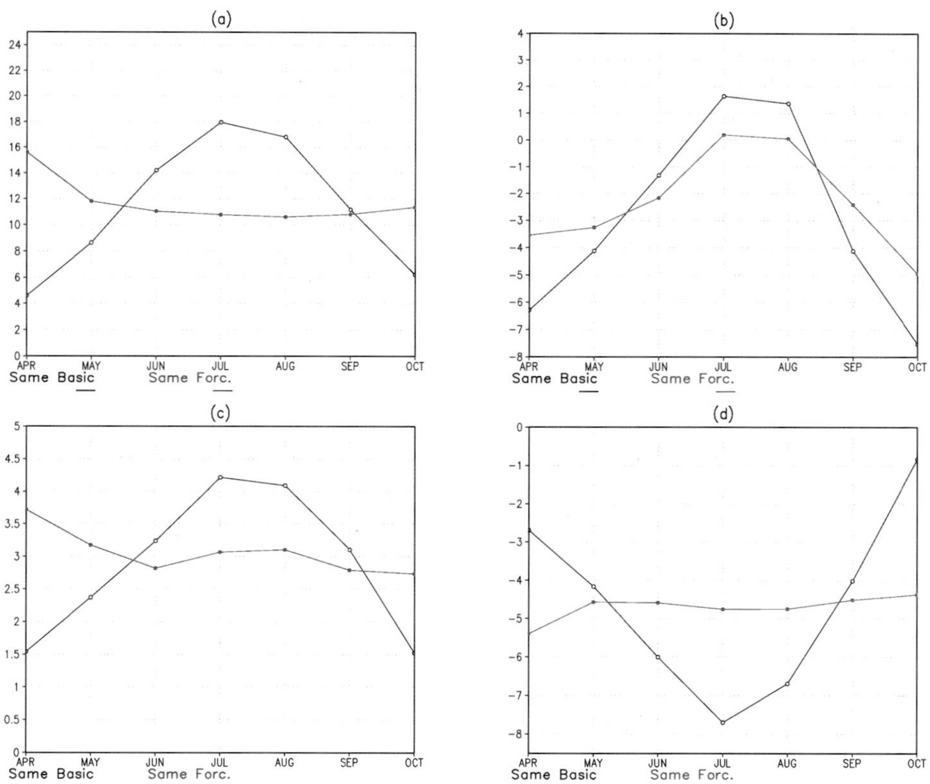

Figure 12. Regional stationary wave indices based on nonlinear model streamfunction responses to monthly varying total forcings but with fixed seven-month averaged basic state (open circle, black), and to fixed seven-month averaged forcings but with monthly varying basic state (solid circle, blue) for Tibetan anticyclone (a), East Asian trough (b), North Pacific subtropical high (c), and monsoon trough (d).

We performed nonlinear model calculations to illustrate the role of the seasonally varying zonal mean state (seasonally varying forcings) by fixing the forcing (zonal mean state) at the seven-month average value and only varying the zonal mean state (forcing) monthly. The results are shown in terms of the stationary wave indices in Fig. 12. Clearly, the seasonally varying zonal mean state with fixed seven-month average forcing (solid circle) cannot explain the seasonal variation of the stationary wave features during Asian monsoon. When varying the forcing with fixed zonal mean state at the seven-month average, however, the seasonal variation of the stationary wave features is comparable to those with both the forcing and basic state varying (Fig. 6). The only exception is the East Asian trough (Fig. 11b), which shows a significant contribution to its seasonal cycle by the effect of varying the basic state as well as that of varying the forcing. Thus the stationary wave seasonal cycle during the Asian monsoon is not directly related to the seasonal variation of the zonal mean basic state, rather it is strongly related to the seasonally varying heating.

5. Summary

The maintenance and seasonal cycle of the stationary wave features during Asian monsoon for the seven-month period, from April to October, are studied using the linear and nonlinear stationary wave models as well as the NCEP/NCAR reanalysis. Four prominent stationary wave features during this period, namely, the Tibetan anticyclone and the East Asian trough at the upper levels and the monsoon low and the subtropical high at the lower level, are examined. All four stationary wave centers go through distinct seasonal cycle during this seven-month period, with the Tibetan anticyclone, the subtropical high and the monsoon trough intensifying from April to July, and weakening from August to October, while the East Asian trough weakens from April to August and re-establishes itself from August to October. While the linear and nonlinear models reproduce the same seasonal cycle, they underestimate the amplitude of the stationary waves due to inaccuracies in the forcing fields. When using the nonlinear model to decompose the stationary waves into those forced by heating, orography and transients alone, the results clearly show that the effect of the heating is the most dominant in contributing to the stationary wave seasonal cycle.

We further separate the heating into three regional centers, the Western Pacific heating, the East Asian monsoon heating, and the Indian monsoon heating. The purpose is to examine if the monsoon heating and/or the heating center associated with western Pacific ITCZ and SPCZ is directly responsible for the seasonal variation of the stationary waves. The results show that the monsoon heating is the one that causes the seasonal variation of the stationary wave centers except the East Asian trough. The heating associated with ITCZ and SPCZ in the western Pacific does contribute to the amplitude of these stationary wave centers, but not the seasonal variation of the amplitude.

The results indicate that the atmospheric circulation and the diabatic heating associated with the Asian monsoon are closely linked. Earlier studies have shown evidence that the interannual variability of the East Asian monsoon rainfall is closely related to the variation in the stationary wave features, such as the Tibetan anticyclone in the upper troposphere, the western Pacific subtropical high and the monsoon low at the lower troposphere (Tao and Chen 1987; Murakami and Matsumoto 1994; Nitta and Hu 1996; Samel *et al.* 1999). The results in this study suggest that the seasonal cycle of these stationary wave features over the Asian monsoon region is a direct response to the monsoon heating, with the exception of the East Asian trough. Similar results were reported in Rodwell and Hoskins (2001) for the subtropical highs and monsoon heating forcing. Thus the relationship between atmospheric circulation and monsoon rainfall can be very complicated due to the highly interactive nature between the two. The responses to regional heating centers also reveal that the effect of Indian monsoon heating and the East Asian monsoon heating are fairly local in their maximum centers, suggesting that the two monsoon systems may vary independently sometimes due to feedback from the regional monsoon heating (Wang *et al.* 2001).

Acknowledgments

The authors would like to thank the support from the following: the Chinese National Science Foundation Young Overseas Chinese Scientists award, the NSF grant ATM 00-00585 and NOAA grant NA06GP0393.

References

Annamalai, H., J. M. Slingo, K. R. Sperber, and H. Hodges, 1999: The mean evolution and variability of the Asian summer monsoon: Comparison of ECMWF and NCEP-NCAR reanalyses. *Mon. Wea. Rev.*, **127**, 1157-1186.

Goswami, B. N., V. Krishnamurthy, and H. Annamalai, 1999: A broad scale circulation index for the interannual variability of the Indian summer monsoon. *Quart. J. Roy. Meteor. Soc.*, **125**, 611-633.

Hoskins, B. J., 1996: On the existence and strength of the summer subtropical anticyclones. *Bull. Amer. Meteor. Soc.*, **77**, 1287-1292.

Huang, R. H., and L. Li, 1989: Numerical simulation of the relationship between the anomaly of the subtropical high over East Asia and the convective activities in the western tropical Pacific. *Adv. in Atmos. Sci*, **6**, 202-214.

Huang, S. S., and T. H. Yue, 1962: On the structure of the subtropical highs and some associated aspects of the general circulation of atmosphere. *Acta Meteorologica Sinica*, **31**, 339-359 (in Chinese).

Kalnay, E., and Coauthors, 1996: The NCEP/NCAR 40-year reanalysis project. *Bull. Amer. Meteor. Soc.*, **77**, 437-471.

Krishnamurti, T. N., 1973: Tibetan high and upper tropospheric tropical circulation during northern summer. *Bull. Amer. Meteor. Soc.*, **54**, 1234-1249.

Lau, K.-M., K.-M. Kim, and S. Yang, 2000: Dynamical and boundary forcing characteristics of regional components of the Asian summer monsoon. *J. Climate*, **13**, 2461–2482.

______, and S. Yang, 1997: Climatology and interannual variability of the southeast Asian summer monsoon. *Adv. in Atmos. Sci*, **14**, 141-162.

Liu, Y. M., G. X. Wu, H. Liu, and P. Liu, 2001: Condensation heating of the Asian summer monsoon and the subtropical anticyclone in the Eastern Hemisphere. *Climate Dyn.*, **17**, 327-338.

Meehl, G. A., 1993: A coupled air-sea biennial mechanism in the tropical Indian and Pacific regions: Role of ocean. *J. Climate*, **6**, 31-41.

______, 1994: Coupled land-ocean-atmosphere processes and South Asian monsoon variability, *Science*, **266**, 263-267.

Murakami, T., and J. Matsumoto, 1994: Summer monsoon over the Asian continent and western north Pacific. *J. Meteor. Soc. Japan*, **72**, 719-745.

Nitta, T., 1987: Convective activities in the tropic western Pacific and their impact on the Northern Hemisphere summer circulation. *J. Meteor. Soc. Japan*, **65**, 373-390.

______, and Z.-Z. Hu, 1996: Summer climate variability in China and its association with 500 hpa height and tropical convection. *J. Meteor. Soc. Japan*, **74**, 425-445.

Qian, W., and S. Yang, 2000: Onset of the regional monsoon over Southeast Asia. *Meteor. Atmos. Physics*, **75**, 29-38.

Rodwell, M. J., and B. J. Hoskins, 2001: Subtropical Anticyclones and Summer Monsoons. *J. Climate*, **14**, 3192-3211.

Samel, A. N., S.-W. Wang, and W.-C. Wang, 1995: A comparison between observed and GCM simulated summer monsoon characteristics over China. *J. Climate*, **8**, 1690-1696.

______, W.-C. Wang, and X.-Z. Liang, 1999: The monsoon rainband over China and relationship with the Eurasian Circulation. *J. Climate*, **12**, 115-131.

Tanaka, M., 1992: Intraseasonal oscillation and the onset and retreat dates of the summer monsoon over the east, southeast and western north Pacific region using GMS high cloud amount data. *J. Meteor. Soc. Japan*, **70**, 613-629.

Tao, S. Y., and L. Chen, 1987: A review of recent research on the East Asia summer monsoon in China. *Monsoon Meteorology*, C. P. Chang and T. N. Krishnamurti, Eds., Oxford University Press, 60-92.

Ting, M., 1994: Maintenance of northern summer stationary waves in a GCM. *J. Atmos. Sci.*, **51**, 3286-3308.

______, and I. M. Held, 1990: The stationary wave response to a tropical SST anomaly in an idealized GCM. *J. Atmos. Sci.*, **47**, 2546-2566.

______, H. Wang, and L. Yu, 2001: Nonlinear stationary wave maintenance and seasonal cycle in the GFDL R30 GCM. *J. Atmos. Sci.*, **58**, 2331-2354.

______, and L. Yu, 1998: Steady response to tropical heating in wavy linear and nonlinear baroclinic models. *J. Atmos. Sci.*, **55**, 3565-3582.

Wang, B., R. Wu, and K.-M. Lau, 2001: Interannual variability of the Asian summer monsoon: Contrasts between the indian and the western north Pacific–East Asian monsoons. *J. Climate*, **14**, 4073–4090.

Wang, H., and M. Ting, 1999: Seasonal cycle of the climatological stationary waves in the NCEP/NCAR reanalysis. *J. Atmos. Sci.*, **56**, 3892-3919.

Wang, P.-K., and D. Zhang, 1992: Recent studies of the reconstruction of east Asian monsoon climate in the past historical literature of China. *J. Meteor. Soc. Japan*, **70**, 423-446.

Webster, P. J., and S. Yang, 1992: Monsoon and ENSO: Selectively interactive systems. *Quart. J. Roy. Meteor. Soc.*, **118**, 877-926.

Yang, S., and W. J. Gutowski, 1992: On the relationship between tropical Chinese rainfall and the Indian summer monsoon. *J. Meteor. Soc. Japan*, **70**, 997-1004.